CAT

Deformation, Fluid Flow, and Reservoir Appraisal in Foreland Fold and Thrust Belts

Edited by

Rudy Swennen
François Roure
James W. Granath

AAPG Hedberg Series, No. 1

Published by

The American Association of Petroleum Geologists

Tulsa, Oklahoma, U.S.A.

ISBN: 0-89181-900-2

AAPG Editor: Ernest A. Mancini
Geoscience Director: James B. Blankenship

This publication is available from:

The AAPG Bookstore
P.O. Box 979
Tulsa, OK U.S.A. 74101-0979
Phone: 1-918-584-2555 or 1-800-364-AAPG (U.S.A. only)
Fax: 1-918-560-2652 or 1-800-898-2274 (U.S.A. only)
E-mail: bookstore@aapg.org
www.aapg.org

About the Editors

Rudy (J.A.) Swennen is Professor of Geology at the Catholic University of Leuven (Belgium) and Vice-head of the department of Geography/Geology. His main present-day research focuses on the reconstruction of fluid flow in foreland fold and thrust belts. In addition his research team is involved in geochemical exploration and diagenetic studies, and the study of interaction processes between bio- and geosphere.

François Roure is an Associate Research Director and Professor at the Institut Français du Pétrole in Rueil-Malmaison, France. His main research interest concerns are the architecture and structural evolution of foreland fold and thrust belts. He is currently involved in geophysical-geological integrated studies for petroleum appraisal of tectonically complex areas.

James W. Granath received B.S. and M.S. degrees in geology from the University of Illinois at Urbana-Champaign, and a PhD in structural geology from Monash University in Australia. Since 1976 he has taught at SUNY Stony Brook and spent 18 years in Conoco in research, international exploration, and new ventures. In 1999 he opened a consulting practice focused on structural geology and tectonics as applied to exploration problems. He is a member of AAPG, AGU, GSA, HGS, GITA, and SEAPEX, and is a certified petroleum geologist (#5512) and a Texas Professional Geologist (#733). Granath's expertise lies in seismic interpretation and integration with structural analysis, regional synthesis, and prospect and play evaluation. Recent projects have focused on the thrust belts of Europe and the Andes, GIS applications to Southeast Asian tectonics and the petroleum basins of Mexico, and the Moroccan passive margin.

Acknowledgments

AAPG wishes to thank the following
for their generous contributions to

Deformation, Fluid Flow, and Reservoir Appraisal in Foreland Fold and Thrust Belts

Instytut Nafty/Gazu—Oil and Gas Institute, Poland
U.S. Geological Survey

Contributions are applied toward the production cost of publication,
thus directly reducing the book's purchase price
and making the volume available to a larger readership.

Table of Contents

Swennen, R., F. Roure, and J. W. Granath, 2004, Introduction, *in* R. Swennen, F. Roure, and J. W. Granath, eds., Deformation, fluid flow, and reservoir appraisal in foreland fold and thrust belts: AAPG Hedberg Series, no. 1, p. 1–2.

Introduction

Deformation, Fluid Flow, and Reservoir Appraisal in Foreland Fold and Thrust Belts

Rudy Swennen
Katholieke Universiteit, Leuven, Belgium

François Roure
Institut Français du Pétrole, Rueil Malmaison, France

James W. Granath
Granath and Associates, Houston, Texas, U.S.A.

INTRODUCTION

This volume constitutes the proceedings of the American Association of Petroleum Geologists–Institut Français du Pétrole Hedberg conference entitled Deformation, Fluid Flow, and Reservoir Appraisal in Foreland Fold and Thrust Belts held on May 14–18, 2002, in Mondello, Sicily, with the participation of 80 earth scientists from the petroleum industry and universities. It comprises 21 papers, selected from among 38 oral and 13 poster presentations.

Reservoir risk remains one of the key concerns to the exploration of deep subthrust prospects in foothill areas of fold and thrust belts. Case studies that have integrated structural geology, geochemistry, and reservoir petrology not only now provide a better understanding of the multiscale processes involved in reservoir damage (particularly small-scale deformation and cementation) but also have begun to identify the critical parameters that are likely to enhance secondary porosity and permeability in deeply buried sandstone and carbonate series. Recent progress in analytical resolution makes it more and more possible to approach the problems on a microscale level and link the evolution of rock properties to the geological history on a broader scale.

Basin-scale numerical simulations now incorporate the kinematics and thermal evolution of fold and thrust belts, as well as reconstructions of the history of fluid flow and pore-fluid pressure. Basin modeling techniques provide quantitative and predictive results based on both thermodynamics and mass-balance calculations of fluid-rock interactions at reservoir scale. Instantaneous fluid-flow values derived from such models can be used as boundary conditions for forward diagenetic simulations applied to a specific reservoir layer, particularly for those episodes of either vertical or horizontal fluid circulation in an open system. In addition to refining burial and chemical compaction models, future developments are expected to add more mechanics in two- and three-dimensional basin-modeling tools. In the near future, it may become routine to include both tectonically driven pressure-solution and compaction processes driven by layer-parallel shortening and hydraulic fracturing in computations of average porosities and permeabilities of reservoirs and seals for each evolutionary stage of any given basin or prospect.

DOI:10.1306/1025682H12607

One of the fundamental problems in any modeling in fold and thrust belts is the determination of the thermal history. The first section of this volume describes the use of two important and distinct paleothermometers (hydrocarbon-bearing fluid inclusions and apatite fission tracks) for reconstructing pressure-temperature (P-T) evolution of subthrust reservoirs.

The second section emphasizes the importance of the link between structural kinematics and thermal modeling. Four chapters trace the burial (P-T) evolution of potential source rocks and reservoirs in three case studies in the southern Apennines, Colombia, and Pakistan.

The third section focuses on analytical results and integrated studies, substantiating the links between deformation and fluid circulation in various fold and thrust belts, such as the Sierra Madre in Mexico, the central Brooks Range and the Arctic in Alaska, the coastal belt in northeastern Spain, and the Ukraine.

Sections 4 and 5 deal with the links between deformation and reservoir characteristics and between fluid flow and diagenesis, respectively, using petrography, integrated case studies, and ultimately, basin modeling mainly for carbonate reservoirs (i.e., in the Apennines and the Canadian Rockies) and also for sandstone reservoirs (i.e., in eastern Venezuela).

Finally, the last section illustrates the impact of thrust-belt evolution on regional petroleum systems and petroleum potential, first, in the Tunisian Atlas and Sicily, close to the venue for the Hedberg conference and postconference field trip, and second, in the Gaspé Appalachians. The former systems are Cretaceous and Tertiary in age, whereas the latter involves a unique example of multiphase Paleozoic deformation and its strong control on the source rock burial history, petroleum generation and migration, and oil charge of the traps.

Paleothermometers

1

Guilhaumou, N., L. Benchilla, P. Mougin, and P. Dumas, 2004, Advances in hydrocarbon fluid-inclusion microanalysis and pressure-volume-temperature modeling: Diagenetic history, pressure-temperature, and fluid-flow reconstruction— A case study in the North Potwar Basin, Pakistan, *in* R. Swennen, F. Roure, and J. W. Granath, eds., Deformation, fluid flow, and reservoir appraisal in foreland fold and thrust belts: AAPG Hedberg Series, no.1, p. 5–20.

Advances in Hydrocarbon Fluid-inclusion Microanalysis and Pressure-volume-temperature Modeling: Diagenetic History, Pressure-temperature, and Fluid-flow Reconstruction— A Case Study in the North Potwar Basin, Pakistan

Nicole Guilhaumou

Museum National d'Histoire Naturelle, Centre National de la Recherche Scientifique, Paris, France

Lakhdar Benchilla

Institut Français du Pétrole, Rueil Malmaison, France

Pascal Mougin

Institut Français du Pétrole, Rueil Malmaison, France

Paul Dumas

Laboratoire pour l'Utilisation du Rayonnement Électromagnétique and Laboratoire de Spectrochimie Infrarouge et Raman, Centre National de la Recherche Scientifique, Orsay, France

ABSTRACT

Several advances have been made for the reconstruction of fluid circulations and diagenetic history in subthrusted petroleum reservoirs because of the combination of the in-situ microanalysis of hydrocarbon fluid inclusions by Synchrotron Fourier transform infrared spectroscopy and PVTX modeling coupled to diagenetic history and tectonic setting. Integrated study has been made in the Eocene Chorgali formation (North Potwar Basin, Pakistan), where the shallow-marine carbonates formed important fractured reservoirs. Hydrocarbon fluid inclusions recognized in authigenic quartz and calcite from

DOI:10.1306/1025683H13110

hydroveins show atypical association of CO_2-rich light oil depleted in H_2O in sulfates-quartz-calcite along simultaneous dissolution recrystallization processes at micrometer scale. Synchrotron Fourier transform infrared spectroscopy analyses, microthermometry, and pressure-volume-temperature modeling led to the beginning of quartz and calcite recrystallization at no more than 75–85°C and 150–180 bar in conditions of sulfate-calcite transformation. Temperatures of 150°C measured in aqueous fluid inclusions from calcite hydroveins are in favor of a thermosulfatoreduction mechanism. Early diagenetic sulfates are reduced by organic acids, and CO_2 comes from organic matter decomposition and/or previous decarbonation. A second phase of quartz growth is evidenced by the homogeneous entrapment in fluid inclusions of more mature oil in 60% CH_4 and a large amount of water at temperatures reaching 150–170°C. This late production of CH_4 agrees with $\delta^{13}C$ depletion (−20 and −36‰) measured in veins and the crystallization of saddle dolomite.

Thrustpack® modeling shows that the onset of hydrofracturing and quartz precipitation at 1.5 km (1 mi) depth and 15–10.8 Ma (middle Siwalik) began when temperatures of 65 ± 10°C were reached at the end of sedimentation in the basin. It lasted until 4–6 km (2.5–4 mi) depth at temperatures as much as 170°C and reached the development of the thrust sheet at 5 Ma. Thus, circulations of hydrocarbon-rich fluids may be considered in thermal equilibrium with host rocks in both cases. The oil could then be derived from source rocks in the deep Mesozoic formation for the first input. The second input originated from the deep part of the basin itself and mixed with tectonic and meteoric water along the circulation pathways. The fluids are mainly driven by tectonics. They are expelled from the hinterland farther to the north and move updip toward the south in the Chorgali conduits, below the Kuldana seals. The potential source rock for organic matter is known as type II and type III kerogens in coal and black shales from the Paleocene.

INTRODUCTION

As migration of oil in reservoirs involves a multiphase mixture of oil, gas, and water (e.g., Rudkiewicz et al., 1994), fluid inclusions entrapped in diagenetic minerals commonly contain hydrocarbons. Hydrocarbons are allowed to migrate in a mixed phase of oil and water, and the oil-bearing fluid inclusions formed during diagenetic events are in the immiscible region and can be made up of pure oil or a mixture of oil and water in any proportion (Rudkiewicz et al., 1994). Since their first description by Murray (1957) in fluorite mineral, several studies involving the composition and densities of these particular fluid inclusions have been made for understanding the pressure-temperature (P-T) conditions of fluid circulations in sedimentary basins and ore deposits (reviewed in Roedder, 1984; McLimans, 1987; Karlsen et al., 1993; Goldstein and Reynolds, 1994; Goldstein, 2001). A combination of trapping temperature, molecular composition, and isochoric pressure-volume-temperature (PVT) modeling of oil-bearing fluid inclusions lead directly to define the P-T conditions of the fluid circulations that occurred during structural events. The microanalyses of several phases associated with hydrocarbons found in these fluid inclusions as dissolved components or daughter minerals aim to define the conditions of diagenesis along these fluid circulations. In subthrusted reservoirs, data from fluid-inclusions lead directly to the reconstruction of the diagenetic history of sediments during syntectonic deformation (Larroque et al., 1996; Guilhaumou et al., 2000; Benchilla et al., 2003).

The trapping temperature is currently measured by microthermometry (Roedder, 1984), and the molecular composition has been tentatively obtained by various destructive techniques involving decrepitation (Barker and Underwood, 1992), crushing, and leachates analyses (George et al., 1998). From exceptionally large fluid inclusions (Guilhaumou et al., 2000), the composition and biomarkers were obtained by individual extraction under the microscope and gas chromatography–mass spectrometry (GCMS) analysis, but the analysis of the C1–C5 fraction is missing. Jones and Macleod (2000) proposed a sophisticated, time-consuming methodology to extract and clean up the sample before crushing for leachate analyses. The main limitations include a possible contamination during the extraction and the mixing of several generations of oil. Additionally, oil-water fluid inclusions are commonly encountered in heterogeneous trapping; the use of hydrocarbon isochores may be inappropriate in such fluid inclusions, because it precludes the linking of the individual homogenization temperture with the real composition for PVT modeling.

In the last 10 yr, advances in molecular infrared microspectrometry (Guilhaumou et al., 1990; Rankin et al., 1990; Pironon et al., 2001) enabled us to perform quantitative, nondestructive molecular analyses in a single fluid inclusion under the microscope. The size of the fluid inclusions analyzed is currently restricted to the wavelength of the incident beam (20 μm). The interfacing of a Synchrotron source to a microscope aims to overcome this limit and to perform analyses and mapping of the components at a few-micrometer scale. Using internal calibration, semiquantitative analyses can be done, and the composition of individual fluid inclusions can be linked directly to its density. At the same time, progress in isochoric modeling using modified equations of state (Aplin et al., 1999; Guilhaumou et al., 2000; Thiery et al., 2000; Benchilla et al., 2003) led us to calculate thermodynamic properties, i.e., the phase envelopes and isochoric P-T path from analytical data of petroleum-bearing fluid inclusions.

In this work, in-situ microanalyses, microthermometric data, and isochoric modeling of petroleum fluid inclusions from cemented fractures of the Eocene Chorgali reservoir (Potwar Basin, Pakistan) have been performed and integrated with kinematic and thermal two-dimensional basin modeling. Then, the diagenetic history could be related to the burial and thermal evolution of the reservoir. Finally, an interpretation on the way these parameters controlled the quality of the reservoir could be deduced (Benchilla, 2003).

GEOLOGIC SETTING

The Potwar Plateau and Salt Range are located in the southwestern foothills of the Himalayas in northern Pakistan. The region extends more than 120 km (75 mi) from the main boundary thrust in the north to the Jhelum River in the south. The Jhelum River limit constitutes an ill-defined border of the Potwar Plateau to the east, whereas its boundary is rather sharply defined by the Mianwali reentrant to the west (Figure 1A).

Summary of Sedimentology and Diagenesis of the Chorgali Formation

Shallow-marine carbonate strata of the Eocene Chorgali formation form an important hydrocarbon-producing horizon in the northern part of the Potwar Basin (northern Pakistan, Figure 1B). Based on limited core samples from the Dakhni gas-condensate field and on data gathered from an outcrop analog located at Chorgali Pass north of this producing field, the sedimentary, early diagenetic, and fracture-related characteristics of this dual-porosity reservoir were studied and compared with fold and thrust belt evolution of Benchilla (2003).

The paragenetic sequences were defined from petrography and microtectonic data. Details may also be found in Benchilla (2003). The Eocene succession represents a shallowing-upward sequence, starting with open-marine, bioclastic, medium-bedded limestones, which grade into fabric-destructive, massive sucrosic, and subsequently, fine-crystalline dolomites. The presence of anhydrite nodules and chicken-wire-shaped calcite pseudomorphs after anhydrite, as well as the existence of dolomitized algal mats, points toward restricted sedimentary conditions. The near-surface sedimentation conditions are also reflected by the presence of flat channel conglomerate and breccia beds in the upper part of the dolomites. The breccias display features typical for early diagenetic evaporite collapse breccias. These dolomites, which are 15 m (50 ft) thick, are overlain by a 32-m (105-ft)-thick shale and marl succession, in which few lacustrine limestone beds are intercalated. Subsequently, a few meters of pedogenetically overprinted marine limestones develop below the regionally recognized Kuldana sealing shales, above which Fateh Jang conglomerates occur. This is followed by shales and fine-grained sandstones of the Miocene Murree Formation that unconformably overlies the Eocene strata. The $\delta^{18}O$ and $\delta^{13}C$ values of the dolomite host rock test for marine dolomites that were recrystallized by meteoric water infiltration. The marine limestones were also affected by this recrystallization and possibly by additional interaction with hot fluids.

Fracturing

Late diagenesis is mainly characterized by hydrofracturing, whereby at least seven distinct fracturing (Figure 2) episodes were recognized. They are well developed in the limestones just below the Kuldana shales.

An attempt was made to characterize individual vein generation as recognized from petrography and by its stable isotopic signature, as described in Figure 1C. In general, there is a major overlap in stable isotopic composition, both in $\delta^{13}C$ and $\delta^{18}O_{PDB}$ between the vein generations V1 and V5. A clear covariant behavior between both parameters is present. Authigenic quartz developed in these hydroveins during mesogenesis as diagenetic crystals with carbonates, but they could not be dated precisely. They are surprisingly rich in primary and pseudosecondary (Roedder, 1984) hydrocarbon-bearing fluid inclusions.

Veins-filling calcite developed prior to quartz in the fractures in which anhydrite and gypsum needles and barite minerals are commonly observed. In some calcite crystals, two generations of small-sized (1–5 μm), primary biphasic oil fluid inclusions are distinguished by

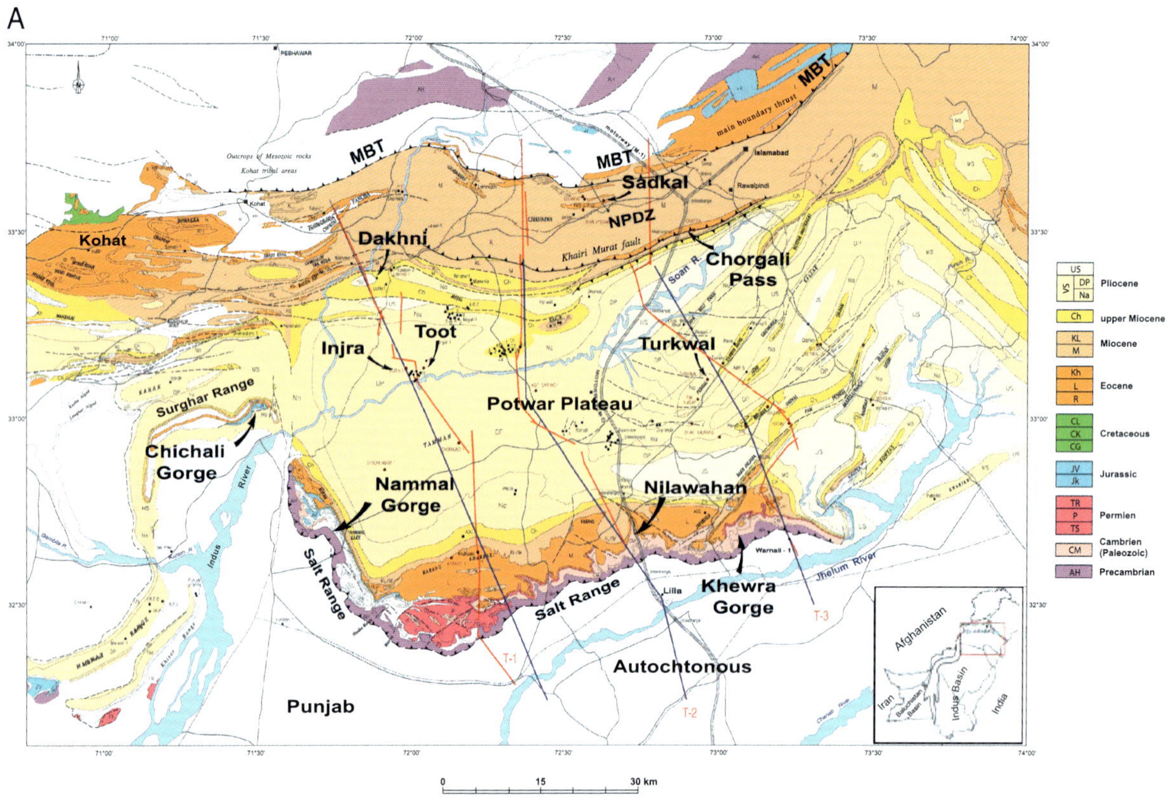

FIGURE 1. (A) Geological setting of the Potwar Plateau and Salt Range (modified after Roure et al., 1999). MBT = Main Brain thrust; NPDZ = near Potwar deformation zone.

ultraviolet (UV) fluorescence, attesting to the fact that they were formed during the migration of two successive oil-bearing solutions. Locally, open fractures are cemented by authigenic quartz rich in hydrocarbon fluid inclusions. Some partly dissolved and recrystallized crystals of carbonates and sulfates are accompanying the quartz crystallization. Several generations of hydrocarbon-bearing fluid inclusions were observed in both diagenetic quartz and calcite. Isolated crystals made up of recrystallized carbonates are common in quartz. They show elongated primary oil inclusions filling cavities, having a fluorescence emission in the light blue region. Some dissolution features are observed, suggesting intensive dissolution-recrystallization processes during the hydrocarbon entrapments; however, there is no indication of fractures or cleavages, supporting the hypothesis of a secondary oil recharge of the primary cavities.

FLUID-INCLUSIONS DATA

Analytical Process

Unmounted, doubly polished sections were prepared for the study of fluid inclusions by optical observations in transmission and polarized light. Fluorescence emission was collected during transmission using a UV source (365 nm). Microthermometric runs were carried out on aqueous and hydrocarbon fluid inclusions to measure the temperatures of the phase transitions (homogenization temperature, T_h; temperature of first melting or eutectic temperature, T_{fm}; and last melting temperature, T_{mi}) using a Linkam THMSG 600 heating-cooling stage coupled to a video system controlled by a computer. Fourier transform infrared molecular microanalyses in point spectroscopy and mapping modes were then acquired using a Synchrotron radiation as excitation source in the midinfrared range on selected samples in the 1000–4000-cm^{-1} frequency domain. We used a 3×3-μm^2 aperture (32× Schwarzchild objectives) and a confocal arrangement. The experimental station consists of an FTIR Thermo-Nicolet Magna 860 spectrometer (spectral range = 650–4000 cm^{-1}) coupled to an optical microscope (Nicplan), beamline Mirage, Laboratoire pour l'Utilisation du Rayonnement Électromagnétique (LURE), Orsay, France. The high brightness of the Synchrotron source allows an aperture setting of 3×3 μm while achieving high signal-to-noise spectra. These aperture dimensions set the probed area close to those obtained by Raman microspectrometry (1×1 μm). A Linkam microthermometric stage was adapted to the microscope by Thierry Marin (LURE), allowing us to perform analyses from room temperature to 400°C. Error

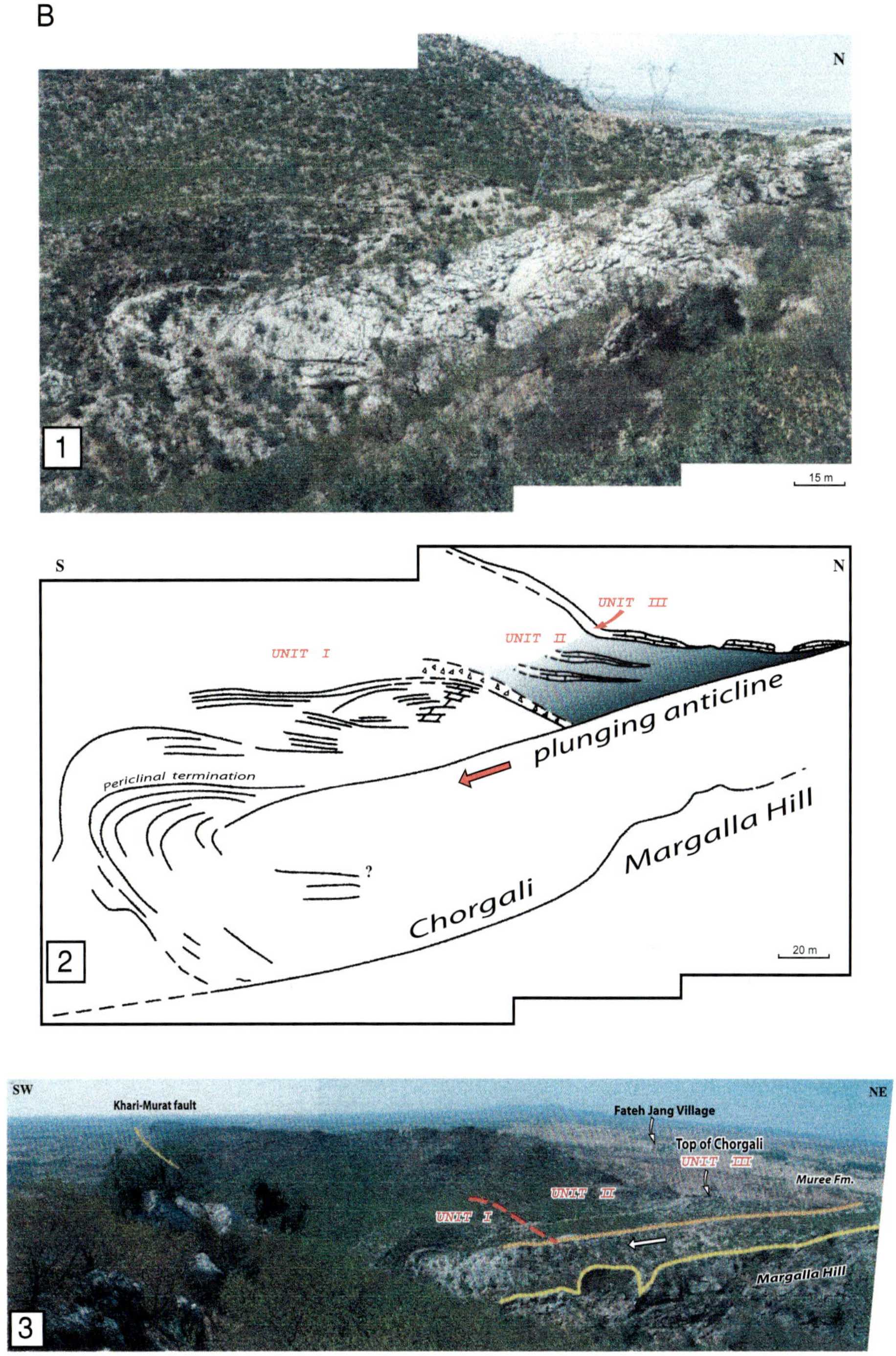

FIGURE 1. (cont.) (B) The Chorgali Pass area: Emplacement of the three units recognized and interpreted cross section of the stratigraphic outcroppings.

is presented at ±1°C. Heating runs in these experiments were performed from 25°C to homogenization temperatures (T_h) using a heating rate of 1°C/min. The liquid/vapor volume ratios of hydrocarbon fluid inclusions have been measured at room temperature through a microscope using a Biorad Radiance confocal laser unit and an Olympus X100 dry objective using the method described in Aplin et al. (1999).

Description of Oil-bearing Fluid Inclusions in Authigenic Quartz and Calcite

In the authigenic quartz formed in veins (Figure 3A), two generations of hydrocarbon fluid inclusions are identified. First are exceptionally large-sized primary fluid inclusions (Figure 3B), ranging from 10 μm to 1.2 mm. They systematically contain a liquid and vapor organic phase and one or several brown bodies. The liquid oil displays a light yellow color and gives a fluorescence emission in the light blue region of UV light, whereas the brown parts do not give any visible fluorescence (Figure 3C, D). Such brown bodies were also encountered in oil-bearing fluid inclusions hosted in fluorite from other areas studied in Pakistan (Guilhaumou et al., 2000; Benchilla et al., 2003). They were commonly observed in fluid inclusions hosted in quartz and fluorite occurring in veins from the Mississippi Valley-type (MVT) deposits and found as ore mineralizations (Richardson and Pinckney, 1984; Roedder, 1984) closely associated to petroleum migration episodes. These bodies are organic phases and are considered to be asphaltenes formed inside fluid inclusions by the precipitation of the heavy-oil fraction by the dissolved CO_2 and/or a decrease in temperature and pressure (Guilhaumou et al., 1990). They should be distinguished from bitumens formed by bacterial sulfate reduction (BSR) and/or thermosulfatoreduction (TSR) (never observed in diagenetic mineral assemblages). In these samples, additional pleochroic minerals are commonly seen in the oil, either inside the fluid inclusions or close to the oil-quartz interface (Figure 3B). They could be daughter minerals and/or mechanically

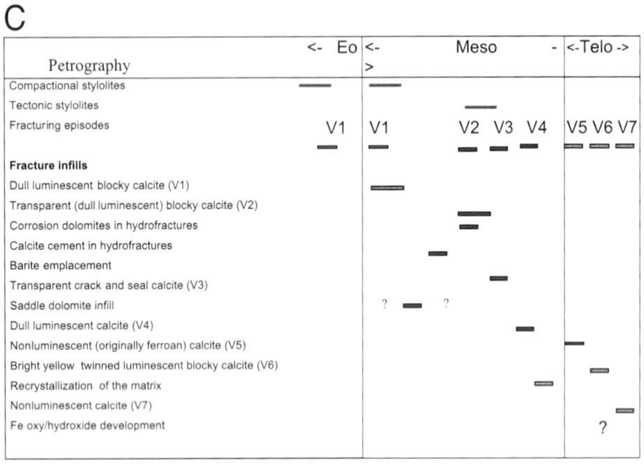

FIGURE 1. (cont.) (C) Fracture succession of the Chorgali Formation in the Chorgali Pass section from eogenesis (Eo) to telogenesis (Telo) (modified from Benchilla, 2003). The calcite in V7 fractures and development of Fe oxides are not dated.

entrapped phases (Roedder, 1984). Numerous, very small, unidentified solid particles ranging from 1 to 5 μm are systematically scattered in the oil phase of some primary cavities (S in Figure 3B). This was never observed before in such oil fluid inclusions and relates probably to a very specific composition of the migrating fluid during the growth of a single crystal.

Another generation of fluid inclusions called pseudosecondary fluid inclusions occurs as polygonal features subparallel to the walls of authigenic crystals (Figure 3E, F). Their entrapment was directly controlled by a lateral overgrowth of quartz crystals. They are smaller than primary fluid inclusions, range from 1 to 10 μm in size, and constitute regular arrangements. Their fluorescence emission under UV light is in the yellow to orange region. The homogenization temperatures of primary hydrocarbon-bearing fluid inclusions in quartz range between 40 and 75°C, with a maximum at 55–60°C. Homogenization temperatures (T_h) of the generation of pseudosecondary fluid inclusions range from 70 to 125°C, with a peak at 70–75°C. This later distribution is similar to the first one, suggesting a continuous increase in temperature during the quartz growth. However, a correction caused by the difference in compositional parameters should be applied before comparison.

Some small-sized authigenic calcite crystals that crystallized inside the authigenic quartz also contain biphasic hydrocarbon fluid inclusions (Figure 4), as revealed by their fluorescence emission in the light blue region. Their small size (less than 5 μm) does not permit the measurement of their homogenization temperatures.

OIL COMPOSITION: SYNCHROTRON FOURIER TRANSFORM INFRARED SPECTROSCOPY MICROANALYSIS

Molecular analyses and mapping were performed by Synchrotron infrared microspectroscopy on the two generations of hydrocarbon fluid inclusions, which are described below in quartz and in primary fluid inclusions in calcite crystals.

Primary Oil Fluid Inclusions in Authigenic in Quartz and Calcite

The absorbance spectra from primary fluid inclusions in quartz (Figure 5A) show the characteristics of the early petroleum generation. These fluid inclusions are mainly composed of linear alkanes that are detected in the whole cavity by their absorption at 2800–3000 cm^{-1} (Colthup et al., 1990). They have a high CO_2 content, now occurring as (1) a phase dissolved in the alkane fraction (absorption at 2336 cm^{-1}; Barres et al., 1987; Guilhaumou et al.,1990, 1998; Berreby, 1983) and as (2) gaseous CO_2 under pressure in the vapor phase (additional typical absorption at 2340 cm^{-1}). The molecular mapping indicates that most of the CO_2 and CH_4 (main absorption at 3010 cm^{-1}) are now concentrated in the vapor phase. This is consistent with the thermodynamic properties of such organic-fluid systems in current pressure-temperature reservoir conditions. Aromatics and water are not detected. Semiquantitative analyses using an internal CH_4 calibration and following the procedure of Pironon et al. (2001) have been performed on several primary fluid inclusions in five different quartz crystals. They lead us to define a typical composition of aliphatic light oil in C_{10}–C_{12} enriched in CO_2, with an average content of 30% CH_4 and free from water. The microanalyses of all fluid inclusions show that this oil

FIGURE 2. Microscopic observations on thin sections showing the main types of veins and cements along the fracturing episode in the Chorgali Pass: (A) Tectonic stylolites layer-parallel shortening (LPS) crosscutting the veins V2. (B) Calcite cementation in "crack and seal" in V3 calcite veins. (C) Several generations of veins are distinguished by their cathodoluminescence emission. (D) The pyrites (Py) are crosscut by V6 and V7 veins. (E) Calcite veins associated to barite crystals. (F) Fibrous anhydrite as cement in dolomites (polarized light). (G) Other examples of fractures filled with anhydrites; some oil droplets may be seen in the intergranular spaces. (H) Bioclastic carbonate wackestone with tectonic stylolites LPS, in transmitted light showing the opening of a secondary porosity (black arrow).

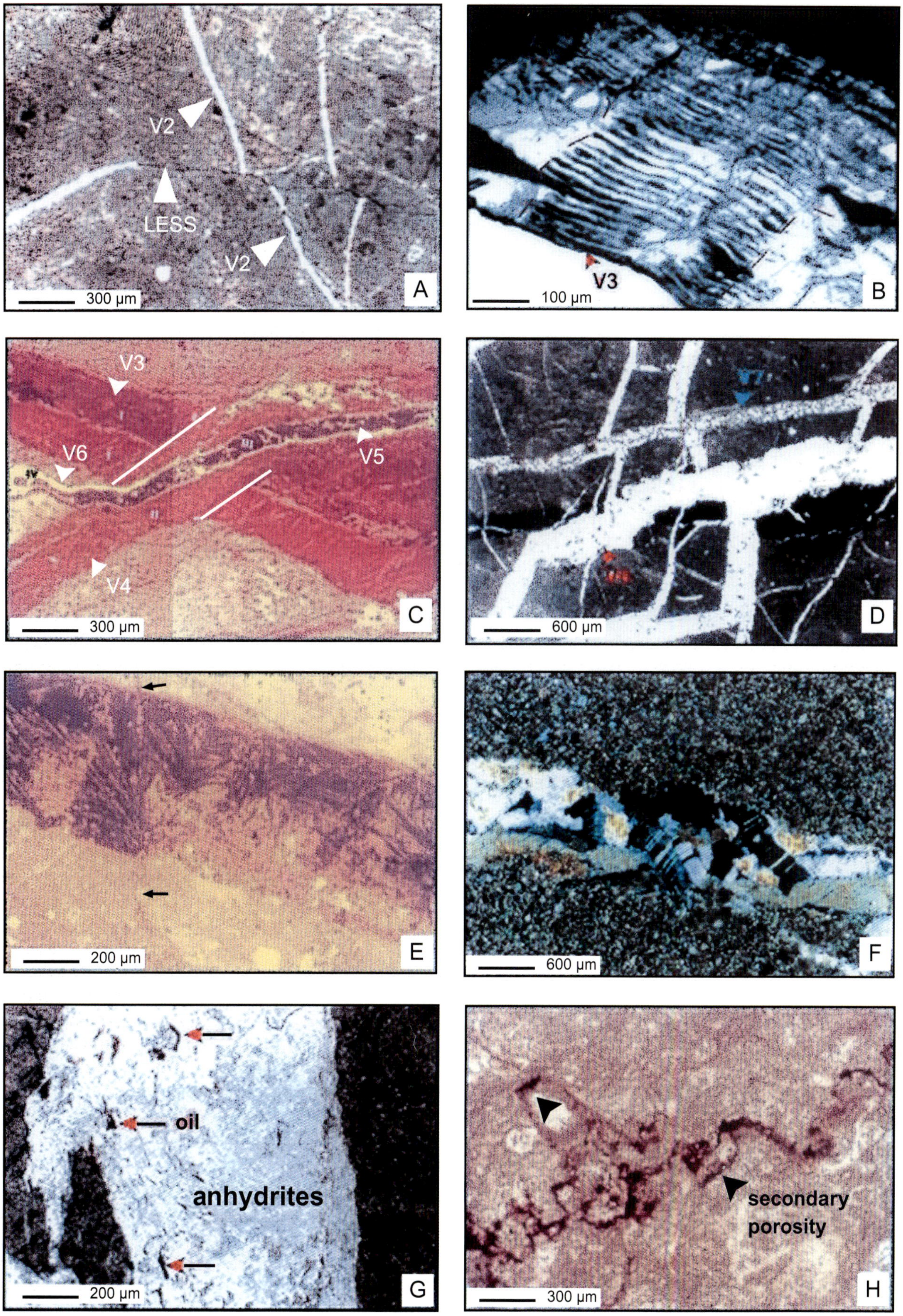
V2
V2
LESS
300 μm
A
V3
100 μm
B
V3
V6
V5
V4
300 μm
C
600 μm
D
200 μm
E
600 μm
F
oil
anhydrites
200 μm
G
secondary porosity
300 μm
H

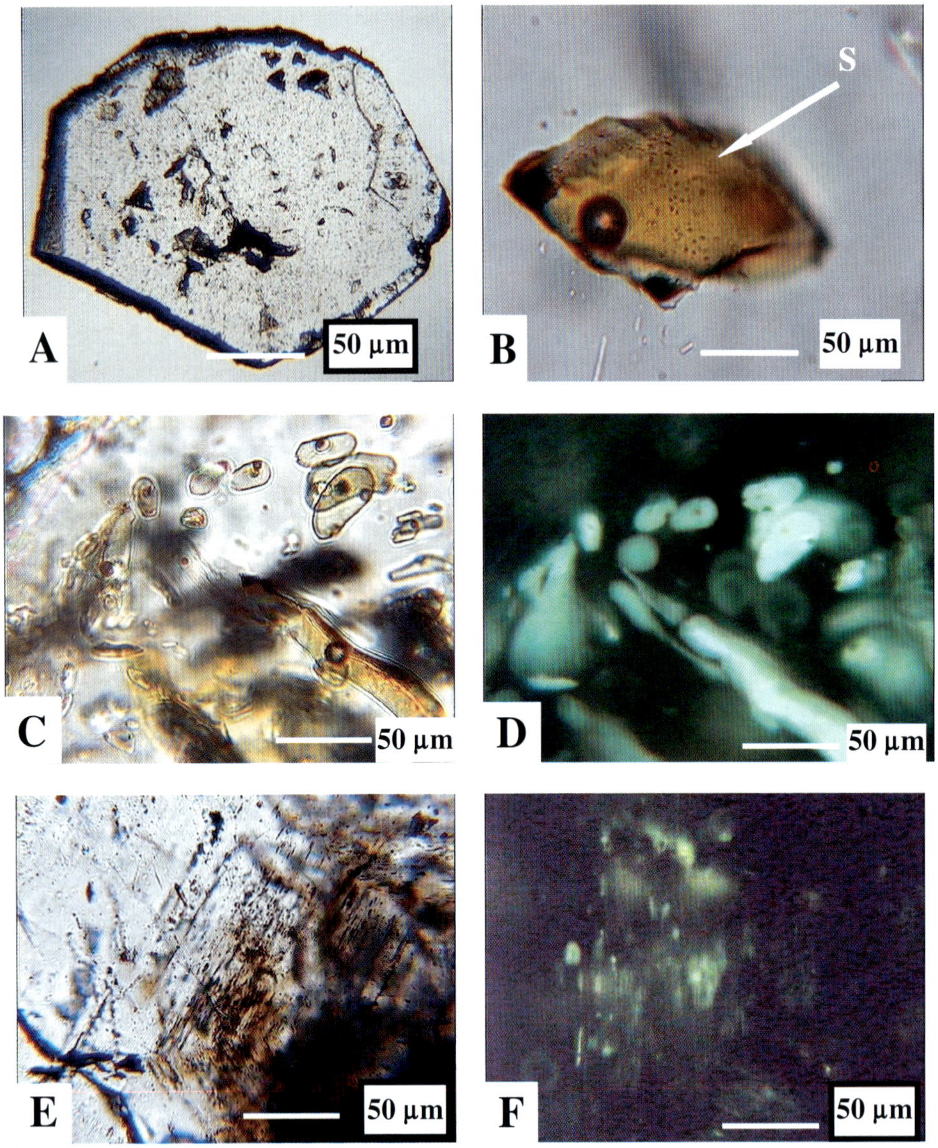

Figure 3. Different types of hydrocarbon fluid inclusions in authigenic quartz. (A) Typical authigenic quartz; (B) primary oil fluid inclusions in small solid phases (S); (C, D) primary oil fluid inclusions in transmitted and UV light; (E, F) pseudosecondary fluid inclusions in external crystal growth in transmitted and UV light.

was entrapped homogeneously at the beginning of the crystal growth (Table 1).

Very small crystals with sizes between 1 and 6 μm are scattered in oil (Figure 5B); they show an additional absorption at 2520 cm^{-1}, which is unambiguously attributed to the CO_3^{2-} bonds and identifies these solid phase as carbonates (Farmer, 1974; Salisbury et al., 1991). Comparison with some spectra recorded from calcite hosted in quartz shows that they are probably small seeds of calcite minerals (Figure 5B). These solids have been observed dissolving at least partly during heating runs. Experiments were performed by heating this type of fluid inclusions using a microthermometic stage from room temperature to homogenization temperature under the Synchrotron IR beam. Some spectra recorded on the same point show the slow decrease of the CO_3^{2-} absorption until its complete disappearance at homogenization temperature (Figure 5B) and its subsequent reappearance during decreasing temperature. This identifies these carbonates as calcite daughter minerals.

The FTIR microanalysis of a large-sized hydrocarbon fluid inclusion (Figure 5C) reveals an exceptional assemblage of minerals entrapped in oil as a key point for diagenesis reconstruction. A round-shaped, highly pleochroic sulfate crystal A (polarized light) similar to some observed inside quartz occurs close to a small crystal B in the center of the fluid inclusions. Crystal A has a high absorption in the OH area, having a main band near 3620 cm^{-1}. This suggests that this mineral is a partly dissolved crystal of hydrated sulfate that could be the hemihydrate calcium sulfate (Mandal and Mandal, 2002). Crystal B shows an absorption near 2520 cm^{-1}, typical of carbonate microcrystals (Legodi et al., 2001). Some polarizing minerals 5–20 μm in size are also identified as carbonates; they occur at the interface between quartz and oil in those fluid inclusions. This assemblage could represent the typical replacement of hydrated calcium sulfates [as gypsum or anhydrite(?)] by calcium carbonate as calcite, the reaction being fossilized because of the formation of primary oil fluid inclusions in authigenic quartz.

Primary oil fluid inclusions in coeval calcites entrapped in quartz (Figure 6) have been tentatively analyzed by Synchrotron Fourier transform infrared spectroscopy (SFTIR). To avoid the effect of the overlapping of alkane absorption to the absorption of carbonates in the 2800–3000-cm^{-1} range, a spectra of the calcite matrix having the same intensity (calibrated on the 2520-cm^{-1} absorption) has been subtracted (Figure 6,

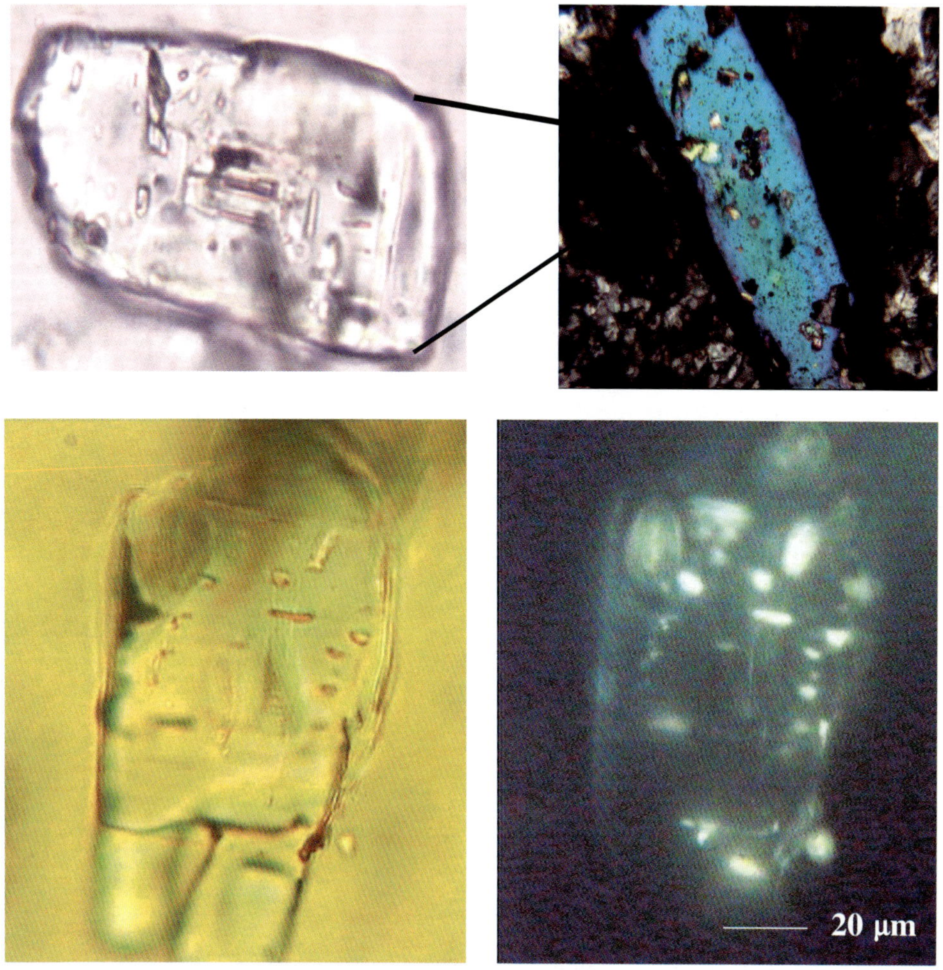

FIGURE 4. Primary hydrocarbon fluid inclusions in calcites recrystallized inside authigenic quartz in transmitted and UV light.

spectrum 4). This led to an oil composition comparable to the composition of the oil fluid inclusions in quartz (Table 1) with no water being detected. These results hereafter would confirm the crystallization of diagenetic quartz and calcite at the same time in the same organic fluid environment enriched in CO_2.

Pseudosecondary Oil Fluid Inclusions in Quartz

The SFTIR analyses of the oil fluid inclusions of the second generation (Figure 7) indicate a water-oil mixture depleted in CO_2 (not detected). Molecular mapping shows that the trapping is homogeneous from one inclusion to another along crystal-growth zones. In particular, the oil phase always appears associated with a noticeable water phase. High detection level of FTIR suggests that the water is not the dominant phase but is just wetting the walls of the cavities. Therefore, such low quantity allows water to be dissolved in oil at least partly during migration. Quantitative estimates led to a more mature hydrocarbon in C_{15}–C_{17}, with a high content of CH_4 reaching 60% (Table 1). These compositions would indicate a more mature and/or partly degraded oil that is consistent with the difference in color (shifted in the red region) observed in the fluorescence emission under UV light (Korashani, 1986; Guilhaumou et al., 1990).

Aqueous Inclusions in Veins

Very few primary aqueous fluid inclusions were found on one quartz crystal only; they are generally not observed associated to oil fluid inclusions in quartz. Their T_h values display temperatures between 60 and 90°C (six measurements). The T_h values of primary fluid inclusions in calcite V3 (detailed in Benchilla, 2003) range from 120 to 150°C; the salinities range between 0.4 and 16.1 equiv. wt.% NaCl. The T_h values of late aqueous inclusions range from 40 to 140°C, with two distinct modes near 70 and 120°C, respectively. Salinities range between 0 and 30 equiv. wt.% NaCl, with a mean value at about 2 equiv. wt.% NaCl. The broad range in T_{mi} in all types of inclusions is best explained by the presence of fluid of different salinities (at least two fluids) that mixed (Parry et al., 1991). Such an observation, together with the occurrence of higher temperatures, has already been described by Muchez and Slobodnik (1996) and interpreted as the result of a complex history of mixing, stretching and/or leakage, and refilling of the cavities. However, the homogeneity in content of the oil fluid inclusion populations and the correlated higher temperatures-higher maturity of the hydrocarbon fluid inclusions (i.e., higher CH_4 content) suggest here an increase in temperature. The succession of several pulses of fluid migration is also attested by the differences in composition observed for the latter oil-bearing fluid inclusions mixed with water and without significant CO_2 at the periphery of quartz crystals. They are in agreement with the large

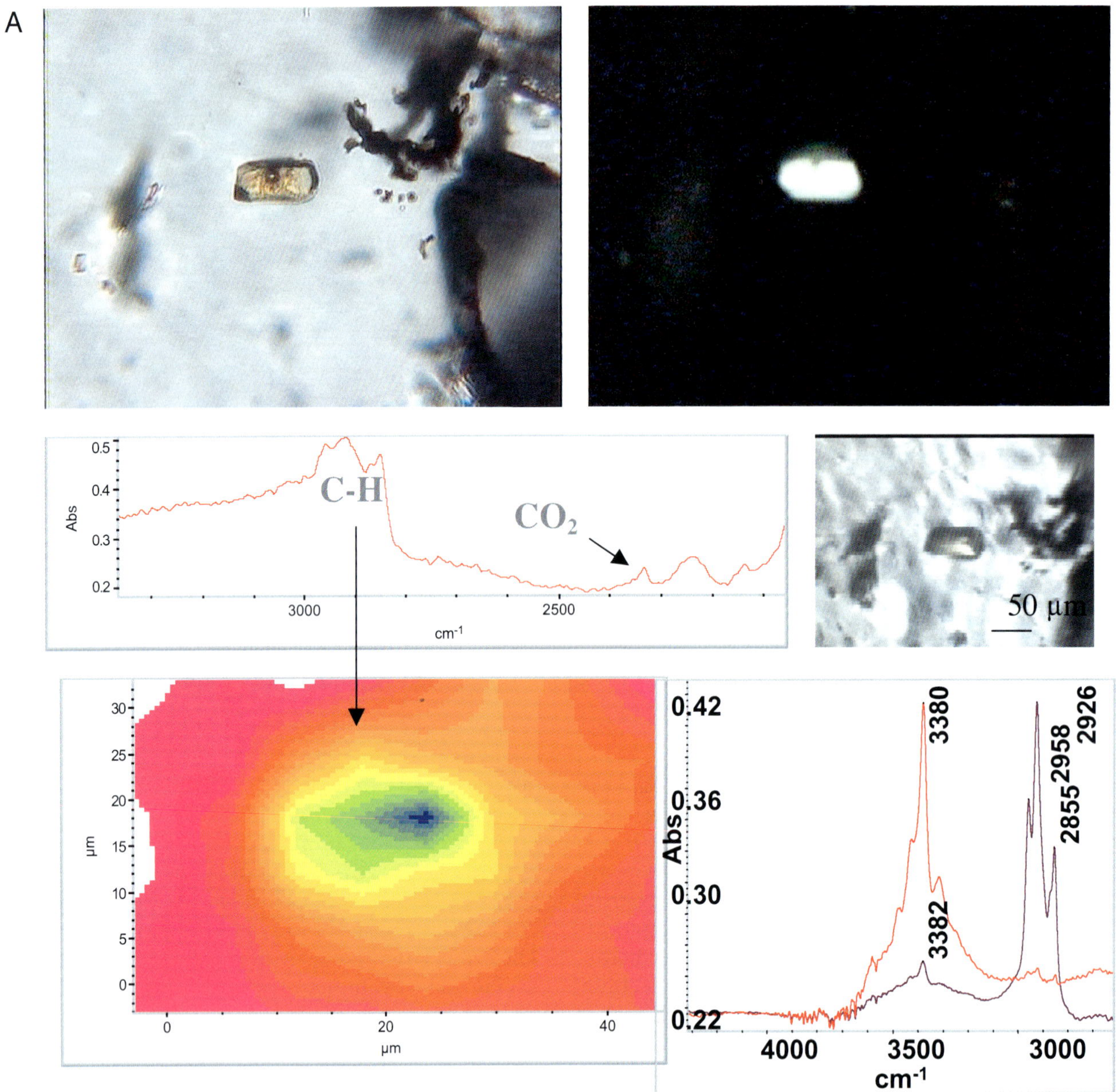

FIGURE 5. Synchrotron Fourier transform infrared spectroscopic microanalyses and mapping of primary hydrocarbon fluid inclusions. (A) Detection of CO_2 and alkanes into liquid oil phase and mapping of the CH_2–CH_3 in primary fluid inclusions in quartz. Water is not detected. On the right side, comparison with a typical spectrum of the quartz matrix shows OH in the defects (3380 cm^{-1}). (B) Fourier transform infrared spectra of the solid microphases S included in the flat primary oil fluid inclusions during the heating experiment. The CO_3^{2-} absorption at 2520 cm^{-1} recorded at room temperature is disappearing slowly until the homogenization temperature of 60°C is reached (T_h). The band reappears at the same point during cooling. (C) Microphothographies of primary oil fluid inclusions (transmitted and polarized light) in quartz and punctual FTIR analyses of a typical association of mechanically entrapped minerals. The bands are characteristics of a sulfate-hydrated mineral, hemihydrate(?) (solid, 1), and calcite (solid, 2) in the 2000–4000-cm^{-1} spectral range. Spectrum 2 in the 2800–3000-cm^{-1} range is composed of both alkane and carbonate absorptions. Absorption of oil nearby in spectrum 3 is shown for comparison in the same spectral range. The 3381-cm^{-1} absorption belongs to OH in the quartz matrix. Abs = absorbance.

Figure 5. (cont.).

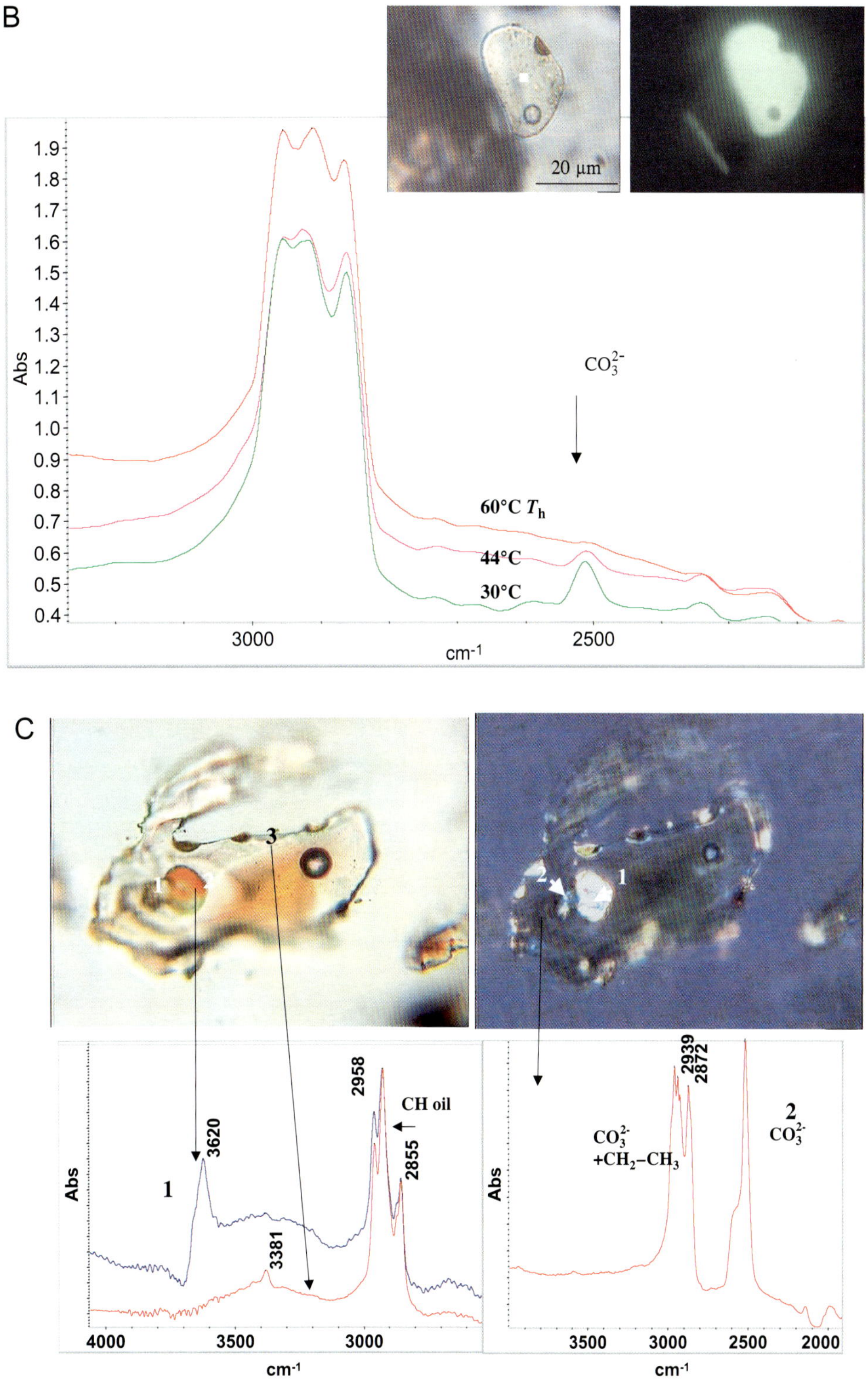

spectrum in isotopic $\delta^{18}O$ and Sr data reported by Benchilla (2003).

DISCUSSION

Two hydrocarbon-rich fluid circulations are identified during veining and diagenetic recrystallization. Both result from a homogeneous trapping of a single fluid system.

The first fluid was entrapped in large cavities in authigenic quartz and calcite cements. It is composed of a light oil enriched in CO_2. Water is surprisingly not detected, but OH is detected in the defects of the quartz matrix. This was soon observed in other samples from different occurrences (Guilhaumou et al., 1996). The injection of the CO_2 in previous oil may account for the asphaltene bodies that are systematically found inside these fluid inclusions only (never found in the host minerals) and are common in oil fluid inclusions from MVT deposits (reviewed in Roedder, 1984; Goldstein, 2001). The composition in mole percent derived from in-situ SFTIR microanalyses is nearly 30% CH_4, 60% alkanes, and 1% CO_2, with an average chain length of C_{10}–C_{12}, which is close to that of normal oil (Montel, 1993). Almost the same oil rich in CO_2 and without water detected is found in fluid inclusions hosted in recrystallized calcite, the overestimate of CH_4 being in the error margin in this case.

Some solid calcite minerals are identified inside primary fluid inclusions as mechanically entrapped and/or daughter minerals. Fourier transform infrared spectroscopy during heating runs until homogenization temperature unambiguously shows that some of these calcite "seeds," which are scattered inside the oil phase, are at least partly disappearing in the liquid phase at homogenization temperature. Some carbonates are also observed to be closely associated with hydrated sulfates, partly dissolved and probably

Table 1. Composition of hydrocarbon fluid inclusions by SFTIR for primary fluid inclusions and pseudosecondary alignments in authigenic quartz.

Reference	*Primary Fluid Inclusions*							*Pseudosecondary*		
	A1 (6)	A2 (8)	Q12 (6)	Q6 (5)	Q5P (5)	IF3 Ca(5)	Q18 (6)	IF4 S1	IF5 S2	IF6 S3
S H_2O	0	0	0	0	0	0	0	++	++	++
S1 CH_3/CH_2	0.65	0.57	0.69	0.70	0.70	0.81	0.58	0.61	0.61	0.63
S2 CH_2/CH_3	1.54	1.74	1.45	1.43	1.43	1.24	1.72	1.65	1.64	1.57
S CH_4/S alkanes	0.11	0.07	0.09	0.08	0.09	0.14	0.09	0.29	0.31	0.32
Chain length equiv.	C10	C10	C10–C12	C12	C12	C10	C12	C17–C19		
CH_4 (mol%)	37.33	30.36	33.00	30.00	31.00	41.00	34.00	62.61	64.13	63.29
CO_2 (mol%)	0.09	1.29	0.00	3.10	5.00	0.50	2.22	0	0	0

preserved in intermediate state by the formation of oil fluid inclusions by heterogeneous trapping during the dissolution-crystallization process. These results are of prime importance, because they identify at micrometer scale the replacement of sulfates by carbonates, which implies a mechanism of sulfatoreduction like BSR and/or TSR, whereby sulfates are reduced by hydrocarbons with a concomitant oxidation of the organic matter. The hypothesis of TSR is supported in this case by several arguments. Temperatures are slightly lower than 100°C, generally considered for the beginning of TSR conditions, but higher than 80°C, which is stated as the upper limit for bacteria in BSR. The TSR has no sharply defined minimum temperature (Worden et al., 1995; and Bildstein et al., 2001, considered that it may be effective at low temperature, e.g., 80°C). In this occurrence, TSR may have occurred in the range of 85–140°C, the upper limit being near the entrapment of the second generation of oil fluid inclusions in quartz (e.g., 140–160°C). The reactants are branched and n-alkanes, which are the main components of the fluid inclusions, instead of organic acid involved in BSR.

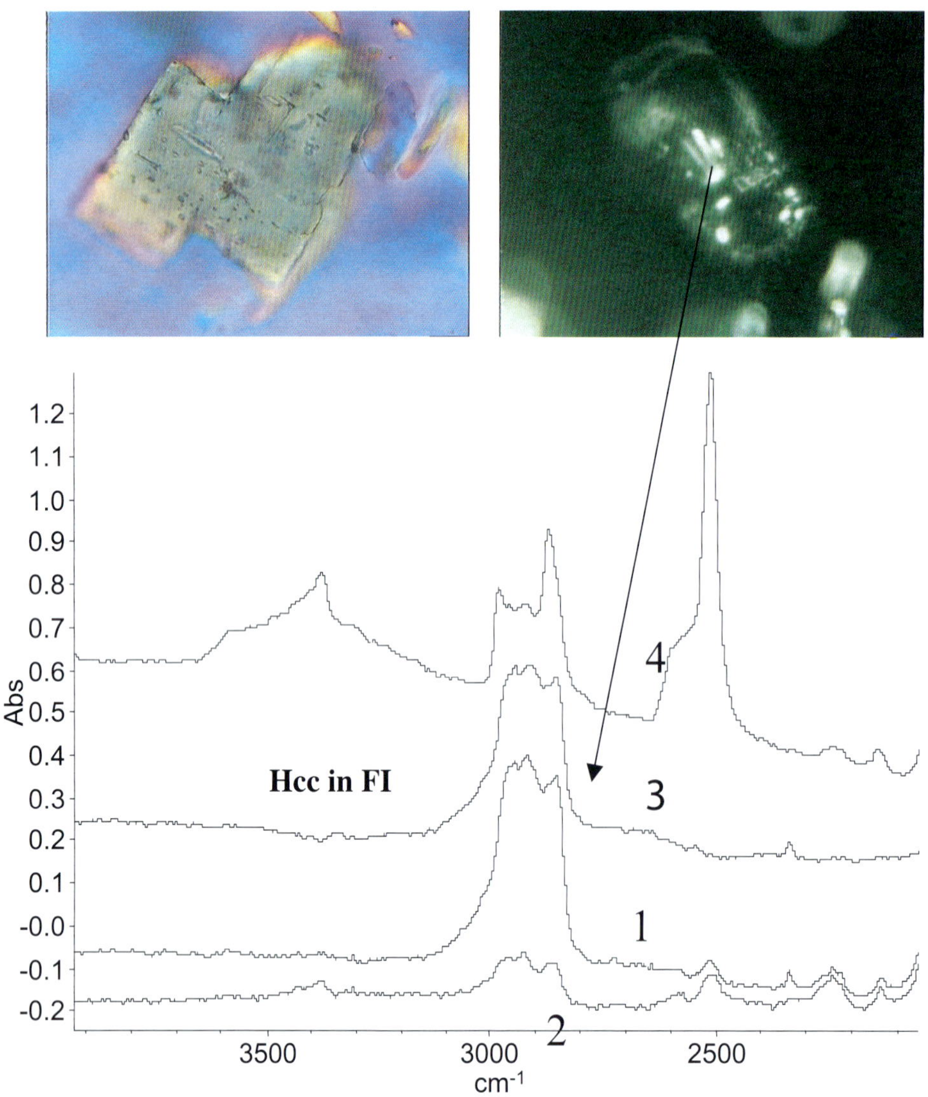

Figure 6. FTIR microanalysis and calculation of CH_2/CH_3 ratio from hydrocarbon fluid inclusions in recrystallized carbonates. The spectra of the matrix at the same intensity (2) are subtracted from the whole spectra (1) to obtain oil spectra in (3). In (4), the spectrum of the carbonate matrix nearby (uncalibrated) is shown. FI = fluid inclusions; Hcc = hydrocarbons.

FIGURE 7. FTIR microanalysis of the pseudosecondary oil fluid inclusions in quartz. Alkanes and a large amount of water are systematically detected. The alkanes mapping show their regular repartition in the quartz matrix.

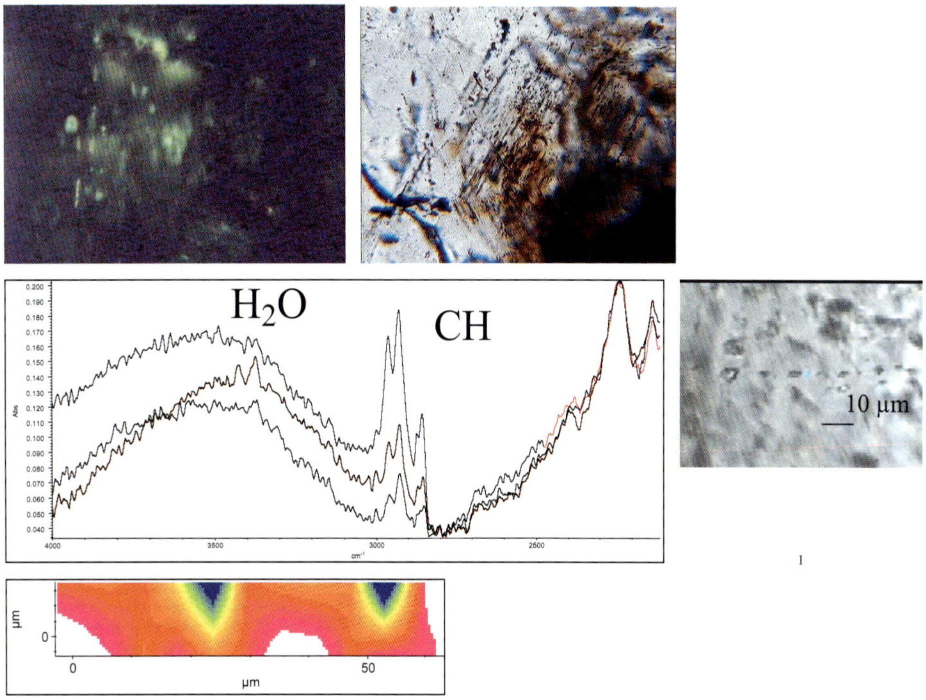

In this sample, water content is very low (undetected by FTIR), indicating that water release during the reaction is negligible (Machel, 2001). In TSR mechanisms, the main organic reactant is considered to be CH_4 (Bildstein et al., 2001). However, we know very little about the chemistry of such possible reactions with low water activity, because most of the TSR studied are located in the gas-water transition zone of petroleum reservoirs (gas cap), where hydrocarbons and large amount of water are present. An explanation could be that rapid fluid circulations induced in fractures along tectonic evolution, as observed in thrust belts, probably change the condition of the chemical reactions. As exploration is now developing in these zones, models considering these parameters will be very helpful.

The second fluid influx consists of a mixed water-oil solution enriched in methane and depleted in CO_2, with a composition of 60% CH_4 and an average chain length of C_{15}, thus indicating a more mature hydrocarbon source. This fluid is entrapped along parallel crystal growth in the outer parts of quartz crystals. It probably migrated when carbonates were stabilized and during a temperature increase of the circulating fluid, as attested by the high CH_4 content, the increase in chain length, and higher homogenization temperatures. The homogeneity of the content for the fluid-inclusion population evidenced from SFTIR mapping in these late hydrocarbon fluid inclusions does not support the hypothesis of a secondary in-situ modification process of fluid inclusions, such as stretching or refilling of old cavities through overheating and/or fracturing (Roedder, 1981; Goldstein and Reynolds, 1994).

PVTX Properties of the Hydrocarbon Fluid Inclusions

The liquid-vapor curve and isochoric P-T path of the oil content in primary hydrocarbon fluid inclusions have been calculated using a software developed at the Institut Français du Pétrole (Ungerer and Batut, 1997) and validated on Pakistani samples from Baluchistan by Guilhaumou et al. (2000) and Benchilla et al. (2003). The software uses a modified Peng Robinson equation of state to do PVT iterative runs to calculate the liquid-vapor curves and the fluid inclusion isochores from the oil composition analyzed by GCMS and the PVT properties of the oil fluid inclusions. These are later defined using the liquid/gas volume ratio and the two fitting parameters α and β controlling the natural hydrocarbon distribution in the database of Montel (1993). More details on the optimization procedure may be found in Benchilla et al. (2003). The first version has been adapted to consider only the semiquantitative composition derived from SFTIR microspectrometry as analytical data (instead of GCMS analyses). The C1 fraction measured by FTIR (Pironon et al., 2001) is used as an additional parameter to the liquid vapor volume ratio at room temperature measured by confocal microscopy (Guilhaumou et al., 2000). The two sets of data were checked and found to be in good agreement. Other models were developed by Aplin et al. (1999), and a similar approach has been developed recently by Thiery et al. (2000). In this new version of the model, we tried to be closer to the experimental data currently obtained by in-situ nondestructive analyses. We introduced the alkane equivalent, i.e., CH_2/CH_3 ratio, the experimental percentage of CO_2 and CH_4, and a fixed aromatic percentage of 25% corresponding to the alkane equivalent greater than C_5 (database, Montel, 1993). The phase envelopes and isochores have been calculated for seven primary hydrocarbon fluid inclusions hosted in

TABLE 2. Experimental data used for isochoric modeling of the primary hydrocarbon fluid inclusions and resulting distribution parameters. The percent of gas is measured by UV confocal microscopy. Homogenization temperatures T_h are also indicated. In inclusion A, the gas volume is overestimated because of the large amount of solid phases inside the fluid inclusions (FI).

Inclusion	*Volume of Gas (μm³)*	*Volume of FI (μm³)*	*Volume of Oil + Bitumen*	*% Gas*	*T_h (°C)*	*CH_4 (%)*	*CO_2 (%)*	*R* CH_2/CH_3 n-Alkanes (equ.)*
E	64	1977	1913	3.35	60	29.7	1.13	1.45 C9–10
C (2)	101	2632	2531	3.99	65	31.8	1.2	1.43 C9–10
G	15	869	854	1.76	58	26	0.4	1.56 C10
D	38	2595	2557	1.49	56	27	0.4	1.62 C10
B	17	1478	1461	1.16	60	30	1.1	1.49 C9–10
A**	35	544	510	6.86	68	36	0.5	1.81 C12
C	47	3703	3656	1.29	55	26	0.8	1.49 C9–10

R = area of CH_2 absorption/area of CH_3 absorption from spectra registered in fluid inclusions. This is expressed after internal calibration (Guilhaumou et al., 1998) as an equivalent of the hydrocarbon chain length because it could be compared with the database made by gas chromatography–mass spectrometry for oil types and compositions by petroleum chemists.

***Sure estimated gas volume.*

different quartz crystals analyzed by SFTIR, with the data reported in Table 2. The volume ratios were measured by confocal microscopy with an accuracy of 0.1%. The α and β distribution parameters obtained from the data above in Chorgali led to a P-T path showing a slightly different phase envelope (cricondenbar) for inclusion A but a similar isochoric evolution for all fluid inclusions in quartz (Figure 8). However, the liquid/vapor ratio and CH_4 amount in case A may be overestimated because of the high asphaltene content.

Conditions of Fluid Circulations in Hydrofractures

Thrustpack® modeling led us to consider that the temperatures of 80 ± 10°C corresponding to trapping temperatures of primary hydrocarbon fluid inclusions were reached during the lower Siwalik (Benchilla, 2003). The burial would indicate corresponding pressures of 150–180 bar. Therefore, the entrapment of this fluid and quartz and calcite precipitation would have occurred at the onset of fluid migration and initial conditions close to 75–85°C and 150–180 bar (first hydrocarbon fluid inclusions entrapment), along with the replacement of sulfate by calcite. The hypothesis of sulfate reduction and calcite precipitation is supported by the presence of solid calcium sulfates closely associated to calcite inside hydrocarbon fluid inclusions. The association of sulfates and organic matter in sediments is thermodynamically unstable, and the branched and n-alkanes are considered as the main reactants for TSR mechanism in all diagenetic environments (Pierre and Rouchy, 1988; Machel, 2001). Along a second input, the fluids reached minimal temperatures of 125°C (T_h of the second hydrocarbon fluid inclusions entrapment) during the second episode of peak oil generation and migration within water influx and late quartz precipitation. Temperatures as much as 150°C measured in aqueous fluid inclusions and low values of carbon and oxygen isotopes signatures for calcite in the hydroveins reported by Benchilla (2003) are in favor of TSR mechanism (Bildstein et al., 2001; Machel, 2001). The conditions agree for the second fluid input, the mechanism of the first input, and the beginning of sulfate-calcite recrystallization process, with very low amount of water presently being not well understood. The origin of CO_2 in primary oil generation may come from either the decarbonation of early calcite or the decomposition of the organic matter itself. High content of CH_4 and CO_2 depletion for the second generation of hydrocarbon fluid inclusions suggest a

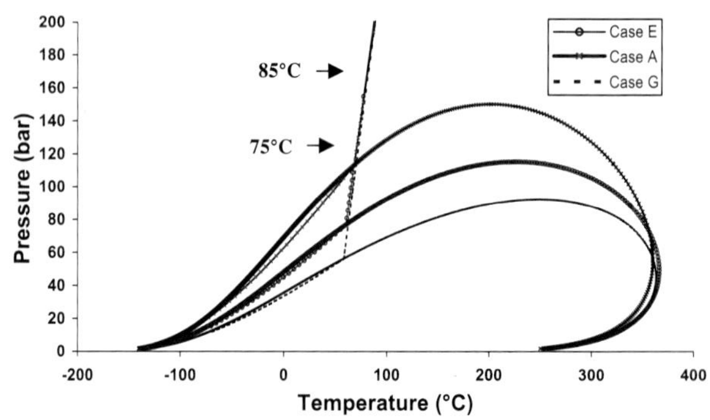

FIGURE 8. Results of the isochoric modeling of primary oil fluid inclusions. Three main groups of fluid inclusions may be defined from their liquid vapor curves calculated from their chemical compositions and volumetric properties (Table 2); the isochores have the same P-T slopes but initiate at different pressures values. The entrapment P-T conditions fall in the range 80 ± 5°C and 165 ± 15 bar.

reducing environment. This is in agreement with low values of $\delta^{18}O$ for the hydraulic veins (−7.8 to −14.1‰), which represent high temperatures of crystallizations for calcite from meteoric water, i.e., 90–150°C. The low values of $\delta^{13}C$ (−7.81 to −14.55‰) in calcite veins reflect the conditions of TSR (Benchilla, 2003).

CONCLUSION

The combination of in-situ micro-FTIR analyses, microthermometry, and PVT modeling of the successive generations of hydrocarbon-rich fluid inclusions entrapped in the hydroveins of the Chorgali formation supports the hypothesis of hydrofracturing and diagenetic veining during the reduction of sulfates and calcite precipitation, along with an increase in temperature. The onset at near 75–85°C and 150–180 bar and final temperatures as much as 150–170°C suggests a TSR mechanism instead of a BSR. However, the two mechanisms may have followed each other along successive inputs of fluids in the hydroveins. The late production of CH_4, the depletion in $\delta^{13}C$ (−20 and −36‰) measured in the veins, and the crystallization of saddle dolomite (Benchilla, 2003) favor the TSR mechanism. The reactants may be mostly n-alkanes during the first input of light oil rich in CO_2, having very low water content, and then a mixture of more mature hydrocarbons enriched in CH_4 and acid waters coming from compaction and meteoric infiltration during the second input at temperatures higher than 140°C. The very low content of water (undetected by FTIR) in quartz cements and the presence of hydrated sulfates partially transformed in carbonates in primary oil fluid inclusions suggest that these reactions and quartz precipitations may occur when oil becomes present in the reservoir. Such quartz precipitations in the presence of organic fluid depleted with water were observed in diagenetic quartz formed in the Terres Noires (southeastern France; Guilhaumou et al., 1990, 1996).

The correlation of these data with Thrustpack® modeling (Benchilla, 2003) shows that hydrofracturing was initiated at 1.5 km (1 mi) depth and was reached at 15–10.8 Ma. It lasted until 4–6 km (2.5–4 mi) depth was reached at −5.7 Ma during the development of the thrust sheet in Dock Patan formation. The fluids appear in both cases in thermal equilibrium with the carbonaceous sediments around and are driven mainly by tectonics from the hinterland farther to the north to the Chorgali Pass. The origin of the oil in former circulation has to be found outside the Paleogene, in which organic matter was not mature. One hypothesis could be the migration of the oil from the deep Mesozoic sediments, where the source rock is known as being mature at near 15–10.8 Ma. The later hydrocarbon fluid entrapment in quartz occurred when the sediments were in the oil window in the whole basin.

ACKNOWLEDGMENTS

We gratefully acknowledge Thierry Marin of the Laboratoire pour l'Utilisation du Rayonnement Électromagnétique for helpful technical assistance, Adriana Dutkiewitz, University of Sydney (Australia), for the help in doing confocal microscopy, and François Roure for the support and fruitful discussions. Tarik Jaswal and Khursheed Akhtar from the Oil and Gas Development Company are acknowledged for the sampling and financial support. Etienne Brosse did a very constructive review of the first draft of the chapter.

REFERENCES CITED

Aplin, A. C., G. Mcleod, S. R. Larter, K. S. Pedersen, H. Sorensen, and T. Booth, 1999, Combined use of confocal laser microscopy and PVT simulation for estimating the composition and physical properties of petroleum fluid inclusions: Marine and Petroleum Geology, v. 16, p. 97–110.

Barker, C., and W. D. Underwood, 1992, Analysis of gases in individual fluid inclusions in natural minerals using a dual mass spectrometer system: The Analyst, v. 117, p. 1407–1411.

Barres, O., A. Burneau, J. Dubessy, and M. Pagel, 1987, Application of micro FTIR spectroscopy to individual hydrocarbon fluid inclusion analysis: Applied Spectroscopy, v. 41, no. 6, p. 1000–1008.

Benchilla, L., 2003, Circulation des fluids et diagenèse des réservoirs carbonates et gréseux dans les fronts des chaînes plissées et leur avant pays: Le cas du Salt Range— Potwar (Pakistan): Ph.D. thesis, University of Leuven, Belgium, 160 p.

Benchilla, L., N. Guilhaumou, P. Mougin, T. Jaswal, and F. Roure, 2003, Paleoburial and pore pressure reconstruction of reservoir rocks in Foothille areas: A sensitivity test in the Hammam Zriba (Tunisia) and Koh-I-Maran (Pakistan) ore deposit: Geofluid, v. 3, p. 103–123.

Berreby, L., 1983: Ph.D. thesis, Paris VI, France, p. 85.

Bildstein, O., R. H. Worden, and E. Brosse, 2001, Assessment of anhydrite dissolution at the rate-limiting step during thermochemical sulfate reduction: Chemical Geology, v. 176, p. 173–189.

Colthup, N. B., L. H. Daly, and S. E. Wiberly, 1990, Major spectral structure correlation by spectral regions, Introduction to infrared and Raman spectroscopy, 3d ed.: San Diego, Academic Press, p. 547.

Farmer, V. C., 1974, The infrared spectra of minerals: London, Mineralogical Society.

George, S. C., P. J. Eadington, M. Lisk, and R. A. Quezada,

1998, Constraining the oil charge history of the South Peper oil field from the analysis of oil bearing fluid inclusions: Organic Geochemistry, v. 29, p. 631–648.
Goldstein, R. H., 2001, Fluid inclusions in sedimentary and diagenetic systems: Lithos, v. 55, p. 159–193.
Goldstein, R. H., and T. J. Reynolds, 1994, Systematics of fluid inclusions in diagenetic minerals: Society for Sedimentary Geology Short Course 31, 199 p.
Guilhaumou, N., N. Szydlowski, and B. Pradier, 1990, Characterization of hydrocarbon fluid inclusions by infrared and fluorescence microspectrometry: Mineralogical Magazine, v. 54, p. 311–324.
Guilhaumou, N., J. C. Touray, V. Parthuisot, and F. Roure, 1996, Paleocirculation in the basin of southeastern France, sub-Alpine Range: A synthesis from fluid inclusions studies: Marine and Petroleum Geology, v. 13, no. 6, p. 695–706.
Guilhaumou, N., P. Dumas, G. L. Carr, and G. P. Williams, 1998, Synchrotron infrared microspectrometry applied to petrography in the micron-scale range: Fluid chemical analysis and mapping: Applied Spectroscopy, v. 52, no. 8, p. 1029–1034.
Guilhaumou, N., N. Ellouz, T. Jaswall, and P. Mougin, 2000, Genesis and evolution of hydrocarbon entrapped in fluorite deposits of Koh-i-Maran (North Kirthar Range, Pakistan): Marine and Petroleum Geology, v. 17, p. 1151–1164.
Jones, D. M., and G. Macleod, 2000, Molecular analysis of petroleum fluid inclusions: A practical methodology: Organic Geochemistry, v. 31, p. 1163–1173.
Karlsen, D., T. Nedkvitne, S. R. Larter, and K. Bjørlykke, 1993, Hydrocarbon composition of authigenic inclusions: Applications to elucidation of petroleum reservoir filling history: Geochimica et Cosmochimica Acta, v. 57, p. 3641–3659.
Korashani, G. K., 1986, Novel development in fluorescence microscopy of complex organic mixtures: Application in petroleum geochemistry: Organic Geochemistry, v. 11, no. 3, p. 157–168.
Larroque, C., N. Guilhaumou, J. F. Stephan, and F. Roure, 1996, Large scale advection of fluids at the front of the Sicilian Neogene subduction complex: Tectonophysics, v. 254, nos. 1–2, p. 41–56.
Legodi, M. A., D. de Waal, J. K. Potgieter, and S. S. Potgieter, 2001, Rapid determination of $CaCO_3$ in mixtures utilizing FTIR spectroscopy: Mineral Engineering, v. 14, no. 9, p. 1107–1111.
Machel, H. G., 2001, Bacterial and thermochemical sulfate reduction in diagenetic settings. Old and new insights: Sedimentary Geology, v. 140, p. 143–175.
Mandal, P. K., and T. K. Mandal, 2002, Anion water in gypsum ($CaSO_4{\cdot}2H_2O$) and hemihydrate ($CaSO_4{\cdot}H_2O$): Cement and Concrete Research, v. 32, p. 313–316.
McLimans, R. K., 1987, The application of fluid inclusions to migration of oil in diagenesis in petroleum reservoirs: Applied Geochemistry, v. 2, p. 585–603.
Montel, F., 1993, Phase equilibria needs for petroleum exploration and production industry: Fluid Phase Equilibria, v. 84, p. 343–367.
Muchez, P., and M. Slobodnik, 1996. Recognition and significance of multiple fluid inclusion generations in telogenetic calcites: Mineralogical Magazine, v. 60, p. 813–819.
Murray, R. C., 1957, Hydrocarbon fluid inclusions in quartz: AAPG Bulletin, v. 41, p. 950–956.
Parry, W. T., D. Hedderly-Smith, and R. L. Bruhn, 1991, Fluid inclusions and hydrothermal alteration on the Dixie Valley fault, Nevada: Journal of Geophysical Research, v. 96, p. 19,733–19,748.
Pierre, C., and J. M. Rouchy, 1988, Carbonate replacements after sulfate evaporites in the middle Miocene of Egypt: Journal of Sedimentary Petrology, v. 58, no. 3, p. 446–456.
Pironon, R., R. Thiery, M. Ayt Ougougdal, S. Teinturier, S. Baudouin, and F. Walgenwitz, 2001, FTIR measurements of petroleum fluid inclusions: Methane, n-alkanes and carbon dioxide quantitative analysis: Geofluids, v. 1, p. 2–10.
Rankin, A. H., B. L. Hodge, and M. Moser, 1990, Unusual oil-bearing inclusions in fluorite from Baluchistan, Pakistan: Mineralogical Magazine, v. 54, p. 335–342.
Richardson, C. K., and D. M. Pinckney, 1984, The chemical and thermal evolution of the fluid in Cave in Rock fluorspar district, Illinois: Mineralogy, paragenesis and fluid inclusions: Economic Geology, v. 79, p. 1833–1836.
Roedder, E., 1981, Origin of fluid inclusions and changes that occur after trapping, *in* L. S. Hollister and M. L. Crawford, eds., Short course in fluid inclusions: Application to petrology: Mineralogical Association of Canada, 644 p.
Roedder, E., 1984, Fluid Inclusions: Reviews in Mineralogy, American Mineralogical Society, v. 12, 644 p.
Roure, F., W. Sassi, E. Lafargue, and M. Vanderbrouke, 1999, Potential source rock and hydrocarbon products of the Salt Range and Potwar Basin (Pakistan): Institut Français du Pétrole internal report, p. F4140.
Rudkiewicz, J. L., O. Brévart, J. Connan, and F. Montel, 1994, Primary migration behaviour of hydrocarbons: From laboratory experiments to geological situation through fluid flow models: Organic Geochemistry, v. 22, nos. 3–5, p. 631–639.
Salisbury, J. W., L. S. Walter, N. Vergo, and D. M. D'Aria, 1991, Infrared (2–15 μm) spectra of minerals: Baltimore, Maryland, John Hopkins University Press, 386 p.
Thiery, R., J. Pironon, F. Walgenwitz, and F. Montel, 2000, PIT (Petroleum inclusion Thermodynamic): A new modelling tool for the characterization of hydrocarbon fluid inclusions from volumetric and microthermometric measurements: Journal of Geochemical Exploration, v. 69–70, p. 701–704.
Ungerer, P., and C. Batut, 1997, Prédiction des propriétés volumétriques des hydrocarbures par une translation de volume améliorée: Revue de l'Institut Français du Pétrole, v. 52, p. 609–623.
Worden, R. H., P. C. Smalley, and N. H. Oxtoby, 1995, Gas souring by thermochemical sulfate reduction: AAPG Bulletin, v. 79, no. 6, p. 853–871.

2

Osadetz, K. G., B. P. Kohn, S. Feinstein, and R. A. Price, 2004, Foreland belt thermal history using apatite fission-track thermochronology: Implications for Lewis thrust and Flathead fault in the southern Canadian Cordilleran petroleum province, *in* R. Swennen, F. Roure, and J. W. Granath, eds., Deformation, fluid flow, and reservoir appraisal in foreland fold and thrust belts: AAPG Hedberg Series, no. 1, p. 21–48.

Foreland Belt Thermal History Using Apatite Fission-track Thermochronology: Implications for Lewis Thrust and Flathead Fault in the Southern Canadian Cordilleran Petroleum Province

Kirk G. Osadetz
Natural Resources Canada, Earth Sciences Sector, Geological Survey of Canada— Calgary, Alberta, Canada

Barry P. Kohn
School of Earth Sciences, University of Melbourne, Victoria, Australia

Shimon Feinstein
Department of Geological and Environmental Sciences, Ben Gurion University of the Negev, Beer Sheva, Israel

Raymond A. Price
Department of Geological Sciences and Geological Engineering, Queen's University, Ontario, Canada

ABSTRACT

Apatite fission-track (AFT) data from rocks above and below Lewis thrust fault lying in the footwall and hanging wall of Flathead normal fault record different thermal-history components, depending on individual structural and stratigraphic positions. Apatite fission-track temperature-history models (THMs) indicate that rapid cooling of the Lewis thrust sheet began at about 75 Ma. This cooling coincided with major displacement on the Lewis thrust. Subsequently, folding of the Lewis thrust sheet by underlying thrust duplex culminations formed the Akamina syncline, and a fossil AFT partial annealing zone was superimposed on the syncline. Apatite fission-track data from east of the Flathead graben record a subsequent cooling event during the middle Eocene onward

DOI:10.1306/1025684H13111

that was coeval with extensional displacement on the Flathead fault and with accompanying uplift and erosion of its footwall. Apatite fission-track data from lower Oligocene sediments in the Flathead graben preserved the temperature history of the sediment source regions in the Lewis thrust sheet without significant subsequent annealing. A set of similar THMs that are consistent with the regional structural history can account for observed variations in AFT parameters at various levels, which are exposed in the Lewis thrust sheet and are penetrated below the thrust sheet by deep wells. From the onset of displacement on the Lewis thrust until the early Oligocene, paleogeothermal gradients in the thrust sheet (8.6–12°C/km) were lower than present values (~17°C/km). The changes in geothermal gradients are attributed to advective heat transfer by tectonically induced, topographically driven, deeply penetrating meteoric water flow. This is a complicated heat-transfer mechanism that can affect organic maturation history and petroleum systems in overthrust belts.

INTRODUCTION

The structural stacking inherent in the formation of fold and thrust belts generates topography, induces tectonic burial, and creates structures in which petroleum may accumulate (Dahlstrom, 1970). Moreover, it influences the hydrodynamic regime and thereby the geothermal field (Majorowicz et al., 1985), which controls chemical reactions governing diagenesis (Ghent and Miller, 1974; Hoffman et al., 1976; Hoffman and Hower, 1979; Hutcheon et al., 1980), coalification of organic matter (Hacquebard and Donaldson, 1974; England, 1984; Bustin and England, 1989), and the evolution of petroleum systems (Faure et al., 2004). The temporal relationship between orogenic deformation and thermally induced chemical reactions controls economically important processes in fold and thrust belts, including the generation (Angevine and Turcotte, 1983; Edman and Surdam, 1984), migration (Furlong and Edman, 1984), accumulation (Dahlstrom, 1970; Fermor and Moffat, 1992), and alteration of petroleum (Orr, 1974; Krause et al., 1988). Petroleum resource variations along the Canadian Cordilleran Foreland belt (Stockmal et al., 2001) (Figure 1) show the important influence of thermotectonic history on petroleum resource potential.

In this study, we have used apatite fission-track (AFT) data from the southern Canadian Cordillera to constrain temperature-history models (THMs) of this prolific petroleum province, in which more than 900 billion m^3 of initial in-place raw gas have been discovered in more than 680 pools (Stockmal et al., 2001). Our study was performed in the vicinity of Waterton field, the largest petroleum field in the southern Canadian foreland belt. Extensive and continuing petroleum exploration in this region has provided a constrained model of the structure and tectonic evolution (Bally et al., 1966; Fermor and Moffat, 1992) (Figure 2).

The study area includes rocks occurring both above and below the Lewis thrust fault in the Flathead Valley–Crowsnest Pass region of the southern Canadian foreland fold and thrust belt (Figure 2). The analysis of AFT data from this region contributes to the resolution of questions regarding times of fault displacement and the linkages among deformation, geothermics, and hydrodynamics and their affect on petroleum systems. Temperature-history models inferred from AFT data indicate that specific rapid-cooling events or intervals affected specific structural elements and stratigraphic levels at specific times. We have inferred that these cooling events or intervals are related to large displacements on major faults and to the associated changes in the hydrodynamic regime and geothermal field. Thermal-history models, combined with the constraints implicit in structural and stratigraphic data and in variations in levels of prethrusting coalification, indicate that there has been a significant decrease in geothermal gradient since thrusting began, evidently because of topographically driven advective heat transfer, a process that is still active today (Majorowicz et al., 1985).

GEOLOGICAL AND GEOTHERMAL SETTING

The Canadian Rocky Mountain foreland fold and thrust belt (Gabrielse and Yorath, 1992) is a northeastward-tapering deformational belt. It consists of Mesoproterozoic, Paleozoic, and Mesozoic strata from several different depositional basins that have been scraped off the underriding North American craton and accreted to the overriding Intermontane terrane during the Late Jurassic to Paleocene convergence between them (Monger and Price, 1979; Price, 1981, 1994; McMechan and Thompson, 1989, 1993). The geological structure in the deformational belt is dominated by thrust faults, most of which are listric and northeasterly or easterly verging. The thrust faults commonly follow long, bedding-parallel detachments that are separated by ramps along which the faults change their stratigraphic level rapidly (Douglas,

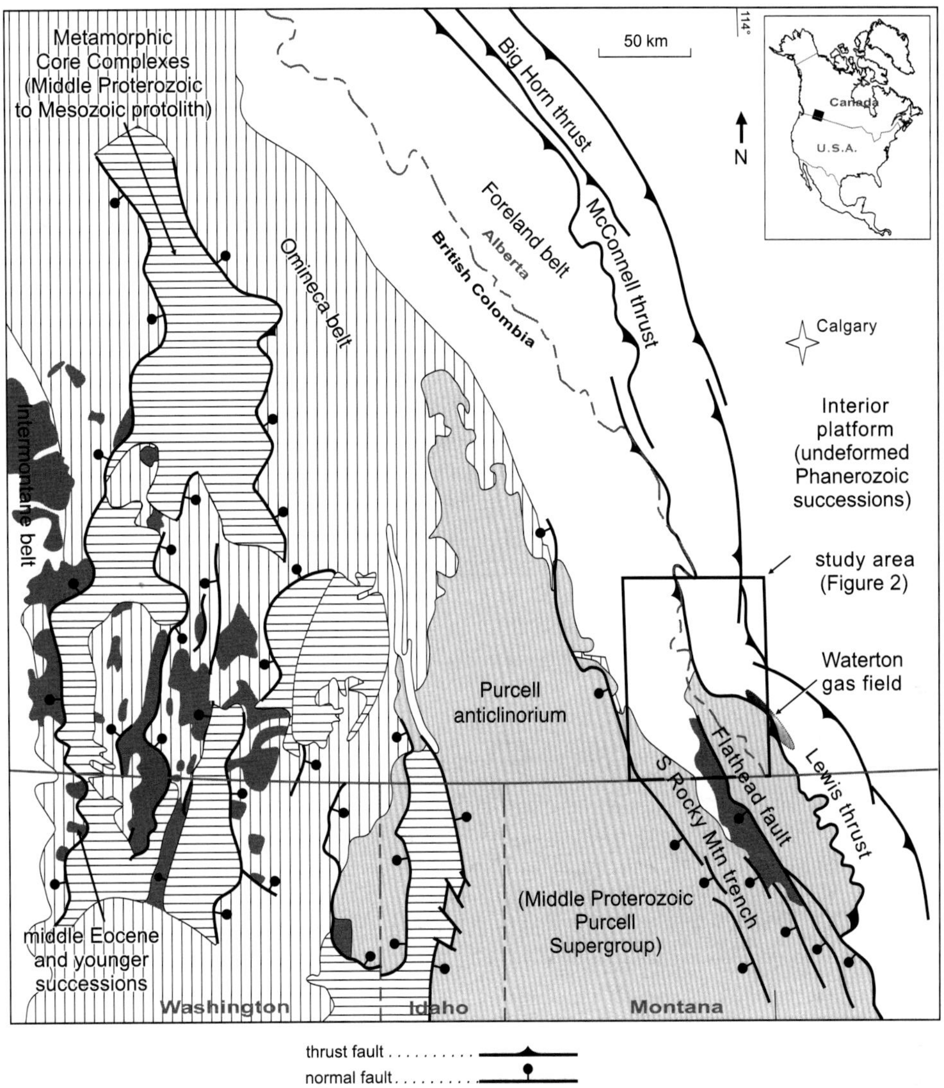

FIGURE 1. A location map that shows the regional tectonic and geologic setting of the study area (outlined by the box).

1950; Bally et al., 1966; Dahlstrom, 1970; Fermor, 1999). The thrusts merge downward with a basal detachment that converges with the contact between the sedimentary rocks and the underlying crystalline basement. The westward-dipping basal detachment extends into the Cordilleran metamorphic core at mid-crustal levels (e.g., Price and Fermor, 1985). The thrust faults generally cut up through the stratigraphic succession in the direction of displacement of the hanging wall, and they generally juxtapose older over younger strata, causing stratigraphic repetitions. The supracrustal cover has been vertically thickened and horizontally shortened.

The total horizontal shortening across the foreland fold and thrust belt is localized mainly on a small number of major thrust faults with large displacement. One of the largest of these is the Lewis thrust (Figures 1, 2). The Lewis thrust is more than 225 km (140 mi) long; it has a maximum displacement of more than 80 km (50 mi); and it juxtaposes a more than 8-km (5-mi)-thick Mesoproterozoic to Cretaceous succession (Figure 3) over Cretaceous and older strata (Price, 1964, 1965; Mudge and Earhart, 1980; Fermor and Moffat, 1992; van der Velden and Cook, 1994; Sears, 2001). Fault bend and detachment folding occurred in conjunction with the thrusting, and fold-related strain varies along strike with stratigraphically controlled changes in mechanical properties of the stratigraphic succession. In the southern foreland belt, folds make a relatively small contribution to the overall horizontal shortening; however, many thrust faults are folded with underlying strata as a result of fault bend and detachment folding. Thrusting and folding in the northeastward-tapering wedge of supracrustal rocks along the western flank of the North American craton increased its thickness by more than 10 km (6 mi). The resulting tectonic load produced a lithospheric flexure in which as much as 8 km (5 mi) of Upper Jurassic, Cretaceous, and Paleocene foreland basin sediments accumulated concurrently with the deformation (Price, 1973; Beaumont, 1981; Mack and Jerzykiewicz, 1989; Peper, 1993). More and more foreland basin deposits were incorporated into the fold and thrust belt as it expanded toward the craton.

Late Cretaceous and Paleocene thrusting in the southern Canadian Rocky Mountains foothills and eastern Front Ranges was associated with oblique, right-hand convergence between the Intermontane terrane and the North American craton (Price, 1994). In the south-central Canadian Cordillera, the oblique right-hand transpression was replaced by late Paleocene to middle Eocene right-hand transtension (Ewing, 1980; Price, 1994), involving conspicuous east–west crustal extension and the tectonic exhumation of midcrustal metamorphic rocks, including Archean and Paleoproterozoic basement rocks (Carr et al., 1987; Parrish et al., 1988; Doughty and Price, 1999, 2000). This change from regional horizontal compression to regional horizontal extension has been inferred to mark the end of both

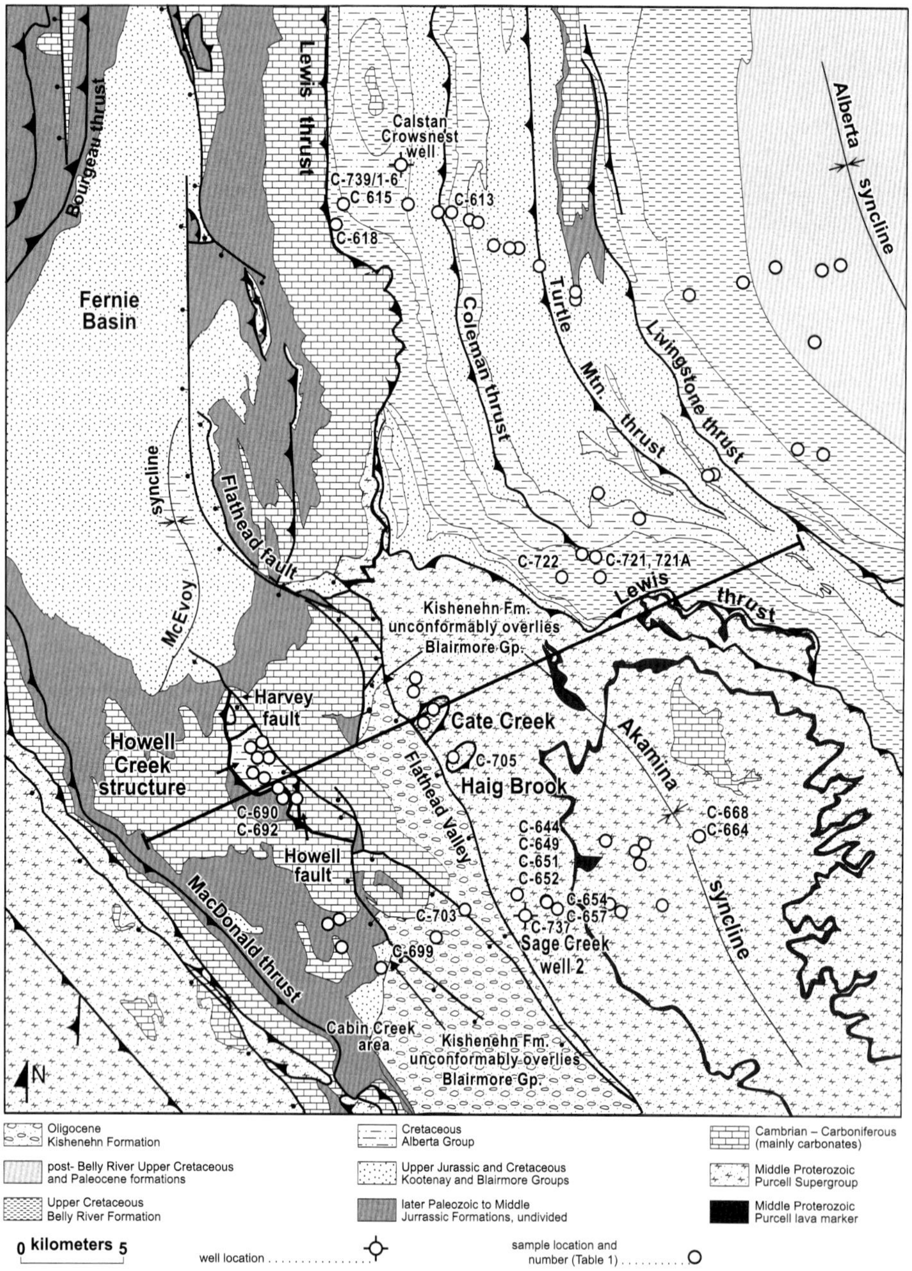

FIGURE 2. Simplified geological map of the study area. Sample locations are shown as circles, some of which are keyed to sample numbers in Table 1. The southeast corner of the map is at long. 114.0°W, lat. 49.0°N. The line of structure sections (Figures 4, 5) is shown as a heavy bar. Simplified stratigraphic units are identified in the legend. The map is a simplification of the geological map (Price, 1962), which should be consulted for details.

thrusting and folding in the foreland belt, and the flexural subsidence of the foreland basin (Price, 1994).

Late Paleocene–middle Eocene and younger crustal extension and the uplift of midcrustal metamorphic core complexes in the Omenica belt (Carr, 1992; Doughty and Price, 1999, 2000; Lorencak et al., 2001; Vanderhaeghe et al., 2003) was accompanied by a westward-increasing erosional exhumation of the foreland fold and thrust belt and the adjacent margin of the undeformed interior platform. The thickness of the strata that eroded increases westward from about 2 to 3 km (1 to 2 mi) at the eastern edge of the fold and thrust belt (Magara, 1976; Hacquebard, 1977; Nurkowski, 1984; England and Bustin, 1986a; Issler et al., 1990; Majorowicz et al., 1990) to more than 8 km (5 mi) in the uplifted footwall of the Flathead normal fault (Figures 1–5). Farther west, in the metamorphic core complexes of the Omineca crystalline belt (Figure 1), isostatic uplift of midcrustal rocks and the accompanying extensional exhumation is commonly more than 25 km (16 mi), and the basal detachment of the Cordilleran foreland fold and thrust is exposed locally at more than 2 km (1 mi) above sea level (Parrish et al., 1988; Carr, 1992).

Upper Cretaceous strata occur in the footwall of the Lewis thrust along most of the eastern edge of the thrust sheet in southwestern Alberta and adjacent Montana (Price, 1962; Mudge and Earhart, 1980) and also in windows eroded through the Lewis thrust adjacent to the Flathead fault at Cate Creek and Haig Brook in southeastern British Columbia (Figure 2) (Price, 1962). Thus, in the area north and east of the Cate Creek and Haig Brook windows, the footwall of the Lewis thrust evidently follows a detachment zone in Campanian–Maastrichtian strata (Fermor and Price, 1987; Fermor and Moffat, 1992; Jerzykiewicz et al., 1996). Northeast of the Flathead fault, the Lewis thrust fault and the strata in its hanging wall have been deformed into a large fault-bend fold (the Akamina syncline) by two antiformally stacked, thrust duplex culminations involving underlying Paleozoic carbonate rocks (Bally et al., 1966; Yin and Kelty, 1991; Boyer, 1992). The duplex structures contain large petroleum accumulations (Fermor and Moffat, 1992). The Akamina syncline, the duplex culminations, and the entrapment of the petroleum

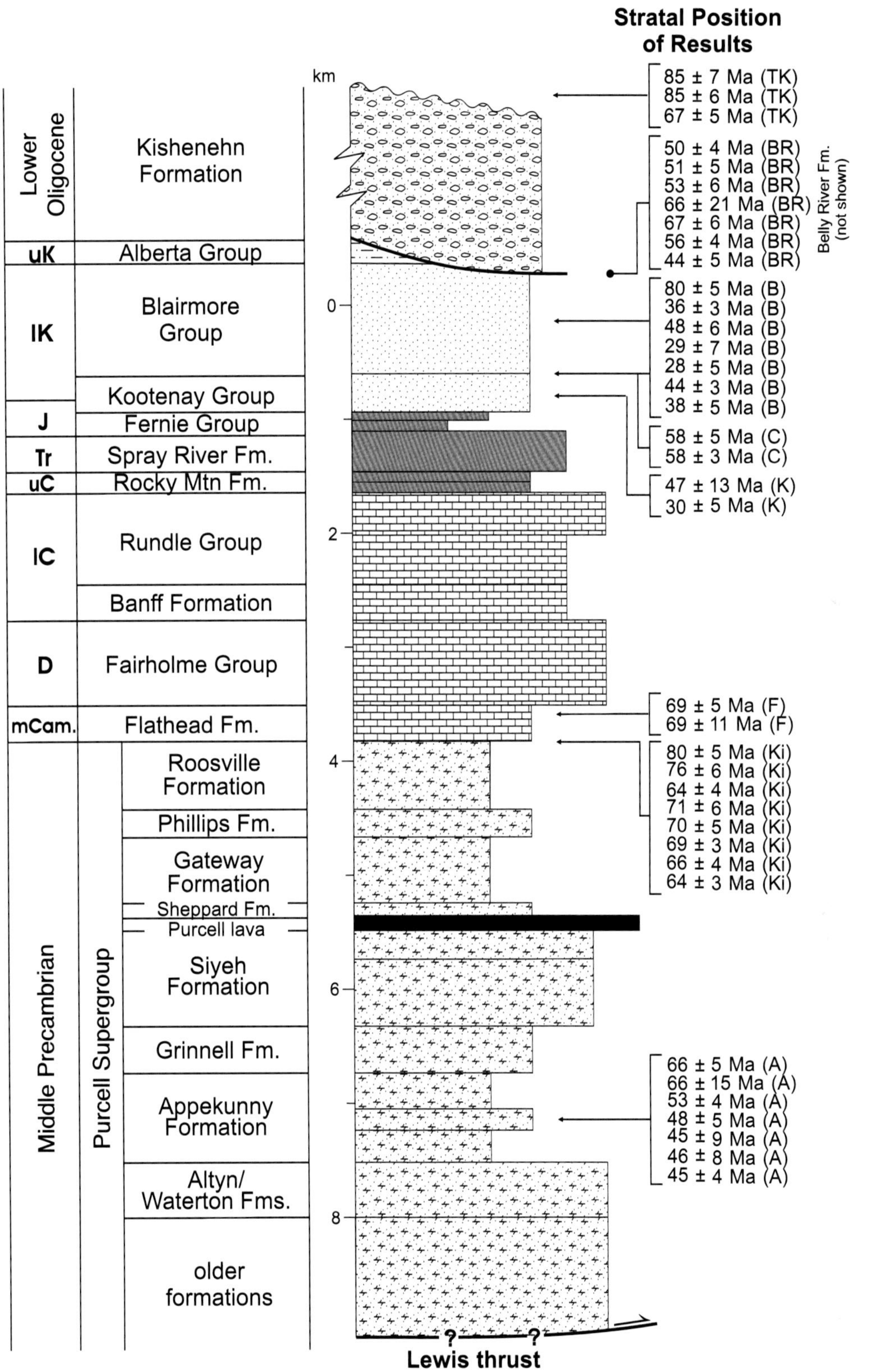

FIGURE 3. Schematic stratigraphic column for Lewis thrust sheet. Thickness data are from the hanging wall (Fairholme Group and younger) and footwall (Flathead Formation and older) of the Flathead fault. Ornamentation of stratigraphic units is as in Figure 2. The stratigraphic positions and calculated AFT ages of samples from Table 1 and Figure 2 are identified. Abbreviated stratigraphic unit names: TK = Kishenehn Formation; BR = Belly River Formation; B = Blairmore Group; C = Crowsnest volcanics; K = Kootenay Group; F = Flathead Formation; Ki = middle Cretaceous hypabyssal alkaline igneous intrusive sample at or near the stratigraphic contact between Proterozoic and Phanerozoic rocks; A = Appekunny Formation. Abbreviated system names: uK = Upper Cretaceous; lK = Lower Cretaceous; Tr = Triassic; uC = Upper Carboniferous; lC = Lower Carboniferous; D = Upper Devonian; mCam = Middle Cambrian; J = Jurassic. The Kishenehn Formation overlies middle Blairmore Group and older rocks; however, the Alberta Group is present in Lewis thrust sheet to the west in Fernie Basin.

within them are inferred to postdate the main displacement on the Lewis thrust. These relationships mean that the thermal history of the Lewis thrust sheet is directly relevant to the analysis of foreland petroleum occurrences: They provide a relative sequence of events leading to the formation of the petroleum pools.

The study area is unique because of the preservation of a Paleogene erosion surface beneath the late Eocene–early Oligocene and younger fanglomerates and tilted lacustrine and fluvial strata of the Kishenehn Formation that accumulated on the hanging wall during displacement on the Flathead fault (Price 1962; McMechan and Price, 1980; McMechan, 1981; Constenius, 1996) (Figure 4). The Flathead fault is a listric

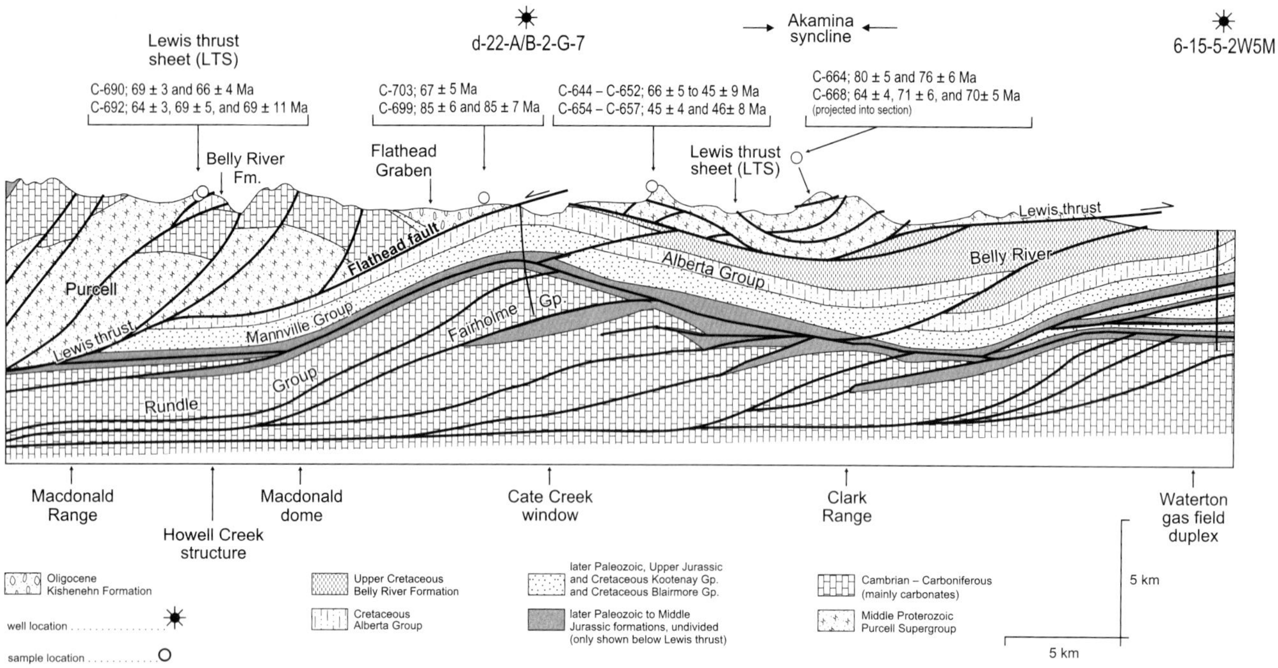

FIGURE 4. Structure section across the Lewis thrust sheet and Flathead graben, simplified from a balanced structure section by Fermor and Moffat (1992). Ornamentation of stratigraphic units is as in Figures 2 and 3. The locations (Figure 2) for selected samples and their calculated AFT ages are shown on or projected into the section. Not all the faults below the Lewis thrust are shown.

southwest-dipping normal fault with an estimated maximum dip slip of about 15 km (9 mi) (Constenius, 1988). It cuts and offsets the Lewis thrust sheet, but it merges with the underlying Lewis thrust without offsetting the strata in its footwall (Bally et al., 1966). The Kishenehn strata form a northeastward-thickening wedge on the flank of a rollover anticline in the hanging wall of the Flathead fault. They were tilted to the northeast during the listric displacement on the Flathead fault (Figures 2, 4) (Price, 1962; McMechan, 1981; Constenius, 1996). Lower Cretaceous strata of the Blairmore Group are preserved beneath the unconformity at the base of the Kishenehn Formation (Figure 2). This shows that in the late Eocene, when the Kishenehn Formation began to accumulate on the hanging wall of the Flathead fault, the Lewis thrust sheet was still stratigraphically at least 7 km (4.3 mi) thick in the vicinity of the Akamina syncline (Figure 5). East of the Flathead fault, where petroleum exploration boreholes penetrate stacked thrust sheets of Upper Cretaceous and older strata below the Lewis thrust, the thrust-faulted Mesozoic and Paleozoic succession is stratigraphically about 4 km (2.5 mi) thick and similar to that in the vicinity of the Waterton gas field thrust duplex (Figure 4). Fault-bounded Upper Cretaceous strata occur at one locality in the Lewis thrust sheet southwest of the Flathead fault, in the vicinity of Howell Creek (Figure 2). These Upper Cretaceous strata have been interpreted as part of the footwall of the Lewis thrust that has been displaced by a minor, out-of-sequence thrust that cuts the Lewis thrust and later offset by normal faults associated with the Flathead fault (Price, 1962; 1965). They have also been interpreted as part of the Lewis thrust sheet that occurs in the footwall of a higher thrust fault and has been downdropped by a Tertiary normal fault (Labrecque and Shaw, 1973) and as large rock slides (Jones, 1977).

By combining the section of the Lewis thrust sheet that is exposed beneath the Kishenehn Formation in the hanging wall of the Flathead normal fault with the section of the Lewis thrust sheet that is exposed above the Cate Creek and Haig Brook windows in the footwall of the Flathead fault and with the section below the Lewis thrust fault that is penetrated by petroleum exploration wells, it is possible to sample a section about 12 km (7 mi) thick through the foreland fold and thrust belt, because it existed prior to the displacement on Flathead fault (Figure 5).

The current geothermal environment in the study region is constrained by temperature measurements associated with petroleum exploration and production activities. Corrected bottomhole logging temperatures indicate a present-day geothermal gradient of about 17°C/km (courtesy of K. Root, Shell Canada Resources Ltd., 1992, personal communication). Organic maturity and inferred maximum temperatures are constrained by vitrinite reflectance studies throughout the region (Hacquebard and Donaldson, 1974; England, 1984; Kalkreuth and McMechan, 1984; Pearson and Grieve, 1985; England and Bustin, 1986a, b; Grieve, 1987; Bustin and England, 1989; Bustin, 1991).

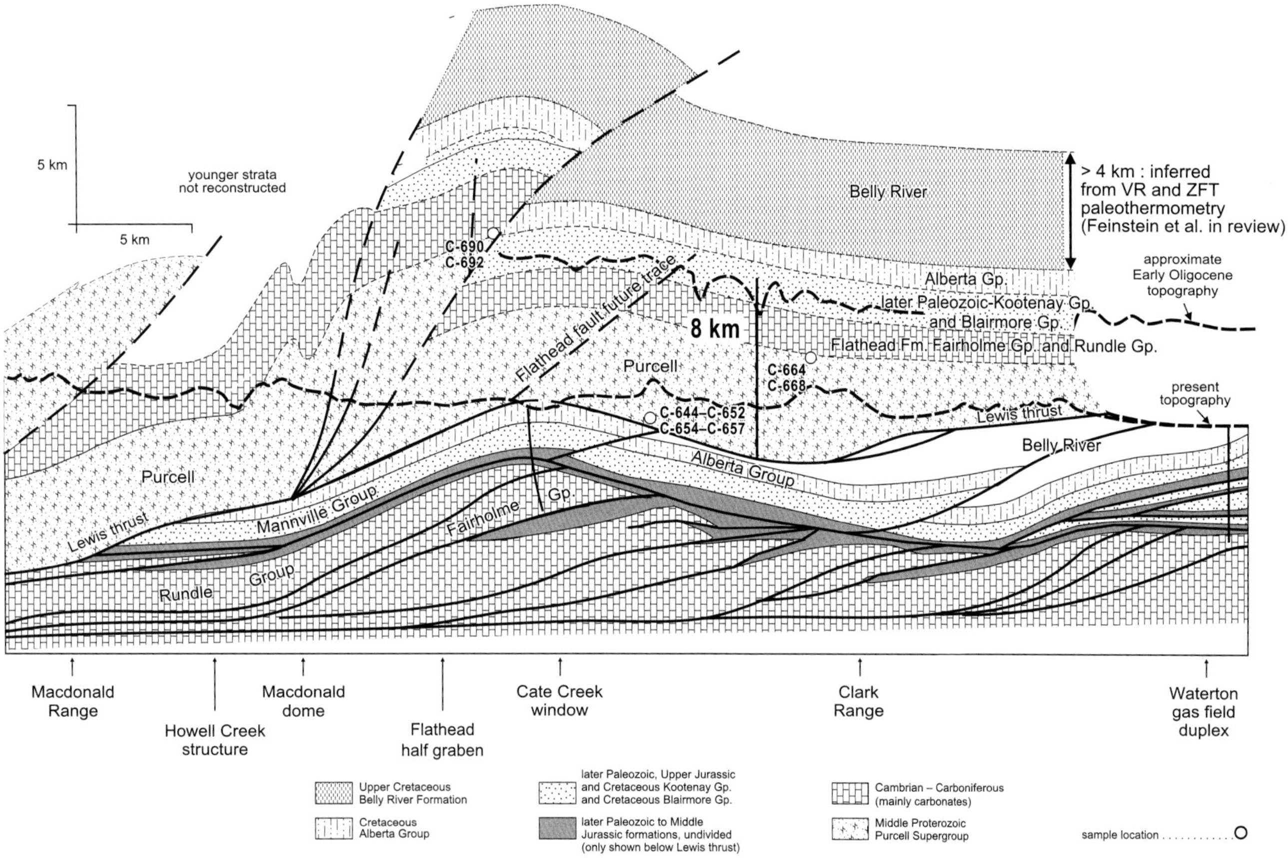

FIGURE 5. Partly restored structure section across the Lewis thrust sheet with the effects of Tertiary crustal extension removed (based on Figure 4). Ornamentation of stratigraphic units is as in Figures 2–4. Preextensional locations of selected AFT samples are shown projected into the section. The inferred approximate early Oligocene erosional surface is based on the observation that lower Oligocene Kishenehn Formation locally lies directly on Lower Cretaceous Blairmore Group strata in the hanging wall of the Flathead fault. Not all the faults below the Lewis thrust are shown.

APATITE FISSION-TRACK THERMOCHRONOLOGY

The temperature sensitivity of fission tracks is used in AFT analysis to estimate maximum paleotemperatures and the time when cooling from maximum paleotemperatures commenced. Wagner and Van den Haute (1992), Gallagher et al. (1998), and Gleadow et al. (2002) provide useful summaries of the principles of AFT dating and annealing. The spontaneous fission of U^{238} in uranium-bearing minerals, such as apatite, produces narrow tubes of intense damage in the crystal lattice continuously throughout geological time. The AFT clock is based on the principle that the density of fission tracks is proportional to the elapsed time since they began to accumulate and the uranium content of the apatite. New tracks have approximately the same initial length ($\sim$16 ± 1 μm in apatite); however, if the mineral is subjected to a sufficiently high temperature for a long enough time, all existing fission tracks will anneal (fade) and eventually disappear, effectively resetting the fission-track clock. The annealing process and its analysis are the key to temperature-history modeling by fission-track thermochronology.

Experimental studies and borehole observations have revealed a wide temperature interval over which gradual annealing occurs. Annealing proceeds more rapidly with increasing temperature, especially at temperatures exceeding about 60°C. With greater annealing, the fission tracks become increasingly shorter. The longer the rock resides in an annealing zone, the greater the proportion of shortened tracks. In apatite, fission-track annealing is sensitive to mineral chemistry, particularly chlorine (Cl) content (Green et al., 1986). Chlorine-rich apatites are more resistant to track annealing than the more commonly found fluorine-rich apatites. In the former, total annealing may occur at higher temperatures between about 125 and 150°C (e.g., Green et al., 1985; Burtner et al., 1994; O'Sullivan and Parrish, 1995).

These observations led to the partial annealing zone (PAZ) concept, in which fission tracks under geological

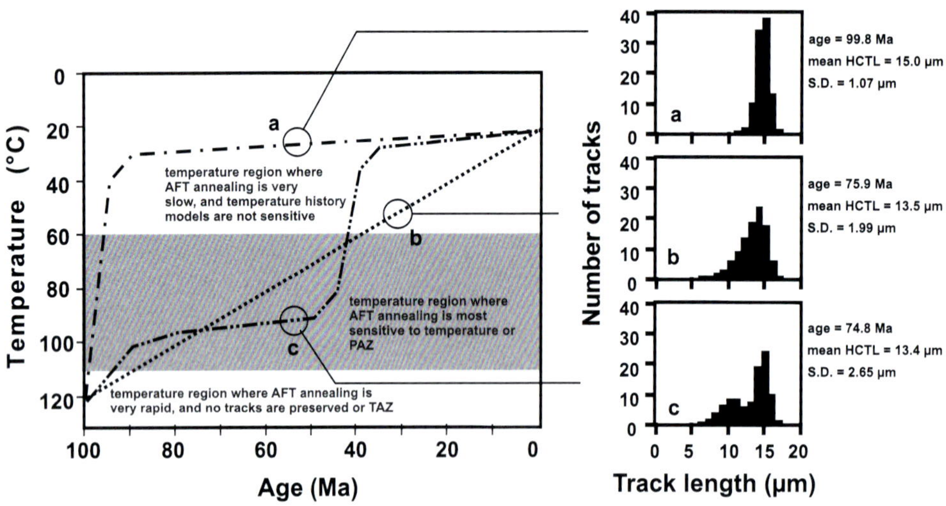

FIGURE 6. Illustration showing the approximate temperature range of the PAZ and the total annealing zone (TAZ) for fission track in fluorapatite. In the PAZ, AFTs are annealed progressively and more rapidly with increasing temperature and annealed to a zero apparent age once temperatures of the TAZ have been attained. Three hypothetical temperature-history paths a, b, and c are shown. These all involve cooling through the PAZ and are characterized by different cooling rates and residence times in the PAZ. The resulting model AFT age and HCTL parameters for each of the three thermal histories, like that calculated by the MonteTrax program (Figure 10), are indicated, as is the different shape of the HCTL distribution for each cooling path. Modified from Gleadow and Brown (2000).

conditions anneal increasingly between ambient surface temperatures and about 110°C. At temperatures below about 60°C, fission tracks in apatite, with the Cl concentration of Durango apatite, anneal at sufficiently slow rates that little apparent AFT age and track length reduction occur (Figure 6). In the region where temperatures are at or below about 60°C, THMs are considered to be less well constrained. The temperature range of about 60–110°C is where AFTs undergo more rapid annealing at a rate that accelerates with increasing temperature (e.g., Gleadow and Fitzgerald 1987; Fitzgerald and Gleadow 1988). This temperature interval was previously referred to as the PAZ, but this term is not now used widely, because partial annealing also occurs at lower temperatures. However, the PAZ concept retains some value with respect to the description of previous temperature history, as in the description of a "fossil or exhumed" PAZ (i.e., a region where AFT parameters retain a record of when a sample previously resided at temperatures between ~60 and 110°C). Near the base of this temperature range, where temperatures reach about 110°C, fission tracks in apatite anneal rapidly, resulting in a zero apparent age; this defines the boundary with the total annealing zone (TAZ; Green et al. 1986). The depths to these critical annealing temperatures depend on the chemical composition of the apatites, the geothermal gradient, and the duration of heating.

The concept of an exhumed apatite PAZ is commonly related to the timing and magnitude of exhumation (e.g., Naeser, 1979; Gleadow and Fitzgerald 1987; Fitzgerald and Gleadow 1988). Cooling of the rock column may have followed an episode where it remained at temperatures between about 60 and 110°C (i.e., in the PAZ), and the AFT ages of the rocks in the PAZ were reduced. Fitzgerald and Gleadow (1988) suggested that the position of the lower break in slope (concave downward) in an apatite age-depth profile represents the base of an exhumed apatite PAZ, which may provide temporal information related to the onset of subsequent cooling. Alternatively, if the rock column had previously been located at paleotemperatures above about 110°C prior to cooling, the upper break in slope (concave upward) in an age-depth profile may indicate the top of an exhumed apatite PAZ.

Because of the numerous possible temperature-history scenarios for a sample, an AFT age alone can be interpreted in several ways (Figure 6). However, more rigorous temperature-history constraints can be placed on an AFT observed age when it is combined with horizontally confined fission-track length (HCTL) data (Figure 6). The analysis of AFT age and length parameters can differentiate between different temperature-history options below temperatures of about 110°C (Gleadow et al., 1986), because AFT data reflect the time over which tracks have been retained as a function of host-material thermal history. Forward models can be used to calculate fission-track age and track length parameters resulting from a given THM using annealing algorithms (e.g., Laslett et al., 1987), and the results of plausible THMs can be tested against the observed fission-track data. This procedure has been automated using a computer program (MonteTrax®, Gallagher, 1995) that provides a quasi-inverse modeling method, which statistically tests the resulting model AFT parameters from numerous temperature-history simulations against observed AFT parameters. The model uses a Monte Carlo approach and a genetic algorithm to converge rapidly on THMs that match the observed AFT data.

SAMPLING STRATEGY, METHODS AND RESULTS

Sampling Strategy

We collected samples to study the thermal history of a current elevation profile in the Rocky Mountain Front Ranges and to compare this with the restored elevation profile prior to middle Eocene extension (Figures 2–5). The Lewis thrust sheet samples were collected both in the footwall (i.e., east) and in the hanging wall (i.e., west) of the Flathead fault. Other samples, all of which were collected at elevations more than 2 km (1 mi) in the Clark Range (Figure 4), provide data at similar elevations from different stratigraphic units in the Lewis thrust sheet (Figure 3). Rocks below the Lewis thrust were collected from outcrops east of the main fault trace, from erosional windows through the Lewis thrust sheet near the Flathead fault, and from petroleum exploration boreholes, e.g., Calstan Crowsnest and Shell Sage Creek wells (Figure 2). East of the Flathead fault, we obtained data from an approximately 3.4-km (2.1-mi)-long composite profile. The top of the profile is in the Lewis thrust sheet in the west limb of the Akamina syncline (+2454 to +1380 m [+8052 to +4528 ft]); it was extended below the Lewis thrust to −906 m (−2973 ft), using samples from the Calstan Crowsnest and Shell Sage Creek No. 2 wells.

The elevation profile from east of the Flathead fault was combined with fission-track data from the Lewis thrust sheet, west of the Flathead fault, to reconstruct a vertical profile that postdates most or all of the displacement on the Lewis thrust but that predates extensional displacement on the Flathead fault. Prior to Tertiary normal faulting, samples from the Lewis thrust sheet in outcrops west of the Flathead fault were higher in the vertical profile than any samples collected east of Flathead fault. Thus, the combined vertical profile that has been restored to remove the effects of the Paleogene normal faulting is more than 7 km (4 mi) long (Figure 5). In addition, we sampled Kishenehn Formation sandstones and conglomerates in the Flathead graben. The conglomerate sample included cobbles and boulders of middle Cretaceous hypabyssal intrusions that were eroded from the Lewis thrust sheet. These samples record the temperature history both of their provenance and during their burial in the graben.

Methods

Mineral separation methods and details of the techniques used to acquire the AFT data followed those described by Kohn et al. (1995), with the exception that for samples from the Calstan Crowsnest well (samples 739/1–739/6; Table 1), thermal neutron fluences were monitored using muscovite attached to the Corning-5 standard glass (Bellemans et al., 1995). Apatite fission-track central ages (Galbraith and Laslett, 1993) were determined using the external detector method (Green, 1985). To ascertain whether compositional effects have influenced the annealing behavior of fission tracks in the apatite studied, we analyzed, by electron microprobe analyzer, the Cl content of individual grains from 11 of the samples. Analytical details are given in Table 1.

Results

Apatite fission-track data are presented in Table 1. Figure 7 illustrates the distribution of sample single-grain AFT ages, both as a histogram and in a radial plot (Galbraith, 1990), as well as sample horizontal confined track length measurements. Figure 8 illustrates the Cl content of studied apatites. The samples from Cretaceous igneous rocks (e.g., samples 668 and 690, Figure 8) and Proterozoic Appekunny Formation (samples 644 and 657) have maximum Cl contents less than that of the Durango apatite standard (dashed vertical lines in Figure 8) with a small compositional variation. In contrast, the Lower Cretaceous Blairmore Formation (739/1–739/6, Figure 8) and Oligocene Kishenehn Formation (sample 699, Figure 8) sandstones have more variable Cl content, and maximum values are far greater than that of Durango apatite. Horizontally confined fission-track length data and apatite composition influence sample selection for THMs that use Durango AFT annealing kinetics. Not all samples have sufficient HCTL data, and for this study, we arbitrarily defined samples with more than 25 measurable track lengths as providing an adequate sample of track length distribution for thermal-history modeling. Likewise, not all samples, especially those with Cl-rich apatites or highly variable apatite compositions, will be reliably modeled using Durango AFT kinetics.

Apatite fission-track age and mean HCTL data from rocks east of Flathead fault, both above and below Lewis thrust, vary with elevation and structural position (Table 1; Figure 9), regardless of stratigraphic position (Figure 3). Apatite fission-track ages decrease from about 68 Ma at about 2500 m (8200 ft) to approximately 30 Ma at about −1000 m (−3300 ft). The range of AFT ages at any specific elevation is large. At the highest elevations east of the Flathead fault, mean HCTLs are about 13.8 μm or longer. These decrease to less than or equal to 12 μm

TABLE 1. Lewis thrust sheet AFT data.

Sample	*Unit*	*Age*	*Elevation (m)*	*Number of Grains*	*Standard Track Density (10^6 cm^{-2})*	*Fossil Track Density (10^6 cm^{-2})*	*Induced Track Density (10^6 cm^{-2})*
		Lewis thrust sheet in the footwall of the Flathead normal fault					
644	Appekunny	Proterozoic	2404	12	1.314 (6107)	0.257 (229)	0.908 (808)
649	Appekunny	Proterozoic	2184	1	1.324 (6107)	0.453 (24)	1.623 (86)
651	Appekunny	Proterozoic	2062	10	1.334 (6107)	0.590 (270)	2.622 (1201)
652	Appekunny	Proterozoic	2015	6	1.265 (3883)	0.375 (104)	1.750 (486)
652	Appekunny	Proterozoic	2015	2	1.344 (6107)	0.392 (32)	2.098 (172)
654	Appekunny	Proterozoic	1397	3	1.218 (3885)	0.616 (50)	2.848 (231)
657	Appekunny	Proterozoic	1382	14	1.289 (3726)	0.504 (140)	2.508 (697)
664	intrusion[†]	middle Cretaceous	2102	31	1.297 (3885)	0.535 (448)	1.540 (1291)
664	intrusion[†]	middle Cretaceous	2102	9	1.364 (6107)	0.319 (265)	1.034 (858)
668	intrusion[††]	middle Cretaceous	2020	20	1.250 (4179)	0.301 (464)	1.044 (1611)
668	intrusion[††]	middle Cretaceous	2020	12	1.320 (4206)	0.296 (221)	0.994 (742)
668	intrusion[††]	middle Cretaceous	2020	12	1.374 (6107)	0.201 (225)	0.703 (787)
		Lewis thrust sheet in the hanging wall of the Flathead normal fault					
690	intrusion[††]	middle Cretaceous	1990	16	1.305 (4016)	0.993 (1013)	3.351 (3419)
690	intrusion[††]	middle Cretaceous	1990	11	1.188 (4206)	0.733 (528)	2.328 (1678)
690	intrusion[††]	middle Cretaceous	1990	16	1.394 (6107)	0.425 (569)	1.639 (2196)
692	Flathead	Middle Cambrian	1993	18	1.332 (4206)	0.514 (243)	1.756 (831)
692	Flathead	Middle Cambrian	1993	3	1.404 (6107)	0.464 (52)	1.670 (187)
		Rocks below Lewis thrust fault east of the Flathead normal fault					
613A	Kootenay	Late Jurassic–Early Cretaceous	1341	11	1.483 (3745)	0.103 (56)	0.888 (483)
615	Crowsnest volcanics	middle Cretaceous	1372	12	1.281 (3885)	0.677 (586)	2.650 (2294)
615A	Crowsnest volcanics	middle Cretaceous	1372	25	1.488 (3745)	0.128 (232)	0.598 (1086)
618	Belly River	Late Cretaceous	1356	9	1.194 (3885)	0.440 (225)	1.877 (960)
705	Belly River	Late Cretaceous	1509	19	1.424 (6107)	0.258 (207)	1.293 (1036)
705/1	Belly River	Late Cretaceous	1509	14	1.397 (4451)	0.251 (189)	1.232 (929)
721	Belly River	Late Cretaceous	1351	3	1.256 (3726)	0.919 (13)	3.111 (44)
721A	Belly River	Late Cretaceous	1351	26	1.498 (3745)	0.195 (239)	0.782 (958)
721A	Belly River	Late Cretaceous	1351	30	1.498 (3745)	0.398 (392)	1.889 (1859)
722	Belly River	Late Cretaceous	1365	10	1.317 (3478)	0.288 (86)	1.556 (465)
737	Kootenay	Late Jurassic–Early Cretaceous	−906	6	1.434 (6107)	0.241 (71)	1.379 (406)
		California Standard: Crowsnest well (6-14-8-5W5)					
739/1	Blairmore	Early Cretaceous	808	25	1.379 (3447)	0.292 (242)	2.095 (1739)
739/2	Blairmore	Early Cretaceous	734	25	1.379 (3447)	0.359 (312)	1.872 (1625)
739/3	Blairmore	Early Cretaceous	189	10	1.379 (3447)	0.142 (20)	1.248 (181)
739/4	Blairmore	Early Cretaceous	−566	20	1.379 (3447)	0.149 (34)	1.406 (322)
739/5	Blairmore	Early Cretaceous	−670	26	1.379 (3447)	0.424 (269)	2.453 (1555)
739/6	Blairmore	Early Cretaceous	−840	25	1.379 (3447)	0.172 (78)	1.159 (525)
		Flathead graben-fill sediments					
699	Kishenehn	early Oligocene	1301	14	1.471 (4451)	0.286 (283)	0.886 (876)
699	Kishenehn	early Oligocene	1301	22	1.414 (61070)	0.453 (568)	1.440 (1804)
703	Kishenehn[‡]	early Oligocene	1318	12	1.260 (327)	0.542 (327)	1.800 (1087)

Brackets show number of tracks or lengths counted. Standard and induced track densities measured on mica external detectors (g = 0.5) and fossil track density on internal surfaces.

**All ages are central ages (Galbraith and Laslett, 1993).*

***Chlorine was determined on a JEOL JXA-5A electron microprobe running at an accelerating voltage of 15 kV, defocused beam size of 15–20 μm, and beam current of 29 nA; samples were calibrated using Durango apatite. Analyses were performed at Geotrack international Pty Ltd. Apatite ages calculated by B. P. Kohn using zeta = 357 for dosimeter glass NIST SRM-612.*

****Apatite age calculated by R. W. Brown using zeta = 350.2 for dosimeter glass NIST SRM-612.*

[†]Middle Cretaceous sill intruded into Middle Cambrian Flathead Formation.

[††]Middle Cretaceous sill or stock intruded into Middle Proterozoic Roosville Formation.

[†††]Apatite age calculated by P. B. O'Sullivan using zeta = 371 for dosimeter glass Corning-5.

[‡]Boulder of middle Cretaceous intrusion in lower Oligocene Kishenehn Formation conglomerate.

TABLE 1. Lewis thrust sheet AFT data (cont.).

Uranium Content (ppm)	*Chi-square Probability (%)*	*Age Dispersion (%)*	*Fission-track Age (±1σ) (Ma)**	*Mean Track Length (μm)*	*Standard Deviation (μm)*	*Cl (wt.%)***
			Lewis thrust sheet in the footwall of the Flathead normal fault			
9	99.4	0.00	66 ± 5	11.70 ± 0.66 (20)	2.95	0.04–0.32
15	–	–	66 ± 15	–	–	–
24	89.9	0.01	53 ± 4	14.02 ± 0.40 (9)	1.21	0.01–0.30
17	86.5	0.00	48 ± 5	–	–	–
19	63.5	0.00	45 ± 9	13.77 (1)	–	–
29	13.5	15.93	46 ± 8	–	–	–
26	99.8	0.00	45 ± 4***	12.63 ± 0.58 (28)	3.06	–
15	59.5	7.53	80 ± 5	–	–	–
9	25.7	7.04	76 ± 6	13.70 ± 0.34 (39)	2.13	0.14–0.38
10	21.0	11.67	64 ± 4	13.79 ± 0.29 (53)	2.18	–
9	55.4	8.65	71 ± 6	–	–	–
6	73.6	0.24	70 ± 5	13.45 ± 0.30 (13)	1.09	0.01–0.16
			Lewis thrust sheet in the hanging wall of the Flathead normal fault			
32	75.7	0.08	69 ± 3	14.40 ± 0.12 (89)	1.17	0.00–0.06
24	60.3	0.03	66 ± 4	–	–	–
15	99.0	0.00	64 ± 3	14.22 ± 0.11 (55)	0.83	–
16	89.6	0.08	69 ± 5	–	–	–
15	86.6	0.00	69 ± 11	–	–	–
			Rocks below Lewis thrust fault east of the Flathead normal fault			
7	15.4	16.8	30 ± 5	11.92 ± 0.48 (29)	2.59	–
27	13.9	9.55	58 ± 3	13.75 ± 0.23 (72)	1.98	–
5	2.6	26.5	58 ± 5	13.04 ± 0.26 (100)	2.59	–
19	98.7	0.00	50 ± 4			
11	7.2	20.13	51 ± 5	12.06 ± 1.14 (6)	2.90	–
11	0.73	29.66	53 ± 6	–	–	–
3	45.6	0.03	66 ± 21	12.26 ± 1.13 (6)	2.90	–
6	11.4	21.08	67 ± 6	–	–	–
16	8.2	16.88	56 ± 4	13.02 ± 0.13 (100)	2.0	–
15	52.2	9.84	44 ± 5	11.96 ± 0.44 (9)	1.32	–
12	0.0	59.68	47 ± 13	13.20 (1)	–	–
			California Standard: Crowsnest well (6-14-8-5W5)			
18	18.5	16.57	36 ± 3†††	13.29 ± 0.16 (101)	1.60	0.00–0.58
16	0.0	39.88	48 ± 6†††	13.13 ± 0.19 (102)	1.90	0.01–0.96
11	39.1	19.74	29± 7†††	12.55 ± 0.40 (29)	2.16	0.01–0.91
12	10.6	24.05	28 ± 5†††	11.31 ± 0.69 (12)	2.39	–
21	48.9	0.44	44 ± 3†††	11.99 ± 0.26 (100)	2.63	0.04–1.12
10	83.6	0.75	38 ± 5†††	11.71 ± 0.22 (92)	2.47	0.00–1.26
			Flathead graben-fill sediments			
7	24.9	9.42	85 ± 6	–	–	–
13	0.01	28.48	85 ± 7	13.85 ± 0.26 (10)	0.83	0.01–1.20
18	98.1	0.00	67 ± 5	13.80 ± 0.27 (74)	2.32	–

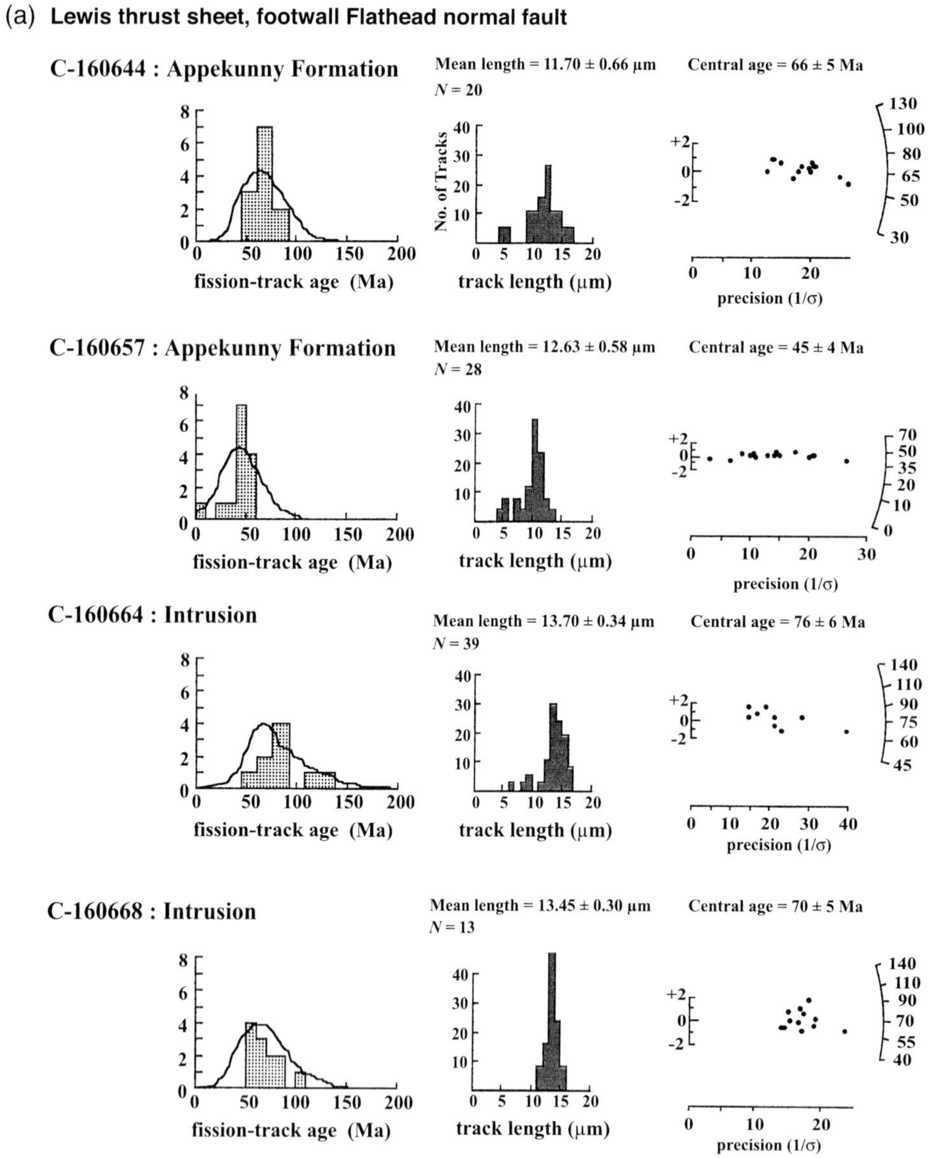

FIGURE 7. Plots showing measured single-grain AFT age data distributions as both histograms and smoothed probability distributions (left hand side), mean horizontal confined length histograms (center), and radial plots (Galbraith, 1990) of single-grain AFT age data (right-hand side) for selected samples from the Lewis thrust sheet east of Flathead fault (a) and west of the Flathead fault and in Kishenehn Formation (b) (keyed to Table 1). The key for numbers appearing above each length histogram is as follows: Mean length = mean HCTL and associated measurement uncertainty; N = number of mean horizontal confined tracks measured. In radial plots, the slope of a straight line from the origin (0) is equivalent to the fission-track age read off radially around the perimeter of the plot. The χ^2 value and the percent relative error (Table 1) measure the precision of each grain age. The further a point plots to the right of the origin, the more precise the individual grain age. If all grains belong to a single age population, then they should scatter within the $\pm 2\sigma$ age range about the central age as shown on the Y-axis. Scatter outside this range indicates a spread of single-grain AFT ages that should also be reflected by the chi-square statistic (Table 1).

at about 1300 m (4300 ft). At about 750 m (2500 ft), longer mean HCTL, more than 13 μm, are observed again. Between elevations of 750 and −1000 m (2500 and −3300 ft), mean HCTL decreases progressively from more than 13 to about 11.5 μm. Lewis thrust sheet samples west of Flathead fault yield AFT ages between 64 and 69 Ma, with mean HCTL between 14.22 and 14.40 μm. The Eocene Kishenehn Formation boulder conglomerate composed of the Lewis thrust sheet lithologies has AFT parameters and F-rich apatites like similar magmatic Cretaceous rocks in the Lewis thrust sheet. A Kishenehn Formation sandstone yields an AFT age of 85 Ma and mean HCTL greater than 13.8 μm. This sample contains two types of apatites, both F-rich and a more Cl-rich source, not seen in the preserved Lewis thrust sheet.

THERMAL HISTORY

Aspects of the Temperature History Inferred Directly from the AFT Data

Apatite fission-track samples above and below the Lewis thrust on different sides of Flathead fault record different components of the temperature history, depending primarily on their structural position. In general, samples with younger observed AFT ages have shorter mean HCTLs. A wide variation in observed AFT age is associated with a given value of mean HCTL (Figure 9). This makes it impossible to infer the temperature history of samples without sufficient HCTL data by comparison to samples with sufficient HCTL data. As a result, we restrict our analysis of temperature history to samples

FIGURE 7. (cont.).

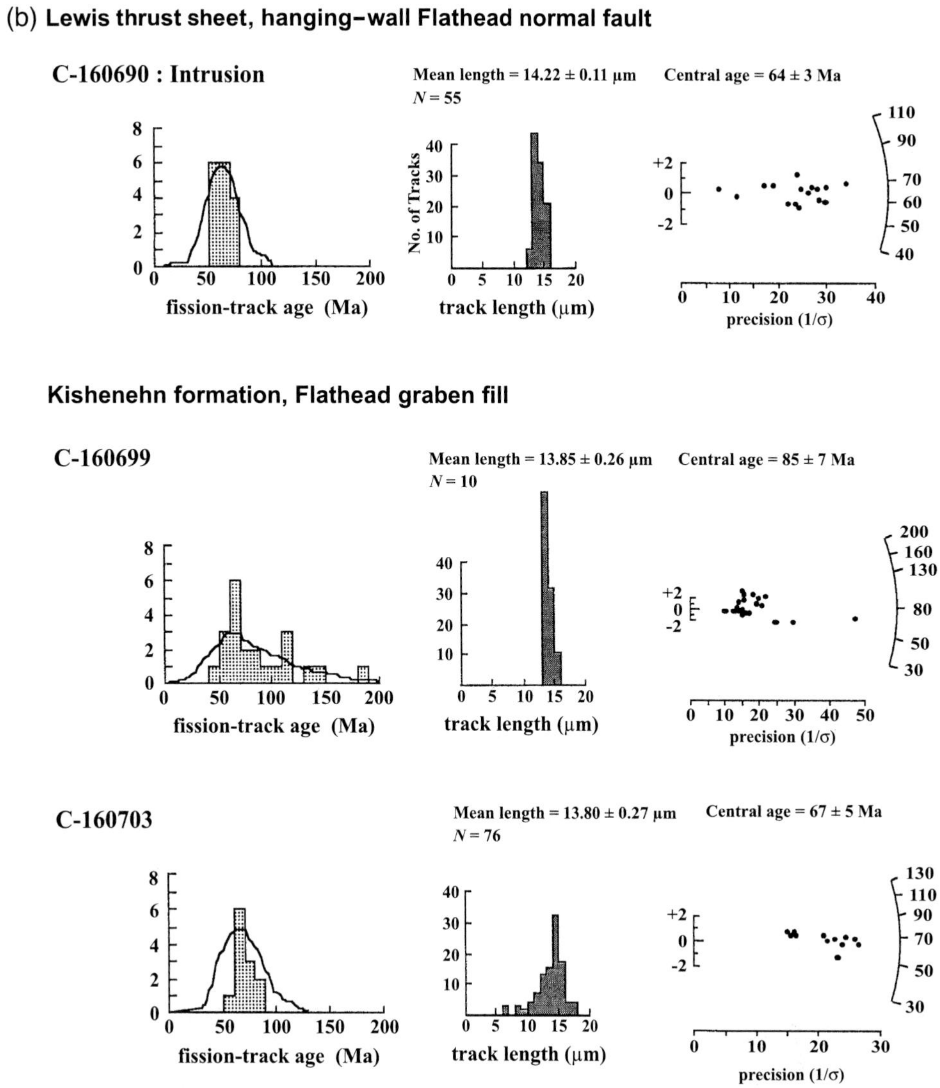

that we consider as those that can be modeled (i.e., >25 HCTL measurements).

The large variation in mean HCTL associated with age is partly attributed to variations in apatite composition. This is especially true among sedimentary samples, and it is illustrated by samples from below Lewis thrust (Table 1; Figure 8). However, the observation that other samples of low Cl apatite composition from different structural elements and stratigraphic positions have similar mean HCTL, but distinctive AFT ages suggest that part of the dispersion in AFT parameters results from temperature-history differences.

Based on the measured AFT parameters and structural position, we identify two sample groups. One group is characterized by F-rich apatites with relatively long mean HCTL (>13.4 μm) and typically older AFT ages (>56 Ma). The second group is characterized by more Cl-rich apatites, with shorter mean HCTL (<13.0 μm), and typically younger observed AFT ages (<56 Ma). The first group occurs in or was eroded from the Lewis thrust sheet. Their long mean HCTL indicates that these samples record relatively rapid cooling (at a time slightly before the observed age) from temperatures above apatite total annealing threshold (more than about 110°C) to temperatures at which subsequent AFT annealing was insignificant to moderate. The second sample group is, with two exceptions, from rocks below Lewis thrust. Their shorter mean HCTL indicates that they record a more complicated temperature history.

Temperature-history Models

Sample THMs were constructed with AFT age and sufficient HCTL measurements using the MonteTrax program. A combination of provenance age and relatively low uranium content in many detrital apatites has resulted in low numbers of measurable HCTL. This has reduced the number of samples suitable for temperature-history modeling. Additional complications result from apatite compositional variations. We present five illustrative THMs that represent the four distinctive structural settings (Table 1; Figure 10). Specifically, we illustrate THMs for

- two samples from the Lewis thrust sheet located in the footwall (east) of the Flathead fault (samples 668 and 657; note the reconstructed sample depths prior to motion on the Flathead fault, Figure 5);
- one sample from the Lewis thrust sheet from in the hanging wall (west) of the Flathead fault (sample 690, Figure 5);
- one sample from below and east of the Lewis thrust (sample 615A, Figure 2); and
- one sample from the Oligocene Kishenehn Formation in the Flathead graben (sample 703, Figure 4).

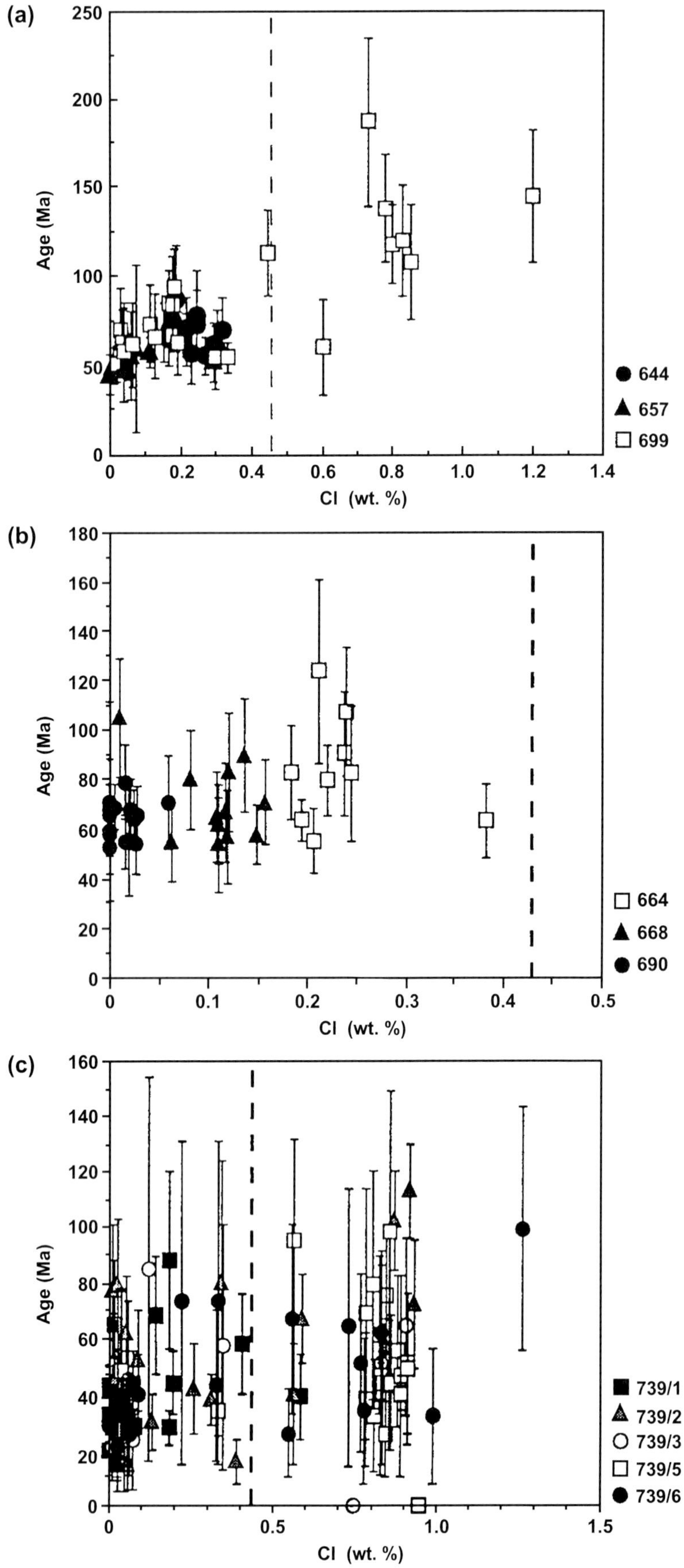

FIGURE 8. Plots of apatite grain chlorine content (wt.%) vs. single-grain AFT age for select samples (Table 1) from (a) Appekunny Formation (samples 644 and 657) and Kishenehn Formation (sample 699) sandstones; (b) middle Cretaceous alkaline igneous intrusive rocks in Lewis thrust sheet; and (c) Blairmore Formation sandstones in the Calstan Crowsnest well. The dashed vertical line indicates the chlorine content of the Durango apatite standard, the annealing parameters for which are used in the MonteTrax temperature-history model. Note the presence of both a distinctive provenance in Kishenehn Formation sandstone (sample 699; Table 1) that is characterized by older AFT ages and higher chlorine contents, as well as the large variation in single-grain AFT age and chlorine content of Blairmore Group samples in the Crowsnest well (739/1–739/6; Table 1).

One sample from each specific structural setting comprises the distinctive alkalic volcanics of the middle Cretaceous Crowsnest Formation volcanics (Pearce, 1970) or its intrusive equivalents (Gordy and Edwards, 1962; Price, 1962). The apatites from this alkalic igneous suite are relatively young, high in uranium, and uniformly low in Cl (Table 1; Figure 8). Differences observed among the AFT parameters of these samples probably result from differences in temperature history alone, because apatites in these samples have similar compositions. The uniformly low Cl content of the accessory apatites in the Cretaceous igneous suite also matches the annealing parameters used in the modeling (Gallagher, 1995). This suite of magmatic apatites was buried below strata in which coal ranks are more than Vitrinite reflectance (VR) = 1.0% R_o equivalent (Hacquebard and Donaldson, 1974). Accordingly, the AFT clocks of these samples were totally reset during burial, prior to thrusting. As a result, four of the models have inferred temperature-history differences that accurately reflect sample thermal-history differences. The fifth model is for a Lewis thrust sheet Middle Proterozoic sandstone sample at an elevation similar to the Crowsnest Formation sample from below the Lewis thrust.

All modeling followed a similar procedure. Initially, upper and lower time-temperature limits were set widely, and 1000 iterations were run using a genetic algorithm, until the model results converged on

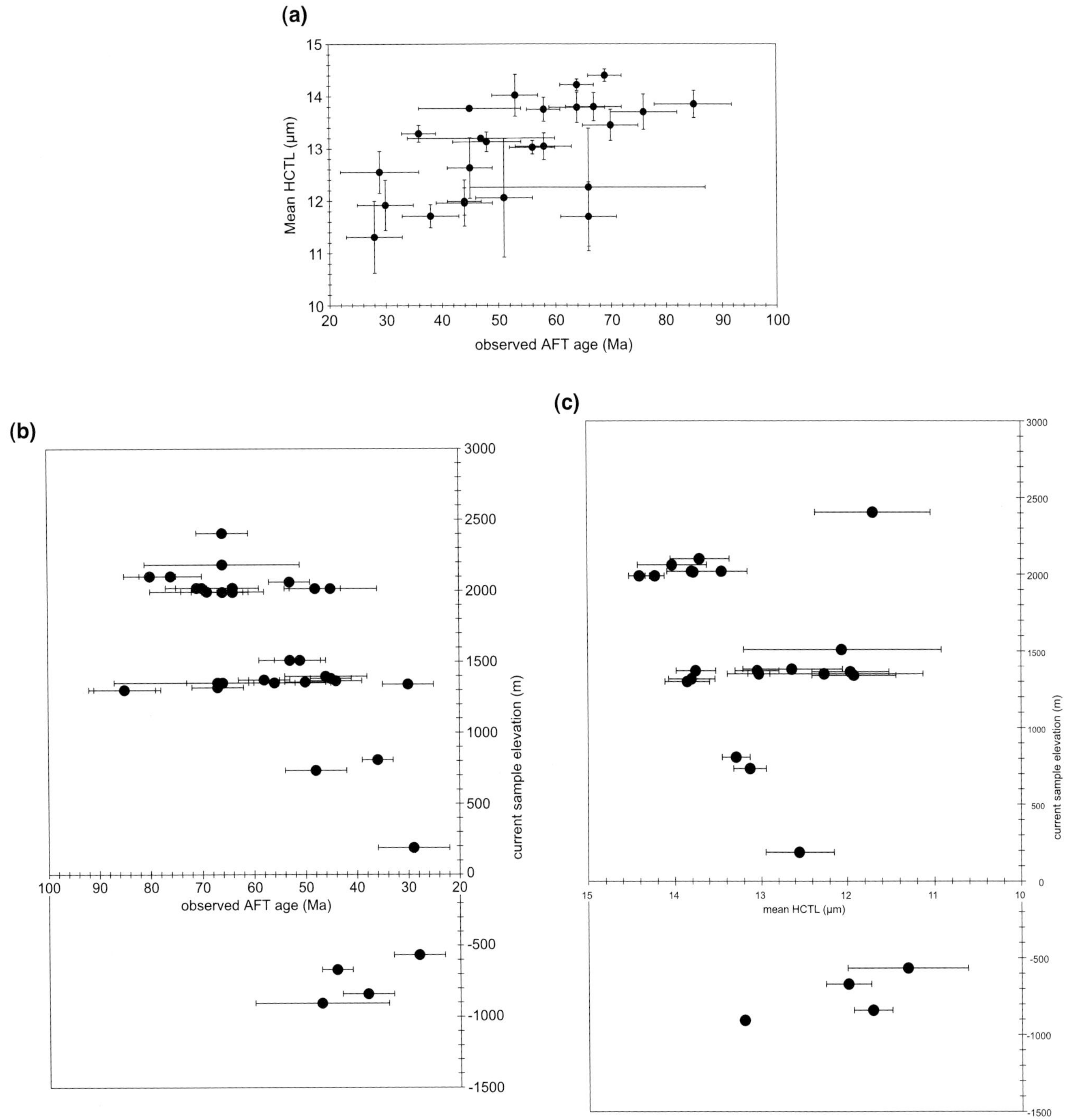

FIGURE 9. Apatite fission-track data plots from Table 1: (a) Mean HCTL vs. AFT age; (b) variation of AFT age with sample elevation; and (c) mean HCTL variation with sample elevation.

"acceptable" temperature histories, which matched the observed AFT parameters. Standard model acceptance criteria were employed (Gallagher, 1995). Stratigraphic and structural constraints were incorporated into the models (Figure 10). The stratigraphic record indicates times when samples lay within a small range of surface temperatures. In contrast, coalification history provides both maximum temperatures to which and maximum ages for when samples were heated.

Consider, for example, the boulders that occur in the Oligocene Kishenehn Formation in the Flathead graben that were eroded from the middle Cretaceous alkaline-intrusive suite in the Lewis thrust sheet east of the Flathead fault. These samples cooled to less than or equal to 80°C after intrusion into Proterozoic and Middle Cambrian strata of the Lewis thrust sheet at about 100 Ma. Subsequently, they reached or exceeded the approximately 145°C temperature (VR = 1.18% R_o) recorded

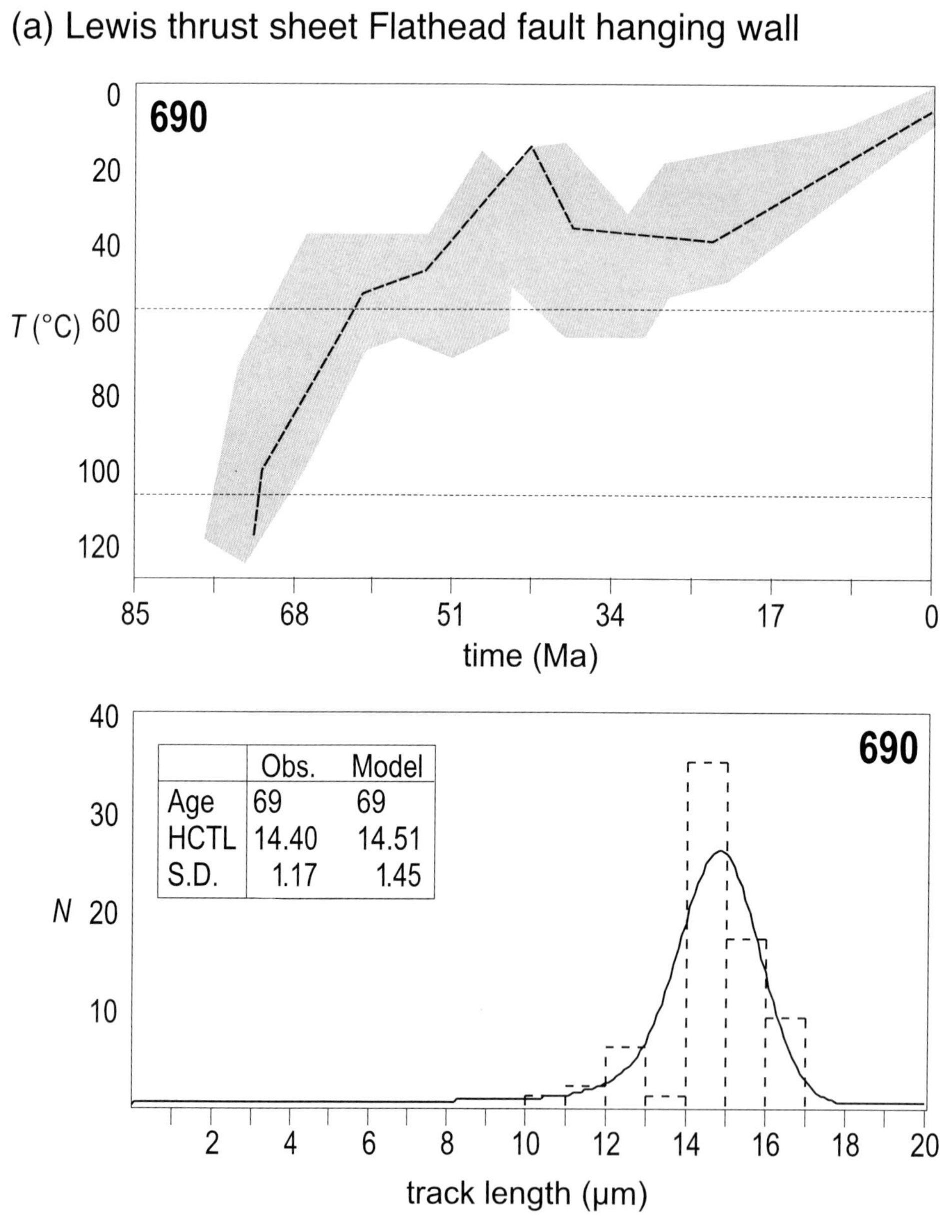

FIGURE 10. Time-temperature plots for five representative samples (see Table 1) derived using observed AFT data and the genetic algorithm forward modeling procedures described by Gallagher (1995). The models assume an initial track length of 16.3 µm and are based on 1000 iterations for each sample. The upper diagram shows the resulting range of possible temperature histories that satisfy both observed AFT age and HCTL parameters in the shaded region. All acceptable models are potential temperature-history solutions. A single model, shown as a heavy dashed line and commonly the "best fit," is also illustrated. On the lower part of the diagram, the observed AFT HCTL data are illustrated as a histogram with model track lengths for the best individual run shown as a smoothed probability distribution. The inset table compares observed and modeled AFT age and HCTL parameters (note that the entire lower part of the diagram shows all the information related to the HCTL data). Irrespective of whether samples pass the chi-square at the 5% probability level, they all show a similar temperature history involving two main cooling phases, one beginning in the Late Cretaceous and a second in the Eocene. Note that the first cooling event is illustrated by the time when apatites begin quantitatively retaining fission tracks, because all the samples resided at temperatures greater than 110°C prior to the first cooling. It should also be noted that the original model formulation of Laslett et al. (1987) using the initial unannealed track length of 16.3 µm does not predict sufficient amounts of annealing at temperatures less than about 50–60°C. Deriving thermal histories using this model commonly leads artificially to the inference of major late Tertiary cooling (Kohn et al., 2002). Hence, in this case, early Miocene cooling from about 60°C for samples (b–d) is considered to be an artifact of the modeling procedure.

Figure 10. (cont.).

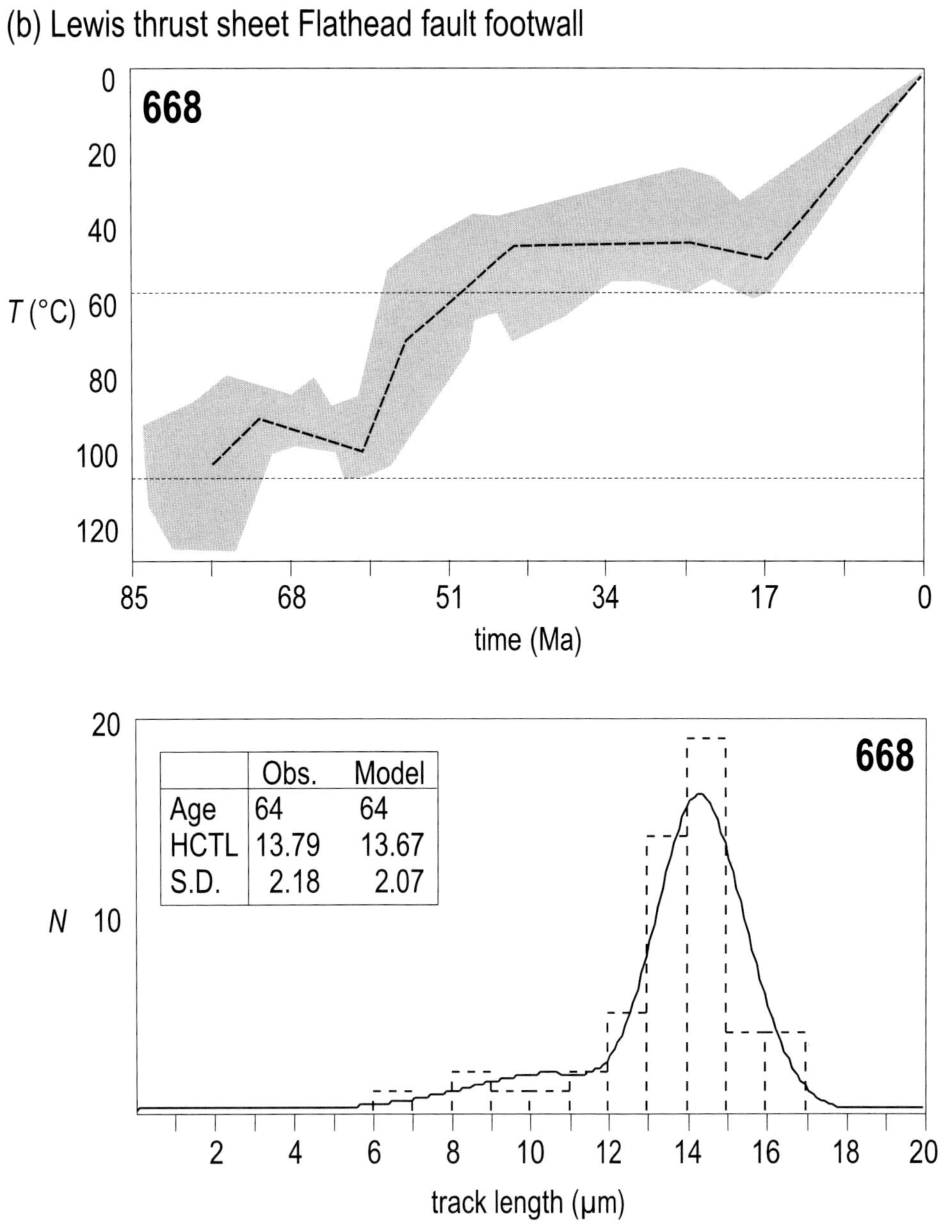

by overlying Kootenay Group coals. From this, we infer that their AFT clocks were totally reset at maximum burial depth in the Late Cretaceous. During the interval from late Campanian to early Oligocene, the AFT clocks in these rocks record a cooling history in the Lewis thrust sheet. Then, when these boulders were eroded from the Lewis thrust sheet and deposited in the Kishenehn Formation in the Flathead graben, they must have cooled to surface temperatures (no later than the early Oligocene). The low level of organic thermal maturity of interbedded lignites and lacustrine oil shales in the Kishenehn Formation (Curiale, 1987) confirms that during burial in the Flathead graben, the boulders containing the apatites were not heated sufficiently to significantly alter their AFT parameters.

All five modeled apatite samples are inferred to have begun accumulating fission tracks between about 75 and 70 Ma, although the inferred subsequent cooling history differs. The similar Cl content of four apatite samples ensures that model temperature-history differences reflect sample temperature-history differences only. Therefore, temperature-history model general characteristics indicate directly that all five samples began to accumulate fission tracks over a short time interval, in the late Campanian–Maastrichtian (Palmer and Geissman, 1999).

LEWIS THRUST SHEET IN THE HANGING WALL OF THE FLATHEAD FAULT

The Lewis thrust sheet sample from the hanging wall (west) of the Flathead fault (sample 690) are from a middle Cretaceous alkaline igneous intrusion into Middle Cambrian sandstones of the Flathead Formation. The sandstones lie above a thrust fault that places Mesoproterozoic and younger strata on Upper Cretaceous strata in the vicinity of Howell Creek. The AFT age is similar to the AFT age of the host rock Middle Cambrian sandstones (sample 692); however, adequate fission-track length measurements for temperature-history modeling could not be obtained from the sandstones. The THMs presented in Figure 10 (sample 690) are probably representative for the other samples collected from the Lewis thrust sheet west of the Flathead fault (Table 1). The THMs for this sample are all similar and very simple. They indicate the onset of rapid cooling from temperatures more than about 110°C at about 75 Ma. The sample cooled rapidly to less than 80°C; by the end of the Cretaceous or slightly earlier, it has

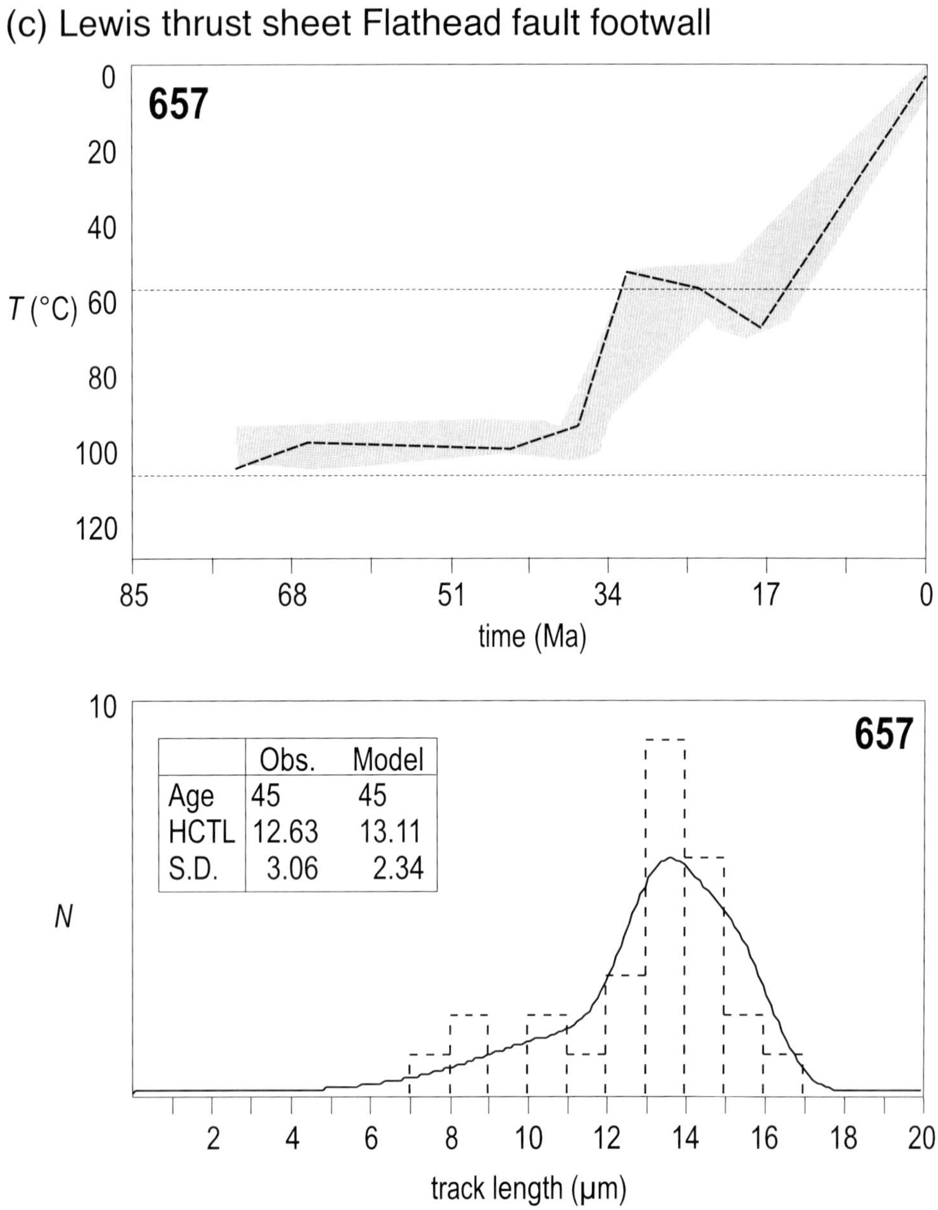

FIGURE 10. (cont.).

remained at temperatures less than about 60°C, where little or no significant annealing occurs since that time. (Note that model temperature histories are not reliably constrained when temperatures are lower than ~60°C.)

LEWIS THRUST SHEET IN THE FOOTWALL OF THE FLATHEAD FAULT

One Lewis thrust sheet sample from the footwall (east) of the Flathead fault (sample 668) is from a middle Cretaceous alkaline igneous intrusion within Mesoproterozoic strata of the Purcell Supergroup, close to the stratigraphic contact with the Paleozoic strata near the hinge of the Akamina syncline (Figure 2). The stratigraphic position of this sample is similar to that of the sample from Lewis thrust sheet in the hanging wall of the Flathead fault (sample 690), and it also shows cooling below about 110°C at about 75 Ma; it remained at temperatures more than 80°C, at which significant track annealing occurred until sometime in the interval from late Paleocene to early Eocene. During the middle Eocene onward, it cooled rapidly to a temperature at which no subsequent significant annealing occurred. The temperature history of this sample suggests that prior to late Paleocene to early Eocene, it was located lower in the deformed structure than sample 690, although it is now some 30 m (90 ft) higher in elevation. This THM is compatible with the structural relationships between samples 668 and 690, only if the initial cooling commenced during the thrusting and folding.

A second Lewis thrust sheet sample (sample 657), from the western limb of the Akamina syncline just east of the Flathead fault in Mesoproterozoic sandstones of the Appekunny Formation (Figure 2), is from the lowest stratigraphic horizon in the Lewis thrust sheet that yielded AFT data. This sample was collected at +1382 m (+4534 ft), approximately the same elevation as a sample of the Crowsnest Formation volcanics (sample 615A) from east of the Lewis thrust at +1372 m (+4501 ft). As with the other samples, THMs for sample 657 indicate that it began to cool and accumulate fission tracks in the Late Cretaceous. However, it remained at temperatures of about 100°C until the late Eocene to early Oligocene, when it cooled rapidly to temperatures below which no further significant annealing occurred. Thus, two cooling events are recorded by this sample. The first began in the late Campanian; the second occurred in the Eocene to

FIGURE 10. (cont.).

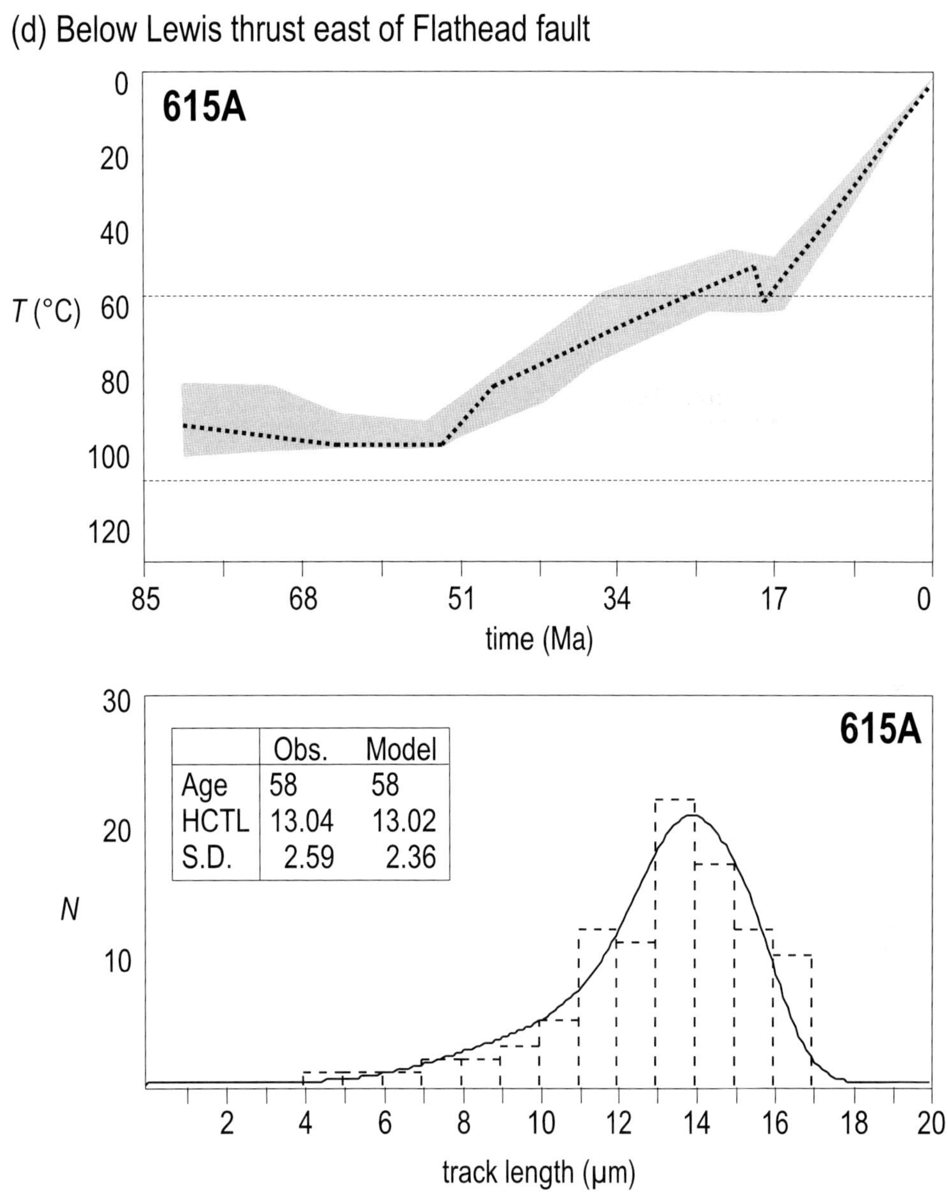

Oligocene. Younger portions of the temperature history are not reliable because of cooling to temperatures below about 60°C during the second cooling event. The difference in observed AFT parameters and THMs between samples 668 and 657 can be attributed to the 640-m (2100-ft) difference in elevation between them, if this produced a temperature difference of about 10–20°C during the interval from the late Campanian to Eocene (Figure 11).

BELOW THE LEWIS THRUST SHEET

The THMs for sub-Lewis thrust rocks in the footwall (east) of Flathead fault is based on samples 615 and 615A from the middle Cretaceous alkaline volcanic rocks of the Crowsnest Formation in the Coleman thrust sheet. Sample 615 is from tuffaceous lahars, and sample 615A is from xenoliths of subvolcanic intrusives. These samples are similar to samples 703, 668, and 690, which also are from middle Cretaceous alkaline igneous rocks. The Crowsnest volcanic samples are from a higher stratigraphic level than any of the samples from the Lewis thrust sheet. The AFT age and HCTL parameters of both samples 615 and 615A are very similar. Their THMs are also effectively identical, and only sample 615A is presented (Figure 10d). The THM for sample 615A is similar to that for sample 657.

Sample 615A began to accumulate fission tracks in the Maastrichtian, when it cooled rapidly from peak temperatures of more than 125°C (VR = 1.0% R_o; Hacquebard and Donaldson, 1974) to about 100°C. The age of this cooling is constrained by the combination of model results and geological relationships, because it is unlikely that cooling predates the deposition of overlying coal-bearing strata (Jerzykiewicz et al., 1996). The sample remained at about 100°C until the early Eocene (about 50 Ma). Subsequently, it began cooling at an accelerated rate until at least the late Oligocene (about 25 Ma), when it cooled to temperatures below 60°C, where no subsequent significant annealing occurred, and the THMs lose sensitivity. This defines the two cooling events recorded by samples 615A and 615. The first, which started the accumulation of tracks in AFT clocks, is Maastrichtian, and the second is an Eocene and younger cooling event. Samples 657 and 615A are from similar present-day elevations but from opposite sides of Lewis thrust, with kilometers of stratigraphic separation. However, both have similar temperature histories. Therefore, prior to the second cooling event, both samples 615A and 657 were at temperatures between 60 and 110°C, following the start of folding of

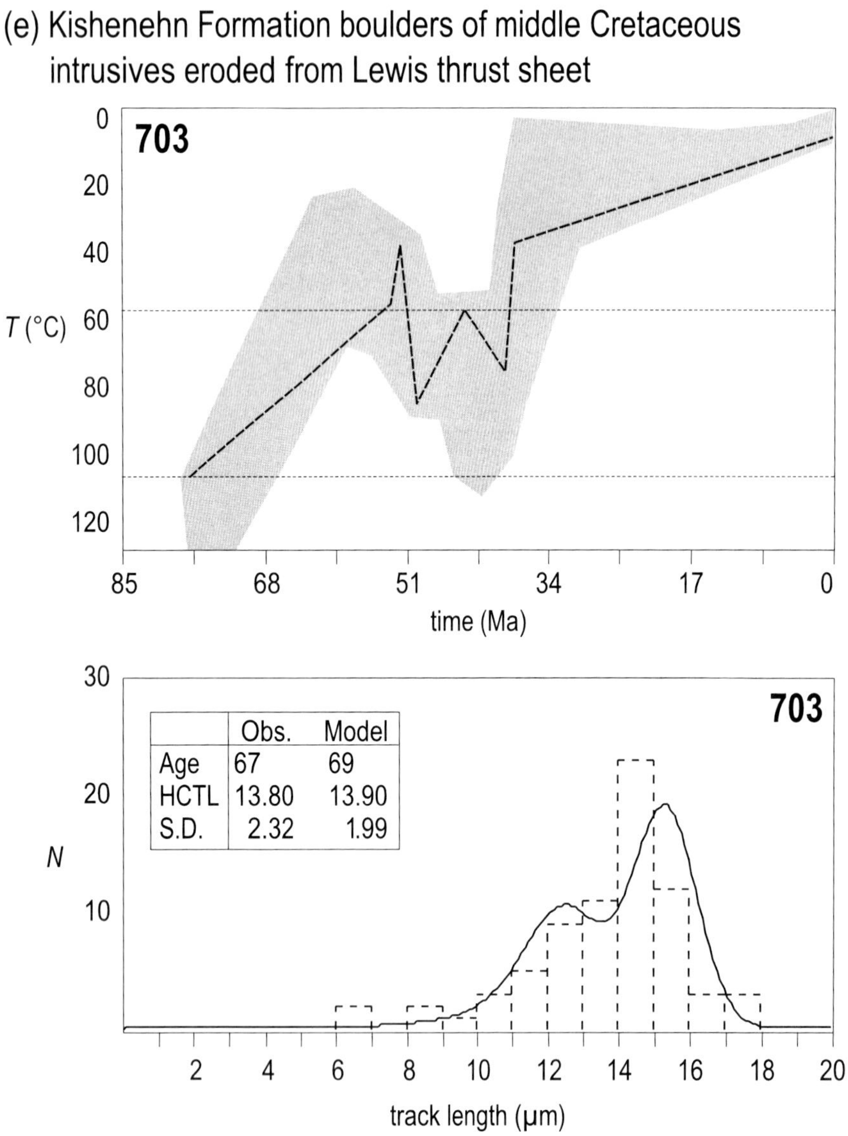

Figure 10. (cont.).

volved a short duration of reheating during the middle and late Eocene to temperatures of about 80°C but not likely exceeding about 90°C. This confirms that the sample cooled relatively rapidly from temperatures of rapid total track annealing at a time slightly prior to the observed FT age during the contractional deformation while the sample was still a part of the Lewis thrust sheet. The THMs indicate that the sample was subjected to only minor reheating subsequent to its deposition in the Kishenehn Formation.

The Eocene and younger temperature history is consistent with the low organic maturity of Kishenehn Formation oil shales (Curiale, 1987). Thus, the THMs of sample 703 are controlled primarily by its provenance. They are roughly comparable to other samples of middle Cretaceous alkaline igneous rocks from the Lewis thrust sheet (samples 668 and 690, Figure 10), until the time when the boulders were eroded from the Lewis thrust sheet and deposited in Kishenehn Formation. The Oligocene and younger segment of the THMs for sample 703 constrain, albeit poorly, the temperature history of both Kishenehn Formation samples (samples 699 and 703) after their deposition. Apatite from sample 699 shows a bimodal distribution of Cl content that is correlated to AFT age, indicating that sample 699 was derived from both the Lewis thrust sheet and another provenance (Figure 8).

Lewis thrust in Akamina syncline (i.e., both samples reached near-current relative structural positions with respect to the basal detachment by the early Eocene).

KISHENEHN FORMATION SEDIMENTS

Sample 703 comprises boulders of middle Cretaceous igneous rock from the late Eocene and Oligocene Kishenehn Formation just west of the Flathead fault (Figure 2). All time-temperature models for sample 703 show a rapid cooling below about 110°C at about 75 Ma. In many models, the sample cooled very rapidly to temperatures less than 70°C, so that prior to or by the end of the Cretaceous, no subsequent significant annealing occurred. Other models suggest that the latter cooling history of sample 703 may have in-

INTERPRETATION

The major cooling events indicated by AFT parameters and THMs can be related to times of major displacements on the Lewis thrust and Flathead normal fault.

Lewis Thrust Sheet Thermal History

Lewis thrust sheet THMs confirm that many samples from the deepest stratigraphic levels of the Lewis thrust sheet cooled rapidly from temperatures more

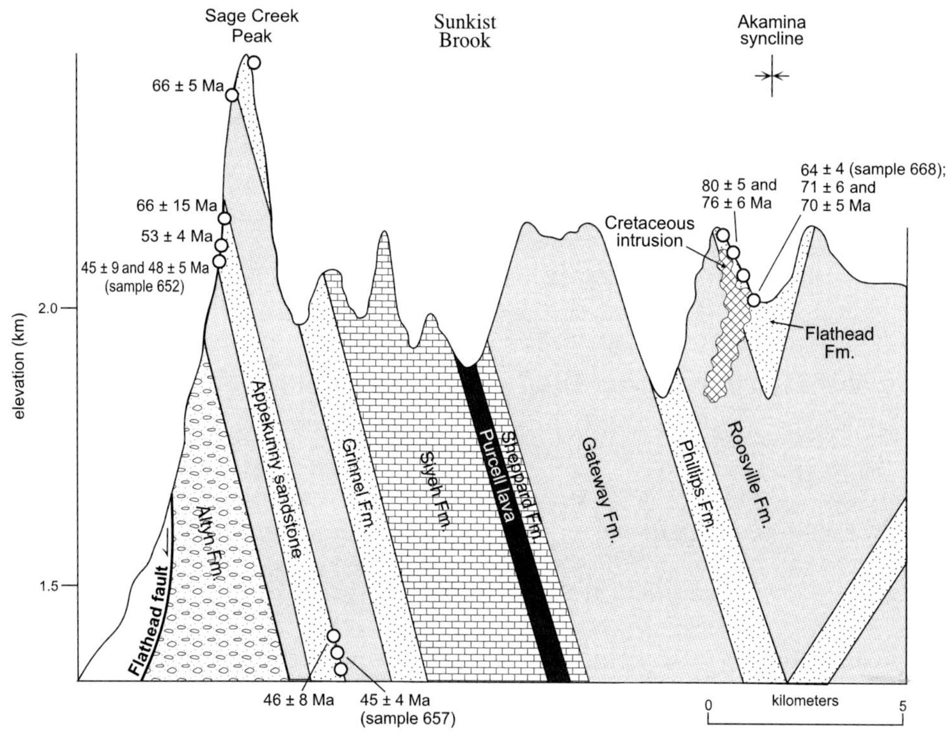

FIGURE 11. Variations of AFT sample ages in the western limb of the Akamina syncline, east of the Flathead fault. Samples near the synclinal hinge have thermal histories illustrated by 668 in Figure 10b. Samples at low elevations have temperature histories like sample 657 in Figure 10c. A band of temperatures between 60 and 110°C (i.e., fossil apatite PAZ) (sample 657) that postdates the folding of Lewis thrust and thrust sheet in Akamina syncline includes these two samples. Note the presence of older (>70 Ma) AFT ages at high elevations in Sage Creek peak and the anomalously younger AFT age (~48 Ma) at a high elevation in Sage Creek peak that is attributed to local hydrothermal activity, similar to current hot-spring activity associated with nearby Flathead fault (Grasby and Hutcheon, 2001).

than 110°C, at a time slightly before their late Campanian (about 75 Ma) measured AFT age. The levels of vitrinite reflectance measurements (VR >0.8% R_o) from Mesozoic strata in the upper parts of the Lewis thrust sheet (Grieve, 1987) show that rocks in the Lewis thrust sheet must have been at temperatures more than 120°C prior to the late Campanian rapid cooling recorded by the AFT clocks. The late Campanian cooling can be correlated with the termination of sedimentary burial and organic maturation of these rocks as they were detached, displaced upward and northeastward along the Lewis thrust fault, and incorporated into the expanding foreland thrust and fold belt. Rapid cooling of the Lewis thrust sheet was underway as the youngest sediments preserved below the Lewis thrust (Maastrichtian–Campanian, Jerzykiewicz et al., 1996) were being deposited.

The late Campanian age (about 75 Ma) for the initiation of major displacement on the Lewis thrust inferred from the apatite fission-track analysis (FTA) is consistent with that for the Lewis–Eldorado–Hoadley thrust faults, lateral equivalents of the Lewis thrust in Montana, inferred from the ages of offset volcanic rocks and crosscutting dikes (Sears, 2001). It is also comparable to the oldest K-Ar radiometric ages obtained from sub-Lewis thrust bentonites (72 Ma, Hoffman et al., 1976). These late Campanian ages differ significantly from the time of faulting (51.5 ± 3.5 Ma) inferred from an "illite age analysis" of fault gouge from the Lewis thrust near the Crowsnest Pass (Van der Pluijm et al., 2001). That study estimated the age of an authigenic illite component in the fault gouge based on the assumption that the authigenic illite was produced by thrust fault displacement and not by other, later processes (e.g., hydrothermal activity). The inferred thrust origin of the illite age implies an age of thrusting for the Lewis fault that is younger than the well-constrained age of the regional crustal extension in the western part of the foreland thrust and fold belt and adjacent Omenica and Interior belts of the southern Canadian Cordillera (Parrish et al., 1988; Carr, 1992; Doughty and Price, 1999, 2000).

The oldest AFT ages and longest mean HCTL in the Lewis thrust sheet occur in samples 690 and 692 (Table 1). This indicates directly that at one time, these samples resided higher in the structure than any of the preserved Lewis thrust sheet east of the Flathead fault. The THM for sample 690 (Figure 10a) indicates only one rapid cooling event to less than 60°C during the Late Cretaceous, which is consistent with a higher structural position than that of sample 668, which preserves a record of higher temperatures in post-Campanian time (Figure 10b).

Samples nearest the Lewis thrust east of the Flathead fault (samples 657 and 615A) have AFT ages (Table 1) and THMs (Figure 10) that are mutually comparable, although they come from opposite sides of the Lewis thrust. In both of these samples, AFT ages are younger than in samples from farther above the thrust (e.g., samples 690 or 668). Generally, the AFT ages in the west limb of

Akamina syncline correlate with present elevation instead of stratigraphic position (Figure 11). From this, we infer that subsequent to the late Campanian cooling event, the Lewis thrust sheet was folded to form the Akamina syncline, and that this happened before both samples 657 and 615 came to reside within a band of temperatures between 60 and 110°C (i.e., an "exhumed" apatite PAZ). The record of this exhumed PAZ is preserved by samples with AFT ages less than 55 Ma at elevations near 1500 m (4900 ft) the age profile (Figure 9b).

Thermal History below the Lewis Thrust

The THMs of sample 615 illustrate the cooling history below the Lewis thrust. Samples from below the Lewis thrust are predominantly from either the Albian Blairmore Group, which includes Crowsnest volcanics, or from the Campanian, and in some places, the lower Maastrichtian parts of the Belly River Formation. Because of structural stacking, Albian samples from below Lewis thrust occur throughout the elevation profile (Table 1; Figure 9). The observed organic maturity of this succession is sufficiently high that all AFT clocks below the thrust were totally reset subsequent to deposition but prior to the recorded cooling history. The THMs show that rocks below the Lewis thrust began to cool during the Maastrichtian, shortly after the youngest preserved strata below Lewis thrust were deposited (Jerzykiewicz et al., 1996, Figure 10). Subsequent to their deformation, rocks below Lewis thrust resided in a band of temperatures between 60 and 110°C (i.e., fossil apatite PAZ), without exceeding their predeformational maximum temperatures, until the second cooling event, generally in the Eocene onward (Figure 10). Temperature-history models for samples from the Calstan Crowsnest and Sage Creek wells (739/1–739/6 and 737) indicate that all apatites below Lewis thrust had begun to accumulate fission tracks before latest Eocene time. However, the range of apatite Cl content and the differences between model and sample apatite annealing parameters limit the interpretation of THMs for these samples.

Implications for the Flathead Fault

Samples 668, 657, and 615A, from east of the Flathead fault, both above and below the Lewis thrust, record the second cooling event that began in the Eocene and resulted in cooling from more than about 80 to less than about 60°C (Figure 10b–d). Prior to this cooling episode, these rocks resided in the PAZ that had been established after the folding of the Lewis thrust. The second cooling event is also recorded by other samples with younger apatite FT ages that occurred lower in the profile, below the Lewis thrust (Figure 10). The second cooling is not recorded in sample 690 from west of the Flathead fault. The beginning of the second cooling event coincides with Eocene rift-fill sedimentation in the Flathead graben in Montana (Constenius, 1996), with displacement on Flathead fault, and with the uplift and erosion of the rocks in its footwall. Extension on the Flathead fault can be correlated with the onset of regional right-hand transtension, crustal extension, and tectonic exhumation of metamorphic core complexes in the interior of the Cordillera (Carr, 1992; Price, 1994; Doughty and Price, 1999, 2000; Lorencak et al., 2001; Vanderhaeghe et al., 2003). Thermal springs occur locally along the Flathead fault (Grasby and Hutcheon, 2001). One FTA sample (sample 652, Figure 11) from the Mesoproterozoic Appekunny Formation at a high elevation (+2015 m; 6610 ft) in the Lewis thrust sheet near the Flathead fault records cooling only during the middle Eocene onward. This sample may have been completely annealed by Eocene hydrothermal activity associated with the Flathead fault zone.

Implications for Kishenehn Formation

The AFT THMs from igneous cobbles of middle Cretaceous alkaline igneous rocks in the Kishenehn Formation in the Flathead graben can be divided into two parts. The younger part records the temperature history since the early Oligocene (33.7 Ma), when these clasts were part of the Kishenehn Formation. The sample also records an earlier temperature history (Figure 10e) during the interval late Campanian to late Eocene, when it was part of the Lewis thrust sheet from which it was eroded. Analysis of the THMs indicate that little reduction in age and track length can be attributed to the early Oligocene and younger portion of the temperature history (sample 703, Figure 10e). Most of the recorded temperature history pertains to the interval late Campanian to late Eocene, when the sample was still part of the Lewis thrust sheet. The thermal history of boulders and cobbles of Cretaceous intrusions preserved the late Campanian cooling without significant subsequent annealing, like samples from the highest preserved elevations and structural positions in the Lewis thrust sheet (e.g., samples 690 and 668). This is consistent with the low organic maturity of Kishenehn Formation oil shale outcrops (Curiale, 1987). From this, we conclude that sample 703 preserves a provenance thermal-history record without significant postsedimentation annealing.

Similarly, from stratigraphic relationships, we can infer that apatite grains from the Kishenehn Formation sandstones (sample 699; Table 1; Figure 8) also record primarily the temperature history of their provenance. However, unlike the middle Cretaceous alkaline igneous rocks from the Lewis thrust sheet, the apatite grains in the sandstones exhibit a much wider range of Cl content and AFT parameters (sample 699, Figure 8a). Generally, the apatite grains in sample 699 either have single-grain Cl contents and AFT ages like the middle

Cretaceous alkaline igneous samples from the Lewis thrust sheet or have more Cl-rich grains that have generally older single-grain ages (compare Figure 8a, b). This suggests that the apatite grains in the Kishenehn Formation sandstone samples come from two distinctive provenances. One provenance has an apatite composition similar to the Lewis thrust sheet samples, whereas another has apatite with higher Cl content and AFT ages more than 100 Ma, which is derived from an unidentified provenance.

Geothermal Gradient Variations

East of the Flathead fault current geothermal gradients are about 17°C/km (K. Root, Shell Canada Resources Ltd., 1992, personal communication). Considering the current geothermal gradient and the geological history of the region, it was unexpected to find apatites in the Lewis thrust sheet with low Cl content (≤0.4 wt.%) having AFT ages more than or equal to 65 Ma and long mean HCTL (≥13.45 μm) (Figure 5). If the current geothermal gradient had prevailed during and immediately after the thrusting and folding, all samples from within and below the Lewis thrust sheet east of the Flathead fault would have been at temperatures more than 110°C prior to displacement on the Flathead fault. Accordingly, geothermal gradients during the period of thrusting and folding must have been lower than the current gradient.

In the Lewis thrust sheet, coal rank variations with respect to stratigraphic position follow Hilt's law (Grieve, 1987). A vitrinite reflectance gradient for the Lewis thrust sheet can be estimated using data from Alberta Group strata in the McEvoy syncline (VR = 0.75% R_o) and the underlying Blairmore Group and Kootenay Group (VR = 1.2% R_o) strata exposed along the edges of the Fernie Basin (Figure 2). This implies that the Lewis thrust sheet geothermal gradient just prior to thrusting and cooling can be estimated (Middleton, 1982) to be about 20°C/km, and that the predeformational geothermal gradient in Lewis thrust sheet was comparable, or slightly higher than, the present temperature gradient. Temperature-history models for rocks from east of the Flathead fault indicate that the Appekunny Formation was at temperatures between about 105 and 95°C between approximately 75 and about 60 Ma (sample 657, Figure 10), and that temperatures at the Proterozoic–Phanerozoic contact were between 100 and 85°C during the same interval (sample 688, Figure 10). The small temperature difference between these two sample localities, which are separated vertically by 638 m (2093 ft), suggests a low paleogeothermal gradient (approximately 6.4 to about 12.8°C), as does the relatively low temperature at the then still deeply buried (>4 km; >2 mi) Precambrian–Cambrian contact. Early Oligocene geothermal gradients can also be estimated using structural and stratigraphic data (Figure 5). Assuming that sample 657 from the Appekunny Formation in the footwall of the Flathead fault was overlain by the middle Blairmore Group strata that now lie below Kishenehn Formation in the hanging wall of the Flathead fault, the early Oligocene geothermal gradient is estimated to have been between approximately 8.6 and about 12°C/km (i.e., [100 to 75°C (sample 657, Figure 10) – 10°C (surface temperature)]/7.5 km [Figure 3]). Therefore, during thrusting and folding, the geothermal gradient east of the Flathead fault was probably between about 8.6 and 12 °C/km.

Mechanisms Affecting the Thermal History

Our results show that changes in temperature and geothermal gradient can be correlated with deep sedimentary burial in the Cordilleran foreland basin, with Late Cretaceous and Paleocene thrust faulting and thrust-related folding, and with Eocene and younger normal faulting. The inferred temperature and geothermal gradient changes are large and rapid in a tectonic setting where changes in basal heat flow are not expected (Beaumont, 1981). In the study area, the present, relatively low geothermal gradient has been attributed to advective cooling by deeply penetrating, topographically driven groundwater flow (Majorowicz et al., 1985; Osadetz et al., 1990). We attribute the regional cooling associated with the thrusting and thrust-related folding to deeply penetrating groundwater flow. Advective heat transport by deeply penetrating meteoric water may be an integral part of the development of a foreland thrust and fold belt. Topographic relief generated during thrusting and folding drives deep, regional penetration of cold meteoric water, whereas the dilation caused by the brittle deformation (e.g., fracturing) that accompanies thrusting and thrust-related folding enhances hydraulic conductivity. Normal faulting also generates topographic relief and enhanced hydraulic conductivity because of similar effects. This mechanism also affects adjacent regions of the hydrodynamic regime (Osadetz et al., 1990; Bustin, 1991). No other possible mechanism accounts for the combination of direction (cooling), range, and rapidity of temperature and geothermal gradient changes.

Implications for Tectonics, Organic Maturation, and Petroleum Systems

Tectonic history affects temperatures and geothermal gradients that in turn control coalification and petroleum systems. It has commonly been assumed that conductive heat transfer dominated overthrust structure thermal evolution models (Oxburgh and Turcotte, 1974; Wygrala et al., 1990). This study and many others

(Majorowicz et al., 1985; Osadetz et al., 1990; Bustin, 1991; Burtner and Nigrini, 1994; Burtner et al., 1994) suggest that rapid, advective cooling caused by deeply penetrating meteoric water may modify the geothermal environment significantly.

Commonly, coalification in overthrust belts is observed to increase with stratigraphic instead of structural depth; the benchmark study was performed in this study region (Hacquebard and Donaldson, 1974). Hacquebard and Donaldson (1974) inferred that peak temperatures in Mesozoic rocks lying below and east of the Lewis thrust were reached simultaneously prior to the deformation. This has been inferred to indicate that petroleum generation occurred before the traps were formed, a temporal relationship considered to be unfavorable for petroleum accumulation (Gautier et al., 1998). However, the very large petroleum accumulation that occurs in the Waterton field (about 3 tcf natural gas recoverable + crude oil, Figure 1) is paradoxical if the relative timing of petroleum generation and trap formation is unfavorable. Temperature-history modeling based on AFT thermochronology can resolve this apparent paradox.

The stratigraphic depth control on coalification is clearly illustrated by the similarity of vitrinite reflectance in the Upper Cretaceous (Alberta Group–Belly River Formation; about VR = 0.8% R_o) and Lower Cretaceous (Kootenay Group, about VR = 1.2% R_o) strata, both above and below the Lewis thrust (Hacquebard and Donaldson, 1974; Pearson and Grieve, 1985; Grieve, 1987; T. Gentzis, K. G. Osadetz, and S. Feinstein, 1989, personal communication). Prior to extension on Flathead fault, there was a structural separation of about 8 km (5 mi) between a given formation in the Lewis thrust sheet and the same formation below it (Figure 5). Still, the coalification of these rocks follows Hilt's law: Coalification increases with respect to stratigraphic position, regardless of structural position (Hacquebard and Donaldson, 1974; Pearson and Grieve, 1985; Grieve, 1987).

However, the AFT temperature histories show that rocks in Lewis thrust sheet were already cooling and deforming (Figure 10) prior to the end of sedimentation and coalification in the succession below Lewis thrust (Jerzykiewicz et al., 1996). Hence, different structural elements reached peak temperatures and observed coalification levels at different times. Either the rocks below Lewis thrust reached maximum paleotemperatures prior to being deformed in response to a sedimentary burial by a succession that is no longer preserved (Hacquebard and Donaldson, 1974), or they reached peak temperatures because of tectonic burial below the cooling and eroding Lewis thrust sheet, in an autochthonous "subthrust" setting, prior to being incorporated into the deformation belt. The second possibility that was not recognized previously exists because of both the large footwall flat in the Lewis thrust in the Front Ranges and foothills (Figure 4) (see Fermor and Moffat, 1992, their figure 6) and the Akamina syncline. Because of the footwall flat in the Lewis thrust, the overthrusting of Mesozoic rocks would result in a "pseudostratigraphic" burial of the subthrust autochthonous footwall by the Lewis thrust sheet, until the autochthonous footwall rocks were incorporated into the deformation belt. The Akamina syncline indicates that displacement on the Lewis thrust was largely complete prior to the deformation of its footwall succession (Constenius, 1996).

In summary, different structural elements reached peak temperatures and organic maturities at different times, as the thrust and fold belt expanded toward the northeast. Both rapid cooling and a major reduction in the geothermal gradient accompanied the displacement of the Lewis thrust sheet. Coalification of the Mesozoic strata that now lie beneath or east of the Lewis thrust occurred in the Maastrichtian or later periods, based on stratigraphic considerations alone, when the Lewis thrust sheet was already moving. Therefore, petroleum generation below the Lewis thrust, where large reserves occur, resulted from either a deep stratigraphic burial by a thick, very rapidly deposited succession that is no longer preserved or in response to tectonic burial by the cooling and eroding Lewis thrust sheet as it overrode its autochthonous footwall (Figure 5). The AFT-derived THMs suggest that the time available for either mechanism is relatively short, reducing the interval between petroleum generation and structural trap formation, which, in turn, reduces the risks associated with petroleum accumulation in the Canadian overthrust belt.

CONCLUSIONS

Apatite fission-track data from rocks above and below the Lewis thrust lying in the footwall and hanging wall of the Flathead fault record different components of the thermal history, depending on their individual structural and stratigraphic position. Apatite fission-track clocks in the Lewis thrust sheet record a rapid cooling that began in the late Campanian (about 75 Ma). This cooling episode, which is observed on both sides of the Flathead fault, coincides with major displacement on the Lewis thrust. During this first cooling event, Mesoproterozoic strata deeper in Lewis thrust sheet, east of the Flathead fault, cooled from temperatures greater than about 110°C to temperatures about 110–60°C. Subsequent to the major displacement on Lewis thrust, the Lewis thrust sheet was folded to form the Akamina syncline. After this folding, a temperature band equivalent to the PAZ was superimposed on the deformed structure, above and below the folded thrust. In rocks east of the Flathead graben, the AFT clocks record a subsequent Eocene and younger cooling event that coincided with the extensional displacements on Flathead fault and accompanying uplift and erosion of its footwall. This second event accelerated the cooling that began in the Late Cretaceous, leading to the commencement of the AFT record below the Lewis thrust.

Sets of similar THMs that are consistent with the regional structural history can account for observed variations in AFT parameters at the various levels that are exposed in the Lewis thrust sheet and that are penetrated below the Lewis thrust by deep wells. Apatite fission-track data from lower Oligocene Kishenehn Formation sediments filling in the Flathead graben preserve the thermal-history record of the sediment source regions without significant post-Oligocene annealing. Some of the sediment sources for the Kishenehn Formation can be directly linked lithologically (Price, 1965; McMechan and Price, 1980; McMechan, 1981) and thermochronologically to the Lewis thrust sheet in the footwall of Flathead fault. These observations and analyses confirm the history of displacement on the Lewis thrust fault, its folding caused by the formation of sub-Lewis antiformal thrust duplexes, and its extension caused by displacements on the Flathead normal fault.

Using AFT thermal histories from the footwall of Flathead fault, we conclude that geothermal gradients in the Lewis thrust sheet were lower (about 8.6–12°C/km) from the onset of thrusting until displacement on the Flathead normal fault than they are now (about 17°C/km). Episodes of cooling in the thrust and fold belt of the southern Canadian Cordillera coincide with times of orogenic deformation that created topographic relief. We infer that the changes in geothermal gradients since the onset of the first cooling in the Upper Cretaceous were caused by tectonically induced, topographically driven, advective heat transfer by deeply penetrating meteoric water flow, a process that still affects the thermal regime in these rocks. This is a complicated heat-transfer mechanism that can affect organic maturation history and petroleum systems in overthrust belts.

ACKNOWLEDGMENTS

We thank Elizabeth Macey for her assistance in the revisions of the figures. The chapter benefited from the thorough and constructive reviews by Th. Moore, G. Stockmal, W. Sassi, and an anonymous reviewer. This study was partially funded by the Earth Sciences Sector, Geological Survey of Canada— Calgary and the Office for Energy Research and Development (Canada); Geological Survey of Canada Contribution no. 2003099. It was also partially funded by the Australian Institute of Nuclear Science and Engineering, the Australian Research Council, and the Natural Science and Engineering Research Council of Canada.

REFERENCES CITED

Angevine, C. L., and L. Turcotte, 1983, Oil generation in overthrust belts: AAPG Bulletin, v. 67, no. 2, p. 235–241.

Bally, A. W., P. L. Gordy, and G. A. Stewart, 1966. Structure, seismic data, and orogenic evolution of southern Canadian Rocky Mountains: Bulletin of Canadian Petroleum Geology, v. 14, p. 337–381.

Beaumont, C., 1981, Foreland basins: Geophysical Journal of the Royal Astronomical Society, v. 65, p. 291–329.

Bellemans, F., F. De Corte, and P. Van den Haute, 1995, Composition of SRM and CN U-doped glasses: Significance for their use as thermal neutron fluence monitors in fission track dating: Radiation Measurements, v. 24, no. 2, p. 153–160.

Boyer, S. E., 1992, Geometric evidence for synchronous thrusting in the southern Alberta and northwest Montana thrust belts, *in* K. R. McClay, ed., Thrust tectonics: New York, Chapman and Hall, p. 377–390.

Burtner, R. L., and A. Nigrini, 1994, Thermochronology of the Idaho–Wyoming thrust belt during the Sevier orogeny: A new calibrated, multiprocess thermal model: AAPG Bulletin, v. 78, p. 1586–1612.

Burtner, R. L., A. Nigrini, and R. A. Donelick, 1994, Thermochronology of Lower Cretaceous source rocks in the Idaho–Wyoming thrust belt: AAPG Bulletin, v. 78, p. 1613–1636.

Bustin, R. M., 1991, Organic maturity in the Western Canada sedimentary basin, *in* W. Kalkreuth, R. M. Bustin, and A. R. Cameron, eds., Recent advances in organic petrology and geochemistry: A symposium honoring Dr. P. Hacquebard: International Journal of Coal Geology, v. 19, p. 319–358.

Bustin, R. M., and T. D. J. England, 1989, Timing of organic maturation (coalification) relative to thrusting in the southeastern Canadian Cordillera: International Journal of Coal Geology, v. 13, p. 327–339.

Carr, S., 1992, Tectonic setting and U-Pb geochronology of the early Tertiary Ladybird leucogranite suite, Thor–Odin–Pinnacles area, southern Omineca belt, British Columbia: Tectonics, v. 11, p. 258–278.

Carr, S. D., R. R. Parrish, and R. L. Brown, 1987, Eocene structural development of the Valhalla complex, southeastern British Columbia: Tectonics, v. 6, p. 175–196.

Constenius, K. D., 1988, Structural configuration of the Kishenehn basin delineated by geophysical methods, northwestern Montana and southeastern British Columbia: The Mountain Geologist, v. 25, p. 13–28.

Constenius, K. D., 1996, Late Paleogene extensional collapse of the Cordilleran foreland fold and thrust belt: Geological Society of America Bulletin, v. 108, no. 1, p. 20–39.

Curiale, J. A., 1987, Steroidal hydrocarbons of the Kishenehn Formation, northwest Montana: Organic Geochemistry, v. 11, p. 233–244.

Dahlstrom, C. D. A., 1970, Structural geology in the eastern margin of the Canadian Rocky Mountains: Bulletin of Canadian Petroleum Geology, v. 18, p. 332–406.

Doughty, P. T., and R. A. Price, 1999, Tectonic evolution of the Priest River complex, northern Idaho and Washington: A reappraisal of the Newport fault with new

insights on metamorphic core complex formation: Tectonics, v. 18, p. 375–393.
Doughty, P. T., and R. A. Price, 2000, Geology of the Purcell Trench rift valley and Sandpoint conglomerate: Eocene en echelon normal faulting and synrift sedimentation along the eastern flank of the Priest River metamorphic complex, northern Idaho: Geological Society of America Bulletin, v. 112, p. 1356–1374.
Douglas, R. J. W., 1950, Callum Creek, Langford Creek and Gap map areas: Geological Survey of Canada Memoir 255, 124 p.
Edman, J. D., and R. C. Surdam, 1984, Influence of overthrusting on maturation of hydrocarbons in Phosphoria Formation, Wyoming–Idaho–Utah overthrust belt: AAPG Bulletin, v. 68, p. 1803–1817.
England, T. D. J., 1984, Thermal maturation of the Western Canadian sedimentary basin in the Rocky Mountain foothills and plains of Alberta south of the Red Deer River: M.Sc. thesis, University of British Columbia, Vancouver, 171 p.
England, T. D. J., and R. M. Bustin, 1986a, Thermal maturation of the Western Canadian sedimentary basin south of the Red Deer River: I. Alberta plains: Bulletin of Canadian Petroleum Geology, v. 34, p. 71–90.
England, T. D. J., and R. M. Bustin, 1986b, Effect of thrust faulting on organic maturation in the southeastern Canadian Cordillera: Organic Geochemistry, v. 10, p. 609–616.
Ewing, T. E., 1980, Paleogene tectonic evolution of the Pacific northwest: Journal of Geology, v. 88, p. 619–638.
Faure, J.-L., K. G. Osadetz, Z. Narimane-Benaouali, F. Schneider, and F. Roure, 2004, Kinematic and petroleum modelling of the Alberta foothills and adjacent foreland west of Calgary: Revue de l'Institut Francais du Petrole, v. 59, p. 81–108.
Fermor, P., 1999, Aspects of the three-dimensional structure of the Alberta foothills and Front Ranges: Geological Society of America Bulletin, v. 111, no. 3, p. 317–346.
Fermor, P. R., and I. W. Moffat, 1992, Tectonics and structure of the Western Canada foreland basin, *in* R. W. Macqueen and D. A. Leckie, eds., Foreland basins and fold belts: AAPG Memoir 55, p. 81–105.
Fermor, P. R., and R. A. Price, 1987, Multiduplex structure along the base of the Lewis thrust sheet in the southern Canadian Rockies: Bulletin of Canadian Petroleum Geology, v. 35, p. 159–185.
Fitzgerald, P. G., and A. J. W. Gleadow, 1988, Fission-track geochronology, tectonics and structure of the Transantarctic Mountains in northern Victoria Land, Antarctica: Chemical Geology (Isotope Geoscience Section), v. 73, p. 169–198.
Furlong, K. P., and J. D. Edman, 1984, Graphic approach to determination of hydrocarbon maturation in overthrust terrains: AAPG Bulletin, v. 68, p. 1818–1824.
Gabrielse, H., and C. J. Yorath, 1992, Geology of the Cordilleran orogen in Canada, Geological Survey of Canada: Geology of Canada, no. 4, 844 p.
Galbraith, R. F., 1990, The radial plot: Graphical assessment of spread in ages: International Journal of Radiation Measurement and Applied Instrumentation, Part D, Nuclear Tracks and Radiation Measurement, v. 17, p. 207–214.
Galbraith, R. F., and G. M. Laslett, 1993, Statistical models for mixed fission track ages: Nuclear Tracks and Radiation Measurement, v. 21, p. 459–470.
Gallagher, K., 1995, Evolving temperature histories from apatite fission track data: Earth and Planetary Science Letters, v. 136, p. 421–435.
Gallagher, K., R. W. Brown, and C. Johnson, 1998, Fission track analysis and its applications to geological problems: Annual Review of Earth and Planetary Sciences, v. 26, p. 519–572.
Gautier, D. L., G. L. Dolton, and E. D. Attanasi, compilers, 1998, 1995 National oil and gas assessment of onshore federal lands: U.S. Geological Survey Open-file Report 95-0075-N, 64 p.
Ghent, E. D., and B. E. Miller, 1974, Zeolite and claycarbonate assemblages in the Blairmore Group (Cretaceous) southern Alberta foothills, Canada: Contributions to Mineralogy and Petrology, v. 44, p. 313–329.
Gleadow, A. J. W., and R. W. Brown, 2000, Fission track thermochronology and the long-term denudational response to tectonics, *in* M. A. Summerfield, ed., Geomorphology and global tectonics: Chichester, John Wiley and Sons, Ltd., p. 57–75,
Gleadow, A. J. W., and P. G. Fitzgerald, 1987, Uplift history and structure of the Transantarctic Mountains: New evidence from fission track dating of basement apatites in the Dry Valleys area, southern Victoria Land: Earth and Planetary Science Letters, v. 82, p. 1–14.
Gleadow, A. J. W., I. R. Duddy, P. F. Green, and J. F. Lovering, 1986, Confined fission track lengths in apatite: A diagnostic tool for thermal history analysis: Contributions to Mineralogy and Petrology, v. 94, p. 405–415.
Gleadow A. J. W., D. X. Belton, B. P. Kohn, and R. W. Brown, 2002, Fission track dating of phosphate minerals and the thermochronology of apatite, *in* J. Hughes, M. Kohn, and J. Rakovan, eds., Phosphates, geochemical, geobiological and materials importance: Reviews in Mineralogy and Geochemistry, v. 48, p. 579–630.
Gordy, P. L., and G. Edwards, 1962, Age of the Howell Creek intrusives: Journal of the Alberta Society of Petroleum Geologists, v. 10, no. 7, p. 369–372.
Grasby, S. E., and I. Hutcheon, 2001, Controls on the distribution of thermal springs in the southern Canadian Cordillera: Canadian Journal of Earth Sciences, v. 38, no. 3, p. 427–440.
Green, P. F., 1985, Comparison of zeta calibration baselines for FT dating of apatite, zircon and sphene: Chemical Geology, v. 58, p. 1–22.
Green, P. F., I. R. Duddy, A. J. W. Gleadow, and P. R. Tingate, 1985, Fission-track annealing in apatite: Track length measurement and the form of the Arrhenius plot: Nuclear Tracks and Radiation Measurement, v. 10, p. 323–328.
Green, P. F., I. R. Duddy, A. J. W. Gleadow, P. R. Tingate, and G. M. Laslett, 1986, Thermal annealing of fission tracks in apatite: 1. A qualitative description: Chemical Geology (Isotope Geoscience Section), v. 59, p. 237–253.
Grieve, D. A., 1987, Coal rank distribution, Flathead coalfield, southeastern British Columbia (82G/2, 82G/7),

in Geological fieldwork, 1986: British Columbia Ministry of Energy, Mines and Petroleum Resources Paper 1987-1, p. 361–364.

Hacquebard, P. A., 1977, Rank of coal as an index of organic metamorphism for oil and gas in Alberta, *in* G. Deroo, T. G. Powell, B. Tissot, and R. G. McCrossan, eds., The origin and migration of petroleum in the Western Canadian sedimentary basin, Alberta—A geochemical and thermal maturation study: Geological Survey of Canada Bulletin, v. 262, p. 11–22.

Hacquebard, P. A., and J. R. Donaldson, 1974, Rank studies of coals in the Rocky Mountains and inner foothills belt, Canada, *in* R. R. Dutcher, P. A. Hacquebard, J. M Schopf, and J. A Simon, eds., Carbonaceous materials as indicators of metamorphism: Geological Society of America Special Paper 153, p. 75–94.

Hoffman, J., and J. Hower, 1979, Clay mineral assemblages as low grade metamorphic geothermometers: Application to the thrust faulted disturbed belt of Montana, U.S.A., *in* P. A. Scholle and P. R. Schluger, eds., Aspects of diagenesis: SEPM Special Publication 26, p. 55–79.

Hoffman, J., J. Hower, and J. L. Aronson, 1976, Radiometric dating of time of thrusting in the disturbed belt of Montana: Geology, v. 4, p. 16–20

Hutcheon, I., A. Oldershaw, and E. D. Ghent, 1980, Diagenesis of Cretaceous sandstones of the Kootenay Formation at Elk Valley (southeastern British Columbia) and Mt. Allan (southwestern Alberta): Geochimica et Cosmochimica Acta, v. 44, p. 1425–1435.

Issler, D. R., C. Beaumont, S. D. Willett, R. A. Donelick, J. Mooers, and A. Grist, 1990, Preliminary evidence from apatite fission-track data concerning the thermal history of the Peace River arch region, Western Canada sedimentary basin: Bulletin of Canadian Petroleum Geology, v. 38A, p. 250–269.

Jerzykiewicz, T., A. R. Sweet, and D. H. McNeil, 1996, Shoreface of the Bearpaw Sea in the footwall of the Lewis thrust, southern Canadian Cordillera, Alberta, *in* Current Research 1996-A: Geological Survey of Canada, p. 15–163.

Jones, P. B., 1977, The Howell Creek structure—A Paleogene rock slide in the southern Canadian Rocky Mountains: Bulletin of Canadian Petroleum Geology, v. 25, p. 868–881.

Kalkreuth, W., and M. E. McMechan, 1984, Regional pattern of thermal maturation as determined from coal-rank studies, Rocky Mountain foothills and Front Ranges north of Grande Cache, Alberta—Implications for petroleum exploration: Bulletin of Canadian Petroleum Geology, v. 32, p. 249–271.

Kohn, B. P., K. G. Osadetz, and R. K. Bezys, 1995, Apatite fission-track dating of two crater structures in the Canadian Williston basin: Bulletin of Canadian Petroleum Geology, v. 43, no. 1, p. 54–64.

Kohn, B. P., A. J. W. Gleadow, R. W. Brown, K. Gallagher, P. B. O'Sullivan, and D. A. Foster, 2002, Shaping the Australian crust over the last 300 million years: Insights from fission track thermotectonic and denudation studies of key terranes: Australian Journal of Earth Sciences, v. 49, p. 697–717.

Krause, H. R., C. A. Viau, L. S. Eliuk, A. Ueda, and S. Halas, 1988, Chemical and isotopic evidence of thermochemical sulphate reduction by light hydrocarbon gases in deep carbonate reservoirs: Nature, v. 333, p. 415–419.

Labrecque, J. E., and E. W. Shaw, 1973, Restoration of basin and range faulting across the Howell Creek window and Flathead Valley of southeastern British Columbia: Bulletin of Canadian Petroleum Geology, v. 21, no. 1, p. 117–122.

Laslett, G. M., P. F. Green, I. R. Duddy, and A. J. W. Gleadow, 1987, Thermal annealing of fission tracks in apatite: 2. A quantitative analysis: Chemical Geology (Isotope Geoscience Section), v. 65, p. 1–13.

Lorencak, M., D. Seward, O. Vanderhaeghe, C. Teyssier, and J. P. Burg, 2001, Low-temperature cooling history of the Shuswap metamorphic core complex, British Columbia: Constraints from apatite and zircon fission-track ages: Canadian Journal of Earth Sciences, v. 38, no. 11, p. 1615–1625.

Mack, G. H., and T. Jerzykiewicz, 1989, Provenance of post-Wapiabi sandstones and its implications for Campanian to Palaeocene tectonic history of the southern Canadian Cordillera: Canadian Journal of Earth Sciences, v. 26, p. 665–676.

Magara, K., 1976, Thickness of removed sedimentary rocks, paleopore pressure and paleotemperature, southwestern part of Western Canada basin: AAPG Bulletin, v. 60, p. 554–565.

Majorowicz, J. A., F. W. Jones, H. L. Lam, and A. M. Jessop, 1985, Terrestrial heat flow and geothermal gradients in relation to hydrodynamics in the Alberta basin, Canada: Journal of Geodynamics, v. 4, p. 265–283.

Majorowicz, J. A., F. W. Jones, M. E. Ertman, K. G. Osadetz, and L. D. Stasiuk, 1990, The relationship between thermal maturation gradients, geothermal gradients and estimates of the thickness of the eroded foreland section, southern Alberta Plains, Canada: Marine and Petroleum Geology, v. 7, p. 138–152.

McMechan, R. D., 1981, Stratigraphy, sedimentology, structure and tectonic implications of the Oligocene Kishenehn Formation, Flathead Valley graben, southeastern British Columbia: Ph.D. thesis, Queens University, Kingston, 327 p.

McMechan, R. D., and R. A. Price, 1980, Reappraisal of reported unconformity in the Palaeogene (Oligocene) Kishenehn Formation: Implications for Cenozoic tectonics in the Flathead valley graben, southeastern British Columbia: Bulletin of Canadian Petroleum Geology, v. 28, p. 37–45.

McMechan, M. E., and R. I. Thompson, 1989, Chapter 4: Structural style and history of the Rocky Mountain fold and thrust belt, *in* B. D. Ricketts, ed., Western Canada sedimentary basin: A case history: Canadian Society of Petroleum Geologists Special Paper 30, p. 47–76.

McMechan, M. E., and R. I. Thompson, 1993, The Canadian Cordilleran fold and thrust belt south of 66 influence on the Western Interior Basin, *in* W. G. E. Caldwell and E. G. Kauffman, eds., Evolution of the Western Interior Basin: Geological Association of Canada Special Paper 39, p. 73–90.

Middleton, M. F., 1982, Tectonic history from vitrinite

reflectance: Geophysical Journal of the Royal Astronomical Society, v. 68, no. 1, p. 121–132.

Monger, J. W. H., and R. A. Price, 1979, Geodynamic evolution of the Canadian Cordillera— Progress and problems: Canadian Journal of Earth Sciences, v. 16, p. 771–791.

Mudge, M. R., and R. L. Earhart, 1980, The Lewis thrust fault and related structures in the disturbed belt, northwestern Montana: U.S. Geological Survey Professional Paper 1174, 18 p.

Naeser, C. W., 1979, Fission-track dating and geologic annealing of fission tracks, *in* E. Jager and J. C. Hunziker, eds., Lectures in isotope geology: Heidelberg, Springer-Verlag, p. 154–169.

Nurkowski, J. R., 1984, Coal quality, coal rank variation and its relation to reconstructed overburden, Upper Cretaceous and Tertiary plains coals, Alberta, Canada: AAPG Bulletin, v. 68, p. 285–295.

Orr, W. L., 1974, Changes in sulfur content and isotopic ratios of sulfur during petroleum maturation: AAPG Bulletin, v. 55, no. 11, part 1, p. 2295–2318.

Osadetz, K. G., D. E. Pearson, and L. D. Stasiuk, 1990, Paleogeothermal gradients and changes in the geothermal gradient field of the Alberta Plains; Current research part D: Geological Survey of Canada Paper 90-1D, p. 165–178.

O'Sullivan, P. B., and R. R. Parrish, 1995, The importance of apatite composition and single grain ages when interpreting fission track data from plutonic rocks: A case study from the Coast Ranges, British Columbia: Earth and Planetary Science Letters, v. 132, p. 213–224.

Oxburgh, E. R., and D. L. Turcotte, 1974, Thermal gradients and regional metamorphism in overthrust terrains with special reference to the eastern Alps: Schweizerische Mineralogische un Petrographische Mitteilungen, v. 54, p. 641–662.

Palmer, A. R., and J. Geissman, 1999, 1999 geologic time scale: Geological Society of America: http://www.geosociety.org/science/timescale/timescl.pdf (accessed July 27, 2004).

Parrish, R. R., S. D. Carr, and D. L. Parkinson, 1988, Eocene extensional tectonics and geochronology of the southern Omineca belt, British Columbia and Washington: Tectonics, v. 7, p. 181–212.

Pearce, T. H., 1970, The analcite-bearing volcanic rocks of the Crowsnest Formation, Alberta: Canadian Journal of Earth Sciences, v. 7, p. 46–66.

Pearson, D. E., and D. A. Grieve, 1985, Rank variation, coalification and coal quality in the Crowsnest coalfield, British Columbia: Canadian Institute of Mining and Metallurgy Bulletin, v. 78, no. 881, p. 39–46.

Peper, T., 1993, Tectonic control on the sedimentary record in foreland basins: Inferences from quantitative subsidence analyses and stratigraphic modeling. Thesis and supplement, Vrije Universiteit Amsterdam, CIP-DATA Doninklijke Bibliotheek, Den Haag, 187 p.

Price, R. A., 1962, Fernie map area, east half, Alberta and British Columbia: Geological Survey of Canada Paper 61-24, 65 p.

Price, R. A., 1964, The Precambrian Purcell system in the Rocky Mountains of southern Alberta and British Columbia: Bulletin of Canadian Petroleum Geology, v. 12, p. 399–426.

Price, R. A., 1965, Flathead map-area, British Columbia and Alberta: Geological Survey of Canada Memoir 336, 221 p.

Price, R. A., 1973, Large-scale gravitational flow of supracrustal rocks, southern Canadian Rockies, *in* K. A. De Jong and R. Scholten, eds., Gravity and tectonics: New York, Wiley and Sons, p. 491–502.

Price, R. A., 1981, The Cordilleran thrust and fold belt in the southern Canadian Rocky Mountains, *in* K. R. McClay and N. J. Price, eds., Thrust and nappe tectonics. Geological Society (London) Special Publication 9, p. 427–448.

Price, R. A., 1994, Chapter 2: Cordilleran tectonics and the evolution of the Western Canada sedimentary basin, *in* G. Mossop and I. Shetsen, compilers, Geological atlas of the Western Canada sedimentary basin: Calgary and Edmonton, Canadian Society of Petroleum Geologists and Alberta Research Council, p. 13–24.

Price, R. A., and P. R. Fermor, 1985, Structure section of the Cordilleran foreland thrust and fold belt west of Calgary, Alberta: Geological Survey of Canada Paper 84-14, 1 p.

Sears, J. W., 2001, Emplacement and denudation history of the Lewis–Eldorado–Hoadley thrust slab in the northern Montana Cordillera, U.S.A.: Implications for steady-state orogenic processes: American Journal of Science, v. 301, p. 359–373.

Stockmal, G. S., K. G. Osadetz, D. Lebel, and P. K. Hannigan, 2001, Structure and hydrocarbon occurrence, Rocky Mountain foothills and Front Ranges, Turner Valley to Waterton Lake— Fieldtrip Guidebook: Geological Survey of Canada Open-file Report 4111, 161 p.

Vanderhaeghe, O., C. Tessier, I. McDougall, and W. J. Dunlap, 2003, Cooling and exhumation of the Shuswap metamorphic core complex constrained by $^{40}Ar/^{39}Ar$ thermochronology: Geological Society of America Bulletin, v. 115, p. 200–216.

Van der Pluijm, B. A., C. M. Hall, P. J. Vrolijk, D. R. Pevear, and M. C. Covey, 2001, The dating of shallow faults in the Earth's crust: Nature, v. 412, p. 172–175.

van der Velden, A. J., and F. A. Cook, 1994, Displacement of the Lewis thrust sheet in southwestern Canada: New evidence from seismic reflection data: Geology, v. 22, p. 819–822.

Wagner, G. A., and P. Van den Haute, 1992, Fission-track dating: Dordrecht, The Netherlands, Kluwer Academic, 285 p.

Wygrala, B. P., M. N. Yalcin, and L. Dohmen, 1990, Thermal histories and overthrusting— Application of numerical simulation technique, *in* B. Durand and F. Behar, eds., Advances in organic geochemistry 1989: Organic Geochemistry, v. 16, p. 267–285.

Yin, A., and T. K. Kelty, 1991, Structural evolution of the Lewis plate in Glacier National Park, Montana: Implications for regional tectonic development: Bulletin of the Geological Society of America, v. 103, p. 1073–1089.

Kinematics and Thermal Modeling

3

Sciamanna, S., W. Sassi, R. Gambini, J.-L. Rudkiewicz, F. Mosca, and C. Nicolai, 2004, Predicting hydrocarbon generation and expulsion in the southern Apennines thrust belt by 2-D integrated structural and geochemical modeling: Part I— Structural and Thermal Evolution, *in* R. Swennen, F. Roure, and J. W. Granath, eds., Deformation, fluid flow, and reservoir appraisal in foreland fold and thrust belts: AAPG Hedberg Series, no. 1, p. 51–67.

Predicting Hydrocarbon Generation and Expulsion in the Southern Apennines Thrust Belt by 2-D Integrated Structural and Geochemical Modeling: Part I— Structural and Thermal Evolution

Simone Sciamanna[1]
Enterprise Oil Italiana, Rome, Italy

William Sassi
Institut Français du Pétrole, Rueil Malmaison, France

Roberto Gambini[2]
Enterprise Oil Italiana, Rome, Italy

Jean-Luc Rudkiewicz
Institut Français du Pétrole, Rueil Malmaison, France

Fausto Mosca[3]
Enterprise Oil Italiana, Rome, Italy

Carlo Nicolai
Enterprise Oil Italiana, Rome, Italy

ABSTRACT

The structural deformation and the petroleum system of the southern Apennines thrust belt (SATB) are studied along a regional cross section traversing the Monte Alpi–Tempa Rossa fields. The SATB is interpreted as a system of three major structural blocks incorporating the basement and the sediments up to the Apulian platform deposits beneath an allochthonous complex. The Thrustpack® software has been used to reconstruct the successive geometries and their progressive burial under foredeep sediments and the allochthonous complex. The bottom of the Apulian platform and the basement are involved in the deformation, and the thickness of the Permian interval, drilled in the foreland, is extended regionally. The timing of the deformation is constrained by the ages of the Pliocene foredeep sediments drilled on top of the Apulian platform. This record was also instrumental to propose a flexure scenario

[1]*Present address:* Repsol-YPF, Buenos Aires, Argentina.
[2]*Present address:* OMV, Vienna, Austria.
[3]*Present address:* Humble Geochemical, Houston, Texas, U.S.A.

DOI:10.1306/1025685H13112

of the migrating foredeep-forebulge system, in which the slope of the topography had to be maintained to a realistic value. These assumptions and boundary conditions were tested by successive, two-dimensional kinematic and thermal reconstructions until a satisfactory match could be obtained with the available temperatures and vitrinite reflectance data. A final good thermal calibration has been obtained for the structural blocks of Monte Alpi and Tempa Rossa. However, the relatively poor quality of the temperature and vitrinite data available for the most hinterland structure questions the conclusions about the validity of our proposed geometry and assumed accumulated thrust displacement. The methodology used in this work is a useful tool in exploration, because it forces one to improve and update structural scenarios and to provide the grounds for highlighting important data gathering to further enhance an evaluation of the hydrocarbon potential at a basin scale. This latter point will be described in a companion chapter.

INTRODUCTION

The southern Apennines thrust belt (SATB) is a prolific petroleum province in Italy (Casero et al., 1991; Mattavelli et al., 1993), where a giant oil field has already been discovered. The main exploration target in the SATB is the Cretaceous part of the Apulian platform carbonates that forms the deepest structures of the SATB, below the outcropping Mesozoic–Cenozoic allochthonous complex, where potential source rocks, reservoir, and traps all reside. The present-day depth below ground level of the main structures is in the range of 1500–5000 m (4900–16,400 ft). Potential source rocks are still a matter of discussion, because only a few direct observations have been made from subsurface and surface samples, and most of the information pertains to oil-to-source correlation. Recent geochemical studies suggest that an oil-prone type I-equivalent kerogen of middle Cretaceous age could be considered the main petroleum source in the study area (Mosca et al., 2004).

This chapter is the first of two companion chapters. In it, the structural evolution and the timing of hydrocarbon generation and expulsion are studied along an almost 130-km (80-mi)-long regional line across the Val d'Agri, where major oil discoveries are located (traversing the Monte Alpi–Tempa Rossa fields). The study was performed using the two-dimensional basin modeling methodology applied with the Thrustpack® software (Roure and Sassi, 1995; Sassi et al., 1998; Sassi and Rudkiewicz, 1999, 2000). Thrustpack was developed in Institut Français du Pétrole to (1) reconstruct the structural evolution and associated deformation of foreland and folded thrust belts using a forward kinematics deformation algorithm; (2) model the evolution of the thermal regime through time and space; (3) monitor the onset of hydrocarbon generation and expulsion from source rocks, resulting from temperature history and kerogen kinetic scheme; and (4) account for changes in the composition of hydrocarbon during generation and expulsion.

Here, the burial and the thermal histories of the potential source rocks are addressed, considering the structural architecture of the sedimentary successions from reflection seismic and well data, as well as both present-day thermal regime and vitrinite reflectance data.

Timing of deformation has been defined integrating several detailed stratigraphic analyses that allow subdivision at the progressive geological deformation into several specific time-step intervals. Of importance here is the youngest siliciclastic foredeep deposition on top of the Apulian platform, which marks the onset of platform subsidence and emplacement of the allochthonous units. However, an important structural problem in any basin-scale reconstruction of the SATB evolution is the amount of shortening in the buried Apulian belt. This question requires determination of the degree of involvement of crystalline basement, which pertains to the style of thrust belt evolution and hence to the evolution of its petroleum system. Despite the large amount of seismic data available in the study area, the magnitude of displacement along major thrusts is still a matter of discussion because of the strong attenuation of seismic resolution in the thrust footwall, where lithological contrast disappears. The bottom of the Apulian platform seems to be involved in the deformation, and displacement along thrusts may range from 1 km (0.6 mi) to a few tens of kilometers. Furthermore, at a lithosphere scale, the migrating foredeep-forebulge system should be reconstructed to properly constrain the maximum burial reached during the thrust belt evolution. In the geological literature, many studies on the dynamics of foreland flexure exist (Jordan, 1981) that quantitatively explain the present-day foredeep geometry (Karner and Watts, 1983; Lyon-Caen and Molnar, 1983; Royden, 1988). Most of these works deal principally with the present-day geometry or with the latest stages of evolution (Waschbusch and Royden, 1992) of the foredeep basins. In particular, Waschbusch and Royden (1992) modeled the southern Apennines upper Pliocene–Pleistocene foredeep along a section roughly parallel to the section presented here.

In this chapter, an attempt is made to use a straightforward approach, extrapolating analytically derived flexural profiles backward in time to late Messinian, following geometric considerations driven by geological data and thermal and maturity modeling.

A sensitivity analysis for the thermal and maturity modeling was carried out to highlight the controlling factors in the temperature history, which was then used to compute a realistic temperature evolution along the cross section. Inconsistencies with R_o data in some structural units enable one to discuss the validity of the proposed structural interpretation and yield a best estimate for the burial rates that are used for compositional modeling of source rock generation and expulsion presented in the accompanying chapter.

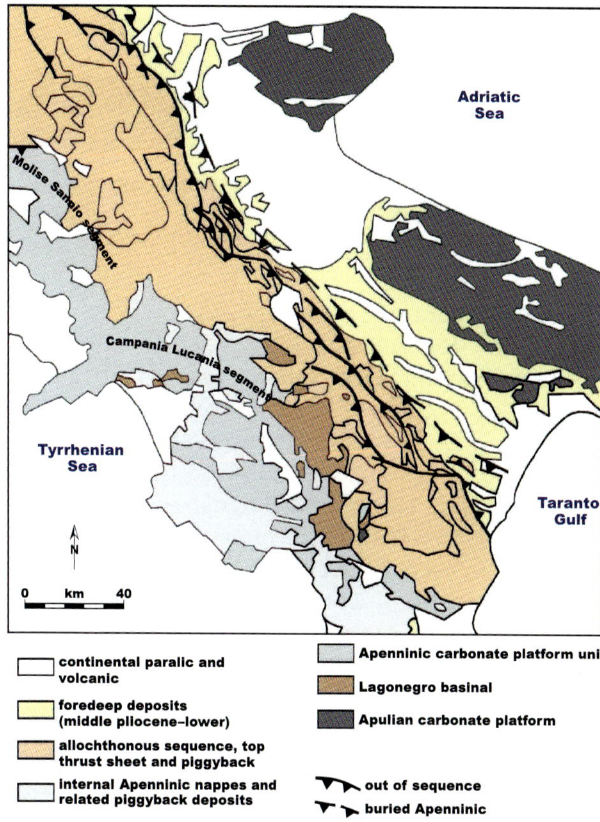

FIGURE 1. Simplified geological map of the southern Apennines thrust belt (SATB) (modified after Patacca et al., 1992).

REGIONAL GEOLOGICAL FRAMEWORK

Almost 40 yr of oil and gas exploration in Italy has resulted in a large amount of subsurface data (Mattavelli et al., 1993). The SATB is roughly a northwest–southeast-trending Neogene postcollision segment of the Mediterranean Africa-verging mountain system that developed in the hanging wall of a west-directed and east-retreating subduction zone (Doglioni et al., 1996). The Adriatic region is the common foreland for the Hellenides and the SATB.

Geodynamic reconstructions of the western Mediterranean show that after a stage of oceanic subduction, the complete closure of the Neotethyan domain was achieved, except along the Calabrian Arc, where oceanic lithosphere is still subducting (Catalano et al., 2001). A suture zone between the northern and southern continental margins was formed in the early Miocene. The continuing convergence related to the counterclockwise rotation of the Corsica–Sardinia block produced compression in the Apennines and widespread calc-alkaline volcanic activity in western Sardinia. Subduction of the western margin of the Adriatic foreland continued during and after this rotation, leading to the SATB evolution.

The SATB is bordered to the east by a foredeep lying all along its Adriatic side and filled by Pliocene–Pleistocene sediments (Casnedi et al., 1981; Balduzzi et al., 1982a, b; Bigi et al., 1992) and to the west by a back-arc basin, the Tyrrhenian Sea (Figure 1). The propagation of the compressive deformation toward the foreland is clearly documented by the development and evolution of a series of eastward-migrating foredeep basins and by the occurrence of several piggyback basins that developed on top of the advancing allochthonous units (Patacca and Scandone, 1989, 2001). Thrust propagation and progressive opening in the Tyrrhenian back-arc basin, since the late Tortonian to Quaternary, appear to have been closely controlled by the rollback of the sinking foreland lithosphere (Malinverno and Ryan, 1986; Patacca et al., 1990; Doglioni, 1991).

From a structural point of view, the SATB is made up of a huge duplex system consisting of the allochthonous complex and the Apulian platform. The allochthonous complex, derived from the deformation of both peritidal carbonate platforms and pelagic basin successions, is completely detached from its original substratum and transported onto the carbonates of the Apulian platform. Figure 2 is a very schematic paleogeographic framework. Because of structural complexity in the SATB, several and often conflicting paleogeographical models have been proposed, sometimes characterized by different numbers of carbonate platforms and pelagic domains. For a detailed description, refer to published papers (D'Argenio et al., 1973, 1975; Mostardini and Merlini, 1986; Sgrosso, 1988; Marsella et al., 1995).

Three major thrust sheets comprise the allochthonous complex lying on the parautochthonous Apulian platform. From the uppermost to the lowermost in the thrust pile and from the west toward the east in the original paleogeography, there are the internal units, the Apenninic platform and the Lagonegro-Molise Basin.

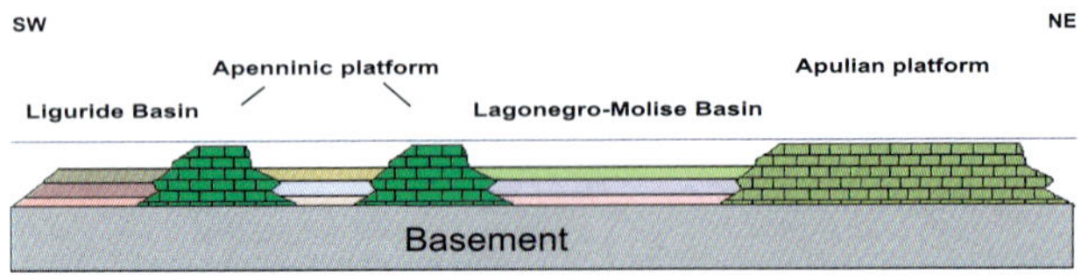

FIGURE 2. Schematic distribution of the paleogeographic units forming the allochthonous complex and the Apulian platform (modified after D'Argenio et al., 1973).

The internal units were deposited on oceanic and transitional crust. These units include sedimentary successions belonging to different domains in the internal zone toward an inferred oceanic ridge in the Neotethyan domain. These locally bear ophiolite rocks, which are affected by a metamorphic overprint of variable grades and types. The structural signature of these units records the effects of the deformation associated to the buildup of an accretionary prism (Knott, 1987).

The Apenninic platform is made up of a few kilometers (as much as 6 km [4 mi]) of Triassic–lower Miocene shallow-water carbonates, overlain by condensed and siliciclastic deposits that mark its subsidence and the onset of the foredeep environment (Patacca et al., 1990). The tectonostratigraphic units derived from this domain crop out extensively in the Calabria–Lucania borderland. These carbonate units are detached from their original substratum along Triassic evaporates or bituminous dolomites. This substratum has never been penetrated in any of the available exploration wells.

The Lagonegro-Molise Basin represents a sedimentary succession deposited in a relatively deep, basinal domain. The nature of the basement (oceanic or thinned continental crust) originally underlying this domain remains uncertain. The tectonic units derived from this domain are made up of a lower succession ("Serie calcareo-silico-marnosa" sensu Scandone, 1967) made up of a 2-km (1-mi)-thick pile of cherty limestones, cherts, and argillites, ranging in age from Late Triassic to Early Cretaceous and of an upper Mesozoic–Tertiary clastic sequences (Scandone, 1967; D'Argenio et al., 1973, 1975).

On top of the accreted tectonostratigraphic units derived from the aforementioned paleogeography domains, several unconformable, diachronous, and space-dependent Miocene–Pliocene–Pleistocene clastic sequences occur. These deposits mark both the polarity of the northeastward-migrating foredeeps and the occurrence of piggyback basins.

The Apulian platform is made up of Mesozoic platform carbonates, with a limited Cenozoic transitional facies and more abundant calcarenites and Quaternary deposits. A Permian volcaniclastic interval has been drilled at the base of this unit in the present-day foreland area. The Apulian platform crops out in the Puglia region (southeastern part of Italy) and constitutes the preorogenic sedimentary cover of the foreland area. The Apulian platform has been identified in the subsurface of the SATB by seismic investigation and explored by several boreholes. It forms the deepest tectonic units of the SATB and is conformably overlain by upper Messinian and Pliocene terrigenous marine foredeep sediments. These were deposited during the flexural subsidence phases before the Apulian platform was included in the thrust belt to form several imbricate units.

REGIONAL CROSS SECTION

The regional cross section traverses the southern part of the SATB, where the major oil discoveries are located and where a large amount of data are available. The present-day structural interpretation and the Apulian platform restored geometry obtained using the Geosec(TM) software are shown in Figure 3. This cross section has been obtained integrating depth-converted composite seismic lines, well data, and detailed surface geological surveys.

The main tectonic units distinguished in Figure 3 are the Allochthon and the Apulian platform. The former is an ensemble of accreted Mesozoic–Cenozoic formations detached from their original basement as described above. It represents the backbone of the southern Apennines Mountain chain. The complex internal geometry of the Allochthon has been simplified for modeling purpose. The implication of this simplification for the thermal and structural forward modeling will be discussed later.

The Apulian platform is clearly dissected by thrust faults. Three main structural trends can be identified from southwest to northeast: the internal trend, Monte Alpi, and Tempa Rossa, respectively (Figure 3). The Monte Alpi structure is a giant onshore field containing very light liquid hydrocarbons, whereas the Tempa Rossa contains a thick column of heavy oil. The internal trend structure has been the target of recent exploration wells.

At a regional scale, a constant thickness can be assigned to the Apulian platform as evidenced by deep wells drilled in the foreland that identify seismic reflectors interpreted as top and bottom of the Apulian platform (D. Scrocca, S. Sciamanna, E. Di Luzio, M. Tozzi, C. Nicolai, and R. Gambini, 2004, personal communication). A constant time-interval thickness, corresponding to 6000–7000 m (19,700–23,000 ft) depending on the average interval velocity used for time-depth conversion, can be assumed. Lateral thickness variations (average magnitude of 500 m [1600 ft]) caused by different paleogeography and/or erosion and nondeposition phases are locally shown by wells located just out of the section plane. These thickness variations are schematically taken into account between Monte Alpi and Tempa Rossa. The internal stratigraphic subdivision is based on regional consideration. The Apulian platform top and bottom are characterized by almost subhorizontal seismic reflectors. A

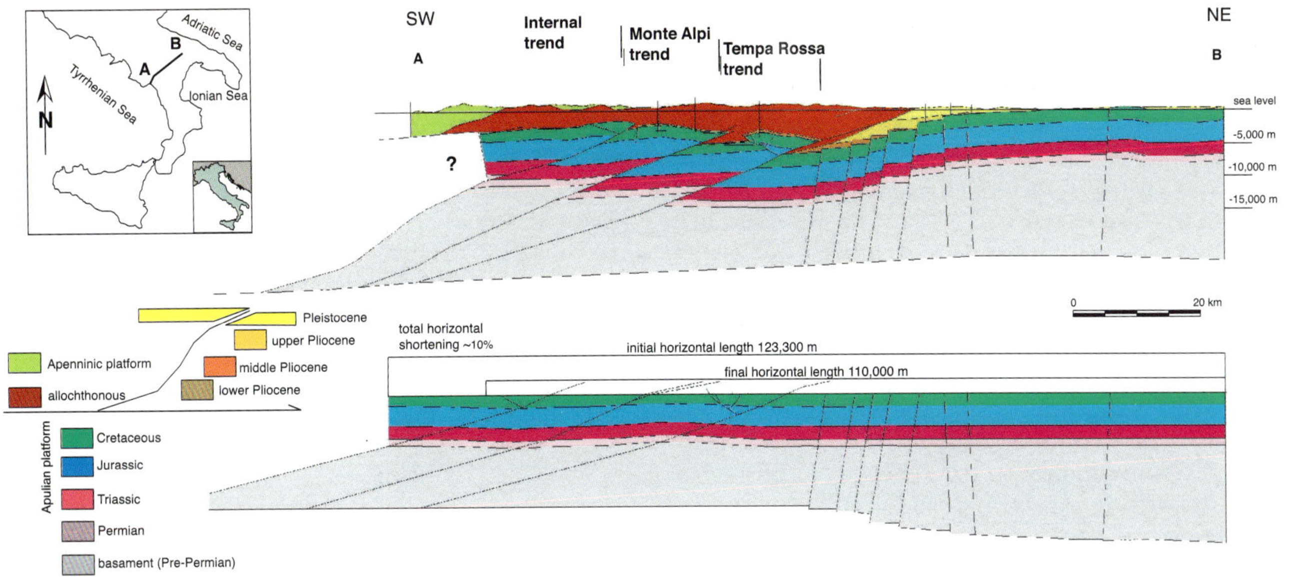

FIGURE 3. Regional geological cross section. Present-day geometry and restored geometry of the Apulian platform.

constant thickness of Permian sediments of about 1000 m (3300 ft) has been placed regionally at the base of the Apulian platform using the information gathered from a well drilled in the foreland area (well Puglia 1).

As mentioned in the Introduction, the displacement along major thrusts may range from 1 km (0.6 mi) to a few tens of kilometers. The Monte Alpi footwall cutoff (Figure 3) has been defined as the minimum admissible displacement following both seismic interpretation and hydrodynamic considerations. The Allochthon deposits in the footwall of Monte Alpi may act as potential fluid barriers and explain the different oil-water contacts between the Monte Alpi and Tempa Rossa oil fields. In the external Tempa Rossa trend, minimum offset can be estimated to be about 7 km (4 mi), distributed between a frontal ramp and a backthrust. Along the internal trend, seismic data do not allow a clear definition of the total displacement. The geometry shown along the regional cross section is very poorly constrained, and a displacement of few kilometers (1–2 km) was considered.

In the structural interpretation shown in Figure 3, it is assumed that thrusts are deeply rooted in the basement and involve the Apulian platform and its base (Permian plus basement). It implies a large discrepancy between the "small" horizontal shortening in the Apulian platform (almost 15 km [9 mi] between the Internal and the Tempa Rossa trends) and the transport of the allochthonous complex. Indeed, the total lateral displacement of the allochthonous complex onto the Apulian platform to its present-day position can be estimated in the range of almost 40 km (25 mi), not considering the extra shortening caused by internal repetition. This difference in shortening could be explained by the gravity sliding of the allochthonous sequences. According to this model, the thrusting occurring in the Apulian platform generated a morphological relief that triggered the sliding of the overlying allochthonous complex. In this way, the allochthonous complex moved a few tens of kilometers in the foredeep. This "uplift and lateral spreading" model has already been presented by Migliorini (1946, 1948), Merla (1951), and Ogniben (1969). Although highly simplified, this deformation mechanism has been considered in the model to calibrate the Apulian platform burial and thermal histories. However, the different shortening in these two overlying tectonic units could be indicative of a hidden shortening in the lower unit. In the western area of the SATB, the seismic quality is generally poor, and the calculated shortening could be highly underestimated. However, this topic goes beyond the scope of the project, and it has not been investigated further.

The results of the structural and thermal-maturity modeling are presented later. The discussion is presented through several paragraphs for the sake of simplicity, but because of the nature of each modeling exercise based on a trial-and-error approach, cross-references exist throughout the discussion.

THERMAL AND MATURITY CALIBRATION DATA

The following data set has been used to constrain both the structural kinematics and the thermal history:

- present-day bottomhole temperatures from well logs
- temperatures from production tests
- vitrinite reflectance data (R_o)

Formation temperatures are used to calibrate the present-day thermal regime and, hence, the basal heat-flux boundary condition in the thermal modeling. The temperatures in boreholes are calculated on each logging run, using a suite of maximum recording thermometers. Because drilling fluids tends to cool the formation, it is necessary to analyze the rate at which formation temperature reverts to its original value after the mud circulation is stopped, so that corrected temperatures can be calculated. The method developed by Fertl and Wichman (1977) has been used. Care was taken not to corrupt the results of modeling through the use of uncorrected temperature data, the use of which was necessary where corrected data were unavailable. Temperatures from production tests have been considered as approximating true formation temperatures.

Temperature data were grouped along the modeled cross section according to the structural trends (i.e., Tempa Rossa, Monte Alpi, or internal trends) to highlight any existing lateral variation of the thermal regime. Apart for the internal trend structure where data are scanty, the other structural trends show similar average thermal gradient as schematically shown in Figure 4a.

Vitrinite reflectance in oil (R_o) is an optical parameter and is the most widely used indicator of maturity of organic material in basin modeling for calibrating the thermal evolution. Because R_o values measured on different vitrinite particles in a sample tend to vary widely, a large number of determinations are averaged to ensure reliable results.

The available R_o data (Figure 4b) display a marked difference in maturity distributions between the Allochthon and the Apulian platform. The Allochthon has essentially a higher maturity, suggesting an inherited complexity in its burial and thermal history prior to thrusting over the Apulian platform. Within the latter, three different maturity trends exist that follow the structural domains as shown in Figure 4a. In the internal trend, only one of the available wells shows reliable maturity data coming from the Mesozoic section of the carbonate platform (Cretaceous) and the siliciclastic Miocene deposits stratigraphically on top of the Apulian platform. These R_o values are generally greater than 1.0. Moving northeastward, the Monte Alpi trend data fall in the mature domain, with values in the range of 0.70–1.0, and the Tempa Rossa trend is in the early to mature domain, with R_o between 0.40 and 0.75. The distribution of these maturity indicators show that the more external, deeper trend has lower maturity in terms of R_o distribution.

STRUCTURAL MODEL, TIMING OF DEFORMATION, AND THERMAL MODELING

Constraints on the Timing of the Deformation

Timing of deformations can be partly constrained using subsurface and surface data. Figure 5 shows a schematic distribution of foredeep deposits across each structural trend. These siliciclastic foredeep sediments deposited on top of the Apulian platform record the timing of emplacement of the Allochthon units. Indeed, these deposits are younger, moving from the internal trend toward the foreland (from southwest toward northeast). The earliest emplacement of the Allochthon onto the Apulian platform cannot be older than late Messinian, which is the age of the drilled siliciclastic sediments overlying the Apulian platform along the internal trend. The Allochthon has progressively buried the Apulian

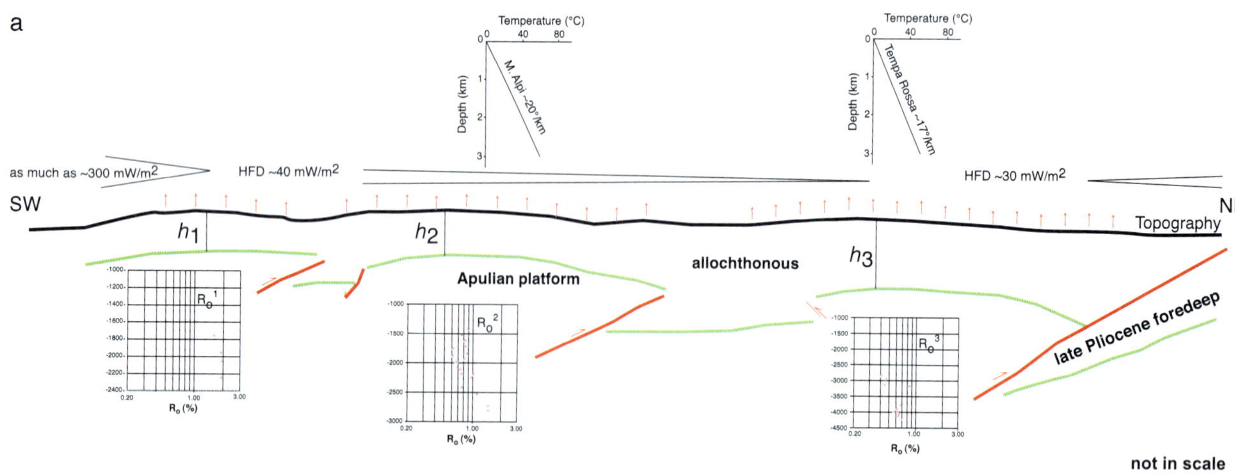

FIGURE 4. (a) Schematic section showing the top of the Apulian platform (green horizon), vitrinite reflectance data, average geothermal gradient, and surface heat-flow distribution. The vitrinite reflectance data are measured in the shallower part of the Mesozoic section and the overlaying siliciclastic deposits. (b) Vitrinite reflectance (R_o) data vs. depth collected in the Allochthon and the underlying Apulian platform. HFD = heat-flow density; TVDSS = true vertical depth subsea.

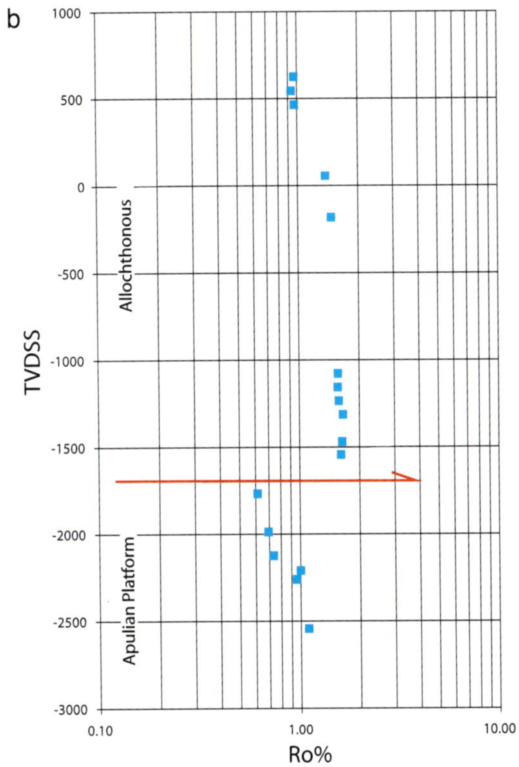

FIGURE 4. (cont.).

platform from the southwest for a period sufficient to reach the observed maturity before denudation processes began. In the present study, the structural trends in the Apulian platform are assumed to have followed a normal forward thrust propagation; that is, the internal trend deforms earlier than Tempa Rossa.

The end of the compression phase was set to early Pleistocene using the findings of Patacca and Scandone (2001), who suggested an upper boundary at 0.66 Ma based on seismic stratigraphy and facies distribution across the northeastern margin of the Sant'Arcangelo basin (Basilicata region), where the authors recognize a clear morphological barrier at depth, locally named Nocara ridge, that shows a precise timing for the nucleation of the structure (see Patacca and Scandone, 2001, for detailed discussions and maps). This area can be interpreted as a southeastward extension of the Tempa Rossa trend, and the timing of deformation can be used to constrain our model.

The latest deformation phase considered in our model is a regional uplift reflected by the elevation of middle Pleistocene marine deposits across the mountain chain (Patacca and Scandone, 2001) and particularly by the late Pleistocene marine terraces as shown in Bordoni and Valensise (1998). Patacca and Scandone (2001) set the beginning of this regional uplift at 0.66–0.65 Ma, when prograding fluviodeltaic systems were established along the outer margin of the thrust belt.

Flexural Subsidence

The flexural geometry of the Apulian platform through time represents a fundamental structural constraint defining the evolution of the Allochthon geometry through time and, hence, the burial history of the Apulian platform (Zoetemeijer et al., 1993). The relationship between flexural angle, burial history, thrust geometry, shortening, and maturity is sketched in Figure 6.

The Apulian platform was progressively buried through time by the allochthonous complex. Figure 6a shows the situation just before activation of thrust faults in the Apulian platform (dashed lines represents the next active thrusts). H_a is the total Allochthon height above the potential source rock area and is a function of the basal surface slope β (flexural subsidence) and the topographic slope α. H_a, in conjunction with the basal heat flow and the timing of deformation, determines the maturation of the organic matter. Figure 6b and c show in different ways the possible relations between shortening and H_a. In Figure 6b, point A will need more horizontal displacement than B to reach the same regional level C, because it has been brought from a greater depth (i.e., higher H_a).

This sketch shows that variable β (flexural angle) determines the overall tectonic uplift; the maximum burial (H_a) undergone prior to deformation and before erosion processes; and the displacement along thrusts, which, in turn, depends on thrust fault geometries (angles) as shown in Figure 6d.

In this example, topography is not a factor, as long as it keeps the same slope. It is quite clear that higher H_a can be achieved by increasing topographic slope, keeping constant the other parameters, but this will affect the tectonic wedge frontal angle. Although the frontal angle can be well constrained for many present-day worldwide accretionary wedges and thrust belts (Davis et al., 1987, and reference therein), it is not well constrained for other geological periods.

To reconstruct flexural profiles through time, the following input was considered:

1) The age and the average thickness of the late Messinian to Pleistocene foredeep deposits obtained from wells and seismic data
2) The present-day thrust geometry as interpreted along seismic lines and its related displacement
3) The present-day bottomhole temperatures
4) The maturity indicators (the vitrinite reflectance data) collected in the shallowest part of the Apulian platform and the overlying Pliocene deposits
5) The present-day average topographic slope across the southern Apennines

The late Messinian–Pliocene deposits on top of the Apulian platform along the section represent the

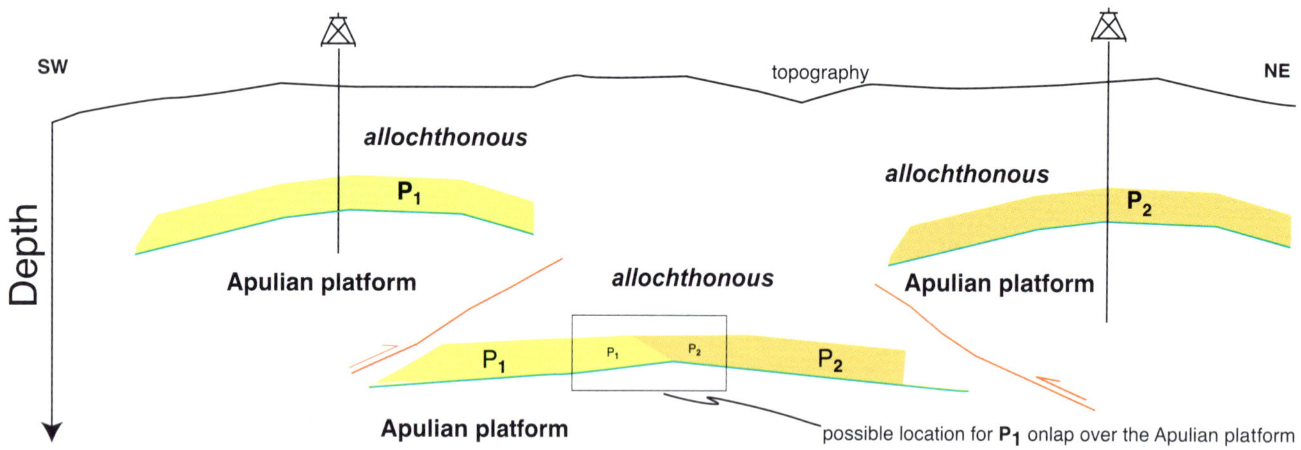

Figure 5. Schematic section showing the distribution of the foredeep sediments drilled on top of the Apulian platform along the regional line. P1 and P2 are Pliocene deposits, with P_1 older than P_2. The geometrical relationships between foredeep deposits, Allochthon, and thrusting of the Apulian platform help in defining the timing of deformation. The allochthonous emplacement onto the Apulian platform occurred after or at least during P_1–P_2 time interval and before the thrust, breaching the Apulian platform. The position of the inferred onlap of P1 above the Apulian platform has been used to reconstruct the migration of the foredeep-forebulge system.

most useful data to reconstruct the flexural geometry of the Apulian platform through time. The lateral subsurface extent of these sediments can be mainly inferred from well data. It would normally be assumed that (between foredeep intervals of different age) stratigraphic onlap should exist somewhere to show that the foredeep depositional center was migrating (Figure 5). Several such onlaps above the Apulian platformApulian platform itself can be used to trace the migration of the peripheral bulge and then to generate a profile of the corresponding foredeep.

Because the thickness of late Messinian–early Pliocene foredeep deposits (between *Globorotalia margaritae* and *Globorotalia puncticulata* planktonic foraminiferal zones in Patacca and Scandone, 2001) can be on an average of 200–400 m (660–1300 ft) and the different time zones are well spread out geographically from the internal to the external areas, there should be a clear onlap shift forelandward. However, the upper Pliocene–Pleistocene foredeep sedimentation is characterized by more than 3000 m (10,000 ft) of sediments whose forelandward pinch-outs are only slightly apart, indicating that the foredeep basin deepened and narrowed through time without a migration of strong foredeep-forebulge system (Waschbusch and Royden, 1992). These evidences suggest that the flexural subsidence through time was not self-similar, and the flexural angle was probably increasing with time as the foredeep-forebulge system not steadily migrating.

The shape of the tectonic wedge, the thrust plane geometry and associated displacements, and a sea level that was assumed almost constant through time represent the input parameters that constrained the flexural angles used to perform the forward modeling. The topographic slope through time has been kept almost constant. The successive profiles of the total flexure through time are shown in Figure 7. These have been calculated using the classical equation for the bend of an elastic beam (Turcotte and Schubert, 1982). Flexural parameter α and elastic rigidity D have been modified to obtain the best flexural profile to fit the input parameters discussed above. This approach is purely geometric; the deepest part of the curves has been calculated using a first-order linear equation to approximate a constant slope.

Kinematic Reconstruction

The structural configuration and timing of deformation imply thrusts propagating in the Apulian platform once the Allochthon was emplaced onto the Apulian platform, and the overlying Pliocene foredeep deposits had accumulated. As a consequence, the Allochthon is modeled as a tectonic wedge accreting and initially sliding over a basal surface above the flexing Apulian platform, until deeper thrusts cut the platform itself to build up the internal trend, Monte Alpi, and Tempa Rossa. At this point, the uplifted structural trends underwent erosion, and further modeling involves thickening and advance of the Allochthon by means of thrust displacement and gravity sliding mechanisms. To reconstruct the geometry of the Allochthon tectonic wedge in time and space, the present-day maximum Allochthon thickness, that is, the thickness above Tempa Rossa, was initially assumed as the minimum allochthonous tectonic wedge's height and then adjusted to draw the evolution of the tectonic wedge

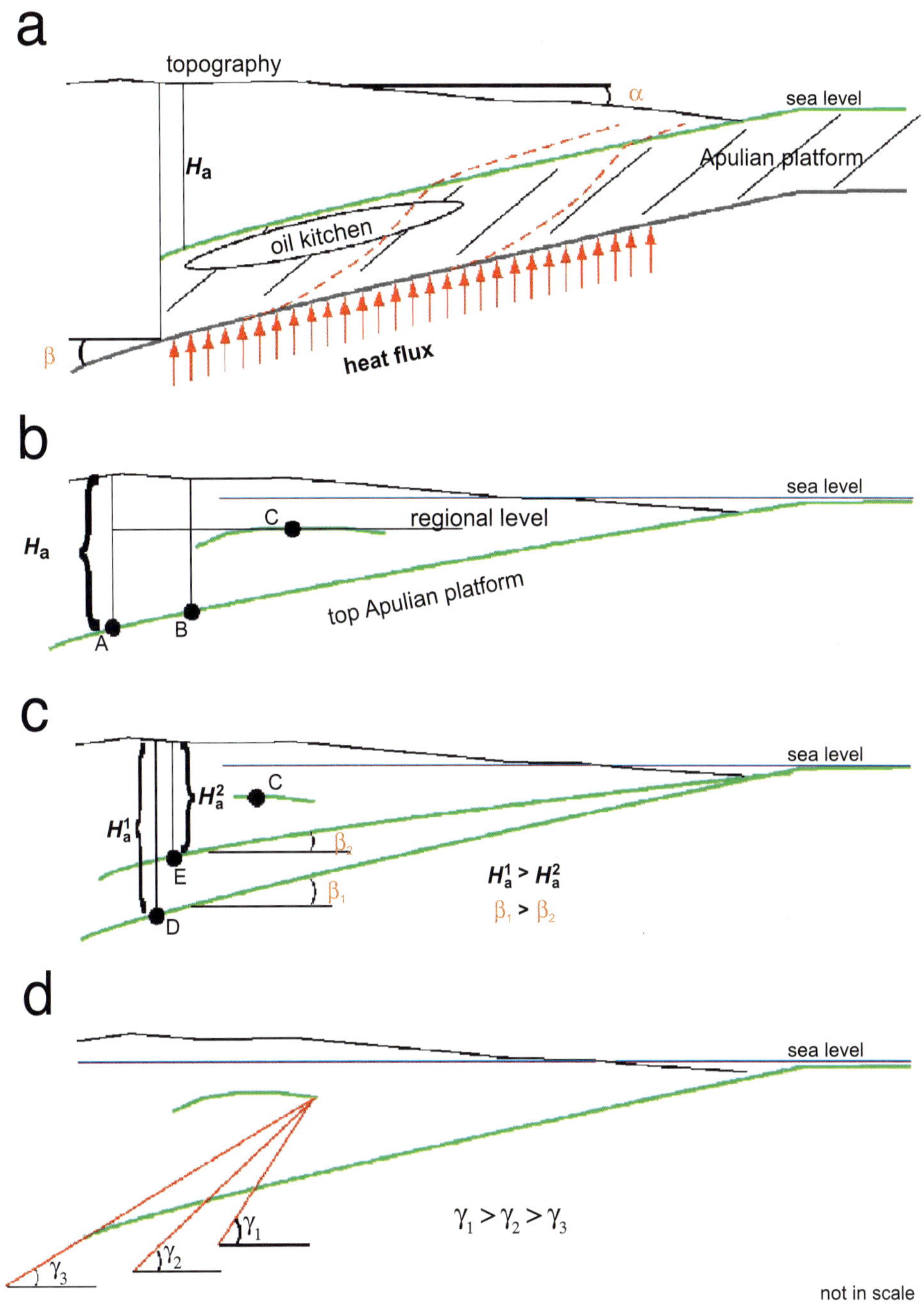

FIGURE 6. Relationships between (a) topographic slope, (b, c) flexural angle, and (d) thrust geometry and their impact on burial history. See text for explanation.

that allowed maturity and thermal calibration. A variation of the Allochthon wedge thickness implied change of the angle of the flexing Apulian platform as described above.

The deformation stages describing the geometric evolution of the SATB along the cross section is shown in Figure 8. An ensemble of accreted units models the allochthonous tectonic wedge emplacement over the flexuring Apulian platform while foredeep sedimentation occurs. The initial geometry is set at Messinian time (7.2 Ma). A regional uplift is modeled during the late Pleistocene. In Bordoni and Valensise (1998), the southern Apennines Arch between the Tyrrhenian and Adriatic Coast is shown, and uplift rates of as much as 1.2 mm/yr (0.04 in./yr) for the last 0.125 Ma are calculated. Following Patacca and Scandone (2001), to define the starting time of this regional uplift back to 0.66–0.65 Ma and assuming the uplift rate of Bordoni and Valensise (1998) to be constant in time, a maximum uplift of almost 800 m (2600 ft) in the central part of mountain chain is obtained. This uplift is modeled as a northeastward tilt between the mountain chain and the Adriatic Coast. Normal faulting in the southwestern part of the modeled section has been partly related with this uplift.

Thermal Modeling: Sensitivity Analysis

To explain (1) the observed maturity distribution in the Apulian platform as a function of depth, (2) the relative lateral homogenous present-day thermal regime along the modeled section, and (3) the very high sedimentation rates, several models have been generated, and a sensitivity analysis was carried out. These initial sensitivity and successive calibration phases allowed reconstruction of the Apulian platform burial-history curves and, hence, the Allochthon thickness variation through time along the section.

The model boundary conditions defined in Thrustpack are the heat flow (uniform or not in space) that is applied at the base of the model (a basement of almost 20 km [12 mi] is considered) and the surface temperature. These parameters can be set as constants through time or can be allowed to vary.

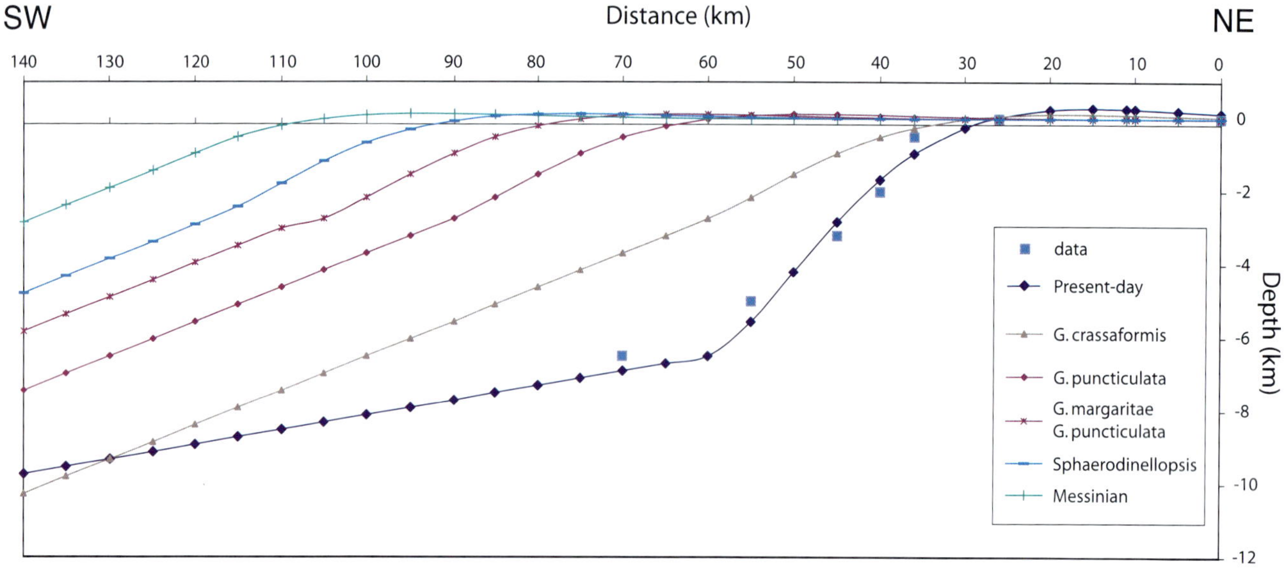

FIGURE 7. Flexural profiles showing the geometry of the Apulian platform at selected ages.

Instead of introducing a considerable thermal anomaly through time to explain their higher maturity, it has been initially assumed that either the Monte Alpi and internal trend could have experienced a maximum burial higher than Tempa Rossa, or a similar maximal burial occurred for a longer period of time.

The present-day Allochthon thickness observed in Tempa Rossa was used as an input parameter, and a series of preliminary attempts to calibrate present-day temperature and maturity data was made using different basal heat-flow values constant in time and space.

The results using a steady state or a transient state were completely different. Figure 9 shows a comparison between maturity distribution obtained in steady and transient states using almost the same basal heat flow. In steady state, it was possible to calibrate present-day temperatures using a basal heat flow of 30 mW/m^2, but neither maturity indicators nor surface heat-flux distributions were in agreement with available data (Mongelli et al., 1989). As soon as the basal heat flow increases to calibrate the observed maturity data, there was no longer a match with present-day temperatures. The discrepancies were between 20 and 40°C. In addition, the calculated Tempa Rossa maturity was always higher than in the Monte Alpi. This preliminary calibration suggested that the SATB should be modeled in transient state to consider the thermal disequilibrium caused by the high sedimentation rates.

A sensitivity analysis was then carried out using a variable basal heat flow in time (but constant in space) and running the models in transient state. These models were designed to test past heat-flow peaks capable of matching both temperature and maturity data. The resulting thermal perturbations, introduced in time by heat-flow peaks, were unable to match the present-day temperature distribution, unless extreme boundary conditions were used during the last step of the model (very low present-day basal heat flow).

THERMAL CALIBRATION AND STRUCTURAL IMPLICATIONS

The sensitivity analysis has shown that the assumption that basal heat flow was constant in time and space, or that this varied in time, permits the model to match temperature or maturity data, but not both at the same time. This problem was not even solved by refining the Apulian platform burial history. To get around this problem, variable basal heat flow in space was introduced, and more attention was paid to model the present-day surface heat-flux distribution.

Heat-flow density data along the Apennines mountain chain are available in the literature (Della Vedova et al., 1984; Mongelli et al., 1989; Pasquale et al., 1993, 1999). Different heat-flow density zones can be distinguished. High values are present in Tyrrhenian Sea and pre-Apenninic belt both in the northern and southern Apennines. Lower values exist, moving toward the central part of the SATB (Figure 10).

For calibration purpose, three different basal heat-flow zones were defined with 40, 60, and 80 mW/m^2, respectively, from northeast to southwest and constant in time. The results of the thermal modeling in terms of surface heat-flow distribution, temperature history, and burial curves are shown in Figure 11. The computed surface heat-flow distribution is worth comparing with the observed data (Figure 10). The high sedimentation rates in the upper Pliocene–Pleistocene foredeep are directly responsible for the lowest heat-flow

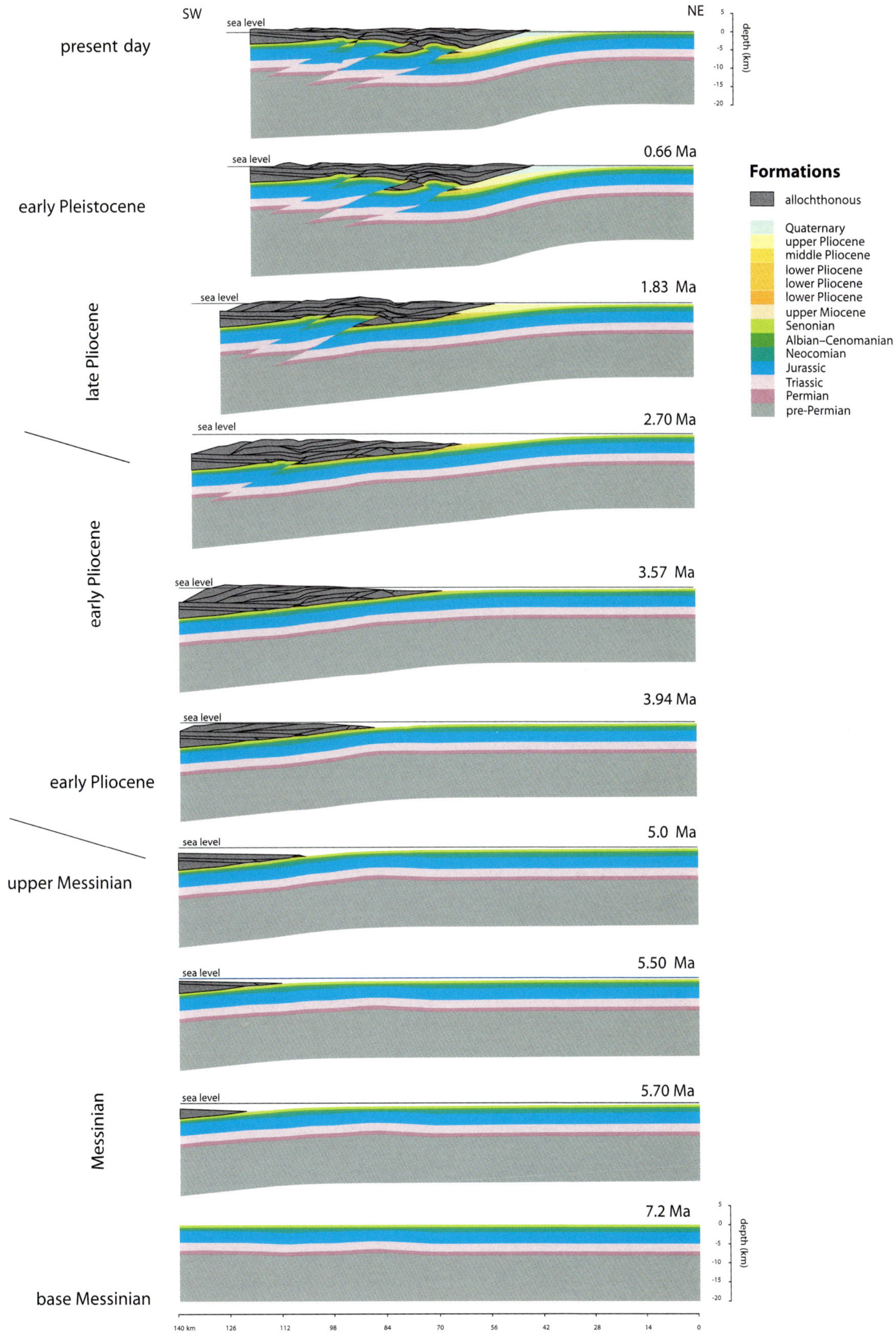

FIGURE 8. Forward structural modeling. Evolution of the SATB geometry since the late Messinian.

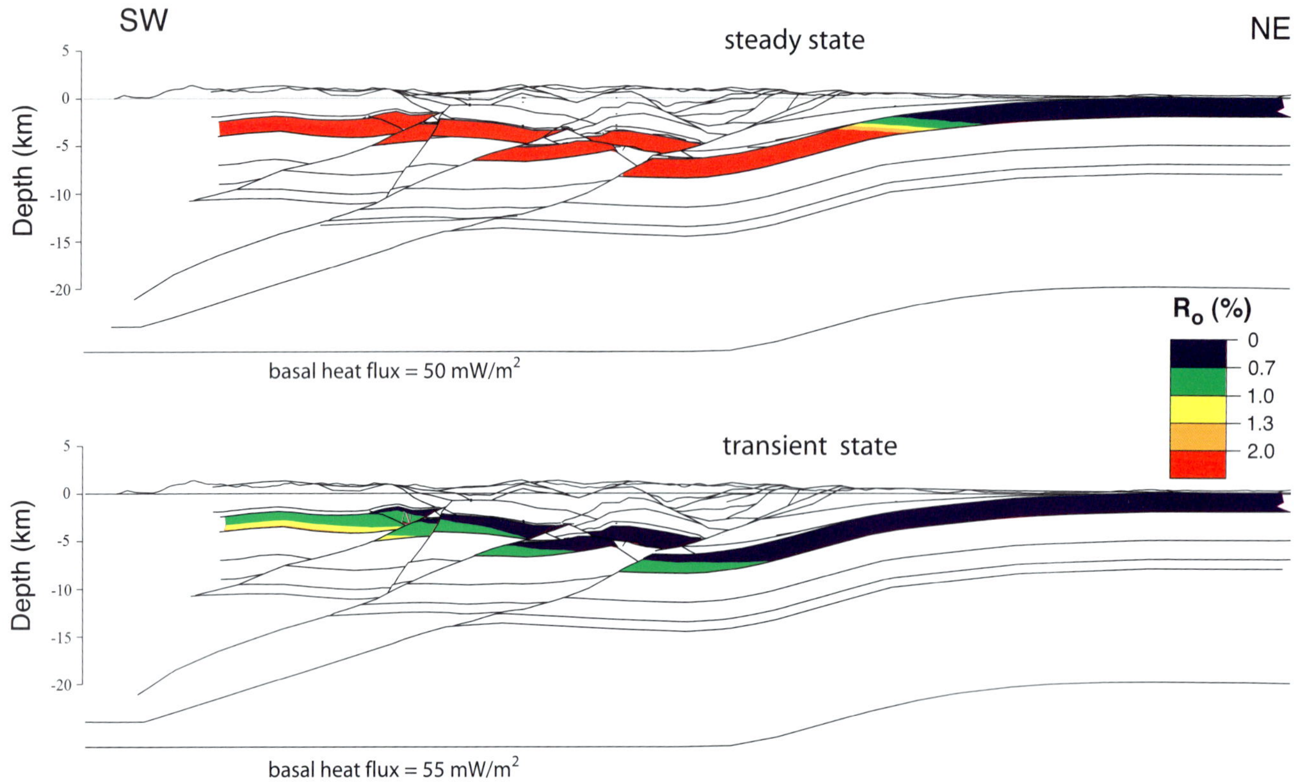

FIGURE 9. Maturity distribution along the regional line computed in transient and steady state using almost the same boundaries conditions (basal heat-flow history).

values (below 35 mW/m^2). Furthermore, one can notice that although all points are being eroded in the later stages, their temperature does not necessarily decrease. This is attributed to the combination of transient thermal regime and the two-dimensional geometry. This pattern could not be reproduced using one-dimensional basin modeling computation. The temperature and maturity calibration is shown in Figure 12. Observation points A and B plus C represent the Tempa Rossa and Monte Alpi area, whereas D represents the internal trend. Pseudo-observation point E was introduced to investigate the burial and thermal histories of more internal areas.

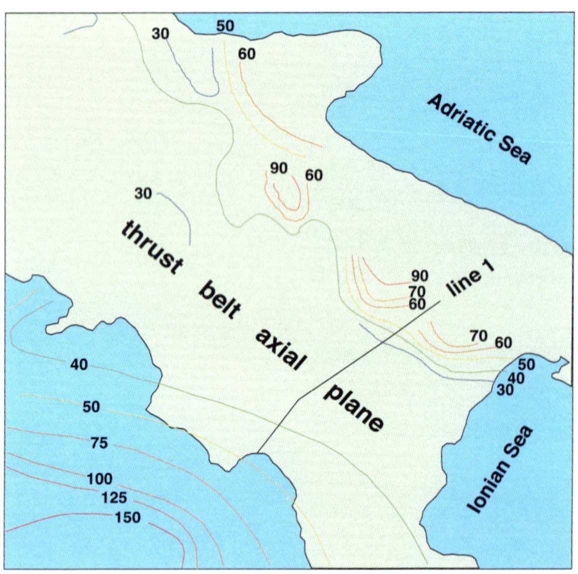

FIGURE 10. Southern Apennines surface heat-flux distribution and modeled cross section location (modified after Mongelli et al., 1989).

Observed vs. calculated temperatures almost fit along the three structural trends. At observation point C, differences are in the order of 5–10°C. Two reasons can be offered to accept this fit. The first reason is model dependence; that is, it is dependent on the way the allochthonous complex is modeled. An average thermal conductivity has been assigned to the Allochthon, because it was outside the scope of this work to study lithology variation in the allochthonous complex in detail. The second reason is related to the fact that present-day temperatures are probably cooled by the presence of very active aquifers in the Allochthon and Apulian platform, a phenomenon that has not been considered in the model.

At observation point D, that is to say in the internal trend, the temperature misfit, in addition to the reasons described above, is caused by data-quality problems. Most of the bottomhole temperatures are not corrected and, moreover, projected into the section from afar.

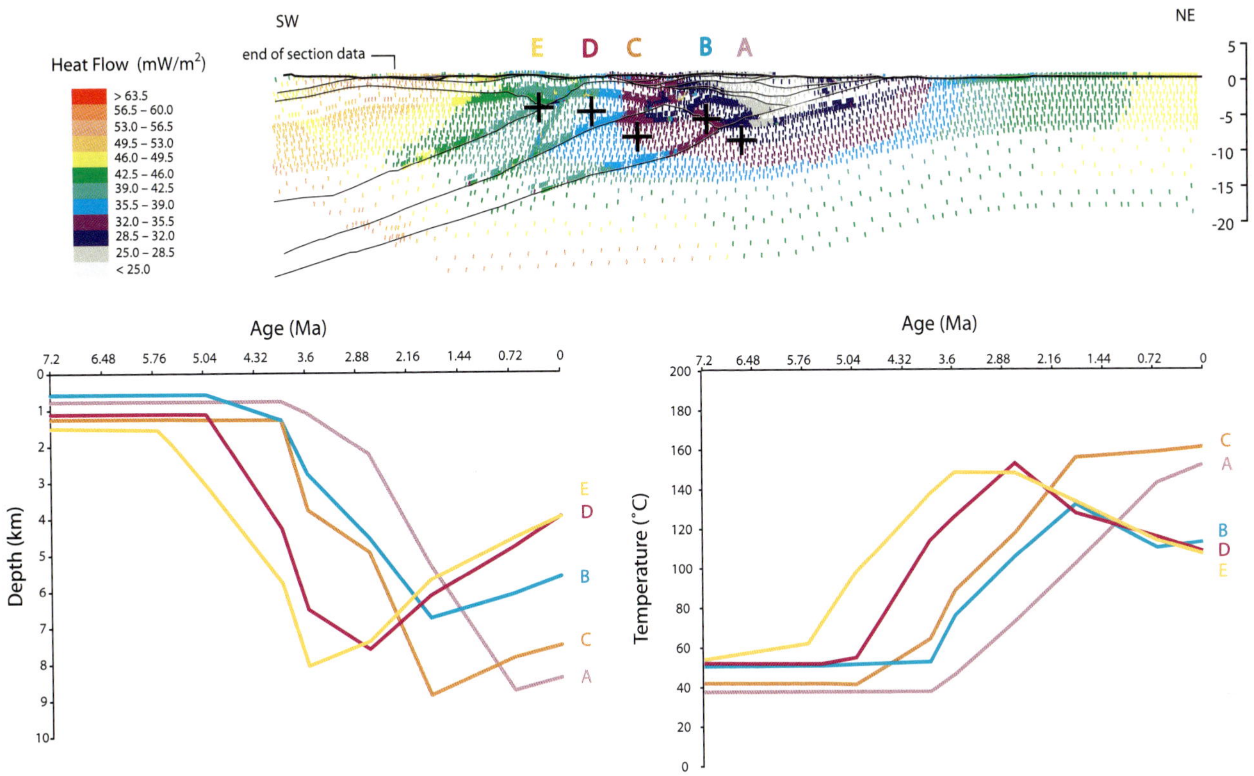

FIGURE 11. Final model's results showing the computed heat-flux distribution within and at the surface of the cross section (to be compared with the surface distribution in Figure 10) and the burial and temperature history graphs for the points A, B, C, D, and E shown on that section.

Main discrepancies between observed and calculated R_O values are shown at observation point D in the internal trend, whereas across Tempa Rossa and Monte Alpi, a good calibration has been achieved (points A, B, and C). As a matter of fact, the calculated maturity along observation point D is in the order of magnitude of Monte Alpi. Instead, the pseudo-observation point E (Figure 12) shows that computed maturity, despite the depth misfit, should be in the range of that observed in the internal trend. Therefore, the model suggests that the short displacement assumed between the Monte Alpi and the internal trend cannot consider the observed maturity, simply because there is not enough space to assume strong lateral thermal- or burial-history variations without also affecting the Monte Alpi trend.

Although the different level of maturity calculated for the internal trend and the pseudo-observation point is primarily caused by the applied basal heat flow as shown in Figure 11, it is evident that the same result would have been achieved by varying the Allochthon thickness evolution in the more internal area, that is, more burial along the internal areas for a longer period of time. The direct consequence would have been a bigger angle of flexure of the Apulian platform according to the aforementioned geometric constrains. This would imply an important change in the geometry of thrust along the internal trend, and a lower detachment and no basement-involved style should be considered to allow longer displacement to match the present-day Apulian platform regional elevation. The model is strongly suggesting that structural reinterpretation could be necessary along the internal trend.

The original structural interpretation between the Tempa Rossa and Monte Alpi area has already been partly revised to honor as much as possible the calibration data. In Figure 13, the Monte Alpi's thrust geometry has been modified, and the total displacement has been increased to improve the Monte Alpi calibration by modifying its burial history. So far, the point C present-day position stays at a greater depth. This revised structural interpretation may have direct implication in terms of where the main active source rock kitchens should be located and how the migration pathways were distributed to explain the hydrocarbon charge toward Tempa Rossa and Monte Alpi.

CONCLUSIONS

Starting from an initial interpretation of the present-day subsurface structures and a restored geometry using balanced cross section techniques, the evolution of the

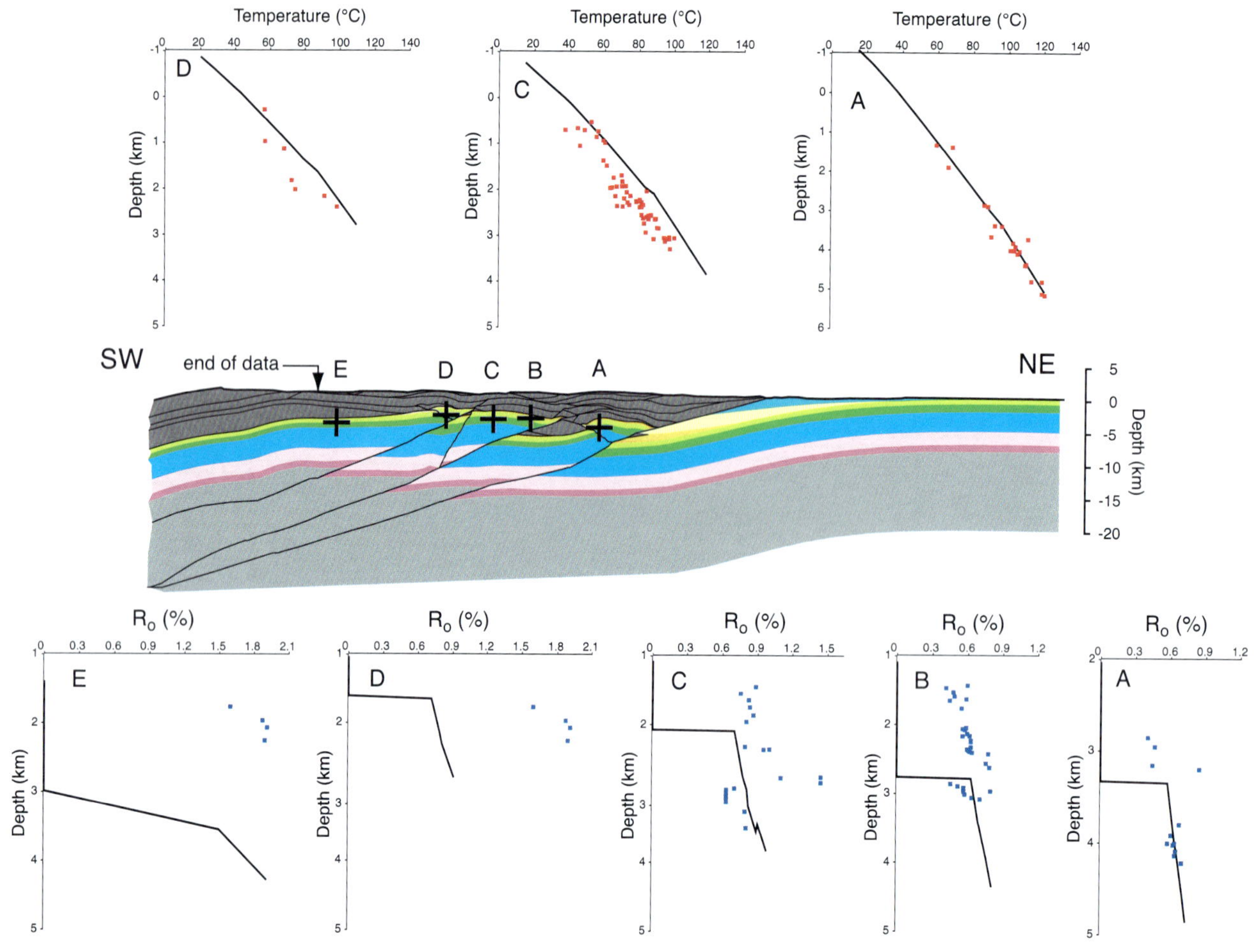

FIGURE 12. Ultimate thermal and maturity calibration obtained in Thrustpack modeling. See text for explanation (legend of the formations as in Figure 8).

SATB has been reconstructed using a trial-and-error procedure on both forward kinematics and thermal computations. This methodology was applied to test and adjust various model parameters and boundary conditions, of which the most important were (1) the chronology and the rate of displacement of the major thrusts, (2) the slope of the topography of the tectonic wedge and that of the steps in the large-scale flexural subsidence through time, and (3) the pattern of the applied surface temperature and basal heat flux across the section and through time. Parameters 1 and 2 are essential in controlling the burial rate history, and parameter 3 controls the corresponding temperature rate variations, the present-day temperatures and surface heat-flux distribution, and the calculated R_o (vitrinite reflectance) profiles.

The good thermal calibration for the structural blocks of Monte Alpi and Tempa Rossa suggests that the proposed interpretation is a plausible scenario. The results in terms of the final conductive thermal regime will be used, as described in the accompanying chapter, to compute the history and composition of hydrocarbon generation and expulsion in the anticipated source rocks of the Apulian platform.

Although the final structural interpretation is viable in the Tempa Rossa and Monte Alpi area, the displacement assumed between Monte Alpi and the internal trend may be too small, thus leading to a new areal-balanced problem that should explain the observed maturity distribution.

Therefore, two main alternatives with different implications for petroleum exploration have to be considered.

1) The structural interpretation driven by the seismic data is valid, and the maturity data needs to be revised; this entails that the internal trend has undergone a geological evolution comparable with that of Monte Alpi oil field area.
2) The maturity indicators are valid, and consequently, the Monte Alpi and the internal trends have undergone different structural evolutions. It can be assumed that these structures were originally (prethrusting

FIGURE 13. The revised structural interpretation between Monte Alpi and Tempa Rossa showing the uncertainty range in the determination of the burial history curves by comparison to the initial structural scenario (continuous line in graphs) for the selected points B, C, D, and E shown in the cross section.

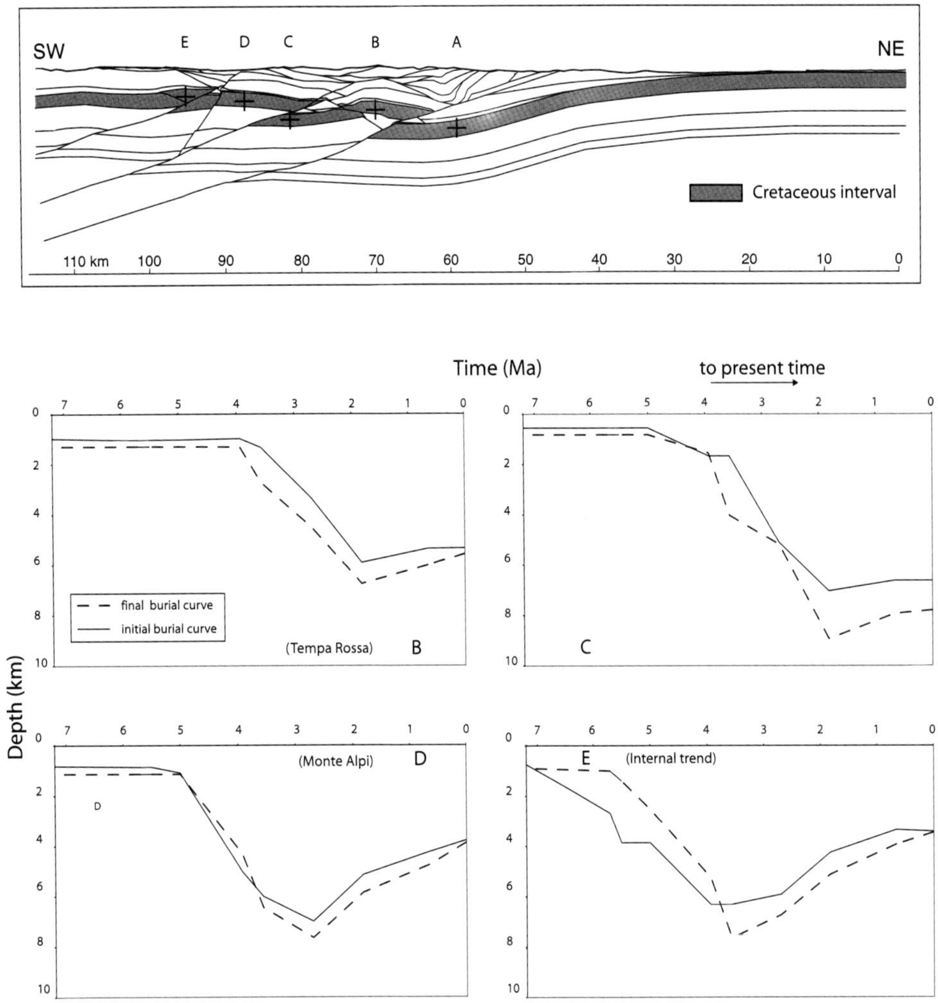

phase) separated by a longer distance, and the present-day setting was achieved because of larger displacement along the thrust that creates the internal trend.

It is worth noting that the original structural interpretation implied a more complex deformation mechanism for the emplacement of the allochthonous tectonic units to compensate for the strong difference in shortening between the Allochthon and the Apulian platform.

Several factors help balance the cross section at all levels:

1) longer displacement along the internal trend
2) more shortening in the Apulian platform than assumed in recent structural interpretations (Mazzoli et al., 2000; Menardi Noguera and Rea, 2000)
3) very little or no involvement of the crystalline basement

Several recent exploration failures along the internal trend show that the petroleum system probably worked differently than in the Monte Alpi and Tempa Rossa oil field areas. If the total shortening and timing of thrusting along the internal part of the SATB can be better constrained by new analysis and subsurface data, the timing of hydrocarbons generation, expulsion, and trap formation, as well as even the existence of the same source rock interval, may be better constrained.

The methodology used in this work is a useful tool in exploration, because it forces one to improve and update structural scenarios. It also can be used to highlight important data gathering to further enhance an evaluation of the hydrocarbon potential at a basin scale. This latter point will be described in the accompanying chapter (Mosca et al., 2004).

ACKNOWLEDGMENTS

The model results presented here have been achieved using mainly unpublished data available in Enterprise Oil Italiana archive and subordinately from available published data.

This research and development project was supported by the European Community THERMIE Program, contract OG/161/98/FR/IT. We acknowledge the Enterprise Oil Italiana S.p.A. and the Institut Français du Pètrole for permission to publish the results of this study.

We thank P. Scandone and E. Patacca for providing unpublished stratigraphic data and for fruitful discussions and D. Scrocca for his contribution to the cross section balancing part and also P. Shiner and V. Zucconi.

We acknowledge J. Granath and R. Zoetemeijer for their suggestions to improve this chapter.

REFERENCES CITED

Balduzzi, A., R. Casnedi, U. Crescenti, and M. Tonna, 1982a, Il Plio-Pleistocene del sottosuolo del Bacino pugliese (Avanfossa appenninica): Geologica Romana, v. 21, p. 1–28.

Balduzzi, A., R. Casnedi, U. Crescenti, F. Mostardini, and M. Tonna, 1982b, Il Plio-Pleistocene del sottosuolo del Bacino lucano (Avanfossa appenninica): Geologica Romana, v. 21, p. 89–111.

Bigi, G., D. Cosentino, M. Parotto, R. Sartori, and P. Scandone, 1992, Structural model of Italy: P. F. Geodinamica, 114. Centro Nazionale delle Ricerche, scale 1:500,000.

Bordoni, P., and L. Valensise, 1998, Deformation of the 125 ka marine terrace in Italy: Tectonic implications: Geological Society (London) Special Publication, v. 146, p. 71–110.

Casero, P., F. Roure, and R. Vially, 1991, Tectonic framework and petroleum potential of the southern Apennines, *in* A. M. Spencer, ed., Generation, accumulation and production of Europe's hydrocarbons: European Association of Petroleum Geoscientists Special Publication 1, p. 381–387.

Casnedi, R., U. Crescenti, C. D'amato, F. Mostardini, and U. Rossi, 1981, Il Plio-Pleistocene del sottosuolo molisano: Geologica Romana, v. 20, p. 1–42.

Catalano R., C. Doglioni, and S. Merlini, 2001, On the Mesozoic Ionian basin: Geophysical Journal International, v. 144, p. 49–64.

D'Argenio, B., T. Pescatore, and P. Scandone, 1973, Schema geologico dell'Appennino meridionale (Campania e Lucania), *in* Moderne vedute sulla Geologia dell'Appennino, Convegno (Roma, February 16–18, 1972): Problemi attuali di Scienza e di Cultura, Quaderno 183, Accademia Nazionale dei Lincei, Roma, p. 49–72.

D'Argenio, B., T. Pescatore, and P. Scandone, 1975, Structural pattern of the Campania–Lucania Apennines, *in* C. Squyres, ed., Geology of Italy: The Earth Science Society of Lybyan Arabic Republic, p. 313–327.

Davis, D., J. Suppe, and F. A. Dahlen, 1987, Mechanics of fold-and-thrust belts and accretionary wedges: Journal of Geophysical Research, v. 88, p. 1153–1172

Della Vedova, B., G. Pellis, J. P. Foucher, and J. P. Réhault, 1984, Geothermal structure of the Tyrrheian Sea: Marine Geology, v. 55, p. 271–289.

Doglioni, C., 1991, A proposal of kinematic modelling for W-dipping subductions— Possible application to the Tyrrhenian–Apennines system: Terra Nova, v. 3, p. 423–434.

Doglioni, C., P. Harabaglia, S. Merlini, F. Mongelli, and G. Zito, 1996, A geodynamic model of the southern Apennines accretionary prism: Terra Nova, v. 8, p. 540–547.

Fertl, W. H., and P. A. Wichman, 1977, How to determine static BHT from well-log data: World Oil, v. 184, no. 1, p. 105–106.

Jordan, T. E., 1981, Thrust loads and foreland basin evolution, Cretaceous, western United States: AAPG Bulletin, v. 65, no. 12, p. 2506–2520.

Karner, G. D., and A. B. Watts, 1983, Gravity anomalies and flexure of lithosphere at mountain ranges: Journal of Geophysical Research, v. 88, p. 10,449–10,477.

Knott, S., 1987, The Liguride complex of southern Italy— A Cretaceous to Paleogene accretionary wedge: Tectonophysics, v. 142, p. 217–226.

Lyon-Caen, H., and P. Molnar, 1983, Constraints on the structure of the Himalaya from an analysis of gravity anomalies and flexural model of the lithosphere: Journal of Geophysical Research, v. 88, p. 8171–8191.

Malinverno, A., and W. B. F. Ryan, 1986, Extension in the Tyrrhenian Sea and shortening in the Apennines as a result of arc migration driven by sinking of the lithosphere: Tectonics, v. 5, p. 227–245.

Marsella, E., A. W. Bally, G. Cippitelli, B. D'argenio, and G. Pappone, 1995, Tectonic history of the Lagonegro domain and southern Apennine thrust belt evolution: Tectonophysics, v. 252, p. 307–330.

Mattavelli, L, M. Pieri, and G. Groppi, 1993, Petroleum exploration in Italy: A review: Marine and Petroleum Geology, v. 10, p. 410–425.

Mazzoli, S., S. Corrado, M. De Donatis, D. Scrocca, R. Butler, D. Di Bucci, G. Naso, C. Nicolai, and V. Zucconi, 2000, Time and space variability of "thin-skinned" and "thick-skinned" thrust tectonics in the Apennines (Italy): Rendiconti della R. Fisiche Accademia dei Lincei, v. 11, p. 5–39.

Menardi Noguera, A., and G. Rea, 2000, Deep structure of the Campanian–Lucanian arc (southern Apennine, Italy): Tectonophysics, v. 324, no. 4, p. 239–265.

Merla, G., 1951, Geologia del'Appennino settentrionale: Bolletino della Societa Geologica Italiana, v. 70, no. 1, p. 92–382.

Migliorini, C., 1946, Osservazioni sulla tettonica dei Monti di Sala Consilina nell'Appennino meridionale: Bolletino della Societa Geologica Italiana, v. 67, p. 29–142.

Migliorini, C., 1948, I cunei composti nell'Orogenesi: Bolletino della Societa Geologica Italiana, v. 65, p. 37–40.

Mongelli, F., G. Zito, N. Ciaranfi, and P. Pieri, 1989, Interpretation of heat flow density of the Apennine chain, Italy: Tectonophysics, v. 164, p. 267–280.

Mosca, F., S. Sciamanna, W. Sassi, J.-L. Rudkiewicz, and R. Gambini, 2004, Predicting hydrocarbon generation and expulsion in the southern Apennines thrust belt by 2-D integrated structural and geochemical modeling: Part II. Geochemical modeling, *in* R. Swennen, F. Roure, and J. W. Granath, eds., Deformation, fluid flow, and reservoir appraisal in foreland fold and thrust belts: AAPG Hedberg Series, no. 1, p. 69–77.

Mostardini, F. and S. Merlini, 1986, L'Appennino centromeridionale. Sezioni geologiche e proposta di modello strutturale: Memorie della Societa Geologica Italiana, v. 35, p. 177–202.

Ogniben, L., 1969, Schema introduttivo alla geologia del confine calabro-lucano: Memorie della Societa Geologica Italiana, v. 8, no. 4, p. 453–763.

Pasquale, V., M. Verdoya, P. Chiozzi, and P. Augliera, 1993, Dependence of the seismotectonic regime on the thermal state in the northern Italian Apennines: Tectonophysics, v. 217, p. 31–41.

Pasquale, V., M. Verdoya, and P. Chiozzi, 1999, Thermal

state and deep earthquakes in the southern Apennines: Tectonophysics, v. 306, p. 435–448.

Patacca, E., and P. Scandone, 1989, Post-Tortonian mountain building in the Apennines. The role of the passive sinking of a relic lithospheric slab, *in* A. Boriani, M. Bonafede, G. B. Piccardo, and G. B. Vai, eds., The lithosphere in Italy. Advances in Earth Science Research: Italian National Committee of the International Lithosphere Programme, Mid-term Conference (Rome, May 5–6, 1987): Atti dei Convegni Lincei, v. 80, p. 157–176.

Patacca, E., and P. Scandone, 2001, Late thrust propagation and sedimentary response in the thrust-belt-foredeep system of the southern Apennines (Plio-Pleistocene), *in* G. B. Vai and I. P. Martini, eds., Anatomy of a mountain: The Apennines and adjacent Mediterranean basins: London, Chapman & Hall, p. 401–440.

Patacca, E., R. Sartori, and P. Scandone, 1990, Tyrrhenian basin and Apenninic arcs: Kinematic relation since late Tortonian times: Memorie della Societa Geologica Italiana, v. 45, p. 425–451.

Patacca, E., P. Scandone, M. Bellatalla, N. Perilli, and U. Santini, 1992, The Numidian-sand event in the southern Apennines: Memorie Scienze Geologiche, v. 18, p. 297–337.

Roure, F., and W. Sassi, 1995, Kinematics of deformation and petroleum system appraisal in Neogene foreland-and-thrust belts: Petroleum Geosciences, v. 1, p. 253–269.

Royden, L., 1988, Flexural behavior of continental lithosphere in Italy: Constraints imposed by gravity and deflection data: Journal of Geophysical Research, v. 93, p. 7747–7766.

Sassi, W., and J.-L. Rudkiewicz, 1999, Thrustpack version 6.2: 2-D integrated maturity studies in thrust areas: Institut Français du Pétrole Report no. 45372, 79 p.

Sassi, W., and J.-L. Rudkiewicz, 2000, Computer modelling of petroleum systems along regional cross-sections in foreland and fold-and-thrust belts (abs.): Proceedings of the Conference on Geology and Petroleum Geology of the Mediterranean and Circum-Mediterranean Basins (European Association of Geoscientists and Engineers Malta, October 1–4, 2000), Extended Abstracts, v. C27, p. 1–4.

Sassi, W., J. L. Rudkiewicz, and R. Divies, 1998, New methods for integrated modeling of deformation and petroleum generation in fold and thrust belts (abs.): AAPG Annual Meeting Program, v. 82, N13 Supplement.

Scandone, P., 1967, Studi di geologia lucana: La serie calcareo-silico-marnosa ed i suoi rapporti con l'Appennino calcareo: Bolletino della Societa dei Naturalisti in Napoli, v. 76, p. 1–175.

Sgrosso, I., 1988, Nuovi elementi per un più articolato modello paleogeografico nell'Appennino centro-meridionale: Memorie della Societa Geologica Italiana, v. 41, p. 225–242.

Turcotte, D. L., and G. Schubert, 1982, Geodynamics: New York, John Wiley, 450 p.

Waschbusch, P. J., and L. Royden, 1992, Spatial and temporal evolution of foredeep basins: Lateral strength variation and inelastic yielding in continental lithosphere: Basin Research, v. 4, p. 179–196.

Zoetemeijer, R., S. Cloetingh, W. Sassi, and F. Roure, 1993, Modelling of piggyback-basin stratigraphy: Record of tectonic evolution, *in* S. Cloetingh, W. Sassi, and F. Horvath, eds., The origin of sedimentary basins: Inferences from quantitative modelling and basin analysis: Tectonophysics, v. 226, nos. 1–4, p. 253–269.

4

Mosca, F., S. Sciamanna, W. Sassi, J.-L. Rudkiewicz, and R. Gambini, 2004, Predicting hydrocarbon generation and expulsion in the southern Apennines thrust belt by 2-D integrated structural and geochemical modeling: Part II—Geochemical modeling, *in* R. Swennen, F. Roure, and J. W. Granath, eds., Deformation, fluid flow, and reservoir appraisal in foreland fold and thrust belts: AAPG Hedberg Series, no. 1, p. 69–77.

Predicting Hydrocarbon Generation and Expulsion in the Southern Apennines Thrust Belt by 2-D Integrated Structural and Geochemical Modeling: Part II—Geochemical Modeling

Fausto Mosca[1]
Enterprise Oil Italiana S.p.A., Rome, Italy

Simone Sciamanna[2]
Enterprise Oil Italiana S.p.A., Rome, Italy

William Sassi
Institut Français du Pétrole, Rueil Malmaison, France

Jean-Luc Rudkiewicz
Institut Français du Pétrole, Rueil Malmaison, France

Roberto Gambini[3]
Enterprise Oil Italiana S.p.A., Rome, Italy

ABSTRACT

The structural deformation and the source rock system evolution of the southern Apennines thrust belt (SATB) are studied along a regional structural profile traversing the Monte Alpi–Tempa Rossa oil fields. In part I (the accompanying chapter), the reconstruction of the structural evolution and the thermal history was addressed to calibrate the burial history of the source rocks along the cross section. Here in part II, the generation and expulsion of hydrocarbons were modeled to test a potential source rock interval and identify geometric factors explaining the observed differences in the nature of the oil found in the three major structural trends. Organic-rich, laminated

[1]*Present address:* Humble Geochemical, Houston, Texas, U.S.A.
[2]*Present address:* Repsol-YPF, Buenos Aires, Argentina.
[3]*Present address:* OMV, Vienna.

DOI:10.1306/1025686H13112

limestones that were penetrated by a few wells in the region represent the best source rock candidate to date. The source interval shows total organic carbon (TOC) values as much as 4% and hydrogen index as much as 632 mg HC/g TOC. This source rock also contains high amounts of sulfur (3–6% in kerogen).

Rock samples and asphaltenes isolated from the oil were analyzed to determine both primary bulk kerogen decomposition and compositional kerogen decomposition products. For the latter, the results include determination of the kinetics of dry gas (C_1), wet gas (C_2–C_5), light oil (C_5–C_{14}), and heavy oil (C_{15+}) components. The southern Apennines Cretaceous source rock behaves as a type I kerogen equivalent, consistent with the distribution of the activation energies dominated by a single activation energy. Most of the predicted generated and expelled hydrocarbons are heavy and light oils. Thermal conditions for secondary cracking of the generated oil into gas could have been reached only in the footwall of the major thrusts. The measured kinetic parameters allow the modeling of a favorable timing of trap formation with respect to hydrocarbon generation and expulsion. When the measured bulk and compositional kinetics are used in the modeling, no oil generation is reached in the Tempa Rossa trend. The model shows that the Tempa Rossa heavy-oil field has been filled by oil that was generated deeper in the surrounding of the structure. Compositional kinetic simulation is consistent with the results of the geochemical analyses performed on several oils from the region. The original oils in the reservoirs should have an API gravity of about 25° API. Only subsequent geological processes (uplift and erosion) provide the pressure-volume-temperature variation responsible for the compositional grading column at the present time.

Finally, kerogen transformation ratio vs. depth shows that the three different transformation ratio-depth zones should be considered to fit the thermal history of the southern Apennines. This two-dimensional information can be used to predict the distribution of potential source rock kitchen areas in the surroundings of the modeled section to guide future exploration.

INTRODUCTION

Success in an exploration play commonly depends on an accurate assessment of the timing of hydrocarbon generation in conjunction with trap formation. The timing of hydrocarbon generation has been shown to vary, depending on kerogen type and related compositional differences, such as variable sulfur and oxygen contents (Tegelaar and Noble, 1993; Jarvie and Walker, 1997; D. M. Jarvie and L. L. Lundell, 2000, personal communication). The goal of a basin-modeling study is to achieve a better understanding of the regional petroleum system potential by integrating the geological and geochemical data.

This chapter focuses on the timing of hydrocarbon generation and expulsion along the regional line already described from a structural evolution point of view in the accompanying chapter (Sciamanna et al., 2004). The modeling has been performed using the two-dimensional (2-D) Thrustpack® software (Sassi and Rudkiewicz, 1999). This tool allows the assessment of the onset of hydrocarbon generation and expulsion from source rocks based on a calibrated thermal regime and a kinetic scheme. An important aspect of the thermal history of thrust belts is that the hanging wall and footwall undergo different pressure and temperature paths: During thrusting, the hanging wall is uplifted, cooled, and generally eroded, whereas the footwall is rapidly buried and warmed. In a simple, one-dimensional approach used in most studies (Angevine and Turcotte 1983; Davy and Gillet 1986; Deming and Chapman 1989; Svara et al., 1998), the thrust faulting is simulated by instantaneous stacking of the structural units. Other papers (Shi and Wang 1987; Wygrala et al., 1990) have discussed the potential of 2-D models by more accurately reconstructing the temperature evolution of thrust sheets. This chapter provides additional evidence of more accurate thermal-history reconstruction by means of a 2-D model like Thrustpack.

The latest version of the code allows kinetic compositional computation during generation (Rudkiewicz et al., 1996; Sassi and Rudkiewicz, 2000).

The study area is known as Val D'Agri, where two giant oil fields have been discovered. Despite recorded knowledge of oil seeps since the days of the Roman

Empire, these giant fields eluded oil finders until the late 1980s. Persistence, along with advances in seismic acquisition and processing technology, are credited with overcoming difficulties imposed by structure and topography (Armour, 1999; Holton, 1999a). The most significant characteristics of these fields are the high flow rates (3000–12,000 bbl/day) and thick oil columns (600–1000 m; 2000–3300 ft), both features lending special consideration to the overall development plan. The total recoverable reserves are estimated at more than 1 billion bbl (Holton, 1999b), making them the largest onshore oil accumulation in Europe.

SOUTHERN APENNINES THRUST BELT: PETROLEUM SYSTEM

Source Rock Definition

The definition and assessment of the potential source rock in the Val D'Agri region is still not fully understood. The fingerprint of the source rock candidate is known, but a few samples are available— no outcrops and samples from Monte Alpi area and good samples only from the Tempa Rossa area (for location, see Sciamanna et al., 2004). Based on these data, it is very challenging to assess the effective source rock properties and, more importantly, the areal distribution.

However, based on recent comprehensive petroleum system studies (D. A. Wavrek, 2000, personal communication; Enterprise Oil Italiana [EOI] proprietary studies) addressing specific questions pertaining to the origin and disparate oil quality in the southern Apennines thrust belt (SATB), some useful indications regarding the potential source rock character can be derived.

The source rocks responsible for generating these oils appear to have been deposited in an anoxic marine carbonate environment in a semiarid climate. A basinal area with a phosphatic or siliceous lithologic composition is also possible. Water depth was relatively shallow (water depth can be increased with good light penetration), because extensive bacterial reworking of the primary biomass is apparent. Independent evidence suggests that the water column was stratified and had a net evaporitic hydrologic character. Stable carbon isotopic data for oils in the southern Apennines, interpreted in the context of analogous global data sets, suggest that the source rock for these oils could be in the Upper Cretaceous or Tertiary (Neogene). These data eliminate the Triassic from being a candidate.

Recently, organic-rich, low-maturity (R_o equivalent 0.6%), laminated limestones (thickness = 400 m [1300 ft]) Cretaceous in age have been drilled by a few wells in the Tempa Rossa. This interval shows good to very good geochemical properties and represents the obvious source rock candidate. The source interval shows total organic carbon (TOC) values ranging from 1 to 4% and hydrogen index ranging from 350 to 650 mg HC/g TOC. This interval contains high amounts of sulfur (3–6% in kerogen), which affects both the thermal thresholds for generation and the distribution of hydrocarbon and nonhydrocarbon fractions. Oil-to-source rock correlations performed by Eni confirmed the Cretaceous interval as the best candidate. However, because of the presence of heavy-oil impregnation (this interval acts at the same time as reservoir and source rock), these results are not conclusive.

Kinetic Study

The transformation of source rock organic matter takes place as temperature increases in the subsurface during burial or other heating events. The temperature at which source rock kerogen is cracked to oil and gas is regionally very different and depends on the source rock type, which is mainly controlled by organic input and preservation during deposition of the source rock layers.

To obtain reliable temperature predictions for source rocks when performing the simulation experiments in the laboratory, it must be ensured that source rock samples are immature or of low maturity to cover the full range of hydrocarbon formation.

In this study, source rock kinetic parameters are derived from a proprietary EOI study performed by the Humble Geochemical Services Division.

Tempa Rossa source rock samples have been selected to assess bulk and compositional kinetics in the modeling exercise. However, some uncertainties exist concerning the maturity level of this interval. The studied intervals correspond to an unusual source rock with types I and II organic matter that is commonly more stable relative to oil formation. The high stability of the source rock organic matter (based on the commonly used maturity parameters) indicates that the selected samples could be in the early phase of the oil generation window.

An alternative solution to this problem could be to use asphaltenes derived from the oils from the study region. Compositional kinetic parameters and yield determinations can be made from oil asphaltene fractions (Di Primio et al., 1999; Jarvie et al., 1999). Oil asphaltenes are remnants of the original kerogenous material that provide a link to the original source when it is not available. Moderate biodegradation does not appear to alter the use of asphaltene fractions for rate and yield assessment (Di Primio et al., 1999).

To this end, two source rock samples and four oil asphaltenes have been selected (Table 1) to determine primary bulk kerogen decomposition and compositional fractions, including dry gas (C_1), wet gas (C_2–C_5), light oil (C_5–C_{14}), and heavy oil (C_{15+}).

TABLE 1. Source rock and oil asphaltene samples average geochemical properties.*

Code	*Type*	*TOC*	*S_1*	*S_2*	*S_3*	*T_{max}*	*HI*	*OI*	*PI*
A1	source rock	2.8	0.4	17.0	0.3	435	600	13	0.03
A2	source rock	5.5	1.9	28.7	0.4	439	525	8	0.06

Code	Type	API (°)	S (%)	% C_{15-}	% C_{15+}	SAT	ARO	RES	ASP
B	asphaltene	19.9	6.1	9.0	91.0	14.8	34.0	4.8	22.5
C	asphaltene	18.3	5.9	16.0	84.0	20.9	47.7	8.2	21.4
D	asphaltene	18.7	4.6	11.0	89.0	39.8	29.1	6.2	19.4
E	asphaltene	21.6	5.5	11.0	89.0	39.8	29.1	6.2	19.4

**HI = hydrogen index; OI = oxygen index; PI = production index; S = sulfur; SAT = saturated; ARO = aromatic; RES = residuum; ASP = asphaltene.*

Modeling of Hydrocarbon Generation and Expulsion

The modeled section crosses the only location where a potential Cretaceous source rock has been sampled. In the model, the entire Cretaceous interval (ranging between 1500–2000 m [5000–6600 ft] at regional scale) has been defined as a potential source rock. This has permitted evaluation of the overall petroleum yield at different depths. The modeling was performed using both the default Institut Français du Pétrole bulk kinetic scheme (type I Green River Shale Formation and type IIS Monterey Formation) and the southern Apennines measured bulk and compositional kinetic parameters.

An average TOC of 2% has been assigned to the entire Cretaceous interval. The results are discussed in terms of timing of hydrocarbon generation and primary expulsion based on an oil-saturation threshold. The oil-saturation threshold assumes that the source rock system is 100% water saturated before any oil generation takes place. To reach a relative permeability to oil (condition for oil migration), a certain amount of oil has to be generated. The amount of oil necessary to achieve this oil-permeability condition is a function of the source rock porosity, TOC, heating rates, etc.

In this study, the expulsion threshold has been set at 20% of oil saturation in the source rock system. Because of the very low porosity assigned to the Apulian Platform (fixed average porosity of 4%), this variable strongly affects the timing of hydrocarbon expulsion.

Bulk Kinetics Results and Implications

A comparison of transformation ratio and expulsion rates at selected locations (about 500 m [1600 ft] below the top of the Apulian Platform) is shown in Figure 1 for bulk type IIS, type I, and Cretaceous source rock kinetics.

Because of the high reactivity of type IIS kerogen along the entire section, most of the hydrocarbon expulsion could have occurred before the formation of the structural traps. A better timing with trap formation exists in the Monte Alpi and the internal trend areas when a type I kerogen or Cretaceous source rock kinetics are used. However, it is worth noting that if the source rock is located only in the Monte Alpi and Tempa Rossa hanging wall (1000 m [3300 ft] from the top of Apulian Platform), the model shows very a low transformation ratio for both type I and Cretaceous source rock kerogens. The most important implication is that transformation ratio values that are consistent with the huge amount of oil generated are reached only in the Monte Alpi and Tempa Rossa footwalls (points A and C in Figure 1) or out of the section plane for a type I or Cretaceous source rock organic matter.

It follows that when default and measured kinetics (the latter from both rock samples and asphaltenes) are compared (Figure 1), the southern Apennines Cretaceous source rock clearly behaves as a type I kerogen equivalent. This result is consistent with the conclusions of the kinetic study. The distribution of the activation energies is dominated by a single activation energy, which indicates that the organic matter in the source rock is very homogeneous and equivalent to algae-enriched type I kerogen.

An important economical implication is that no hydrocarbons, including heavy oils (API gravity ranging from 10 to 20°), could have been generated in the Tempa Rossa area. The amount of original oil in place in the Tempa Rossa field implies that a larger fetch area should be considered.

Furthermore, high-resolution C_7 Gas Chromatography-Flame Ionization Detection analyses of Monte Alpi and

FIGURE 1. Comparison between default kerogen kinetic (IIS and I) and bulk kinetic scheme from the Tempa Rossa rock sample. Kerogen type I and Tempa Rossa bulk kinetics show a better relation between petroleum expulsion and timing of trap formation. A = Tempa Rossa footwall; B = Tempa Rossa; C = Monte Alpi footwall; D = Monte Alpi; E = internal trend.

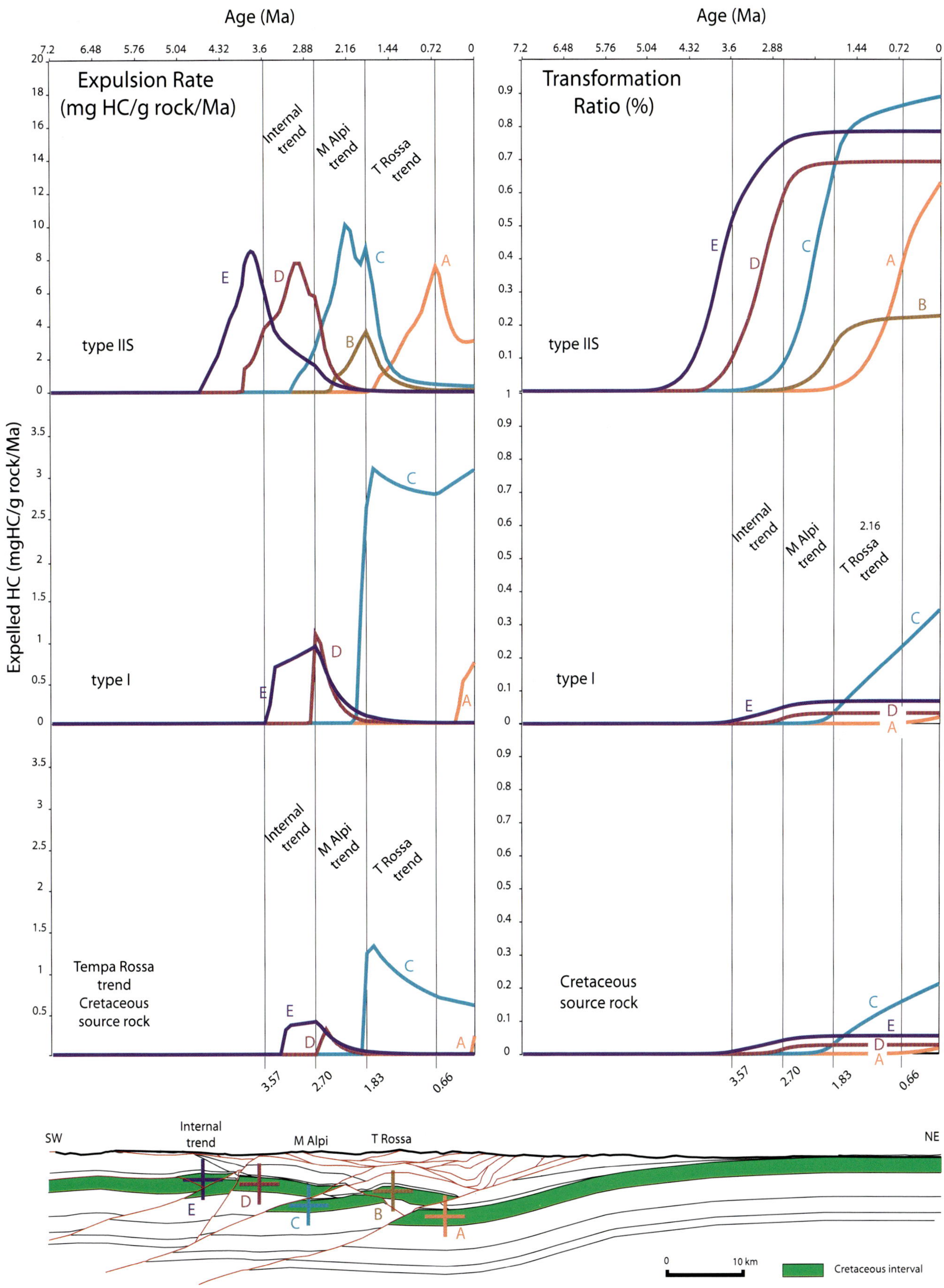
Age (Ma)
Expulsion Rate (mg HC/g rock/Ma)
Transformation Ratio (%)
Expelled HC (mgHC/g rock/Ma)
Internal trend
M Alpi trend
T Rossa trend
type IIS
type I
Tempa Rossa trend Cretaceous source rock
Cretaceous source rock
3.57
2.70
1.83
0.66
SW
NE
Internal trend
M Alpi
T Rossa
10 km
Cretaceous interval

Tempa Rossa crude oil samples performed on the southern Apennines oils during different EOI proprietary studies (D. A. Wavrek, 2000, personal communication) provide important information. A useful application of the high-resolution C7 analysis is to quantify the temperature at which hydrocarbon generation starts (Mango, 1997). As advocated by Mango, the generation temperature is calculated via the equation °C = 140 + 15(ln [24 dimethylpentane (DMP)/23DMP]), which is based on steady-state catalytic theory and calibrated with laboratory heating experiments. The results for the SATB oils show a general range in generation temperature between 110 and 115°C and a total range of 105–120°C. A similar range of generation temperatures is obtained when bulk kinetics relative to a type IIS kerogen is used in the model. In this case, the model shows a temperature of 115°C when 10% of the transformation ratio is achieved.

When we apply the Cretaceous source rock bulk kinetic parameters, the model shows temperatures at 10% transformation ratio in the range of 137–155°C. A similar range of generation temperatures is obtained when bulk kinetics relative to a type I kerogen is used.

It is important to note that in oils that are generated under high levels of thermal stress and/or have high gas-oil ratio (GOR) values (D. A. Wavrek, 2000, personal communication), this form of the Mango's equation tends to underestimate the generation temperature.

The southern Apennines thrust belt oils show low GOR, but most of them have been generated under high levels of thermal stress (thrust event). Furthermore, given the evidence established for a secondary hydrocarbon charge of light oil (D. A. Wavrek, 2000, personal communication), it is logical that the lighter oils would have a higher generation temperature.

Compositional Kinetics Results and Implications

Using compositional kinetics derived from the Cretaceous source rock samples and from the asphaltenes of Monte Alpi and Tempa Rossa oils (Table 1), an assessment of the quality and quantity of the generated hydrocarbons has been performed. The results shown in Figure 2 are expressed as dry gas (C_1), wet gas (C_2–C_5), light oil (C_5–C_{14}), heavy oil (C_{15+}), and a quantity of hydrocarbons (milligrams of hydrocarbons per gram of rock) that are generated at three key points: the Monte Alpi oil field, the Monte Alpi footwall, and the internal trend. The three selected points are located at about 1000 m (3300 ft) below the top of the seismic reflector that is regionally interpreted as the top of the Apulian Platform. Basically, in terms of oil composition and quantity, the results using different samples look very similar. In particular, the amount of volatile fraction is very low and consistent with the low amount of gases found so far in the area. Only the volatile fraction generated by Monte Alpi trend asphlatenes is slightly higher than the relative amount generated by Tempa Rossa trend asphaltenes and Cretaceous source rock samples. The amount of light and heavy oil generated and expelled is approximately the same and occurs simultaneously. Based on these results, we should expect an oil composition in the reservoir that reflects this mixing. If we assume a light-oil average composition of 35° API and a heavy-oil average composition of 20° API, we can estimate an average API gravity of more or less 25° API. Unfortunately, the API gravity in the reservoir ranges from 20 to 42° in the Monte Alpi field, which is a range too wide to imply a match between model and reservoir. Apparently, these results seem to be in contradiction with the oil quality in reservoir, unless we consider secondary physical processes that affected the oil column after the filling of the reservoir.

The physiochemical rationale for the disparate oil quality in this reservoir has been attributed to compositional grading as described in an EOI proprietary report (D. A. Wavrek, 2000, personal communication). Basically, a single oil is thought to have arrived in the reservoir during the main migration phase and subsequently separated into light and heavy oils because of a gravity segregation process. Furthermore, this study also established circumstantial evidence for a subsequent minor light-oil charge from the same source rock as the primary oil charge. Based on these considerations, this modeling exercise provides support for this final oil-quality assessment. The 2-D simulation shows that the physical uplift and concomitant reduction in pressure and temperature experienced by Monte Alpi trend is responsible for a huge pressure-volume-temperature (PVT) variation. A rough estimate shows a decrease of reservoir pressure between 5400 and 7700 psi and reservoir temperature 40–60°C over the past 2 m.y. These numbers have to be considered as indicative of the PVT variation and need to be checked by a pressure modeling calculation through time. The amount of uplift experienced by the Tempa Rossa trend is insignificantly different from modeling result. In this case, the lack of any effective PVT variation, together with the low maturity of the source rock, can explain why only heavy oil occurs in the Tempa Rossa field. Finally, to predict the extent of source rock maturity (transformation ratio) as a function of temperature and depth, the modeling exercise has been applied to the entire Apulian Platform. The compositional kinetics parameters determined from the Tempa Rossa asphaltenes have been used in the simulation. Figure 3 shows the burial histories and cumulative transformation ratios for four horizons at each of the three locations along the cross-solution. The modeling exercise gives the relation between source rock depth and transformation ratio that can be extrapolated to the surrounding area using the same boundary

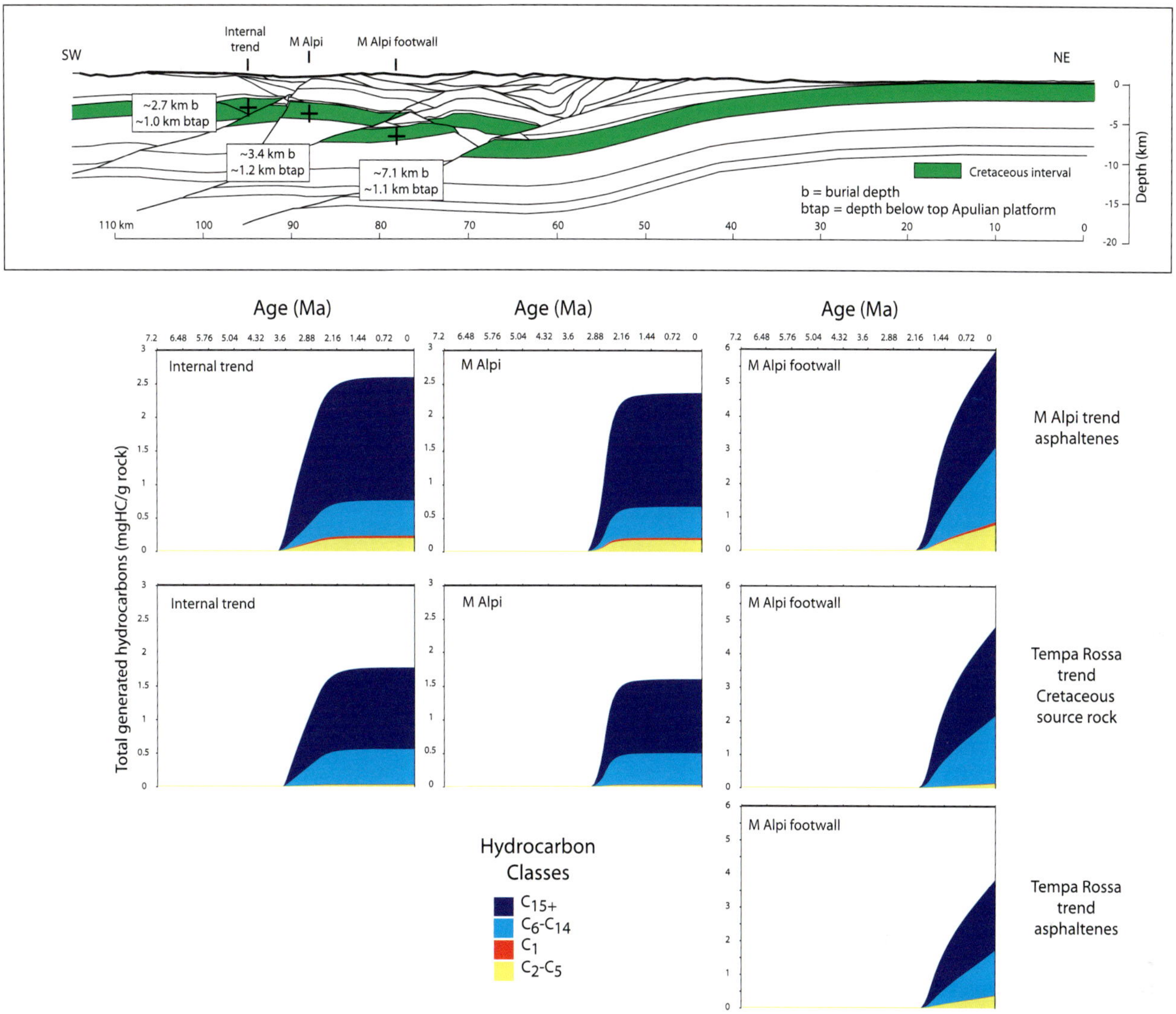

FIGURE 2. Compositional modeling performed using the kinetics parameters determined from the Tempa Rossa rock sample and asphaltenes of Monte Alpi and Temp Rossa oils. Most of the total expelled hydrocarbons are light and heavy oils. The calibrated thermal history does not allow measurable oil expulsion in the Tempa Rossa in the modeled Cretaceous interval (almost 1500 m [5000 ft] thick).

conditions used to calibrate the section (Sciamanna et al., 2004). Laterally, at the same depth of burial, the transformation ratio varies because of the different timing of deformation, which is younger moving northeastward. Therefore, at least three different transformation ratio-depth zones should be considered to fit the thermal history of the southern Apennines and to predict the distribution of potential source rock kitchen areas for future exploration targets.

CONCLUSIONS

This 2-D basin modeling study of the SATB presents the first quantitative investigation of the hydrocarbon compositional generation and expulsion in an attempt to match a real case situation of interest to petroleum companies in Italy.

This modeling exercise has shown the following:

- The southern Apennines Cretaceous source rock clearly behaves as a type I kerogen, which is consistent with a distribution of the activation energies dominated by single activation energy.
- Oil-generation temperatures, calculated at transformation ratio (TR) = 10%, are higher than generation temperatures calculated on the oil samples.
- Most of the generated and expelled hydrocarbons are heavy and light oils. Thermal conditions for secondary cracking of the oil into gas could have been reached only in the footwall of the main structures.
- Compositional kinetic simulation is consistent with the results of the geochemical analyses performed

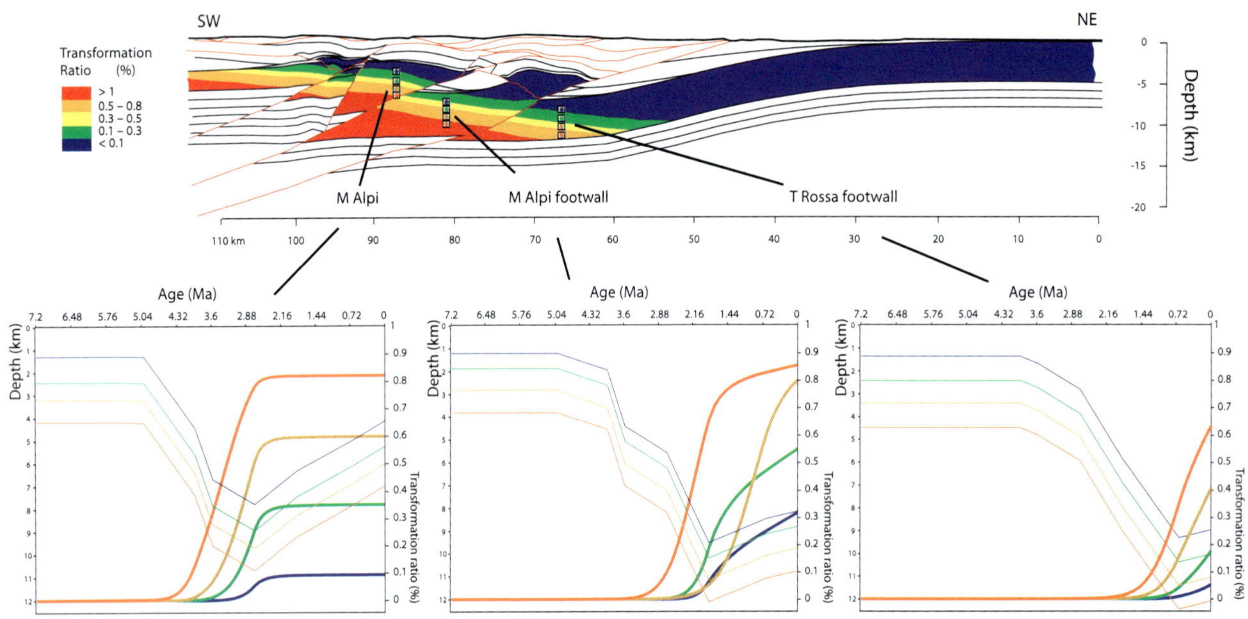

FIGURE 3. Transformation ratio vs. depth. The compositional kinetics of the Tempa Rossa asphaltenes was used to define potential source rock levels even deeper than the Cretaceous interval to evaluate the source rock maturity as a function of the modeled thermal regime. At least three main transformation ratio vs. depth zones (from southwest toward northeast) should be created to fit the calibrated thermal state of the SATB.

on several oils from the region. The original oils in the reservoirs should have an API gravity of (uplift and erosion) provide the PVT variation that is responsible for the compositional grading column at the present time.

- Oil generation and expulsion is mainly achieved in the footwall of the major structures. The measured kinetic parameters allow modeling a favorable timing of trap formation with respect to hydrocarbon generation and expulsion.
- When the measured bulk and compositional kinetics are used in the modeling, no oil generation is reached in the Tempa Rossa trend. The model shows that the Tempa Rossa heavy-oil field has been filled by oil that is generated deeper in the surrounding of the structure.

Finally, kerogen transformation ratio vs. depth shows that three different transformation ratio-depth zones should be considered to fit the thermal history of the southern Apennines. This 2-D information can be used to predict the distribution of potential source rock kitchen areas in the surroundings of the modeled section to guide future exploration.

ACKNOWLEDGMENTS

The model results presented here have been achieved using mainly unpublished data available in Enterprise Oil Italiana archive and subordinately from available published data.

This research and development project was supported by the European Community THERMIE Program, contract OG/161/98/FR/IT. We acknowledge the Enterprise Oil Italiana S.p.A and the Institut Français du Pètrole for permission to publish the results of this study.

We thank D. Wavrek (Petroleum Systems International Inc., Salt Lake City, Utah) and D. Jarvie (Humble Geochemical Services, Houston, Texas) for providing fruitful discussion and for their contribution to the interpretation of the geochemical data.

We acknowledge J. Granath and B. Wygrala for their helpful review.

REFERENCES CITED

Angevine, C. L., and D. L. Turcotte, 1983. Oil generation in overthrust belts: AAPG Bulletin, v. 67, no. 2, p. 235–241.

Armour, A., 1999, Southern European thrust belt exploration— Persistence pays off (abs.): AAPG Annual Meeting Program Extended Abstracts, v. 83, p. 20.

Davy, P. H., and P. H. Gillet, 1986, The stacking of thrust slices in collision zones and its thermal consequences: Tectonics, v. 5, no. 6, p. 913–929.

Deming, D., and D. S. Chapman, 1989, Thermal histories and hydrocarbon generation: Example from Utah–Wyoming thrust belt: AAPG Bulletin, v. 73, no. 12, p. 1455–1471.

Di Primio, R., B. Horsfield, M. A. Guzman-Vega, and D. Jarvie, 1999, Petroleum asphaltenes reflect source rock stability at time of liquid phase generation (oral presentation), *in* 19th International Meeting on Organic Geochemistry, Istanbul, Turkey, September 6–10, 1999.

Holton, J., 1999a, Southern Italy; (1) S. Apennines success bodes well for potential off southern Italy: Oil & Gas Journal, v. 97 (December 20, 1999), p. 65–68.

Holton, J., 1999b, Southern Italy; (2) Four geologic settings dominate oil, gas fields of Italy, Sicily: Oil & Gas Journal, v. 97 (December 20, 1999), p. 81–84.

Jarvie, D. M., and P. R. Walker, 1997, Correlation of oils and source rocks in the Williston basin using classical correlation tools and thermal extraction high-resolution C7 gas chromatography (oral presentation), *in* 18th International Meeting on Organic Geochemistry, September 22–26, 1997, Maastricht, Netherlands.

Jarvie, D. M., B. Leroux, W. D. Weldon, P. R. Walker, and D. A. Wavrek, 1999, Compositional kinetic parameters and yields from oil and gas prone source rocks and oil asphaltene fractions, *in* 19th International Meeting on Organic Geochemistry, Istanbul, Turkey, September 6–10, 1999.

Mango, F. D., 1997, The light hydrocarbons in petroleum: A critical review: Organic Geochemistry, v. 26, p. 417–440.

Rudkiewicz, J.-L., F. Schneider, F. Behar, J.-M. Gaulier, J. Wendebourg, and W. Sassi, 1996, Challenges and achievements in basin modeling: Predicting the nature and the volume of hydrocarbons, *in* M. E. Gomez Luna and A. Martinez Cortés, eds., Geoquimica Organica— 5e congreso Latino-American, Cancun, October 6–10, 1996, Resumenes, p. 380.

Sassi, W., and J.-L. Rudkiewicz, 1999, Thrustpack version 6.2: 2-D integrated maturity studies in thrust areas: Institut Français du Pétrole Report no. 45372, 79 p.

Sassi, W., and J.-L. Rudkiewicz, 2000, Computer modelling of petroleum systems along regional cross-sections in foreland and fold-and-thrust belts (abs.), *in* Proceedings of the Conference on Geology and Petroleum Geology of the Mediterranean and Circum-Mediterranean Basins (European Association of Geophysicists and Engineers Malta, October 1–4, 2000), Extended Abstracts.

Sciamanna, S., W. Sassi, J.-L. Rudkiewicz, R. Gambini, F. Mosca, and C. Nicolai, 2004, Predicting hydrocarbon generation and expulsion in the southern Apennines thrust belt by 2-D integrated structural and geochemical modeling: Part I— Structural and thermal evolution, *in* R. Swennen, F. Roure, and J. W. Granath, eds., Deformation, fluid flow, and reservoir appraisal in foreland fold and thrust belts: AAPG Hedberg Series, no. 1, p. 51–67.

Shi, Y., and C. Y. Wang, 1987, Two-dimensional modeling of the P-T-t paths of regional metamorphism in simple overthrust terrains: Geology, v. 15, p. 1048–1051.

Svara, P., A. De Poli, K. Leischner, F. Mosca, and C. Turrini, 1998, Modelling in thrusted areas— Sensitivity analysis and application: First Break, v. 16, no. 1, p. 9–20.

Tegelaar, E., and R. A. Noble, 1993, Kinetics of hydrocarbon generation as a function of the molecular structure of kerogen as revealed by pyrolysis-gas chromatography, *in* N. Telnaes, G. van Graas, and K. Oygard, eds., Advances in organic geochemistry, v. 22, p. 543–574.

Wygrala, B. P., M. N. Yalcin, and L. Dohmen, 1990, Thermal histories and overthrusting— Application of numerical simulation technique: Organic Geochemistry, v. 16, nos. 1–3, p. 267–285.

5

Toro, J., F. Roure, N. Bordas-Le Floch, S. Le Cornec-Lance, and W. Sassi, 2004, Thermal and kinematic evolution of the Eastern Cordillera fold and thrust belt, Colombia, *in* R. Swennen, F. Roure, and J. W. Granath, eds., Deformation, fluid flow, and reservoir appraisal in foreland fold and thrust belts: AAPG Hedberg Series, no. 1, p. 79–115.

Thermal and Kinematic Evolution of the Eastern Cordillera Fold and Thrust Belt, Colombia

Jaime Toro

Department of Geology and Geography, West Virginia University, Morgantown, West Virginia, U.S.A.

François Roure

Institut Français du Pétrole, Rueil Malmaison, France

Nathalie Bordas-Le Floch

Schlumberger, Aberdeen, United Kingdom

Sophie Le Cornec-Lance

Institut Français du Pétrole, Rueil Malmaison, France

William Sassi

Institut Français du Pétrole, Rueil Malmaison, France

ABSTRACT

We modeled the kinematic evolution of two regional-scale transects through the Eastern Cordillera fold and thrust belt of Colombia and then calculated the conductive thermal state of key steps of the kinematic history using Thrustpack® 4.0. The models were constrained by well, seismic, apatite fission-track, and thermal-maturity data.

The main compressional structures in the Cordillera are controlled by Jurassic–Early Cretaceous normal faults of the Bogotá, Cocuy, and the paleo-Magdalena basins. The location of these Mesozoic extensional features strongly influenced thermal evolution. Although shortening and basin inversion started in the early Tertiary, the bulk of the deformation occurred during the Miocene to Holocene Andean orogeny.

Rocks in different structural positions in the thrust belt have distinct thermal and maturation histories that determine the timing of hydrocarbon source rock maturation and the quality of sandstone reservoirs. The internal part of the Cordillera had high heat flow, with peak sedimentary burial and peak maturation during the Oligocene flexural phase. Local structures formed during this time and were followed by major uplift and denudation during the Andean orogeny. Hydrothermal circulation of basinal fluids,

DOI:10.1306/1025687H13114

which was probably expulsed at the onset of structural inversion, led to extensive cementation of Albian reservoirs. In contrast, the Llanos foreland is characterized by continued flexural subsidence and syntectonic sedimentation up to the present time. Thermal maturation results from the combination of syntectonic sedimentation and tectonic burial. Quartz cementation appears to be linked to the appearance of abundant silica in the system from pressure solution during Andean shortening. The thermal regime of the western flank of the Cordillera is cooler than the interior of the range, whereas the structural history is more complex. Along our transect, an active kitchen is located in the west-vergent thrust belt of the Eastern Cordillera. In the Magdalena Valley, there are local kitchens only where a thick stratigraphic section is preserved. The main limitations of our thermal models are (1) the lack of constraints on the thickness and timing of deposition of the Eocene–Oligocene flexural deposits, which are sparsely preserved in the Eastern Cordillera; (2) the paucity of good-quality thermochronologic data to constrain the timing of erosion and rates of fault motion; and (3) the difficulty in modeling the effects of fluid circulation over this large and structurally complex region.

INTRODUCTION

Understanding the thermal evolution of prospective areas in fold and thrust belts is critical in reducing exploration risk. It is now evident that thermal maturation of the source area controls hydrocarbon charge history, and that tectonic and sedimentary events determine the thermal history. However, it is less widely appreciated that both thermal and tectonic processes exert important controls on the cementation and dissolution processes in the reservoirs (Bordas-Le Floch, 1999; Roure et al., 2003). Unfortunately, in areas with complex geological histories, such as most fold and thrust belts, it is difficult to make decisions early enough in the exploration process, because a large body of data is required to constrain the many geological parameters involved.

We integrated such a large geological data set, including both surface and subsurface information provided by a consortium of oil companies, to evaluate the links between the thermal history of the rocks and the evolution of the sandstone reservoirs that are most productive in the fold and thrust belts of the Eastern Cordillera of Colombia. Our objective was to use this specific case to draw general conclusions about the evolution of the source rock and reservoirs to help guide exploration here and in other similar settings.

The thermal history of rocks follow distinct paths in different settings in a thrust belt determined by the interaction of (1) variation of basal heat flow, (2) tectonic and sedimentary burial, (3) erosion, and (4) fluid flow. Both the timing of maturation of source rocks and the quartz cementation history are very different in the internal part of a fold and thrust belt relative to either the foothill belts or the foreland.

To evaluate the effect of the first three factors listed above (heat flow, burial, and erosion), we carried out two-dimensional thermal and kinematic modeling of two transects across the Eastern Cordillera of Colombia, and we coupled this with petrographic, fluid inclusion, and anisotropy of the magnetic fabric studies of reservoir rocks collected both from cores and from outcrop samples. The methodology we used evolved from previous studies carried out in Venezuela (Roure et al., 2003). Insights from regional thermal and kinematic modeling of the Colombian Andes are presented in this chapter. Results from the reservoir studies are also discussed in Roure et al. (2003).

REGIONAL FRAMEWORK

In southern Colombia, the Andes branch into three ranges of different geological character and age (Figure 1). The area west of the Romeral fault, including the Choco block, the Western Cordillera, and part of the Central Cordillera, is composed of accreted rocks of oceanic affinity added to South America since the middle Cretaceous (Barrero, 1979; McCourt et al., 1984; Bourgois et al., 1985). The bulk of the Central Cordillera, which is the site of the modern volcanic arc, is mostly underlain by Paleozoic metamorphic rocks and Mesozoic to Cenozoic plutons. In contrast to the other branches of the Colombian Andes, a large portion of the Eastern Cordillera is made up of deformed but generally unmetamorphosed Cretaceous marine sedimentary rocks (Figure 2). This area of high topography resulted from inversion of a Mesozoic extensional basin system culminating in the Andean orogeny that is still active today.

The evolution of the Andes of Colombia is determined by plate interactions at a triple junction located at the northwest corner of South America (Burke et al., 1984). This triple junction, which originated during the

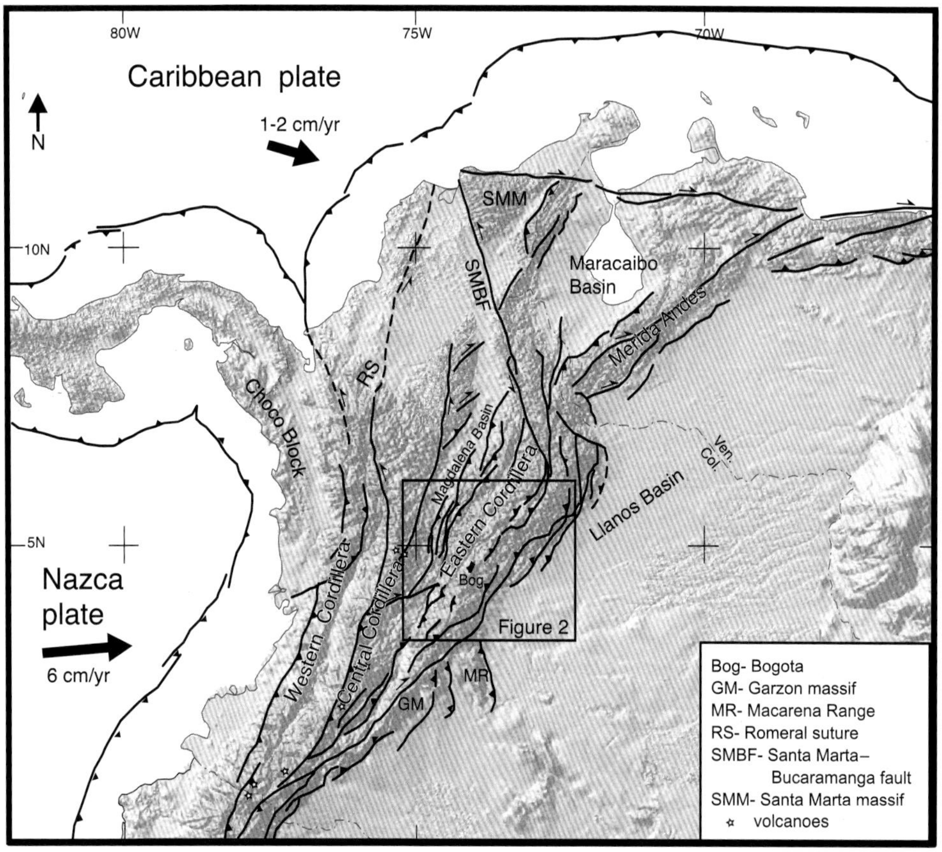

FIGURE 1. Tectonic map of the northern Andes, modified from Taboada et al. (2000). Shaded relief from U.S. Geological Survey GTOPO30 DEM.

Triassic–Jurassic breakup of Pangea, initially involved the North American, Pacific, and South American plates. The extensive belt of arc-related magmatic rocks of Mesozoic to Holocene age that exists in the Central Cordillera of Colombia and in the northern part of the Eastern Cordillera attests to a long history of subduction along the paleo-Pacific–South America boundary. During the opening of the North Atlantic, rifting and intraplate extension affected both the northern margin of South America and the paleo-Pacific back-arc region, creating a system of northeast-trending basins at the site of the present-day Eastern Cordillera, Magdalena Basin, and the Merida Andes of Venezuela. The origin of these basins has been attributed either to a failed rift (Pindell and Dewey, 1982; Jaillard et al., 1990) or to back-arc extension (McCourt et al., 1984; Pindell and Tabbutt, 1995; Sarmiento, 2002), but with the available data, it is difficult to choose between the two hypotheses.

Accretion of terranes of oceanic affinity starting in the middle Cretaceous along the paleo-Pacific margin produced episodes of compressional and transpressional deformation that propagated into the South American plate, creating the Colombian and Venezuelan Andes (Suarez et al., 1983; Aspden and McCourt, 1986). Today, the northwest corner of South America is a broad zone of dextral, transpressional deformation at the junction between the Nazca, Caribbean, and South American plates driven by convergence between South America and the Nazca plate (Figure 1).

EVOLUTION OF THE EASTERN CORDILLERA

The structure and evolution of the Eastern Cordillera were the subject of several classic studies (i.e., Van der Hammen, 1961; Campbell and Bürgl, 1965; Bürgl, 1967; Julivert, 1970), but the discovery of major oil fields during the last two decades caused a revival of interest (Fabre, 1987; Colletta et al., 1990; Dengo and Covey, 1993; Cooper et al., 1995; Roeder and Chamberlain, 1995), culminating with recent regional syntheses (Villamil, 1999; Taboada et al., 2000; Sarmiento, 2002).

Jurassic–Early Cretaceous Extension

Throughout the Eastern Cordillera and the Magdalena Valley, there are isolated exposures of Late Triassic–Jurassic red beds, volcanoclastic and volcanic rocks believed to have been deposited in elongated rift basins. The normal faults that bounded these rift basins are sparsely preserved as such; instead, the faults were inverted during the Tertiary–Holocene Andean compressional deformation (Colletta et al., 1990; Cooper et al., 1995). The existence of the main, basin-bounding normal faults is documented by the distribution of the synrift deposits. For example, in the central part of the Eastern Cordillera (northern part of Figure 2), thick Jurassic units (Arcabuco, La Rusia, Palermo, and Bata formations) are exposed on the hanging wall of the Boyacá thrust along the Arcabuco anticline, whereas directly to the east, thin, Early Cretaceous rocks rest directly on the Paleozoic basement of the Floresta massif.

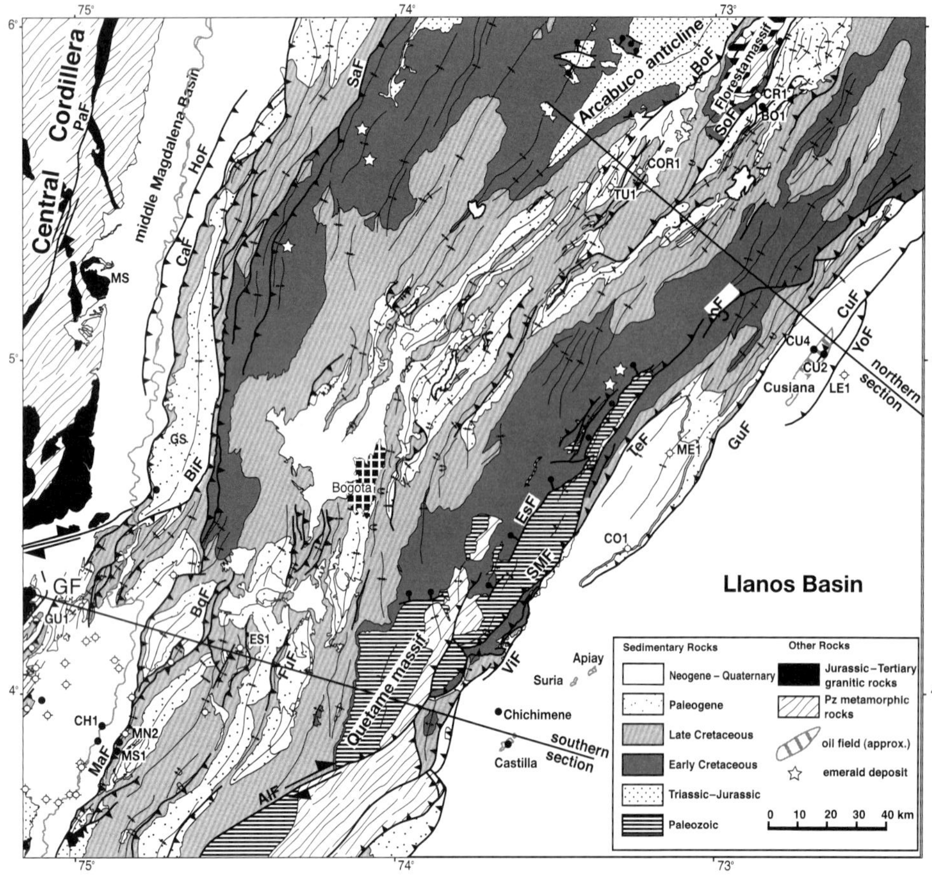

FIGURE 2. Simplified geological map of the central part of the Eastern Cordillera compiled from the 1:100,000 Ingeominas quadrangles. See Plate 1 for a color version and complete references. Abbreviations: GF = Girardot fold belt; GS = Guaduas syncline; MS = Mariquita stock. Faults: AlF = Algeciras; BiF = Bituma; BqF = Boqueron; BoF = Boyacá; CaF = Cambao; CuF = Cusiana; EsF = Esmeraldas; FuF = Fusa; GuF = Guaicaramo; HoF = Honda; MaF = Magdalena; PaF = Palestina; Sa = Salina; SMF = Santa Maria; SoF = Soápaga; TeF = Tesalia; ViF = Villavicencio; YoF = Yopal. Wells: BO = Bolivar; CH = Chenche; CO = Coporo; COR = Cormichoque; CR = Corrales; CU = Cusiana; ES = Estratigrafico; GU = Gualanday; LE = Leticia; ME = Medina; MS = Matachines Sur; TU = Tunja; MN = Matachines Norte; Pz = Paleozoic.

Extension and rapid subsidence during the Neocomian led to the deposition of thick marine strata in the large intracratonic Bogotá basin that covered most of the area now occupied by the Eastern Cordillera and the middle Magdalena Valley (Figure 3) (Fabre, 1987). The northern part of the Bogotá basin was divided into two rapidly subsiding subbasins separated by a basement high (the Floresta–Santander massif) (Figure 2) (Fabre, 1987). The northeastern arm of the Bogotá basin known as the Cocuy basin was bounded on the east by the paleo-Guaicaramo system of normal faults, now inverted (Fabre, 1987). Evidence for the paleo-Guaicaramo fault is that in the eastern flank of the Cordillera, latest Jurassic and Neocomian rocks (Macanal, Las Juntas, Fómeque, and equivalent formations) are more than 5000 m (16,500 ft) thick, whereas in wells of the eastern foothills, rocks of this age are absent. Normal faults in the footwall of the Guaicaramo system have also been recognized in seismic data (Linares, 1996; Sarmiento, 2002). Although good-quality seismic data is lacking along the western margin of the Cocuy subbasin, it appears that the basin was a half graben, because Early Cretaceous subsidence rates were faster along the paleo-Guicaramo system than in the vicinity of the Floresta massif (Sarmiento, 2002). The Tablazo subbasin, which formed the northwestern arm of the Bogotá basin, extended over much of the western flank of the Eastern Cordillera and the Magdalena Valley. Southeast of Bogotá, Neocomian strata decrease in thickness abruptly across an east–west-trending lineament (Figures 2, 3). Fabre (1987) has proposed that this transverse structure (the paleo-Gutierrez fault) marked the southern boundary of the Bogotá basin and extended to the Central Cordillera in the west.

Rapid Early Cretaceous subsidence has been explained by the thinning of the lithosphere using the model of McKenzie (1978). Fabre (1987) proposed that a stretching factor (β) of 2 could account for the subsidence of the Cocuy basin. More detailed subsidence analysis and modeling based on more than 100 wells and surface sections across the Eastern Cordillera shows that crustal stretching for the Early Cretaceous event ranged from about 1.1 to 3.0 (Sarmiento, 2002). Maximum stretching was concentrated in an elongated trough parallel to the paleo-Guaicaramo fault system. An important consequence of lithospheric thinning is an increase of heat flow in the crust because of the rise of the mantle isotherms. Numerous, small basaltic intrusions that cut the Early Cretaceous units (Fabre and Delaloye, 1982) and the high level of thermal maturity of Early Cretaceous units are evidence of elevated heat flow during that period.

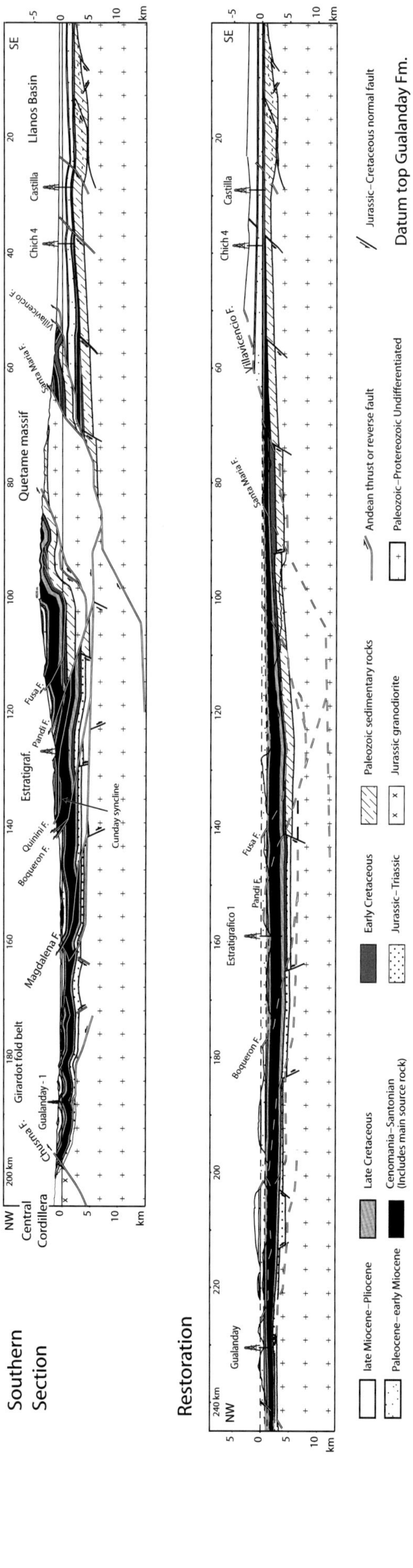

FIGURE 3. Cross section through the Eastern Cordillera south of Bogotá with restoration flattened on the Eocene Gualanday Formation. See Figure 2 for location and Plate 4 for a larger scale color version.

Late Cretaceous–Paleocene Subsidence

The Bogotá basin continued to subside through the Cretaceous as the lithosphere returned to thermal equilibrium. In this chapter, we treat this period, which extended from the Albian to the early Paleocene, as the thermal subsidence phase of the basin (Figure 4), although Sarmiento (2002) identified two localized episodes of accelerated subsidence that may reflect renewed extensional deformation during the Aptian–early Albian and Cenomanian.

Sedimentation was characterized by a transgressive-regressive megacycle that started with progradation of sandy deltaic and coastal facies of the Albian Une, Caballos, and equivalent units (Figure 4). These include some of the potential petroleum reservoir intervals in the basin. During the Turonian, the basin reached its maximum extent; even the basement high that separated the Tablazo and Cocuy subbasins was submerged, and sedimentation prograded into the Llanos Basin to the east and the flank of the Central Cordillera to the west (Etayo et al., 1969; Villamil et al., 1999). The main source rock intervals of the petroleum systems of northern South America were deposited during this period as anoxic marine shales. These are found in the La Luna, Villeta, Gachetá, and Chipaque formations of mainly Cenomanian to Santonian age. This sedimentary megacycle was closed by a relative drop of sea level in the Campanian to Maastrichtian that resulted in the deposition of sandstones of the Guadalupe Group (Villamil, 1999), which are productive reservoirs both in the Llanos and in the upper Magdalena Valley areas. Coal-bearing deltaic to continental deposits of the Guaduas Formation of Maastrichtian to Danian age mark the end of the thermal subsidence phase. A regional unconformity caps the thermal subsidence phase of the Bogotá basin and signals the onset of compressional tectonics and basin inversion east of the Central Cordillera. A detailed sequence-stratigraphic interpretation of the Late Cretaceous sedimentation in the Bogotá basin can be found in Villamil (1999).

Early Tertiary Basin Inversion

During the Late Cretaceous–early Tertiary, the Caribbean plate, characterized by anomalously thick oceanic crust, separated from the Farallon plate, began to move east relative to South America (Burke et al., 1984), and was eventually subducted along the northern margin of Colombia (Pennington, 1981; Taboada et al., 2000). In addition, an oceanic terrane, perhaps part of the Caribbean plate itself, was accreted to the continental margin in the Western Cordillera of Colombia (McCourt et al., 1984). South of the Caribbean–South America–Farallon triple junction, active subduction of

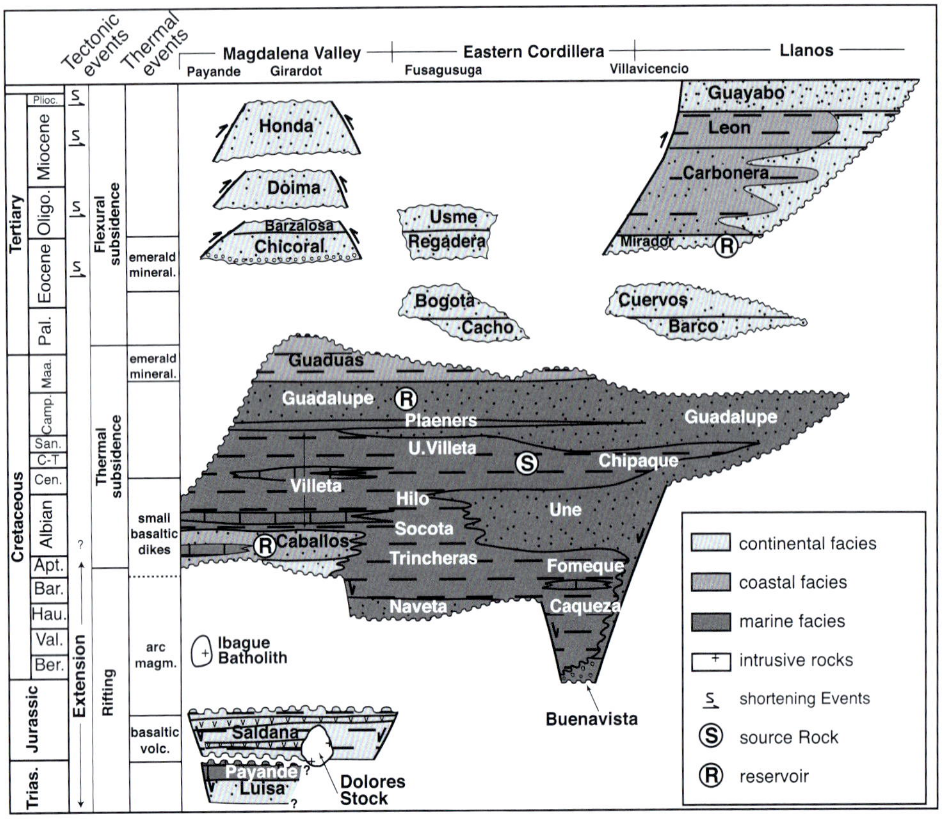

FIGURE 4. Chronostratigraphic diagram for the southern cross section showing the main sedimentary sequences of the Eastern Cordillera and adjacent basins.

the Farallon plate continued along the paleo-Pacific margin. As a result of these plate boundary interactions, transpressional and compressional deformation led to the tectonic uplift of the Central Cordillera and the onset of inversion in the Bogotá basin.

Beginning in the Paleocene, continental sedimentation established itself over most of the Eastern Cordillera domain. Uplift and denudation of the Central Cordillera and parts of the Santander massif are evidenced by the appearance of chert and lithic fragments in the upper Paleocene sandstones of the Cocuy basin (Socha formation; Fabre, 1981), as well as in the early Tertiary sandstones of the southern part of the Bogotá basin (Hoyón formation; Sarmiento, 1989), and by the progradation of coarse clastics to the east from the Central Cordillera source area in the middle Magdalena region, which have been interpreted as synflexural series (Gómez, 2002). In the upper Magdalena Valley, a major angular unconformity overlain by Eocene(?) conglomerates and sandstones of the Gualanday Group truncates earlier folds and thrust faults. The clasts in the Gualanday Group clearly point to a source in the Central Cordillera (Guillande, 1988). This unconformity is also recognized throughout the middle Magdalena Valley and is thought to indicate major deformation in the Central Cordillera before the middle Eocene (Gómez, 2002). Granodiorite of the Mariquita stock, located on the flank of the Central Cordillera 110 km (70 mi) north of our transect (Figure 2), yielded a 32 ± 6 Ma apatite fission-track age, indicative of late Eocene–early Oligocene uplift and denudation (Gómez, 2002).

In general, the Paleogene sedimentary basin was asymmetrical, with its depocenter close to the Central Cordillera, suggesting that it was a flexural basin resulting from tectonic loading in the Central and Western Cordilleras (Cooper et al., 1995). However, it appears that during the early Tertiary, some structural relief also developed in the Eastern Cordillera as the result of the inversion of Mesozoic normal faults (Van Houten and Travis, 1968; Roeder and Chamberlain, 1995; Sarmiento, 2002), so the pre-Andean basin may be more accurately described as a broken foreland basin. Detailed flexural modeling shows that the pattern of early Tertiary subsidence across the Eastern Cordillera and Magdalena Valley requires both thrust loading in the Central Cordillera and local loads in the Eastern Cordillera attributed to the uplift of inverted blocks (Sarmiento, 2002). In addition, areas thinned during Mesozoic rifting underwent greater Neogene flexural subsidence because the effective elastic thickness of the lithosphere had also been reduced.

Isotopic and structural studies in the emerald deposits of the Eastern Cordillera provide some constraints on the timing of deformation. Emerald mineralization in the Eastern Cordillera was produced by hydrothermal activity associated with the expulsion of basinal fluids (Cheilletz et al., 1994; Giuliani et al., 1995). Emerald mines are located along the two flanks of the Eastern Cordillera, close to the borders of the Early Cretaceous Bogotá basin (Figure 2). Emeralds are found in calcite-dolomite-pyrite veins in Lower Cretaceous shales (Macanal and Paja formations). $^{40}Ar/^{39}Ar$ ages of micas from the mineralized veins are 65 ± 3 Ma (Maastrichtian) in the eastern-flank district (Cheilletz et al., 1997) and are associated with extensional structures linked with the dissolution of evaporites (Branquet et al., 1999). In contrast, the ages of the western-flank district are 32–38 Ma (middle to late Eocene; Cheilletz et al., 1994) and occur in compressional structures (Branquet et al., 1999).

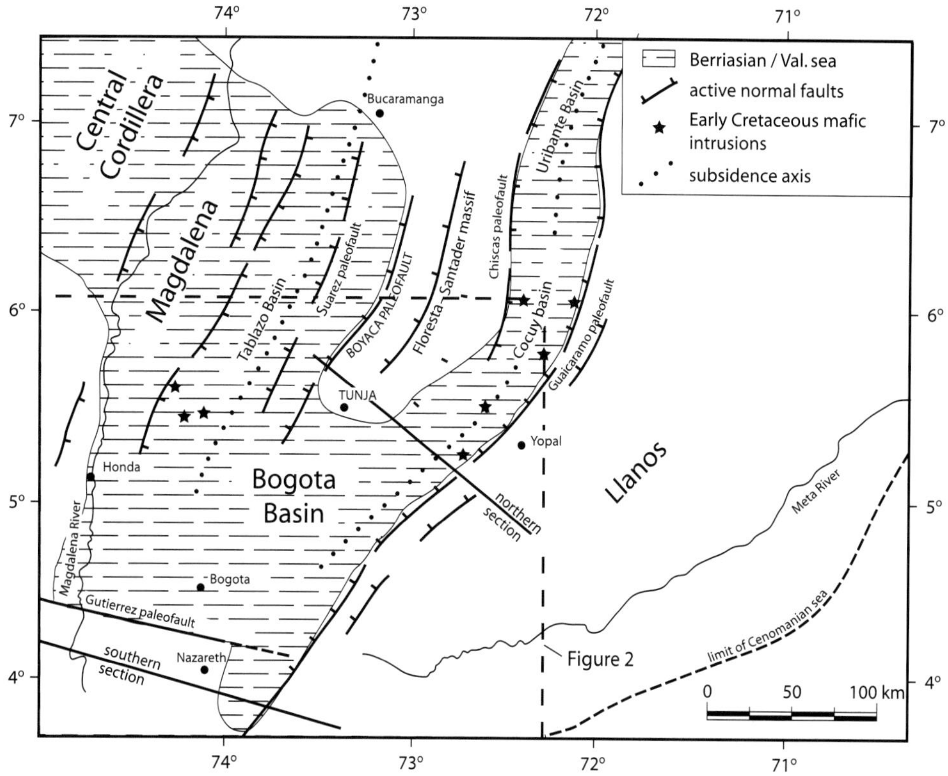

FIGURE 5. Paleogeographic map of the Early Cretaceous Bogotá basin modified from Fabre (1987). Present-day rivers are shown for geographic reference. Val = Valanginian.

Expulsion of hydrothermal fluids during the middle to late Eocene mineralization event probably occurred in response to regional, compressional deformation. The hydrothermal fluids reached a temperature of 290–360°C (Cheilletz et al., 1994) and may have been an important factor in the cementation of the Albian sandstones, as discussed below. The radiometric ages from the emerald deposits show that the history of fluid flow in the Eastern Cordillera is complex and may be associated with diverse tectonic regimes.

The Oligocene–early Miocene sedimentary records of the Llanos Basin (Carbonera Formation) and the northern part of the Eastern Cordillera (Concentración Formation) attest to rapid subsidence and relative tectonic quiescence. Subsidence during this period can be attributed to continued tectonic loading in the Central Cordillera (Cooper et al., 1995). However, there is evidence of Oligocene deformation throughout the region. In the upper Magdalena Valley, Oligocene sandstones (Doima formation) lie unconformably on folded Jurassic to Eocene rocks. Reliable apatite fission-track ages from the flanks of the Guaduas syncline, located in the western foothills of the Eastern Cordillera (Figure 2), are 26 ± 4 Ma, indicating relatively rapid cooling and denudation during the middle Oligocene (Gómez, 2002). In the central part of the Eastern Cordillera, north of Tunja, the shaly Concentración Formation (Oligocene), which is preserved only east of the Floresta massif, contains in its lower part conglomeratic sandstone horizons with fossiliferous limestone and chert clasts of Cretaceous age (Reyes and Valentino, 1974). This can be taken as evidence that deep erosion had already occurred in the Floresta massif or somewhere to the west during the Oligocene, but the relief may date back to Eocene time. Late Oligocene–early Miocene apatite fission-track ages from the Floresta massif (Toro, 1990) and southern Santander massif (Shagam et al., 1984) suggest that uplift and deformation had already begun by that time in the northern part of the Eastern Cordillera. Seismic data from the Llanos foothills in the Cusiana and Medina areas (Figure 2) show that minor thrusting of basement blocks occurred prior to deposition of the Carbonera Formation (Linares, 1996; Rathke and Coral, 1997).

The evidence presented above suggests that by Oligocene time, inversion had already begun in the Eastern Cordillera (Gómez, 2002; Sarmiento, 2002), and a complex paleogeography is likely to have existed. Because the Oligocene sedimentary record is very incomplete in the Cordillera, it is difficult to reconstruct the coeval basin geometry. Our thermal models, discussed below, show that the deepest part of this Oligocene basin was located along the axis of the Eastern Cordillera, instead of along the foothills of the Central Cordillera, as was proposed by Cooper et al. (1995).

Andean Deformation

Major shortening and formation of the high topography of the Eastern Cordillera began during the late Miocene as a response to the accretion of the Panama arc against the western margin of Colombia (Duque, 1990; Kellogg and Vega, 1995; Taboada et al., 2000). The thinned crust of the Bogotá basin was a zone of weakness in the South American plate, where a large amount of shortening could be accommodated. The Early Cretaceous basin was inverted by the reactivation of the old normal faults and by shortcuts on basement structures, producing the doubly vergent orogen that we see today (Figure 5) (Colletta et al., 1990; Cooper et al., 1995).

TABLE 1. Comparison of shortenings proposed for cross sections of the Eastern Cordillera through the Tunja area.

Authors	*Shortening (km)*	$(L_0-L_1)/L_0$ (%)
Cooper et al. (1995)	68	14
Colletta et al. (1990)	105	30
This study (northern section)	60 (partial section)	30
Taboada et al. (2000)	120	33
Dengo and Covey (1993)	150	40
Roeder and Chamberlain (1995)	230	64

Thick synorogenic clastics (Guayabo Formation) filled the new flexural basin of the Llanos and also the intermontane Magdalena Basin (Honda and Real formations). Andean range front faults are still seismically active (Taboada et al., 2000), and even the youngest sedimentary units are involved in the deformation, thus attesting to the youth of the mountain belt.

Several competing structural models have been published recently for the Eastern Cordillera (Colletta et al., 1990; Dengo and Covey, 1993; Cooper et al., 1995; Roeder and Chamberlain, 1995). The main issue is whether the deformation in the Cordillera was accomplished by motion on low-angle faults and the magnitude of shortening is great or by high-angle faults, which would require only moderate shortening to achieve the same structural relief.

A wide range of shortening estimates have been published for the Eastern Cordillera (Table 1). Two end-member interpretations exist: one where the major thrust faults are low-angle detachments that cut the entire crust (Roeder and Chamberlain 1995) and one where the main faults root into either reactivated Jurassic–Early Cretaceous normal faults or new thrust faults dipping approximately 30° (Cooper et al.,1995). Dengo and Covey (1993) proposed a mixed model of the Cordillera, where an early phase of thin-skinned deformation was followed by crosscutting, high-angle reverse faults. To balance the shortening in the cover and the basement, they postulate the existence of a major detachment extending west beneath the Central Cordillera into the subduction zone. However, the Bouguer gravity anomaly associated with the Cordillera indicates that it has a lower crustal root that must be accounted for in the structural model (Colletta et al., 1990; Toro, 1990). We prefer a simpler solution similar to that of Colletta et al. (1990), where the shortening in the cover can be balanced by local crustal thickening of the basement. Our northern section requires 60 km (37 mi) of shortening (30%) to restore the structures of the east-vergent half of the Cordillera. Undoubtedly, there are multiple detachment levels in the stratigraphic section of the Cordillera as pointed out by Dengo and Covey (1993). This leads to disharmonic folding at different structural levels, but there is no need to propose large values of displacement along these detachments. Our southern section is located where the Eastern Cordillera has narrowed to half the width found further north. Along that transect, 50 km (31 mi) of total shortening (25%) is sufficient to restore the structures. Details of the Andean structures are discussed below in the context of our structural transects.

KINEMATIC MODELING

Methods

The cross sections through the Eastern Cordillera in this study (Plate 1; Figures 3, 6) were constructed on the basis of the published geological maps of Ingeominas at 1:100,000 scale locally supplemented by more detailed unpublished maps. We also used stratigraphic data derived both from wells and from the literature and seismic sections provided by Ecopetrol and oil companies belonging to the Subthrust Reservoir Appraisal (SUBTRAP) consortium of Institut Français du Pétrole. The seismic data provided valuable control on the structure of the Llanos and Magdalena foothills. However, in the internal portion of the Cordillera, seismic coverage is discontinuous, and the quality of the existing lines is commonly poor. In particular, the deep architecture of the orogen remains speculative, because deep-crustal seismic data is lacking. We used Bouguer gravity data (Bermudez et al., 1985) to constrain the basement and cover geometry where the seismic data was lacking.

A simplified version of the initial state of the balanced cross sections was input into the software Thrustpack® 4.0 to model the kinematic and thermal history of the Cordillera. The modeling was an iterative process where the Thrustpack 4.0 results lead to modifications of the cross sections until we obtained a good match to the present-day structures, a deformation history that fits all the timing constraints, and a thermal history in agreement with the thermal-maturity data. The deformation algorithm used in Thrustpack 4.0 (Sassi and Rudkiewicz, 1996) is based on the kink-line method of Suppe (1983). Only foreland-verging fold-bend folds could be modeled by Thrustpack 4.0.

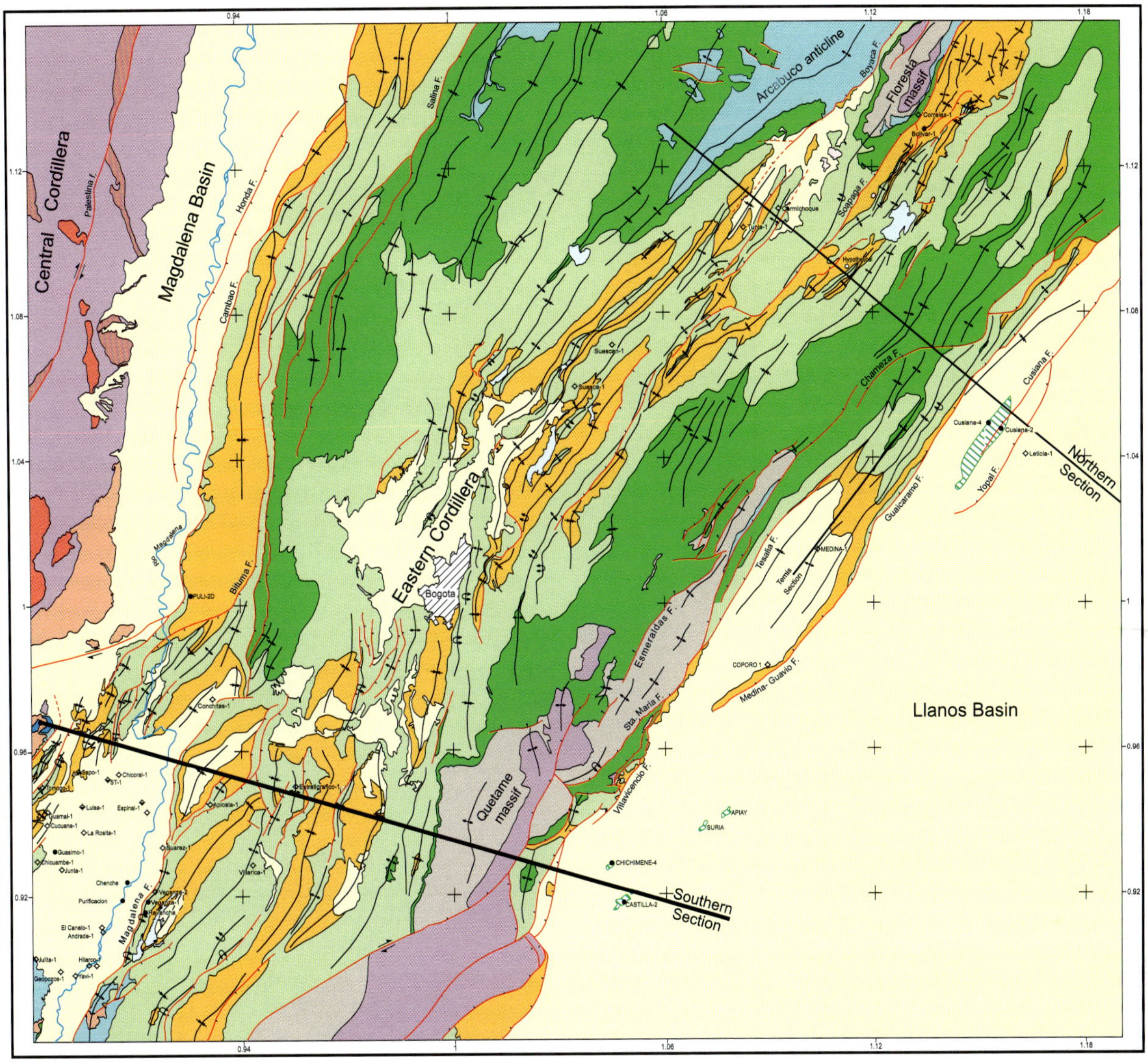

Sedimentary Rocks

- Neogene–Quaternary
- Paleogene
- Late Cretaceous
- Early Cretaceous
- Late Jurassic
- Triassic
- Paleozoic

Igneous and Metamorphic Rocks

- Tertiary volcanic rocks
- Tertiary pluton
- Cretaceous pluton
- Jurassic pluton
- Paleozoic crystalline basement

oil field (approx.)

0 10 20 30 40 km

Map References

Arce, M., and D. H. McLaughlin, 1969, Mapa geologico del cuadrangulo K-11, Zipaquira, scale 1:100,000, Ingeominas 1 sheet.

Barrero, D., and C. Vega, 1976, Mapa geologico del cuadrangulo K-9, Armero, parte sur del J-9, La Dorada, scale 1:100,000, Ingeominas 1 sheet.

Champetier de Ribes, G., G. Weeksteen, and R. R. Quintero, 1961, Mapa geologico de la plancha K-10, Villeta, scale 1:200,000, Ingeominas, Bogotá 1 sheet.

McLaughlin, D. H. Jr., et al., 1969,Mapa geologico del cuadrangulo K- 11 (Zipaquira, Colombia), scale 1:100,000, Ingeominas 1 sheet.

Renzoni, G., 1968, Geología del macizo de Quetame. Geologia Colombiana, v. 5, p. 75–127.

Renzoni, G., 1981, Geología del cuadrangulo J-12, Tunja (1967). Boletín. Geologoco, Ingeominas, v. 24, no. 2, p. 31-48.

Renzoni, G., 1992, Mapa fotogeologico, Yopal (1985), scale 1:100,000, plancha 193. Ingeominas.

Renzoni, G., H. Rosas, C. Ospina, and F. Etayo, 1969, Geología del cuadrangulo J-12. Primer Congreso Colombiano de Geología.

Renzoni, G., et al., 1983, Mapa geologico, plancha 191, Tunja, (1967), scale 1:100,000, Ingeominas 1 sheet.

Segovia, A., 1963, The geology of plancha L-12, Peralonso-Medina: Ph.D. thesis, Pennsylvania State University, 202 p.

Segovia, A., 1967, Geology of Plancha L-12, Colombia, South America. A, Reconnaissance: Geological Society of America Bulletin, v. 78, p. 1007–1028.

Suarez, M., 1985, Mapa fotogeologico de la plancha 174, Paz de Ariporo, Ingeominas, Informe. 1964, Bogotá.

Ulloa, C. E., 1973, Geología del cuadrangulo J-13, Sogamoso, scale 1:100,000, Ingeominas 1 sheet.

Ulloa, C. E., and E. Rodriguez, 1984, Mapa geologico preliminar de la plancha 170, Velez, scale 1:100,000 (1978), Ingeominas 1 sheet.

Ulloa, C. E., and E. Rodriguez, 1991, Mapa geologico de Colombia, plancha 190, Chiquinquira. Memoria explicativa, Ingeominas, 26 p.

Ulloa, C. E., and M. E. Rodriguez, 1978, Mapa geologico de la plancha 190, Chinqinquira, scale 1:100,000, Ingeominas 1 sheet.

Ulloa, C. E., E. Rodriguez, and R. Escovar, 1973a. Mapa geologico, plancha 192, Laguna de Tota, scale 1:100,000, Ingeominas 1 sheet.

Ulloa, C. E., E. Rodriguez, and R. Escovar, 1973b, Mapa geologico, plancha 171, Duitama, scale 1:100,000, Ingeominas 1 sheet.

Ulloa, C. E., E. Rodriguez, and A. Guerra, 1973c, Mapa geologico, plancha 172, Paz del Rio, scale 1:100,000, Ingeominas 1 sheet.

Ulloa, C. E., R. Escovar, and A. H. Pacheco, 1976a, Mapa geologico, plancha 230, Monterrey, scale 1:100,000, Ingeominas 1 sheet.

Ulloa, C. E., R. Escovar, and A. H. Pacheco, 1976b, Mapa geologico, plancha 211, Tauramena, scale 1:100,000, Ingeominas 1 sheet.

Ulloa, C. E., et al., 1975, Mapa geologico del cuadrangulo K-12, Guateque, Colombia, scale 1:100,000, Ingeominas 1 sheet.

Gauss Conformal Projection
Origin Bogota X = 1,000,000 m North
Y = 1,000,000 m East

Plate 1. Geological map of the central Eastern Cordillera.

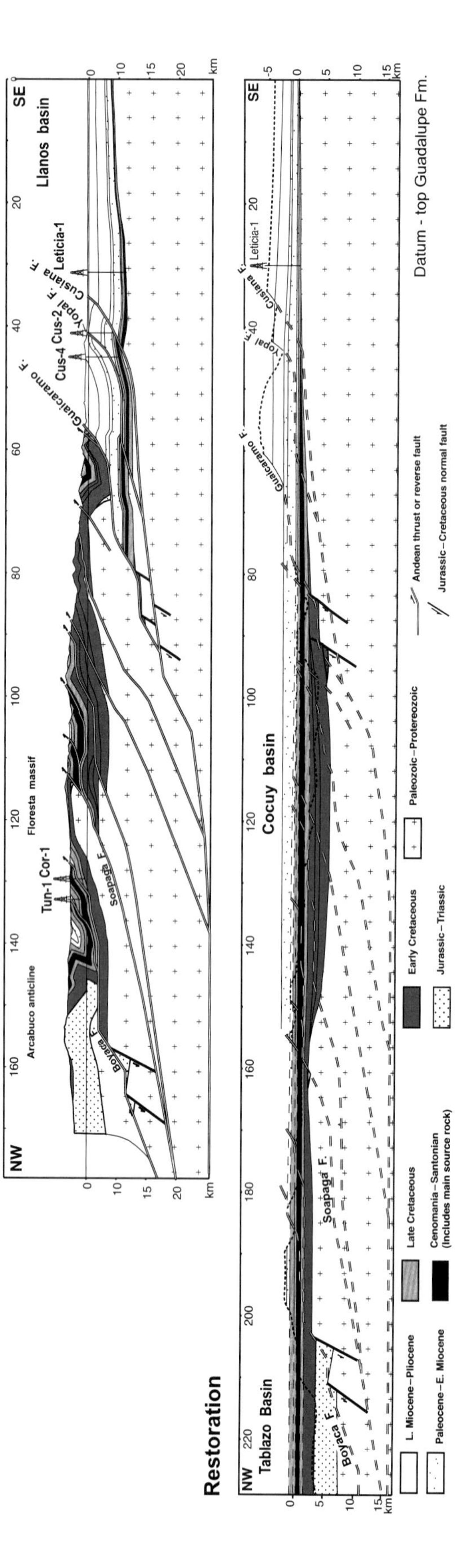

FIGURE 6. Cross section of the eastern flank of the Eastern Cordillera through the Tunja area and its restoration flattened at the top of the Late Cretaceous Guadalupe Formation. The restoration shows the Early Cretaceous rift structures that control the location of the main Andean thrusts. See Figure 2 for location and Plate 2 for a large-scale color version.

Inversion structures, ductile deformation, fault-propagation folds, and back thrusts were not possible with that early version of the program. Additional limitations were that faults could not cut each other, that faults had to begin with a flat, and that faults steeper than 30° were problematic. In the case of the Eastern Cordillera, these restrictions proved not to be a major obstacle, and a satisfactory fit to the hand-drawn cross section could be obtained as long as the east-vergent and the west-vergent flanks are modeled independently. The limitations outlined above have been overcome in more recent versions of Thrustpack (Sassi et al., 1998).

Structural Transects

The two transects described here characterize two distinct areas of the Eastern Cordillera. The northern transect is only a partial cross section located in the northern part of the high plateau known as the Altiplano Cundiboyacese and extends from the Arcabuco anticline in the west to the Llanos foreland in the east (Figures 2, 6; Plate 1). It is located in the widest part of the Cordillera in an area that, in terms of the Early Cretaceous paleogeography, corresponds to the southern extremity of the Floresta paleohigh and to the Cocuy subbasin. In contrast, the southern transect, which extends from the Magdalena Valley to the Llanos, is located south of the southern boundary of the Bogotá basin, where Early Cretaceous strata are relatively thin.

The Eastern Cordillera is a doubly vergent orogen, with thrust faults directed toward both the Llanos Basin and the Magdalena Valley. However, the structure is not symmetrical in cross section as is evident in the southern transect (Figure 3; Plate 1). The greatest structural relief and largest horizontal displacement are on the faults of the eastern foothills, which is on the Guaicaramo system. On a map view, the system is composed of a set of large en echelon, right-stepping thrust faults (Figure 1). In the study area, these faults are the Santa Maria, Tesalia, and Guavio–Guaicaramo faults. These appear to be the master structures of the Eastern Cordillera. The en echelon pattern of the frontal faults is probably caused by a component of right-lateral shear of the Andean deformation. Right-lateral transpressional deformation is the result of the oblique angle between the range-bounding faults and the direction of convergence of the Nazca plate to South America (Figure 1) (Colletta et al., 1990). This oblique angle is inherited from the geometry of the Jurassic–Early Cretaceous rift system.

Northern Transect

The restoration of the northern transect (Figure 6; Plate 1) shows that before the Andean deformation, there were two subbasins bounded on the east by

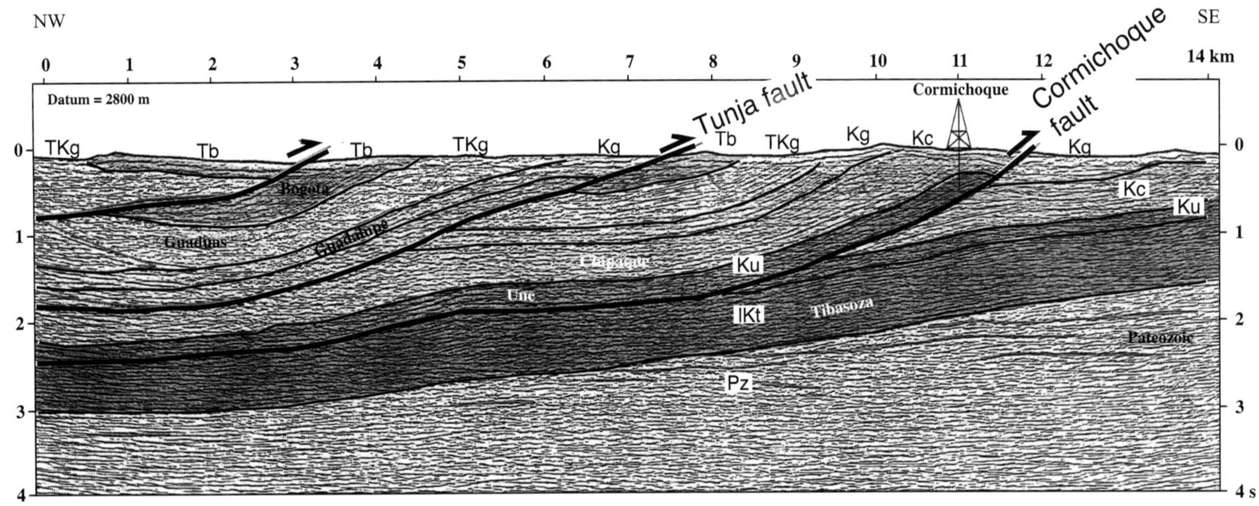

FIGURE 7. Seismic line in the internal part of the Eastern Cordillera located on the footwall of the Boyacá fault. Abbreviations: Tb = Paleocene Bogotá Formation; TKg = Maastrichtian–Paleocene Guaduas Formation; Kg = Upper Cretaceous Guadalupe Formation, Chipaque Formation; Kc = Upper Cretaceous Chipaque Formation; Ku = Albian Une Formation; lKt = Aptian–Barremian Tibazoza Formation; Pz = Paleozoic sedimentary and metamorphic rocks.

normal faults and separated by the Floresta massif. We show the basins as simple half grabens, but a more complex geometry is possible. Each of the two basin-bounding normal fault systems was inverted during the Andean deformation. The geometry of the Arcabuco anticline, which is located at the western end of the section, suggests that the paleo-Boyacá fault was inverted by a shortcut and not by direct reactivation of the normal faults (Figure 6). At the location of our transect, the Boyacá fault does not breach the surface. Instead, its displacement is transferred to the Tunja and Cormichoque thrusts that are beautifully expressed in the seismic data and have been drilled (Figure 7). These two thrusts follow detachments in the shales of the Chipaque and Tibasosa formations. The top of the Paleozoic basement dips gently to the northwest over most of the block comprised between the Boyacá and Soápaga faults.

The Soápaga fault brings Paleozoic basement to the surface in the core of the Floresta massif. This fault can be traced to the north for more than 40 km (25 mi), until it links to the Santa Marta–Bucaramanga left-lateral strike-slip fault (Figure 1). The Soápaga and Boyacá faults may be transferring a large component of the strike-slip displacement of the Santa Marta–Bucaramanga fault into shortening (Toro, 1990). In its northern part, the Soápaga fault thrusts Paleozoic basement over the Oligocene Concentración Formation. In that area, there is a large, basement-cored, recumbent anticline on the hanging wall paired with a recumbent syncline (the Beteitiva syncline) in the footwall. This geometry suggests that the Soápaga fault developed by fault propagation and culminated with a synclinal breakthrough (Toro, 1990). We have interpreted the Soápaga fault as a moderate-angle thrust that joins the same midcrustal detachment as the Boyacá fault. Some of the displacement on the Soápaga fault is transferred forward into a set of smaller thrust faults detached in the middle Cretaceous shales that form complicated, short-wavelength folds. Several back thrusts occur in this area. However, in the regional context, the amount of shortening represented by these west-vergent structures is minor. Because of the limitations of the software, we have neglected them in the kinematic modeling.

In our model, more than 24 km (15 mi) of horizontal motion on the Guaicaramo fault are required to produce the observed structural relief. As we have already stated, the major ramps and flats of the Guaicaramo fault are probably controlled by the location of Early Cretaceous normal faults (Figure 6; Plate 1). The hanging wall of the Guaicaramo fault is dominated by a folded but generally continuous section of Lower Cretaceous sedimentary rocks. In this area, we propose several faults of lesser importance detached in the Lower Cretaceous shales. Fifty kilometers along strike to the south of our transect, the underlying Paleozoic and Jurassic rocks crop out. These pre-Cretaceous block is bounded on the west by the Esmeraldas fault, which appears to be a Jurassic–Early Cretaceous normal fault decapitated and transported to the east by the Guaicaramo fault.

We have presented a very simple interpretation of the range front structures where most of the shortening occurs on a single branch of the Guaicaramo fault. A more complicated geometry probably exists, given the along-strike variations of structures (Estrada et al., 1999), but we lack the seismic data necessary to propose a more accurate model. The Yopal and Cusiana faults are at the present-day deformation front. To have a coherent kinematic model, we have linked the Yopal and

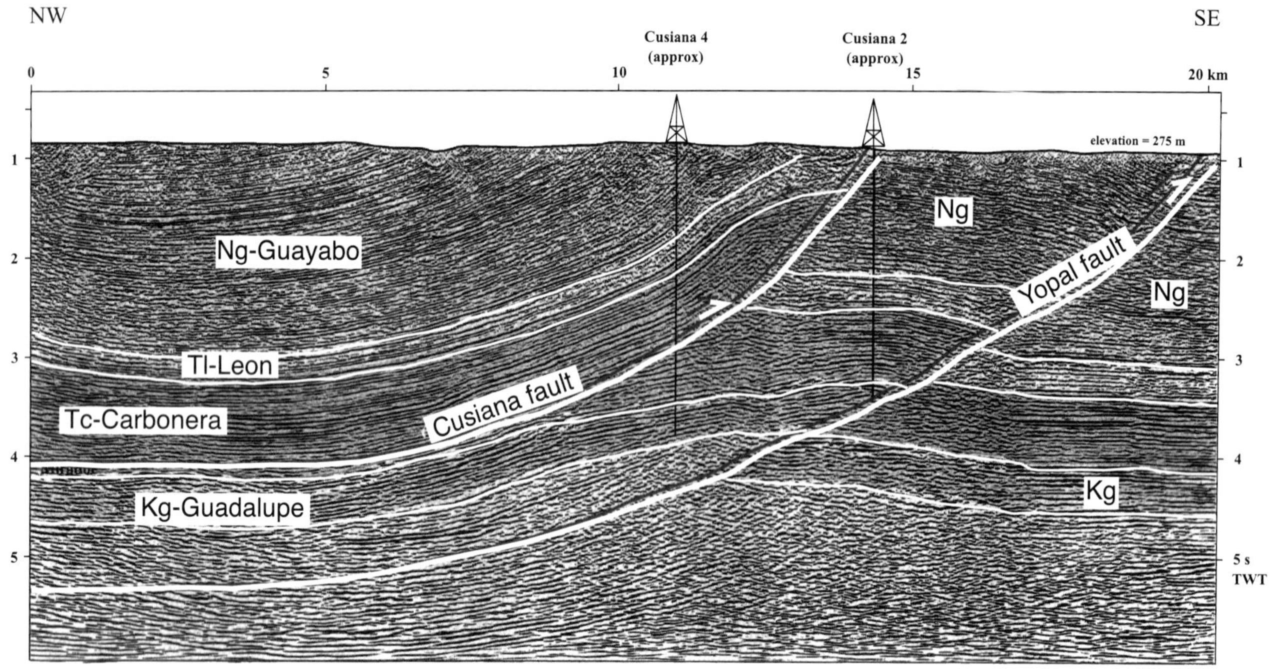

FIGURE 8. Seismic line through the Cusiana field of the Llanos foothills located at the eastern end of our northern section (Plate 2). Folding of the Guayabo Formation (Ng) without fanning of strata indicates that the Cusiana fault is late to post-Neogene. See Cazier et al. (1995) for a complete description of the Cusiana field. Ng = Neogene Guyabo Formation; Tl = Tertiary Leon Formation; Tc = Tertiary Carbonera Formation; Kg = Upper Cretaceous Guadalupe Formation.

Cusiana faults back to the Guaicaramo system following detachments in the lower Carbonera shales and in the Paleozoic basement.

Timing of Deformation

The available constraints on the exact age of motion on the different fault systems of the Eastern Cordillera are poor. This is one of the main sources of uncertainty in the kinematic model. Apatite fission-track data from the Floresta massif show that unroofing in the core of the Cordillera began at 20–25 Ma (Toro, 1990). The track-length distributions suggest a slow cooling rate; however, the mediocre quality of the data does not permit an unambiguous interpretation. In contrast, unpublished fission-track ages from the eastern flank and the foothills indicate rapid cooling between 10 and 5 Ma. The coarse clastic sediments of Guayabo Formation (late Miocene–Pliocene) of the Llanos foreland are thought to be associated with active thrusting in the Eastern Cordillera. However, in the Cusiana area, the Guayabo reflectors are parallel, generally isopachous, and are folded in the hanging walls of the Cusiana and Yopal faults (Figure 8). No evidence exists for syndepositional deformation in the Guayabo Formation. Therefore, the frontal faults of the Cordillera postdate the entire Guayabo section. Seismicity along the Guaicaramo fault indicates that it remains active today (Taboada et al., 2000). Paleobotanical data coupled with radiometric ages of volcanic ashes from the late Miocene–Holocene sediments of the Bogotá area (Helmens, 1990) show that a tropical lowland environment persisted until the earliest Pliocene (about 5 Ma). The lowland floras were replaced by lower sub-Andean forests and then by upper sub-Andean forests. These paleoenvironmental changes were interpreted as the result of uplift of the Bogotá region from less than 500 m (1600 ft) above sea level at 5 Ma to more than 2200 m (7300 ft) beginning at 3 Ma (Helmens, 1990).

In summary, deformation had begun in the Floresta massif by 20 Ma, but the main pulse of shortening and uplift occurred in the last 5 Ma. In the kinematic model presented below, we introduce minor motion on the Boyacá and Soápaga faults before 20 Ma, then the deformation propagated eastward. The large displacement of the Guaicaramo fault requires that it remains active from 16 Ma to the present to maintain moderate displacement rates. In our model, about 20 km (13 mi) of total shortening had already occurred by 12 Ma. This is difficult to reconcile with the paleobotanical data from the Bogotá plateau but agrees with the apatite fission-track constraints from the foothills. It must be remembered that the paleoflora is first a paleoclimatic indicator and only indirectly a measure of paleoelevation, and that we have control neither on the rates of fault motion nor on the deep geometry of the faults. If we assume higher fault velocities (which are possible) or steeper faults, it would be permissible to concentrate more on the deformation in the Pliocene–Holocene period.

Kinematic Model (Northern Transect)

To reduce computational problems in Thrustpack 4.0, we carried out the modeling of the northern section in two segments: (1) an internal segment, including all the faults comprised between the Boyacá fault and the Guaicaramo fault in the east (Plate 2; Table 2), and (2) a frontal segment, including only the Chámeza, Guaicaramo, Cusiana, and Yopal faults (Plate 3; Table 3). Below, we highlight the important features of the kinematic model of the internal segment of the cross section. The initial and final stages of the model are shown in Figure 9. The complete model, as well as a model of the frontal structures, is shown in Plate 3.

34 Ma: Initial state. The initial structure is characterized by the Cocuy and Tablazo Jurassic–Cretaceous rift basins.
20 Ma: Early Carbonera deposition. We assume that rocks of the Carbonera–Concentración Formation extended continuously over the Cordillera. Today, the Concentración Formation is only preserved along a narrow band on the footwall of the Soápaga fault, and there is great uncertainty about its age range and original thickness. Fission-track data from the Floresta massif indicate that denudation began at 25–20 Ma in that area, so we activate the Soápaga and Tunja faults (faults 1 and 3) and begin to erode the highlands.
16 Ma: Late Carbonera–Leon deposition. In this period, there is additional deposition of Carbonera-equivalent rocks in the Cocuy basin. Motions on the Soápaga and Boyacá fault systems (faults 1–3) continue leading to more erosion of the highlands.
12 Ma: Early Guayabo deposition. Displacement continues on the Boyacá and Soápaga fault systems (faults 1–5), and deformation propagates to the east (faults 6–8). Rapid accumulation of Guayabo sediments begins.
8 Ma: Middle Guayabo deposition. Motion ceases in the Boyacá and Soápaga fault systems (faults 1–5). The faults of the eastern part (faults 6–8) continue to be active. Guayabo sedimentation continues, accompanied by erosion of the highlands.
4 Ma: Late Guayabo deposition. Displacement continues on the eastern faults (faults 6–8) and is accompanied by erosion. Within the Cordillera, Tertiary rocks are only preserved in the synclinoria. Rapid sedimentation continues in the Llanos.
0 Ma: Holocene. Only the Guaicaramo fault (fault 8) is active. erosion is present along the entire section. The final geometry is a good match to the hand-drawn structural cross section, but the uplift of the Arcabuco anticline is not sufficient. The west-vergent fault system of the Magdalena flank, which is neglected in this model, is necessary to achieve the correct amount of uplift of the internal part.

Southern Transect

Below, we describe the principal structures of the southern transect from east to west. Unlike the northern transect, there are two small basement-involved anticlines in the foreland east of the range front (Figure 3; Plate 4). Both structures, which host the Chichimene and Castilla oil fields, are bounded by high-angle reverse faults that cut the late Miocene–Pliocene strata. In the Chichimene anticline, the Cretaceous units are thicker in the hanging wall, suggesting that the structure formed by reactivation of a Cretaceous normal fault (Kluth et al., 1997). The Castilla anticline is broader and lacks clear evidence of tectonic inheritance. However, the similarity in age, character, and setting between the two structures suggests that they had similar origins.

The main frontal faults of the Eastern Cordillera along the southern transect are the Villavicencio and Santa Maria faults. Although the Santa Maria fault links to the south with the Algeciras right-lateral strike-slip fault (Figures 1, 2), seismic data from 30 km (19 mi) north of the location of our transect (Figure 10) demonstrate that it is relatively low-angle thrusts. In our kinematic model, the Santa Maria and Villavicencio faults have combined horizontal displacement of about 28 km (18 mi); thus, these two faults account for about half the total shortening along the southern section (Table 4). The Lower Cretaceous units exposed in the hanging wall of the Villavicencio fault (Caqueza Group) are much thicker than in the foreland, confirming that the Andean front is controlled by Early Cretaceous normal faults.

On the basis of surface geology (Cáceres et al., 1970), the structure of the Quetame massif and the internal part of the Cordillera appears to be rather simple. It is a large block, tilted to the west by the Santa Maria–Villavicencio faults, with a complete lower Paleozoic to Paleocene stratigraphic section exposed at the higher elevations (Figure 3; Plate 4). In contrast, the Magdalena flank of the Cordillera is cut by numerous west-vergent thrust faults. The most important ones are the Fusa, Pandi, Quinini, Boquerón, and Magdalena faults (Figure 3; Plate 4). The Fusa fault was not recognized by Cáceres et al. (1970), but it is required in the seismic interpretation (Figure 11). Generally, the west-vergent thrusts that are imaged in the seismic lines are low angle, have moderate displacements, and involve most of the Cretaceous sedimentary section (Figure 11). In our interpretation, we have linked most of the faults into a single detachment in pre-Cretaceous rocks at about 5 km (3 mi) below sea level. This interpretation is speculative;

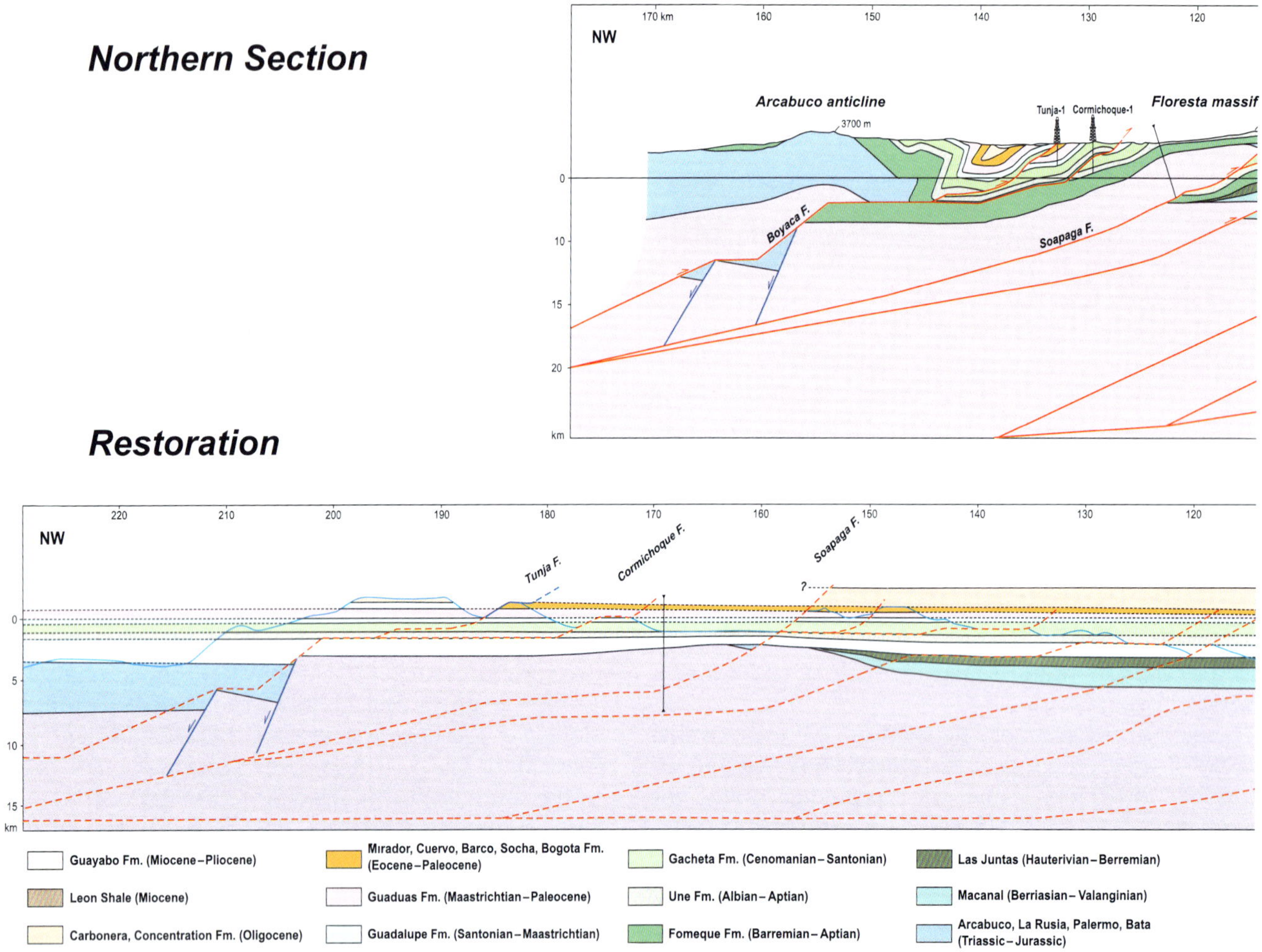

Plate 2. Structural cross section of the eastern flank of the Eastern Cordillera in the Tunja–Cusiana area. See Plate 1 for location.

in fact, we have not fully addressed the problem of the interaction between the east-vergent master fault and the west-vergent thrusts.

The distribution of pre-Cretaceous rocks is not well known. Devonian and Carboniferous clastic rocks underlie the Early Cretaceous Fómeque formation in the Quetame massif but are not exposed further west (Figure 2; Plate 1). In the Magdalena Valley, thick, rift-related Jurassic red beds and felsic to intermediate volcanic rocks of the Saldaña Formation exist. In our interpretation, we have used the conceptual model that most ramps of the Andean faults are controlled by Jurassic normal faults, but in general, we lack direct evidence of this relationship. A notable exception is the Espinal high, a large, symmetrical anticline located in the middle of the Magdalena Valley, which has a geometry typical of inversion structures (Figure 3; Plate 4).

West of the Estratigrafico 1 well, there are some irregularities in the thrust belt. The Quinini fault strikes to the north-northwest, in contrast to the dominant northeast strike of the rest of the structures (Figure 2; Plate 1). This fault overrides the Cunday syncline almost at 90° to the fold axis. The Quinini fault has a strong component of motion out of the plane of the transect (Figure 11). In the Thrustpack 4.0 model presented below, we neglected the Quinini fault because the software cannot deal with out-of-plane effects. Analog model experiments suggest that oblique structures such as these can form where the stratigraphic section abruptly changes thickness parallel to the direction of thrusting (B. Colletta, 1998, personal communication). In this case, the oblique structures are located at the southern limit of the Bogotá basin, where the Early Cretaceous strata thin over the paleo-Gutierrez fault.

The Magdalena range front is formed by a resistant ridge of Guadalupe Formation thrusted over Miocene–Pliocene sediments of the Honda Formation by the Magdalena fault (Figure 12). However, the Magdalena fault is also linked to a deeper triangle structure that deforms and uplifts the Oligocene–Pliocene section of the Magdalena Valley. This triangle zone is laterally extensive and is productive in the Matachines Norte, Matachines Sur, and Chenche fields, where the Guadalupe Formation is preserved in the triangle zone.

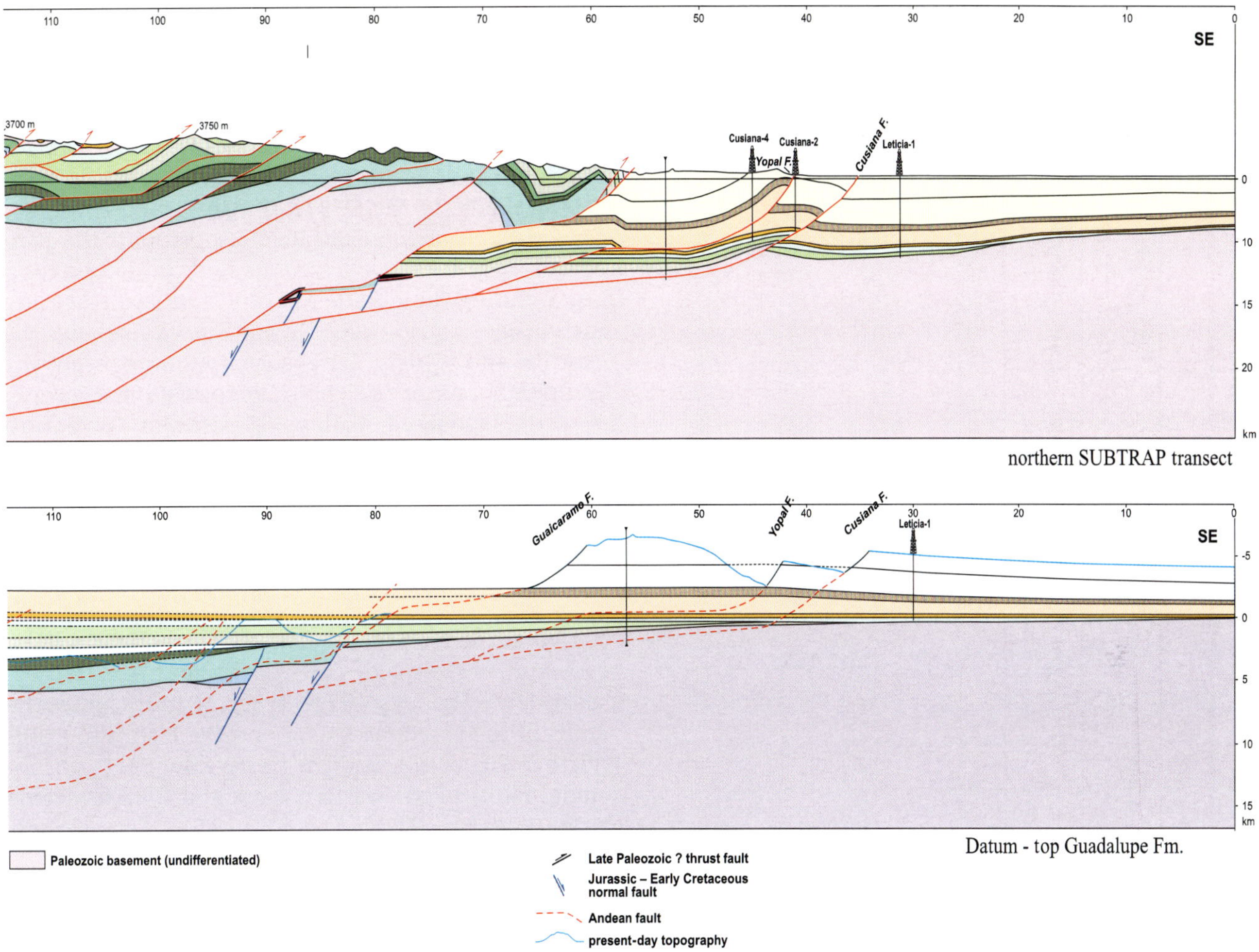

Plate 2. (cont.).

Additional compressional structures are exposed west of the Espinal high. These folds have shorter wavelengths than most of the folds in the Cordillera, indicating that they are caused by deformation on shallower detachment levels. Along our transect, the Chusma fault forms the range front of the Central Cordillera and the western boundary of the Magdalena Basin. This east-vergent thrust brings to the surface Paleozoic schists of the Central Cordillera overlain by Triassic to Jurassic rocks of the Luisa and Payande formations and intruded by the Ibague batholith (Late Jurassic–Early Cretaceous).

Sequence of Deformation

The timing constraints along the southern transect are consistent with those further north. As discussed above, there is ample evidence for pre-Eocene deformation in the Central Cordillera and Magdalena Basin. Notably, the thickness of the Guaduas Formation (Maastrichtian–Paleocene) ranges between 1.8 and 0 km (1.1 and 0 mi) below the pre-Eocene unconformity in the wells of the Girardot subbasin. Apatite fission-track ages from the Guaduas syncline (Gómez, 2002) and the timing of mineralization in the emerald deposits of the western flank of the Eastern Cordillera (Cheilletz et al., 1994) indicate that early Tertiary shortening also affected the Eastern Cordillera along the southern transect.

An additional deformational event is evidenced by the development of the Espinal high (Figure 12), where folded Jurassic through Eocene rocks are unconformably overlain by coarse clastics of the Oligocene(?) Doima formation. These rocks are in turn covered by claystones and shales of the La Cira Formation that signal a period of subsidence and relative tectonic quiescence. The Doima–La Cira interval is roughly equivalent to the Carbonera Formation of the Llanos Basin.

Five good-quality apatite fission-track ages from Tertiary and Cretaceous rocks in the flanks of the Guaduas syncline, located north of our transect on the hanging wall of the Cambao fault, cluster between 12 and 5 Ma (Gómez, 2002). These ages suggest that the main cooling and denudation episode on the Magdalena flank of the Cordillera was late Miocene to Pliocene. This is consistent with apatite fission-track data

Table 2. Fault displacement model of the internal segment of the northern section. Faults are numbered from left to right.

Stage	Time (Ma)	Fault 1 Tunja km	Fault 1 Tunja km/m.y.	Fault 2 Cornichoque km	Fault 2 Cornichoque km/m.y.	Fault 3 Soápaga km	Fault 3 Soápaga km/m.y.	Fault 4 km	Fault 4 km/m.y.	Fault 5 km	Fault 5 km/m.y.	Fault 6 km	Fault 6 km/m.y.	Fault 7 km	Fault 7 km/m.y.	Fault 8 Guaicaramo km	Fault 8 Guaicaramo km/m.y.	Sedimentation
1	34																	after Mirador
2	34–20	2	0.14			2	0.14											lower Carbonera
3	20–16	2	0.5	2	0.5	2	0.5											upper Carbonera and Leon
4	16–12			2	0.5	2	0.5	1.5	0.37	2	0.5	1.6	0.4	1.6	0.4	6	1.5	early Guayabo
5	12–8											1.6	0.4	1.6	0.4	6	1.5	middle Guayabo
6	8–4											1.6	0.4	1.6	0.4	6	1.5	late Guayabo
7	4–0															6	1.5	uplift and erosion
	Total	4		4		6		1.5		2		4.8		4.8		24		total = 51 km

from the Garzon massif, which is located in the Eastern Cordillera about 100 km (630 mi) to the south of our transect. There, two episodes of uplift are recorded: one at about 12 Ma (middle Miocene) and a second one between 6 and 3 Ma (late Miocene–Pliocene; Van der Wiel and Andriessen, 1991). Proprietary fission-track data available to the SUBTRAP consortium along the eastern flank of the Cordillera also support the conclusion that the main phase of Andean deformation began during the middle Miocene and peaked during the Pliocene to Holocene. In the Llanos foreland, the Castilla and Chichimene anticlines fold even the youngest strata of the Guayabo Formation. Likewise, in the Magdalena Valley, the Honda Formation (Miocene–Pliocene) is steeply folded in the footwall of the Magdalena fault. Therefore, it is clear that the deformation has been active until very recently.

Kinematic Model

We carried out the Thrustpack 4.0 modeling of the east- and west-vergent parts of the southern transect separately. The east-vergent segment includes the faults of the Quetame massif and the Llanos foothills (Plate 5; Figure 13), and the west-vergent segment includes faults of the western flank and the Magdalena Basin (Plate 6).

Eastern Flank (Chichimene Section)

The Castilla and Chichimene faults are interpreted as inversions of small normal faults in the foreland. The dominant structures of the model are the Villavicencio and Santa Maria faults. The fault displacements used in this model are shown in Table 5.

34 Ma: Initial state. The model begins after the deposition of the San Fernando formation (Mirador equivalent). The Cretaceous units (mainly, the Une and Villeta formations) thicken toward the center of the Bogotá basin. We show an Early Cretaceous half graben to account for the thickness of the Caqueza Group present along the eastern flank but absent farther east. Flexure during this stage is presumably caused by tectonic loading in the Central Cordillera and Magdalena Valley.
16 Ma: Lower Carbonera deposition. Thermal-maturity data of the Estratigrafico 1 well in the western flank of the Cordillera require burial by thick Tertiary deposits. We propose that Carbonera-equivalent rocks covered the Cordillera domain.
12 Ma: Leon deposition. At this stage, deformation begins with motion on the Santa Maria and Villavicencio faults.
8 Ma: Early Guayabo. Displacement continues on the Santa Maria and Villavicencio faults. Deposition of the syntectonic foreland sequence begins.

Plate 3. Thrustpack 4.0 kinematic model of a section of the Eastern Cordillera in area between Tunja and the Llanos foothills. See Plate 1 for location, Table 2 for the fault displacement model and the text for discussion of each stage of the model. Both vertical and horizontal scales in kilometers.

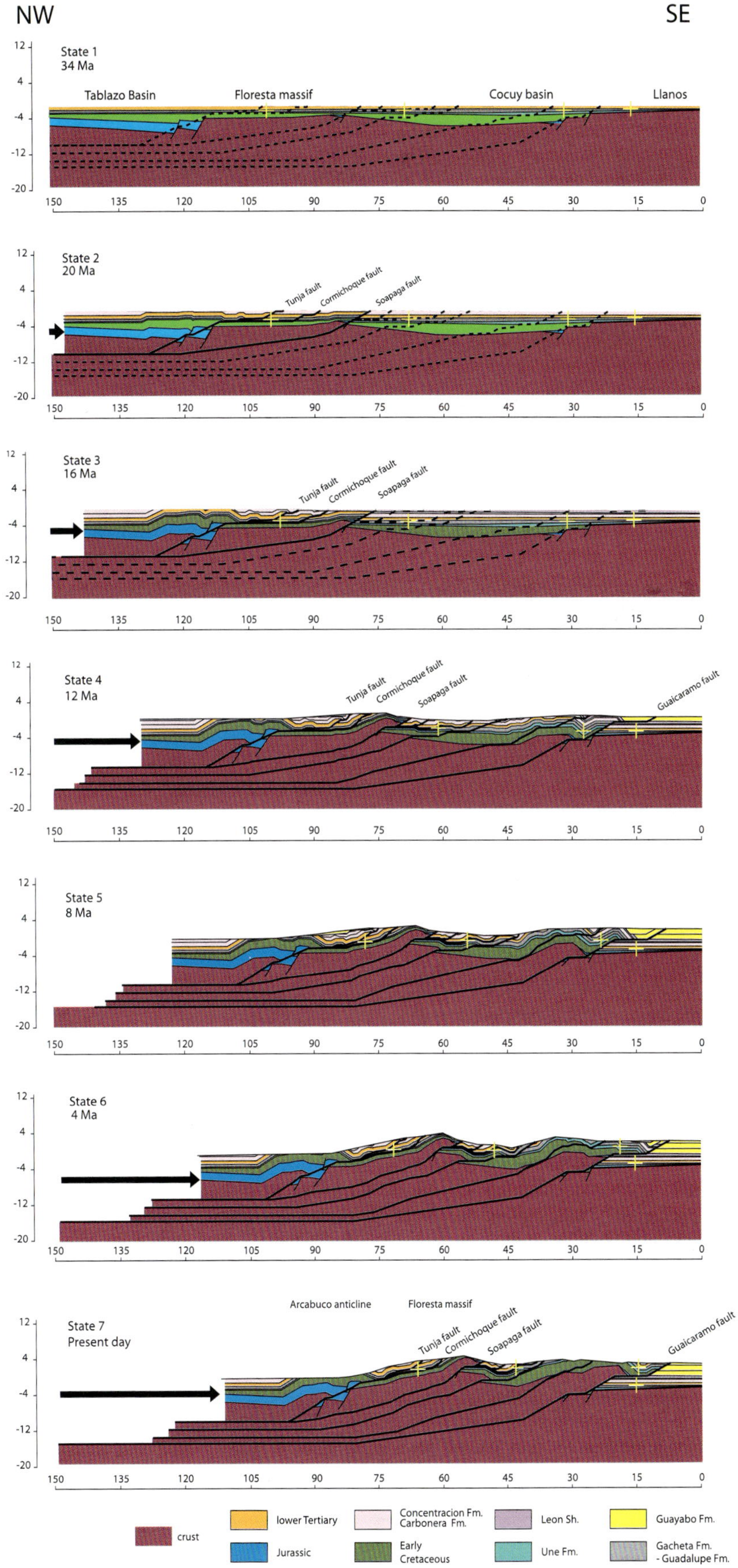

4 Ma: Late Guayabo. Further displacement occurs on the Santa Maria and Villavicencio faults and important unroofing of the highlands as indicated by apatite fission-track data.
0 Ma: Holocene. Deformation continues on the Santa Maria and Villavicencio faults, and the inversion of the normal faults in the foreland takes place, forming the Castilla and Chichimene anticlines. The Paleozoic basement is unroofed in the Quetame massif.

Western Flank (Magdalena Section)

In this kinematic model (Figure 13; Plate 6), we neglect the structures west of the Espinal high, such as the Gualanday 1 anticline and parts of the Girardot fold belt. Fault displacements for this model are shown in Table 5.

34 Ma: The kinematic model begins after deposition of the Gualanday Group. We do not model the pre-Eocene deformation, because it is unclear which specific structures in the Cordillera were active in that episode. In the initial stage, we present a hypothetical set of Jurassic grabens. In effect, the distribution of the Saldaña Formation and the structures associated with it are not well known.
31 Ma: Inversion of Espinal high and Doima deposition. To simulate the inversion of a Jurassic graben in Thrustpack 4.0, we employed a ramp-flat geometry linked to the regional detachment (fault 5 in Table 5). This geometry greatly exaggerates the horizontal shortening at this stage of the model and produces some uplift in the Cordillera. Both effects could be regarded as artifacts. To preserve the Gualanday Group rocks on both flanks of the Espinal high, the Doima formation

Table 3. Kinematic model of the frontal section. Faults are labeled from left to right. See Plate 3.

Stage	*Time (Ma)*	*Fault Displacements*								*Sedimentation*
		Fault 1 Chameza Fault		*Fault 2 Guaicaramo Fault*		*Fault 3 Yopal Fault*		*Fault 4 Cusiana Fault*		
		km	*km/m.y.*	*km*	*km/m.y.*	*km*	*km/m.y.*	*km*	*km/m.y.*	
1	16–12	3	0.75	6	1.5					early Guayabo
2	12–8	3	0.75	6	1.5					middle Guayabo
3	8–4			6	1.5					late Guayabo
4	4–0			6	1.5	3	0.75	2	0.5	uplift and erosion
	Total	6		24		3		2		total = 35 km

must be deposited in the incised paleo-Magdalena Valley.

20 Ma: La Cira–Carbonera deposition. To account for the thermal maturity of the Villeta Formation in the Estratigrafico 1 well of the western flank of the Cordillera, we propose that Carbonera-equivalent rocks covered the Cordillera domain. Because the thickness of Oligocene rocks required by the thermal-maturity data is greater in the Cordillera than in the Magdalena Valley, it appears that the depocenter of the Oligocene basin was located away from the Central Cordillera. This can be explained either by the presence of local, tectonic loads already in place in the Eastern Cordillera or by the effect of zones of reduced flexural rigidity inherited from the Early Cretaceous extensional event.

16 Ma: Upper Carbonera–Leon. Deposition continues in the Eastern Cordillera–Llanos domain.

12 Ma: Lower Honda. West-vergent thrusting begins in the interior of the Cordillera with motion on the Pandi and Boqueron faults. The Honda Formation begins to fill the Magdalena Basin.

8 Ma: Middle Honda. Displacement continues on the Pandi and Boqueron faults and starts on the Magdalena fault, accompanied by additional deposition of the Honda Formation.

4 Ma: Upper Honda. The same pattern continues, accompanied by erosion of the highlands.

0 Ma: Holocene. All the faults in the Cordillera remain active, and a triangle zone develops under the Magdalena range front, uplifting the Honda beds (Figure 12). Regional uplift and erosion are present.

Summary of Structural Framework

The main conclusions regarding the structural and kinematic framework of the Eastern Cordillera can be summarized as follows. The Andean structures are strongly controlled by the Jurassic–Early Cretaceous extensional features of the Bogotá basin and the paleo-Magdalena Basin. This control is manifested both by the inversion, commonly through shortcuts, of the Jurassic–Cretaceous normal faults and by the determinant role of the thickness and rheology of the Cretaceous stratigraphy on the structural style. The present-day geometry can be satisfactorily modeled using moderate amounts of shortening by considering the large variations of stratigraphic thickness. Although the width of the fold and thrust belt diminishes considerably from north to south (Figure 2), the magnitude of strain is fairly consistent. Along our northern transect, the shortening of the section is 30%, whereas along the southern transect, it is 25%. The difference between these two values is probably less than the uncertainty inherent in the cross sections. The structure of the eastern flank of the Cordillera along the northern transect is dominated by three major fault blocks that involve basement. In contrast, along the southern transect, the structure is dominated by one major east-vergent fault system along which most of the shortening is concentrated. The west-vergent faults can be regarded as back thrusts that accommodate only a lesser part of the overall shortening. Evidence exists for three compressional tectonic events in the Eastern Cordillera: (1) pre-Eocene, (2) pre-Oligocene, and (3) middle Miocene to Holocene.

Thermal Modeling

Thermal History Constraints and One-dimensional Models

We had available for this study an extensive database of confidential thermal-maturity data from surface and wells of the Eastern Cordillera and its foothills. Although we cannot present the raw data in detail, in this section, we describe the main features of the thermal maturation pattern. Along the northern transect, the Early Cretaceous rocks that are exposed east of the Floresta massif and west of the Guaicaramo fault system attained a very high level of maturity (supermature range, vitrinite reflectance > 3.5) that is

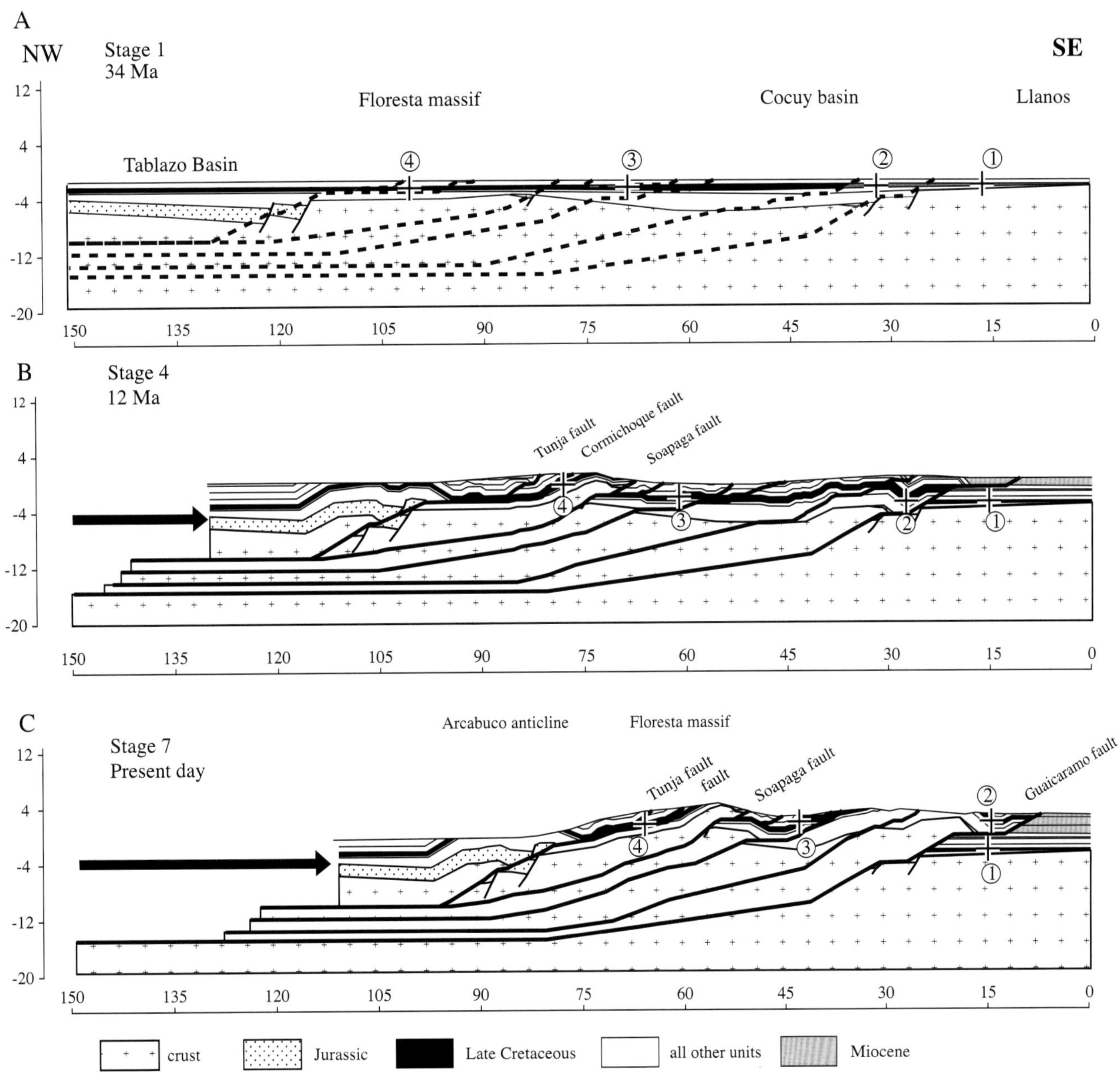

Figure 9. Three key stages of the kinematic model of the internal part of the northern section. See Plate 3 for the complete model. Both vertical and horizontal scales in kilometers. (A) Initial state of the model (late Eocene) showing the Jurassic–Early Cretaceous Cocuy and Tablazo rift basins buried in the foreland basin of the Central Cordillera. (B) Middle Miocene state after thrusting in the Floresta massif had ceased, the internal part of the Cordillera was undergoing erosion, thrusting had propagated to the frontal structures, and sedimentation had commenced in the foreland basin of the Eastern Cordillera. (C) The final state where most of the deformation is concentrated along the frontal thrusts, and erosion dominates the core of the Cordillera. The hinterland of the model does not show sufficient uplift because of the missing west-vergent part of the Cordillera that lies to the southwest and the lack of isostatic uplift. This model neglects the Yopal and Cusiana faults of the Llanos foothills, which were modeled independently. The thermal histories of the four numbered points are shown in Figure 16 and discussed in the text.

also manifested in the degree of transformation of clay minerals into phyllosilicates (Hébrard et al., 1987). In contrast, Cenomanian rocks on both sides of the Floresta massif are in the mature range, with vitrinite reflectance of about 1.0 (i.e., in the Tunja well, Figure 14). Even west of the Floresta massif, where the Early Cretaceous section is thin, there is rapid increase of the thermal-maturity gradient in the Early Cretaceous units relative to the overlying middle Cretaceous rocks. We attribute this maturation pattern to the thermal effects of active Early Cretaceous rifting in the Bogotá basin. Simple models of rifting show that thinning of the

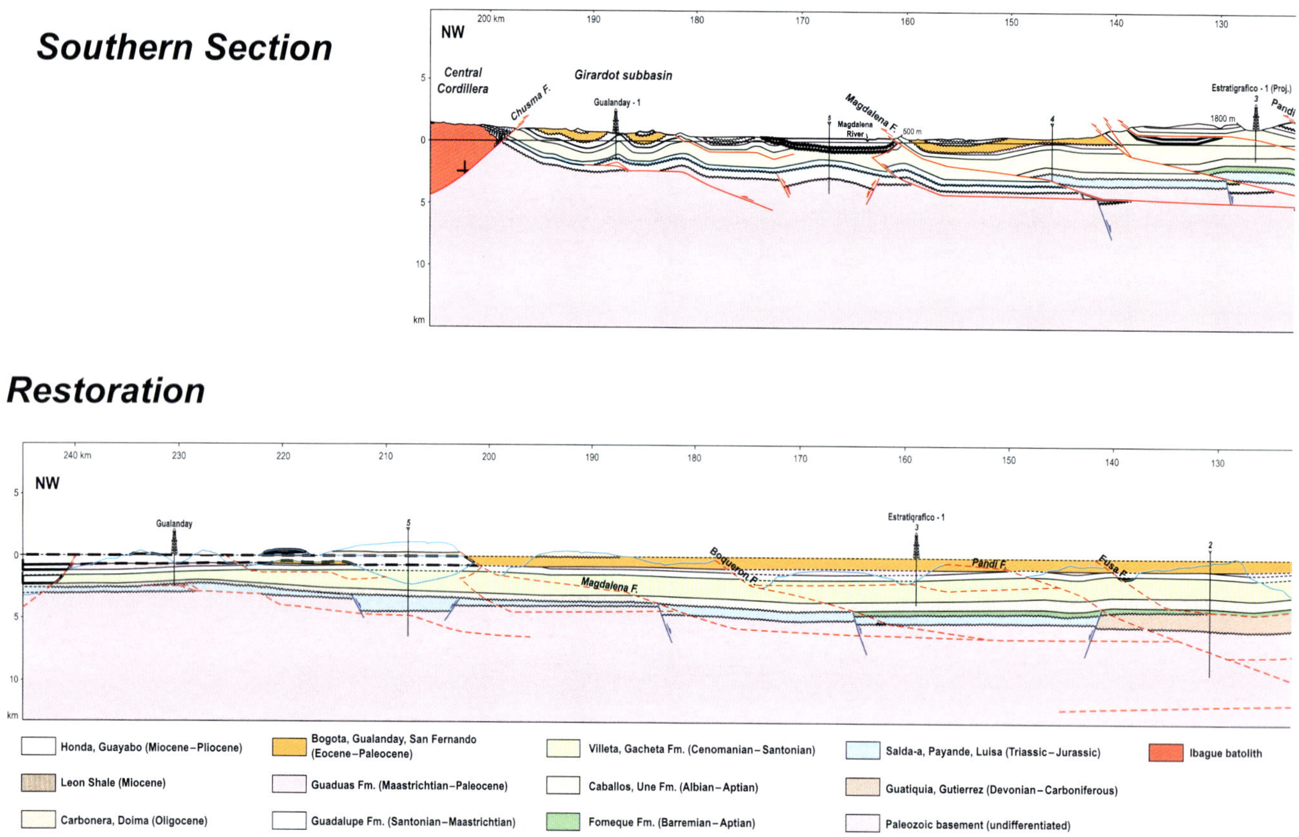

PLATE 4. Structural cross section of the Girardot to Castilla area. See Plate 1 for location.

lithosphere leads to an increase in surface heat flow as a function of the stretching factor (β), followed by exponential decay of heat flow over the next 100 Ma or so as the lithosphere reequilibrates thermally (McKenzie, 1978). We first estimated the value of lithospheric stretching (β) by applying a simple McKenzie (1978) model to the backstripped subsidence pattern for each well or section studied. The values of stretching during Early Cretaceous rifting in the Tunja area ranged from 1.5 to 2.0. We then estimated the values of basal heat flow in the internal part of the Cordillera using one-dimensional thermal models carried out using the Genex-Gentect® software (Beicip-Franlab, 1995) and constrained by the well and surface thermal-maturity data. The best fitting models simulated a rifting event using a modified McKenzie (1978) model that incorporates linear changes of heat flow during the rifting event, followed by thermal decay during the thermal subsidence phase (Figure 14A). The best fitting models using constant basal heat flow (no rifting) require between 25 and 45 mW/m^2 of basal heat flow, depending on the specific location.

Fluid inclusions in microcracks that cut both quartz grains and cement in sandstone of the Albian Une Formation in the vicinity of our northern transect indicate that the trapped fluid equilibrated at temperatures greater than 235°C (Bordas-Le Floch, 1999). Our one-dimensional thermal models demonstrate that the Une Formation could not achieve this high temperature by sedimentary and tectonic burial alone (Figure 14B); therefore, we hypothesize that hydrothermal circulation is responsible for these high temperatures. The effects of hydrothermal circulation have only been documented previously in the vicinity of the emerald deposits of the flanks of the Eastern Cordillera (e.g., Cheilletz et al., 1994). However, we believe that hydrothermal circulation was regional in nature and affected the entire basin. Paleothermometric data from the emerald deposits reveal temperatures as high as 300°C in the Early Cretaceous host shales (Cheilletz et al., 1994). Using Genex, we simulated a late Eocene hydrothermal event by introducing a 300°C temperature spike in the Neocomian units during the period 38–32 Ma according to the timing of emerald mineralization in the western flank of the Cordillera (Cheilletz et al., 1994). This is sufficient to elevate the temperature of the overlying Une Formation to more than 250°C purely by conduction (Figure 14D). Once the hydrothermal event ceases, temperature returns close to the normal burial temperature in a few million years. This simplistic model shows that it is possible to attain the temperatures recorded by the fluid inclusions by a short period of hydrothermal circulation in the underlying rocks.

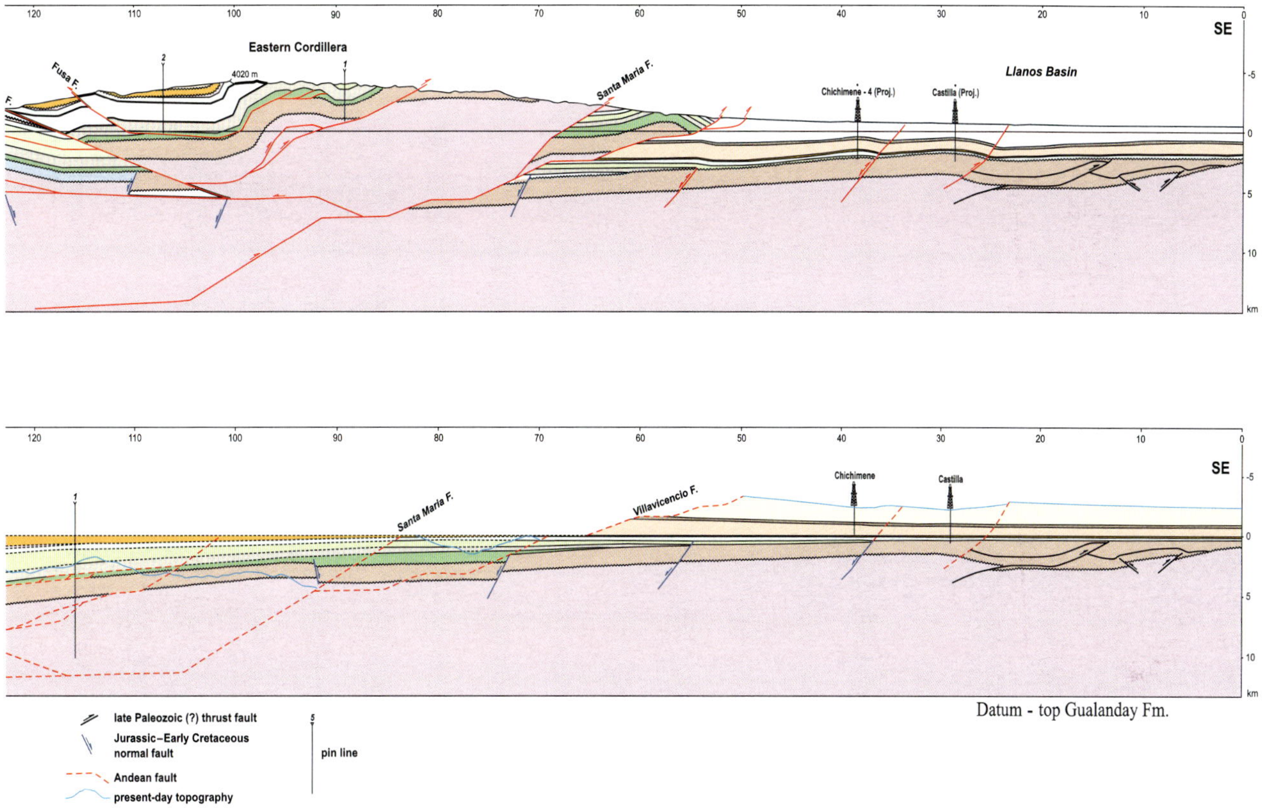

Plate 4. (cont.).

The vitrinite reflectance profile of wells located on the Llanos foothills (Medina 1, Chichimene 1, and Coporo 1) can be satisfied with basal heat flow in the range of 10–22.5 mW/m^2 without incorporating a rifting event. Likewise, wells in the Magdalena Valley in the vicinity of our southern section (Chenche 1, Matachines Norte 1, and Matachines Sur 1) were successfully modeled using basal heat flow of less than 20 mW/m^2.

In summary, it is clear from our one-dimensional thermal models that the thermal regime in the central part of the Bogotá basin, which corresponds to the interior of the Eastern Cordillera, was hotter than in the flanks. We will explore this further with two-dimensional thermal models discussed below. One manifestation of high heat flow in the internal part of the Cordillera is late Miocene to Pliocene high-potassium rhyolitic plugs that are found in the vicinity of the Laguna de Tota (Figure 2). The age and composition of these igneous bodies suggest that they are caused by melting of the lower crust of the Cordillera, which was thickened during Andean shortening.

While carrying out the thermal models of wells in various settings in the Eastern Cordillera, it became clear that the present-day temperature profile of many of those wells could not be explained by simple, conductive heating. In most cases, the geothermal gradient implied by the thermal modeling constrained by the maturation indicators is higher than the present-day gradient observed in the wells. In some extreme cases (Corrales 1), temperature actually decreases downhole. In other wells where a detailed temperature profile is available (Coporo 1), there are inflections in the slope of the thermal gradient that indicate that it is in disequilibrium. It is clear that circulation of meteoric water that was driven by the great topographic gradients across the mountain range is an important factor in determining the present-day subsurface temperature and may have affected the maturation of hydrocarbons in the past. Regional-scale hydrodynamics have also been documented in the Llanos Basin (Bachu et al., 1995). Despite this limitation, we can derive useful insights from the conductive models discussed below.

Two-dimensional Thermal Models

Thrustpack 4.0 can be used to calculate the thermal state of every stage in the kinematic model. The inputs are the surface temperature profile, the basal heat-flow profile, and the thermal properties of the materials that make up the section (conductivity, heat capacity, porosity, density, and radioactive content). Thrustpack 4.0 grids the section and solves the conductive heat equation in two dimensions for each point of the grid. Because we do not have direct measurements of the thermal properties of the rocks of the Eastern Cordillera, we used values for the average lithology of each stratigraphic

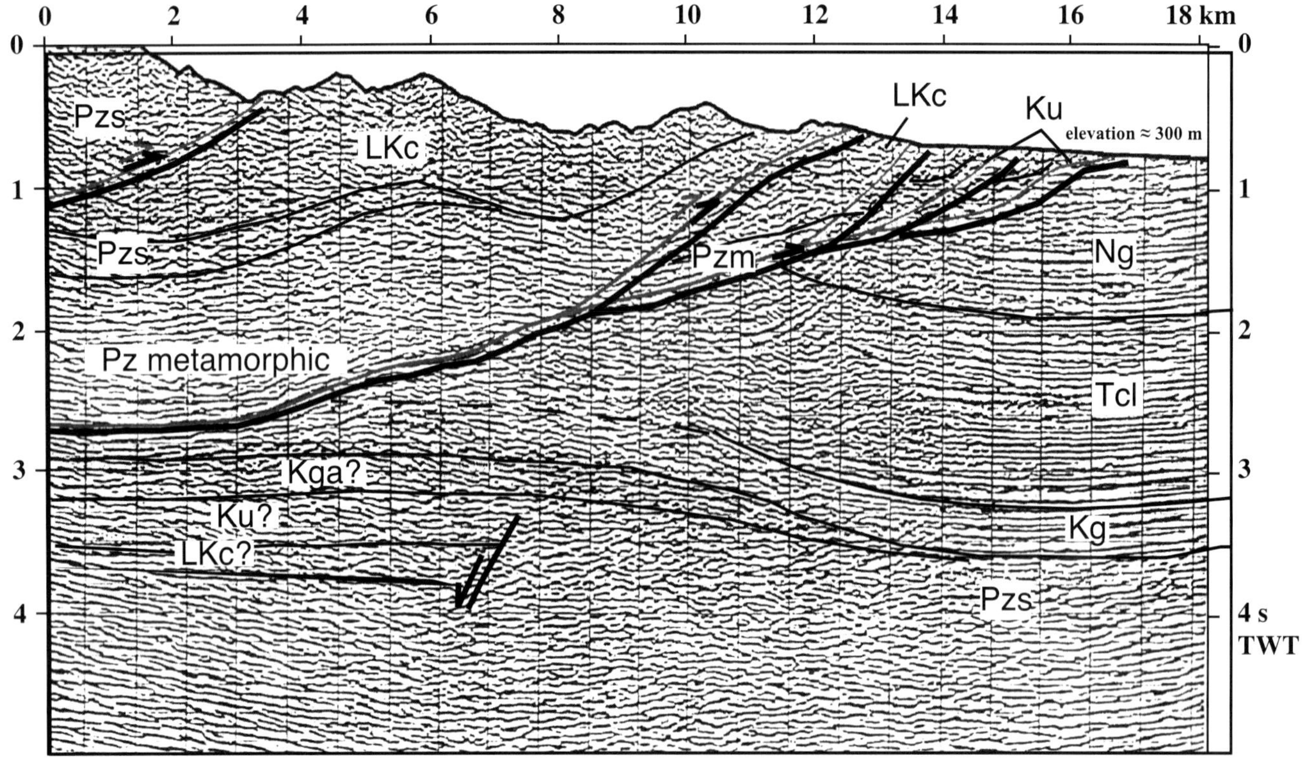

FIGURE 10. Seismic line through the Llanos foothills located 20 km (13 mi) north of the southern section. The Villavicencio–Santa Maria fault is a moderate-angle thrust with more than 20 km (13 mi) of horizontal displacement. It places Paleozoic metamorphic (Pzm) and sedimentary rocks (Pzs) and a thick section of Lower Cretaceous shales of the Caqueza formation (LKc) on Pliocene and Tertiary strata of the Guayabo (Ng) and Leon–Carbonera (Tcl) formations. Albian sandstone of the Une Formation (Ku) is present in the hanging wall and perhaps in half grabens in the footwall of the frontal thrust. Kga = Upper Cretaceous Guaduas Formation; Kg = Upper Cretaceous Guadalupe Formation.

TABLE 4. Fault displacement model of the east-vergent segment of the southern section (Chichimene section). Faults are numbered from west to east.

State	*Time (Ma)*	*Fault Displacements*										*Sedimentation*
		Fault 1 Sta. Maria		*Fault 2 Villavicencio*		*Fault 3*		*Fault 4 Chichimene*		*Fault 5 Castilla*		
		km	*km/m.y.*	*km*	*km/m.y.*	*km*	*km/m.y.*	*km*	*km/m.y.*	*km*	*km/m.y.*	
1	34											after Mirador Formation
2	34–16											Carbonera Formation
3	16–12	2	0.5	5	1.25							upper Carbonera and Leon formations
4	12–8	2	0.5	5	1.25							lower Guayabo Formation
5	8–4	1.6	0.4	5	1.25							middle Guayabo Formation
6	4–0	3.2	0.8	5	1.25	1	0.25	1	0.25	1	0.25	upper Guayabo and erosion
	Total	8.8		20		1		1		1		total = 31.8 km

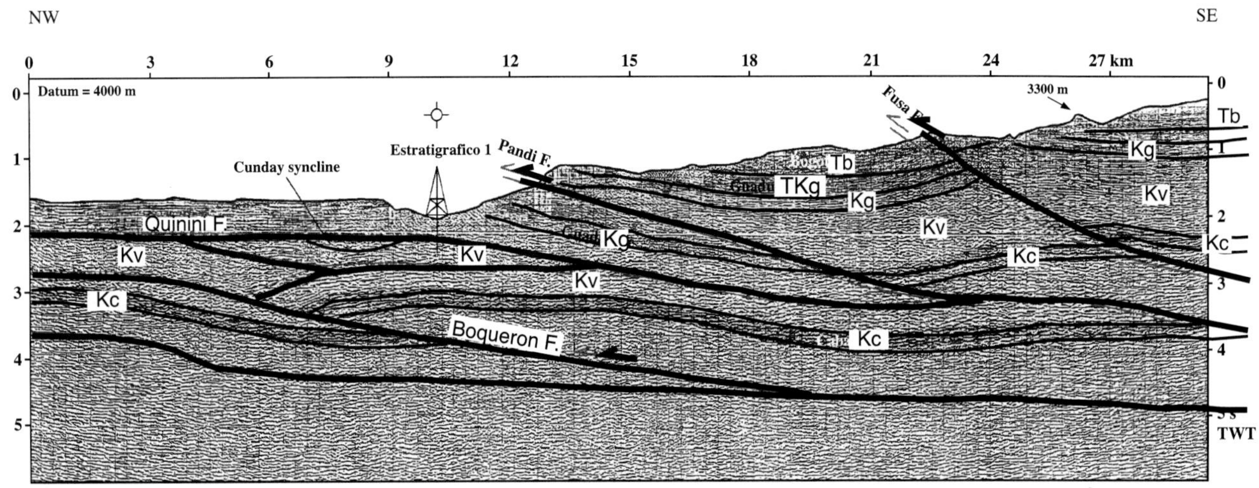

FIGURE 11. Seismic line crossing the western flank of the Eastern Cordillera along the southern section. See Plate 4 for approximate location. The motion of the Quinini fault is predominantly out of the plane of this section. Rocks below the Caballos Formation (Kc) are probably Jurassic to late Paleozoic sedimentary rocks. Abbreviations: Tb = Paleocene Bogotá Formation; TKg = Maastrichtian–Paleocene Guaduas Formation; Kg = Campanian–Maastrichtian Guadalupe Formation; Kv = Albian–Cenomanian Villate Formation; Kc = Albian–Aptian Caballos Formation.

unit that are consistent with those used in the one-dimensional thermal modeling package Genex-Gentect (Beicip-Franlab, 1995). The variation of porosity with depth is calculated using empirical curves appropriate for each lithology. Surface temperature was approximated as a linear function of elevation (20°C near sea level, 10°C at 2500 m [8300 ft]). The basal heat flow was adjusted to obtain a match to one-dimensional thermal models of wells along the transect carried out with Genex-Gentect, although the values of basal heat flow that used Thrustpack 4.0 are higher, because to facilitate the modeling, we did not incorporate radiogenic heating into the crust in Thrustpack 4.0. All the heat flow was imposed at the base of the crust. After the temperature grid had been calculated for each stage of the cross section, we also computed the maturity of the organic matter for these successive stages. We had no data on the kinetic parameters of the kerogen found in

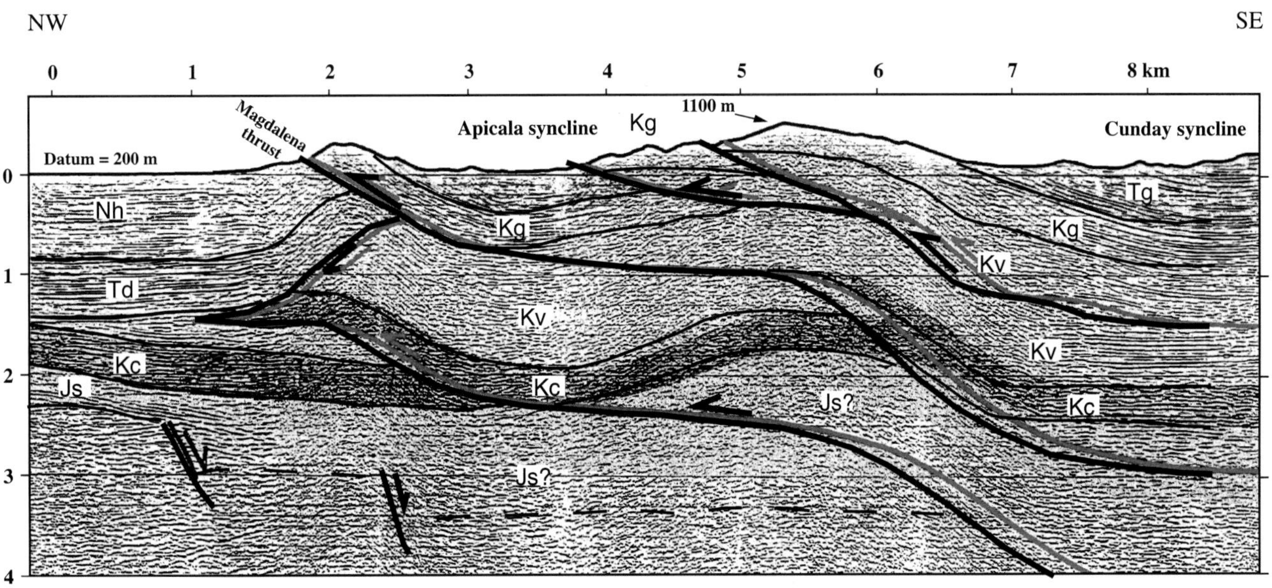

FIGURE 12. Interpreted seismic line across the Magdalena range front located 15 km (9 mi) south of the southern section (Plate 4). The Magdalena fault is an out-of-sequence thrust that cuts the triangle zone that forms the actual structural front. Abbreviations: Nh = Neogene Honda Formation; Td = Oligocene Doima formation; Tg = Eocene Gualanday Formation; Kg = Late Cretaceous Guadalupe Formation; Kv = middle Cretaceous Villeta Formation; Kc = Albian–Aptian Caballos Formation; Js = Jurassic Saldaña Formation.

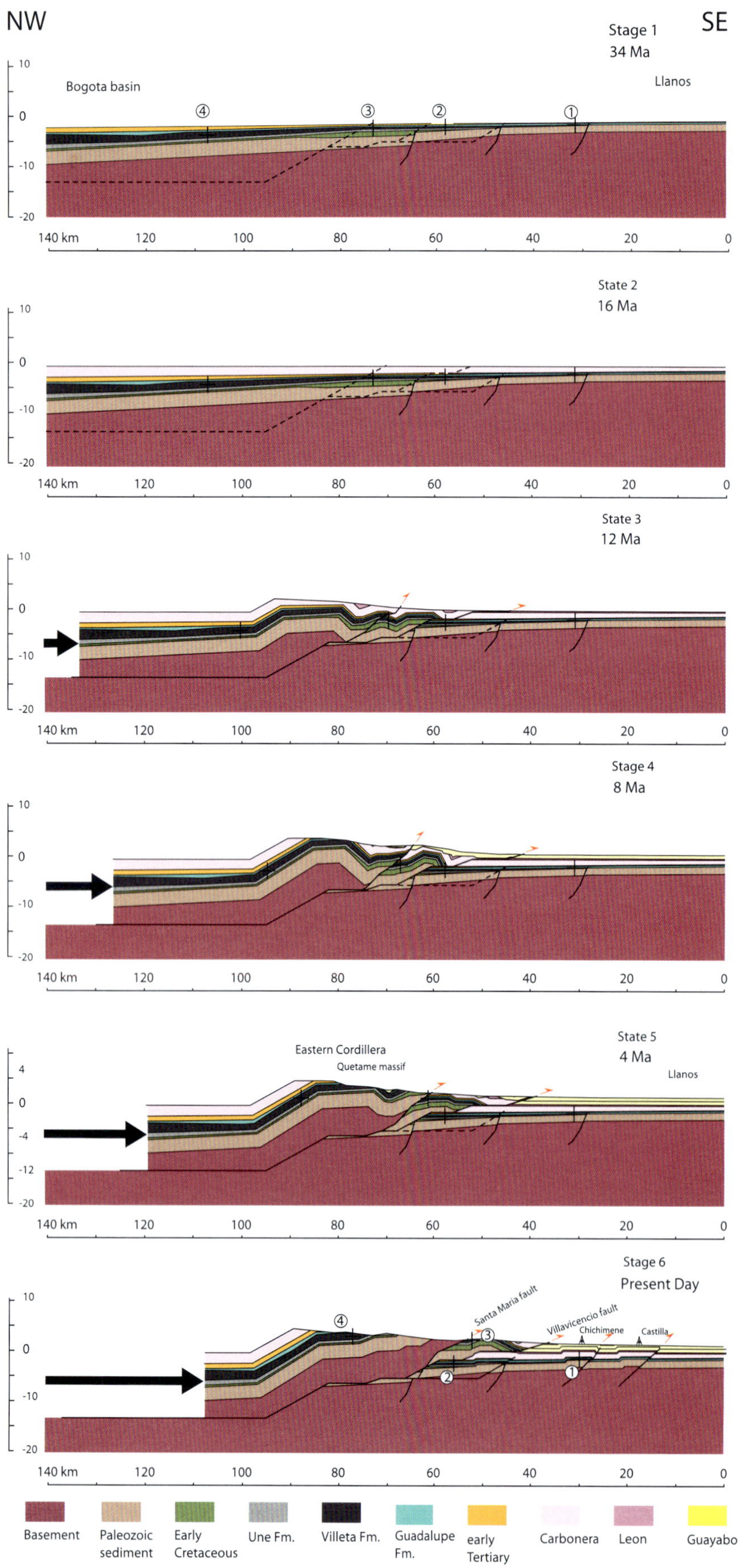

PLATE 5. Thrustpack 4.0 kinematic model of the east–west part of the Eastern Cordillera along our southern section. See Table 4 for fault displacement and the text for the details about each stage of the model. Vertical scale in kilometers.

the Eastern Cordillera source rocks. In our models, we used default parameters typical of types II and III organic matter and assigned these to the Cretaceous and Tertiary units, respectively.

The two-dimensional models are most useful for understanding the influence of sedimentation, deformation, and erosion on the thermal profile at the regional scale. We have not attempted to account for the effect of hydrothermal circulation or circulation of meteoric water.

Thermal Model, Northern Section, Internal Segment

This set of thermal models was carried out using a surface temperature of 10°C independent of elevation and heat flow at the base of the model that varied from 35 mW/m^2 under the foreland to 45 mW/m^2 under the Cordillera (Plate 7; Figure 15).

34 Ma: Initial state. The rocks in the deep part of the Cocuy basin are already overmature at the beginning of the model. The geothermal gradient is higher in the Cocuy basin than in the foreland. This results in the upward deflection of the isotherms and is caused by the combined effects of the higher basal heat flow and the blanketing effect of low conductivity of the shaly rocks that fill the Cocuy basin.

20 Ma: Early Carbonera (Concentración) deposition. Deposition of the Carbonera (Concentración) leads to greater burial and higher temperatures for the source rocks of the Cocuy basin. The Albian Une Formation enters the oil window in the internal part.

16 Ma: Late Carbonera–Leon (Concentración) deposition. Along most of the section, the Chipaque Formation

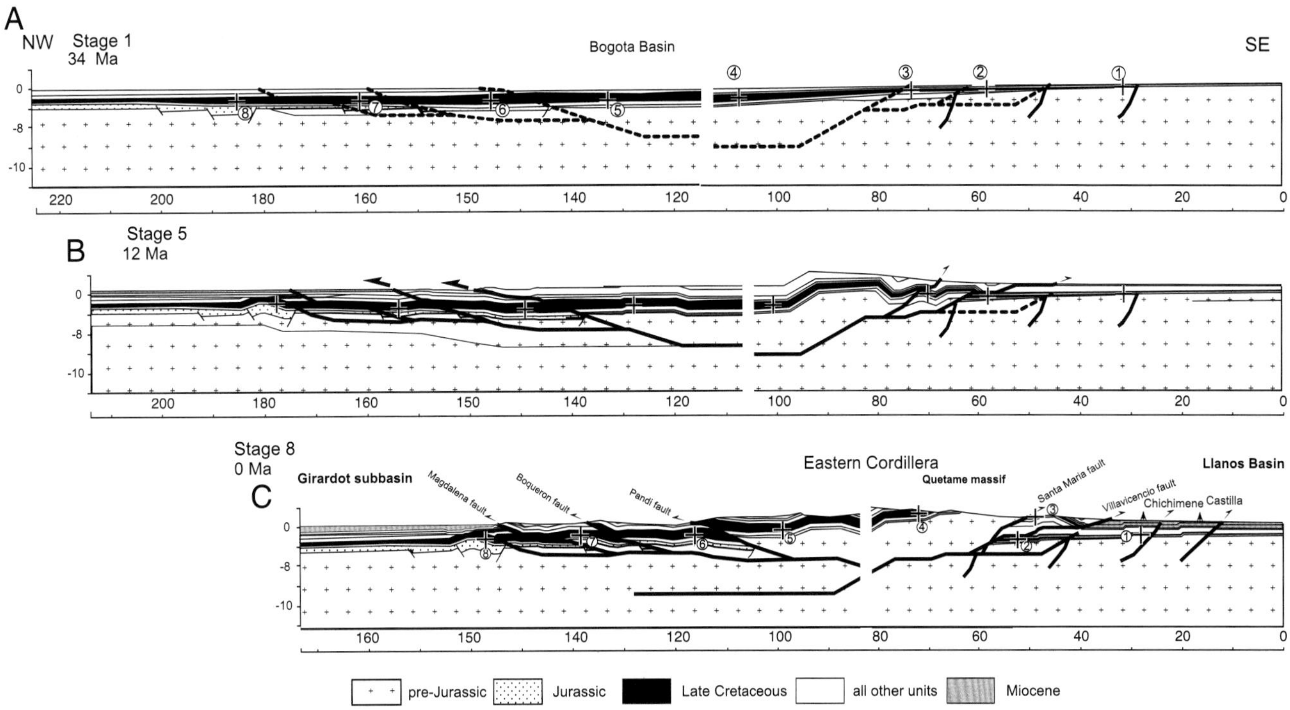

FIGURE 13. Three key stages of the kinematic model of the southern section. The east- and west-vergent parts of the Cordillera were modeled independently, so the two fault systems are not fully coupled in this composite figure. See Plates 5 and 6 for the complete models. Most of the basin fill in the initial state of the model (late Eocene) (A) represents the thermal subsidence phase of the Bogotá basin, although there are Jurassic grabens in the Magdalena Valley, and small Early Cretaceous half grabens along the basin boundary on the Llanos side. By the middle Miocene (B), inversion of some Jurassic grabens had occurred in the Magdalena Basin and the uplift of the Cordillera. Through the late Miocene to Holocene (C), thrusting in the Cordillera continued and was dominated by displacement on the eastern range front to produce the structural fan observed today. The thermal histories of the points marked by crosses are shown in Figure 19. Both vertical and horizontal scales in kilometers.

remains in the oil window. The onset of erosion over the Arcabuco and Floresta anticlines stops maturation. The base of the Lower Cretaceous reaches temperatures in excess of 250°C in the deep part of the Cocuy basin.

12 Ma: Early Guayabo deposition. The beginning of uplift on the Guaicaramo fault stops maturation of the frontal anticline. Tectonic burial and rapid syntectonic deposition cause maturation to increase in the footwall of the Guaicaramo fault.

8 Ma: Middle Guayabo deposition. During the final stages, erosion dominates the model except in the frontal part, where tectonic and sedimentary burial continue.

4 Ma: Late Guayabo deposition. The same patterns continue.

0 Ma: Holocene. In the present-day configuration, most of the Cretaceous source rocks of the Cordillera are overmature; however, areas that underwent uplift during the early Miocene, for example, the hanging wall of the Soápaga fault, retain some generating potential. The Tertiary rocks (Guaduas Formation and younger) are immature for the most part because of the slower kinetics of type III kerogen.

Discussion of the Thermal History of the Northern Section

Figure 16 shows the temperature and maturation history of four points along the Tunja section. The points are located in the Upper Cretaceous Gachetá Formation, which is the main source rock in the area. Point 1, located in the footwall of the Guaicaramo fault, follows a path of increasing temperature because of progressive sedimentary and tectonic burial. In contrast, the points that are within the Cordillera proper (points 2–4) are heated during deposition of the Carbonera–Concentración section, but with the onset of deformation, temperature ceases to increase, and as uplift and erosion dominate the system, the points begin to cool. Following our hypothesis that thrusting began in the vicinity of the Floresta massif and propagated to the east, the most internal points (points 3 and 4) begin to cool earlier.

Maturation is also linked to burial by the Carbonera–Concentración Formation, but the difference in geothermal gradient in the interior of the Cordillera relative to the foreland causes the more internal points

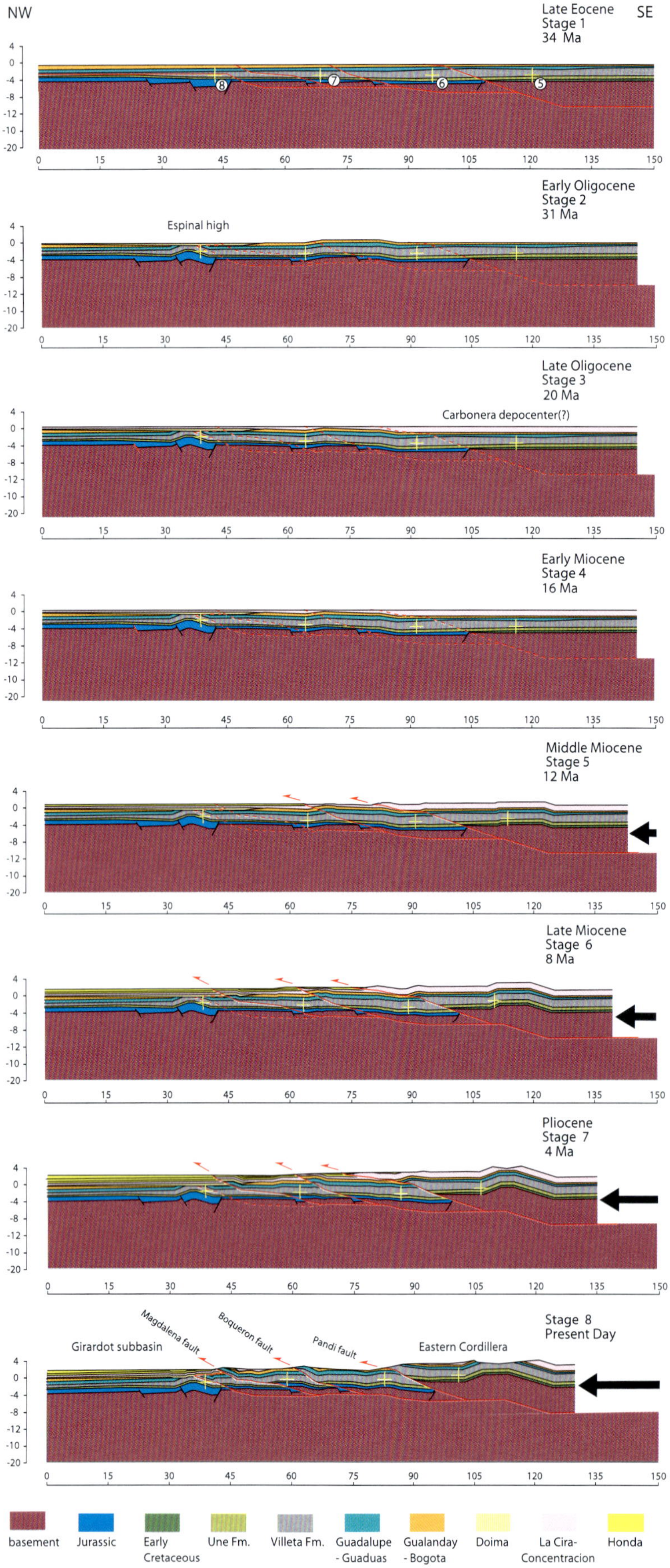

Plate 6. Thrustpack 4.0 kinematic model of the west–vergent portion of the Eastern Cordillera along our southern section. See Table 5 for the fault displacement model and the text for details about each stage of the model. Both vertical and horizontal scales in kilometers.

(points 3 and 4) to enter the oil window first (Figure 17). The exact timing of the maturation depends largely on the hypothesis on the thickness, age, and rate of sedimentation of the Tertiary units that is introduced into the model. The timing of the onset of deformation determines when maturation will stop in the internal portion and whether some hydrocarbon potential still remains. Oil seeps and shows in wells in the areas surrounding the Floresta massif suggest that pockets of active source rock are locally preserved. Because the Tertiary stratigraphy has been largely eroded from the Cordillera, and the paleontological dating of the preserved section is poor, the uncertainty of the model is great. However, the thermal histories shown in Figure 16 characterize each structural region of the Cordillera, even if the detailed timing may not be exact.

The two-dimensional thermal models of the frontal segment of the northern transect (Figure 18) show that at the initial stage, before significant inversion of the Guaicaramo fault system, the deep part of the Cocuy basin was already overmature. Any hydrocarbon generated would have migrated toward the Llanos foreland unimpeded. However, once the Guaicaramo system became active, the foreland was isolated from the Cordillera domain. Prospects in the Llanos foothills are only viable if they are connected to active hydrocarbon-generating zones in the tails of the frontal thrust sheets as shown in Figure 18.

THERMAL AND MATURITY MODEL OF THE SOUTHERN TRANSECT

Eastern Flank

The thermal model of the eastern flank (Figure 17; Plate 8) was done using heat flow at the base of the model that varied from 40 mW/m^2 under the foreland to 55 mW/m^2

TABLE 5. Fault displacement model of the Magdalena section. Faults are numbered from east to west.

Stage	*Time (Ma)*	*Fault Displacements*										*Subsidence/ Uplift (slope)*	*Sedimentation*
		Fault 1 Pandi Fault		*Fault 2 Boqueron Fault*		*Fault 3 Magdalena Fault*		*Fault 4 Triangle Zone*		*Fault 5 Espinal High*			
		km	*km/ m.y.*	*km*	*km/ m.y.*	*km*	*km/ m.y.*	*km*	*km/ m.y.*	*km*	*km/ m.y.*		
1	34												Gualanday Formation
2	34–31									4	1.3		Doima formation
3	31–20											−0.005	Carbonera–La Cira–Barsaloza
4	20–16												
5	16–12	1.0	0.25	1	0.25								lower Honda Formation
6	12–8	1.6	0.4	1	0.25	1.0	0.25					0.002	middle Honda Formation
7	8–4	2.4	0.6	1	0.25	1.6	0.4					0.006	upper Honda Formation
8	4–0	0.8	0.2	1	0.25	2.4	0.6	1.2	0.3			0.005	uplift and erosion
Subtotal		5.8		4		5		1.2					total = 16 km

under the core of the Cordillera. These values were calibrated by fitting the maturity profile of the Estratigrafico 1 well in the Cordillera, the bottomhole temperatures of the Chichimene well in the Llanos, and limited vitrinite reflectance data in the Magdalena Valley. Type II kerogen was assigned to the Lower and Upper Cretaceous, a type III to the Tertiary units.

34 Ma: Initial state. At the initial state, the entire sedimentary section of the Llanos is immature. Because of the thicker Cretaceous stratigraphy and the higher thermal gradient, the Early Cretaceous units are mature in the Eastern Cordillera domain.
16 Ma: Lower Carbonera deposition. With the deposition of the Carbonera, the Upper Cretaceous rocks become mature to overmature in the Cordillera domain.
12 Ma: Leon deposition. With the onset of deformation and the coeval uplift and erosion, maturation in the hanging wall of the Santa Maria fault is arrested. Tectonic burial, coupled with burial by the syntectonic sediments, causes the maturation front to migrate toward the Llanos. The Chipaque–Villeta source rocks enter the oil window in the hanging wall of the Villavicencio fault and also in the footwall. At this stage, the mature source rocks of the Cordillera are no longer in communication with the Llanos.
8–4 Ma: Guayabo. The same patterns continue to develop.
0 Ma: Holocene. In the final stage, there is a wedge of mature source rock in the footwall of the Villavicencio fault capable of feeding the Chichimene and Castilla traps. The rocks in the Cordillera itself are mostly overmature and are isolated from the traps of the Llanos.

Western Flank

The thermal model of the Magdalena section was carried out using a surface temperature of 20°C. The heat flow at the base of the model varied from 45 mW/m^2 under the Magdalena Valley to 55 mW/m^2 under the core of the Cordillera in all the stages. Type II kerogen was assigned to the Lower and Upper Cretaceous, a type III to the Tertiary units (Figures 17, 18; Plate 9).

34 Ma: Initial state. The entire section is immature except for the basal part of the Cretaceous in the Cordillera domain.
31 Ma: Inversion of Espinal high and Doima deposition. Because almost the entire section is immature, the inversion has little effect on the maturity.
20 Ma: La Cira–Carbonera deposition. Deposition of the Oligocene sediments leads to rapid increase of the maturity in the Cordillera domain.
16 Ma: Upper Carbonera–Leon. The same pattern continues. The lower Villeta Formation becomes overmature

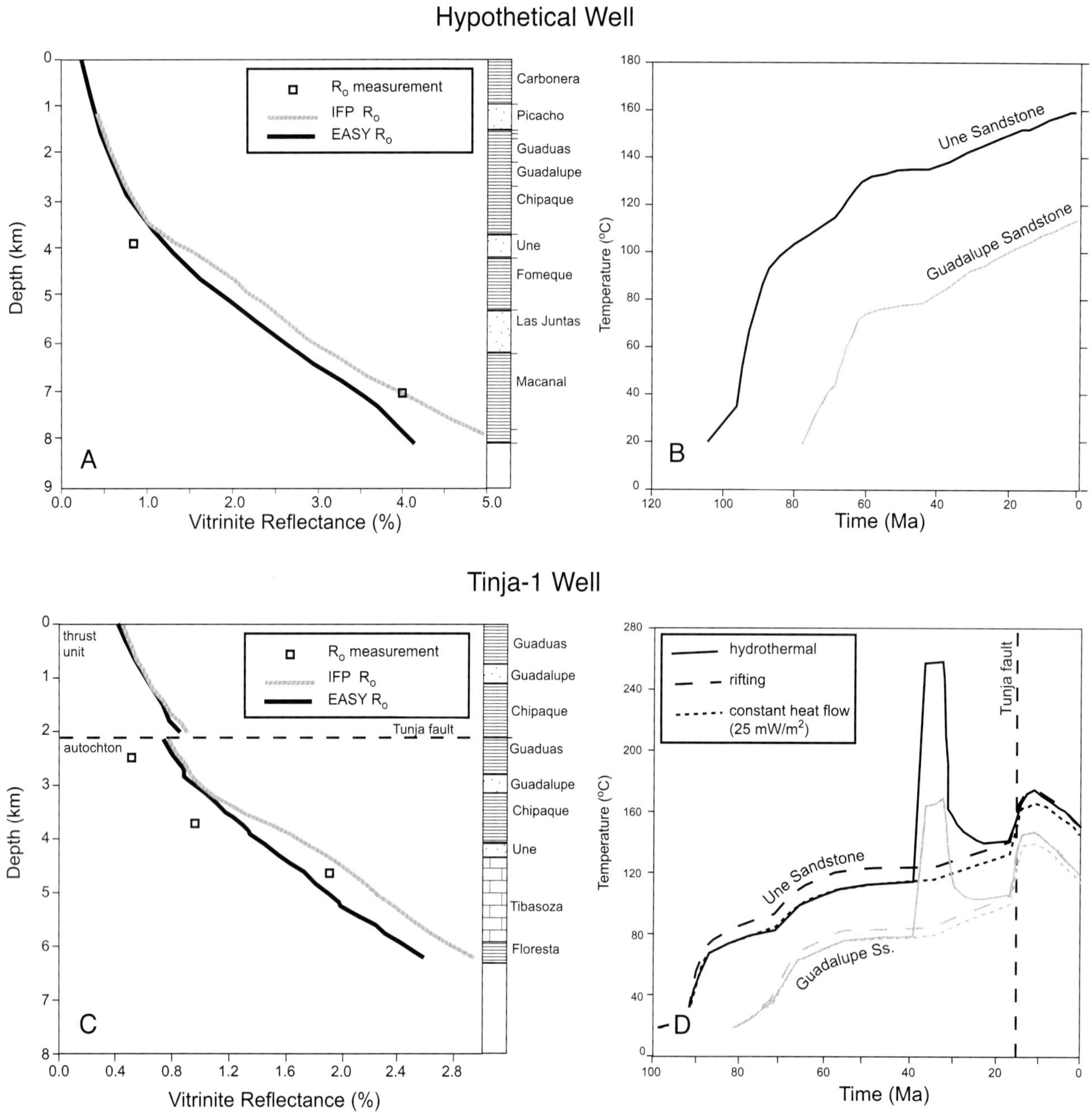

FIGURE 14. One-dimensional conductive thermal models carried out using Genex-Gentect of a hypothetical well located in the east of the Floresta massif (A and B) and the Tunja well (C and D). See Figure 2 or Plate 1 for the well locations. The models are constrained by surface and well vitrinite and T_{max} measurements. The vitrinite reflectance profiles were calculated using a McKenzie-type rifting model, with rifting between 140 and 110 Ma and a stretching factor of 2. IFP R_o and EASY R_o are two alternate kinetic models of vitrinite maturation (Beicip-Franlab, 1995). In the hypothetical well (B), the maximum temperature at the level of the Albian Une sandstone is about 160°C. The models shown in (D) attempt to account for the temperatures in excess of 235°C found in fluid inclusions in the quartz cement of the Une Formation. A temperature spike of 300°C was introduced into the Neocomian rocks at 38–32 Ma, and the overlying units heat up by conduction alone.

in the Cordillera domain, but it is still immature in the Magdalena Basin.

12 Ma: Lower Honda. With the beginning of the deposition of the Honda Formation, the lower part of the Villeta Formation enters the oil window in the footwall of the Magdalena fault.

8 Ma: Middle Honda. Further sedimentation and tectonic burial drive the maturation front toward the Magdalena Basin.

4 Ma: Upper Honda. The same pattern continues.

0 Ma: Holocene. In the present stage, the Villeta Formation is within the oil window or overmature in all

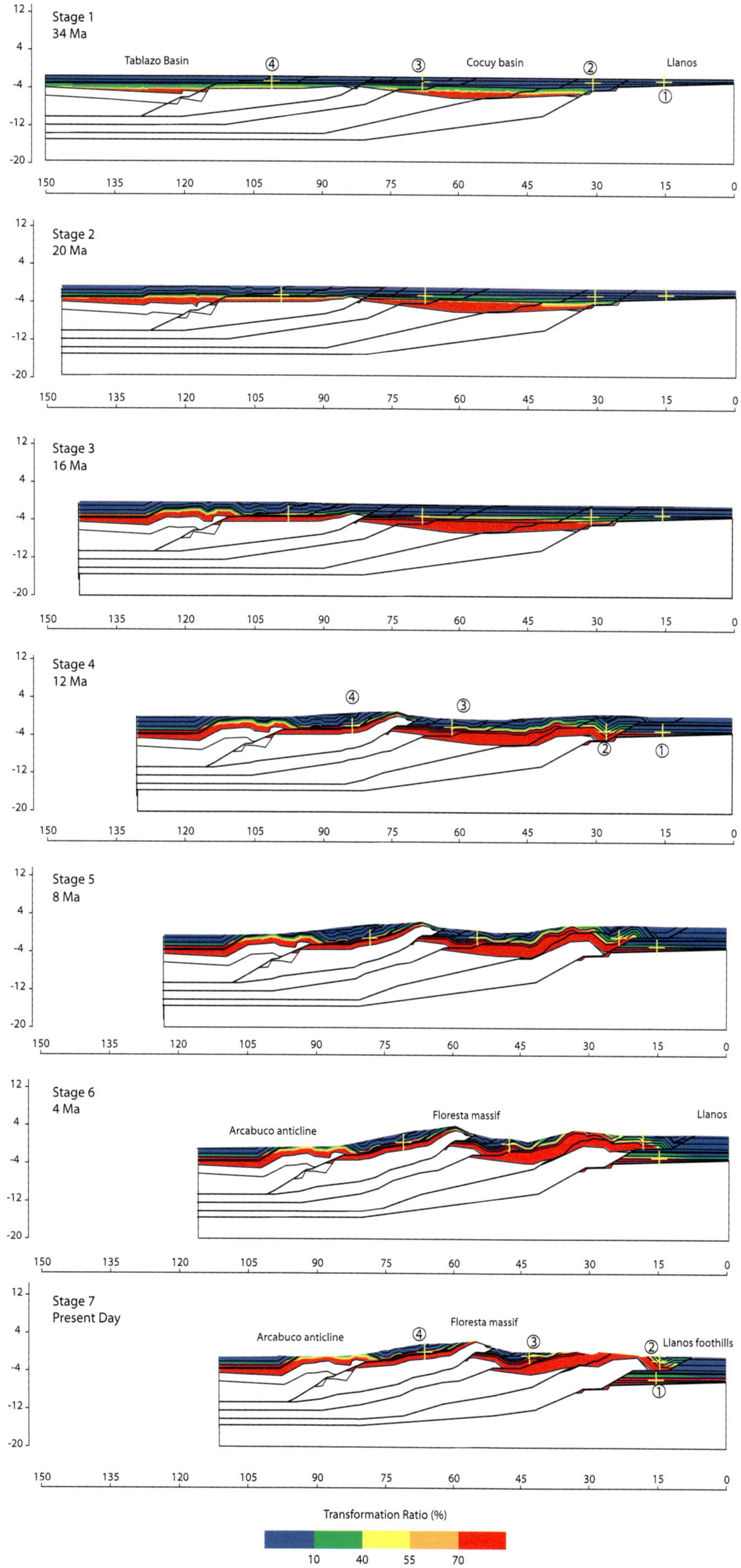

Plate 7. Two-dimensional Thrustpack 4.0 thermal model of the internal part of our northern section of the Eastern Cordillera. See Plate 1 for location. The yellow crosses represent points whose thermal histories are tracked in Figure 17. Both vertical and horizontal scales in kilometers.

the thrust-bounded blocks of the cross section. In the Girardot fold belt, the Villeta is mostly immature along the transect.

Discussion of the Thermal History of the Southern Section

We have tracked the thermal and maturation histories of eight points across the southern transect (Figures 17, 19). These points are located approximately at the same stratigraphic level in the source rocks of the Villeta–Gachetá formations. Points 1 and 2, which are located in the foreland, experienced continuous burial and increasing temperature throughout the entire history (Figure 19c). Point 2, located in the footwall of the Villavicencio fault, suffered greater tectonic and sedimentary burial and reached the highest present-day temperature. In contrast, points 3 and 4, which are located on the hanging wall of the frontal thrusts, experienced an initial period of sedimentary burial by the pre-Andean foreland basin units, followed by uplift and unroofing, leading to rapid cooling (Figure 19c). The difference in the initial temperatures between point 4 (120°C at 35 Ma) and points 1–3 (less than 80°C at 35 Ma) is caused by both the thicker Cretaceous–Eocene sedimentary section and the higher basal heat flow of the Cordillera domain relative to the Llanos foreland.

The maturation history of these points is also very characteristic of their setting in the fold and thrust belt (Figure 19d). Because of the prevailing higher heat flow, the interior of the Cordillera (point 4) matures

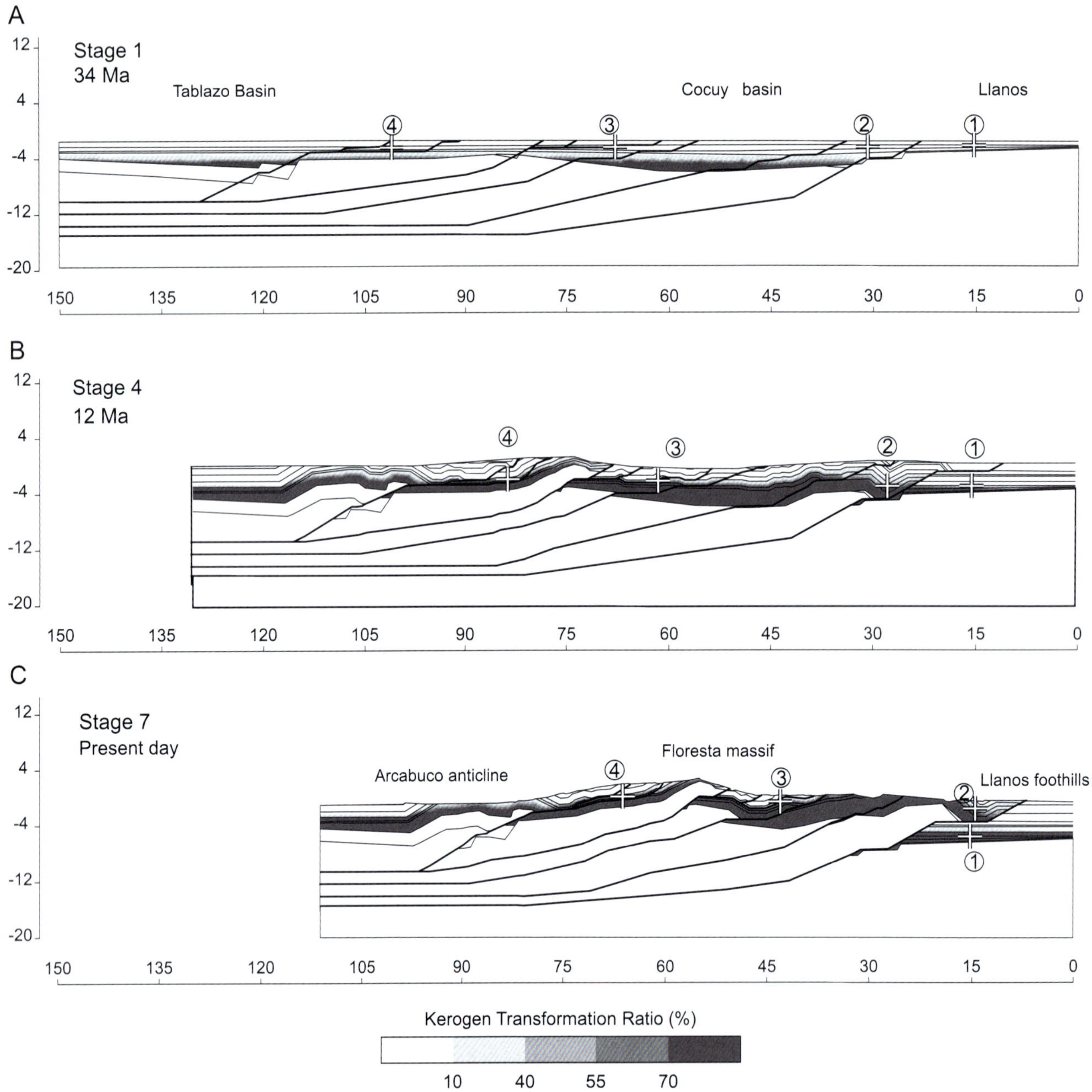

FIGURE 15. Three key stages of the two-dimensional thermal model of the internal part of the northern section. See Plate 7 for the complete model and the text for details. (A) Initial state of the model (late Eocene) caused by the high heat flow in the Cocuy basin and the thick sedimentary pile, Early Cretaceous units in the rift basins were already overmature before the main phase of Andean deformation. (B) Burial by Oligocene units of the pre-Andean foreland further increased the maturity of Cretaceous rocks, but by the middle Miocene, this process was arrested by the onset of thrusting and erosion in the core of the Cordillera. (C) Today, exhumation dominates the high part of the Cordillera, whereas tectonic and sedimentary burial of the footwall of the Guaicaramo fault causes the wedge of mature source rock to prograde toward the foreland. The thermal histories of the four numbered crosses are shown in Figure 17 and discussed in the text. Both vertical and horizontal scales in kilometers.

during the early Oligocene, before the main phase of Andean deformation; however, with the onset of uplift and erosion, maturation ceases in the Miocene. The maturity of the footwall of the frontal faults is directly linked to the onset of Andean deformation and the deposition of the syntectonic series (Figure 18). The maturation of point 2 is the most relevant to the hydrocarbon prospects of the Llanos, because it is the active kitchen capable of feeding the Andean-age traps. The foreland itself (point 1) is a relatively cold area, where a large amount of burial is required to initiate the maturation.

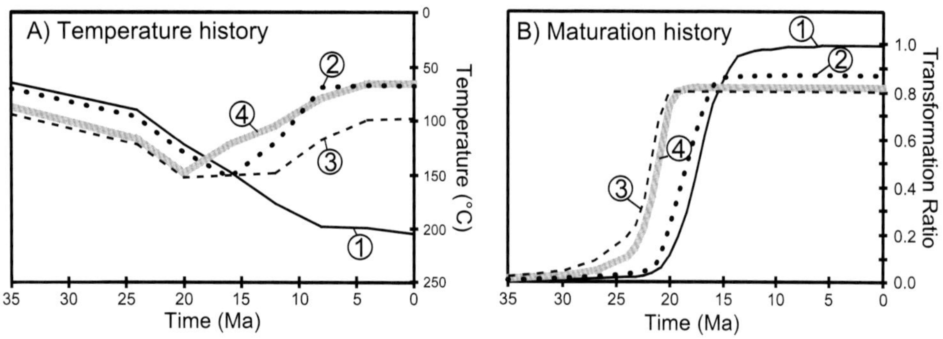

FIGURE 16. (A) Teperature and (B) maturation history of four points at the stratigraphic level of the Late Cretaceous source rock of the northern section of the Eastern Cordillera. See Figures 9 and 15 for location of the points. Because of higher heat flow and greater burial under the pre-Andean foreland basin deposits, points in the internal part of the Cordillera (3 and 4) undergo earlier maturation relative to the foreland (points 1 and 2). However, the onset of thrusting in the core of the range at about 20 Ma leads to denudation and stops the maturation process. The footwall of the frontal thrust (point 1) undergoes continuous burial, whereas the hanging wall also undergoes denudation and cooling once deformation propagates eastward in the Miocene. The exact timing of the maturation episodes is not well constrained because of lack of accurate biostratigraphic and thermochronologic data, but the overall patterns are well defined.

The differences in thermal histories of points in the western flank of the Cordillera are more subtle than in the eastern flank (curves 5–7 in Figure 19a), and therefore, the model is very sensitive to the input parameters. The timing of maturation depends on the sedimentation of the Carbonera-equivalent units over the Cordillera domain, yet these rocks have been completely stripped during Andean denudation. In general terms, our model predicts that the maturation front moved westward with time until the onset of uplift and

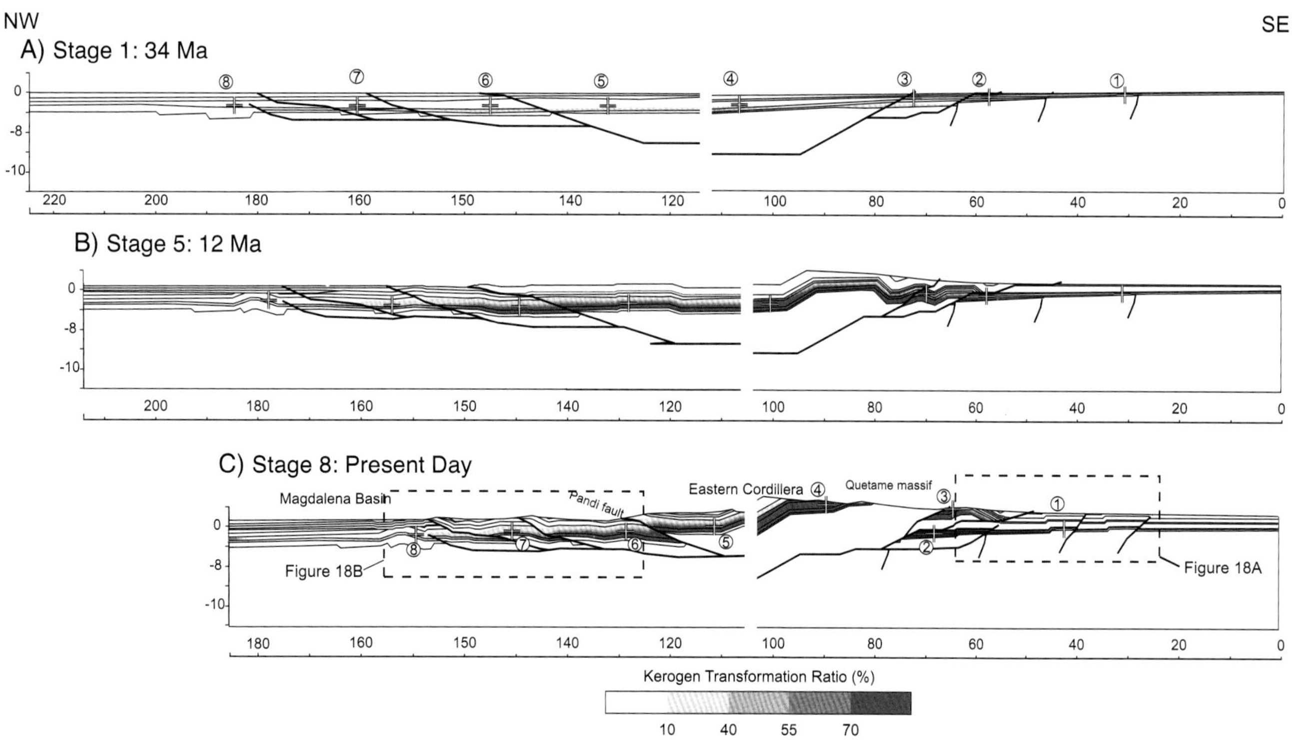

FIGURE 17. Three key stages if the two-dimensional Thrustpack thermal model of our southern section calculated using 45 mW/m^2 of basal heat flow for the Magdalena and Llanos basins and 55 mW/m^2 for the Cordillera. See Plates 8 and 9 for the complete models. The eight numbered crosses represent thermal history points in the Late Cretaceous Gachetá–Villeta source rock, which are plotted in Figure 16. Both vertical and horizontal scales in kilometers.

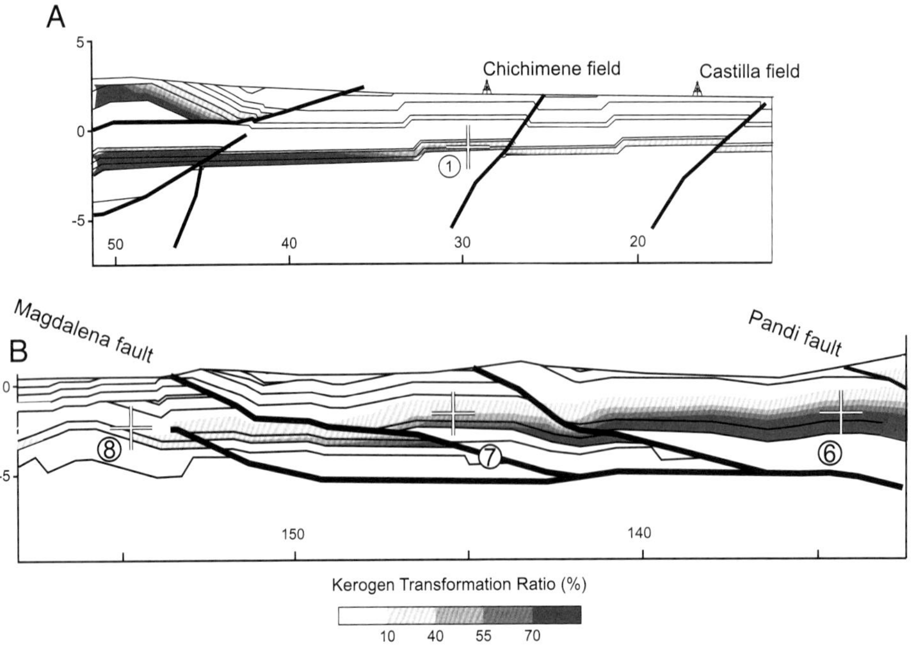

FIGURE 18. Detail of the final stage of the maturation model of (A) the Llanos foothills and (B) the Magdalena foothills of the Eastern Cordillera along the southern transect. The yellow crosses are located at the level of the Villeta–Gachetá source rock. See Figure 17 for location within the full model. Maturation in the Llanos front is controlled by burial by Tertiary units of the pre-Andean and the Andean foreland sequences. Fields in the Llanos foreland are fed by a wedge of mature source that progrades to the east with progression of the thrust front. Reservoirs in that setting are affected by cementation because of silica influx triggered by pressure solution driven by tectonic stresses linked to thrusting episodes. Maturation in the Magdalena foothills (B) was initiated by burial by the Oligocene–early Miocene pre-Andean foreland sequence sourced in the Central Cordillera. Prospects in the foothills require connection to the mature source rock in the thrust sheets within the Cordillera or in deep parts of the Magdalena Basin (not present along this profile). Both vertical and horizontal scales in kilometers.

erosion in the Cordillera. Because of the thinner stratigraphy and lower heat flow in the Magdalena Valley, maturation was lower in that area (curve 8 in Figure 19b). The source rocks can only reach thermal maturity through tectonic burial in the footwall of the Magdalena fault or in the depocenter of the Magdalena Basin, where the Miocene to Holocene sections are thickest.

CONCLUSIONS

We differentiate four distinct structural settings in the Eastern Cordillera fold and thrust belt that had characteristic thermal histories. These thermal histories lead to different timing of maturation of source rocks and different histories of cementation of the sandstone reservoirs.

Eastern Cordillera Interior

The thermal evolution of the Cordillera was strongly influenced by the location of Mesozoic extensional basins. Thermal gradients were higher in the interior of the Bogotá basin than in the flanks. During the Early Cretaceous, high heat flow developed because of the effects of ongoing rifting. Higher thermal gradient persisted in the basin during the Tertiary because of the long time required for thermal reequilibration of the lithosphere plus the blanketing effect of the thick, low-conductivity Cretaceous sedimentary pile. As a result, the maturation values at equivalent depths are higher in the interior of the Cordillera than in the margins of the Llanos Basin. Peak sedimentary burial occurred in the middle Tertiary during the pre-Andean foreland basin stage, leading to an early phase of hydrocarbon maturation and migration that predated the Andean traps. In addition, the timing of mineral deposits and high temperatures recorded by fluid inclusions in the Albian sandstones record the expulsion of hot basinal fluids probably at the onset of compressional deformation. Silica transported during this episode effectively plugged some of the potential sandstone reservoirs. Uplift and erosional denudation started at about Oligocene time, arresting the maturation process. The subsequent history has been of continued deformation, uplift, and erosion. Some hydrocarbon kitchens were preserved by early uplift. Exploration in the interior of the Cordillera is risky because of the combined effects of structural complexity, early hydrocarbon maturation, and poor reservoir conditions.

PLATE 8. Two-dimensional Thrustpack 4.0 thermal model of the Llanos foothills along the southern section. See Plate 1 for location. Only the maturation of Cretaceous rocks was computed. The yellow crosses represent points whose thermal histories are tracked in Figure 19. Both vertical and horizontal scales in kilometers.

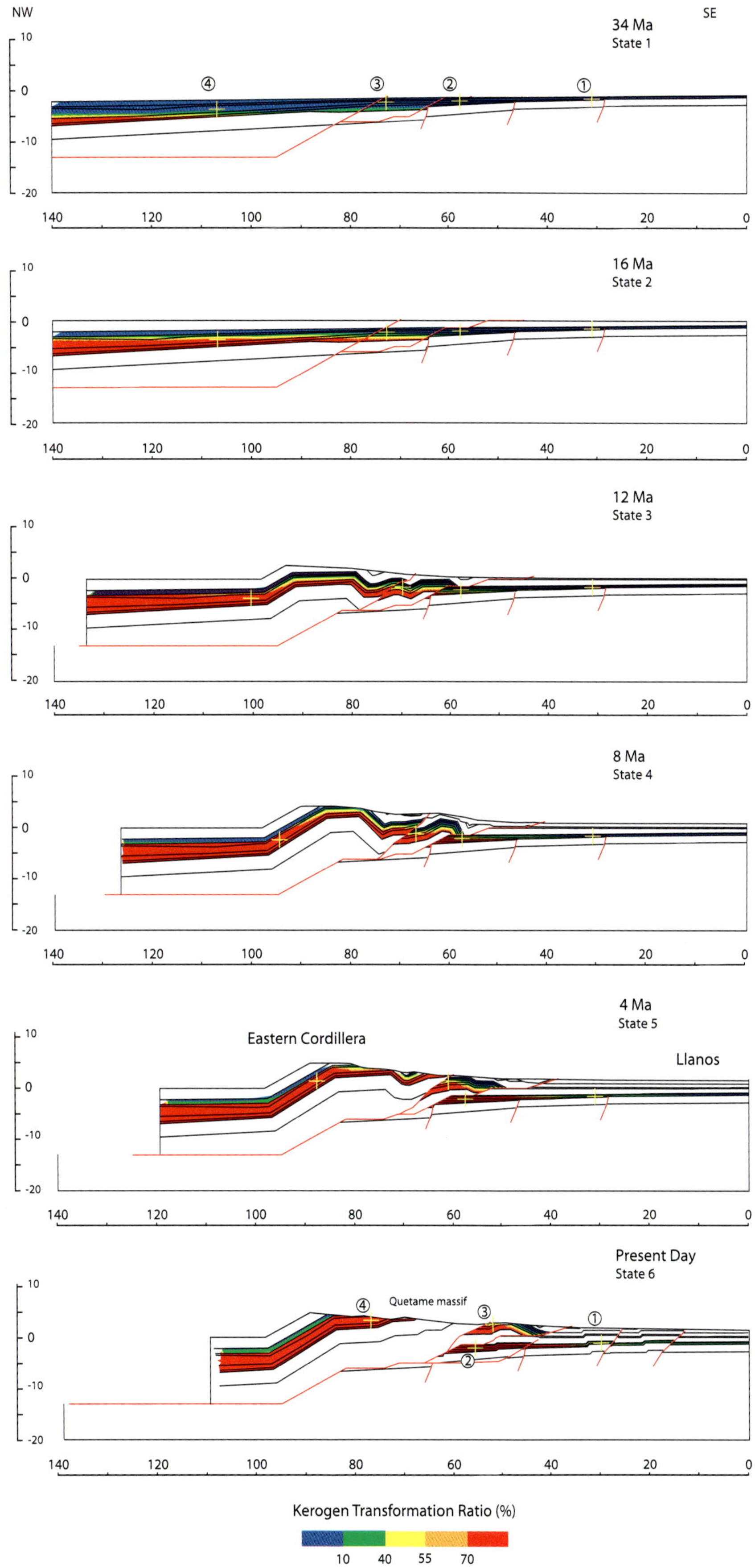

Llanos Foothills

The thermal evolution of this area was characterized by continued flexural subsidence and syntectonic sedimentation through the Miocene and Pliocene, until the onset of displacement of the frontal thrusts. Because both the thermal gradient and the thickness of pre-Miocene stratigraphic units were less than in the central part of the Bogotá basin, thermal maturity of the source rocks was delayed relative to the interior of the Cordillera. Burial of rocks on the hanging wall of the frontal thrusts (i.e., Cusiana, Yopal, and Guavio faults) ceased when these faults became active. Prospects in the foothills are viable when connected to active kitchens in the underthrust zone. The combination of fluid-inclusion studies and thermal modeling indicates that quartz cementation in the foothills prospects is tied with the onset of Andean deformation (Roure et al., 2003). This is probably caused by the influx of silica supplied by tectonically driven pressure solution of shale units that surround the reservoir (Roure et al., 2003). However, details of the cementation process are complicated by local effects, such as the petrography of the reservoirs and the chemistry of the fluids present.

Llanos Foreland

As in the foothills, the thermal regime has been cooler than in the Cordillera. Today, the basin is at maximum burial; however, because the thickness of the stratigraphic section decreases rapidly away from the thrust

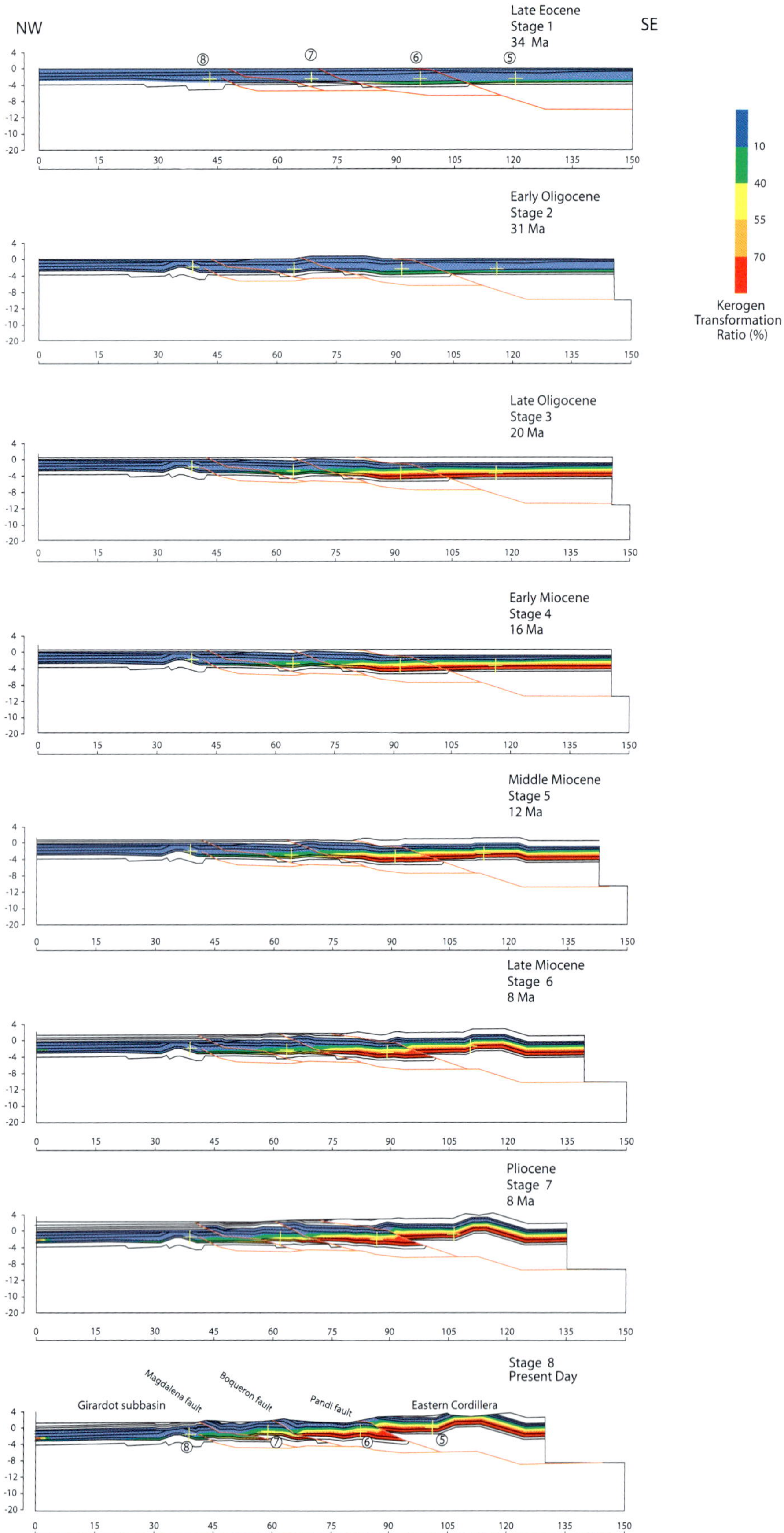

Plate 9. Two-dimensional Thrustpack 4.0 thermal model of the Magdalena flank of the southern section of the Eastern Cordillera. Only the maturation of Cretaceous rocks was computed. The yellow crosses represent points whose thermal histories are tracked in Figure 17. Both vertical and horizontal scales in kilometers.

front, the source rocks are, for the most part, immature. Traps in the Llanos require long-distance migration, but they may preserve the best reservoir conditions.

Magdalena Valley

The thermal regime of the western flank of the Cordillera is also cooler than the interior of the range, but the structural history is more complex. Along our southern transect, an active kitchen is located in the underthrust zone of the west-vergent thrust belt of the Cordillera Oriental. In the Magdalena Valley, there are local kitchens only where a thick section of syntectonic units is present in basin lows. The timing of hydrocarbon generation is strongly influenced by local conditions.

Although our models help us understand the large-scale thermal history of the Eastern Cordillera fold and thrust belt and make useful generalizations, it is important to emphasize their limitations. The critical parameters controlling thermal maturation of source rocks in the interior of the Cordillera are the amount of burial by middle Tertiary units and the timing of erosional denudation. Geological constraints on these two parameters are very poor. Middle Tertiary rocks have been, for the most part, stripped from the Cordillera, and where they are preserved, their dating is very poor. Further study of these units using nonconventional dating methods, such as

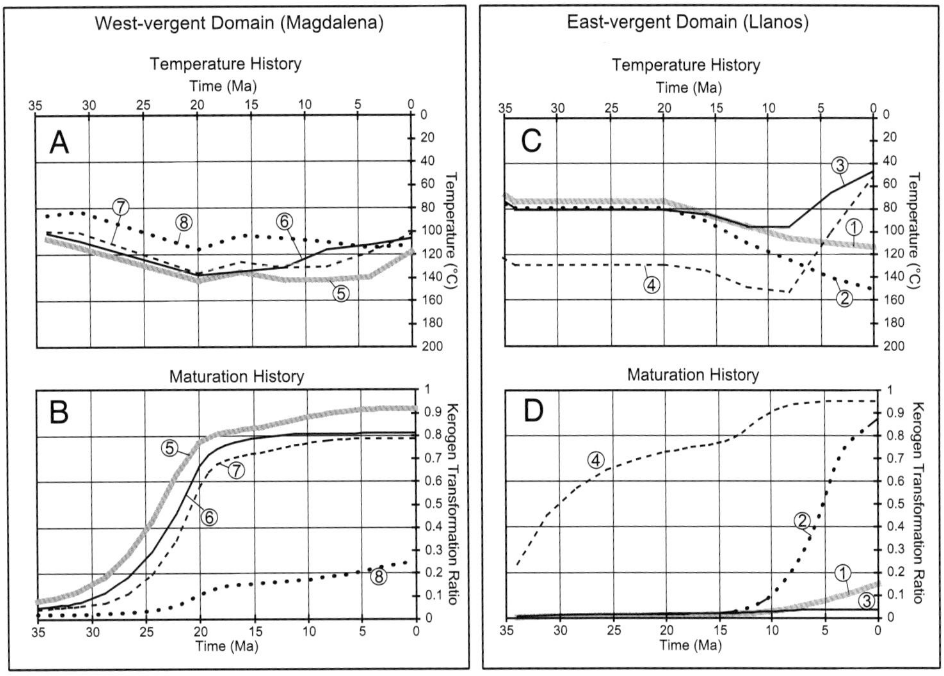

FIGURE 19. Thermal (A and C) and maturation histories (B and D) for eight points of the two-dimensional thermal models of the southern cross section of the Eastern Cordillera. See Figures 13 and 17 and Plates 8 and 9 for the location of each point. The models were calculated using a basal heat flow of 45 mW/m^2 in the Magdalena and Llanos and 55 mW/m^2 in the Cordillera. This, coupled with greater sedimentary burial prior to the Andean deformation, leads to higher initial temperatures (A) and greater thermal maturation of the more internal points (B). In the Llanos foothills (C and D), the onset of thrusting on the Santa Maria and Villavicencio faults in the late Miocene leads to uplift and cooling of the hanging-wall rocks (points 3 and 4), whereas the footwall rocks undergo continuous burial (points 1 and 2).

magnetostratigraphy and radiometric dating of ashes, are critical in improving our understanding of the evolution of the Eastern Cordillera. The thermochronological data available in the Cordillera are also very sparse and commonly of poor quality; therefore, we have very poor control on the timing and rates of erosion and, consequently, on rates of fault motion. In some areas, such as the Magdalena flank of our southern section, the thermal model is very sensitive to small changes in the initial assumptions. New apatite fission-track data, coupled with modern kinetic modeling of track-length distributions, or U-Th-He thermochronology would be other important contributions to future progress.

ACKNOWLEDGMENTS

This work was only possible due access to proprietary data contributed by the members of the SUBTRAP consortium organized by François Roure at the Institut Française du Pétrole (IFP) and by their financial support. We are also grateful for their permission to publish the results shown in this chapter. These companies included BP, ChevronTexaco, ConocoPhillips, Ecopetrol, ExxonMobil, Lasmo, Petroleos de Venezuela S.A., and Total-Fina-Elf. Finally, we thank the staff at the drafting department at IFP for their help in producing the figures. The manuscript was greatly improved by careful reviews by Luis Fernando Sarmiento, Rod Graham, and Tomas Villamil.

REFERENCES CITED

Aspden, J. A., and W. J. McCourt, 1986, Mesozoic oceanic terrane in the Central Andes of Colombia: Geology, v. 14, p. 415–418.

Bachu, S., J. C. Ramon, M. E. Villegas, and J. R. Underschultz, 1995, Geothermal regime and thermal history of the Llanos Basin, Colombia: AAPG Bulletin, v. 79, p. 116–129.

Barrero, D., 1979, Geology of the central Western Cordillera, west of Buga and Roldanillo, Colombia: Ingeominas, Publicación Geológica Especial, v. 4, 75 p.

Beicip-Franlab, 1995, Genex User Guide, Beicip-Franlab, Ruiel-Malmaison, France.

Bermudez, A., R. Acosta, and M. Garzón, 1985, Mapa gravimetrico de anomalias simples de Bouguer: Bogotá, Ingeominas, scale 1:1,500,000, 1 sheet.

Bordas-Le Floch, N., 1999, Diagenèse, compaction et déformation des réservoirs gréseux dans les chaînes plissées: Ph.D. thesis, University of Paris VI, Paris, 298 p.

Bourgois, J. et al., 1985, Les ophiolites des Andes de Colombie: Évolution structurale et signification géodynamique, *in* A. Mascle, ed., Géodynamique des Caraïbes: Paris, Editions Technip, p. 475–493.

Branquet, Y., B. Laumonier, A. Cheilletz, and G. Giuliani, 1999, Emeralds in the Eastern Cordillera of Colombia: Two tectonic settings for mineralization, Geology, v. 27, p. 597–600.

Bürgl, H., 1967, The orogenesis of the Andean system of Colombia: Tectonophysics, v. 4, p. 429–443.

Burke, K., C. Cooper, J. F. Dewey, P. Mann, and J. Pindell, 1984, Caribbean tectonics and relative plate motions, *in* W. E. Bonini, R. B. Hardgraves, and R. Shagam, eds., The Caribbean–South American plate boundary and regional tectonics: Geological Society of America Memoir 162, p. 31–64.

Cáceres, C., F. Etayo, R. Llinas, M. Rubiano, and L. J. Perez, 1970, Mapa geológico del cuadrangulo L-10, Fusagasugá: Bogotá, Ingeominas, scale 1:100,000, 1 sheet.

Campbell, C. J., and H. Bürgl, 1965, Section through the Eastern Cordillera of Colombia, South America: Geological Society of America Bulletin, v. 76, p. 567–590.

Cazier, E. C., A. B. Hayward, G. Espinosa, J. Velandia, J. F. Mugniot, and W. G. Leel, 1995, Petroleum geology of the Cusiana field, Llanos Basin foothills, Colombia: AAPG Bulletin, v. 79, p. 1444–1463.

Cheilletz, A., G. Féraud, G. Giuliani, and C. T. Rodríguez, 1994, Time-pressure and temperature constraints on the formation of Colombian emeralds: An $^{40}Ar/^{39}Ar$ laser microprobe and fluid inclusion study: Economic Geology, v. 89, p. 361–380.

Cheilletz, A., G. Giuliani, Y. Branquet, B. Laumonier, M. A. J. Sanchez, F. Rueda, G. Féraud, and T. Arhan, 1997, Datation K-Ar and $^{40}Ar/^{39}Ar$ à 65 M.a. des gisements d'émeraudes du district Chivor-Macanal: Argument en faveur d'une déformation précoce de la Cordillère orientale de Colombie: Comptes Rendu de l'Academie des Sciences, Paris, série II, Paris, v. 324, no. 5, p. 369–377.

Colletta B., F. Hébrard, J. Letouzey, P. Werner, and J.-L. Rudkiewicz, 1990, Tectonic style and crustal structure of the Eastern Cordillera (Colombia) from balanced cross-sections, *in* J. Letouzey, ed., Petroleum and tectonics in mobile belts: Paris, Editions Technip, p. 81–100.

Cooper, M. A. et al., 1995, Basin development and tectonic history of the Llanos Basin, Eastern Cordillera, and middle Magdalena Valley, Colombia: AAPG Bulletin, v. 79, p. 1421–1443.

Dengo, C. A., and M. C. Covey, 1993, Structure of the Eastern Cordillera of Colombia: Implications for trap styles and regional tectonics: AAPG Bulletin, v. 77, p. 1315–1337.

Duque, H., 1990, The Chocó block in the northwestern corner of South America: Structural, tectonostratigraphic, and paleogeographic implications: Journal of South American Earth Sciences, v. 3, p. 71–84.

Estrada, C., J. Martínez, J. Hossack, and R. Herbert, 1999, Structural evolution of the Piedemonte triangle zone, Eastern Cordillera, Colombia (abs.): AAPG Annual Meeting Abstracts, v. 1999, p. A39.

Etayo, F., G. Renzoni, and D. Barrero, 1969, Contornos sucesivos del mar Cretaceo en Colombia: I. Congreso Colombiano de Geología: Bogota, Asociación Colombiana de Geólogós, Memorias, p. 217–253.

Fabre, A., 1981, Estratigrafia de la Sierra Nevada del Cocuy, Boyacá y Arauca, Cordillera Oriental, Colombia: Geologia Norandina, v. 4, p. 3–12.

Fabre, A., 1987, Tectonique et génération d'hydrocarbures: Un modèle d'évolution de la Cordillère orientale de Colombie et du bassin des Llanos pendant le Crétacé et le Tertiaire: Archive Science Genève, v. 40, p. 145–190.

Fabre, A., and M. Delaloye, 1982, Intrusiones basicas cretacicas de la Cordillera Oriental: Geología Norandina, v. 6, p. 19–28.

Giuliani, G., A. Cheilletz, C. Arboleda, V. Carillo, F. Rueda, and J. Baker, 1995, An evaporitic origin of the parent brines of Colombian emeralds: Fluid inclusion and sulphur isotope evidence: European Journal of Mineralogy, v. 7, p. 151–165.

Gómez, E., 2002, Tectonic controls on the Late Cretaceous to Cenozoic sedimentary fill of the middle Magdalena Valley basin, Eastern Cordillera, and Llanos Basin, Colombia: Ph.D. thesis, Cornell University, New York, 619 p.

Guillande, R., 1988, Evolution Méso-Cénozoïque d'une vallée intercordilléraine andine: La Haute Vallée du Rio Magdalena (Colombie): Ph.D. Tesis, Université Pierre et Marie Curie, v. 88-26, 358 p.

Hébrard, F., M. Leikine, J. Bourgois, R. Ricateau, and P. Charitat, 1987, Métamorphisme et diagenèse des séries crétacées et tertiaires du versant est de la Cordillère orientale de Colombie (abs.): Un exemple d'évolution thermique liée à l'ouverture d'un bassin au Crétacé: II Simposio Bolivariano Exploracion Petrolera en las Cuencas Subandinas, Bogota.

Helmens, K. F., 1990, Neogene–Quaternary geology of the high plain of Bogotá, Eastern Cordillera, Colombia (stratigraphy, paleoenvironments and landscape evolution): Dissertationes Botanicae, v. 163, 202 p.

Jaillard, E. P., P. Solar, G. Carlier, and T. Mourier, 1990, Geodynamic evolution of the northern and central Andes during early to middle Mesozoic times: A Tethyan model: Journal of the Geological Society, v. 147, p. 1009–1022.

Julivert, M., 1970, Cover and basement tectonics in the Cordillera Oriental of Colombia, South America, and a comparison with some other folded chains: Geological Society of America Bulletin, v. 81, no. 12, p. 3623–3646.

Kellogg, J. N., and V. Vega, 1995, Tectonic development of Panama and the Colombian Andes: Constraints from gravity and global positioning system, geodetic studies, *in* P. Mann, ed., Geologic and tectonic development of the Caribbean plate boundary in southern Central America: Geological Society of America Special Paper 295, p. 75–90.

Kluth, C., R. Ladd, M. De Armas, L. Gómez, and N.

Tilander, 1997, Different structural styles and histories of the Colombian foreland basin, Castilla and Chichimene oil field areas, east-central Colombia (abs.), *in* VI Simposio Bolivariano, Exploracion Petrolera en las Cuencas Subandinas, Cartagena de Indias, September 14–17, 1997, Asociación Colombiana de Geólogós y Geofisicos del Petróleo, p. 18.

Linares, R., 1996, Structural styles and kinematics of the Medina area, Eastern Cordillera, Colombia: M.S. thesis, University of Colorado, Colorado, 104 p.

McCourt, W. J., T. Feininger, and M. Brook, 1984, New geological and geochronological data from the Colombian Andes: Continental growth by multiple accretion: Journal of the Geological Society (London), v. 141, p. 831–845.

McKenzie, D., 1978, Some remarks on the development of sedimentary basins: Earth and Planetary Science Letters, v. 40, p. 25–32.

Pennington, W. D., 1981, Subduction of the eastern Panama basin and seismotectonics of northern South America: Journal of Geophysical Research, v. 86, no. B11, p. 10,753–10,770.

Pindell, J. L., and J. Dewey, 1982, Permo-Triassic reconstruction of western Pangea and the evolution of the Gulf of Mexico–Caribbean region: Tectonics, v. 1, p. 179–211.

Pindell, J. L., and K. D. Tabbutt, 1995, Mesozoic–Cenozoic Andean paleogeography and regional controls on hydrocarbon systems, *in* A. J. Tankard, R. Suarez-Soruco, and H. J. Welsink, eds., Petroleum basins of South America: AAPG Memoir 62, p. 101–128.

Rathke, W. W., and M. G. Coral, 1997, Cupiagua field, Colombia: Interpretation case history of a large, complex thrust belt gas condensate field, *in* VI Simposio Bolivariano, Exploracion Petrolera en las Cuencas Subandinas, Cartagena: Bogota, Asociación Colombiana de Geólogós, p. 9.

Reyes, I., and M. T. Valentino, 1974, Geología del yacimineto y variabilidad de las características geoquímicas del mineral de hierro en la región de Paz Vieja (Municipio de Paz del Río, Departamento de Boyacá): Primer Congreso Colombiano de Geología, Separata, p. 267–324.

Roeder, D., and R. L. Chamberlain, 1995, Eastern Cordillera of Colombia: Jurassic–Neogene evolution, *in* A. J. Tankard, R. Suarez-Soruco, and H. J. Welsink, eds., Petroleum basins of South America: AAPG Memoir 62, p. 633–645.

Roure, F., N. Bordas-Le Floch, J. Toro, C. Aubourg, N. Guillaumou, E. Hernandez, S. Le Cornec-Lance, C. Riveiro, P. Robion, W. Sassi, 2003, Petroleum systems and reservoir appraisal in the Sub-Andean basins (Eastern Venezuela and Eastern Colombian foothills), *in* C. Bartolini, R. T. Buffler, and J. Blickwede, eds., The Circum-Gulf of Mexico and the Caribbean: Hydrocarbon habitats, basin formation, and plate tectonics: AAPG Memoir 79, CD-ROM, p. 750–775.

Sarmiento, L. F., 1989, Stratigraphy of the Cordillera Oriental west of Bogotá, Colombia: Master's thesis, University of South Carolina, South Carolina, 102 p.

Sarmiento, L. F., 2002, Mesozoic rifting and Cenozoic basin inversion of the Eastern Cordillera, Colombian Andes: Inferences from tectonic models: Ph.D. thesis, Vrije Universiteit, Amsterdam, 295 p.

Sassi, W., and J.-L. Rudkiewicz, 1996, Thrustpack version 3.3: 2-D integrated maturity studies in thrusts areas: Institut Française du Pétrole Report 43393, 113 p.

Sassi, W., J.-L. Rudkiewicz, and R. Divies, 1998, New methods for integrated modelling of deformation and petroleum generation in fold and thrust belts: AAPG Bulletin, v. 82, no. 10, p. 1883–1984.

Shagam, R., B. P. Kohn, P. O. Banks, L. E. Dasch, R. Vargas, G. I. Rodriguez, and N. Pimentel, 1984, Tectonic implications of Cretaceous–Pliocene fission-track ages from rocks of the circum-Maracaibo Basin region of western Venezuela and eastern Colombia, *in* W. E. Bonini, R. B. Hardgraves, and R. Shagam, eds., The Caribbean–South American plate boundary and regional tectonics: Geological Society of America Memoir 162, p. 385–412.

Suarez, G., P. Molnar, and C. B. Burchfiel, 1983, Seismicity, fault planes solutions, depth of faulting and active tectonics of the Andes of Peru, Ecuador and southern Colombia: Journal of Geophysical Research, v. 88, no. B12, p. 10,403–10,428.

Suppe, J., 1983, Geometry and kinematics of fault-bend folding: American Journal of Science, v. 283, p. 684–721.

Taboada, A., L. A. Rivera, A. Fuenzalida, A. Cisternas, H. Philip, H. Bijwaard, J. Olaya, and C. Rivera, 2000, Geodynamics of the Northern Andes: Subduction and intracontinental deformation (Colombia): Tectonics, v. 19, no. 5, p. 787–813.

Toro, J., 1990, The termination of the Bucaramanga fault in the Cordillera Oriental, Colombia: Master's thesis, University of Arizona, Arizona, 60 p.

Van der Hammen, T., 1961, Late Cretaceous and Tertiary stratigraphy and tectonogenesis of the Colombian Andes: Geologie en Mijbouw, v. 40, p. 181–188.

Van der Wiel, A. M., and P. A. M. Andriessen, 1991, Precambrian to Recent thermotectonic history of the Garzon massif (Eastern Cordillera of the Colombian Andes) as revealed by fission-track analysis, *in* A. M. Van der Wiel, ed., Uplift and volcanism of the SE Colombian Andes in relation to Neogene sedimentation of the upper Magdalena Valley: Ph.D. thesis, Free University, Amsterdam Wageningen, 208 p.

Van Houten, F. B., and R. B. Travis, 1968, Cenozoic deposits, upper Magdalena Valley, Colombia: AAPG Bulletin, v. 52, p. 675–702.

Villamil, T., 1999, Campanian–Miocene tectonostratigraphy, depocenter evolution and basin development of Colombia and western Venezuela: Palaeogeography, Palaeoclimatology, Palaeoecology, v. 153, p. 236–275.

Villamil, T., C. Arango, and W. W. Hay, 1999, Stratigraphic and structural constraints on Campanian–Oligocene tectonostratigraphic development of northern South America, *in* E. Berrera and C. C. Johnson, eds., Evolution of the Cretaceous ocean-climate system: Geological Society of America Special Publication 332, p. 110–121.

6

Fowler, J. N., R. Graham, W. Sassi, J. D. Smewing, and J. Warburton, 2004, Two-dimensional kinematic modeling of the southern Kirthar fold belt, Pakistan, *in* R. Swennen, F. Roure, and J. W. Granath, eds., Deformation, fluid flow, and reservoir appraisal in foreland fold and thrust belts: AAPG Hedberg Series, no. 1, p. 117–131.

Two-dimensional Kinematic Modeling of the Southern Kirthar Fold Belt, Pakistan

J. N. Fowler,[1] R. Graham,[2] J. D. Smewing,[3] J. Warburton[4]
Eni-LASMO Oil Pakistan Ltd., Karachi, Pakistan

W. Sassi
Institut Français du Pétrole, Rueil-Malmaison, France

ABSTRACT

A study of the southern Kirthar fold belt in Pakistan was undertaken to elucidate the hinterland structure and hydrocarbon prospectivity. Interpretation of structure and stratigraphy is difficult because of suboptimal seismic data, a lack of hinterland well data, and a transition from shelfal to basinal stratigraphy. An interpretation of two cross sections was made using outcrop and seismic data and well data from foreland discoveries. The Institut Français du Pétrole Thrustpack® software was used to validate the structural model and provide data on the maturity of the source rock.

The Kirthar fold belt is dominated by open and symmetrical folds that are driven by inversion of basement-involved Jurassic extensional faults. Thrusts have been interpreted with two detachments, thrusts with a shallow detachment in the Eocene mudstones and thrusts with a deeper detachment in the Lower Cretaceous source rock interval that involve the reservoir during deformation.

The major mountain-building episode is interpreted as late Pliocene–Pleistocene, but there is evidence for earlier inversion dating from the late Paleocene associated with the emplacement of the Bela ophiolite and constrained by maturity data obtained from outcrop. Early inversion and uplift impacts the burial curve and, thus, the prospectivity of the area.

INTRODUCTION

The Kirthar fold belt (KFB) lies in southern Pakistan and is part of the Himalayan chain, marking the collision between the Indo-Pakistan and Eurasian plates (Figure 1). The KFB is the focus of much attention for hydrocarbon and mineral exploration, and in recent years, there have been several large gas discoveries in the mountain front and foreland. Commercial gas reserves were discovered in Bhit during 1997 in the Upper

[1]*Present address:* Agip KCO, Den Haag, The Netherlands.
[2]*Present address:* Amerada Hess, London, United Kingdom.
[3]*Present address:* Earth Resources Ltd., Swansea, United Kingdom.
[4]*Present address:* Executive Insight Pty. Ltd., Paddington, Sydney, Australia.

DOI:10.1306/1025688H13112

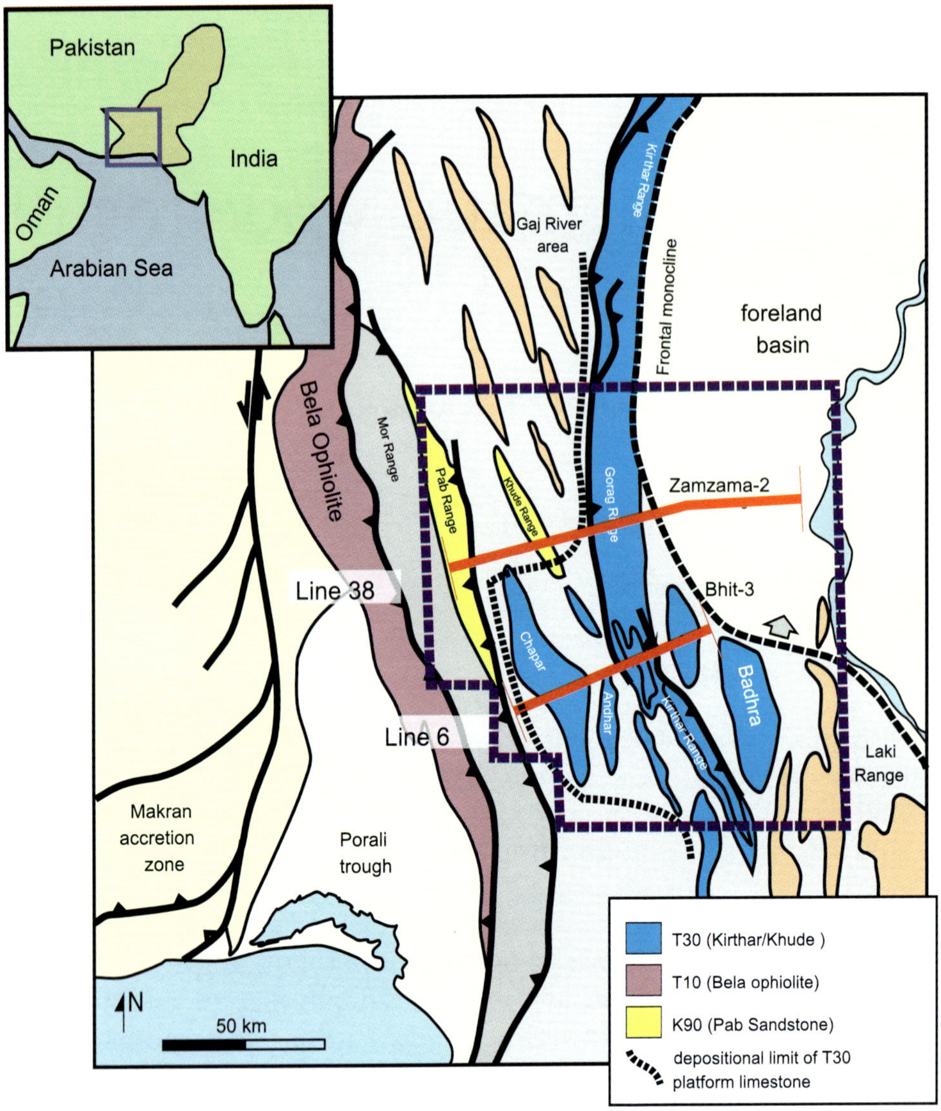

FIGURE 1. A simplified surface geology map of the southern Kirthar fold belt, Pakistan. The study area is outlined in a dashed blue box.

Cretaceous Pab sandstone, and further gas discoveries were made in 1998–1999 at Zamzama and Badhra in the Pab and Mughal Kot sandstones, respectively. Approximately 4–5 tcf of gas reserves has been found to date in the Eni-LASMO area of interest (Hedley et al., 2001).

Exploration success in the foreland led to activity in the hinterland of the fold belt, where many large structures remain undrilled. Understanding the type of structure and the involvement, or noninvolvement, of the reservoir within structural closure is critical. The deduction of structural style, sequence, and timing also aids the prediction of source rock maturity and timing of hydrocarbon generation.

Because of the poor seismic imaging and lack of well ties in the hinterland, little is known about the subsurface structure and sedimentary architecture. This is compounded by the presence of a shelf-to-basin-floor transition from the foreland to the hinterland, making it difficult to correlate stratigraphy. To a large extent, this difficulty has been overcome by the application of a sequence-based stratigraphic scheme of the southern KFB (Smewing et al., 2002a, b). This scheme has been adopted in the present study and is illustrated in Figure 2, which shows the depositional sequences and their chronostratigraphic equivalent.

This chapter presents the results of work carried out to address the uncertainties associated with exploration in the southern KFB. The study began with the interpretation of a two-dimensional (2-D) seismic data set over the Eni-LASMO licenses in the study area (Figure 3), and two regional lines were selected for restoration and thermal modeling of the source rock.

Roure and Sassi (1995) introduced a methodology that uses Thrustpack® to assess the petroleum potential of foreland fold and thrust belts. The method employs the use of forward modeling and cross section balancing of intermediate geometries (in addition to initial and final geometries), coupled with thermal modeling, allowing the user to constrain the burial history and estimate the timing of hydrocarbon generation. To accurately reconstruct the burial history in the hinterland and thus determine source rock maturity and reservoir characteristics, the amount of overburden eroded must be ascertained. Maturity and fission-track data can be incorporated during the modeling to estimate the initial thickness of eroded overburden (see the Appendix for more details about Thrustpack).

The ability to integrate a diverse data set when modeling with Thrustpack led to its use during the study to investigate the following questions:

1) Structure: Is the deformation style thick skinned or thin skinned? What level are the detachments, and is the reservoir involved during deformation? Does the modeling confirm the deduction made

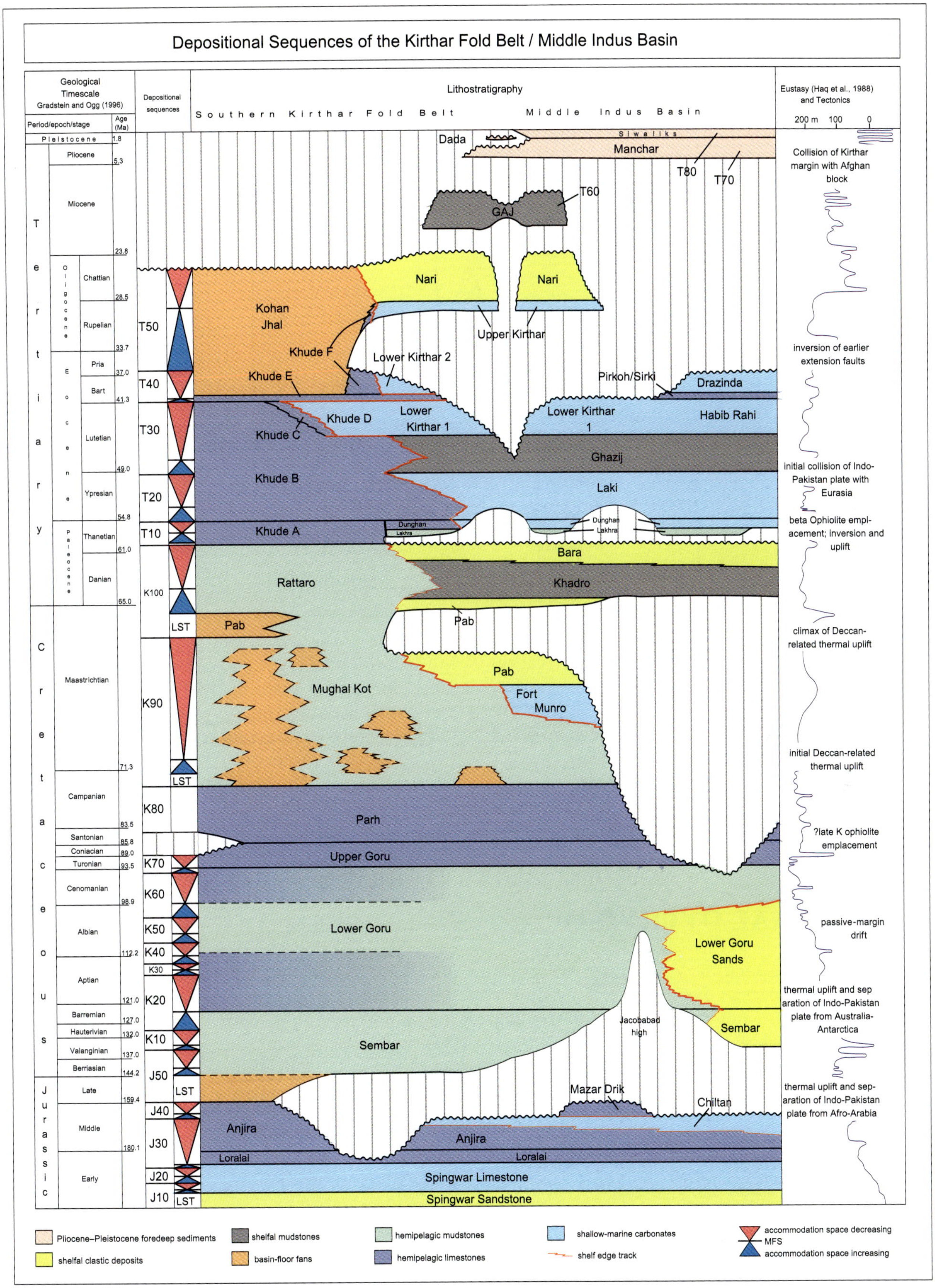

FIGURE 2. Chronostratigraphy and sequence stratigraphy of the Kirthar fold belt modified after Smewing et al. (2002a). The alphanumeric naming system has been employed in this study.

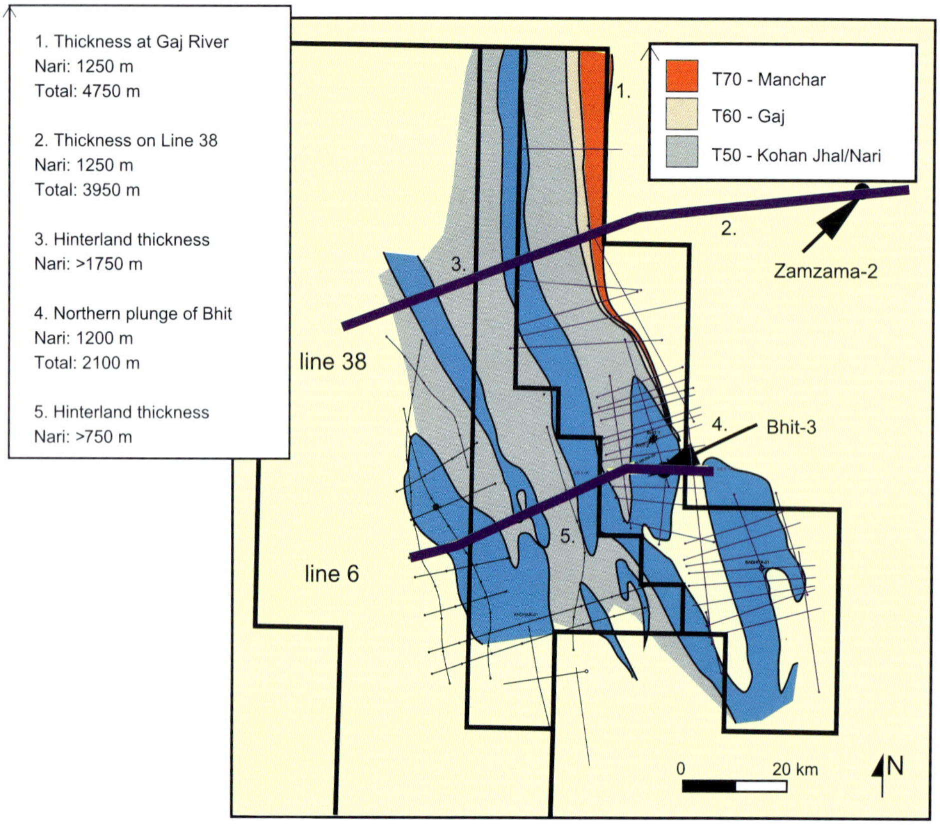

FIGURE 3. Simplified surface geological map of the study area showing the thickness of the foreland basin megasequence taken from outcrop. The black outline represents the extent of the Eni-LASMO licenses during the study. The 2-D seismic data that were used are also shown.

from field and seismic evidence that the elevation of the Kirthar Range is essentially thick skinned, and that thin-skinned detachments are minor and structurally dependent on the basement-involved deformation. Finally, does the modeling predict the detachment levels that we observe?

2) Source rock: What is the maturity of the source rock? What is the timing of trap generation vs. hydrocarbon generation?

REGIONAL OVERVIEW

A series of plate tectonic maps illustrating the formative tectonic episodes of the fold belt are shown in Figure 4. Figure 4A shows the position of the KFB during the Early Cretaceous (K10–K70). After rifting and drifting away from the Pangean supercontinent at the end of the Jurassic, Pakistan formed part of a northwest-facing passive margin throughout the Cretaceous. Large deltas prograded from the emergent Indian continent, depositing the Goru and Sembar formations in the middle Indus Basin (MIB) and shedding turbidites into the KFB area. The Goru formation acts as reservoir for the middle Indus Basin gas discoveries, and the Sembar Formation is thought to be the source rock in southern Pakistan (Figure 4A). Scant data on the source rock characteristics of the Sembar Formation in the study area (see below) are available.

In southern Pakistan, the late Campanian–Maastrichtian (sequence K90; Figure 2) is a period of relative sea level fall (Hedley et al., 2001). The drop in relative sea level is accompanied by the deposition of the Mughal Kot turbidites in the present-day hinterland of the fold belt (Figure 4B). The large clastic input into the basin is attributed to uplift and denudation of the Indian landmass as it passed over a mantle hot spot (Eschard et al., 2004). The deepwater turbidite sandstones are exposed in the Pab and Khude Ranges and constitute the primary reservoir target in the hinterland (Figure 1). A subsequent rise in relative sea level shifted the depocenter onto the shelf (present-day foreland), leading to the deposition of the Pab formation in the latest Maastrichtian. The braided-fluvial Pab delta forms the reservoir for the Bhit and Zamzama gas discoveries.

During the Paleocene (sequence T10), the Indo-Pakistan plate collided with an intraoceanic subduction zone, resulting in the emplacement of the Bela ophiolite (Gnos et al., 1998; see Figure 4C). The emplacement resulted in basinwide deepening and appears to have inverted structures in parts of the study area, a point investigated by the modeling. The early inversion impacts local paleogeography and, thus, the petroleum system. Inversion continued into the Eocene, affecting the development of the Ghazij and Khude B slope deposits and the Kirthar and Khude D platform limestone (sequence T30).

During early collision of the Indo-Pakistan and Eurasian plates in the Eocene–Oligocene, the erosional products of the nascent Himalaya infilled the Kirthar basin (sequences T40–T70). This influx of clastic sediment represents the start of the foreland basin megasequence proper. At its base are the Kohan Jhal and Nari, the former a deep-water flysch turbidite succession and the latter a shallow-water molasse (Figure 4D).

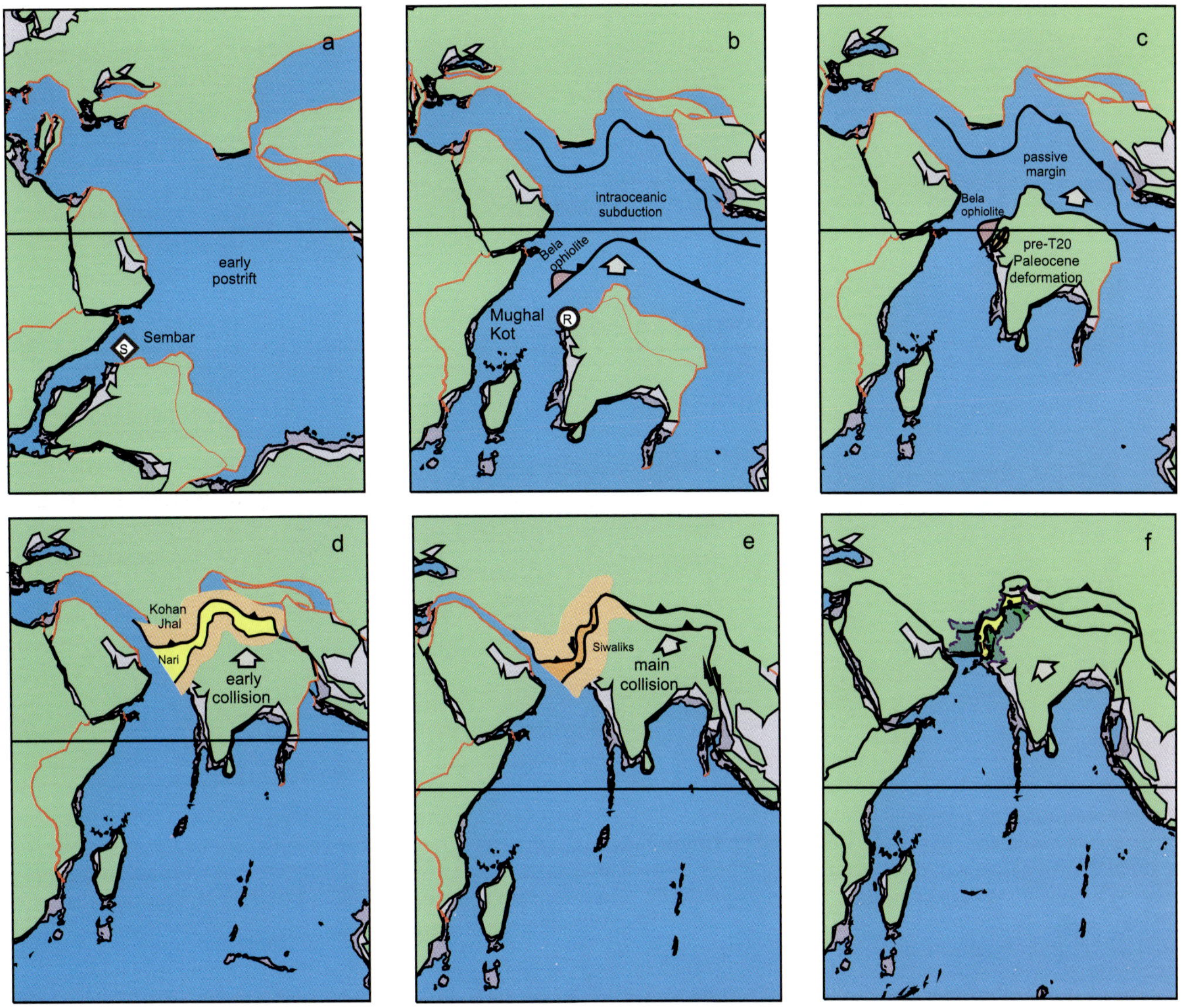

FIGURE 4. Plate tectonic maps illustrating the transition of the Kirthar margin on the west of the Indo-Pakistan plate during (A) K10–K70, (B) K50, (C) T10, (D) T40–T70, (E) T80, and (F) present day (modified after Hedley et al., 2001).

Younger basinal deposits than the T40–T50 (Kohan Jhal Formation) are not evident in the hinterland because of erosion or nondeposition. Deposition of shallow-marine molasse in the foreland continued with the Nari, passing gradually upward into the Gaj and Manchar (Figure 4E). Estimates of the amount of eroded overburden (T40–T70 sequences) are needed to determine the burial history of the study area.

The final sequence (T80) is represented by the terrestrial Dada conglomerate and the Siwaliks, which onlap the underlying marine clastics. The base of this sequence appears to represent the major mountain-building phase in the study area (Figure 4F) and is dated as late Pliocene–Pleistocene (Hunting Survey Corporation Ltd., 1960).

DATA SET

Two cross sections were interpreted for Thrust-pack modeling, with each line highlighting differences in structure, structure timing, paleogeography, and sedimentary architecture. The lines were interpreted using 2-D seismic data (Figure 3), well data from mountain front gas discoveries, and a proprietary surface geological map covering an area of approximately 8000 km^2 (3100 mi^2).

In the north, line 38 was interpreted using a composite seismic line, which ties into Zamzama-2 (Figures 5, 6). Seismic data coverage ceases at the western side of Gorag Ridge. The geological map was used to extend the section westward over the Khude, Zarro, and Pab ranges.

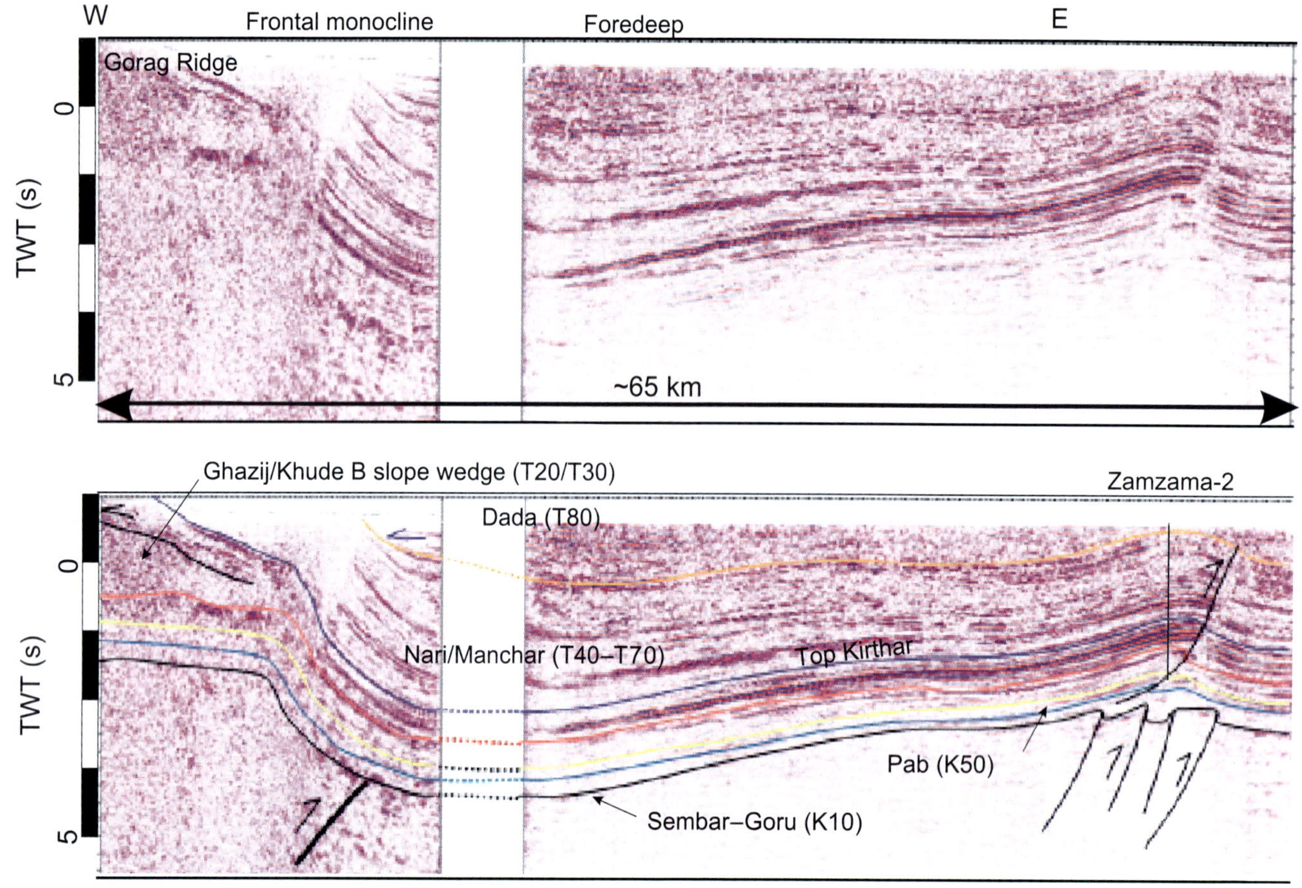

FIGURE 5. Composite seismic line used to interpret line 38. The Zamzama structure is interpreted as a fold driven by inversion of basement-involved extensional faults, modified by a thrust with a shallow detachment in the K10. Note the degradation in seismic quality from the foreland to the hinterland. The two seismic lines do not overlap additional seismic data were used to correlate the horizons. TWT = two-way traveltime.

The southern cross section, line 6, ties into Bhit-3 (Figure 1). Seismic data and geological maps are available along the entire line, covering the Kirthar Range and Chapar (Figures 6, 7).

Vitrinite reflectance data obtained from well cuttings were used to calibrate the heat flow, along with corrected bottomhole temperatures and paleotemperature data obtained through fluid-inclusion analysis (FIA) of the K90 sandstones. Further vitrinite reflectance and FIA data were gathered from outcrops in the west of the study area to calibrate the thickness of eroded overburden.

STRUCTURE OF THE SOUTHERN KFB

The KFB is dominated by large, broadly symmetrical, open anticlines. They commonly have steep or oversteepened eastern limbs that may be broken by thrusts, but the thrusts are generally small displacement structures, and the folds are the significant structural element. Thrusts are apparent on surface geological maps but with one very notable exception; the stratigraphic separations are negligible (the Kirthar Range, see Figure 1). The most significant stratigraphic duplications exist on a small scale in the T30 sequence of the Hab Valley and beneath the T30 carbonate platform on the western slopes of the Gorag Ridge. These Gorag thrusts are westward verging and form part of the major back-thrust system described by Banks and Warburton (1986) that is visible on all the maps and Landsat images of the KFB.

The Kirthar Range shows a combination of thick-skinned and thin-skinned deformation. "Thick skinned" is the term for contractional deformation (folding) that is associated with steep reverse faults that penetrate the basement. Commonly, but not invariably, such faults were once extensional. Because rift- and passive-margin extensional faults do not, by definition, enter the postrift section, inversion of these structures creates only folding in these sequences. Where faulting is present, it is associated with local strain. "Thin-skinned"

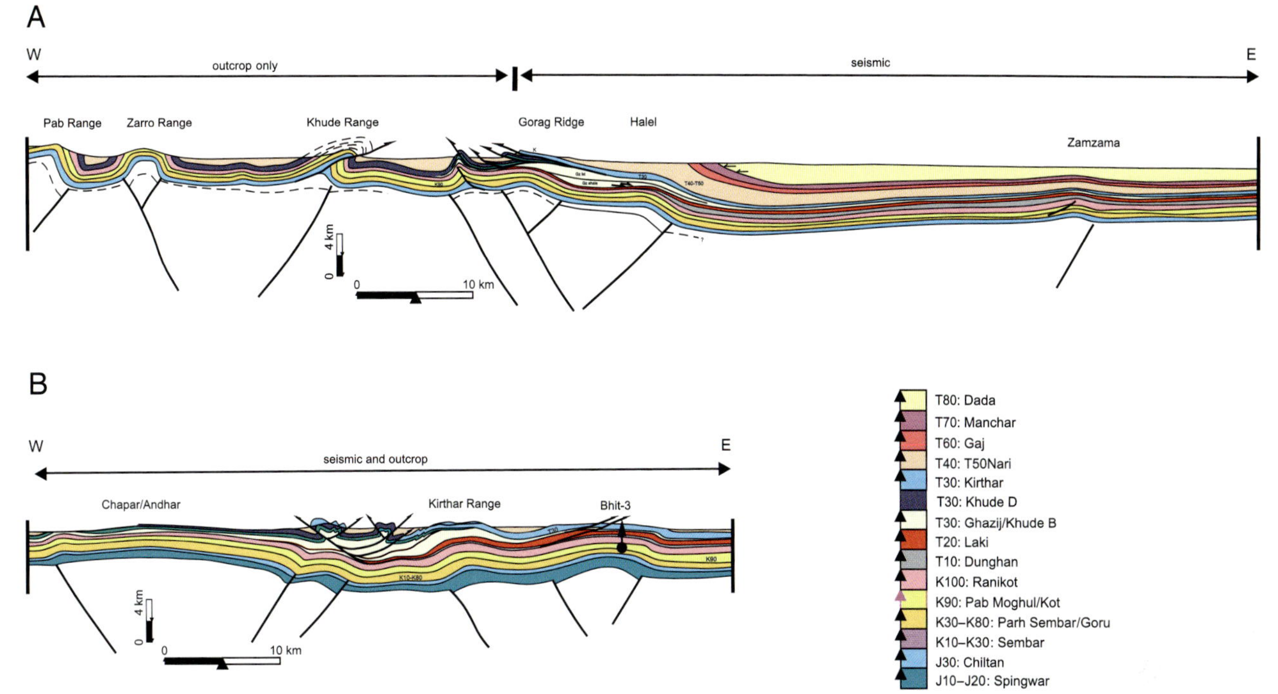

FIGURE 6. The cross sections used for Thrustpack modeling are displayed at the same scale: (A) Line 38 and (B) line 6.

deformation refers to shortening associated with displacement surfaces that mostly lie bedding parallel but ramp upsection locally to produce fault-related folds.

Locally, in the KFB, thin-skinned deformation is clear and unequivocal, for example, on the west flank of Gorag Ridge. Commonly, surface structure and lack of a break in deep reflections strongly suggest thin-skinned deformation (Figure 7), for example, the rootless thrusts of the Hab Valley, which probably relate to expulsion from a tight syncline (out-of-the-syncline thrusts). However, the simplicity of the KFB folds and lack of seismic evidence of significant, thin-skinned displacement at depth suggest that they have some other origin.

On several seismic lines, the division between the pre- and synrift section is recognized by a change from irregularity to regularity. Seismic data occasionally indicate faults, and in the Badhra structure (Figure 1), thickening of the postrift section to form a local sag basin is evident. Direct continuation of fold geometry below the base of the postrift section is not observed, so the notion of large-scale buckling is not favored, and the KFB structures, such as Chapar and Bhit, are interpreted as thick-skinned inversion structures (Figure 6).

The leading limbs of these folds are commonly broken by thrusts, but they are small displacement thrusts that are not responsible for the major elevation and formation of the structures. These minor thrusts result from oversteepening of the fold limbs and are local strain phenomena. It is likely, although not proven to date, that the controlling faults are Mesozoic extensional faults, which do not break into the postrift section in any coherent fashion. When the structures were inverted during regional contraction, the postrift and foreland basin megasequences above them were folded.

The Isle of Wight and Lulworth structures in southern England are typical examples of this sort of structure (Underhill and Paterson, 1998); these folds are asymmetrical because they are inverted half grabens. The folds in the study area are more symmetrical and probably represent an inverted graben, a situation that requires the physical displacement of one or both of the controlling faults. Interestingly, the leading limb of the Lulworth anticline is broken by a small thrust; the analogy with the leading limbs of the Kirthar structures is close. Examples of these sorts of structure, perhaps more closely analogous to the southern KFB because of their larger scale and more extreme inversion-related elevation, can be found throughout the Andes and Central Asia. The Eastern Cordillera of Colombia is a "popped out" (inverted), symmetrical, back-arc graben, with limited displacement thrusts on the margin (Colletta et al., 1990; Cooper et al., 1995). Inspection of any published geological map shows that the Andean mountain front from Putomayo to the Maranon is essentially a monocline, with a leading limb some 5 km (3 mi) or more in height. On the Pongo de Manseriche (the mountain front monocline west of the Maranon Basin of Peru), there is a back thrust whose limited displacement seems to complement the elevation of the frontal monocline, just as the back-thrust system on the Gorag Ridge seems to accommodate the elevation of the Kirthar Range. Even more spectacular analogs for the Kirthar are the Merida Andes of Venezuela, which, on its north side, elevates the basement 12 km (8 mi) across a 10–15-km

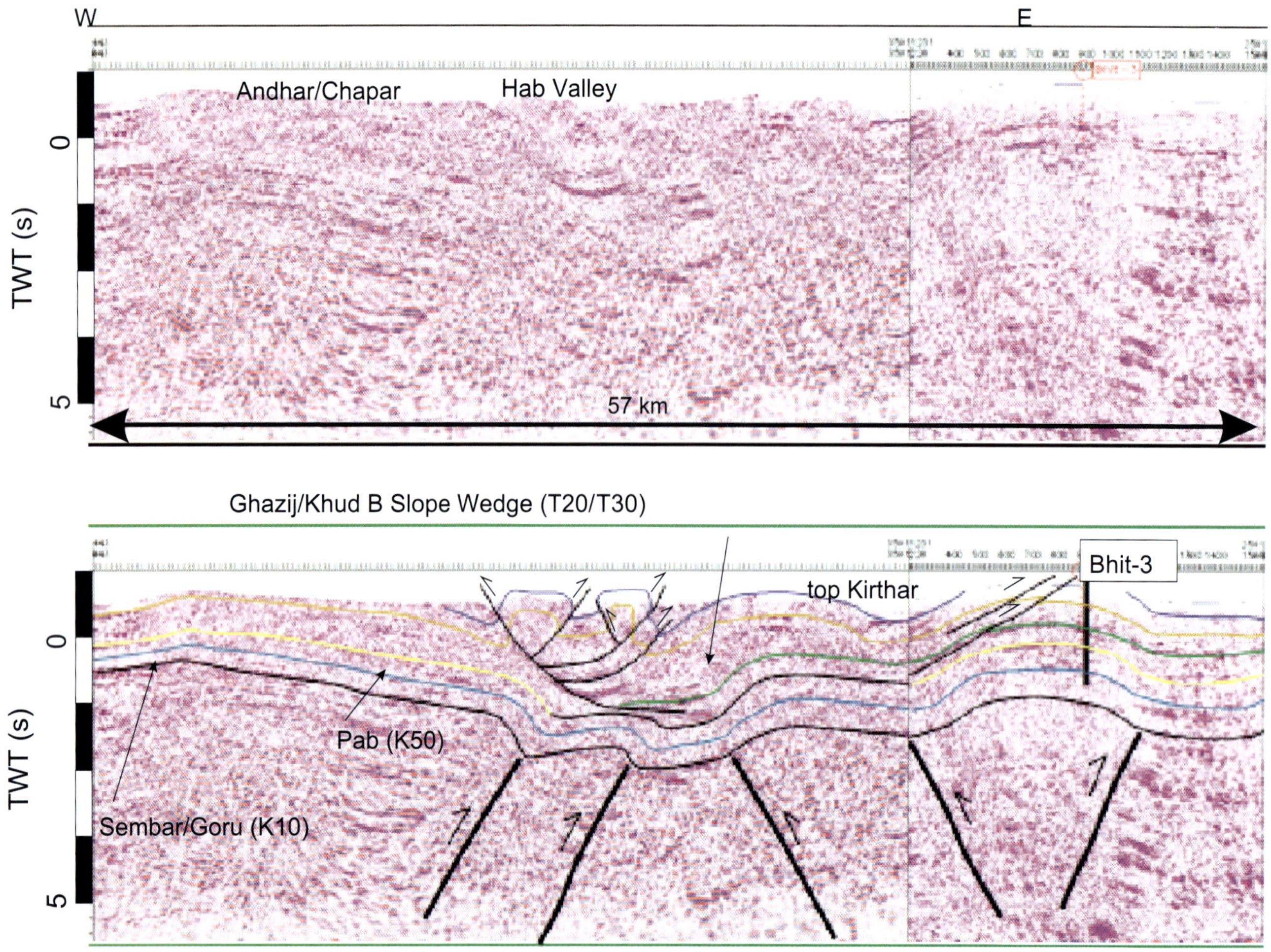

FIGURE 7. Composite seismic line used to interpret line 6. Note the poor quality of seismic data in the hinterland relative to the foreland seismic data shown in Figure 4. TWT = two-way traveltime.

(6–9-mi) zone with very limited shortening. The Tien Shan, Altyn Tag, and the Great Caucasus (the northern margin is a complex monocline with as much as 8 km [5 mi] of elevation) are all large, thick-skinned uplifts. Compared to them, the inversion in the southern Kirthar Range is small. Although subvertical strike or oblique slip folds may occur internally in these ranges, there is hardly any evidence of strike slip on the structures in the frontal zones. Most inversions are, of course, oblique. We can hardly expect the amazing coincidence of extension and contraction in exactly the same direction. However, the strain is normally partitioned. In the southern Kirthar fold belt, the strike-slip component that was necessitated by the global transpressive kinematic state is localized on the Chaman fault zone; the Kirthar foreland is closer to pure compressive inversion.

Because of the relatively poor quality of seismic data over the hinterland (Figure 7; subsequently improved through reprocessing), much of the early interpretation was based on outcrop data and therefore reliant on the projection of surface structures at depth using conventional geometric techniques. Modeling the subsurface this way introduces uncertainty about the involvement (or noninvolvement) of reservoir and source rock during deformation and the size of trap closure. Conversely, seismic data in the foreland are of better quality and therefore provide valuable information about the structural style of the fold belt. The Zamzama structure in Figure 5 shows a thin-skinned thrust climbing upsection from a bedding-parallel detachment in the K10–K70 sequences (the source rock interval). Zamzama has only been modified and not created by this thrust. The footwall of the thin-skinned thrust is elevated by a fold structure that extends beyond it. The geometry suggests association with a steeper fault. The thin-skinned thrust apparently loses displacement upward and does not break the surface, implying movement synchronous with deposition of the T40–T70 sequences. The excellent imaging of the Zamzama structure has led to its use as an analog during the integrated interpretation of the hinterland. The interpretation of the Khude Range (Figure 6) is derived directly from the Zamzama analogy.

Timing of Inversion

Constraints on the timing of inversion are needed to determine the burial history. Early inversion and uplift (pre-Pliocene–Pleistocene) could reduce maximum burial, which would impact the source rock maturity. Evidence for early inversion on some structures in the southern KFB, listed below, has been obtained using stratigraphic data primarily from outcrop and also from seismic and maturity indicators.

1) From north to south, the depocenter of the T20–T30 slope deposits shifts westward (Figure 6). Prograding over these shales is the T30 Kirthar Formation and its distal equivalent, the Khude D Member, the former a proximal platform limestone and the latter a distal platform limestone. The margin of the carbonate platform shifts dramatically westward, and thus basinward, on line 6 (Figure 1). To enable deposition of the carbonate platform in the photic zone, a substrate on which the limestone can prograde is needed. Basement inversion in the south was invoked, uplifting the Chapar and Andhar sequences in the hinterland. Inversion in the Eocene, possibly initiated in the Paleocene during ophiolite obduction, may also account for differing geometries of the slope wedge, which is thinner in the south.
2) Multiple erosion surfaces associated with the Laki fault in the east Laki Range (Figure 1) provide evidence for Paleocene inversion in the south of the study area that continued throughout the Tertiary (Smewing et al., 2002b). Multiphase inversion resulted in four erosion surfaces in the hanging wall, eventually resulting in exposure of the K90 sequence. Further evidence in the Laki Range for pre-Pleistocene uplift is found in the Maastrichtian Pab sandstone (K90), which contains reworked Campanian Parh Limestone clasts (K80), indicating inversion as early as the Maastrichtian.
3) The T40–T70 sequences record inversion from the Oligocene to Pleistocene. Outcrop data from the mountain front show all formations of the T40–T70 sequence thinning onto the nose of the Bhit anticline: In the north, the total thickness is about 4.75 km (3 mi), and in the south, it is about 2.1 km (1.3 mi) (Figure 3). The stratigraphic thinning implies uplift prior to and synchronous with the deposition of T40–T70 (from Eocene to Oligocene onward). The projection of these overburden sequences into the hinterland is problematic, because only the base of the T40 sequence is recorded; therefore, only thickness minimums are available. The identification of early inversion and uplift in the south of the study area will reduce accommodation space for T40–T70 deposits, so a "layer cake" projection of stratigraphy from the foreland would not be suitable. Maturity data constrain T40–T70 thicknesses in the hinterland to about 2 km (1.2 mi) on line 38 and about 4 km (2.5 mi) on line 6, further evidence of early uplift in the south of the study area.

The evidence outlined above indicates inversion from the Paleocene onward, with a phase of inversion during the Maastrichtian in the south of the study area along an east–west trend from Chapar to the Laki Range. Differences in the orientation of the rift architecture across the accommodation zone may control the timing of uplift in the south of the study area, with the strike of the extensional faults, relative to Sigma 1, conducive to early inversion. During the Oligocene to Pliocene, the structures in the north of the study area were inverted, as recorded by a northward shift of the depocenter for the T40–T70 sequences.

Outcrop data do not indicate rapid uplift prior to and during T30 deposition. These data indicate subaerial exposure in parts of the fold belt, and inversion has continued throughout the Eocene to Miocene. However, the main mountain-building phase is interpreted as late Pliocene–Pleistocene, when the Dada conglomerates were deposited. The base of the Dada represents a major structural unconformity and a facies change from marine to nonmarine deposition. The Dada Formation possesses reworked deposits of the T40–T70 sequences and constrains uplift to approximately 1.6 m.y. ago (Hunting Survey Corporation Ltd., 1960). The timing of uplift remains uncertain, and more work is needed to properly constrain the mountain-building phase in the hinterland.

Line 38: Thrustpack Modeling

The inverted basement-involved extensional faults are poorly imaged by seismic data and are therefore difficult to interpret. The faults are therefore interpolated at the inflection point of folds, with the orientation determined by the fold vergence (Figure 8). Although Thrustpack was designed to simulate deformation on low-angle fault planes, introduction of steep faults (>50°) is possible, provided that they are not closely spaced (see the Appendix for details regarding the modeling of steep faults using the Foldis algorithm). As a result, many of the antithetic faults present in the initial interpretation (Figure 6) could not be simulated. Symmetric folds could still be generated by "popping up" minigraben-producing anticline-syncline pairs in the cover sequence, K10 and younger deposits (Figure 8).

The introduction of a basal detachment, a Thrustpack requirement, together with predicted steep, basement-involved faults, enabled the generation of the large uplift with minimal thrusting (total horizontal shortening is approximately 18 km [11 mi]). Figure 6 indicates as much as 4 km (2.5 mi) of uplift with the

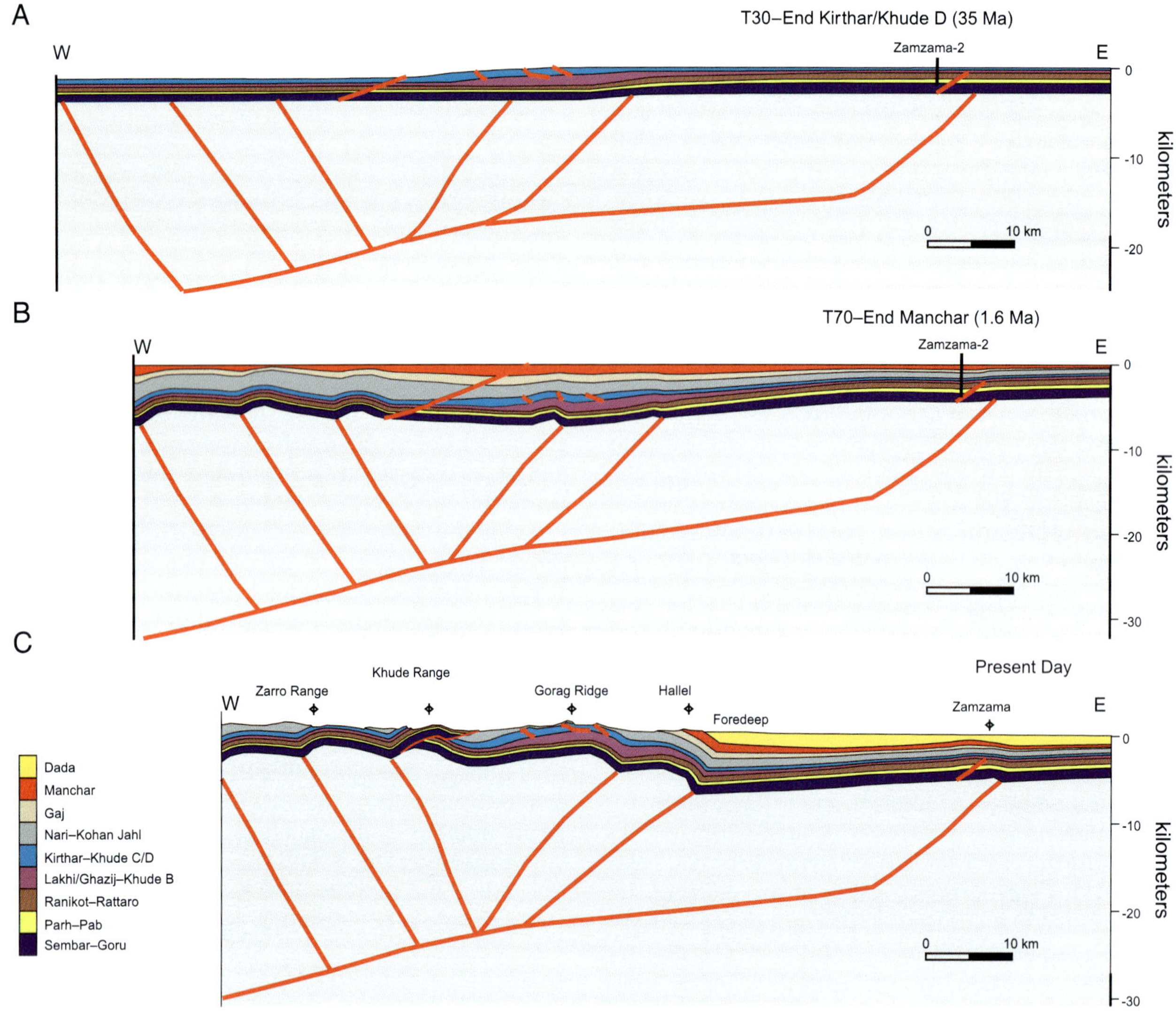

FIGURE 8. Thrustpack output for line 38 showing the Kirthar Range during (A) Eocene prior to flexure, (B) postflexure and prior to the major phase of mountain building, and (C) present day.

Kirthar Range maintaining a common horizontal level behind the frontal monocline.

The back-thrust zone, west of the frontal monocline, was interpreted as passive accommodation because of the elevation of the Gorag Ridge on the inverted basement faults. The initial interpretation of 5–10 km (3–6 mi) of displacement on the back thrusts could not be reconstructed in the model, an issue discussed below.

Line 6: Thrustpack Modeling

The southern line, although of relatively limited, lateral extent by comparison with line 38, records significant changes in structure and stratigraphy (Figure 9). In the basement, we interpret a change in the polarity of the basement faults along strike. On line 38, the Khude fault dips east. Moving along strike to line 6, the Andhar fault dips west. The configuration of basement faults was a product of trial and error in the modeling process. The polarity, cutoff, and number of basement faults finally adopted served to recreate the geometries observed on seismic and in map view. The change in polarity of the basement faults from line 38 to line 6 is attributed to inherited rift architecture, with an accommodation zone separating the Khude fault from the Andhar fault. The change in basement fault configuration from north to south is also reflected by the change in position of the mountain front, which steps eastward in the region of line 6. Surface anticlines break eastward in an en echelon fashion, possibly reflecting a similar fabric in the inherited rift architecture (Figure 1).

The imbricate back-thrust system on line 38 is not seen on line 6 (compare Figure 9 with Figure 6). In the south, a thin-skinned system displaying eastward vergence was initially interpreted. These out-of-the-syncline thrusts, interpreted as expulsion from a tight syncline, were difficult to replicate using Thrustpack (see below).

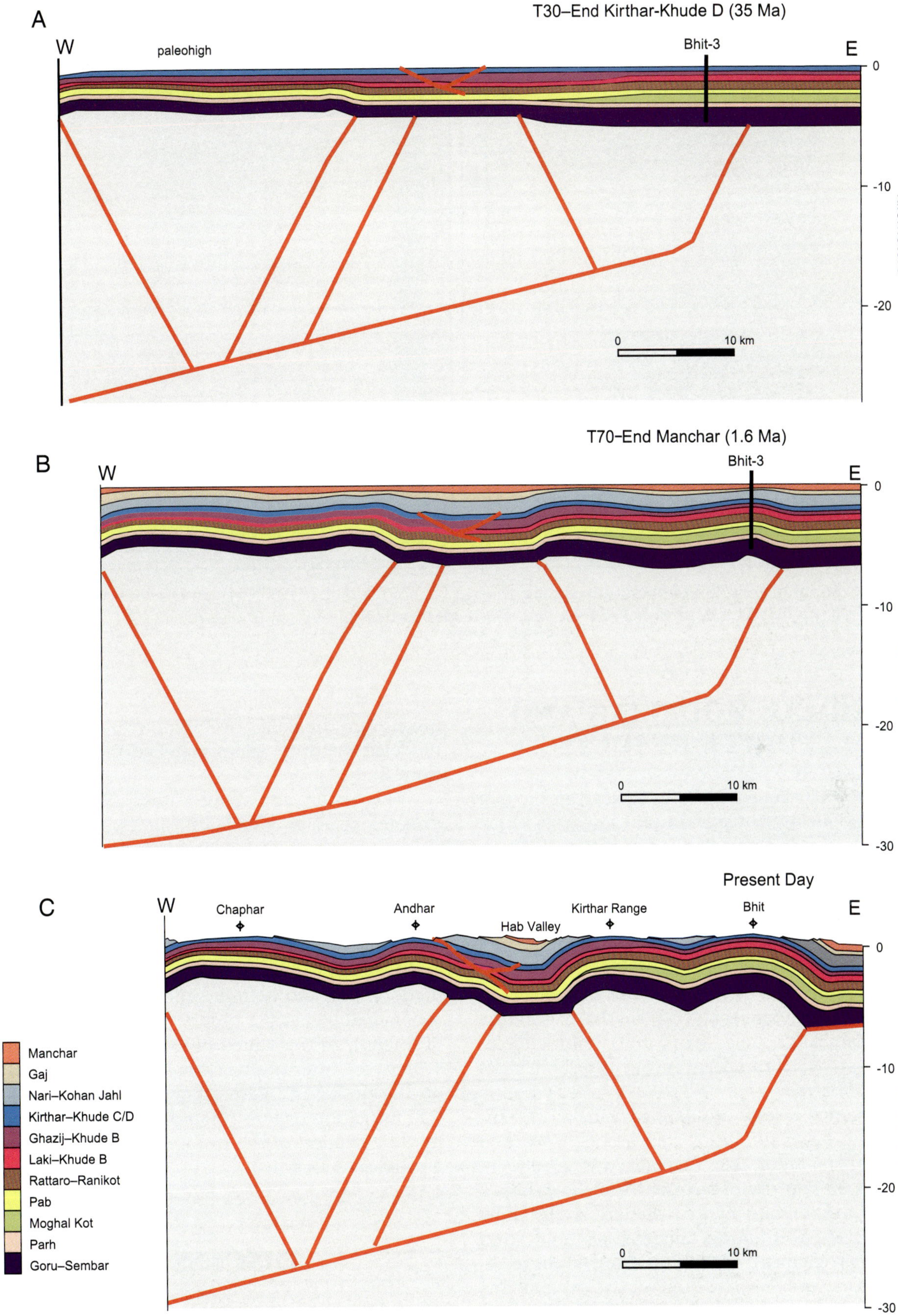

FIGURE 9. Thrustpack output for line 6 showing the Kirthar Range during (A) Eocene prior to flexure and during early inversion, (B) postflexure and prior to the major phase of mountain building, and (C) present day.

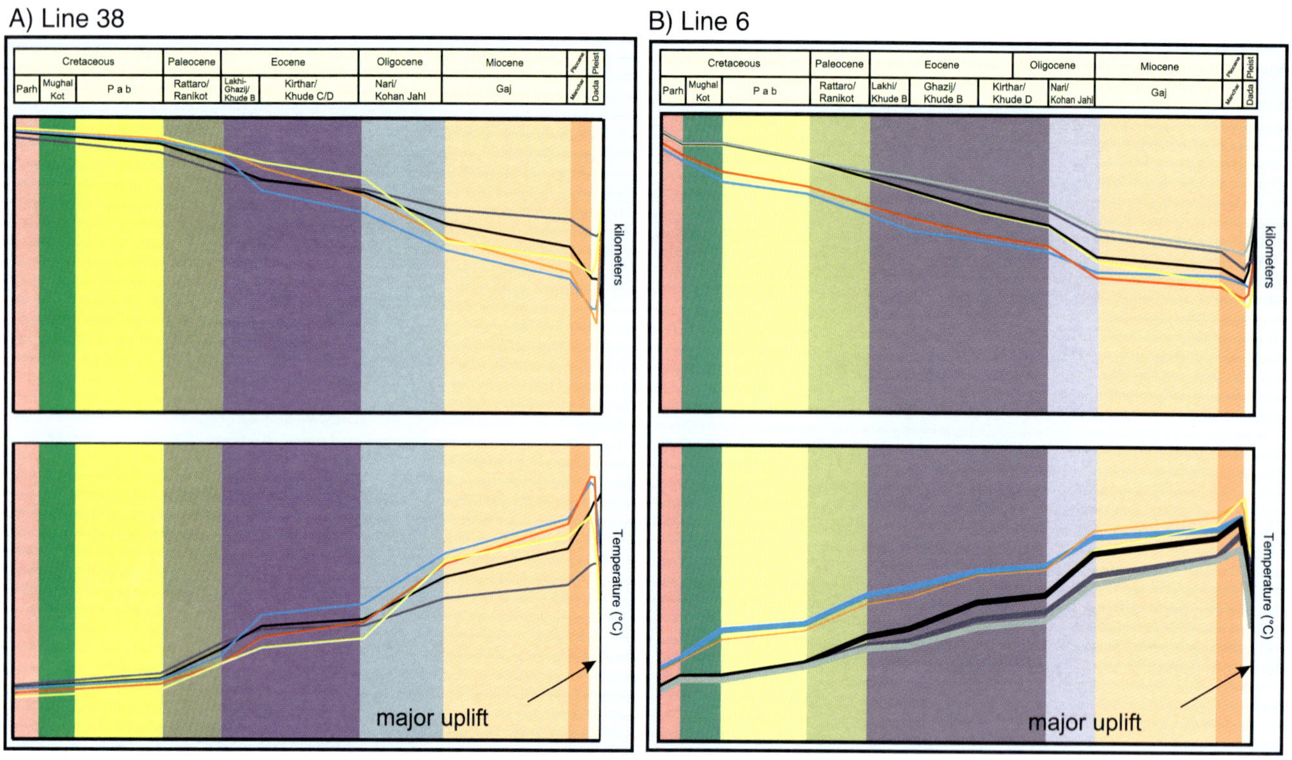

FIGURE 10. Burial history and temperature data for the Sembar source rock (K10). Each curve relates to a history point inserted in the modeled cross sections (not shown). (A) relates to line 38 and (B) to line 6.

THERMAL MODELING AND MATURITY RESULTS

The K10–K60 sequences (Goru–Sembar) are thought to be the producing source rock in the foredeep, which is typically assigned a type III or a mixed type II/III kerogen. No direct measurement of Goru/Sembar source rock quality in the study area is available. To the west in the Mor Range and to the north in Gaj River, the measured potential of the source rock is less than 2% total organic carbon (TOC) (Figure 1).

The quantity and quality of source rock in the license area are therefore equivocal. For this reason, the Goru and Sembar (modeled as one unit) were ascribed a TOC value of 2%. The models were run using both types II and III kerogen.

The thermal history was defined using a constant (steady-state) heat flow. Steady-state heat flow was used because sedimentation rates were not considered rapid enough to warrant use of transient heat flow. Bhit-3 served as the calibration point for present-day heat flow using corrected bottomhole temperatures (indicating a geothermal gradient of 27–30°C/km), vitrinite reflectance data, and FIA data (from quartz overgrowths in the Pab sandstone). A value of 50 mW/m^2 was obtained using an average surface temperature of 20°C. Using present-day heat flow and overburden estimates taken from the flanks of Bhit anticline (Figure 3), the modeled maturity values matched the data obtained from Bhit-3.

The western end of line 6 was calibrated using FIA paleotemperatures from Eocene limestone (used as temperature minimums) and vitrinite reflectance data from outcrop. The calibrated heat-flow value from line 6 was used for line 38. The eroded overburden was constrained using FIA maximum burial temperatures taken from quartz overgrowths in the Pab sandstone.

A record of burial history, temperature, transformation ratio, and vitrinite reflectance (among others) is obtained by placing history points in the formation of interest. In this study, history points placed in the source rock interval, the Goru–Sembar, provided data on the timing of hydrocarbon generation relative to trap formation (Figure 10; for reasons of confidentiality, the location of the history points are not shown on the cross sections).

DISCUSSION

Comparison of line 38 from Thrustpack (Figure 8C) with that from the initial interpretation (Figure 6A) reveals a difference in the amount of displacement on the thin-skinned imbricate thrusts of the Gorag Ridge. The initial interpretation estimated 5–10 km (3–6 mi) of shortening along these back-thrust faults as a response to shortening and uplift in the basement. The displacement could not be recreated using the Foldis algorithm, which estimated about 2 km (1.2 mi) of total displacement on the Gorag Ridge faults. To balance the section, excess stratigraphic bed length in the cover was accommodated by the Khude Range fault. According to outcrop data, the modeled displacement on this fault is not reasonable. Shortening in the cover therefore needs to be redistributed. The imbricate back thrusts may have a deeper detachment in the K10 sequence. The seismic data are equivocal and do not refute this interpretation, but outcrop data indicate a Ghazij detachment. Alternatively, the Zarro Range and Pab Range modeled as folds may have a thrust component, thus accommodating additional shortening.

The back-thrust system of the Gorag Ridge is one of the few areas in the southern KFB where low-angle reverse faults are observed, but there is insufficient evidence of major shortening across the ridge, as proposed in the passive roof duplex model of Banks and Warburton (1986). Outcrop data along the western flank of the ridge indicate that the Eocene shelf margin can be reconstructed through the various thrust sheets without invoking shortening in excess of about 10 km (6 mi). The low-angle back thrusts of the Gorag Ridge pass southward into high-angle reverse faults that are more typical of structures elsewhere in the southern KFB.

Warburton (2000) revised the interpretation of the Gorag Ridge area as a passive roof duplex system. The chapter concludes that the back thrusts have a small displacement, in the order of kilometers and not tens of kilometers, and shortening is dominantly caused by folding instead of thrust faulting. This interpretation is consistent with outcrop data and the results of the Foldis kinematic modeling, which indicate minimal shortening along these faults (~2 km [~1.2 mi] and as much as 10 km [6 mi]).

The tight folds in the Hab Valley on line 6 proved difficult to replicate using the current discretization of structural blocks in Thrustpack (compare Figure 6B with Figure 9C). This may be caused by an underestimation of sediment thicknesses or modeled shortening, and uplift was not sufficient to produce out-of-the-syncline thrusts. Alternatively, the structural interpretation may be incorrect; the steep limb dips and short wavelength of these structures may be the product of oblique slip. Further modeling runs are needed to determine which of these factors control the generation of these structures. It must be noted that tight folds such as these (short wavelength and high amplitude) are difficult to replicate without a more refined geometric description of the basin architecture.

Of fundamental difference between lines 38 and 6 is the estimation of eroded overburden (T40–T70 sequences). The maturity data indicate an overburden thickness of no greater than 2 km (1.2 mi) on line 6 and about 4 km (2.4 mi) on line 38, resulting in greater burial depths in the north of the study area (Figure 10). These values, obtained using Thrustpack, contrast the initial interpretation of overburden thickness of as much as 6 km (4 mi) prior to uplift, which would have resulted in overmature source rock throughout the area. The maturity data, in combination with the stratigraphic and structural data outlined previously, indicate early uplift in the south of the study area. The early inversion of structures around line 6, and possibly further south, may therefore have trapped early formed hydrocarbons.

CONCLUSIONS

- Through integration of all available surface and subsurface data, Thrustpack modeling has helped develop a structural model and quantify source rock risk in the study area. Two east–west transects selected for modeling revealed two distinct structural provinces in the study area: a southern province, where inversion was initiated during the Paleocene, possibly because of Bela ophiolite emplacement, and a northern province, where inversion occurred later, from Oligocene onward.
- The fold belt was generated by inversion on steep basement-involved extensional faults, although more evidence is needed to corroborate this interpretation. Symmetric folds were generated by inversion of synthetic and antithetic extensional fault pairs. Small displacement thrusts occasionally displace the frontal limbs, but folds are the dominant structure.
- Structures with steep limb dips and short wavelength have shallow detachments either in the Eocene (T20–T30) or Lower Cretaceous (K10) mudstones. Thrustpack modeling indicates that thrusts detaching in the T30 have small displacements (~2 km; ~1.2 mi) when using a model of fold generation, the result of basement inversion. Thrusts detaching in the K10 sequence are dominant and involve the reservoir during deformation.
- Paleocene inversion in the southern province reduced the accommodation space for the deposition of the T40–T70 sequences, which reached thicknesses of about 2 km (1.2 mi). Inversion in the north started in the Oligocene, and the T40–T70 deposits reached thicknesses of about 4 km (2.4 mi).
- Structures generated in the south of the study area during Paleocene inversion may have trapped early formed hydrocarbons.

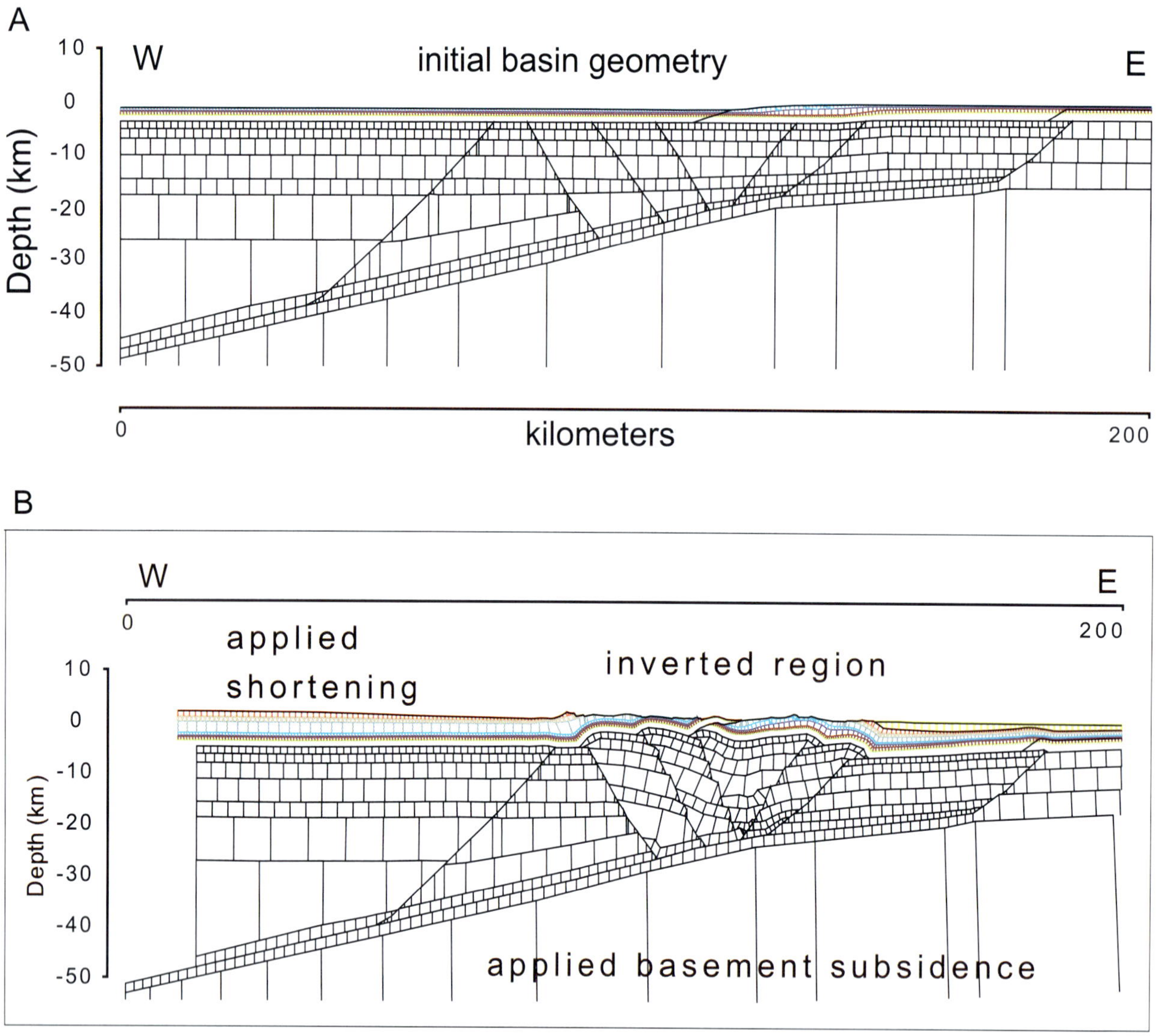

FIGURE 11. A section from Thrustpack showing the Foldis mesh for line 38; see text for discussion. (A) Before deformation and (B) present day.

- Thrustpack allows the user to interrelate and verify interpretations of stratigraphy, paleogeography, structure, and structure timing and to predict hydrocarbon generation. The result is a clearer understanding of the basin-scale sedimentary architecture and structure.

ACKNOWLEDGMENTS

The authors thank BHP Pakistan for permission to publish their data. The contributions of Mujahid Ali, Nazir ul Haq, Simon Beswetherick, Jonathan Craig, and Alastair Sharp are gratefully acknowledged. The management of Eni-LASMO Pakistan, Ltd., are thanked for permission to publish this chapter. The chapter benefited from reviews by Jean-Louis Mugnier and John Bell.

APPENDIX

Thrustpack

Thrustpack is a forward structural modeling and thermal modeling program based on geometric constraints and a conductive thermal regime. The program uses the Foldis algorithm (Divies, 1997) to simulate folding by flexural slip and mechanical compaction caused by burial increase (Sassi et al., 1998; Sassi and Rudkiewicz, 1999, 2000).

The Foldis algorithm simulates the kinematics of folding. Key to the modeling is the discretization of a geological cross section into several structural blocks. Each block is subdivided into several sedimentary layers that are free to slide at their base, and both their length and thickness remain constant during deformation. In this modeling approach, the strata are subdivided into juxtaposed rectangular cells, for which the area is preserved (Figure 11). When a given layer moves along a fault trajectory, the cells are translated over the flat portions of the fault segments and subjected to deformation only around the region of fault-bend points (anticlines or synclines). The geometry of the cells at a fold axis is slightly modified, i.e., the constraints on thickness is relaxed, but this is done in a way that preserves the overall length of the entire layer. The principle of mass preservation using constant layer length and thickness is the basis of the fault-bend fold theory of Suppe (1983), but in the Foldis algorithm, no kink axes are constructed. The geometric reconstruction using Foldis will converge with that of Suppe's kink model when a larger number of layers are used. The use of constant bed length and thickness has limitations: When the fault-strata angle is greater than 30°, the Foldis model predicts back-shearing. Therefore, when applying the model to inverted, steeply dipping normal faults, it is necessary to discretize the graben into numerous horizontal beds. The back-shear effect can then be counteracted using differential offsets between strata to generate the geometry of the anticline and syncline transmitted into the overlying overburden. Finally, in Thrustpack, the mechanical compaction caused by porosity changes with depth can be superimposed on the kinematic deformation, and large-scale basin subsidence or uplift may be applied using vertical shear at the base of the substratum.

Source rock modeling considers TOC and burial history. The degradation of organic matter is simulated using a compositional kinetic scheme, and the hydrocarbons generated and expelled are replicated using temperature histories obtained from thermal modeling (Sassi and Rudkiewicz, 1999; Husson and Moretti, 2002).

REFERENCES CITED

Banks, C. J., and J. Warburton, 1986, "Passive-roof" duplex geometry in the frontal structures of the Kirthar and Sulaiman mountain belts, Pakistan: Journal of Structural Geology, v. 8, nos. 3–4, p. 229–237.

Colletta, B., F. Hebrard, J. Letouzey, P. Werner, and J. L. Rudkiewicz, 1990, Tectonic style and crustal structure of the Eastern Cordillera (Colombia) from a balanced cross-section, *in* J. Letouzey, ed., Petroleum and tectonics in mobile belts: Proceedings of the Institut Français du Pétrole Exploration and Production Research Conference: Paris, Editions Technip, p. 81–100.

Cooper, M. A. et al., 1995, Basin development and tectonic history of the Llanos Basin, Eastern Cordillera, *in* A. J. Tankard, R. Suarez-Soruco, and H. J. Welsink, eds., Petroleum basins of South America: AAPG Memoir 62, p. 659–665.

Divies, R., 1997, FOLDIS: Un modèle cinématique de bassins sédimentaires par éléments discrets associant plis, failles, érosion/sédimentation et compaction: Doctoral thesis, University of Grenoble I, Grenoble, France, 220 p.

Eschard, R., E. Albouy, F. Gaumet, and A. Ayub, 2004, Comparing the depositional architecture of basin floor fans and slope fans in the Pab Sandstone, Maastrichtian, Pakistan, *in* S. A. Lomas and P. Joseph, eds., Continued turbidite systems: Geological Society (London) Special Publication 222, p. 159–185.

Gnos, E., M. Khan, K. Mahmood, A. S. Khan, N. A. Shafique, and I. M. Villa, 1998, Bela oceanic lithosphere assemblage and its relation to the Reunion hotspot: Terra Nova, v. 10, p. 90–95.

Hedley, R., J. Warburton, and J. Smewing, 2001, Sequence stratigraphy and tectonics in the Kirthar fold belt, Pakistan, *in* Proceedings of the Annual Technical Conference of the SPE—Pakistan Association of Petroleum Geologists, Islamabad, Pakistan, November 7–8, 2001, p. 61–72.

Hunting Survey Corporation Ltd., 1960, Reconnaissance geology of part of west Pakistan: Ottawa, Canada, Miracle Press, 550 p., 331 figures, 9 plates, 4 maps, 29 geological maps, scale 1:25,000.

Husson, L., and I. Moretti, 2002, Thermal regime of fold and thrust belts. An application to the Bolivian sub-Andean zone: Tectonophysics, v. 345, p. 253–280.

Roure, F., and W. Sassi, 1995, Kinematics of deformation and petroleum system appraisal in Neogene foreland fold-and-thrust belts: Petroleum Geoscience, v. 1, p. 253–269.

Sassi, W., and J.-L. Rudkiewicz, 1999, Thrustpack version 6.2: 2-D integrated maturity studies in thrust areas: Institut Français du Pétrole Report no. 45372, 79 p.

Sassi, W., and J.-L. Rudkiewicz, 2000, Computer modelling of petroleum systems along regional cross-sections in foreland and fold and thrust belts (abs.): European Association of Geoscientists and Engineers Annual Meeting, Malta, *in* Proceedings of the Conference on Geology and Petroleum Geology of the Mediterranean and Circum-Mediterranean Basins (European Association of Geophysicists and Engineers Malta, October 1–4, 2000) Extended Abstracts Paper C-27, 4 p.

Sassi, W., J.-L. Rudkiewicz, and R. Divies, 1998, New methods for integrated modelling of deformation and petroleum generation in fold and thrust belts (abs.): AAPG Annual Meeting Program, v. 7 May 17–18.

Smewing, J. D., J. Warburton, T. Daley, P. Copestake, and N. ul-Haq, 2002a, Sequence stratigraphy of the southern Kirthar fold belt and middle Indus Basin, Pakistan, *in* P. D. Clift, D. Kroon, C. Gaedicke, and J. Craig, eds., 2002, The tectonic and climatic evolution of the Arabian Sea region: Geological Society (London) Special Publication 195, p. 273–299.

Smewing, J. D., J. Warburton, A. Cernuschi, and N. ul-Haq, 2002b, Structural inheritance in the southern Kirthar fold belt: Society of Petroleum Engineers-Pakistan Association of Petroleum Geologists, p. 26–34.

Suppe, J., 1983, Geometry and kinematics of fault-bend folding: American Journal of Science, v. 283, p. 684–721.

Underhill, J. R., and S. Paterson, 1998, Genesis of tectonic inversion structures: Seismic evidence for the development of key structures along the Purbeck— Isle of Wight disturbance: Journal of the Geological Society (London), v. 155, p. 975–992.

Warburton, J., 2000, Passive roof duplex model and future fold belt exploration in Pakistan, *in* Proceedings of the Annual Technical Conference of the SPE—Pakistan Association of Petroleum Geologists, Islamabad, Pakistan, November 9–10, 2000, p. 67–80.

Deformation and Regional Fluid Flow

7

Ferket, H., R. Swennen, S. Ortuño-Arzate, M. C. Cacas, and F. Roure, 2004, Vein formation in Cretaceous carbonates in the Laramide foreland fold and thrust belt of eastern Mexico, *in* R. Swennen, F. Roure, and J. W. Granath, eds., Deformation, fluid flow, and reservoir appraisal in foreland fold and thrust belts: AAPG Hedberg Series, no. 1, p. 133–156.

Vein Formation in Cretaceous Carbonates in the Laramide Foreland Fold and Thrust Belt of Eastern Mexico

H. Ferket
Fysico-chemische Geologie, K. U. Leuven, Heverlee, Belgium

R. Swennen
Fysico-chemische Geologie, K. U. Leuven, Heverlee, Belgium

S. Ortuño-Arzate
Instituto Mexicano del Petróleo, México D. F., Mexico

M. C. Cacas
Institut Français du Pétrole, Rueil-Malmaison, France

F. Roure
Institut Français du Pétrole, Rueil-Malmaison, France

ABSTRACT

A structural analysis and petrographic investigation has been performed on veins in Cretaceous carbonates in the Cordoba Platform in eastern Mexico, which is part of the Laramide foreland fold and thrust belt (FFTB). This chapter focuses on the different episodes of vein formation, vein morphology, and possible mechanisms of vein formation. Vein (fracture) formation is interpreted in relation to the kinematic evolution of the FFTB. Evidence for the development of hydrofractures during this evolution is given.

This study documents veins (fractures) related to Laramide FFTB development in the Cordoba Platform. These veins (fractures) are related to the kinematic evolution of the area and the inferred paleostress conditions. The kinematic evolution can be split up into three major stages: a precompression phase with platform development; a Laramide compressional stage, during which the FFTB developed; and finally, a late Basin and Range-related extension phase.

Compound veins and densely spaced microveins record multiple fracturing events in a cyclic stress field during burial, most probably caused by changes in fluid pressure. They are interpreted in relation with early foreland flexuring. With rising compressional stress, less well-oriented veins and breccia veins develop because of a lowered differential stress in the prefolding stage. Progressive layer-parallel shortening (LPS) leads to a

DOI:10.1306/1025689H12965

caterpillar-type scenario of fluid migration toward the foreland, eventually causing hydrofracturing, succeeded by pressure solution and development of vertical stylolitic planes. These LPS stylolites have the potential to be reopened during subsequent folding of the strata. In addition, older LPS-parallel planes and extrados fractures may open in anticlinal hinges. Shear-associated, shallow-dipping veins develop after LPS development, possibly because of bedding-parallel shear and/or thrust migration. Other post-LPS veins are steeply dipping and commonly reuse older vein orientations. Dark, banded veins, which are filled with a silt-sized and clay-sized material and lack significant cementation, are interpreted to reflect fracture planes along which recrystallization of matrix occurred. Many post-LPS dissolution-enlarged veins and breccias relate to telogenetic karstification. Post-LPS multiple brecciation just above a major thrust plane in the buried tectonic front area is interpreted to reflect the damage zone of that fault.

INTRODUCTION

Veins in foreland fold and thrust belts (FFTBs) may record (part of) the deformation and fluid-flow history of the area. However, to deduce the deformation–fluid-flow history from the structural, petrographic, and geochemical studies of these veins, a good understanding of vein formation mechanisms and conditions is needed. In the past decade, the widely used reference to some classical models has been subjected to discussion, and alternative mechanisms were proposed. Several studies focused on the parameters that control typical vein morphologies and modeled the inferred processes (e.g., Bons and Jesell, 1997; Bons, 2001a; Hilgers et al., 2001; Hilgers and Urai, 2002). Several of these authors warned of rapid conclusions on vein formation mechanisms based only on typical vein morphologies. The discussion concentrated mainly on the applicability of the "crack-seal" mechanism as defined by Ramsay (1980), the distinction between "vein formation without fracturing" and "cemented fractures," as well as the driving mechanism for vein opening. Hydrofracturing is commonly invoked as the responsible mechanism for vein formation, but it is not easy to prove directly. Many questions about vein formation still remain unanswered. Therefore, more case studies that combine both a structural and a diagenetic approach are needed. The latter help to deduce ancient deformation and fluid flow and allow to document the occurrence of typical vein morphologies and geochemical characteristics at different tectonic stages.

This chapter focuses on several episodes of vein and stylolite formation in Cretaceous carbonates in the Cordoba Platform in eastern Mexico and describes vein morphology, orientation, and relative timing of vein formation or reactivation and stylolite development. These microstructures develop at different deformation stages and are placed in the kinematic evolution and related to the inferred paleostress conditions. The aim of this study is to gain some insights into the deformation–fluid-flow history during FFTB development. Hereby, the discussion on fracture vs. vein, on the possible fluid-flow mechanisms, and on the evidence of hydrofracturing, is essential to estimate the importance of several vein formation episodes for exploration-directed research, because the fracturing and fluid-flow history also affect reservoir characteristics and hydrocarbon migration.

The area has previously been studied to deduce the diagenetic history and geochemical characteristics of paleofluids (Ferket et al., 2000, 2003a, b; Ortuño-Arzate et al., 2003). However, to really interpret the processes during FFTB development, a combined structural and diagenetic approach is essential. Moreover, a better understanding of how petrographic observations can be used to interpret vein formation processes is needed to avoid overgeneralizations of certain mechanisms. These negligences comprise the use of "crack-seal vein" as a descriptive term and the interpretation of all "breccia veins" as hydrofractures based only on their outline. This study also focuses on breccia veins, which are commonly reported to occur in the compressional stage of FFTB areas (e.g., Swennen et al., 2003). The study area forms an excellent case for this type of veins, because they developed here in different tectonic stages.

GENERAL CONCEPTS ON VEIN FORMATION

For a concise but thorough discussion on the formation, morphology, and associated terminology of veins, the reader is referred to Bons (2000). Here, in three introductory sections, we focus on some issues that are relevant for this study.

Fluid Transport Mechanisms

Veins can form as cemented fractures (e.g., Knipe and McCaig, 1994; Knoop et al., 2002), or they can form without fracturing (Toriumi and Hara, 1995; Bons and

Jesell, 1997; Wiltschko and Morse, 2001). These different modes of vein formation may involve different fluid transport mechanisms.

Vein formation without fracturing is associated with diffusional mass transfer (Bons and Jesell, 1997), commonly in relation with pressure solution (Wiltschko and Morse, 2001). These veins are then intimately associated with stylolites (Knipe and White, 1979; Smith, 2000). Because fluid movement is limited, these veins can only account for small-scale transport. The infilling cement phases are typically host rock buffered. Because of dissolution in the host rock and precipitation at another favorable site, a net decrease in porosity takes place. These veins consequently have a negative effect on reservoir development. Diffusional transport may, however, also occur along fractures. The pressure difference between small pores and the open fracture space may then establish a thermodynamic drive for dissolution and precipitation (Bjørlykke, 1994).

Larger distance transport can occur along open fractures by Darcian or advective flow. The fluid can, to a variable degree, interact with the host rock, depending on the flow velocity, which is in turn related to fracture opening and on the permeability and lithology of the host rock. Numerous examples of veins as "ancient fluid migration pathways" have been described (e.g., Callan and Spooner, 1998; Travé et al., 1998; Knoop et al., 2002).

Other transport mechanisms comprise the concept of "mobile hydrofractures" (Bons, 2000, 2001b), which implies that fracture and fluid move together at high speed but in discrete events. This mechanism accounts for large volumes of fluid transported over large distances without substantial fluid-rock interaction. Another fast but localized transport mechanism is turbulent flow that occurs in larger open spaces, such as wide fractures and karst channels (White, 2002). The fast transport prevents the fluid from chemical equilibration with the host rock. Transport along narrow microstructures (e.g., stylolites) can be driven by capillary forces (Bethke et al., 1991). Finally, the formation of a vein can involve a combination of fluid transport mechanisms. For example, both diffusive and advective flow can contribute to vein formation, or the change in opening rate of a fracture through time may change the operating flow systems.

Processes of Vein Formation and Related Morphology

The interpretation of ancient fluid transport mechanisms and vein formation processes is based on morphological and geochemical characteristics. Unfortunately, vein morphology does not always allow an unambiguous interpretation of the responsible mechanisms. Table 1 gives an overview of some morphological observations and their interpretation toward vein formation and fluid transport mechanisms.

The crack-seal mechanism, which was first described by Ramsay (1980), has commonly been referred to as the responsible mechanism for the formation of fibrous and elongate-blocky veins (Bons, 2000). Crack-seal involves the opening of a crack as an extension fracture and subsequent filling of the open space by crystalline material. The process may be repeated several times resulting in compound veins. The generally used diagnostic criterion for crack-seal is the occurrence of inclusion bands parallel to the vein wall. However, recent studies have shown that the presence of such inclusion bands does not necessarily relate to a crack-seal process (Hilgers et al., 2001; Wiltschko and Morse, 2001; Hilgers and Urai, 2002).

Alternatively, Caputo and Hancock (1999) proposed a crack-jump mechanism to explain dense sets of hairline microveins. When, after sealing of the fracture, the elastic parameters of the rock are recovered, stresses can rebuild to a critical level. A new fracture can be formed either in the rock mass (crack-jump) or at the vein wall-matrix contact (crack-seal), depending on which is mechanically weakest.

Driving Mechanisms for Vein Formation

Veins can form from precipitation of a cement phase in an open fracture. Fractures can form as hydrofractures in overpressured conditions, or they can form under normal fluid pressure. Extension fractures form under low differential stress ($\Delta\sigma$) and low confining pressure (p_c) if not driven by an increased fluid pressure. They consequently form in extensional regimes in the shallow subsurface, for example, in karst areas. Shear fractures form under higher $\Delta\sigma$ and p_c. Tension gashes may be associated with shear fracture development. Petit et al. (1999) describe a crack-seal, slip mechanism to explain sheared calcite veins along a growing shear fault, although this process requires a suprahydrostatic fluid-pressure buildup. Far-field compressive stress can also induce local tensile stresses around flaws, leading to the development of extension fractures in a compressional regime (Lorenz et al., 1991; Twiss and Moores, 1992). The effect of anisotropic far-field stresses may be significant for the interpretation of fractures developing in the foreland of fold and thrust belt areas (Lorenz et al., 1991).

Hydrofracturing is perhaps the most popular mechanism to explain fracture formation and is commonly reported to occur in the compressional stage of FFTB development (Swennen et al., 2003). Natural hydrofracturing is the brittle failure of a material under elevated fluid pressure (Hubbert and Willis, 1957; Secor, 1965). Several studies regarding the mechanics of hydrofracture development have been published during the past decades and especially recently (e.g., Ramsay, 1980; Sibson, 1994; Miller, 1995; Cosgrove, 1997; Simpson, 2000; Bons, 2001b; Gudmundsson and Brenner, 2001). These studies demonstrate a wide variety of conditions under

	Observation	Interpretation	References
Macroscopic	isolated veins in 3-D	fluid from matrix, not via fracture	Hilgers and Urai, 2002
	fracture-shaped (continuous) veins	shape relates to fracturing, but not diagnostic for vein formation process	Bons, 2000
	close relation with stylolite seams (e.g., hyperactivation of stylolites)	diffusion	Smith, 2000
	cemented breccia	advective flow, (hydro-) fracturing, near hydrostactic $\Delta\sigma$	Ramsay and Huber, 1987; Lorenz et al., 1991
Microscopic	blocky cement	growth in open space; fast fracture opening (or recrystallization of preexisting cement)	Bons, 2000; 2001a Hilgers and Urai, 2002
	drusy fabric	growth in open space	
	elongate-blocky or stretched cement (CPO)	relatively slow opening of fracture, no nucleation, no large supersaturation (typical for crack-seal mechanism)	
	fibrous cement (symmetric antitaxial vein)	slow opening of vein, permanent contact between walls and cement, typical for diffusion	
	growth competition	relatively large opening	
	continuous inclusion bands parallel to vein wall and growth competition	crack-seal mechanism	
	(parallel) inclusion bands	not necessarily crack-seal (chemical oscillations may cause banding)	
Geochemical	host rock buffering	diffusion and/or advection with host-rocklike fluid	Bradbury and Woodwell, 1987; Hilgers and Urai, 2002
	different geochemistry between vein and host rock	fluid flow, except for veins with diffusion and chemical oscillation	

Table 1. Table summarizing some petrographic observations that can be used to help interpret vein and fracture formation conditions.

which hydrofractures can develop. They form in shallow (Cosgrove, 1997), intermediate, or deep (Sibson, 1994; Davies, 1999; Bons, 2001b) conditions in compressional, extensional, or strike-slip regimes. They can be related to active faults (Sibson, 1994; Gudmundsson, 2001) or caused by permeability changes in the matrix (Miller, 1995). Hydraulic fracturing results in extension fractures, if the differential stress ($\Delta\sigma$) is smaller than four times the tensile strength of the rock ($\sigma_1 - \sigma_3 < 4T$) and in shear fractures when the differential stress exceeds $4T$ (Secor, 1965). Furthermore, the smaller the differential stress (and in the case of $\Delta\sigma < 4T$), the less preferred orientations of extension fractures occur. A variety of fracture styles exist between straight linear and more randomly oriented fractures until the extreme situation of zero differential stress (i.e., a hydrostatic stress field) without preferred orientations and, consequently, the formation of breccia (Cosgrove, 1997).

The recognition of paleohydrofracturing is difficult, because the morphology of the fractures is not always indicative for the process of formation nor for the driving force. Moreover, hydrofractures that form at shallow depth are easily healed and unlikely to be preserved (Cosgrove, 2001). At greater depths, hydrofractures can be active without leaving any traces (Bons, 2001b). Some clear indications of hydraulic fracturing are the occurrence of sedimentary dikes (Cosgrove, 2001; Jolly and Lonergan, 2002), Hill-type fault and fracture meshes (Sibson, 1994), and cemented wall-rock breccias (Ramsay and Huber, 1987; Lorenz et al., 1991; Sibson, 1994; Cosgrove, 1997). Other features that are generally interpreted to relate to high fluid pressures are multiple crack-seal veins (Ramsay, 1980; Callan and Spooner, 1998; Caputo and Hancock, 1999), although they do not necessarily require high fluid pressure to form. Lee and Wiltschko (2000) described a mechanism for coarse, crack-seal calcite veins that formed along a continuously slipping fault in a porous rock with normal fluid pressure. Wiltschko and Morse (2001) proposed a model for banded veins in which the opening mechanism is attributed to crystallization pressure instead of internal fluid pressure.

In the same way, wide veins with large, euhedral crystals are believed to have formed in a fluid-filled, large, open space. Other evidence for high pore pressures can be found in the paleopressures deduced from fluid inclusions in the respective veins (Roedder, 1984; Kenis et al., 2002).

Various causes for overpressures that lead to hydrofracturing in sedimentary basins are discussed in the literature (e.g., Sibson, 1994; Osborne and Swarbrick, 1997). The development of seals and pressure compartments is extensively discussed in Ortoleva (1995). The main controlling factors are stress, permeability changes, and diagenesis, which are commonly interdependent. Compaction during rapid burial can lead to dewatering-related overpressure. Additionally, tectonic compaction associated with layer-parallel shortening (LPS) also affects the lateral permeability, especially in carbonate rocks. If impermeable barriers are present, pore fluid can be forced into a smaller rock volume and become overpressured or migrate along some tectonic gradient and induce overpressures at some other site (Ramsay and Huber, 1987). Burial and tectonic compactions also lead to pressure solution, with the development of stylolitic seams perpendicular to σ_1. These features will in turn influence the permeability of the rock. Faults and fractures can either store or expel fluids, depending on the stress field and fluid pressure. Sibson (1994) underlines the importance of stress cyclicity. Diagenetic processes can also cause an increase in fluid pressure, for example, smectite-illite dehydration, gypsum-anhydrite transformation, or hydrocarbon maturation. Aquathermal pressuring can increase pore pressures under elevated temperatures, for example, because of burial.

Vein formation without fracturing can be driven by force of crystallization (Lee and Wiltschko, 2000; Wiltschko and Morse, 2001; Hilgers and Urai, 2002), which replaces the function of pore-fluid pressure in the opening and propagation of the vein. Crystallization is controlled by supersaturation, which in turn may be controlled by pressure solution in the host rock. A geochemical self-organization is achieved, which can explain vein development without fluid flow or increased fluid pressure, although the latter processes are not excluded by the mechanism. They can help in the process of crack formation and propagation.

GEOLOGICAL SETTING AND KINEMATIC EVOLUTION

The study area is the Cordoba Platform, a Cretaceous carbonate platform, which is situated in the Sierra Madre Oriental fold and thrust belt of the Laramide (or Hidalgoan) orogen in Mexico (Figure 1). The platform strata are Lower to Upper Cretaceous in age (Figure 2) and range from a shallow platform environment, with deposition of limited amounts of evaporites, to a stable platform, with (partly dolomitized) limestones. The Lower Cretaceous evaporitic layers served as decollement surface (Ortuño-Arzate et al., 2003). In the western part of the area, middle and Upper Cretaceous reservoir lithologies crop out in antiformal crests of fault-related folds in thin-skinned thrust sheets. Upper Cretaceous and Paleocene flysch sediments cover the platform strata and represent the last phases of deposition in the western part of the area. To the east, the tectonic front strata are buried below a thick foreland basin of Tertiary–Quaternary age, i.e., the Veracruz Basin (Figure 3). All producing oil fields are located in Cretaceous or Tertiary reservoirs in this tectonic front setting.

The kinematic evolution of the area started during the Middle to Late Jurassic with the opening of the Gulf of Mexico and the development of several basins and rift systems (Ortuño-Arzate, 1991; Sedlock et al., 1993). During the Late Jurassic, a series of roughly north–south to N160°-oriented basins opened in the region that is now eastern Mexico. The general opening direction could be partly inherited from Paleozoic basement structures (Ortuño-Arzate, 1991). During the Lower Cretaceous, several transgressive phases led to the development of platforms (e.g., Cordoba Platform) on the basement highs. From Late Cretaceous to middle Eocene (Mossman and Viniegra, 1976), the area was affected by tectonic compression along a N60–65° direction (Ortuño-Arzate, 1991). Deformation prograded through the foreland from west to east (de Cserna, 1989; Sedlock et al., 1993; Ortuño-Arzate et al., 2003). From late Eocene to Oligocene onward, an intraplate extension phase affected the area and reactivated older Laramide structures (Ortuño-Arzate, 1991).

Oil migration, which has been calculated from thermal modeling (Ortuño-Arzate et al., 2003), started after Laramide orogeny and Tertiary subsidence in the Veracruz Basin. This burial was necessary for source rocks (Kimmeridgian and Lower and middle Cretaceous argillitic limestones) to reach the oil window.

DEFORMATION HISTORY

This section provides an overview of the different deformation stages and the resulting microstructures through the evolution of the Cordoba Platform. The structural and diagenetic evolution has been split up into three major stages that roughly coincide with the main kinematic stages. This division is based on cross-cutting relationships between veins and stylolitic seams. A set of bedding-parallel stylolites (BPS) formed during burial compaction and a set of stylolites perpendicular or oblique to bedding developed during LPS (Andrews and Railsback, 1997). Layer-parallel shortening stylolites

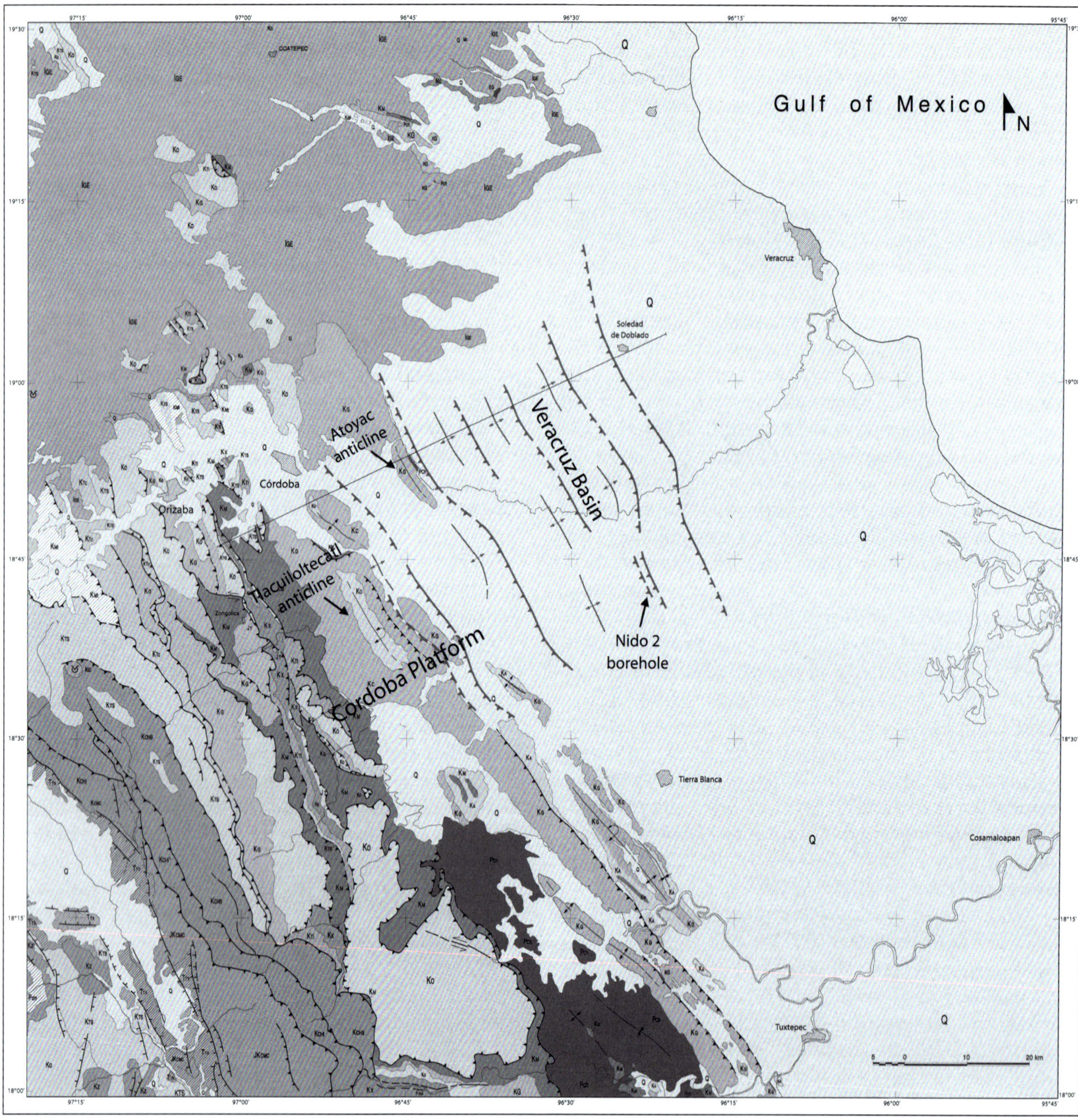

FIGURE 1. Geological map of the Cordoba Platform and adjacent Veracruz Basin, with indication of the sampling locations mentioned in the text. The map has been worked out at the Institut Français du Pétrole based on available seismic, surface, and Pemex borehole data.

(LPSS) typically develop prior to folding and thrusting of the strata in fold and thrust belts (Tavarnelli, 1997; Storti and Salvini, 2001) as the LPS deformation front migrates through the foreland (Groshong, 1988; Gray and Mitra, 1993; Ortuño-Arzate et al., 2003).

A careful sampling strategy has been followed, whereby crucial crosscutting relationships were sampled with a hand drill. Strike and dip of these plugs were measured to determine the orientation of all sampled (micro-) veins and stylolites. In total, about 150 microvein and macrovein directions have been measured in the field and determined from oriented plug samples. Additionally, cemented fractures from field and core samples have been studied petrographically.

The Pre-BPS to Syn-BPS Stage

An early set of fracture-shaped veins, which are crosscut by BPS planes (Plate 1A), has been recognized in fine-grained limestones and partly dolomitized

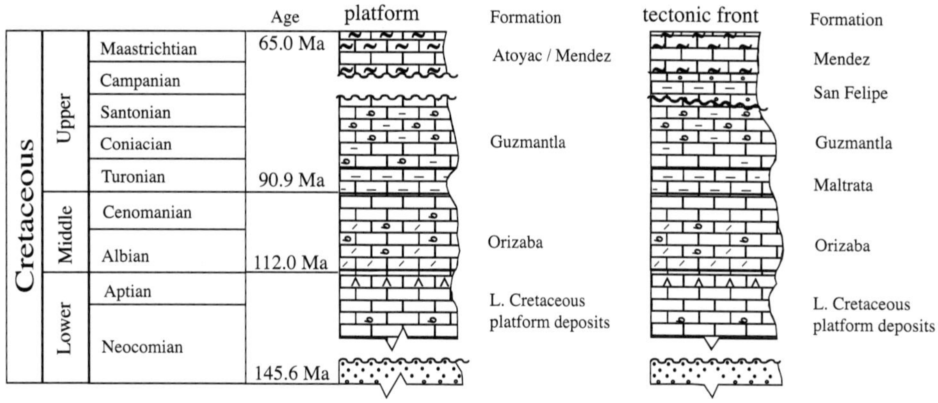

FIGURE 2. Stratigraphic log of the Cretaceous platform strata in the study area.

limestones (mudstone to wackestones) at the top of the Orizaba and Maltrata formations (upper part of the middle Cretaceous, Figure 2). These early veins are formed in the Tlacuiloltécatl anticline (Figure 1) and in core samples from the buried tectonic front area. They occur either in densely spaced (centimeter-scale to micrometer-scale) sets of microveins that have a typical width of 5–150 μm (Plate 1B), or they may form thicker (as much as 1.5 cm [0.6 in.]) compound veins (Plate 1C). In some thin sections, more than 100 different phases of microveins have been observed. The microveins can converge into a thicker compound vein, or reversely, thicker veins can branch into several microveins.

Under the microscope, some veins have a breccia-vein outline because of the incorporation of host rock slivers, where nearby veins locally formed nonparallel (Plate 1D). Breccia veins are veins that are internally composed of fragments of host rock material cemented together. The pre-BPS to syn-BPS breccia veins can be described as cemented crackle float breccia in the classification system of Morrow (1982). Most veins have a typical serrated outline (Plates 1B, E) because of curving of the vein around heterogeneities such as microfossils. Bands of solid and/or fluid inclusions, parallel to the vein wall and continuous through different crystals, are common and occur with a typical spacing of 10–35 μm.

The infilling calcite cement is of elongate-blocky or blocky type (Plate 1D), and an increase in crystal size is sometimes observed toward the center of the vein. The relationship between host rock and vein cement is difficult to assess in a micrite-dominated matrix, but where local intragranular cement occurs (e.g., in foraminifera), the vein cement has grown in optical continuity with the matrix cement. The vein cement has grown also in optical continuity throughout several adjacent microveins and inclusion bands (Plate 1E). The veins opened syntaxially, and the cement growth direction is toward the center of the vein. Based on mineralogic and isotopic evidence, the pre-BPS to syn-BPS veins reflect host rock buffering (Ortuño-Arzate et al., 2003).

The thicker veins have a dark brown appearance and show evidence of strong crystal-plastic deformation with intense calcite twinning (Plate 1F) and unduleous extinction and dynamic recrystallization with the formation of subgrains. This deformation destroys (partly) the original (elongate) blocky texture.

The orientations of the pre-BPS fractures have been statistically processed, and the strike directions have been plotted in a rose diagram (Figure 4a) and on a conceptual block diagram relative to horizontal bedding (Figure 5). The dip of these fractures is approximately perpendicular to bedding. Two main groups of strike orientations can be distinguished, a north–south to N20° and a N30–40° direction.

The Post-BPS and Pre-LPSS to Syn-LPSS Stage

Two types of veins that cut through BPS planes, but which are cut by LPSS planes, occur in outcrops and cores. They are fracture-shaped veins and breccia veins.

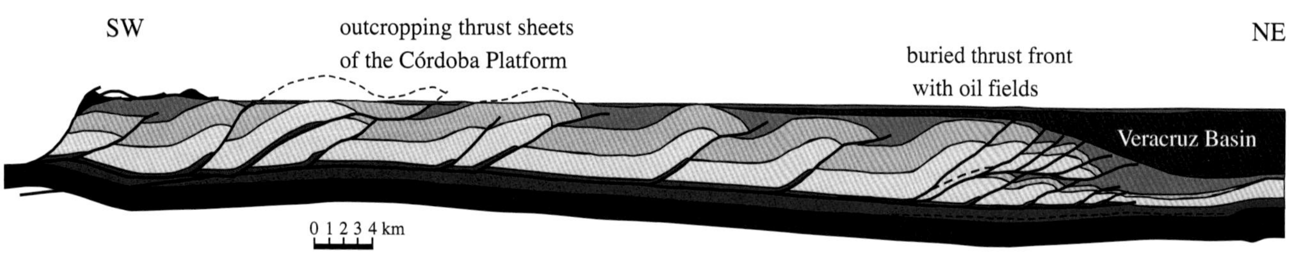

FIGURE 3. Cross section through the foreland fold and thrust belt along the indicated transect on Figure 1. The cross section has been worked out at the Institut Français du Pétrole based on available seismic, surface, and Pemex borehole data.

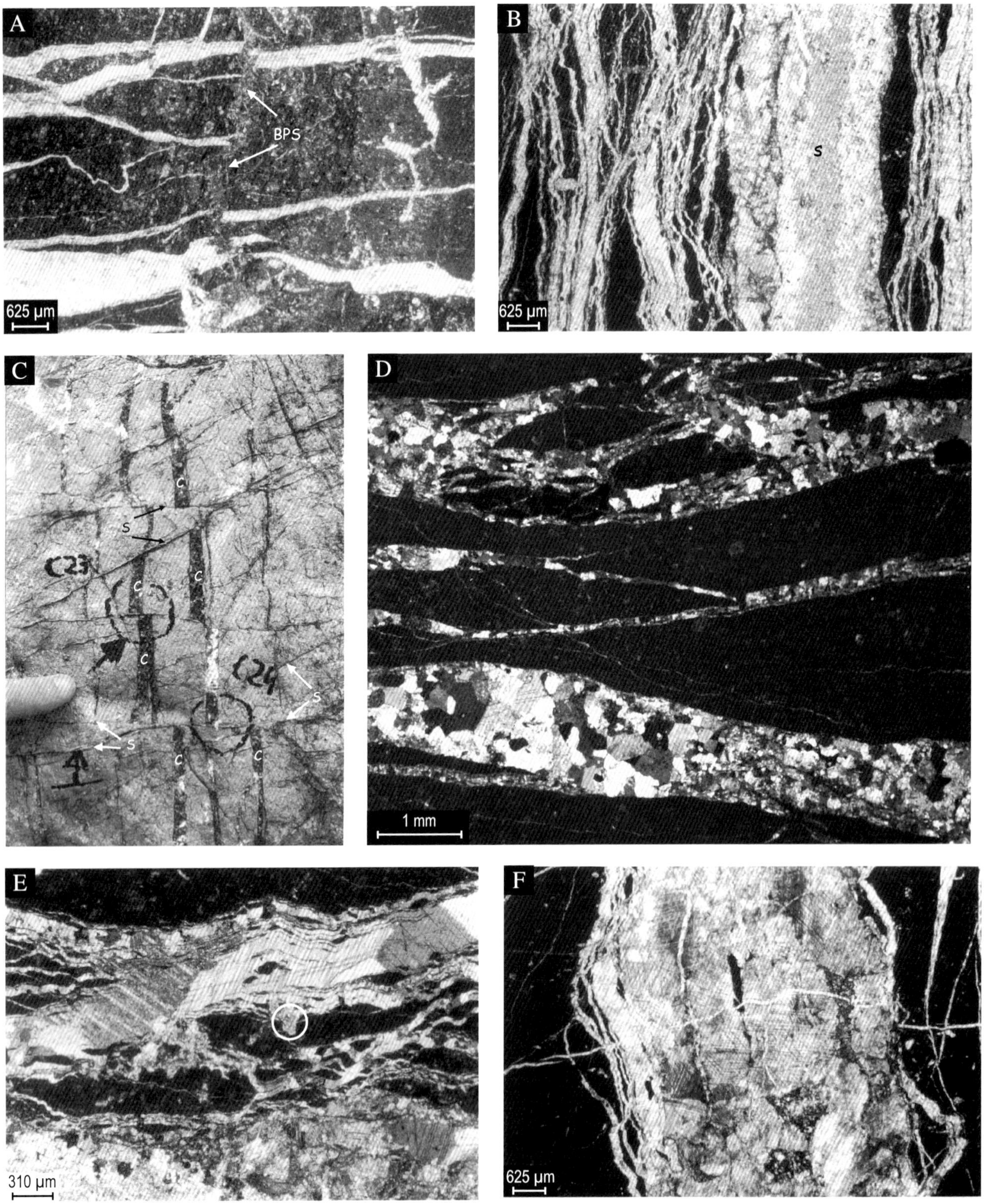

Plate 1. (A) Serrated calcite veins cut by a BPS plane. (B) Pre-BPS to syn-BPS serrated, calcite microveins converging into a compound vein, which has been reactivated by a later karst-related vein with vadose-silt infill (s). (C) Outcrop view of a vertical wall with pre-BPS to syn-BPS thick compound veins (C), which are cut by subhorizontal and shallow-dipping shear veins (S). (D) Pre-BPS to syn-BPS vein, composed of several branching and converging microveins, with a breccia-vein outline and filled with blocky calcite cement. View under crossed polars. (E) Pre-BPS to syn-BPS serrated compound vein taken under crossed polars. Note the optical continuity in crystal growth through different bands (upper part) and the highly deformed crystals in the thicker vein (lower part). (F) Pre-BPS to syn-BPS thick compound vein with calcite cement that displays intensely twinned crystal planes. Note the serrated outline of the thinner side veins.

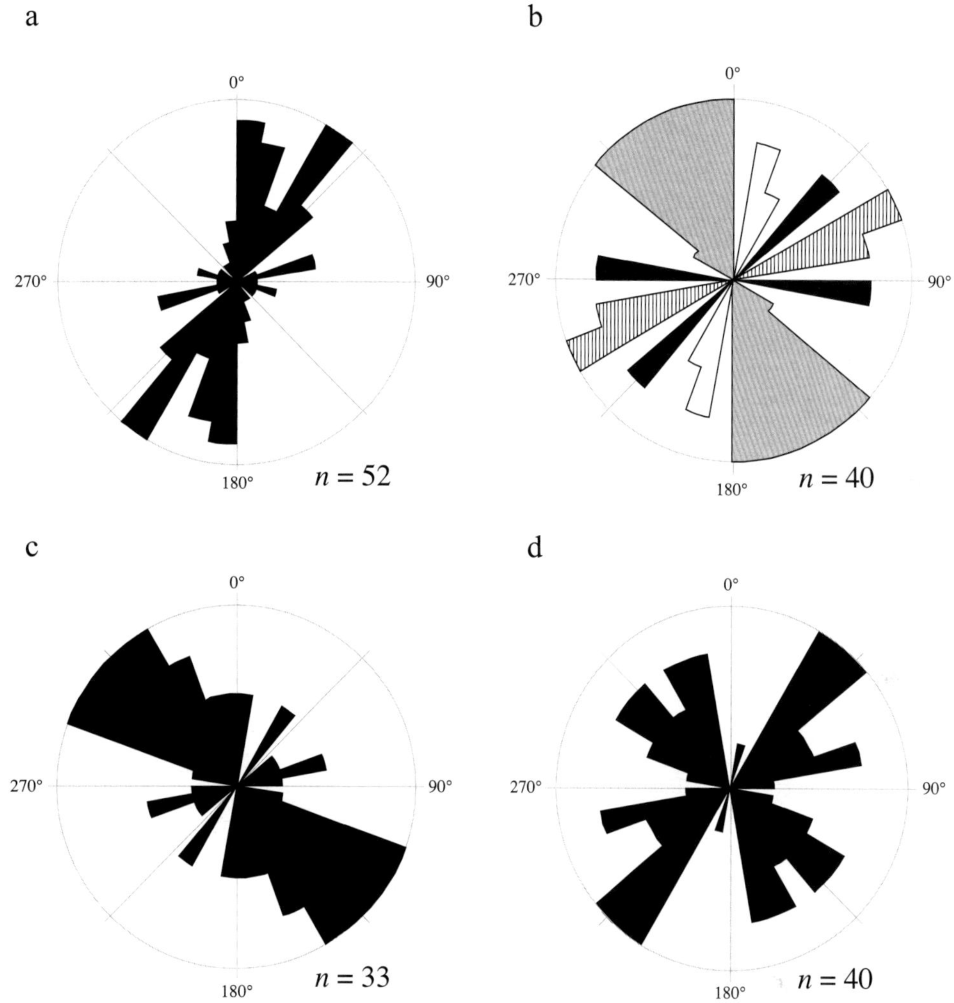

FIGURE 4. Rose diagrams showing strike of (a) pre-BPS to syn-BPS veins measured in subhorizontal stratification and therefore not corrected for bedding orientation. (b) Post-BPS and pre-LPSS to syn-LPSS veins, the hatchings only serve to distinguish orientation groups. (c) LPSS planes in the Tlacuiloltécatl anticline. (d) Post-LPSS (steeply dipping) veins.

The fracture-shaped veins developed under a high angle to bedding and commonly consist of a main vein with several branching and converging, smaller side veins (Plate 2A). Vein widths range between 300 μm and 1.5 cm (0.6 in.). Smaller side veins are generally composed of elongate-blocky calcite cement (Plate 2A), which grew in optical continuity with the cement of the main vein when converging. Thicker veins are generally composed of different phases of calcite cement (Plate 2B). A rim of thin calcite crystals (10–180 μm) lines both vein walls. This is followed by larger elongate-blocky (crystal long axis between 120 μm and 3 mm) or blocky calcite cement growing symmetrically from both vein walls toward the center of the vein. Finally, blocky calcite crystals filled the remaining space in the center of some veins. Growth competition of the elongate-blocky cement also caused an increase in crystal size from wall to center (Plate 2B). The vein-filling cement grew in optical continuity with local matrix cement (Plate 2C). Bands of solid and/or fluid inclusions parallel to the vein walls and continuous throughout different crystals are common (Plates 2A, C). These bands have a typical spacing of 10–20 μm. The filling cement grew in optical continuity through a series of parallel bands (Plate 2A).

Other post-BPS and pre-LPSS to syn-LPSS veins have a breccia-vein morphology (Plates 2D–G). The post-BPS and pre-LPSS to syn-LPSS breccia veins can be described as cemented crackle to mosaic float breccia in the classification of Morrow (1982). In two dimensions, the fragments seem to float into the vein cement (Plate 2D). In three dimensions, an anastomosing vein network that has a mean preferred orientation developed. The total breccia-vein width can be several centimeters thick. These breccia veins are observed in limestones of the Guzmantla and Orizaba formations in several outcrops and cores. In dolomite and evaporitic lithologies, these breccia veins evolve to more extensive breccia without any preferred orientation. Similar breccia veins also develop before LPSS formation in other fold and thrust belt settings worldwide (Van Geet et al., 2002; Swennen et al., 2003).

Apart from host rock slivers, a geopetally arranged, silt-sized carbonate granulate material occurs in some breccia veins (Plate 2E). The cement filling the space between fragments and vein wall is blocky to elongate-blocky (Plate 2F) and developed syntaxially. Around the silt-sized material, many small crystals nucleated. An increase in crystal size away from the vein wall or host rock fragments is caused by growth competition and drusy fabric in wider vein branches. The crystals are commonly rich in fluid inclusions.

Two stages of cement precipitation with different geochemical characteristics can be distinguished in

a

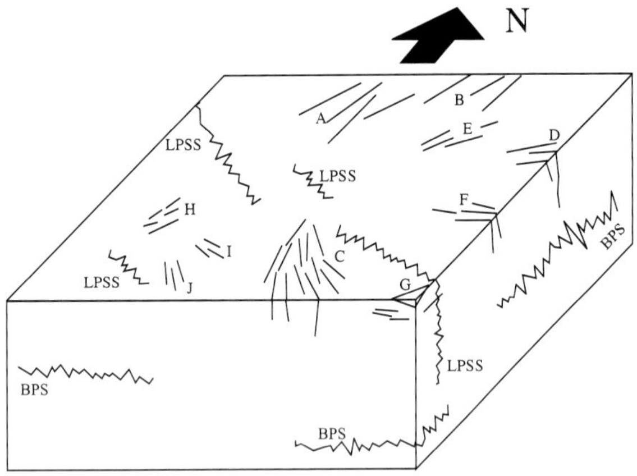

b

Group	Timing	Type of Veins	Orientation
A	Pre-BPS or syn-BPS	fracture-shaped, syntaxial, (elongate-) blocky veins microveins or compound veins with parallel inclusion bands serrated and intensive crystal-plastic deformation	NS-N20° N30-40°
B C D E F	Post-BPS and pre-LPSS or syn-LPSS	fracture-shaped, syntaxial,elongate-blocky veins, anastomosing patterns, parallel inclusion bands breccia veins with a crackle-mosaic texture and blocky cement	N10-30° reuse of A N130-180° N60-80° N40-50° N90-100°
G	Post-LPSS	shear-associated, irregular veins composed of (stacked) lenticular and sigmoidal segments and filled with a blocky calcite cement	shallow dipping
H I J		dark, banded veins lacking significant cementation fracture-shaped, dissolution-enlarged veins and breccia-veins (cemented and particulate, mosaic float breccia and pack breccia) often with a geopetally arranged silt infill and/or blocky calcite cement with a drusy texture	N30-50° N120-130° N150-160°

Figure 5. (a) Block diagram showing the orientations of the studied vein sets and stylolites relative to horizontal bedding. (b) Table summarizing the different vein groups with their relative timing, the type of veins, and their typical orientations.

the diagenetic evolution of this post-BPS and pre-LPSS to syn-LPSS period. An early phase of cementation is mineralogically host rock buffered (e.g., calcite cement in limestone fractures and dolomite cement in dolomite fractures). A second phase of cementation that fills veins displays different mineralogies (e.g., fluorite, saddle dolomite; Plates 2E, G). This is also attested by a slight $\delta^{18}O$ depletion and similar to host rock $\delta^{13}C$ values for the first phase, in contrast to a stronger $\delta^{18}O$ and a slight $\delta^{13}C$ depletion for the later phase (Ortuño-Arzate et al., 2003). The late-stage veins that are filled with saddle dolomite and fluorite sometimes predate LPSS development and, in other cases, cut through these planes. It is therefore argued that the timing of the latter veins coincides with LPS.

Reuse of older (pre-BPS to syn-BPS) veins is commonly observed, with generally less deformed cement phases that develop across BPS planes, whereas the older, intensely twinned crystal planes stop against BPS planes (Plate 2H). Reused older calcite veins sometimes contain dolomite or fluorite.

The orientations of the post-BPS and pre-LPSS to syn-LPSS veins have been measured in dolomites and limestones of the Orizaba Formation (middle Cretaceous) and mainly in limestones of the Guzmantla Formation (Upper Cretaceous) in the Atoyac and Tlacuiloltécatl anticlines (Figure 1). The vein data are represented in Figures 4b and 5, and the orientations of LPSS planes (Figure 4c) are given as an indication for the Laramide tectonic stress direction, because LPSS planes develop mainly perpendicular to the main compressive stress. The LPSS orientations range chiefly between N110 and 150° (Figure 4c), with subvertical dips relative to horizontal bedding. The veins that reuse older (pre-BPS) veins display a N10–30° orientation. The remaining post-BPS and pre-LPSS to syn-LPSS veins fall into four groups. A wide scatter in vein orientations exists between N130 and 180°. A set between N60 and 80° is at a high angle to this N130–180° group and bisects two other sets of conjugate vein orientations at about 25°, with an orientation of N40–50° and N90–100°, respectively. Crosscutting relationships indicate that the oldest veins are in the N130–180° group, and that they are cut by N60–80° fractures that are, in turn, cut by N40–50° and N90–100° fractures. However, these relationships are not always unambiguous. The two different types of veins (fracture-shaped veins and breccia veins) can be found in a similar group, and younger veins can follow the direction of previously formed veins. Therefore, plotting the strike data for both vein

Plate 2. (A) Thick calcite vein with smaller side veins filled with elongate-blocky calcite cement. Note the bands of solid inclusions parallel to the vein wall (left part of the main vein). View under crossed polars. (B) Vein filled with a thin rim of calcite cement along the vein walls and elongate-blocky calcite cement in the center of the vein. Note the crystal growth in different directions and the larger development of some well-oriented crystals, which indicate growth competition. View under crossed polars. (C) Two converging side veins filled with elongate-blocky (E) and blocky (B) calcite cement. The vein cement has syntaxially overgrown adjacent matrix cement (arrows). Note also the microveins and inclusion bands (visible below the lowest arrow). View under crossed polars. (D) Post-BPS and pre-LPSS to syn-LPSS breccia vein with pieces of host rock material floating into the vein cement and a LPSS crosscutting the vein (marked). (E) Breccia vein in dolomite with geopetally infilled silt-sized material (arrows) and blocky calcite cement. (F) Breccia vein with blocky calcite cement, taken under crossed polars. (G) Hand specimen of a post-BPS and pre-LPSS to syn-LPSS breccia vein with dolomite fragments in a calcite cement (C). (H) Early calcite veins (E), which stop against a weathered BPS filled with silt and cement, whereas a later vein (L) cuts through the filled BPS. Part of an older vein (left) and of the BPS fill have been damaged because of sample preparation.

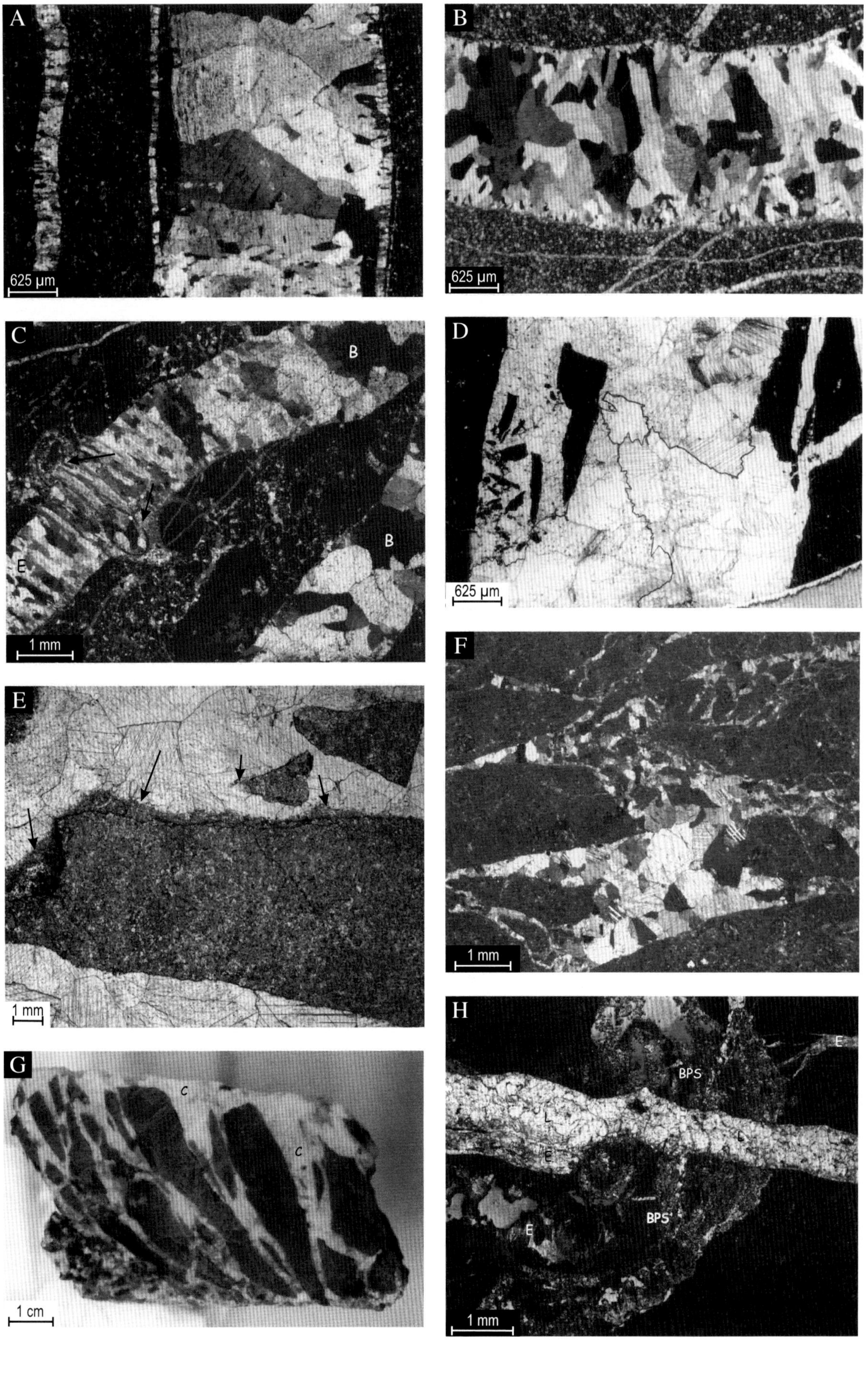
A
625 μm
B
625 μm
C
B
B
E
1 mm
D
625 μm
E
1 mm
F
1 mm
G
c
c
1 cm
H
E
BPS
L
L
E
BPS
E
1 mm

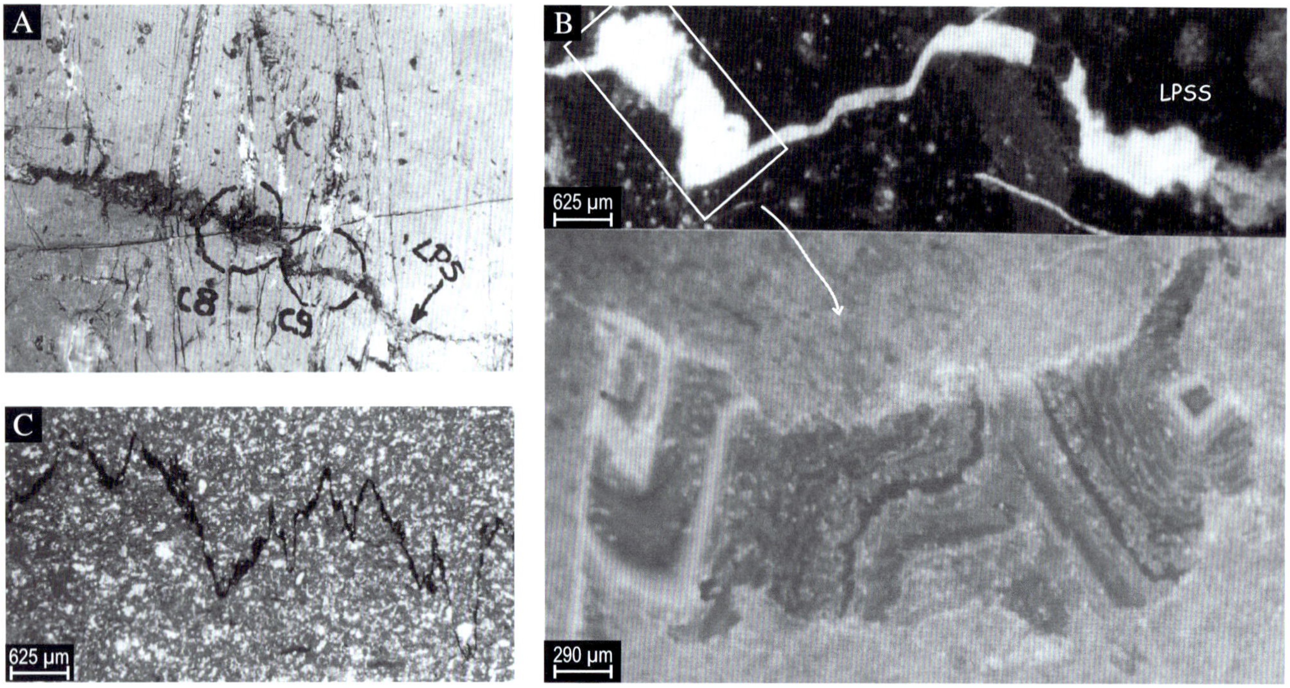

PLATE 3. (A) Outcrop view of a dissolution-enlarged LPS stylolite with vadose-silt infill and calcite cement. (B) Upper part: Transmitted light view of an LPS stylolite filled with calcite cement. Lower part: Cathodoluminescence view of a detail of the cemented stylolite with a cauliflowerlike bright- and dull-luminescent zoned calcite cement. (C) Oil-stained stylolite in an Upper Cretaceous wackestone.

morphologies separately is considered meaningless. Consequently, three vein families were distinguished purely based on their orientation (Figure 4b).

The Post-LPSS Stage

Field and petrographic observations of stylolites, which are filled with a karst-related infill (Plate 3A) or a cement phase (Plate 3B), and of oil-stained stylolites (Plate 3C) give evidence of the opening of LPSS planes and fluid flow along these planes after formation. It means that stylolites can have two functions, initially as impermeable boundaries and later during diagenesis as fluid pathways (see also Van Geet et al., 2002).

Subhorizontal to 45°-dipping veins, which crosscut and displace LPSS planes and older veins (Plate 1C), have been observed mainly in the Tlacuiloltécatl anticline (Figure 1), where their orientation has been measured. Because of the shallow dips and the inherent difficulty of precise and more variable strike measurements, the poles to the planes have been plotted (Figure 6) instead of the strike data in a rose diagram. The mean value is perpendicular to a plane with orientation N42° and dipping 10° to the south. When these veins cut through older veins, intensely deformed and bent twin lamellae locally develop in crystals of the older veins (Plate 4A). The shallow-dipping veins typically possess an irregular morphology (Plate 4A). Their thickness varies along strike, and they branch sometimes into side veins. In cross section, the veins are commonly composed of a

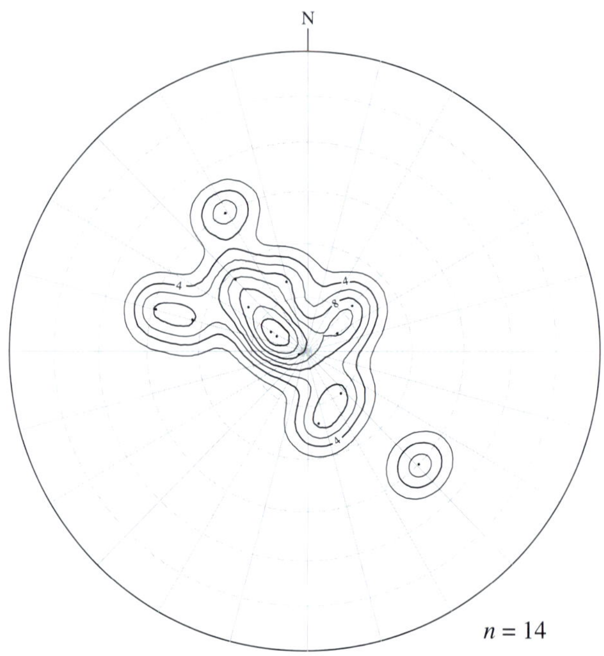

FIGURE 6. Density pole stereogram for post-LPSS shear veins.

succession of lenticular, sheetlike, or sigmoidal segments that are partly stacked on top of each other (Plate 4B). Between different segments, thin bands of siltlike (recrystallized) matrix material or discontinuous inclusion bands are present (Plate 4A). The vein cement is formed of blocky to elongate-blocky calcite crystals, which may be perpendicular or oblique to the vein walls (Plate 4C). Generally, a thin rim of calcite cement lines a segment and is followed by precipitation of larger calcite crystals. The increase in crystal size toward the center of a segment reflects growth competition. The growth direction of the vein cement is syntaxial, and where matrix cement is present, syntaxial overgrowths are sometimes observed. Where several vein segments are stacked on top of each other, intensely twinned planes and subhedral crystals are formed because of deformation (Plate 4B). In subhorizontal strata, like in anticlinal hinges, these veins are commonly closely associated with BPS planes. They tend to follow partly the BPS planes, and sometimes, BPS traces can be found in these veins. These BPS planes are locally reactivated, and also, the plane between two vein segments may be stylolitized (Plates 4B, C).

Other post-LPSS veins are mostly steeply dipping, and the measured strikes plot in a wide range of directions (Figures 4d, 5). The main groups are N30–50°; N120–130°, and N150–160°.

Dark veins, which lack significant cement precipitation between (some of) their successive bands, occur at several localities and at different diagenetic timings but are most commonly observed as post-LPSS features in the Tlacuiloltécatl anticline (Figure 1). They form along joint planes and as fracture-shaped veins. These veins possess a banded structure, which is the result of planar accumulations of clay-sized material, separated from each other by a granular, silt-sized material (Plates 4D, E). This material is a kind of microsparite, which is a typical recrystallization product of micrite matrix (Bathurst, 1975). Where vein cement occurs between the bands, only thin and inclusion-rich, granular crystals are developed without preferential orientation.

These veins and other preexisting veins are commonly affected by a new fracturing and/or dissolution phase (Plates 4D, E) that leads sometimes to breccia formation (Plate 4F). The type of breccia is a cemented and particulate mosaic, float and pack breccia (according to Morrow, 1982). The resulting porosity locally is filled with a geopetally organized sedimentary silt-infill (Plates 1B, 4F). This sedimentary silt-infill is succeeded by a dissolution phase and/or by blocky calcite cement precipitation, commonly with a drusy texture (Plate 4F). The same diagenetic phases are also found in post-LPSS veins that typically possess an irregular, vuggy (dissolution-enlarged) outline. The silt infill and blocky cement form the latest phases in the paragenetic sequence in the western, outcropping part of the study area and relate to telogenetic karst development (Ferket et al., 2000, 2003b; Ortuño-Arzate et al., 2003).

A post-LPSS multiple brecciation has affected dolomite strata from the Orizaba Formation in the Nido 2 borehole in the buried tectonic front area (Figure 1). This breccia occurs just above a major thrust plane. Dolomite host rock has been brecciated, partly dissolved (with biomold formation and dissolution-enlarged fractures) and subsequently cemented by blocky dolomite cement with drusy texture. The resulting texture is a cemented crackle float breccia (Morrow, 1982). Another crackle brecciation and dissolution event subsequently affects the rock and is followed by precipitation of euhedral dolomite rhombs (Plate 4G). A microbrecciation event has subsequently cut through previous cements and has created a fine, cataclastic material. The resulting texture is a particulate rubble pack breccia or a cemented rubble float breccia, where it has been extensively cemented. The infill material consists of sharp pieces of recrystallized host rock material and fragments of dolomite cement (Plate 4G). A concentric, "cauliflower" type of zoned calcite cement, which occurs also in opened LPS stylolites (Plates 3B, 4G), fills up (part of) the remaining porosity. Traces of oil are found in the breccias and in other fractures in this borehole.

DISCUSSION

The Pre-BPS to Syn-BPS Stage

The pre-BPS to syn-BPS veins are fracture shaped and formed as dense sets of individual microveins that may branch and converge and result in thicker, banded compound veins. Nearby, open cracks cannot coexist, and therefore, these veins must reflect multiple phases of crack formation. The (elongate) blocky texture also indicates crystal growth in an open space. The pre-BPS to syn-BPS veins can consequently be regarded as cemented fractures. The typical serrated outline is caused by bending of the fracture surface around local heterogeneities such as microfossils. The breccia-vein outline is the result of subparallel fractures that branch and converge, whereby host rock slivers become isolated.

Crystal-growth morphology and syntaxial overgrowth of matrix cement indicate syntaxial opening of the fractures. The crystal growth may be toward the center of the vein in bitaxial veins (Plate 1D) or from one side toward the other in unitaxial veins that maintain crystal width and crystallographic orientation throughout the different microcrack events (Plate 1E). The bitaxial syntaxial veins reflect cracking in the center of the vein, whereas the unitaxial syntaxial veins reflect cracking along a vein-wall contact. The syntaxial overgrowth and the evidence of host rock buffering (Ferket et al., 2003b; Ortuño-Arzate et al., 2003) in these veins

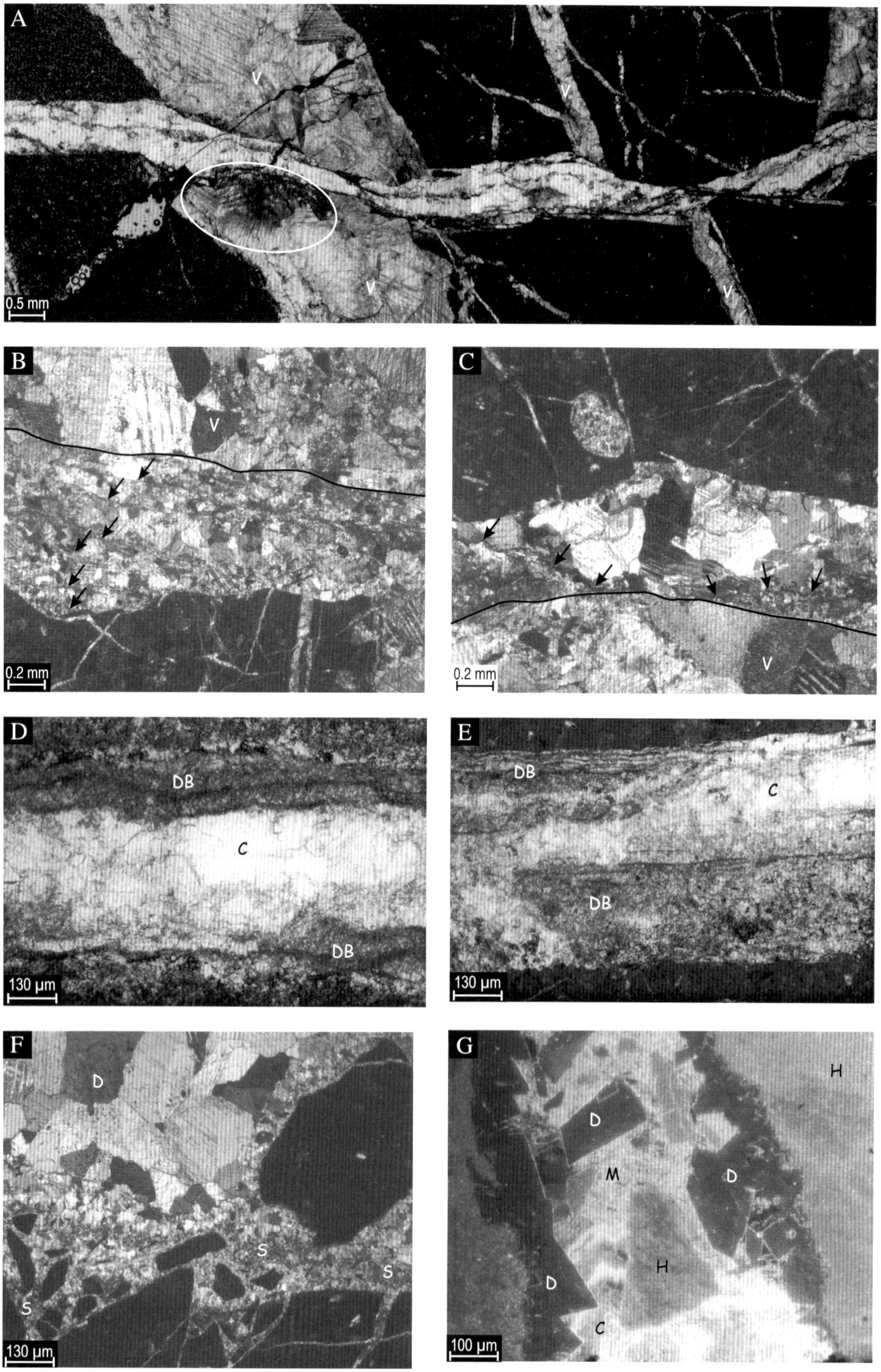
A
V
V
V
V
0.5 mm
B
V
0.2 mm
C
V
0.2 mm
D
DB
C
DB
130 µm
E
DB
C
DB
130 µm
F
D
S
S
S
130 µm
G
H
D
M
D
H
D
C
100 µm

suggest the involvement of a host-rocklike fluid transported by diffusion and/or advection.

The local occurrence of dense sets of microveins or thicker compound veins, in contrast to locations that almost lack vein formation, shows that these veins occur clustered. Moreover, their formation requires multiple fracturing events (Ramsay, 1980; Pollard and Segall, 1987; Caputo and Hancock, 1999). Nearby, parallel, open fractures are in contradiction with the extension failure criterion, where a critical stress must build up at least equal to the tensile strength of the rock. Even with the fracture clustering mechanism of Simpson (2000), the observed spacing cannot be explained by a single fracturing phase, because in this model, a new fracture is forced to develop at a distance of two fracture lengths away from a nearby fracture. The only apparent possibility is to invoke multiple fracturing stages and, consequently, a cyclic stress behavior (Sibson, 1994). This cyclic behavior can be caused by variations in local stresses or by oscillations in fluid pressure.

The coexistence of thicker compound veins and single hairline veins can be explained by the degree and rate of cementation of a fracture before the opening of a new nearby fracture. If cementation is uncompleted, a new fracture will be forced to develop at the weakest contact, which is the preexisting fracture surface. When cementation of the preexisting fracture is completed, a new fracture may be formed elsewhere in the host rock or along a preexisting fracture, depending on the mechanical strength of both. These veins can thus be regarded as crack-seal (sensu Ramsay, 1980) and crack-jump veins (sensu Caputo and Hancock, 1999).

The pre-BPS to syn-BPS veins must thus have formed in multiple, syntaxial fracturing events in a cyclic stress field. Key questions relate to the cause of this cyclic stress field and to the mechanisms that control the opening and cementation of the fractures. Therefore, the timing and the kinematic framework of the pre-BPS to syn-BPS fracture formation must be considered.

The pre-BPS to syn-BPS vein set also affects top strata of the Orizaba Formation and thus must be younger than middle Cretaceous. They must have been formed prior to or synchronous with BPS development (i.e., during burial). Compaction stylolites or dissolution seams in carbonates can start to form at depths of 800–1500 m (2600–5000 ft) (Railsback, 1993) or even less (300–600 m [1000–2000 ft]; Machel, 1990), depending on a variety of parameters like microfacies, diagenesis, degree of cementation, water chemistry, and temperature. The presence of monophase fluid inclusions in these calcite veins indicates precipitation temperatures lower than 50°C (Sabouraud et al., 1980). Considering a low geothermal gradient between 15 and 25°C/km (24 and 40°C/mi) (on the assumption that the Cretaceous gradient was not very different than the present-day gradient, calculated from bottom-hole temperatures), a maximum burial depth of 800–1700 m (2600–5600 ft) can be defined. The overlying Guzmantla Formation comprises a thickness of about 1500 m (5000 ft) of Upper Cretaceous (Turonian–Santonian) rocks and is followed by a period of local erosion and karst (Ferket et al., 2000, 2003b) before deposition of Late Cretaceous flysch. Therefore, the period of vein development can be constrained between the Cenomanian and Campanian. Vein formation probably occurred at a burial depth of about 1–1.5 km (0.6–1 mi).

Tensile stresses may be expected at depths of no more than 1 km (0.6 mi) (Twiss and Moores, 1992), and tensile effective stresses below 1 km (0.6 mi) can exist with increased pore-fluid pressures. Furthermore, an early burial origin is corroborated by the crystal-plastic deformation of the veins, probably because of later burial and tectonic compaction.

The orientation of these veins fits with the general direction of basin and platform development in the area west of the Gulf of Mexico rift, which is around NS–N160° and which is at least partly inherited from Paleozoic basement structures (Ortuño-Arzate, 1991). However, the Gulf of Mexico rift cannot be the driving mechanism for the opening of these fractures, because rifting ended during the Callovian (Late Jurassic). The fractures are generally subvertical and most probably developed perpendicular to the least principal stress, in line with classical extension fracturing.

The fractures mainly developed in mudstones to wackestones that virtually lack intergranular (i.e.,

Plate 4. (A) A post-LPSS irregular vein cuts through pre-BPS to syn-BPS veins (V), thereby causing the bending of twin planes (white ellipse). The post-LPSS vein consists of lenticular segments stacked on top of each other and separated by inclusion bands. (B) View under crossed polars of a post-LPSS shear vein cutting through and displacing an older vein (V). The different segments (indicated with arrows) are filled with a blocky calcite cement and stacked on top of each other, thereby causing crystal-plastic deformation. (C) A post-LPSS shear vein cuts through and displaces an older vein (V). Note the presence of a stylolitic plane in the shear vein (indicated by arrows). (D and E) A post-LPSS dark, banded vein (DB) with clay-sized and silt-sized granular cement has been reopened, partly dissolved, and filled with a white, blocky calcite cement (C). (F) View under crossed polars of a crackle to mosaic breccia filled with vadose silt (S) and blocky calcite cement with a drusy texture (D). (G) Cathodoluminescence view of a brecciated dolomite host rock (H), which has subsequently been filled with zoned dolomite cement (D) with free faces toward the vein center. A next microbrecciation event has caused locally a pulverizing of the host rock into a micriticlike material (M) that partly fills open fractures. A cauliflower type of zoned calcite cement (C) also partly fills up open spaces.

between the grains) matrix cements. Because these veins cut through intragranular (i.e., in microfossils) matrix cement and the vein cement has grown in optical continuity with these cements, it can be concluded that fracture formation postdates matrix cementation. The sediment, consequently, was at least partly consolidated before fracture formation, and fracturing cannot be related to dewatering of a nonconsolidated sediment pile. In carbonate systems, the early diagenetic history is complex to understand because of the different porosity-permeability evolution for each lithofacies. Early cementation and compaction in some lithofacies type and retardation of these processes (e.g., because of framework-stabilizing cementation; Van Geet et al., 2002; Ferket et al., 2003a) in others have been recognized. These contrasts in early diagenetic processes in different lithofacies can differentiate aquifers that can store interstitial water. However, to cause episodic dewatering of these aquifers, a triggering mechanism is required. Tectonic causes are a likely candidate to force fluid expulsion before BPS development, although burial compaction by itself might be the driving force for multiple fracturing and cementation (Ramsay, 1980; Ramsay and Huber, 1987; Miller, 1995).

Layer-parallel shortening in the study area did not begin until the end of the Cretaceous or the beginning of the Paleocene and consequently cannot be the triggering mechanism. However, pre-Paleocene foreland flexuring and associated fluid flow could explain the pre-BPS to syn-BPS episodic fracturing events. On a regional scale, during the late Early Cretaceous to early Late Cretaceous, subduction of the Farallon plate below North America occurred in the west, while final accretion of terranes was going on, together with arc magmatism (Sedlock et al., 1993). Tectonic activity started at that time in the western hinterland and could be responsible for foreland flexuring in the study area. Local karst development in the Atoyac area during the Campanian (Ferket et al., 2000, 2003b) gives also evidence of the development of an early forebulge. This foreland flexuring may be accompanied by the migration of overpressured fluids. Closure of a volume of pores at one site may expel fluids and force them to migrate to another site, along existing permeability pathways.

Stage 1 in Figure 7 represents the inferred paleostress situation for the pre- to syn-BPS period. A stable platform has developed by that time, and compression has not yet started in the Cordoba Platform. At a burial

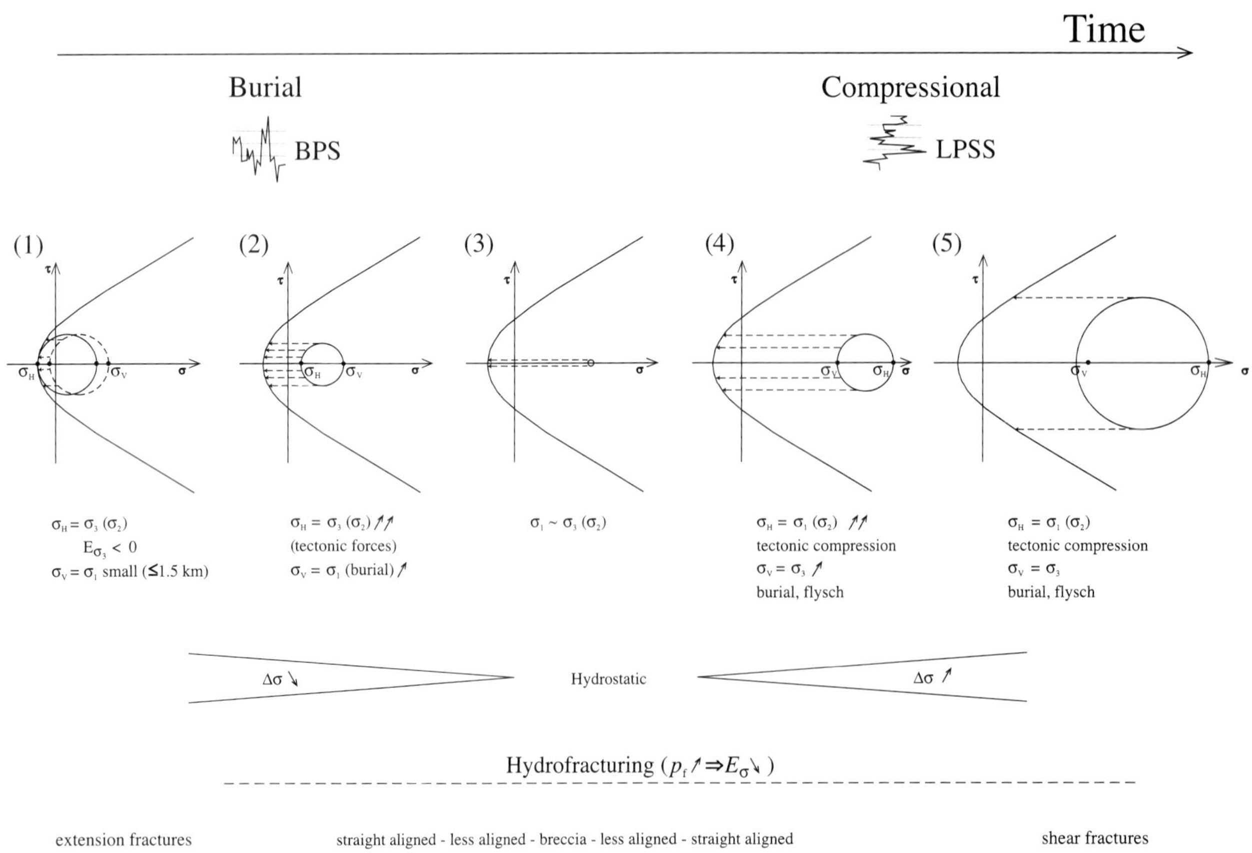

FIGURE 7. Schematic evolution of the paleostress conditions during the growing compressional regime in the Cordoba Platform.

depth of less than 1.5 km (1 mi), the vertical stress (most probably σ_1) is small, and the least principal stress (σ_3) might be tensile, or only a small fluid pressure would be required to reach the failure envelope for extension fracturing. The compressive stress causing foreland flexuring is a far-field stress, which, however, may have served to force fluid migration, to induce local tensile stresses around flaws (Lorenz et al., 1991; Twiss and Moores, 1992) in the study area, and to align fractures.

Whether compaction by itself can be the driving force for the repeated crack-seal and crack-jump cycles can, however, be questioned. Miller (1995) proposed a concept for the development of hydraulic shrinkage fractures, whereby compaction leads to overpressure and fracture development, but in which case, the orientation of the fractures may still be determined by the regional stress field and the host rock mechanical properties. The pre-BPS to syn-BPS veins in the study area grew during burial and indicate host rock buffering. This is in agreement with a pressure solution-related origin for cementation. Because these fractures preferentially develop in fine-grained lithologies lacking important intergranular cement precipitation and compaction is preferentially developed in such lithologies, compaction may have been a major factor in the process of vein formation.

Most probably, foreland flexuring triggered fluid migration, with development and/or propagation of fractures. Pulses in fluid-pressure buildup and release with fracturing can explain the numerous crack-seal and crack-jump cycles. The cementation mechanism most probably relates to pressure solution instead of the migration of large volumes of fluid along the fractures.

The Post-BPS and Pre-LPSS to Syn-LPSS Stage

The post-BPS and pre-LPSS to syn-LPSS veins can be straight and fracture shaped, and many are less well oriented and occur as main veins with branching side veins, in anastomosing patterns and as breccia veins. The less preferential alignment most probably relates to a decrease in differential stress ($\Delta\sigma$) between two (or all) principal stresses.

The fracture-shaped veins have an (elongate-) blocky texture and show strong evidence of growth competition (Plate 2B). The veins must consequently have grown as open fractures in which cement precipitated. The syntaxial overgrowth of matrix cement and the increase in crystal size toward the center of the veins indicate syntaxial opening of the fractures.

Branching and converging side veins and the typical presence of continuous inclusion bands parallel to the vein walls suggests that the veins developed in different phases of crack formation and sealing by cement precipitation (crack-seal cycles).

The breccia veins, displaying a crackle or mosaic texture, can be explained by anastomosing veins with an (elongate-) blocky texture, indicating growth in a relatively small open space. The mosaic type, however, contains more chaotically organized host rock slivers and commonly a geopetally organized, siltlike material in a drusy, blocky cement. This texture indicates host rock-derived, cataclastic material and cementation in a large, open space. This requires a mechanism for cataclastic fracture opening and a mechanism to keep the fracture open after formation. Hydrofracturing, in combination with a low $\Delta\sigma$, is a plausible mechanism for cataclastic breccia-vein development.

Post-BPS veins in the strata of the middle Cretaceous Orizaba Formation must have been formed during Upper Cretaceous or later, whereas post-BPS veins in the upper part of the Upper Cretaceous Guzmantla Formation must be Paleocene or younger. During the end of the Cretaceous and lower Paleocene, the first tectonic pulses of Laramide deformation affected the western part of the study area (de Cserna, 1989). From late Paleocene to Eocene, thrusts started to propagate eastward throughout the platform (de Cserna, 1989; Sedlock et al., 1993; Ortuño-Arzate et al., 2003). Fractures formed between BPS and LPSS development are consequently associated with a changing stress regime. The horizontal stress increased because of tectonic forces, and as a result, the main principal stress evolved from vertical to horizontal, and the least principal stress evolved from horizontal to vertical. The Laramide compressional regime had a regional N60–65° main principal stress (σ_1) component (Ortuño-Arzate, 1991). The fold hinge lines of the Atoyac and Tlacuiloltécatl anticlines (Figure 1) trend N145°. The measured LPSS planes (N110–150°) in both anticlines indicate a main compressive stress direction between N20 and 60°.

The N60–80° group of veins is oriented close to the main Laramide compression direction (N60°). Moreover, these fractures bisect the acute angle between two conjugate fracture sets at N40–50° and N90–100°. The spatial relationship of these fractures, compared with the tectonic stress direction, may reflect a genetic relationship (Figure 8). The veins (fractures) consequently formed parallel to the maximum principal compressive stress σ_1 and perpendicular to the least principal compressive stress σ_3, and the conjugate sets at 25° of σ_1. This fracture configuration is typical of prefolding compression (Twiss and Moores, 1992).

The largest scatter in strike orientation is found in the N130–180° group, which overlaps partly with the LPSS group. The parallelism of these veins with the LPSS structures (N130–150°) and with the fold hinge lines of the anticlines (N145°) suggests that fracture formation relates to the process of LPS (Figure 8). Because

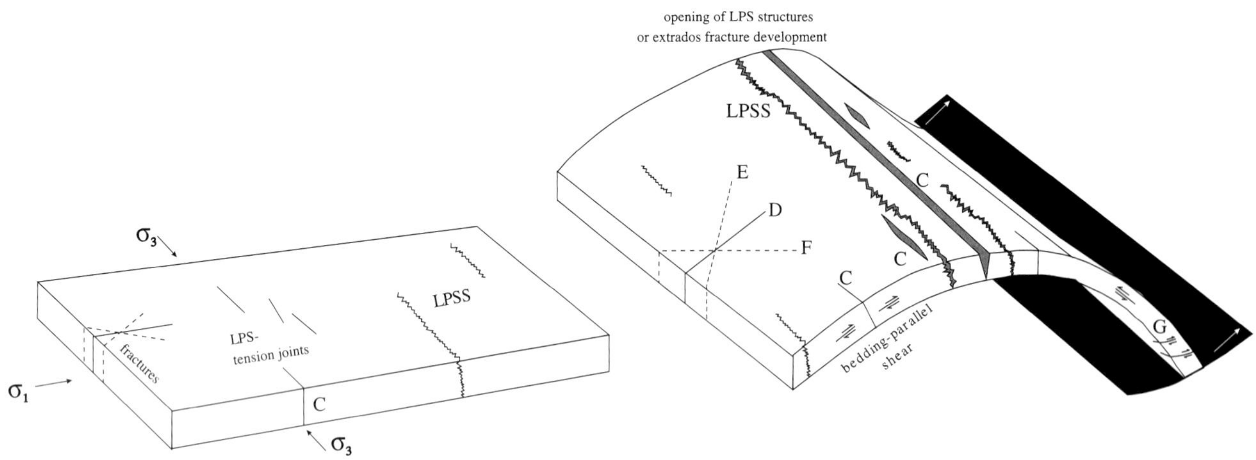

FIGURE 8. Sketch visualizing the development of prefolding, synfolding, and thrust-related structures in an anticlinal structure. The letters refer to the vein generations in Figure 5.

the post-BPS fractures contain cement generations that sometimes predate and sometimes postdate LPSS formation, fracturing must start before LPSS development but can be reactivated or continue after LPSS formation. Therefore, it is likely that at least some of these structures originated as LPS tension joints prior to folding, and that they have been opened later during subsequent folding of the strata (Figure 8 and see the discussion below). For joint formation, smaller stresses are required, and therefore, they develop during a prefolding or postfolding stage instead of during the main compressive phase (Ramsay and Huber, 1987). The wide variation in orientations could be explained by deformation of a rock with an inherited north–south anisotropy from older fractures or alternatively by a lowered differential stress and, thus, development of less well-oriented fractures (Cosgrove, 1997). Alternatively, LPSS-parallel fractures may also form as extrados structures in the hinges of antiformal structures during folding of the strata.

The different intermediate stress stages during this growing compressional regime are schematically represented in Figure 7. The following stages are differentiated. Stage 1 represents the pre-BPS to syn-BPS regime (see above). Stage 2 corresponds to the period of increasing burial, with σ_V (or σ_1) increasing proportionally. Because of the onset of tectonic forces, σ_H (or σ_3) must rise significantly, and consequently, the differential stress ($\Delta\sigma = \sigma_1 - \sigma_3$) decreases. Stage 3 represents the period of growing tectonic stress, until an intermediate stage is reached, where σ_H equals σ_V (i.e., a hydrostatic stress field) and where σ_H approaches zero. Stage 4 corresponds to the period where the tectonic stress rises further, and the horizontal stress exceeds the vertical stress to become the main principal stress ($\sigma_H = \sigma_1$). The vertical stress becomes the least principal stress ($\sigma_V = \sigma_3$) and increases proportionally with burial. Finally, stage 5 represents the period where the differential stress increases and reaches a maximum value simultaneously with σ_1 during the main compressional stage.

In summary, the changing stress field caused first a decrease and afterward an increase in $\Delta\sigma$. Note that in this evolution, the importance of the intermediate principal stress (σ_2) has been neglected, because the real changes between the three principal stresses are not known. However, the general trend with decreasing and subsequently increasing $\Delta\sigma$ remains valid (e.g., between σ_1 and σ_2 or between σ_2 and σ_3).

The decrease in differential stress ($\Delta\sigma$) can explain the less well-oriented fractures that develop during this period. The lower the difference in principal stresses, the less fractures are forced to develop along preferred orientations, and as a consequence, the more chaotic fracture patterns develop (Cosgrove, 1997). This pattern also depends on the lithology, with more aligned fractures in limestone and more breccia development in dolomite and evaporitic rocks.

During stages 2–4, the confining pressure rises, whereas the differential stress decreases, making fracturing without increased fluid pressures impossible according to the failure criteria. For shear fracturing, a larger differential stress is needed with increasing confining pressure, and for extension fracturing, which is favored by decreasing differential stress, a smaller, least principal stress is needed. The only way to explain fracturing under these conditions is to invoke a high internal fluid pressure (p_f) to achieve lower effective stresses (E_σ). Hydrofracturing must be the driving force for fracturing during this period, as was already proposed based on some breccia-vein morphologies. With increasing differential stress (stage 5), the failure criterion predicts the formation of shear fractures instead of extension fractures even at high fluid pressures. This was also found in this study, where extension (hydro-) fractures and tension joints develop prior to and at the same time as LPSS development and which are succeeded by the development of shear fractures.

The geometry of floating matrix fragments in vein cement suggests that an open structure was preserved after hydrofracturing, in which the fragments were mechanically stabilized in three dimensions. This texture could partly be explained by rapid cementation after fluid pressure drop. Because of pressure drop, the solubility of CO_2 in the fluid decreases, and calcite precipitates.

Cementation in the fracture-shaped crack-seal veins and in the breccia veins can be explained either by the migration of large volumes of fluids (advection) or by a local continuous source of ions from pressure solution (diffusion) or a combination of both. Cathodoluminescence and isotopic characteristics of the infilling cements indicate some host rock buffering and also a trend toward higher temperatures (Ferket et al., 2000, 2003b). However, later cement generations in these fractures indicate a change in mineralogy, cathodoluminescence pattern, and isotopic characteristics (see post-LPSS discussion). A caterpillar type of LPS propagation can provide a plausible explanation. During early LPS, fluids are progressively forced toward the foreland. High fluid pressures can cause hydrofracturing, and subsequent compression of the rock volume leads to pressure solution with cement precipitation in open spaces. If this deformation progrades with distinct pulses, the hydrofracture cementation process can be repeated several times and give rise to multiple crack-seal fractures.

The Post-LPSS Stage

In this section, the opening of LPSS planes will be addressed before focusing on fracture and breccia development.

Several mechanisms may be responsible for the opening of LPSS planes after their formation. They may be opened (1) during subsequent folding of the strata or (2) during a later extensional stress regime, when the planes are oriented perpendicular to σ_3. Alternatively, (3) they may be opened because of fluid pressure (i.e., hydrofracturing) or (4) because of crystallization pressure. The porosity-permeability along these planes may be further enhanced by dissolution of adjacent host rock (5) when undersaturated or aggressive fluids circulate along these stylolites. An evaluation of each of these mechanisms is given below. It is not possible to determine one major controlling mechanism. Different mechanisms might occur developing under different circumstances and at different positions and times.

1) LPSS planes form early during the compressional regime, prior to folding and thrusting of the strata (Tavarnelli, 1997; Storti and Salvini, 2001). These stylolites form perpendicular to the main principal stress (σ_1) and consequently chiefly develop perpendicular to bedding. Their orientation makes them sensitive to be opened during subsequent folding of the strata, at least in anticlinal hinges, where the rock is locally in extension. This mechanism, therefore, is especially plausible to explain the sediment infilled (Plate 3A) and cemented stylolites (Plate 3B) in anticlinal hinges, whereby the sediment infill relates to telogenetic (i.e., karst-related), diagenetic processes.
2) Opening of stylolitic planes in an extensional stress field necessarily depends on the orientation of the stylolites and of the principal stresses and also on the mechanical strength of the rock and its heterogeneities. Layer-parallel shortening stylolite planes, similarly to fractures, can form mechanically weaker boundaries caused by loss of cohesion (compare with the anticrack model of Fletcher and Pollard, 1981). Moreover, the concentration of organic material and eventual precipitation of cements along the stylolites influence the mechanical strength of these structures. Opening of LPSS planes will be favored, if the stylolites are oriented perpendicularly to σ_3 and if they are mechanically weaker than the host rock and nearby fractures, for example, because of minor porosity along LPSS planes, or if appreciable insoluble residue is concentrated along these planes.

During Oligocene, an intraplate extension with a roughly N50–60° direction, related with the Basin and Range phase of extension, affected the study area (Sedlock et al., 1993). When we compare the orientation of LPSS planes (Figure 4c) with the later σ_3 component of the extensional stress field (N50–60°), stylolitic planes with a N140–150° orientation should be most sensitive for this opening mechanism.

3) Opening of LPSS planes by hydrofracturing (e.g., because of upward-migrating hydrocarbons) is difficult to assess. Their low initial porosity-permeability makes these stylolites not very suitable as fluid conduits. However, hydrocarbon migration along stylolites is common, as attested by many oil-stained stylolites (Plate 3C). The latter, however, can also be explained by hydrocarbon migration caused by capillary forces (Bethke et al., 1991) or by late-stage migration after opening or creation of secondary porosity along the stylolites.
4) Crystallization pressure may be a driving force for opening LPSS planes along which cements precipitated. However, an initial (small) opening by another mechanism is needed before pressure buildup in relation to cementation can occur. This mechanism depends merely on the confining pressure

and will be more important with lower confining pressures. It is therefore likely to be restricted to the shallow subsurface and outcrops. It may have been a significant factor in the opening of stylolites in the western part of the study area, with emerged thrust sheets subjected to erosion and karstification.

5) Further opening of LPSS planes can also be attributed to dissolution of host rock adjacent to the stylolites by undersaturated or aggressive fluids. This can be important in karstic areas, where meteoric water infiltrates the strata as attested in the study area or when acidic fluids precede oil migration.

The same mechanisms may be responsible for the opening of LPS tension joints. Opening of LPS-related structures during subsequent folding of the strata can enhance vertical fluid migration and contribute to the observed changes in mineralogy, cathodoluminescence pattern, and isotope characteristics of the early post-LPSS cement infills. Bedding-parallel shear during folding of the strata can also contribute to some of these changes, like the precipitation of fluorite. Shear in the Lower and middle Cretaceous evaporitic layers can cause dissolution of evaporites, and consequently, saline fluids can migrate along open fractures and cause precipitation of dolomite and fluorite in the overlying middle and Upper Cretaceous limestones.

The slightly dipping post-LPSS veins displace and deform older veins (or other markers), which they crosscut and which consequently reflect shear parallel to the vein length. Moreover, the close relationship of some of these veins with BPS planes and especially the occurrence of BPS traces in some veins indicate slip along BPS planes or bed-parallel shear. This is especially the case in anticlinal hinges, where these veins form subparallel to bedding. In anticlinal limbs, these veins may form under a higher angle with bedding. The vein and crystal morphology shows that these veins are composed of different lenticular, sigmoidal, or flat segments filled with a blocky cement. This suggests that the veins originated as tension gashes or pull-apart structures in a shear zone, instead of by continuous growth during shearing, in which case slickenfibers would form. These different segments are stacked on top of each other and separated by solid inclusion bands, whereby crystals are deformed and local pressure solution occurs. This indicates that shear also affected the newly formed veins. However, in a classical shear zone, tension gashes develop typically under an angle of 30–45° with respect to the shear direction. The studied veins, on the contrary, are subparallel to the shear direction. A similar vein texture has been explained by Petit et al. (1999), a "crack-seal, slip" mechanism, whereby a preexisting (micro-) vein serves to align the newly formed shear vein. In this model, the vein forms by a cyclic crack-seal process, which is succeeded each time by a slip event and which requires overpressure to reopen the preexisting vein. The development of these shear fractures might relate to thrust propagation (Figure 8), which follows a discontinuous, cyclic deformation path with small slip events restricted in space and time (Twiss and Moores, 1992).

The steeply dipping post-LPSS veins display a large scatter in vein orientations (Figure 4d), which are difficult to interpret, because these fractures may have formed in relation with the Paleocene to Eocene compressional regime, during the transitional stage or in relation with the Oligocene or younger extensional regime. Older directions are reactivated, as can be inferred when Figure 4d is compared with Figures 4a and b. The heterogeneity in post-LPSS fracture orientation can, at least partly, be explained by the mechanical anisotropy of the deformed rock after Laramide FFTB development. During Oligocene intraplate extension, extension fractures are expected to form perpendicular to σ_3 (N50–60°), i.e., with a N140–150° orientation, which is an old compression-related direction. Old fractures that have an orientation close to N140 or N150 are also likely to be reactivated. Different fracture criteria and, consequently, also different orientations exist for anisotropic media (Twiss and Moores, 1992). Therefore, the fracture strength of the deformed crust is weaker than expected.

Dark, banded veins which are filled with silt-sized and clay-sized material and which lack significant cement precipitation (Plates 4D, E) could reflect planes along which opening was not sufficient for crystal growth or along which the circumstances were not favorable for cement precipitation. The infilling material could be interpreted as a kind of gouge material, implicating movement along the plane without increasing the porosity-permeability. A shear (strike-slip) fracture instead of an extension fracture would then develop and act as a sealing structure instead of as a conduit for fluids. When fracture roughness provides a small opening, small-sized cement can precipitate. Repeated reactivation of the fracture planes could explain the banded structure. Alternatively, the observed vein morphology could be explained by a cyclic process of small fracture opening and migration of fluids by capillarity, followed by recrystallization of host rock along the fracture plane and concentration of insoluble residue along the plane. The silt-sized material is indeed similar to vadose microsparite, which is typical of recrystallized micrite. The latter explanation is therefore preferred.

Although these veins might have acted as sealing structures, they have certainly been reactivated and have been used as fluid conduits afterwards. Most of these fractures have been affected by dissolution and fracturing and have subsequently been filled with a blocky cement.

The dissolution-enlarged fractures, eventually leading to breccia development (Plate 4F); the geopetal silt

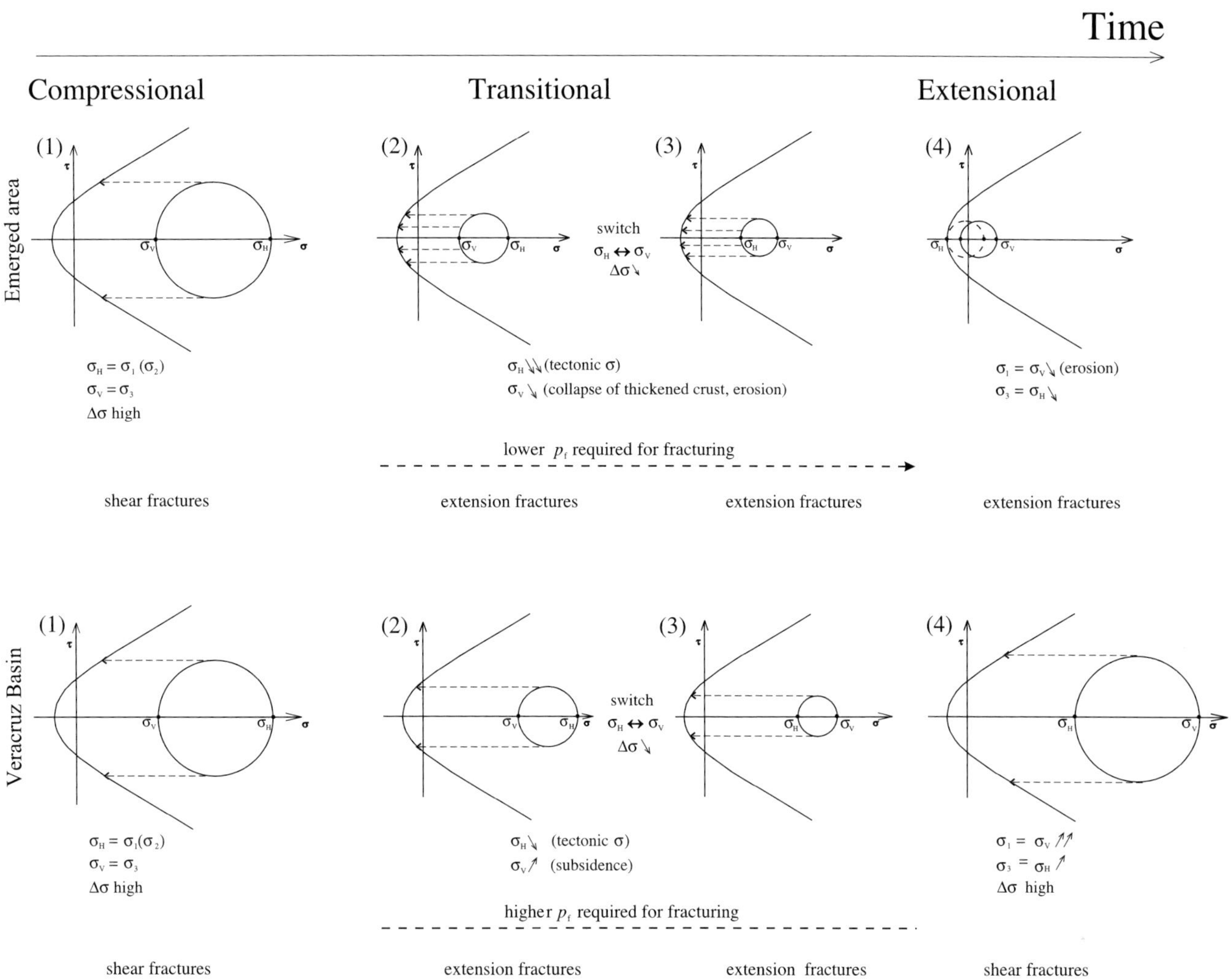

FIGURE 9. Schematic evolution of the paleostress conditions in the study area after LPSS development.

infill, which is typical for the vadose zone (Moore, 2001); and the precipitation of blocky calcite with a drusy texture testify to a stage with appreciable secondary porosity development. The infilling cement shows isotopic signatures indicative of karst processes (Ferket et al., 2000, 2003b; Ortuño-Arzate et al., 2003). Evidence for telogenetic karst development is also found in macroscale cavities and terra rossa infills in the field. Many preexisting veins are affected by these telogenetic karst processes (e.g., Plate 1B).

The kinematic timing of these karst-related veins must postdate the compressional regime (Figure 9, upper part). After Laramide compression, the tectonic stress (σ_1) decreased, and after collapse of the thickened crust and erosion of the emerged areas, σ_3 was also reduced. Before Oligocene extension, a transitional stress stage with a lowered confining pressure and differential stress prevailed. Under these circumstances, a high fluid pressure would be required to form extension fractures. During the subsequent extensional regime, vertical stress exceeded horizontal stress. In the emerged areas subjected to erosion, σ_3 could even become tensile, allowing fracturing without requiring a high fluid pressure.

Diagenetic investigation of the karst-related veins indeed points toward an open, meteoric fluid circulation. Advection was probably the dominant fluid transport mechanism and was related to gravity-driven flow of meteoric water from recharge zones in the topographically elevated areas or to meteoric water that directly infiltrated from the surface into open fractures.

After Laramide deformation, the tectonic front area was buried below the subsiding Veracruz Basin. The vertical stress increased to become the main principal stress (Figure 9, lower part), and with increasing confining pressure, higher fluid pressure becomes a requirement for fracturing. Because of an increasing differential stress, the development of shear fractures is predicted for the buried Cretaceous carbonates during the late Tertiary and Quaternary.

Post-LPSS breccia is only found in the buried tectonic front area. The timing of formation can be constrained to be post-upper Paleocene, i.e., the time of

LPSS development according to coeval flysch sedimentation (de Cserna, 1989). Moreover, breccia formation predates oil migration, which was calculated to start from early Eocene according to thermal modeling (Ortuño-Arzate et al., 2003). The breccia can therefore be assumed to have formed during the early Eocene, which is the period of final thrust emplacement in the tectonic front position. This timing is in line with the occurrence of the breccia just above a major thrust plane. The mechanism responsible for brecciation therefore probably relates to fault development. This breccia is consequently interpreted to reflect the damage zone of the thrust fault (Gudmundsson, 2001). The morphology of the rubble float breccia and of the infilling carbonate cements indicates appreciable porosity development and crystal growth in large, open spaces. The responsible mechanism for the opening of the fracture network is therefore most probably hydrofracturing (Sibson, 1994; Gudmundsson, 2001).

CONCLUSIONS

Pre-BPS to syn-BPS multiple crack-seal and crack-jump veins are interpreted to have formed during burial in a cyclic stress field related to foreland flexuring. The responsible mechanism for fracture opening and propagation is most probably hydrofracturing, whereas cement precipitation likely relates to pressure solution.

Post-BPS, vertical, fracture-shaped veins and breccia veins are interpreted in relation to a caterpillar type of LPS propagation, whereby fluids are progressively forced toward the foreland. Cyclic hydrofracturing caused by repeated compression pulses and subsequent cementation caused by LPS give rise to multiple, cemented crack-seal fractures. The less well-oriented veins and breccia-vein morphology indicate a lowered differential stress, which is in agreement with the changing stress field in that period.

Prefolding LPS structures may be opened during subsequent folding of the strata and create vertical migration pathways. This results in a change in the geochemistry of the infilling cements in post-LPSS veins. Stylolites can consequently have two functions, initially as impermeable barriers and later as conduits for fluids.

Post-LPSS veins that are indicative of shear are composed of multiple segments. They are interpreted to relate to bed-parallel shear and/or thrust migration.

Dark, banded veins reflect small fracture openings along which recrystallization of the host rock occurred. Although these fractures may have acted initially as sealing structures, they have commonly been used as fluid conduits later in the diagenetic evolution.

Dissolution-enlarged fractures and breccia filled with geopetal vadose silt and blocky, drusy calcite cement testify of telogenetic karst processes, which have also affected older veins.

Damage zone-related multiple brecciation has been recognized in the buried tectonic front area and was accompanied by appreciable porosity development.

ACKNOWLEDGMENTS

This study has been realized thanks to the technical and professional assistance of Pemex, Exploración y Producción in the Veracruz district. We are grateful to Miguel Espinoza Nava, Juan Toriz Gama, Luis Ricardo Caraveo Miranda, and Martín Martínez Medrano for their help in organizing the field work in the Cordoba–Veracruz area and for access to data of cores and maps that were necessary to construct the geological map and cross sections of the area. We also warmly thank H. Nijs for the careful thin-section preparation. The constructive and critical comments of Professor Manuel Sintubin (University of Leuven, Belgium) and of the reviewers Gary Gray (Exxon-Mobil) and José M. Grajales Nishimura (Instituto Mexicano del Petróleo) were greatly appreciated and have helped in improving the manuscript.

REFERENCES CITED

Andrews, L. M., and L. B. Railsback, 1997, Controls on stylolite development: Morphologic, lithologic, and temporal evidence from bedding-parallel and transverse stylolites from the U.S. Appalachians: Journal of Geology, v. 105, p. 59–73.

Bathurst, R. C. G., 1975, Carbonate cements and their diagenesis: Developments in Sedimentology, v. 12, 658 p.

Bethke, C. M., J. D. Reed, and D. F. Oltz, 1991, Long-range petroleum migration in the Illinois basin: AAPG Bulletin, v. 75, p. 925–945.

Bjørlykke, K., 1994, Fluid-flow processes and diagenesis in sedimentary basins, *in* J. Parnell, ed., Geofluids: Origin, migration and evolution of fluids in sedimentary basins: Geological Society (London) Special Publication 78, p. 127–140.

Bons, P. D., 2000, The formation of veins and their microstructures: Journal of the Virtual Explorer, v. 2.

Bons, P. D., 2001a, Development of crystal morphology during unitaxial growth in a progressively widening vein: I. The numerical model: Journal of Structural Geology, v. 23, p. 865–872.

Bons, P. D., 2001b, The formation of large quartz veins by rapid ascent of fluids in mobile hydrofractures: Tectonophysics, v. 336, p. 1–17.

Bons, P. D., and M. W. Jessell, 1997, Experimental simulation of the formation of fibrous veins by localised dissolution-precipitation creep: Mineralogical Magazine, v. 61, p. 53–63.

Bradbury, H. J., and G. R. Woodwell, 1987, Ancient fluid flow within foreland terrains, *in* J. C. Goff and B. P. J.

Williams, eds., Fluid flow in sedimentary basins and aquifers: Geological Society (London) Special Publication 34, p. 87–102.

Callan, N. J., and E. T. C. Spooner, 1998, Repetitive hydraulic fracturing and shear zone inflation in an Archean granitoid-hosted, ribbon banded, Au-quartz vein system, Renabie area, Ontario, Canada: Ore Geology Reviews, v. 12, p. 237–266.

Caputo, R., and P. L. Hancock, 1999, Crack-jump mechanism and its implications for stress cyclicity during extension fracturing: Journal of Geodynamics, v. 27, p. 45–60.

Cosgrove, J. W., 1997, Hydraulic fractures and their implications regarding the state of stress in a sedimentary sequence during burial, *in* S. Sengupta, ed., Evolution of geological structures in micro- to macroscales: London, Chapman, and Hall, p. 11–25.

Cosgrove, J. W., 2001, Hydraulic fracturing during the formation and deformation of a basin: A factor in the dewatering of low-permeability sediments: AAPG Bulletin, v. 85, p. 737–748.

Davies, J. H., 1999, The role of hydraulic fractures and intermediate-depth earthquakes in generating subduction-zone magmatism: Nature, v. 398, p. 142–145.

De Cserna, Z., 1989, An outline of the geology of Mexico, *in* A. W. Bally and A. R. Palmer, eds., The geology of North America— An overview: Boulder, Colorado, Geological Society of America, The geology of North America, v. A, p. 233–264.

Ferket, H., F. Roure, R. Swennen, and S. Ortuño-Arzate, 2000, Fluid migration placed into the deformation history of fold-and-thrust belts: An example from the Veracruz Basin (Mexico): Journal of Geochemical Exploration, v. 69–70, p. 275–279.

Ferket, H., S. Ortuño-Arzate, F. Roure, and R. Swennen, 2003a, Lithologic control on matrix porosity in shallow-marine Cretaceous reservoir limestones: A study of the Peñuela reservoir outcrop analogue (Cordoba Platform, southeastern Mexico), *in* C. Bartolini, R. T. Buffler, and J. Blickwede, eds., The Circum-Gulf of Mexico and the Caribbean: Hydrocarbon habitats, basin formation, and plate tectonics: AAPG Memmoir 79, CD-ROM, p. 283–304.

Ferket, H., R. Swennen, S. Ortuño-Arzate, and F. Roure, 2003b, Reconstruction of the fluid flow history during Laramide foreland fold and thrust belt development in eastern Mexico: Cathodoluminescence and $\delta^{18}O$-$\delta^{13}C$ isotope trends of calcite-cemented fractures: Journal of Geochemical Exploration, v. 78–79, p. 163–167.

Fletcher, R. C., and D. D. Pollard, 1981, Anticrack model for pressure solution surfaces: Geology, v. 9, p. 419–424.

Gray, M. B., and G. Mitra, 1993, Migration of deformation fronts during progressive deformation— Evidence from detailed structural studies in the Pennsylvania anthracite region, U.S.A.: Journal of Structural Geology, v. 15, p. 435–449.

Groshong, R. H., 1988, Low-temperature deformation mechanisms and their interpretation: Geological Society of America Bulletin, v. 100, p. 1329–1360.

Gudmundsson, A., 2001, Fluid overpressure and flow in fault zones: Field measurements and models: Tectonophysics, v. 336, p. 183–197.

Gudmundsson, A., and S. L. Brenner, 2001, How hydrofractures become arrested: Terra Nova, v. 13, p. 456–462.

Hilgers, C., and J. L. Urai, 2002, Microstructural observations on natural syntectonic fibrous veins: Implications for the growth process: Tectonophysics, v. 352, p. 257–274.

Hilgers, C., D. Koehn, P. D. Bons, and J. L. Urai, 2001, Development of crystal morphology during unitaxial growth in a progressively widening vein: II. Numerical simulations of the evolution of antitaxial fibrous veins: Journal of Structural Geology, v. 23, p. 873–885.

Hubbert, M. K., and D. G. Willis, 1957, Mechanics of hydraulic fracturing, *in* T. D. Cook, ed., Underground waste management and environmental implications: AAPG Memoir 18, p. 239–257.

Jolly, R. J. H., and L. Lonergan, 2002, Mechanisms and controls on the formation of sand intrusions: Journal of the Geological Society, v. 159, p. 605–617.

Kenis, I., M. Sintubin, P. Muchez, and E. A. J. Burke, 2002, The "boudinage" question in the High-Ardenne slate belt (Belgium): A combined structural and fluid-inclusion approach: Tectonophysics, v. 348, p. 93–110.

Knipe, R. J., and A. M. McCaig, 1994, Microstructural and microchemical consequences of fluid flow in deforming rocks, *in* J. Parnell, ed., Geofluids: Origin, migration and evolution of fluids in sedimentary basins: Geological Society (London) Special Publication 78, p. 99–111.

Knipe, R. J., and S. H. White, 1979, Deformation in low grade shear zones in the O.R.S. from S.W. Wales: Journal of Structural Geology, v. 1, p. 53–66.

Knoop, S. R., L. A. Kennedy, and G. M. Dipple, 2002, New evidence for syntectonic fluid migration across the hinterland-foreland transition of the Canadian Cordillera: Journal of Geophysical Research, v. 107 (B4), no. 2071, p. 1–26.

Lee, Y. J., and D. V. Wiltschko, 2000, Fault controlled sequential vein dilatation: Competition between slip and precipitation rates in the Austin Chalk, Texas: Journal of Structural Geology, v. 22, p. 1247–1260.

Lorenz, J. C., L. W. Teufel, and N. R. Warpinski, 1991, Regional fractures: I. A mechanism for the formation of regional fractures at depth in flat-lying reservoirs: AAPG Bulletin, v. 75, no. 11, p. 1714–1737.

Machel, H. G., 1990, Burial diagenesis, porosity and permeability development in carbonates, *in* G. R. Bloy and M. G. Hadley, eds., The development of porosity in carbonate reservoirs: Canadian Society Petroleum Geologists Short Course Notes 2, p. 1–18.

Miller, T. W., 1995, New insights on natural hydraulic fractures induced by abnormally high pore pressures: AAPG Bulletin, v. 79, p. 1005–1018.

Moore, C. H., 2001, Carbonate reservoirs: Porosity evolution and diagenesis in a sequence stratigraphic framework: Developments in Sedimentology, v. 55, 444 p.

Morrow, D. W., 1982, Descriptive field classification of sedimentary and diagenetic breccia fabrics in carbonate rocks: Bulletin of the Canadian Petroleum Geology, v. 30, no. 3, p. 227–229.

Mossman, R. W., and F. Viniegra, 1976, Complex fault structures in the Veracruz Province of Mexico: AAPG Bulletin, v. 60, p. 379–388.

Ortoleva, P. J., 1995, Basin compartments and seals: AAPG Memoir 61, 447 p.

Ortuño-Arzate, S., 1991, Contribution de la sédimentologie et de la télédetection a la reconstitution de l'évolution du Bassin de Zongolica (Mexique centro-oriental). Place dans la dynamique du couple Golfe du Mexique/systeme cordillerain: Ph.D. thesis, Université de Pau et des Pays de l'Adour, France, 289 p.

Ortuño-Arzate, S., H. Ferket, M.-C. Cacas, R. Swennen, and F. Roure, 2003, Late Cretaceous carbonate reservoirs in the Cordoba Platform and Veracruz Basin (eastern Mexico), *in* C. Bartolini, R. T. Buffler, and J. Blickwede, eds., The Circum-Gulf of Mexico and the Caribbean: Hydrocarbon habitats, basin formation, and plate tectonics: AAPG Memoir 79, CD-ROM, p. 476–514.

Osborne, M. J., and R. E. Swarbrick, 1997, Mechanisms for generating overpressure in sedimentary basins: A reevaluation: AAPG Bulletin, v. 81, p. 1023–1041.

Petit, J.-P., A. J. Wibberley, and G. Ruiz, 1999, 'Crack-seal', slip: A new fault valve mechanism?: Journal of Structural Geology, v. 21, p. 1199–1207.

Pollard, D. D., and P. Segall, 1987, Theoretical displacements and stresses near fractures in rock: With applications to faults, joints, veins, dikes and solution surfaces, *in* B. K. Atkinson, ed., Fracture mechanics of rock: New York, Academic Press, p. 277–359.

Railsback, L. B., 1993, Contrasting styles of chemical compaction in the upper Pennsylvanian Dennis Limestone in the mid-continent region, U.S.A.: Journal of Sedimentary Petrology, v. 63, p. 61–72.

Ramsay, J. G., 1980, The crack-seal mechanism of rock deformation: Nature, v. 284, p. 135–139.

Ramsay, J. G., and M. I. Huber, 1987, Folds and fractures, *in* The techniques of modern structural geology: London, Academic Press, v. 2, 700 p.

Roedder, C. H., 1984, Fluid inclusions: Reviews in Mineralogy, v. 12.

Sabouraud, C., J. C. Macquar, and H. Ravier, 1980, Les inclusions fluides, témoins et faux-témoins des conditions de dépôt. Quelques exemples pris dans les minéralisations Pb, Zn, Ba, F du Sud du Massif Central français: Mineralia Deposita, v. 15, p. 211–230.

Secor, D. T., 1965, Role of fluid pressure in jointing: American Journal of Science, v. 263, p. 633–646.

Sedlock, R. L., F. Ortega-Gutiérrez, and R. C. Speed, 1993, Tectonostratigraphic terranes and tectonic evolution of Mexico: Geological Society of America Special Paper 278, 153 p.

Sibson, R. H., 1994, Crustal stress, faulting and fluid flow, *in* J. Parnell, ed., Geofluids: Origin, migration and evolution of fluids in sedimentary basins: Geological Society (London) Special Publication 78, p. 69–84.

Simpson, G. D. H., 2000, Synmetamorphic vein spacing distributions: Characterisation and origin of a distribution of veins from NW Sardinia, Italy: Journal of Structural Geology, v. 22, p. 335–348.

Smith, J. V., 2000, Three-dimensional morphology and connectivity of stylolites hyperactivated during veining: Journal of Structural Geology, v. 22, p. 59–64.

Storti, F., and F. Salvini, 2001, The evolution of a model trap in the central Apennines, Italy: Fracture patterns, fault reactivation and development of cataclastic rocks in carbonates at the Narni anticline: Journal of Petroleum Geology, v. 24, p. 171–190.

Swennen, R., H. Ferket, L. Benchilla, F. Roure, R. Ellam, and Subthrust Reservoir Appraisal team, 2003, Fluid flow and diagenesis in carbonate dominated foreland fold-and-thrust belts: Petrographic inferences from field studies of late-diagenetic fabrics from Albania, Belgium, Canada, Mexico and Pakistan: Journal of Geochemical Exploration, v. 78–79, p. 481–485.

Tavarnelli, E., 1997, Structural evolution of a foreland fold and thrust belt: The Umbria—Marche Apennines, Italy: Journal of Structural Geology, v. 19, p. 523–534.

Toriumi, M., and E. Hara, 1995, Crack geometries and deformation by the crack-seal mechanism in the Sambagawa metamorphic belt: Tectonophysics, v. 245, no. 3–4, p. 249–261.

Travé, A., F. Calvet, A. Soler, and P. Labaume, 1998, Fracturing and fluid migration during Paleogene compression and Neogene extension in the Catalan coastal ranges, Spain: Sedimentology, v. 45, no. 6, p. 1063–1082.

Twiss, R. J., and E. M. Moores, 1992, Structural geology: New York, Freeman and Company, 532 p.

Van Geet, M., R. Swennen, C. Durmishi, F. Roure, and P. H. Muchez, 2002, Paragenesis of Cretaceous to Eocene carbonate reservoirs in the Ionian foreland fold-and-thrust belt (Albania): Relation between tectonism and fluid flow: Sedimentology, v. 49, p. 697–718.

White, W. B., 2002, Karst hydrology: Recent developments and open questions: Engineering Geology, v. 65, no. 2–3, p. 85–105.

Wiltschko, D. V., and J. W. Morse, 2001, Crystallization pressure versus "crack seal" as the mechanism for banded veins: Geology, v. 29, p. 79–82.

8

Moore, T. E., C. J. Potter, P. B. O'Sullivan, K. L. Shelton, and M. B. Underwood, 2004, Two stages of deformation and fluid migration in the west-central Brooks Range fold and thrust belt, northern Alaska, *in* R. Swennen, F. Roure, and J. W. Granath, eds., Deformation, fluid flow, and reservoir appraisal in foreland fold and thrust belts: AAPG Hedberg Series, no.1, p. 157–186.

Two Stages of Deformation and Fluid Migration in the West-Central Brooks Range Fold and Thrust Belt, Northern Alaska

Thomas E. Moore
U.S. Geological Survey, Menlo Park, California, U.S.A.

Christopher J. Potter
U.S. Geological Survey, Lakewood, Colorado, U.S.A.

Paul B. O'Sullivan
Apatite to Zircon, Inc., Moscow, Idaho, U.S.A.

Kevin L. Shelton
Department of Geological Sciences, University of Missouri-Columbia, Missouri, U.S.A.

Michael B. Underwood
Department of Geological Sciences, University of Missouri-Columbia, Missouri, U.S.A.

ABSTRACT

The Brooks Range is a north-directed fold and thrust belt that forms the southern boundary of the North Slope petroleum province in northern Alaska. Field-based studies have long recognized that large-magnitude, thin-skinned folding and thrusting in the Brooks Range occurred during arc-continent collision in the Middle Jurassic to Early Cretaceous (Neocomian). Folds and thrusts, however, also deform middle and Upper Cretaceous strata of the Colville foreland basin and thus record a younger phase of deformation that apatite fission-track data have shown to occur primarily during the early Tertiary (~60 and ~45 Ma). A structural and kinematic model that reconciles these observations is critical to understanding the petroleum system of the Brooks Range fold and thrust belt.

New interpretations of outcrop and regional seismic reflection data indicate that from the modern mountain front northward to near the deformation front under the coastal plain, the basal thrust detachment for the orogen is located in the Jurassic and Lower Cretaceous Kingak Shale in the upper part of the regionally extensive, gently

DOI:10.1306/1025690H13116

south-dipping, north-derived Mississippian to Early Cretaceous Ellesmerian sequence. The frontal part of the orogen lies in middle Cretaceous foreland basin strata and consists of a thin-skinned fold belt at the deformation front and a fully developed passive-roof duplex to the south. Near the mountain front, the orogen is composed of a stacked series of allochthons and thrust duplexes and associated Neocomian syntectonic deposits that are unconformably overlain by proximal foreland basin strata. The foreland basin strata and underlying deformed rocks are truncated by a younger generation of folds and thrusts. Vitrinite reflectance and stable isotope compositions of veins provide evidence of two fluid events in these rocks, including an earlier higher temperature ($\sim$250–300°C) event that was buffered by limestone and a younger, lower temperature ($\sim$150°C) event that had distinctly lower $\delta^{13}C$ values as a result of oxidation of organic matter and/or methane. Zircon fission-track data from the host rocks of the veins show that the higher temperature fluid event occurred at 160–120 Ma, whereas the lower temperature event probably occurred at about 60–45 Ma.

It is proposed that the Brooks Range consists of two superposed contractional orogens that used many of the same mechanically incompetent stratigraphic units (e.g., Kayak Shale, Kingak Shale) as sites of thrust detachment. The older orogen formed in a north-directed arc-continent collisional zone that was active from 160 to 120 Ma. This deformation produced a thin-skinned deformational wedge that is characterized by far-traveled allochthons with relatively low structural relief, because it involved a thin (1–4-km [0.6–2.5-mi]-thick) stratigraphic section. Deeper parts of the deformational wedge are envisioned to have contained relatively high-temperature fluids that presumably migrated from or through limestone-rich source areas in the underlying autochthon or from deeper parts of the orogen. The younger orogen, which formed initially at about 60 Ma and reactivated at 45 Ma, produced a thrust belt and frontal triangle zone with low amounts of shortening and relatively high structural relief, because it involved a structural section 5–10 km (3–6 mi) thick. Fluids associated with this deformation were relatively of lower temperature and suggest that hydrocarbon migration occurred at this time.

We conclude that hydrocarbon generation from Triassic and Jurassic source strata and migration into stratigraphic traps occurred primarily by sedimentary burial principally at 100–90 Ma, between the times of the two major episodes of deformation. Subsequent sedimentary burial caused deep stratigraphic traps to become overmature, cracking oil to gas, and initiated some new hydrocarbon generation progressively higher in the section. Structural disruption of the traps in the early Tertiary released sequestered hydrocarbons. The hydrocarbons remigrated into newly formed structural traps, which formed at higher structural levels or were lost to the surface. Because of the generally high maturation of the Colville basin at the time of the deformation and remigration, most of the hydrocarbons available to fill traps were gas.

INTRODUCTION

Interest in the petroleum potential of the Brooks Range foreland fold and thrust belt has grown in recent years because of new proposals for gas pipeline construction in northern Alaska and because of hope that exploration strategies that have proven successful in generally similar reservoir lithologies and structural settings in the foreland fold and thrust belt of the Rocky Mountains of Alberta and British Columbia can be employed with equal success in the Brooks Range. This interest is warranted because of the presence of subeconomic discoveries of oil and gas in structural traps in the northern foothills of the Brooks Range and by the recent U.S. Geological Survey assessment that estimates a mean value of more than 32 tcf of gas in the National Petroleum Reserve, Alaska (NPRA) (Bird and Houseknecht, 2002a, b). Because the focus of

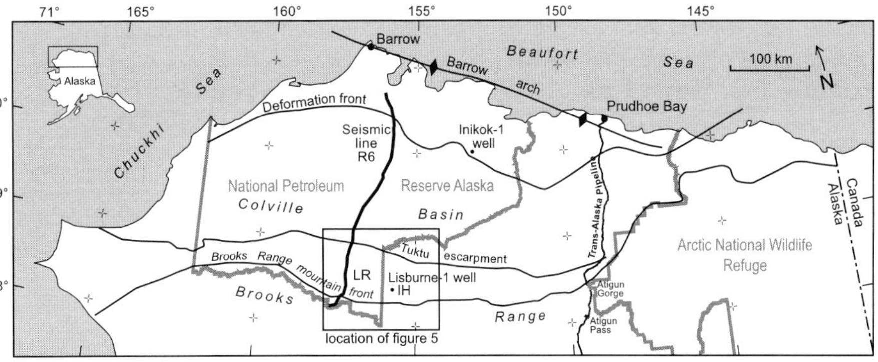

FIGURE 1. Index map of northern Alaska showing locations and features discussed in the text. The North Slope petroleum province encompasses the region north of the mountain front of the Brooks Range. Abbreviations: IH = Ivotuk Hills; LR = Lisburne Ridge.

exploration moved northward into stratigraphic traps in relatively undeformed strata of the Arctic platform and Barrow arch following the discovery of the Prudhoe Bay oil field in 1967, the Brooks Range fold and thrust belt remains a frontier province, with only about 35 separate structures having been tested by drilling during the entire history of exploration in northern Alaska.

New discoveries in the Brooks Range fold and thrust belt will depend, in part, on advances in understanding its structural development and improved understanding of the timing of petroleum generation and migration. The Brooks Range has long been regarded as the northerly continuation of the Late Cretaceous and Tertiary Rocky Mountain system from a geomorphic viewpoint (Wahrhaftig, 1965). Investigations of the geology of the Brooks Range over the past several decades, however, have focused on high-displacement, north-vergent folding and thrusting and emplacement of allochthons that occurred in the Middle Jurassic to Early Cretaceous (e.g., Roeder and Mull, 1978; Mull, 1985; Mayfield et al., 1988; Moore et al., 1994). Structures that deform middle Cretaceous and younger strata in the frontal part of the orogen were regarded as evidence that this orogenic episode persisted into the Late Cretaceous or early Tertiary (Mull, 1985). More recently, O'Sullivan (1996) and O'Sullivan et al. (1997) determined from apatite fission-track dating that major regional episodes of cooling and tectonic-related uplift occurred in the early and middle Tertiary, which led them to propose that the modern Brooks Range was constructed principally in the Tertiary during a separate, younger orogenic event. Although the fission-track evidence for Tertiary deformation is compelling, the structures on which the uplift occurred generally have not been widely recognized, particularly in the axial part of the orogen.

The purposes of this study are to summarize the structural framework in the foreland part of the fold and thrust belt in the west-central Brooks Range, to identify structures that might have formed in the Tertiary, to propose a conceptual model for the origin of the fold and thrust belt, and to identify the fluids that were expelled from the deformed belt through time (e.g., Oliver, 1986). Our analysis was developed during the course of assessing the petroleum potential of structural plays in NPRA in the central North Slope (Moore and Potter, 2003; Potter and Moore, 2003) and uses outcrop studies, seismic reflection investigations, wire-line-log, thermal-maturity, and fission-track data. We conclude that the central and western Brooks Range fold and thrust belt is the product of superposition of two distinct, north-vergent orogenic events, one formed in the Late Jurassic and Early Cretaceous and the other in the early Tertiary. Stable isotope data from calcite and quartz veins that were collected over the study area indicate that each orogenic episode was associated with fluids with distinct compositions and support the interpretation that the younger deformation caused the remigration of hydrocarbons from early-formed stratigraphic traps into structural traps associated with the deformation.

REGIONAL SETTING

The Brooks Range is a north-directed contractional orogen that forms the southern boundary of the North Slope petroleum province in northern Alaska (Figure 1). To the north of the Brooks Range is the Colville basin, the foreland basin of the orogen. The Colville basin is asymmetric, with a deformed, steep southern flank resting on allochthonous rocks of the Brooks Range and a gently dipping northern flank resting on the south-dipping Arctic platform, the autochthon for the orogen. The northern margin of the Colville basin is located at the Barrow arch, where the Arctic platform has subsided northward because of the rifting that resulted in the formation of the Canada basin and the Arctic Ocean. The basin is filled with middle Cretaceous and younger clastic strata derived from the orogenic belt.

Deformation in the Brooks Range in the Jurassic and Early Cretaceous is attributed to collision of an island arc with a south-facing Devonian to Jurassic passive margin, whose northern, proximal part is the Arctic platform autochthon (Roeder and Mull, 1978; Box, 1985; Mayfield et al., 1988; Moore et al., 1994). Them collision resulted in the northward emplacement of Jurassic ophiolite, imbrication of distal parts of the passive margin into extensive allochthons, and development of blueschist facies metamorphic rocks and penetrative strain in continental lithologies in the internal part of the orogen. The deformation began in the Middle Jurassic, with emplacement of ophiolitic allochthons, and continued through the Neocomian (i.e., the pre-Aptian part of the Early Cretaceous), with the successive imbrication and emplacement of allochthons composed of progressively more proximal parts of the passive-margin succession (Moore et al., 1994). The timing of the deformation is constrained by paleontologic dating of synorogenic clastic rocks and is supported by isotopic age data from metamorphic rocks at the base of the ophiolitic klippen and from the metamorphic rocks in the core of the Brooks Range (e.g., Moore et al., 1994; Gottschalk and Snee, 1998).

Following this period of contractional deformation, the hinterland was subjected to orogen-scale exhumation from about 105 to 96 Ma (Vogl et al., 2002). The exhumation has been interpreted to be the result of extensional tectonism and down-to-the-south normal faulting along the southern flank of the Brooks Range (Miller and Hudson, 1991). Sedimentary filling of much of the Colville basin occurred approximately at the same time, suggesting that these deposits were derived from tectonic unroofing in the hinterland of the orogen (Vogl et al., 2002).

Evidence for deformation of the Colville basin and the northeastern Brooks Range in the Late Cretaceous and Tertiary is indicated by stratigraphic relations and fission-track data. In the Colville basin in the central North Slope, strata as young as the Maastrichtian are folded into broad gentle synclines with fault-cored anticlines. Fission-track data from these structures indicate that kilometer-scale uplift-related cooling occurred principally at about 60 Ma (O'Sullivan, 1996; Cole et al., 1997; O'Sullivan et al., 1997) and at about 45 Ma (O'Sullivan et al., 2002). In the northeastern Brooks Range, structural relief is much higher, and metamorphosed early Paleozoic and Precambrian rocks are exposed over broad areas in the mountains and in the cores of folds near the mountain front. Near the thrust front in the coastal area of northeastern Alaska, strata as young as the Miocene are involved in the deformation (e.g., Potter et al., 2004). Apatite fission-track data from these areas indicate that uplift-related cooling of these rocks occurred at about 60, 46, 35, and 24 Ma (O'Sullivan et al., 1997). These observations suggest that the northeastern Brooks Range forms a structural salient that was constructed by prolonged, deep-seated shortening that occurred episodically throughout much of the Tertiary (Potter et al., 2004) and resulted in the denudation of as much as 10 km (6 mi) of structural section (O'Sullivan et al., 1993; Moore, 1999; O'Sullivan and Wallace, 2002).

STRATIGRAPHY OF THE CENTRAL BROOKS RANGE AND THE NORTH SLOPE

The stratigraphy of the Brooks Range and the North Slope is complex and rich in detail and has been organized to emphasize source area and genetic relations (Lerand, 1973; Hubbard et al., 1987). We simplify the discussion of the stratigraphy of this region for the purpose of this study by dividing it into three major tectonic elements, including (1) autochthon, (2) Late Jurassic and Early Cretaceous deformational wedge, and (3) Colville (foreland) basin (Figure 2). This organization provides a useful framework for analyzing the structural development of the region while still being compatible with the stratigraphic divisions of Lerand (1973) and Hubbard et al. (1987).

The autochthon consists of Neocomian and older strata of the Ellesmerian sequence of Lerand (1973) and underlying pre-Mississippian strata and local granitic rocks (Franklinian sequence of Lerand, 1973). The pre-Mississippian rocks were deformed and mildly metamorphosed during the Early Devonian. The thickness of the pre-Mississippian rocks is undetermined, but seismic data and field relations suggest that it must be substantial. The pre-Mississippian rocks are overlain by the Ellesmerian sequence, which is typically 2–4 km (1.2–2.5 mi) thick and consists of carbonate rocks and compositionally mature sediments that were shed from northern source areas (Lerand, 1973). The Ellesmerian sequence rests on a regional unconformity that developed across the pre-Mississippian rocks during rifting and the onset of passive-margin deposition in the early Mississippian. The lowermost part of the sequence consists of nonmarine and marine clastic deposits that are thickest in local extensional basins (Devonian and Mississippian Endicott Group) and overlying carbonate strata (Lisburne Group) that transgressed from south to north during the Mississippian and Pennsylvanian. These strata are overlain by Permian and Lower Triassic deltaic clastic strata that thin and fine southward (Sadlerochit Group) and organic-rich carbonate and fine-grained clastic strata of the Upper Triassic Shublik Formation and Sag River Sandstone. The upper part of the Ellesmerian sequence (Beaufortian sequence of Hubbard et al., 1987) consists of a southward-thinning depositional wedge of Jurassic and Lower Cretaceous fine-grained clastic strata (Kingak Shale). This succession is capped by the regional Lower Cretaceous unconformity that cuts more deeply into underlying rocks to the north and an overlying thin, blanketlike sequence of marine clastic deposits and shale (pebble shale unit) that marks the end of Ellesmerian sedimentation. The upper part of the Ellesmerian sequence was shed from an uplifted region caused by rifting along the northern margin of Alaska, with the Lower Cretaceous unconformity marking the

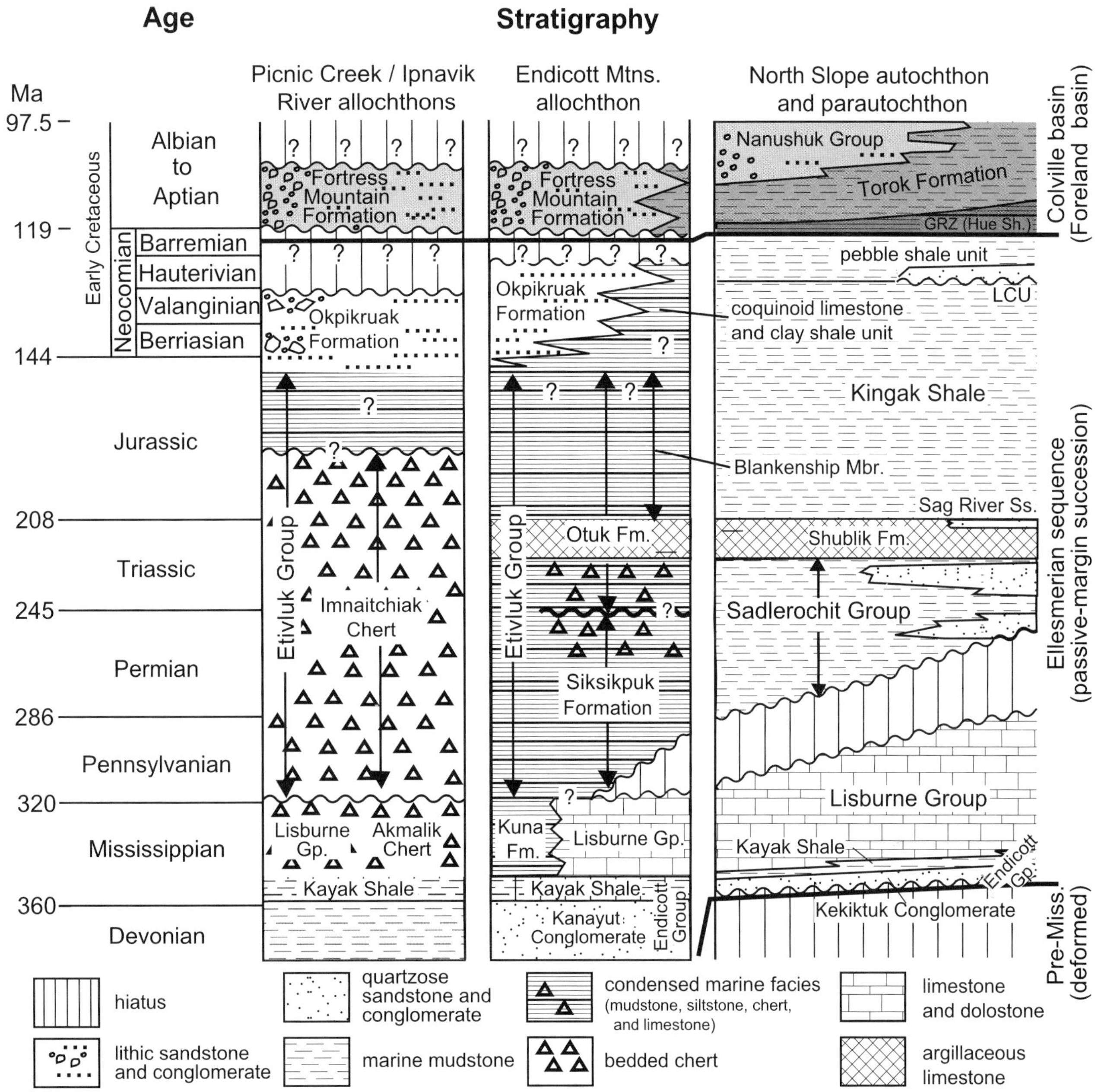

FIGURE 2. Stratigraphic ages, names, and rock types of allochthonous strata in the Brooks Range and northern foothills compared to time-equivalent strata in the autochthon in North Slope subsurface. Brooks Range and foothills stratigraphy shown is composite of Picnic Creek and Ipnavik River (higher) allochthons (left column) and the Endicott Mountains allochthon (center column) (Mull, 1985). Abbreviations: GRZ = gamma-ray zone; LCU = Lower Cretaceous unconformity; Miss. = Mississippian.

time of opening of the Canada basin and formation of the Arctic Ocean (Grantz et al., 1990). Although in as much as 1.5 km (1 mi) thick in the northern part of the central North Slope, the rift-related deposits thin and fine markedly southward in the autochthon into less than 300 m (1000 ft) of shale (Houseknecht, 2003; Houseknecht and Bird, 2004) that rests directly on similar clastic rocks of the Shublik Formation in the southern North Slope.

The Late Jurassic and Early Cretaceous deformational wedge consists of a stacked succession of thin, allochthonous stratigraphic sequences and accompanying synorogenic clastic rocks that characterize the stratigraphy of the northern Brooks Range (Mull, 1985; Mayfield et al., 1988; Moore et al., 1994). The allochthonous stratigraphic successions consist of imbricated clastic, carbonate rocks and chert of Late Devonian to Early Cretaceous age that are correlative with the Ellesmerian sequence and thought to represent the distal, outboard part of a passive-margin sequence. In the central Brooks Range, the structurally higher imbricates consist of chert-rich successions assigned to the Picnic Creek, Ipnavik River, and Nuka Ridge allochthons, each of which is less than 1 km (0.6 mi) in thickness (Mayfield et al., 1988). Allochthons of mafic and ultramafic rocks, including ophiolite, lie at even higher structural levels. Because of their low thicknesses and high structural positions, these imbricates are here collectively referred to as "higher allochthons."

At structurally lower levels, allochthonous rocks display distinctly different stratigraphic features, including (1) a thick, coarse-grained Upper Devonian and early Mississippian succession of clastic strata (Upper Devonian Hunt Fork Shale, Upper Devonian Noatak Sandstone, Upper Devonian and lower Mississippian(?) Kanayut Conglomerate, and lower Mississippian Kayak Shale) that represents a southward-thinning clastic wedge shed from the north principally in the Late Devonian (Nilsen, 1981; Moore and Nilsen, 1984); (2) a lateral facies transition from carbonate platform into basinal shale and chert in Mississippian strata (Lisburne Group); and (3) shale, siliceous shale, and carbonate strata of Permian to Jurassic age. These strata are as much as 4 km (2.5 mi) thick near the Trans-Alaska pipeline in the central Brooks Range (Figure 1) and are collectively referred to as the Endicott Mountains allochthon.

The assembly of the higher allochthons and Endicott Mountains allochthon onto the autochthon occurred on a thrust system that developed on a detachment at the base of the passive-margin sequence. The age of synorogenic clastic deposits (Jurassic and Lower Cretaceous Okpikruak Formation) and stratigraphic relations indicate that the emplacement of the higher allochthons occurred in the Berriasian and Valanginian and was followed by the emplacement of the Endicott Mountains allochthon in the Valanginian to no later than the early Albian (Mull et al., 1997). At the highest structural levels of the range, ophiolite of Middle Jurassic age is present in isolated synforms and is thought to represent preserved remnants of the island arc system, beneath which the Ellesmerian passive-margin succession was underthrust (Box, 1985; Moore et al., 1994). New zircon fission-track data (O'Sullivan et al., 2002; see below) support the interpretation that thrusting occurred in the Middle Jurassic to Early Cretaceous.

The Colville basin is filled with lithic sediments (Brookian sequence of Lerand, 1973) shed from the south and west from source areas in the Brooks Range orogenic belt. Cole et al. (1997) calculated that subsidence reached a maximum in the Barremian, probably caused in part by tectonic loading related to the emplacement of the Endicott Mountains allochthon (Cole et al., 1997). Sedimentary fill in the basin, however, is dominated by wedges of progradational clastic sediments, whose deposition postdated the emplacement of the allochthons. The earliest and most extensive depositional wedge is composed of Aptian and Albian strata (Torok and Nanushuk formations; nomenclature of Mull et al., 2003, is used herein) that form a 10-km (6-mi)-thick progradational sequence covering most of the central and western North Slope. Seismic and well data show that lowermost strata in the wedge are bottomset beds composed of sandstone turbidites (lower Torok Formation). Strata as old as upper Hauterivian could be present at the base of the bottomset interval (Cole et al., 1997). Overlying units consist of clinoforms that are composed of slope mudstone deposits (upper Torok Formation) and topset beds that are composed of shelfal and nonmarine sandstone deposits (Nanushuk Formation) (Mull et al., 2003). Cole et al. (1997) considered the dominance of lateral progradation over vertical aggredation in these strata to be evidence that little or no subsidence occurred in the Colville basin in the Albian. The eastern part of the Colville basin was filled by a succession of younger (Upper Cretaceous and Tertiary) depositional wedges (Bird and Molenaar, 1992).

Regional seismic data show that strata in the Aptian–Albian depositional wedge onlaps the south-dipping strata of the autochthon along the northern margin of the Colville basin, but their relation to the southern margin of the basin is poorly resolved in the seismic data. In the southern part of the northern foothills of the Brooks Range, Colville basin deposits are represented by coarse- to fine-grained, shallow-marine and alluvial deposits of Aptian(?)–Albian age (Fortress Mountain Formation). Although the basal contact of these deposits on deformed rocks of the Late Jurassic and Early Cretaceous deformational wedge is commonly covered or uncertain, a variety of field relations indicate that they rest unconformably on the older rocks (Mull, 1985, their figure 9) (Figure 3). The relatively low thermal maturity of the Fortress Mountain Formation shows that these strata were deposited high on the southern flank of the Colville basin (Cole et al., 1997).

Petroleum source rock units in northern Alaska are rich, abundant, and widespread and include multiple source rock units at several stratigraphic levels in each of the three major tectonic-stratigraphic units. Bird (1991,

FIGURE 3. Photos showing unconformable relation of Aptian(?)–Albian Fortress Mountain Formation on deformed strata of the Early Cretaceous deformational wedge (Endicott Mountains allochthon) in Brooks Range foothills, northeast DeLong Mountains. (A) Westward view of thumbprint syncline defined by Fortress Mountain Formation; (B) eastward view along contact on southern flank of syncline in (A). Bedding in Fortress Mountain Formation (Kfm) dips northward (to left) above south-dipping previously deformed strata of Tingmerkpuk Sandstone (Valanginian) (Kt) and unnamed lithic clastic deposits (Hauterivian and/or Barremian) (Kl) below unconformity.

Kl
Kfm
A

Kl
Kfm
Kt
B

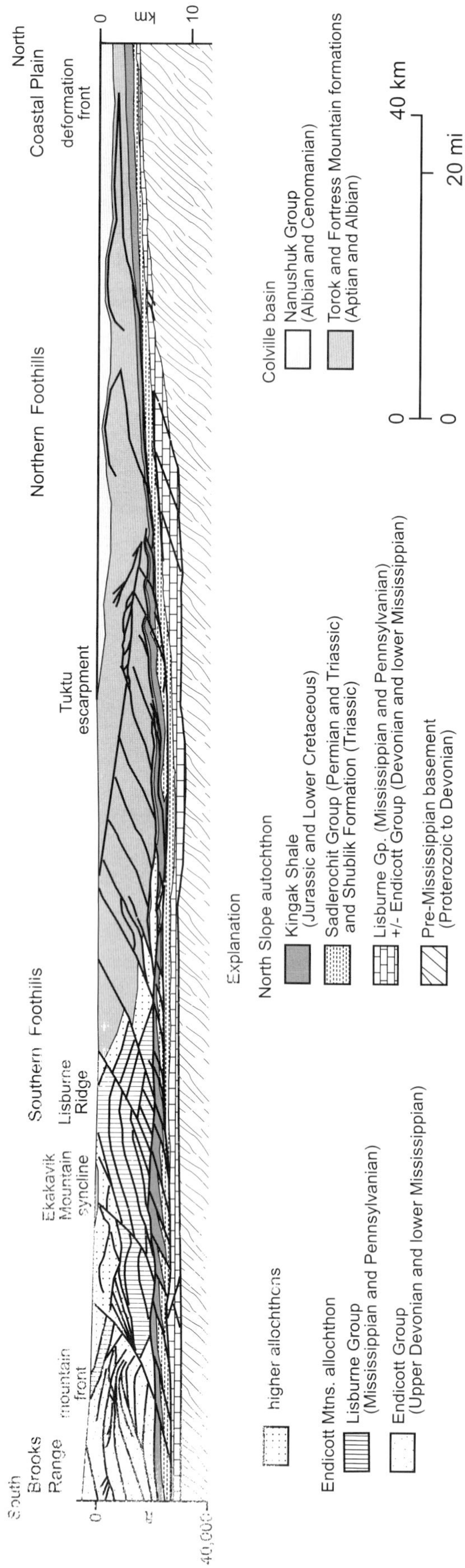

FIGURE 4. Regional cross section based on seismic line R6 (see Figure 1 for location). See text for discussion.

2001) considered the most important source rocks to be marine shales of Mesozoic age, including the Shublik Formation, Kingak Shale, and pebble shale unit of the Ellesmerian sequence in the autochthon and condensed deposits at the base of the Colville basin (Hue Shale and equivalents). Basinal deposits in the Lisburne Group are also potential source rocks for petroleum. Other source rocks include gas-prone deltaic and prodelta deposits at the base of the Ellesmerian sequence of the autochthon and in the Colville basin. The Mesozoic units of the Ellesmerian sequence thin and fine southward and are represented in allochthonous strata of the deformational wedge by the Triassic and Jurassic Otuk Formation, which contains strata with total organic carbon commonly 4–10% and locally as high as 38% in outcrop (P. Lillis, 2002, personal communication; C. G. Mull, E. E. Harris, P. Peapples, 2003, personal communication). Basinal deposits in the Lisburne Group are also abundant in the deformational wedge and could have been an important source rock unit. The structural position, kinematic history, and thermal history of each of these units are important factors for exploration in the fold and thrust belt.

STRUCTURAL ARCHITECTURE OF WEST-CENTRAL BROOKS RANGE FOOTHILLS

A cross section showing our interpretation of relations from the foreland to the Brooks Range is shown in Figure 4. The interpretation in this cross section is based on seismic line R6 and other seismic data collected in the late 1970s and early 1980s and reprocessed by Miller et al. (2000, 2001). The interpretation is also based on the geologic maps of Mull and Werdon (1994), Mull et al. (1994), C. G. Mull (2000, personal communication), and C. G. Mull, D. W. Houseknecht, G. H. Pessel (2003, personal communication), as well as our own examination of key structures in the field (Figure 5). The section has been generalized to depict outcrop and map relations over a large region adjacent to the seismic line.

The cross section portrays the northern part of the Brooks Range as a thin-skinned, north-directed fold and thrust belt, with a basal detachment located in the Jurassic and Lower Cretaceous Kingak Shale near the top of the autochthon across most of the deformational belt. The deformation front for the orogen lies beneath the coastal plain at the leading edge of a thin, frontal, intercutaneous wedge formed in mudstone of the upper Torok Formation. The frontal wedge was formed where the basal detachment ramps upward from the Kingak into the upper part of the Torok Formation and structurally thickened this unit over a distance of about 20 km (13 mi) northward from the ramp, forming a broad, weltlike fold in strata of the Nanushuk Formation.

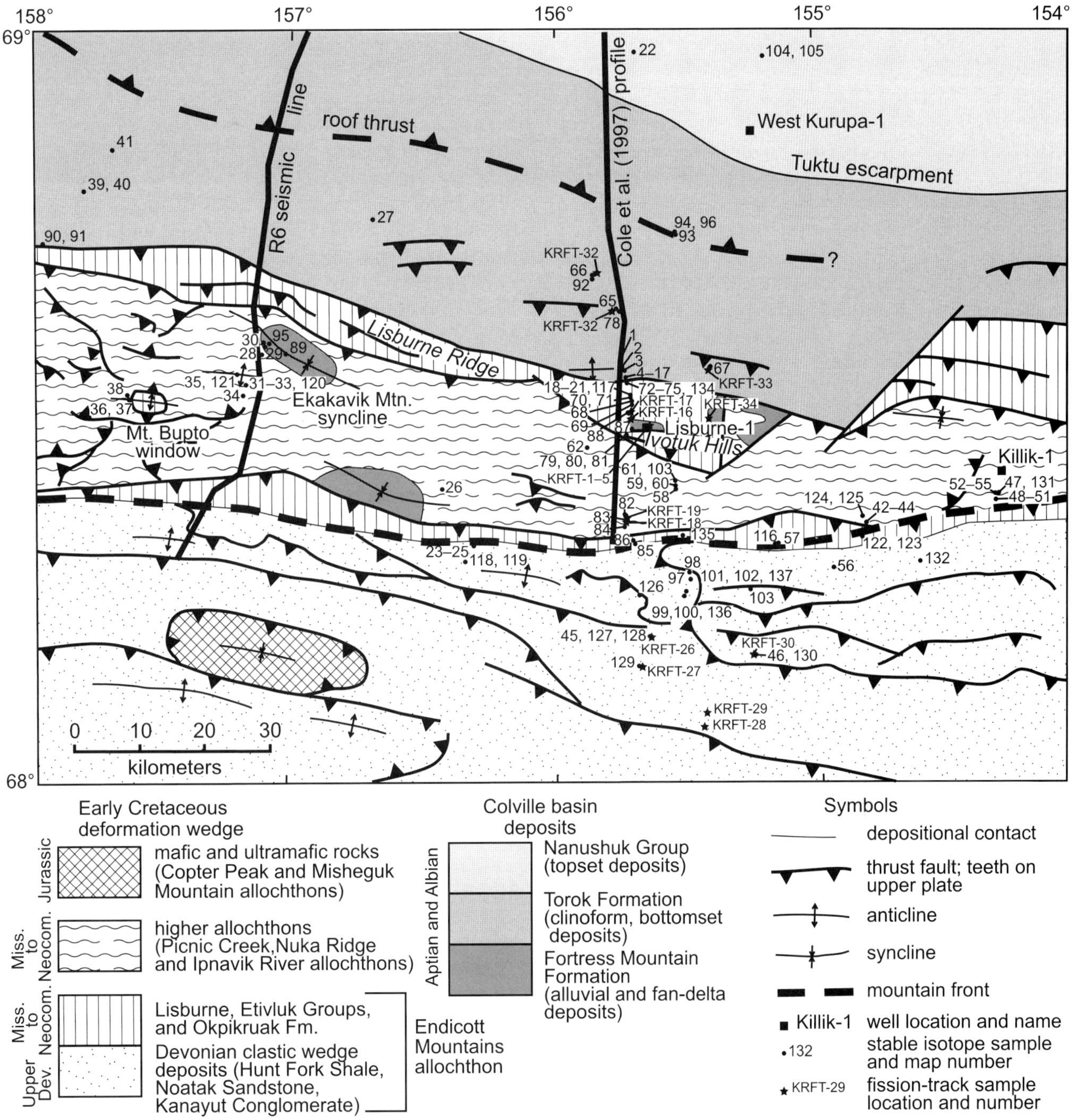

FIGURE 5. Generalized interpretive geologic map showing fission-track and stable isotope sample locations. Sources of map data: Mull and Werdon (1994); Mull et al. (1994); C. G. Mull (2000, personal communication).

Behind the deformation front, the Nanushuk Formation displays a valley and ridge topography that is controlled by upright open folds constructed above thrusts that ramp upward from the Kingak into the upper Torok Formation. Seismic data indicate that the folds developed above the ramps are detachment folds in mudstone of the upper Torok. Steep reverse faults are commonly present in the cores of these folds at the surface, but the faults do not display a consistent pattern of structural transport. In some folds, south-vergent faults are present at the base of the Nanushuk and define local roof thrusts in the cores of the anticlines. The roof thrusts apparently do not continue from anticline to anticline across intervening synclines, indicating that shortening is entirely resolved by thickening in the cores of the anticlines without the development of a regional roof thrust.

The transition from slope deposits in the upper Torok to shelf and nonmarine deposits of the Nanushuk Formation is exposed along the Tuktu Escarpment. The regional gentle northward dip in the Nanushuk Formation and overlying strata in the northern part of the cross section steepens southward in this area, reflecting the presence of a regional passive-roof duplex in the Torok Formation at depth. Surface exposures of folds and faults near the top of the Torok south of the Tuktu Escarpment are consistently south directed and become more numerous southward. The roof thrust is located in the middle part of the upper Torok and intersects the surface but is exposed only locally in the field. Its location is most easily delineated by regional changes from north dips and south-directed folds and thrusts above the roof thrust to south dips and north-directed folds and thrusts below the roof thrust in mudstone deposits of the upper Torok Formation.

The duplex is developed primarily in sandstone turbidites that compose the lower Torok Formation. In seismic data, the sandstone turbidites form a thick, competent unit caught between a floor thrust in the incompetent shale of the Kingak at depth and the roof thrust in the upper part of the Torok Formation. Outcrop-scale folds and faults in the duplex consistently are north directed. The spacing and displacement of linking thrusts in the duplex are difficult to estimate because of poor seismic resolution and exposure but can be located in some places by exposures of imbricates of sandstone turbidites of the lower Torok.

Where it is intersected by the profile, the Brooks Range foothills north of the mountain front is characterized by a regional antiform that is composed of the Endicott Mountains allochthon to the north and a synform containing a klippen of higher allochthons to the south. The principal component of the antiform is a duplex of Lisburne Group carbonate rocks that developed between detachments in the Kayak Shale at the base of the Endicott Mountains allochthon and in the Etivluk Group above the Lisburne. Deformed syntectonic strata of the Okpikruak Formation may represent a structurally higher duplex at the top of the allochthon. The regional synform north of the mountain front contains remnants of higher allochthons. In general, the structural style of the strata in the higher allochthons varies from tectonically mixed melange to broken formation, although duplexes of chert or lithic sandstone units can be identified locally in these rocks.

The structurally and stratigraphically highest strata in this part of the cross section are nonmarine and shelf sandstone and conglomerate of the Aptian(?)–Albian Fortress Mountain Formation. This unit forms broad map-scale "thumbprint" synclines that overlap both the Endicott Mountains and higher allochthons (Figures 3, 5). The Fortress Mountain is generally not preserved in intervening antiforms, but structural patterns in underlying rocks suggest that it was folded into sharp-crested, fault-cored anticlines that were asymmetric to the north.

At the longitude of our profile, the mountain front of the Brooks Range is an abrupt physiographic feature caused by the exposure of resistant, coarse-grained clastic rocks of the Upper Devonian clastic wedge. The Upper Devonian clastic strata (part of the Endicott Mountains allochthon) are structurally imbricated and form a duplex that has been mapped by Mull and Werdon (1994) and is well imaged in seismic data in the subsurface near the mountain front. In a few places, thin Lisburne Group strata are preserved above the Upper Devonian rocks south of the mountain front, but in most parts of the central Brooks Range, the mountain front sharply divides the Upper Devonian strata to the south from less competent, stratigraphically higher rocks to the north.

The Endicott Mountains allochthon was interpreted by Wallace et al. (1997) to comprise a forward-tilted, multistoried stacked succession of duplexes. In this interpretation, the duplex in the Upper Devonian strata is separated from a duplex in strata of the overlying Lisburne Group by a regional detachment in shale of the lower Mississippian Kayak Shale. The regional forward tilt resulted from multiple hanging-wall ramps in the basal detachment that cut upward successively through the Kanayut, Lisburne, and Etivluk groups and probably the Okpikruak Formation and were emplaced on a regional footwall flat in the Kingak Shale. The forward tilt produces a northward shingling of the duplexes and a regional, north-directed wedgelike shape for the Endicott Mountains allochthon (Wallace et al., 1997, their figure 9). The base of the Endicott Mountains allochthon is constrained by gentle, south-dipping seismic reflectors in the autochthon that can be traced at least as far south as the modern mountain front.

A similar wedgelike geometry is inferred for the higher allochthons emplaced on the Endicott Mountains allochthon, but the northward limit of these wedges is uncertain because of erosion and poor seismic resolution. In this study, the composite structural wedge is referred to as the "Early Cretaceous deformational wedge," in recognition of map and stratigraphic relations that show that all of the allochthons were emplaced before the Aptian(?)–Albian, and that the overall configuration of the overlapping wedges is itself a wedge.

Wallace et al. (1997) recognized that the multistoried duplexes in the Endicott Mountains allochthon are locally truncated by breaching thrusts that transect the duplex stratigraphy. Examples of breaching thrusts include steeply south-dipping reverse faults at the mountain front and faults that define the northern limit of exposure of deformed rocks of the deformational wedge. These faults are significant, because they juxtapose structurally lower parts of the deformational wedge against higher parts of the wedge and, locally, the deformational wedge against younger Colville basin deposits to the north. The latter relation indicates that the breaching thrusts are relatively late faults that were active after deposition of the Colville basin deposits in the Aptian–Albian and are thus probably unrelated to thrusts on which the Early Cretaceous deformational wedge was constructed.

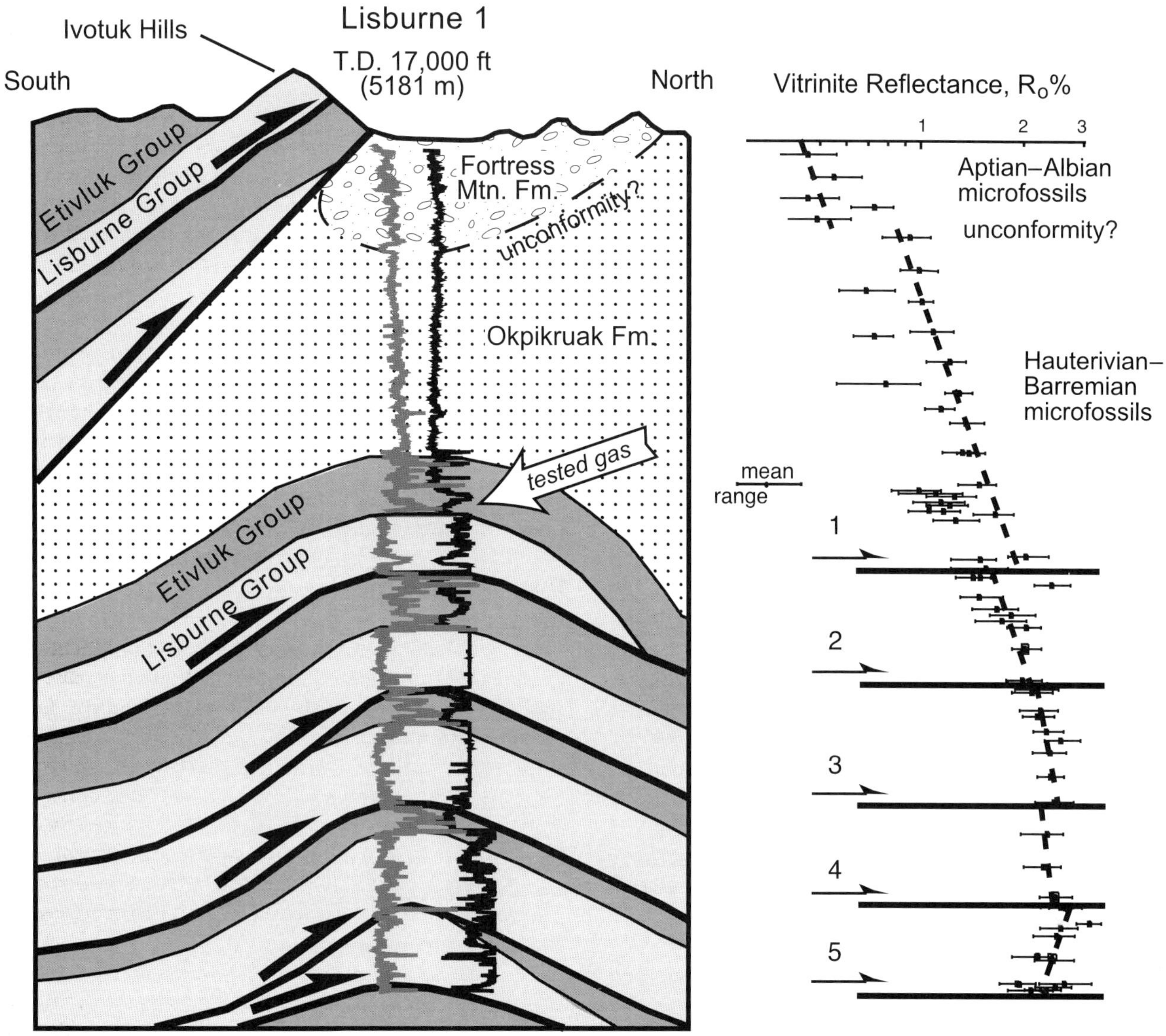

FIGURE 6. Well logs and vitrinite reflectance data from Lisburne-1 well plotted against interpreted cross section (modified from Cole et al., 1997; Dumoulin and Bird, 2002). Five repetitions of the Lisburne and Etivluk groups were encountered in the lower part of the well and two others project above the well from the Ivotuk Hills to the south. Microfossil data indicate that clastic rocks in upper part of the well consist of Okpikuak Formation (here Hauterivian and Barremian) and Fortress Mountain Formation [Aptian(?)–Albian]. The contact between the two units at about 650-m (2100-ft) depth is inferred to be an unconformity, an interpretation supported by vitrinite reflectance data shown on the right. These stratigraphic relations require that the fault at the base of the Ivotuk Hills is a breaching fault of post-Aptian(?)–Albian age, although the Lisburne Group–Etivluk Group duplex low in the well probably formed in the Neocomian. Apatite fission-track thermochronology from the Ivotuk Hills and from the Fortress Mountain Formation in the well suggest that the breaching fault is early Tertiary (~60 Ma). T.D. = total depth.

The effect of breaching thrusts on the structural stacking is evident at the Lisburne-1 well (Legg, 1983) (Figure 6), which is located in the regional foothills antiform east of our regional profile. The structural, stratigraphic, and thermal-maturity data from this well were described by Cole et al. (1997), who showed that the lower 3500 m (11,500 ft) of the 5400-m (17,700-ft)-deep well consists of five repetitions of the Lisburne and Etivluk groups of the Endicott Mountains allochthon. They noted that surface exposures in the Ivotuk Hills near the well indicate that two other imbricates composed of the Lisburne and Etivluk groups project to positions above the top of the well. The upper 2000 m (6600 ft) of the well, however, penetrated Lower Cretaceous clastic rocks, which must lie at structural levels below the exposed imbricates composed of the Lisburne and Etivluk groups. Cole et al. (1997) reported that coarse-grained clastic rocks in the upper 600 m (2000 ft) of the well yielded microfossils of Aptian(?)–Albian age, whereas underlying finer grained clastic rocks yielded Hauterivian to Barremian ages. These results suggest that the well may have been spudded in a synform of the Fortress Mountain Formation and penetrated an unconformity at a depth of about 600 m (2000 ft) into the underlying Hauterivian and Barremian strata of the Okpikruak Formation, an interpretation that is suggested by the vitrinite reflectance data from the well (Figure 6). The presence of the Aptian(?)–Albian strata in the uppermost part of the well with strata of the Lisburne and Etivluk groups above shows that the thrust beneath Lisburne strata in the Ivotuk Hills must be a breaching thrust that was active in postmiddle Cretaceous time. This relation strongly suggests that breaching faults core the anticlines that fold the Fortress Mountain Formation in the foothills area.

Apatite fission-track data from pre-Cretaceous strata in the Ivotuk Hills south of the well yield early Tertiary ages, whereas an apatite age from Aptian(?) to Albian strata in the well is only slightly reduced from the sample's original depositional age (Cole et al., 1997; see below). These data indicate that the structurally highest imbricates south of the well did not cool below about 110°C until the early Tertiary, suggesting that the breaching thrusts on which they are emplaced are early Tertiary. These data confirm the interpretation of Cole et al. (1997) that the Albian(?)–Aptian Fortress Mountain Formation is the structurally highest unit in the foothills region and indicate that these rocks did not experience either stratigraphic or tectonic burial sufficient to become fully annealed since their deposition.

On the basis of these observations, we conclude that at least two episodes of thrusting are present in the foothills belt as has been called for by some previous workers (Cole et al., 1997; Mull et al., 1997; O'Sullivan et al., 1997). The earliest episode is represented by the thrusts on which the Endicott Mountains and higher allochthons were emplaced during the Late Jurassic and Early Cretaceous, as discussed by other workers (Mull, 1985; Mayfield et al., 1988; Moore et al., 1994). This episode of thrusting led to the construction of the Early Cretaceous deformational wedge. Following the deposition of Colville basin strata in the middle Cretaceous, another episode of thrusting occurred in the early Tertiary on thrusts that characteristically breach the structural stratigraphy and juxtapose different structural levels (e.g., Endicott Mountains allochthon against Colville basin strata). Because seismic data indicate that the autochthon is not involved in the younger deformation, the breaching thrusts must root downward into the same basal detachment as those that emplaced the Endicott Mountains allochthon. These relations are shown in the foothills part of Figure 4 as a relatively younger group of steeper thrusts that transect a relatively older group of low-angle thrusts.

FISSION-TRACK THERMOCHRONOLOGY

Apatite fission-track thermochronology (AFTT) provides a record of the thermal history of the host rock below temperatures of about 110°C, which makes this technique ideal for constraining the timing of deformation and uplift of upper crustal rocks. At surface temperatures over geologic timescales, fission tracks in apatite have a fairly constant mean length of about 16 μm (Gleadow et al., 1986). However, following their formation, fission tracks in apatite will progressively shorten (anneal) at a rate that depends primarily on temperature. Increased annealing results in shorter tracks, reduced track density, and a reduction in the fission-track age. Total annealing results in the reduction of the fission-track age to zero.

It has also been recognized that the annealing behavior of fission tracks in apatite is sensitive to chemical composition, and that chlorine-rich apatites are more resistant to track annealing than fluorine-rich apatites, as summarized by Donelick et al. (1999). This property suggests that the closure temperature for individual grains in a sample may vary significantly in response to variations in chlorine content.

Based on 160 microprobe analyses of apatites from eight of the samples analyzed for this study, the overwhelming majority of apatites present are fluorine rich (~0.0–0.02 wt.% Cl) with about 15% of grains ranging between 0.05 and 0.08 wt.% Cl. These data suggest that at temperatures below about 50–60°C, fission tracks in these apatite grains will be annealed but at a slow enough rate that little reduction in the AFTT age occurred, whereas at temperatures of about 110°C or more, tracks in these apatites annealed rapidly, resulting in total resetting of the AFTT age.

Zircon fission-track thermochronology (ZFTT) is useful for constraining a sample's higher temperature thermal history. Recent work by Brandon et al. (1998) indicates a range in the zircon fission-track closure temperature of between about 225 and 260°C for rates of cooling between 1 and 100°C/m.y., respectively. In this study, we therefore assume that reset ZFTT ages indicate that the sampled unit must have been exposed to elevated paleotemperatures of at least about 225°C, whereas unreset ZFTT ages indicate that the sampled unit was not exposed to such elevated paleotemperatures.

Interpretation of AFTT Results

A total of 18 AFTT analyses have been generated from 13 surface samples collected along the southern end of the cross section of Cole et al. (1997), as well as from 5 subsurface samples from the Lisburne well. A preliminary interpretation of these results was first presented in Cole et al.

(1997), and a revised interpretation is presented below. Locations for the samples discussed are shown in Figure 5.

The AFTT results from three samples from the Aptian(?)–Albian Fortress Mountain Formation, including Killik River Fission Track (KRFT)-5 from the Lisburne well and KRFT-16 and KRFT-34 collected just north of the Ivotuk Hills, indicate that these rocks were never heated sufficiently to have totally reset their AFTT ages. This result is generally consistent with vitrinite reflectance data for these samples (Cole et al., 1997). However, the AFTT results from Neocomian and older sedimentary rocks and the Aptian and Albian Torok Formation collected throughout the region, including KRFT-1 to KRFT-4 from the Lisburne well and KRFT-17 to KRFT-19 and KRFT-26 to KRFT-33, give apatite ages significantly less than their original depositional ages. The track-length distributions for most of the surface samples are indicative of rapid cooling at the time suggested by their age. Interpretation of the results from these older rocks suggest that (1) they were exposed to paleotemperatures above about 110°C following deposition, which resulted in the total resetting of the AFTT age; (2) they then cooled rapidly at the time recorded by the AFTT age; and (3) they have subsequently remained at cool temperatures.

Based on the AFTT results, the region experienced at least three distinct cooling and denudation events since the rocks were exposed to their maximum postburial temperatures. The first recorded event occurred during the middle Cretaceous at about 100 ± 5 Ma and, although of regional extent in the axial part of the Brooks Range (O'Sullivan et al., 1997), was only recorded by a few of the samples discussed herein. The two major recorded cooling and denudation events to affect the entire study area occurred during the early Tertiary at about 60 ± 5 and about 45 ± 5 Ma. Furthermore, because the apatite grains in some of the samples varied in composition based on the proven analyses and etch pit parameters (dpar) (e.g., Donelick et al., 1999), many of these samples recorded multiple cooling events. In some cases, the more retentive grains (either chlorine rich or large dpar) with higher blocking temperatures recorded either middle Cretaceous (i.e., KRFT-26) or Paleocene (i.e., KRFT-16) cooling and denudation, whereas less-retentive grains (either fluorine rich or smaller dpar) with lower blocking temperatures recorded either the younger Paleocene or Eocene cooling and denudation. Importantly, there was generally good correlation between etch pit parameters and apatite grain age retentiveness, whereas the correlation between chlorine composition and apatite grain age retentiveness was not always clear. Finally, the data indicate that the samples have experienced protracted cooling to present-day surface temperatures following cooling and denudation during the early to middle Tertiary.

In at least two cases, the distribution of AFTT ages indicates that the region has experienced a complicated structural history. First, AFTT data from KRFT-26 and KRFT-27 record rapid cooling and denudation at about 100 ± 5 Ma, whereas AFTT data from KRFT-28, KRFT-29, and KRFT-30 indicate that rapid cooling and denudation were not recorded until the Eocene at about 45 ± 5 Ma, although all five samples are collected from the Noatak Sandstone and at similar elevations. Similarly, AFTT results from the three Fortress Mountain Formation samples located to the north indicate that they have never been exposed to sufficient paleotemperatures to significantly reduce their AFTT ages. Uplifted Etivluk Group samples just to the south and at higher elevations record rapid cooling and denudation from higher paleotemperatures during the Paleocene at about 60 ± 5 Ma.

Interpretation of ZFTT Results

A total of 14 ZFTT analyses have been generated from 12 surface samples originally collected for AFTT analysis by Cole et al. (1997) along the southern end of their transect and from 2 subsurface samples from the Lisburne well. These results are the first ZFTT results published from the central and western Brooks Range. The sample data and locations are shown in Table 1 and in Figure 5.

The ZFTT results from the Colville basin (Fortress Mountain Formation and Torok Formation, samples KRFT-5, KRFT-16, KRFT-31, KRFT-33, and KRFT-34) yielded distinct populations of zircon single-grain ages older than the stratigraphic age of the sampled units. Samples from the Fortress Mountain Formation yielded a single, distinct population of provenance ages between 140 and 120 Ma, whereas samples from the Torok Formation yielded two distinct populations of provenance ages between 140 and 120 Ma, as well as between 160 and 150 Ma. These results also indicate that this part of the Colville basin was never heated sufficiently to have reset ZFTT ages, a result consistent with AFTT data and vitrinite reflectance data from the same rocks presented previously. These results, however, do indicate that these deposits were eroded from source areas that must have cooled from temperatures above about 225°C in the Late Jurassic and Early Cretaceous, a result that is consistent with derivation from rocks that were exhumed during arc-continent collision and allochthon emplacement in the Brooks Range. Samples from Neocomian syntectonic deposits of the Okpikruak Formation (KRFT-4 and KRFT-17) yield similar results and indicate a similar provenance area for this unit. Coupled with the AFTT data, the ZFTT data indicate that the Okpikruak saw temperatures at or above 110°C but below 225°C, consistent with vitrinite reflectance data for these rocks (Cole et al., 1997).

In contrast, the ZFTT results from samples from Devonian and lower Mississippian rocks south of the present-day mountain front yielded ages younger than the stratigraphic age of the units sampled. Zircon fission-track thermochronology ages from these samples

Table 1. Fission-track results: Lisburne well region.

Lab Number (Unit)*	Field Number	Type (Ap/Zi)	Number of Grains	Latitude N	Longitude W	Elevation (+m) or Depth (−m)	χ^2 Probability (%)	Fission-track Age (Ma)	Uranium (ppm)	Track Length (μm)	Standard Deviation (μm)
Outcrop											
KRFT-16	FC-47C	Ap	25	68.489	155.714	540	0.0	72.2 ± 9.9	10.8	13.28 ± 0.27	1.95
Kfm		Zi	25				91.6	124.6 ± 5.3	198.4		
KRFT-17	FC-51B	Ap	25	68.499	155.712	540	29.0	51.9 ± 3.4	25.7	12.32 ± 0.29	2.26
Ko		Zi	25				96.8	136.8 ± 9.1	147.8		
KRFT-18	FC-84	Ap	25	68.349	155.728	810	0.90	81.2 ± 6.3	22.0	12.47 ± 0.20	2.04
Mk		Zi	30				0.0	267.6 ± 18.9	117.6		
KRFT-19	FC93-86	Ap	25	68.357	155.726	810	0.10	51.6 ± 5.3	26.6	12.09 ± 0.17	1.73
Mk		Zi	17				0.0	252.1 ± 12.6	135.8		
KRFT-26	DH93-95C	Ap	25	68.196	155.633	1510	1.20	97.6 ± 7.5	22.4	13.01 ± 0.20	2.06
Dh		Zi	20				93.7	161.8 ± 6.5	115.8		
KRFT-27	DH93-96B	Ap	25	68.155	155.668	1370	1.40	75.4 ± 5.2	17.1	13.31 ± 0.18	1.86
Dn		Zi	19				55.7	157.2 ± 7.9	165.7		
KRFT-28	DH93-98A	Ap	25	68.067	155.436	1240	58.6	44.7 ± 3.5	20.7	13.90 ± 0.17	1.14
Dn		Zi	25				69.2	119.3 ± 4.1	124.9		
KRFT-29	DH93-99A	Ap	25	68.082	155.415	860	79.8	50.0 ± 4.4	20.5	13.93 ± 0.14	1.47
Dn		Zi	25				61.9	125.1 ± 4.3	148.1		
KRFT-30	DH93-100C	Ap	7	68.171	155.248	1670	91.3	48.9 ± 10.0	6.8	–	–
Dn		–	–				–	–	–		
KRFT-31	DH93-118A	Ap	27	68.639	155.769	405	0.50	59.8 ± 6.3	17.2	13.75 ± 0.19	1.98
Kto		Zi	25				0.0	151.6 ± 10.8	126.8		
KRFT-32	DH93-120D	Ap	25	68.684	155.859	380	91.3	52.6 ± 7.6	7.9	14.86 ± 0.28	0.48
Kto		–	–				–	–	–		
KRFT-33	DH93-122D	Ap	25	68.560	155.408	500	69.8	54.7 ± 5.1	10.6	14.03 ± 0.18	1.57
Kto		Zi	25				61.8	142.3 ± 4.6	165.8		
KRFT-34	DH93-124	Ap	26	68.486	155.434	705	7.5	91.5 ± 11.1	7.0	13.30 ± 0.34	2.55
Kto		Zi	25				42.9	129.9 ± 3.9	218.7		
Lisburne Unit 1 Well											
KRFT-5	89POS-06A	Ap	20	68.485	155.743	−634	5.1	92.1 ± 10.3	7.6	142.22 ± 0.27	0.96
Kfm		Zi	20				17.5	139.8 ± 10.4	106.4		
KRFT-4	89POS-05A	Ap	26	68.485	155.743	−1190	24.0	65.6 ± 6.2	8.9	13.01 ± 0.40	3.29
Ko		Zi	25				55.3	155.6 ± 7.6	98.5		
KRFT-3	89POS-04A	Ap	8	68.485	155.743	−1896	18.0	36.8 ± 10.7	7.4	10.67 ± 2.11	4.22
JTro		–	–				–	–	–		
KRFT-2	89POS-03A	Ap	10	68.485	155.743	−2663	91.4	16.1 ± 5.9	6.8	–	–
JTro		–	–				–	–	–		
KRFT-1	89POS-01A	Ap	6	68.485	155.743	−4148	62.7	20.5 ± 15.2	6.4	–	–
Ps		–	–				–	–	–		

**Kfm = Cretaceous Fortress Mountain Formation; Kto = Cretaceous Torok Formation; Ko = Cretaceous Okpikruak Formation; JTro = Triassic and Jurassic Otuk Formation; Ps = Permian Siksikpuk Formation; Mk = Mississippian Kurupa Sandstone; Dh = Devonian Hunt Fork Shale; and Dn = Devonian Noatak Sandstone.*

Ages for apatite samples (Ap) were calculated using $\zeta = 352.7 \pm 4$ for dosimeter glass SRM612 or $\zeta = 379.2 \pm 3$ for dosimeter glass CN5 (analyst: P. O'Sullivan). Ages for zircon samples (Zi) were calculated using $\zeta = 87.8 \pm 7.5$ for dosimeter glass U3 (analyst: P. O'Sullivan). Errors quoted at $\pm 1\sigma$.

(KRFT-18 and KRFT-19 from the Picnic Creek allochthon and KRFT-26–30 from the Endicott Mountains allochthon) range from as old as 162 ± 13 to as young as 119 ± 8 Ma, indicating that all of these rocks cooled from temperatures above about 225°C in the Late Jurassic and/or Neocomian. Most vitrinite reflectance estimates of peak rock temperatures are above 225°C for these rocks (see below). Because the stratigraphic thickness of pre-Cretaceous strata that overlie these rocks was insufficient to generate such high temperatures by stratigraphic burial alone, these results suggest instead that the rocks were tectonically buried by allochthons that were emplaced during arc-continent collision and exhumed in the Late Jurassic and Early Cretaceous while contractional deformation was still ongoing. Tectonic burial is consistent with the observation of deformed rocks and allochthons, including mafic and ultramafic rocks of the highest structural levels, at structural positions above the sampled units, as well as the provenance ages observed in the Okpikruak Formation and Colville basin deposits. Interestingly, samples KRFT-18 and KRFT-19, which were collected from strata that were thought to represent the lowest stratigraphic unit of the Picnic Creek allochthon, indicate that rocks even in some higher structural positions saw temperatures above 225°C.

The ZFTT results suggest that most if not all of the Devonian clastic strata exposed south of the present day mountain front experienced temperatures above 225°C prior to cooling in the Late Jurassic and Neocomian. These data further indicate that these rocks did not see temperatures above 225°C after the Early Cretaceous, although cooling below about 110°C was not accomplished until the early Tertiary as indicated by the AFTT results from the same rocks. Rocks at stratigraphically high and more forward structural position in the Early Cretaceous deformational wedge (e.g. Okpikruak Formation), however, never experienced such high temperatures, presumably because they were not buried as deeply by tectonic processes in the Late Jurassic and Early Cretaceous.

STABLE ISOTOPE STUDIES

Many studies have demonstrated the utility of combining geochemical and structural data to decipher the complex thermal and fluid history of fold and thrust belts (e.g., Muchez et al., 1994, 2000; Pagel et al., 1996; Shelton et al., 1996; Underwood et al., 1996, 1999; Travé et al., 1998; Ferket et al., 2000; van Geet et al., 2002). To determine the spatial and stratigraphic patterns of thermal maturity and fluid-migration history, carbon and oxygen isotope compositions were determined regionally for calcite veins ($N = 95$), quartz veins ($N = 32$), and limestone matrix and fossil components ($N = 8$) in Devonian to Lower Cretaceous strata of the Early Cretaceous deformational wedge and the Colville basin on the northern flank of the central Brooks Range. Calcite was reacted with 103% phosphoric acid at 25°C (McCrea, 1950). Quartz was reacted with ClF_3 using a modification of the Clayton and Mayeda (1963) method. Isotopic data are reported in standard δ notation relative to Vienna standard mean ocean water for oxygen and the Peedee belemnite (PDB) standard for carbon (Table 2). The standard error of each analysis is ±0.2‰ for $\delta^{18}O$ values of quartz and less than ±0.1‰ for $\delta^{13}C$ and $\delta^{18}O$ values of calcite.

Oxygen Isotopes

The $\delta^{18}O$ values of calcite are 13.5–32.2‰ (mean = 20.1‰; Figure 7). The $\delta^{18}O$ values of quartz range from 9.3 to 26.1‰ (mean = 16.7‰; Figure 8). Quartz and calcite are not in isotopic equilibrium with each other. This fact is not surprising, because textures in most veins indicate sequential precipitation instead of coprecipitation of the two minerals.

There appears to be a relationship between $\delta^{18}O$ values of veins and the degree of thermal maturity (metamorphic grade) of the enclosing host rocks. Vitrinite reflectance data from Cole et al. (1997) and Johnsson et al. (1999) show that rocks south of the present mountain front have markedly higher thermal maturity than those to the north in the foothills of the Brooks Range (Figure 9). The higher grade rocks south of the mountain front have calcite veins with $\delta^{18}O$ values generally between 16 and 20‰; those in lower grade rocks north of the mountain front have $\delta^{18}O$ values typically in the range of 17–30‰. Quartz veins in higher grade rocks of the Devonian clastic wedge have $\delta^{18}O$ values of 9.3–21.2‰ (mean = 14.8‰), whereas quartz veins in lower grade Mississippian to Cretaceous rocks have $\delta^{18}O$ values of 16.7–26.1‰ (Table 2; Figure 8).

The $\delta^{18}O$ values of vein-filling minerals are a function of both the $\delta^{18}O$ values of the fluids that deposited the minerals and the temperatures at which mineral deposition occurred. To determine whether differences between $\delta^{18}O$ values of veins hosted in Devonian rocks and of veins hosted in younger lithologies reflect temperature of deposition or fluid composition, we calculated equilibrium $\delta^{18}O_{water}$ values of vein-forming fluids (Figure 10; Table 2). For calculation, we used paleotemperature estimates based on vitrinite reflection data for wall rocks adjacent to individual veins (Table 2), the calcite-water oxygen isotope fractionation equation of O'Neil et al. (1969), and the quartz-water equation of Clayton et al. (1972).

The calculated $\delta^{18}O_{water}$ values for fluids that deposited calcite veins in Upper Devonian clastic deposits are 9.5–11.2‰ (mean = 10.2‰); those for fluids that deposited quartz veins in these rocks are 5.3–8.7‰ (mean = 6.5‰). In contrast, the calculated $\delta^{18}O_{water}$

Table 2. Stable isotope and vitrinite reflectance data.

Map Location	Sample Number	Sample Description	Stratigraphic Unit*	Latitude (N)	Longitude (W)	$\delta^{13}C$ PDB**	$\delta^{18}O$ SMOW**	Mean R_o (%)	Temperature (°C)	$\delta^{18}O_{water}$	Tectonic Unit
1	DH-93-2A	calcite	Ko	68°34′4″	155°45′16″	−11.5	21.5	0.8	125	5.9	Neocomian syntectonic deposits
2	DH-93-5C	calcite	Ko	68°33′38″	155°44′50″	−13.1	21.1	0.5	75	1.1	Neocomian syntectonic deposits
3	DH-93-7B	calcite	Ko	68°33′19″	155°44′11″	−2.8	19.8	0.8	125	4.2	Neocomian syntectonic deposits
4	DH-93-8	black calcite	Ko	68°32′46″	155°43′52″	−0.7	22.1				Neocomian syntectonic deposits
5	DH-93-8A	Buchia (limestone)	Kcl	68°32′46″	155°43′52″	0.9	24.1				Neocomian syntectonic deposits
6	DH-93-8A	matrix calcite	Kcl	68°32′46″	155°43′52″	1.1	30.6				Neocomian syntectonic deposits
7	DH-93-8B	matrix calcite	Kcl	68°32′46″	155°43′52″	−0.3	22.6				Neocomian syntectonic deposits
8	DH-93-8C	milky calcite	Kcl	68°32′46″	155°43′52″	0.1	21.4				Neocomian syntectonic deposits
9	DH-93-8C	clear calcite	Kcl	68°32′46″	155°43′52″	0.1	21.7				Neocomian syntectonic deposits
10	DH-93-8D	Buchia (limestone)	Kcl	68°32′46″	155°43′52″	1.8	31.1				Neocomian syntectonic deposits
11	DH-93-8D	first-generation calcite	Kcl	68°32′46″	155°43′52″	−1.1	22.5				Neocomian syntectonic deposits
12	DH-93-8D	second-generation calcite	Kcl	68°32′46″	155°43′52″	0.5	21.9				Neocomian syntectonic deposits
13	DH-93-8E	calcite	Kcl	68°32′46″	155°43′52″	0.9	20.9				Neocomian syntectonic deposits
14	DH-93-8E	Buchia (limestone)	Kcl	68°32′46″	155°43′52″	1.4	27.1				Neocomian syntectonic deposits
15	DH-93-8F	Buchia (limestone)	Kcl	68°32′46″	155°43′52″	1.5	29.1				Neocomian syntectonic deposits
16	DH-93-8F	clear calcite	Kcl	68°32′46″	155°43′52″	−4.2	22.4				Neocomian syntectonic deposits
17	DH-93-8F	black calcite	Kcl	68°32′46″	155°43′52″	−1.3	22.4				Neocomian syntectonic deposits
18	DH-93-9D	calcite	JTro or Kcl	68°32′18″	155°43′34″	−0.1	23.4				Neocomian syntectonic deposits
19	DH-93-9D	Buchia (limestone)	Kcl	68°32′18″	155°43′34″	1.5	28.8				Neocomian syntectonic deposits
20	DH-93-9F	calcite	JTro or Kcl	68°32′18″	155°43′34″	22.7	33.5				Neocomian syntectonic deposits
21	DH-93-9F	calcite	JTro or Kcl	68°32′18″	155°43′34″	1.1	28.6				Neocomian syntectonic deposits
22	DH-93-29A	calcite	Kn	68°59′5″	155°42′0″	−6.8	16.6	0.6	95	−1.5	Aptian–Albian Colville basin
23	DH-93-39B	calcite	Mk	68°19′16″	156°17′51″	−7.7	17.2	2.5	240	9.5	Upper Devonian clastic wedge
24	DH-93-39D	calcite	Mk	68°19′16″	156°17′51″	−1.9	17.4	2.5	240	9.7	Upper Devonian clastic wedge
25	DH-93-39E	calcite	Mk	68°19′16″	156°17′51″	−1.1	18.2	2.5	240	10.5	Upper Devonian clastic wedge
26	DH-93-40B	calcite	Ko	68°23′41″	156°24′55″	−13.3	13.5	1.1	160	1.7	Neocomian syntectonic deposits
27	DH-93-52C	calcite	Kto upper	68°45′24″	156°41′32″	−5.3	18.2	0.9	135	4.2	Aptian–Albian Colville basin
28	DH-93-55C	calcite	Ko	68°34′18″	157°6′10″	−17.5	17.2	0.7	120	2.1	Neocomian syntectonic deposits
29	DH-93-57D	calcite	Kfm	68°34′56″	157°5′26″	−13.3	16.7	0.9	135	2.7	Aptian–Albian Colville basin
30	DH-93-58B	calcite	Kfm	68°35′18″	157°5′50″	−22.7	17.4	0.9	135	3.4	Aptian–Albian Colville basin
31	DH-93-59C	calcite	Ko	68°31′48″	157°9′43″	−20.6	18.9	0.8	125	3.2	Neocomian syntectonic deposits
32	DH-93-59D	calcite	Ko	68°31′48″	157°9′43″	−21.7	17.7	0.8	125	2	Neocomian syntectonic deposits
33	DH-93-59D	calcite	Ko	68°31′48″	157°9′43″	−21.9	19.3	0.8	125	3.6	Neocomian syntectonic deposits
34	DH-93-60C	calcite	Ko	68°30′57″	157°10′21″	−19.3	17.2	0.6	95	−0.9	Neocomian syntectonic deposits
35	DH-93-62C	calcite	Ko	68°32′40″	157°11′50″	−8.2	19.7	0.8	125	4.1	Neocomian syntectoni depositsc
36	DH-93-64C	milky calcite	Ko	68°30′35″	157°35′13″	−11.9	25.6				Neocomian syntectonic deposits
37	DH-93-64D	light brown calcite	Ko	68°30′35″	157°35′13″	−12.1	24.8				Neocomian syntectonic deposits
38	DH-93-66	brown calcite	Ko	68°30′45″	157°36′14″	−9.4	20.3	0.9	135	6.3	Neocomian syntectonic deposits
39	DH-93-74C	milky calcite	Kto	68°47′8″	157°47′28″	−10.1	16.1	1.0	150	3.4	Aptian–Albian Colville basin
40	DH-93-74C	clear calcite	Kto	68°47′8″	157°47′28″	−9.5	15.9	1.0	150	3.2	Aptian–Albian Colville basin

Table 2. Stable Isotope and Vitrinite Reflectance Data (cont.).

Map Location	Sample Number	Sample Description	Stratigraphic Unit*	Latitude (N)	Longitude (W)	$\delta^{13}C$ PDB**	$\delta^{18}O$ SMOW**	Mean R_o (%)	Temperature (°C)	$\delta^{18}O_{water}$	Tectonic Unit
41	DH-93-77C	white calcite	Kto	68°50′32″	157°41′9″	−2.1	17.5				Aptian–Albian Colville basin
42	DH-93-84D	milky calcite	Ko	68°20′52″	154°49′12″	−17.3	17.6	1.8	210	8.4	Neocomian syntectonic deposits
43	DH-93-84E	light brown calcite	Ko	68°20′52″	154°49′12″	−13.8	22.2	1.6	200	12.5	Neocomian syntectonic deposits
44	DH-93-84 I	white calcite	Ko	68°20′55″	154°49′5″	−10.6	18.3	1.9	215	9.3	Neocomian syntectonic deposits
45	DH-93-95G	siderite	Dhf	68°11′45″	155°38′0″	−4.1	17.3	3.7	285	11.2	Upper Devonian clastic wedge
46	DH-93-100D	calcite	MDk	68°10′17″	155°14′52	−9.2	16.6				Upper Devonian clastic wedge
47	DH-93-102B	white calcite	Ma	68°23′6″	154°19′50″	−5.9	21.6				Mississippian–Jurassic passive margin
48	DH-93-103B	black calcite	Ko	68°22′35″	154°19′46″	−16.4	14.4	2.0	220	5.7	Neocomian syntectonic deposits
49	DH-93-103D	white calcite	Ko	68°22′35″	154°19′46″	−17.7	19.8	2.0	220	11.1	Neocomian syntectonic deposits
50	DH-93-103D	yellow calcite	Ko	68°22′35″	154°19′46″	−24.2	19.4	2.0	220	10.7	Neocomian syntectonic deposits
51	DH-93-103E	calcite	Ko	68°22′35″	154°19′46″	−22.4	21.6	2.0	220	12.9	Neocomian syntectonic deposits
52	DH-93-103F	white calcite	Ko	68°22′35″	154°19′46″	−24.2	19.2	2.0	220	10.5	Neocomian syntectonic deposits
53	DH-93-103F	white-gray calcite	Ko	68°22′35″	154°19′46″	−20.2	19.3	2.0	220	10.6	Neocomian syntectonic deposits
54	DH-93-103F	white-yellow calcite	Ko	68°22′35″	154°19′46″	−21.5	19.9	2.0	220	11.2	Neocomian syntectonic deposits
55	DH-93-103G	white calcite	Ko	68°22′35″	154°19′46″	−23.8	19.4	2.0	220	10.7	Neocomian syntectonic deposits
56	DH-93-106B	calcite	Mk	68°17′17″	154°56′28″	−5.6	19.8	2.1	225	10.8	Upper Devonian clastic wedge
57	DH-93-107B	dolomite	PMl	68°19′19″	155°9′13″	2.5	27.9				Mississippian–Jurassic passive margin
78	DH-93-108B	brown calcite	Ko	68°23′43″	155°32′32″	−11.4	20.3	0.5	80	0.8	Neocomian syntectonic deposits
79	DH-93-108D	yellow calcite	Ko	68°23′59″	155°32′32″	−16.6	19.5	0.8	125	4.7	Neocomian syntectonic deposits
60	DH-93-108D	white calcite	Ko	68°23′59″	155°32′32″	−16.6	19.8	0.8	125	5	Neocomian syntectonic deposits
61	DH-93-109A	calcite	Ko	68°24′24″	155°32′22″	−9.6	32.2				Neocomian syntectonic deposits
62	DH-93-111B	calcite	Mn	68°27′4″	155°52′27″	−5.2	22.7				Mississippian–Jurassic passive margin
63	DH-93-113	calcite	PMl	68°17′40″	153°6′53″	−20.9	28.7				Mississippian–Jurassic passive margin
64	DH-93-113C	white calcite	PMl	68°17′40″	153°6′53″	2.9	−2.5				Mississippian–Jurassic passive margin
65	DH-93-118H	calcite	Kfm	68°38′20″	155°46′10″	3.6	27.3	0.6	95	9.3	Aptian–Albian Colville basin
66	DH-93-120B	calcite	Kto upper	68°41′1″	155°51′34″	−8.2	30.2				Aptian–Albian Colville basin
67	DH-93-122E	calcite	Kto	68°33′37″	155°24′27″	−3.5	17.7	0.6	95	−0.3	Aptian–Albian Colville basin
68	FC-93-1A	calcite	Ko	68°30′34″	155°42′25″	−11.6	20.3	0.8	125	5.5	Neocomian syntectonic deposits
69	FC-93-2A	white calcite	JTro	68°30′50″	155°42′25″	−22.1	28.4				Mississippian–Jurassic passive margin
70	FC-93-5B	calcite	Ko	68°31′13″	155°42′42″	0.9	27.8	0.9	135	13.9	Neocomian syntectonic deposits
71	FC-93-5J	calcite	Ko	68°31′13″	155°42′42″	−9.9	23.1	0.9	135	9.2	Neocomian syntectonic deposits
72	FC-93-6B	white calcite	Ko	68°31′20″	155°42′50″	21.6	30.4	0.9	135	16.5	Neocomian syntectonic deposits
73	FC-93-6B	transparent calcite	Ko	68°31′20″	155°42′50″	−7.1	19.4	0.9	135	5.5	Neocomian syntectonic deposits
74	FC-93-6C	white calcite	Mn	68°31′20″	155°42′50″	−4.5	22.9	0.9	135	9	Neocomian syntectonic deposits
75	FC-93-6D	white calcite	Ko	68°31′20″	155°42′50″	−14.6	24.1	0.9	135	10.2	Neocomian syntectonic deposits
78	FC-93-8	calcite	Kfm	68°38′18″	155°45′20″	−8.9	21.8				Aptian–Albian Colville basin
79	FC-93-20A	white calcite	Ko	68°28′5″	155°43′33″	−18.2	18.8	0.6	95	0.8	Neocomian syntectonic deposits
80	FC-93-20H	calcite	Ko	68°28′5″	155°43′33″	−15.1	20.5	1.0	150	7.8	Neocomian syntectonic deposits
81	FC-93-20 I	calcite	Ko	68°28′5″	155°43′33″	−21.4	19.5	1.0	150	6.8	Neocomian syntectonic deposits
82	FC-93-25B	white calcite	Ko	68°21′42″	155°44′4″	−3.8	23.7	0.6	100	6.2	Neocomian syntectonic deposits
83	FC-93-26B	calcite	Ma	68°21′15″	155°43′54″	−14.2	20.9	1.5	190	10.7	Mississippian–Jurassic passive margin
84	FC-93-27B	calcite	Ps	68°20′41″	155°44′5″	−21.8	20.6				Mississippian–Jurassic passive margin

TABLE 2. Stable Isotope and Vitrinite Reflectance Data (cont.).

Map Location	Sample Number	Sample Description	Stratigraphic Unit*	Latitude (N)	Longitude (W)	$\delta^{13}C$ PDB**	$\delta^{18}O$ SMOW**	Mean R_o (%)	Temperature (°C)	$\delta^{18}O_{water}$	Tectonic Unit
85	FC-93-28B	white calcite	Mk	68°19′19″	155°41′40″	−7.1	17.8	2.2	230	9.6	Upper Devonian clastic wedge
86	FC-93-29B	calcite	Mku	68°19′34″	155°42′5″	−6.9	18.1	2.5	240	10.4	Mississippian–Jurassic passive margin
87	FC-93-39A	calcite	Ps	68°28′35″	155°42′20″	−1.9	24.9	0.5	80	5.4	Mississippian–Jurassic passive margin
88	FC-93-51	calcite	Ko	68°29′55″	155°42′42″	−11.6	18.9				Neocomian syntectonic deposits
89	FC-93-66C	calcite	Kfm	68°34′24″	157°0′43″	−10.2	17.9				Aptian–Albian Colville basin
90	FC-93-128B	calcite	Kfm	68°42′45″	157°56′30″	−20.7	19.8				Aptian–Albian Colville basin
91	FC-93-128C	white calcite	Kfm	68°42′45″	157°56′30″	−20.2	19.3				Aptian–Albian Colville basin
92	MU-93-2B	calcite	Kto upper	68°40′42″	155°51′24″	−8.3	17.8	0.8	125	2.2	Aptian–Albian Colville basin
93	MU-93-6B	calcite	Kto upper	68°44′19″	155°32′25″	−7.5	16.7	0.7	110	0.4	Aptian–Albian Colville basin
94	MU-93-7B	calcite	Kto upper	68°44′27″	155°32′33″	−11.4	17	0.7	110	0.7	Aptian–Albian Colville basin
96	MU-93-7E	calcite	Kto upper	68°44′27″	155°32′33″	−5.5	16.2	0.7	110	−0.1	Aptian–Albian Colville basin
95	MU-93-8C	calcite	Kfm	68°35′11″	157°4′37″	2.3	20.4	1.9	215	11.4	Aptian–Albian Colville basin
97	TM-93-55C	calcite	Dn	68°16′41″	155°29′21″	−0.6	17.7				Upper Devonian clastic wedge
98	TM-93-56A	calcite	Dn	68°16′57″	155°29′27″	−0.5	17.5				Upper Devonian clastic wedge
99	TM-93-60A	calcite	Dn	68°15′6″	155°30′37″	−2.6	18.3				Upper Devonian clastic wedge
100	TM-93-60C	calcite	Dn	68°15′6″	155°30′37″	−1.5	16.2				Upper Devonian clastic wedge
101	TM-93-62B	gastropod	Dhf	68°15′41″	155°30′24″	1.9	18.3				Upper Devonian clastic wedge
102	TM-93-62C	calcite	Dhf	68°15′41″	155°30′24″	−1.9	15.1				Upper Devonian clastic wedge
103	TM-93-72A	calcite	Dhf	68°15′36″	155°15′37″	−3.7	21.1				Upper Devonian clastic wedge
104	UM-93-1A	calcite	Kn	68°58′45″	155°12′0″	−5.5	15.2				Aptian–Albian Colville basin
105	UM-93-1B	calcite	Kn	68°58′45″	155°12′0″	−6.9	17.3				Aptian–Albian Colville basin
106	AL-19	quartz	NR	NR	NR		14.9				Upper Devonian clastic wedge
107	AL-75	quartz	NR	NR	NR		13.1				Upper Devonian clastic wedge
108	AL-75	breccia quartz	NR	NR	NR		14.2				Upper Devonian clastic wedge
109	AL-78A	quartz	NR	NR	NR		13.4				Upper Devonian clastic wedge
110	AL-78B	quartz	NR	NR	NR		10.9				Upper Devonian clastic wedge
111	AL-100A	quartz	NR	NR	NR		10.8				Upper Devonian clastic wedge
112	AL-100B	quartz	NR	NR	NR		14.2				Upper Devonian clastic wedge
113	AL-QUZ-2A	quartz	NR	NR	NR		15.4				Upper Devonian clastic wedge
114	AL-QUZ-2B	early quartz	NR	NR	NR		13.6				Upper Devonian clastic wedge
115	AL-QUZ-2B	late quartz	NR	NR	NR		13.7				Upper Devonian clastic wedge
116	ALM-93-15D	quartz	PMl	68°19′15″	155°9′50″		22.2	2.2	230	11.7	Mississippian–Jurassic passive margin
117	DH-93-9F	quartz	JTro/Kcl/Ko	68°32′18″	155°43′34″		26.1				Mississippian–Jurassic passive margin
118	DH-93-38A	quartz	Dn	68°17′46″	156°19′38″		18.1				Upper Devonian clastic wedge
119	DH-93-38B	quartz	Dn	68°17′46″	156°19′38″		15.7				Upper Devonian clastic wedge
120	DH-93-59E	quartz crystal	Ko	68°31′48″	157°9′43″		18.3	0.8	125	1.1	Neocomian syntectonic deposits
121	DH-93-62F	quartz	Ko	68°32′40″	157°11′50″		21.1	0.9	135	4.5	Neocomian syntectonic deposits
122	DH-93-83B	white quartz	Ps	68°19′47″	154°50′10″		21.8	1.7	205	9.8	Mississippian–Jurassic passive margin
123	DH-93-83B	clear quartz	Ps	68°19′47″	154°50′10″		21.1	1.7	205	9.1	Mississippian–Jurassic passive margin
124	DH-93-85C	quartz crystal	Dhf-p	68°21′26″	154°49′55″		21.2	1.6	195	8.7	Upper Devonian clastic wedge
125	DH-93-85E	clear quartz	Dhf-p	68°21′26″	154°49′55″		18.7	1.6	195	6.2	Upper Devonian clastic wedge
126	DH-93-87G	white quartz	Dhf	68°15′8″	155°40′36″		14.9				Upper Devonian clastic wedge

TABLE 2. Stable Isotope and Vitrinite Reflectance Data (cont.).

Map Location	*Sample Number*	*Sample Description*	*Stratigraphic Unit**	*Latitude (N)*	*Longitude (W)*	*$\delta^{13}C$ PDB***	*$\delta^{18}O$ SMOW***	*Mean R_o (%)*	*Temperature (°C)*	*$\delta^{18}O_{water}$*	*Tectonic Unit*
127	DH-93-95G	white quartz	Dhf	68°11′45″	155°38′0″		13.3	3.7	285	5.3	Upper Devonian clastic wedge
128	DH-93-95G	quartz crystal	Dhf	68°11′45″	155°38′0″		14.1	3.7	285	6.1	Upper Devonian clastic wedge
129	DH-93-96D	quartz crystal	Dn	68°9′19″	155°40′5″		15.3				Upper Devonian clastic wedge
130	DH-93-100B	white quartz	MDk	68°10′17″	155°14′52″		13.9				Upper Devonian clastic wedge
131	DH-93-102B	quartz	Ma	68°23′6″	154°19′50″		23.1				Mississippian–Jurassic passive margin
132	DH-93-104B	quartz crystal	MDk	68°17′44″	154°37′6″		16.5	2.2	230	6	Upper Devonian clastic wedge
133	DH-93-109A	late quartz	Ko	68°24′24″	155°32′22″		23.7				Neocomian syntectonic deposits
134	FC-93-6B	quartz crystal	Ko	68°31′20″	155°42′50″		19.9				Neocomian syntectonic deposits
135	FC-93-17C	white quartz	Mku	68°19′46″	155°30′57″		16.7	2.0	220	5.7	Mississippian–Jurassic passive margin
136	TM-93-60C	quartz	Dn	68°15′6″	155°30′37″		9.3				Upper Devonian clastic wedge
137	TM-93-62A	quartz	Dhf	68°15′41″	155°30′24″		13.7				Upper Devonian clastic wedge

NR = sample location information not recorded.
**Kfm = Cretaceous Fortress Mountain Formation; Kn = Cretaceous Nanushuk Group; Kto = Cretaceous Torok Formation; Ko = Cretaceous Okpikruak Formation; Kcl = Cretaceous coquinoid limestone unit; JTro = Triassic and Jurassic Otuk Formation; Ps = Permian Siksikpuk Formation; PMl = Mississippian and Pennsylvanian Lisburne Group; Mku = Mississippian Kuna Formation; Ma = Mississippian Akmalik Chert; Mn = Mississippian Nuka Formation; Mk = Mississippian Kayak Formation; MDk = Mississippian(?) and Devonian Kanayut Conglomerate; Dn = Devonian Noatak Sandstone; Dhf = Devonian Hunt Fork Shale; Dhf-p = Devonian Hunt Fork Shale of Picnic Creek allochthon.*
***All delta values are in per mil.*

values for fluids that deposited calcite veins in younger lithologies (Mississippian–Jurassic passive-margin deposits, Neocomian syntectonic deposits, and Aptian–Albian Colville basin deposits) range from −1.5 to 16.5‰ (mean = 6.1‰); those for fluids that deposited quartz veins in these younger rocks range from 1.1 to 11.7‰ (Table 2; Figure 10).

The narrow ranges of calculated water values for quartz and calcite veins in Upper Devonian clastic deposits, near 10.2 and 6.5‰, respectively (Figure 10), are indicative of fluids whose $\delta^{18}O$ values were controlled by oxygen isotope exchange with rocks at relatively high temperatures and low water-to-rock ratios (Shelton et al., 1996). The much larger range of calculated water values for quartz and calcite veins in younger rocks (Mississippian–Jurassic passive-margin deposits, Neocomian syntectonic deposits, and Aptian–Albian Colville basin deposits), extending to $\delta^{18}O_{water}$ values as low as −1.5‰, is indicative of fluids whose oxygen isotope compositions were achieved by exchange with rocks at low temperatures and highly variable water-to-rock ratios (Underwood et al., 1999). The extremely low $\delta^{18}O_{water}$ values of some of these fluids may reflect the presence of relatively unexchanged meteoric waters in these younger rocks. It is important to note that the Upper Devonian clastic deposits were apparently unaffected by these low-$\delta^{18}O$ fluids.

Carbon Isotopes

The $\delta^{13}C$ (PDB) values of calcite veins range from −24.2 to +3.6‰ (mean = −9.9‰; Figure 11). Higher grade rocks of the Upper Devonian clastic wedge contain calcite veins with $\delta^{13}C$ values closer to 0‰, comparable to values of limestones in the stratigraphic section (Figure 12), indicating host rock buffering of carbon isotopes. In contrast, calcite veins in Mississippian to Cretaceous rocks have highly variable, commonly more negative $\delta^{13}C$ values, indicating a limited degree of host rock buffering of carbon isotopes.

Anomalously low $\delta^{13}C$ values for some calcite veins (−16.5 to −24.2‰) in Mississippian to Cretaceous strata may be the result of oxidation of organic matter and/or methane and appear to coincide spatially with areas having evidence of hydrocarbon fluxes along fractures (as determined by crushing veins in the field). The $\delta^{13}C$ signature of these calcite veins thus may be a potential guide to areas of methane venting from source rocks deeper in the frontal Brooks Range.

Interpretation of Stable Isotope Data

When combined with paleotemperatures based on vitrinite reflectance data in adjacent wall rocks, stable isotope compositions of calcite and quartz veins are consistent with at least two types of fluids affecting

Paleozoic–Lower Cretaceous rocks of the foothills and northern Brooks Range. Earlier, higher temperature (~250–300°C) fluids were responsible for vein emplacement in rocks of the Upper Devonian clastic wedge. The fluids responsible for formation of these veins had nearly constant $\delta^{18}O$ values, reflecting isotopic exchange in fluid source regions at low water-to-rock ratios. These fluids also had relatively constant $\delta^{13}C$ values, that were likely buffered by limestone lithologies. Younger (at least as young as the Aptian–Albian deposits of the Colville basin), lower temperature (~150°C) fluids deposited veins in Mississippian to Neocomian strata of the deformational wedge and in Colville basin deposits. These vein-forming fluids had more highly variable and significantly lower $\delta^{18}O$ values than the fluids that deposited veins in Upper Devonian strata, likely reflecting shallower fluids that exchanged isotopically with rocks at low temperatures and highly variable water-to-rock ratios. These fluids also had distinctly lower $\delta^{13}C$ values as a result of oxidation of organic matter and/or methane.

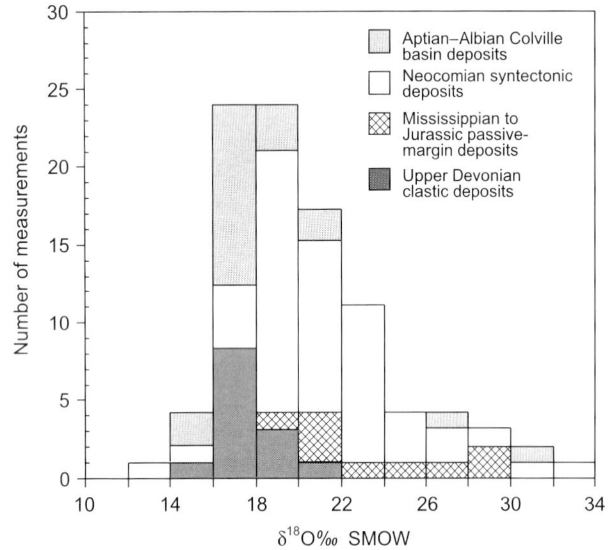

FIGURE 7. Histogram of $\delta^{18}O$ values (SMOW) of calcite veins hosted in Devonian to Cretaceous rocks from Brooks Range fold and thrust belt. Note that Devonian rock-hosted veins have lower values than veins hosted in younger rocks. SMOW = standard mean ocean water.

TIMING OF FLUID EVENTS

The ages of the two fluids can be estimated by comparing the fission-track data and thermal-maturity data from the same rocks. The earlier, higher temperature veins, estimated to have formed at temperatures as much as 285°C (Table 2), are hosted by Upper Devonian clastic rocks south of the modern mountain front. Johnsson et al. (1999), who compiled thermal-maturity data for Alaska, reported that the sampled region has thermal maturities of $R_o = 2–3.6$. The calibration of vitrinite reflectance to maximum temperature is not exact and is probably time dependent, but temperatures of 190–250°C are required to achieve this level of maturity (Barker and Pawlewicz, 1994). The regional distribution of the thermal-maturity data south of the modern mountain front suggests that these temperatures reflect ambient conditions instead of thermal anomalies of local extent.

The time at which temperatures as much as 285°C were active can be constrained by the zircon fission-track data from the same rocks. Fission-track ages in Upper Devonian clastic rocks south of the mountain front confirm that these rocks were subjected to temperatures of at least 225°C, because ZFTT ages record the time of cooling through the temperature range of about 260–225°C. This suggests that the earlier, high-temperature fluid that produced the veins migrated through the rocks at or before the times when the ZFTT cooling ages were set (160–120 Ma) at the time of thrusting and maximum tectonic burial that occurred during arc-continent collision in the Late Jurassic and Early Cretaceous. Moreover, these rocks could not have been reheated to temperatures of more than 225°C after 120 Ma, because such temperatures would have been recorded by the ZFTT data.

The age of the relatively younger, lower temperature fluid is constrained both by the fission-track ages and by stratigraphic relations. The lower temperature veins are hosted by Mississippian to Cretaceous rocks of the deformational wedge and by Aptian–Albian Colville basin deposits that overlie them. Their presence in Aptian–Albian strata thus requires a post-Albian age. The temperature of about 150°C that is estimated for these fluids from vitrinite reflectance data in adjacent

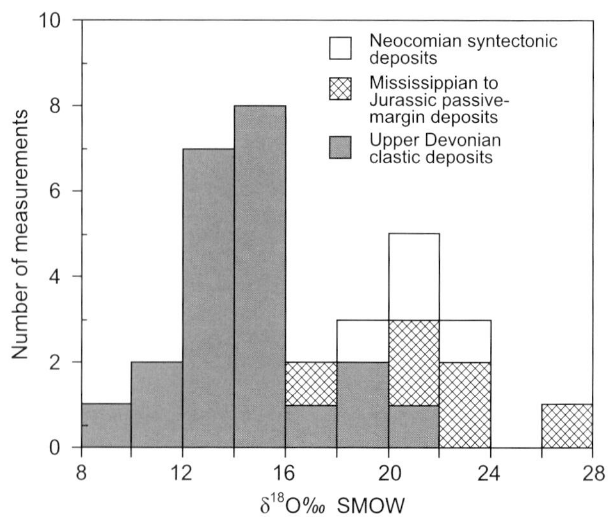

FIGURE 8. Histogram of $\delta^{18}O$ values (SMOW) of quartz veins hosted in Devonian to Cretaceous rocks from Brooks Range fold and thrust belt. Note that Devonian rock-hosted veins have lower values than veins hosted in younger rocks. SMOW = standard mean ocean water.

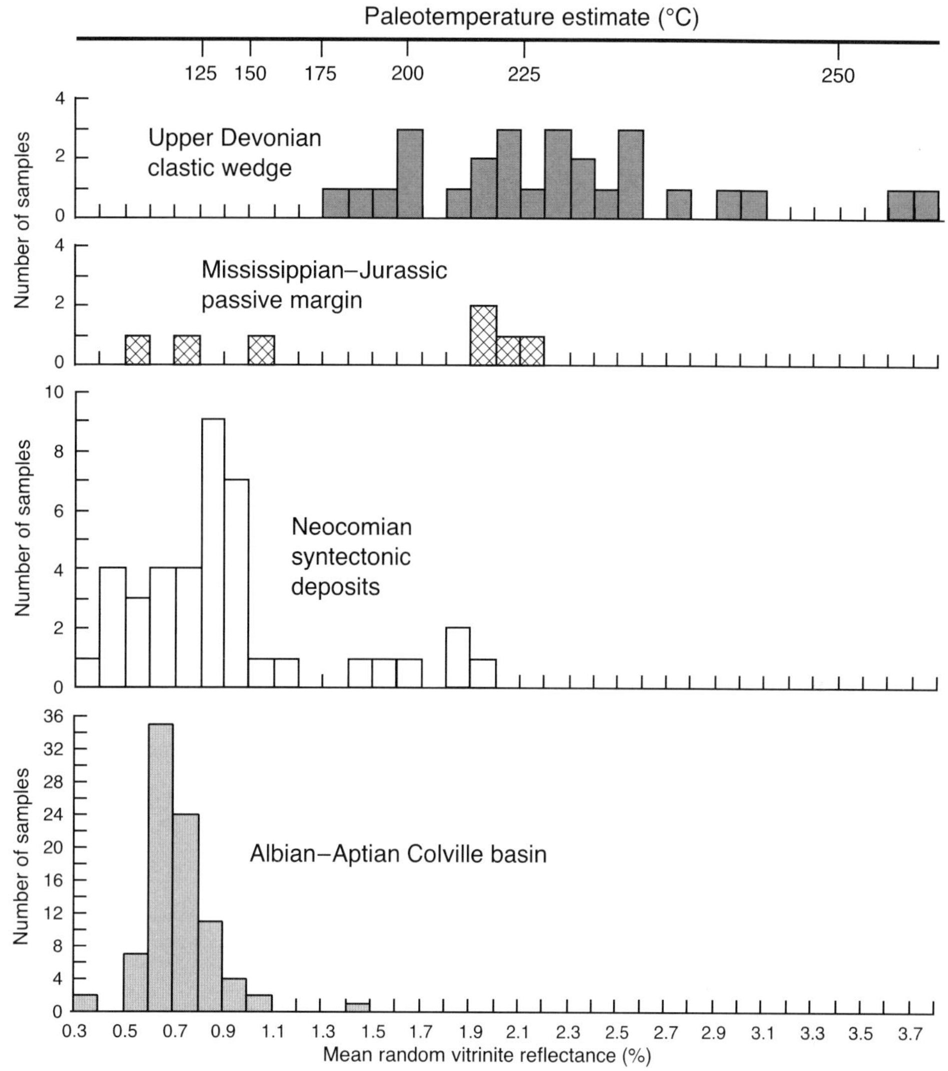

FIGURE 9. Histograms of vitrinite reflectance values (Cole et al., 1997; Johnsson et al., 1999) and paleotemperature estimates displayed by tectonic-stratigraphic unit for samples from the study area. At least 10 vitrinite particles were measured in each sample. The paleotemperature scale is based on the equation of Barker (1988). Note that all samples of the Upper Devonian clastic wedge and the four samples of the Mississippian to Jurassic passive margin with $R_o > 1.2$ were collected at or south of the present-day mountain front.

rocks (Table 2) is consistent with fission-track data that show that these rocks were heated sufficiently to reset AFTT ages (~110°C) but not sufficient to reset ZFTT ages (~225°C). Therefore, cooling from maximum temperatures of about 150°C for these rocks is probably recorded by the AFTT ages of 60 and 45 Ma (early Tertiary).

It is important to note that many aspects of the fluid history remain unresolved by the data presented herein. Vein samples for this study were collected during a reconnaissance investigation prior to the development of the structural framework concepts presented in this study. Most samples were taken from tectonic features, such as faults and joints, that probably were formed and open to fluid flow at the time of deformation, such that the timing of deformation approximates the time of fluid migration; additional work that identifies fluid pathways and ties their fluid history to specific structural features is needed to demonstrate this relation (e.g., Parris and Burruss, 2001). Although unrecognized in our data, relatively younger, low-temperature fluids may have been present in the Devonian clastic wedge, possibly along specific pathways (e.g., Tertiary faults) that were not sampled during this study. Likewise, low-temperature fluids that are coeval with the older, high-temperature fluids at deep levels in the deformational wedge may have been present at high structural positions in Mississippian to Neocomian strata in the frontal part of the wedge. Although such fluids were not identified in this study, scenarios involving a more complicated history of multiple fluid events than that considered here, especially at paleotemperatures of 110–225°C, remain plausible and should be considered in future studies.

TWO-STAGE MODEL OF DEFORMATION

Important structural features observed in the Brooks Range fold and thrust belt include (1) exposure of progressively deeper structural levels to the south in the Brooks Range foothills despite generally southward dips observed in most outcrops; (2) fission-track evidence for at least two periods of contractional deformation and uplift-related cooling in the Late Jurassic–Early Cretaceous and early Tertiary; (3) southward change in

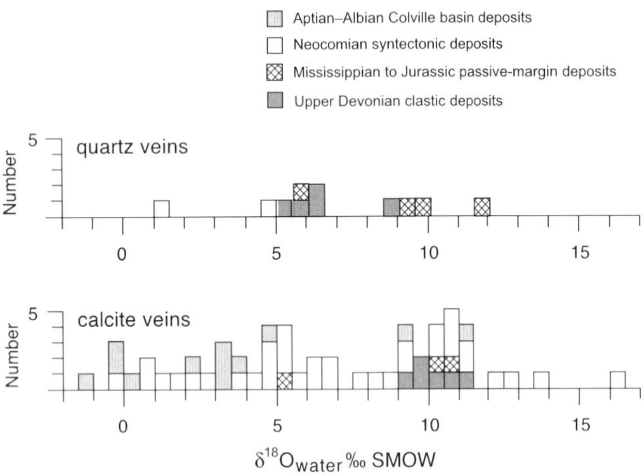

FIGURE 10. Histograms of calculated equilibrium $\delta^{18}O$ values (SMOW) of waters depositing calcite and quartz veins in Devonian to Cretaceous rocks from the Brooks Range fold and thrust belt. Note that Devonian rocks have narrower ranges of water compositions than all younger lithologies. SMOW = standard mean ocean water.

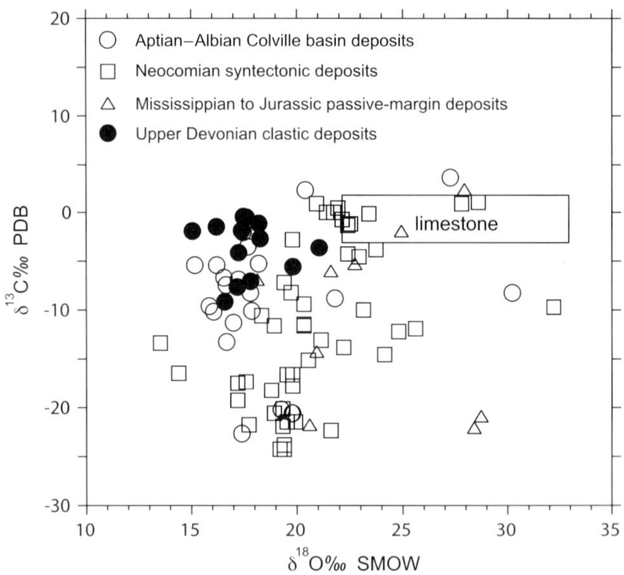

FIGURE 12. Carbon vs. oxygen isotope diagram for calcite veins hosted in Devonian to Cretaceous rocks from Brooks Range fold and thrust belt. SMOW = standard mean ocean water.

structural style, ranging from simple, upright folds in the north to very complex folds and thrusts, including broken formation and melange in the south; and (4) rapid northward thickening and facies changes in Colville basin deposits. We propose that these features are the products of two superposed north-directed deformational belts: (1) a thin-skinned fold and thrust belt formed in the Late Jurassic(?) to Early Cretaceous

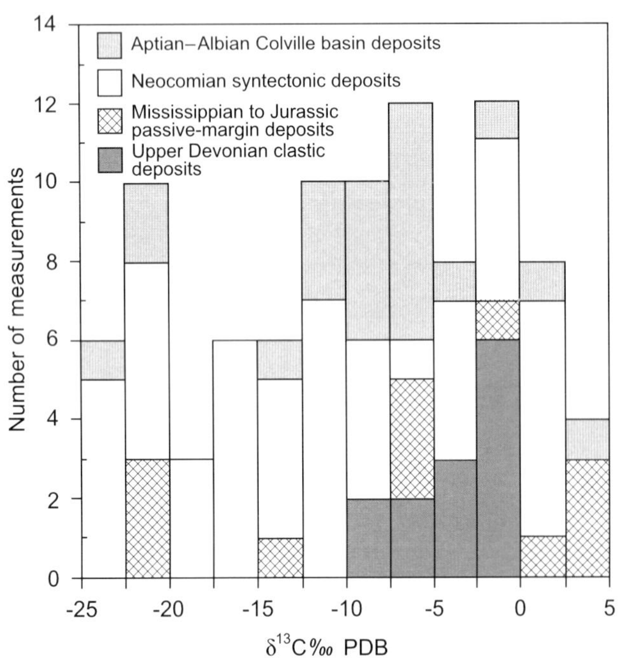

FIGURE 11. Histogram of $\delta^{13}C$ values (PDB) of calcite veins hosted in Devonian to Cretaceous rocks from Brooks Range fold and thrust belt.

(Neocomian) and (2) a thin-skinned fold and thrust belt that developed in the early Tertiary, following sedimentary burial in the Colville basin.

The first episode of thrusting in northern Alaska was related to the Late Jurassic and Early Cretaceous north-facing arc-continent collision, which emplaced the Endicott Mountains and higher allochthons beneath a Jurassic oceanic arc (e.g., Roeder and Mull, 1978; Box, 1985; Moore et al., 1994). During the deformation, allochthons of distal facies of the Ellesmerian and Beaufortian sequences were internally deformed and detached from their continental substrate and assembled by northward thrusting into shingled, north-thinning deformational wedges (Figure 13). Within the composite deformational wedge, the structural style was stratigraphically controlled, with mechanically competent stratigraphic units imbricated in duplexes and less competent units exploited as detachments that separated the duplexes as discussed by Wallace et al. (1997). Zircon fission-track data suggest that the deformational wedge was emplaced onto the south-dipping autochthon from 160 to about 120 Ma, which is consistent with the age of thrusting determined from synorogenic deposits.

By the Aptian, the thrusting resulted in the construction of a structurally and topographically high-standing region, which became the south flank of the Colville basin (Figure 13). Tectonic loading produced by the thrusting caused subsidence of the basin by the Barremian (Cole et al., 1997). The basin was filled with voluminous sediments of Aptian–Albian age shed from source areas south and west (Molenaar, 1988). Vogl et al.

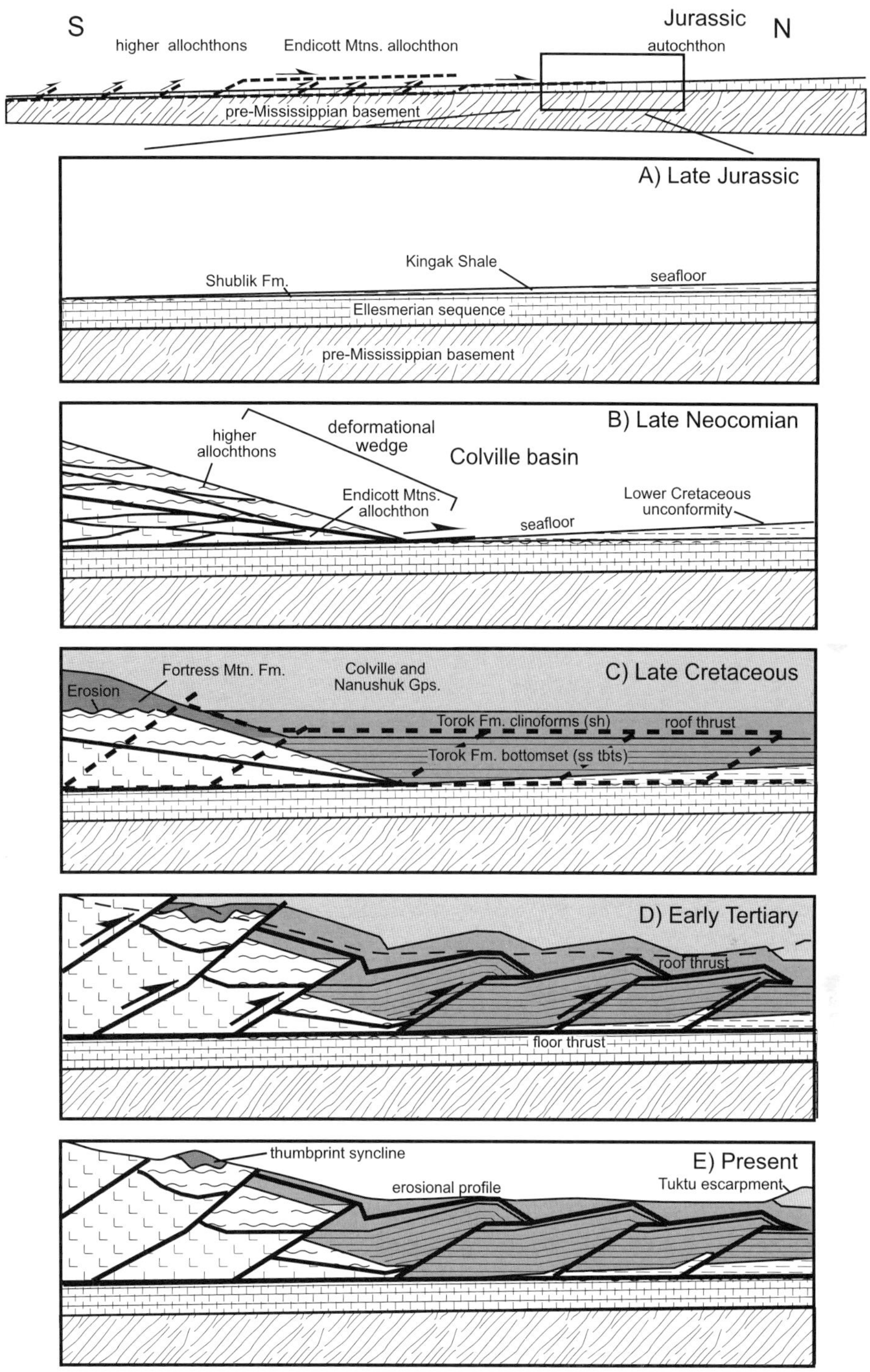

FIGURE 13. Conceptual diagram showing development of Colville basin. Top: Location of diagrams (A–E) on south-facing passive margin. Note restored positions of higher allochthons and Endicott Mountains allochthon. (A) In the Late Jurassic, the location of the diagram was in undeformed foreland with north-vergent, thin-skinned deformation occurring outside the diagram to the south. (B) By the late Neocomian, a deformational wedge consisting of Endicott Mountains allochthon and overlying higher allochthons formed a constructional high that became the south flank of the Colville basin and caused basin subsidence because of tectonic loading. Allochthons are composed of a stacked series of duplexes and broken formation. (C) In the Aptian, Albian, and Late Cretaceous, voluminous sediments that were shed from sources to the west and south prograded longitudinally along the basin from the west and onlapped the deformational wedge to the south (internal duplex structure in deformational wedge shown only schematically). By the end of the Cretaceous, more than 8 km (5 mi) of clastic sediments were deposited. (D) In the early Tertiary, renewed north-vergent deformation formed a duplex with a floor thrust in Kingak Shale of autochthon and a roof thrust in the clinoform strata of the upper Torok Shale. To the south, the Early Cretaceous deformational wedge was deformed by linking thrusts in the duplex, whereas to the north, only Colville basin strata were deformed by linking thrusts. A triangle zone developed near the present Tuktu Escarpment. Dashed line indicates the position of the modern erosional profile. (E) Erosion in the early Tertiary removed about 3 km (2 mi) of strata and exhumed the upper part of duplex.

(2002) suggested that the Colville basin sediments were derived from an uplifted region in the southern Brooks Range formed by extensional uplift active from about 105 to 96 Ma. Thermal heating and subsequent cooling associated with this extensional event may have extended northward into the northern Brooks Range, as evidenced by AFTT ages of about 100 Ma in the southern part of our study area.

The early Tertiary episode of north-directed thrusting occurred principally at about 60 and 45 Ma (O'Sullivan et al., 1997). This thrusting was probably generated by plate interactions in southern Alaska that resulted in northward thrusting in northern Alaska (Grantz et al., 1991). The thrusting occurred above a very deep detachment under the southern Brooks Range, which ramps up to the Kingak Shale beneath the northern flank of the Brooks Range (Fuis et al., 1997, their plate 3). In the southern Brooks Range, where deformation was thick skinned, the shortening caused the formation of a crustal-scale duplex (Oldow et al., 1987; Fuis et al., 1997; O'Sullivan et al., 1998). In the northern foothills, where the deformation was thin skinned, thrusting resulted in formation of reverse faults that truncated and juxtaposed the deformational wedge against Colville basin deposits overlying the wedge. This thrusting extended northward into Colville basin deposits, where it produced the passive-roof duplex and a triangle zone at the Tuktu Escarpment, as well as thrust-related folds and an intercutaneous wedge at the present deformation front (Figure 13).

THERMAL EVOLUTION AND FLUID MIGRATION

Prior to Jurassic and Early Cretaceous thrusting, the thermal structure of the Ellesmerian sequence was controlled by simple stratigraphic burial. Assuming a geothermal gradient typical for a long-standing passive margin (~30°C/km), the depth to the oil window (80–120°C) was greater than the thickness of the entire Ellesmerian sequence (<3 km) in most areas (Magoon and Claypool, 1983, their figure 7). Thus, the top of the oil window must have been in pre-Mississippian strata, below the Ellesmerian sequence in deformed and mildly metamorphosed pre-Mississippian rocks prior to the Late Jurassic. No hydrocarbons could have been generated at this time, except possibly in the deepest parts of isolated extensional basins, where the Ellesmerian sequence attains its largest thickness.

Levels of thermal maturity sufficient for hydrocarbon generation in the Ellesmerian sequence could have been attained in the Late Jurassic and Neocomian by tectonic burial beneath thrust sheets of allochthonous strata (Figure 14, top left). Thermal-maturity and ZFTT data presented in this study demonstrate that temperatures in excess of about 225°C were attained in deeper parts of the deformational wedge. Our isotopic data from veins of this age indicate that fluids present in this part of the Late Jurassic and Early Cretaceous deformational wedge were buffered by limestone. The fluids may have been derived from the underlying autochthon, where a thick, widespread succession of caronate rocks is located, (Lisburne Group) and/or from more rearward parts of the deformational wedge, where a thick, imbricated succession of lower Paleozoic carbonate rocks is present (Baird Group, e.g., Dumoulin and Harris, 1994). Either possibility suggests that fluid migration occurred along thrust faults from deeper levels in the orogenic belt.

The temperatures indicated by the vitrinite and fission-track data suggest that source rocks present in higher and more forward parts of the deformational wedge were cooler but probably had sufficient maturity to generate hydrocarbons at the time of thrusting. If present, these hydrocarbons could have migrated along faults and fractures, and they could have been trapped in reservoir strata involved in thrust-related folds in the frontal part of the deformed belt as suggested by Oliver (1986) (Figure 14).

North of the Neocomian deformation front, stratigraphic thicknesses in most areas were insufficient to generate hydrocarbons from Mississippian and younger strata until after the onset of rapid deposition and infilling of the Colville basin in the Albian (Figure 14, top right). As sediments of the Brookian sequence began to accumulate in the basin, the stratigraphic section subsided through the oil window. With increasing stratigraphic burial, hydrocarbon generation moved upward from the Ellesmerian sequence into Colville basin deposits. Seismic profiles, well data, and structural restorations indicate that as much as 10 km (6 mi) of sediment was deposited in the deepest parts of the Colville basin in the Albian and Cenomanian. By the time of maximum burial just prior to Tertiary deformation, the oil window was entirely within Colville basin deposits in the axial parts of the basin, and the major source units in the Ellesmerian sequence at depth were overmature (Figure 14, middle). Burial-history calculations (Magoon and Claypool, 1983; Burns et al., 2003) confirm that peak generation for these units occurred at about 100–90 Ma in the central and western parts of the basin, indicating that Ellesmerian petroleum systems were active while the basin was in the early stages of being filled. Hydrocarbons generated during this period mostly likely migrated into nearby stratigraphic traps in Colville basin strata.

A key consideration is that isotherms present during filling of the Colville basin were generally concordant with stratigraphy, because the thermal maturity of the basin sediments resulted from sedimentary burial alone (Figure 14, middle). Vitrinite reflectance data from well data suggest that the top of the oil window at the time of maximum burial was located in the lower part of the Nanushuk Formation, and that its base was located in the lower part of the upper Torok Formation (Cole et al., 1997). Where Colville basin strata onlap the deformational wedge along the southern flank of the basin, however, these isotherms transected the basin margin and extended southward into deformed strata of the deformational wedge. For this reason, the upper part of the deformational wedge did not experience temperatures sufficient to cause hydrocarbon generation

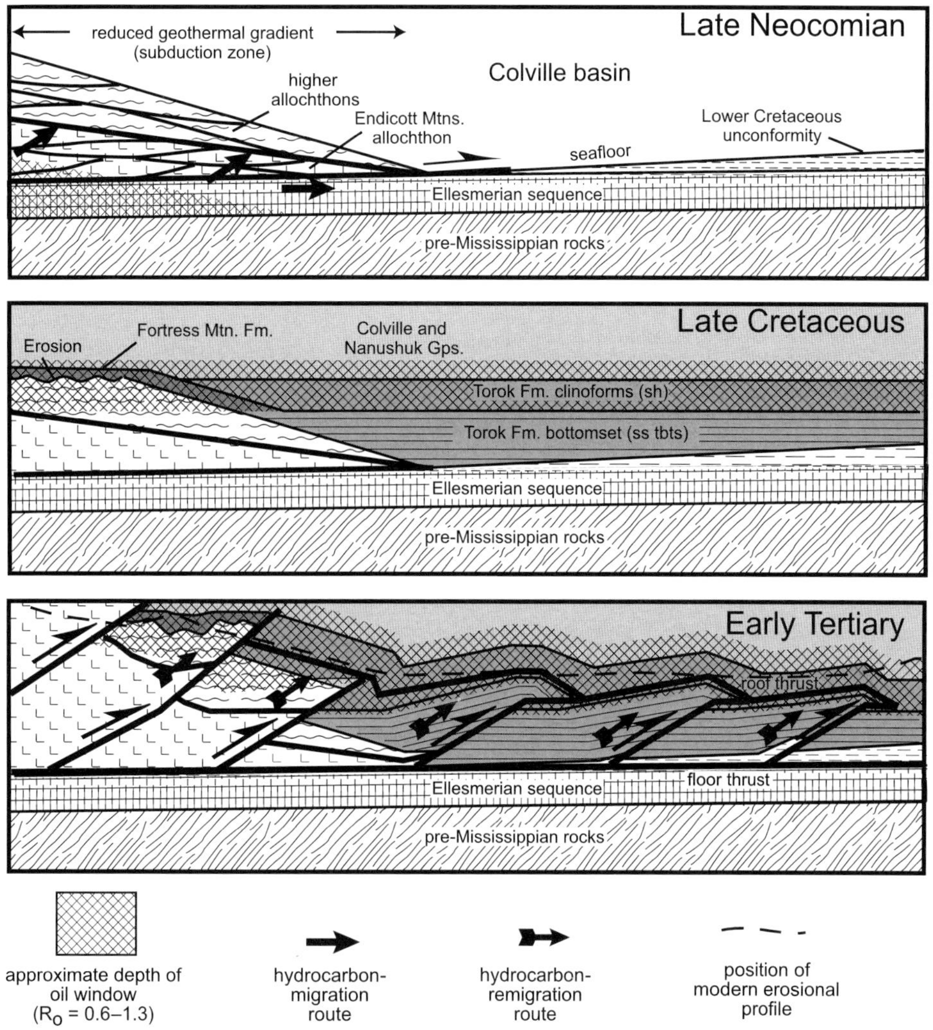

FIGURE 14. Thermal analysis and relation to hydrocarbon generation of the structural model developed in Figure 13. Approximate position of rocks in the oil window (shown by hatchures) superposed on cross sections of Figure 13B–D. Top: In the late Neocomian, the oil window was located beneath the deformational wedge in the thrust belt. Hydrocarbon migration at this time is toward the frontal part of the deformational wedge and up the depositional dip in the underlying autochthon. Ellesmerian sequence in the foreland is too thin to cause maturation by sedimentary burial in most areas (underlying pre-Mississippian rocks are previously deformed and metamorphosed). Middle: During the Late Cretaceous, the position of the oil window moved upward through the section as sedimentary burial under Torok and Nanushuk formations of the Colville basin increased. To the south, the oil window intersected the deformational wedge because of sedimentary onlap of Colville basin strata. Oil migration at this time is local, with accumulations forming principally in stratigraphic traps and preexisting structural traps. Bottom: The oil window that developed at the time of maximum burial was deformed and uplifted by thrusts during early Tertiary deformation. At depth, early formed hydrocarbon traps, now overmature, are breached by faults, releasing hydrocarbons that migrated upward along active faults into newly formed traps.

until much of the basin was filled. This interpretation is supported by relatively low thermal maturity in the Neocomian and older strata in the Brooks Range foothills (Cole et al., 1997; Johnsson et al., 1999).

Apatite fission-track ages across the deformational belt in southern NPRA show that cooling from maximum burial temperatures began at about 60 Ma as a result of thrust-related uplift in the early Tertiary (O'Sullivan, 1996; O'Sullivan et al., 1997). Because the isotherms in the Colville basin were generally concordant with Colville basin stratigraphy prior to thrusting, they probably were faulted and folded with the stratigraphy that contained them (Figure 14, bottom). This suggests that the hydrocarbons and aqueous fluids that were generated from rocks in the oil window at the time of maximum burial near the onset of thrusting could have migrated into adjacent antiformal traps that formed during early stages of the deformation, both in the basin and in the Neocomian deformational wedge along its southern flank. Additionally, the seals for many hydrocarbon accumulations sequestered in stratigraphic traps at depth in the Colville basin probably were ruptured by the deformation. The hydrocarbons released by this process would have remigrated upward into structurally higher traps along faults or other fluid pathways or were lost to the surface (Figure 14, bottom). Fluid migration at this time was from rocks at relatively low temperatures and highly variable water-to-rock ratios, as suggested by the stable isotope data from Mississippian to Cretaceous rocks of the deformational wedge and the Colville basin. These characteristics are consistent with derivation of the fluids from in-situ sources in the basin and in the deformational wedge. Because the deformational wedge had been through an earlier cycle of deformation and dehydration, however, it probably did not represent a major source of expelled aqueous fluids. The relatively low $\delta^{13}C$ values associated with some of these fluids support the interpretation that the fluid pathways were also pathways for remigration of hydrocarbons.

Although source rocks at the surface in the deformational wedge and Colville basin have thermal maturities indicative of heating to oil window temperatures, hydrocarbon generation probably largely ceased with the onset of deformation at 60 Ma. More than 3 km (2 mi) of section have been removed by erosion from much of the deformation belt in the Brooks Range foothills following thrusting in the early Tertiary (O'Sullivan et al., 1997), indicating that temperatures have decreased because of exhumation after the onset of deformation. Any tectonic burial resulting from thrusting in the early Tertiary would have occurred at very deep levels, where strata were already thermally overmature and generation was largely completed.

CONCLUSION

Our analysis of field, seismic, thermal-maturity, fission-track, and stable isotope data support the interpretation that the Brooks Range was the site of at least two north-directed contractional orogenic belts. The earliest was a thin-skinned fold and thrust belt active from about 160 to about 120 Ma. This belt formed in an arc-continent collisional zone that produced blueschist facies and other metamorphic rocks at deep structural levels in the orogen and deformed a 1–4-km (0.6–2.5-mi)-thick, passive-margin stratigraphic succession and accompanying synorogenic deposits in the frontal part of the orogen. In its frontal part, the orogen was characterized by structural stacking of internally deformed but regionally extensive allochthonous sequences. An imbricate fan structural style was probably present near the deformation front, because Jurassic and Lower Cretaceous strata are thin and are present in the autochthon beneath allochthonous rocks. The magnitude of total shortening required by stratigraphic and structural relations was very high (100–500 km; 60–300 mi), although total structural relief in its frontal part was probably relatively low, because only relatively thin stratigraphic sequences were involved in the thrusting. Tectonic burial produced temperatures in deeper parts of the fold and thrust belt in excess of 225°C and contained fluids that may have migrated upward along the thrust faults from strata that included carbonate rocks, such as the thick carbonate successions located in the autochthon or in deeper parts of the orogen. This orogenic belt is probably the northward expression of approximately coeval subduction accretion events that occurred along the western margin of North America in the Jurassic and Early Cretaceous (Grantz et al., 1991).

The second orogenic belt was active at about 60 Ma and was thick skinned in the hinterland and thin skinned in the frontal zone. This deformation produced thrusts that breached the older structures and formed a passive-roof duplex and fold belt near the deformation front in Colville basin deposits. A 10-km (6-mi)-thick succession consisting of rocks of both the preexisting Jurassic–Early Cretaceous deformational wedge and subsequent posttectonic Colville basin deposits were involved in the thrusting. Although total shortening was probably relatively low (perhaps 10–50 km [6–30 mi]), the structural relief was probably high because of the large thickness of the stratigraphic section involved in the deformation. Fluids were cooler than about 150°C and probably hydrocarbon-bearing in the foreland part of the orogen. This belt probably formed as a north-vergent retroarc thrust belt that was linked to northward subduction beneath southern Alaska and may be viewed as the northward expression of the Rocky Mountains deformational belt.

The two orogenic belts were separated by an episode of extensional tectonism that affected the southern Brooks Range and interior Alaska during the period of about 105–96 Ma. The extensional tectonism produced more than 8 km (5 mi) of exhumation in the southern Brooks Range and consequent depositional infilling of the Colville basin as discussed by Vogl et al. (2002). The middle Cretaceous extensional event is important, because it separates the two major episodes of contractional deformation in time and shows that they did not form as part of a single, protracted orogenic event.

As noted by O'Sullivan et al. (1993, 1997), subsequent periods of uplift-related cooling are indicated by AFTT at 45, 35, and 25 Ma. Apatite fission-track thermochronology ages of 45 Ma are widespread north of the Brooks Range and may provide evidence that the second episode of contractional deformation persisted from 60 to 45 Ma. Alternately, a third discrete episode of contractional deformation may have occurred at 45 Ma. The subsequent episodes of contractional deformation at 35 and 25 Ma are most prevalent in the northeastern Brooks Range and indicate that the northeastern salient of the range was constructed in the middle and late Tertiary (O'Sullivan et al., 1993, 1997; O'Sullivan and Wallace, 2002; Potter et al., 2004). Evidence of cooling at 25 Ma is also locally present to the west, indicating that the associated deformation may have extended across much of the foreland. The deformational style of these events in the northeastern Brooks Range is very similar to the thin- to thick-skinned structures on which the deformation at 60 Ma occurred and probably reactivated the older 60-Ma structures.

Hydrocarbon generation from the principal source rocks of the North Slope (Shublik Formation, Kingak Shale, and Otuk Formation) and migration into stratigraphic traps occurred by sedimentary burial principally between 100 and 90 Ma, between the two major episodes of contractional deformation. Subsequent sedimentary burial caused deep stratigraphic traps to become overmature, cracking oil to gas, and some new generation to begin progressively higher in the section. Structural disruption of the traps in the early Tertiary released sequestered hydrocarbons and caused remigration into newly formed structural traps formed at higher structural levels or release to the surface. Because of the generally high maturation of the Colville basin at the time of the deformation and remigration, most of the hydrocarbons available to fill traps were gas.

This analysis suggests that the time of the principal period of hydrocarbon generation in the foreland region of the Brooks Range was not coincident with either Late Jurassic–Early Cretaceous or early Tertiary deformation, although some generation probably did occur near the time of both events. For this reason, the timing of generation and trap development was not ideal. Nonetheless, significant accumulations of hydrocarbons might be expected because of the richness and abundance of source rock strata in northern Alaska and because of the abundance of stratigraphic traps in the foreland basin that could have sequestered hydrocarbons until deformation released hydrocarbons and allowed their remigration into new-formed traps in the early Tertiary.

ACKNOWLEDGMENTS

This study benefited from thought-provoking discussions with Ken Bird, Fran Cole, David Houseknecht, David Howell, Gil Mull, and Wesley Wallace. We thank Mark Pawlewicz for vitrinite reflectance analysis and Cole and Howell for providing samples and field notes. M. B. Underwood and K. L. Shelton thank ARCO and British Petroleum for their support of vein analyses. The manuscript was greatly improved by the thoughtful reviews, comments, and insights of Bird, Mull, Michael McGroder, and Philippe Muchez.

REFERENCES CITED

Barker, C. E., 1988, Geothermics of petroleum systems; implications of the stabilization of Kerogen thermal maturation after a geologically brief heating duration at peak temperature, *in* L. B. Magoon, ed., Petroleum systems of The United States: U.S. Geological Survey Bulletin, 1870, p. 26–29.

Barker, C. E., and M. J. Pawlewicz, 1994, Calculation of vitrinite reflectance from thermal histories and peak temperatures: A comparison of methods, *in* P. K. Mukhopadhhay and W. G. Dow, eds., Vitrinite reflectance as a maturity parameter: Applications and limitations: American Chemical Society Symposium Series 570, p. 216–229.

Bird, K. J., 1991, North Slope of Alaska, *in* J. J. Gluskoter, D. D. Rice, and R. B. Taylor, eds., Economic geology, U.S.: Boulder, Colorado, Geological Society of America, The geology of North America, v. P-2, p. 447–462.

Bird, K. J., 2001, Alaska: A twenty-first-century petroleum province, *in* M. W. Downey, J. C., Threet, and W. A. Morgan, eds., Petroleum provinces of the twenty-first century: AAPG Memoir 74, p. 137–165.

Bird, K. J., and D. W. Houseknecht, 2002a, U.S. Geological Survey 2002 petroleum resource assessment of the National Petroleum Reserve in Alaska (NPRA): U.S. Geological Survey Fact Sheet 045-02, 6 p.

Bird, K. J., and D. W. Houseknecht, 2002b, U.S. Geological Survey 2002 petroleum resource assessment of the National Petroleum Reserve Alaska (NPRA): Play maps and technically recoverable resource estimates: U.S. Geological Survey Open-file Report 02-207, 18 p.

Bird, K. J., and C. M. Molenaar, 1992, The North Slope foreland basin, *in* R. W. Macqueen and D. A. Leckie, eds., Foreland basins and fold belts: AAPG Memoir 55, p. 363–393.

Box, S. E., 1985, Early Cretaceous orogenic belt in northeastern Alaska: Internal organization, lateral extent, and tectonic interpretation, *in* D. G. Howell, ed., Tectonostratigraphic terranes of the circum-Pacific region: Circum-Pacific Council for Energy and Mineral Resources (AAPG), Earth Science Series, no. 1, p. 137–145.

Brandon, M. T., M. K. Roden-Tice, and J. I. Garver, 1998, Late Cenozoic exhumation of the Cascadia accretionary wedge in the Olympic Mountains, northwest Washington State: Geological Society of America Bulletin, v. 110, p. 985–1009.

Burns, M. W, D. O. Hayba, D. W. Houseknecht, and E. L. Rowan, 2003, Timing of hydrocarbon generation in the Colville basin of northern Alaska from burial and thermal history modeling (abs.): AAPG Annual Meeting Program, v. 12, p. A22.

Clayton, R. N., and T. K. Mayeda, 1963, The use of bromine-pentafluoride in the extraction of oxygen from oxides and silicates for isotopic analyses: Geochimica et Cosmochimica Acta, v. 27, p. 43–52.

Clayton, R. N., J. R. O'Neil, and T. K. Mayeda, 1972, Oxygen isotope exchange between quartz and water: Journal of Geophysical Research, v. 77, p. 3057–3067.

Cole, F., K. J. Bird, J. Toro, F. Roure, P. B. O'Sullivan, M. Pawlewicz, and D. G. Howell, 1997, An integrated model for the tectonic development of the northern Brooks Range and Colville basin 250 km west of the Trans-Alaska crustal transect: Journal of Geophysical Research, v. 102, p. 20,645–20,684.

Donelick, R. A., R. A. Ketcham, and W. D. Carlson, 1999, Variability of apatite fission-track annealing kinetics: II. Crystallographic orientation effects: American Mineralogist, v. 84, p. 1224–1234.

Dumoulin, J. A., and K. J. Bird, 2002, Lithofacies and stratigraphy of the Lisburne and Etivluk groups in the Lisburne 1 well and adjacent outcrops, central Brooks Range, Alaska (abs.): AAPG Bulletin, v. 86, no. 6 , p. 1142.

Dumoulin, J. A., and A. G. Harris, 1994, Depositional framework and regional correlation of pre-Carboniferous metacarbonate rocks of the Snowden Mountain area, central Brooks Range, northern Alaska: U.S. Geological Survey Professional Paper 1545, 73 p.

Ferket, H., F. Roure, R. Swennen, and S. Ortuno, 2000, Fluid migration placed into the deformation history of fold and thrust belts: An example from the Veracruz Basin (Mexico): Journal of Geochemical Exploration, v. 69–70, p. 275–279.

Fuis, G. S., J. M. Murphy, W. J. Lutter, T. E. Moore, K. J. Bird, and N. I. Christensen, 1997, Deep seismic structure and tectonics of northern Alaska: Crustal-scale duplexing with deformation extending into the upper mantle: Journal of Geophysical Research, v. 102, no. B9, p. 20,873–20,896.

Gleadow, A. J. W., I. R. Duddy, P. F. Green, and J. F. Lovering, 1986, Confined fission track lengths in apatite: A diagnostic tool for thermal history analysis: Contributions to Mineral Petrology, v. 94, p. 405–415.

Gottschalk, R. R., and L. W. Snee, 1998, Tectonothermal evolution of metamorphic rocks in the south-central Brooks Range, Alaska: Constraints from $^{40}Ar/^{39}Ar$ geochronology, *in* J. S. Oldow and H. G. Avé Lallemant, eds., Architecture of the central Brooks Range fold and thrust belt, Arctic Alaska: Geological Society of America Special Paper, v. 324, p. 225–251.

Grantz, A., S. D. May, and P. E. Hart, 1990, Geology of the Arctic continental margin of Alaska, *in* A. Grantz, L. Johnson, and J. F. Sweeney, eds., The Arctic Ocean region: Boulder, Colorado, Geological Society of America, The geology of North America, v. L., p. 257–288.

Grantz, A., T. E. Moore, and S. M. Roeske, 1991, Continent–ocean transect A-3: Gulf of Alaska to Arctic Ocean: Geological Society of America, scale 1:500,000, 3 sheets.

Houseknecht, D. W., 2003, Beaufortian stratigraphic plays in the National Petroleum Reserve in Alaska (NPRA): U.S. Geological Survey Open-file Report 03-039, 47 p. plus tables and figures, http://pubs.usgs.gov/of/2003/of03-040/ (accessed February 4, 2003).

Houseknecht, D. W., and Bird, K. J., 2004, Sequence stratigraphy of The Kingak Shale (Jurassic–Lower Cretaceous), National Petroleum Reserve in Alaska: AAPG Bulletin, v. 88, no. 3, p. 279–302.

Hubbard, R. J., S. P. Edrich, and R. P. Rattey, 1987, Geologic evolution and hydrocarbon habitat of the Arctic Alaska microplate, *in* I. L. Tailleur and P. Weimer, eds., Alaskan North Slope geology: SEPM Pacific Section, Bakersfield, California, Book 50, p. 797–830.

Johnsson, M. J., K. R. Evans, and H. A. Marshall, 1999, Thermal maturity of sedimentary rocks in Alaska: Digital resources: U.S. Geological Survey Digital Data Series DDS-54, version 1.0.

Legg, G. W., 1983, Geological report on the Lisburne test well no. 1: Husky Oil National Petroleum Reserve Operations Inc., http://nerslweb.cr.usgs.gov/NPRAWeb/PUBS/HUSKYREPORTS.htm (accessed October 8, 2004).

Lerand, M., 1973, Beaufort Sea, *in* R. G. McCrossam, ed., The future petroleum provinces of Canada—Their geology and potential: Canadian Society of Petroleum Geology Memoir 1, p. 315–386.

Magoon, L. B., and G. E. Claypool, 1983, Petroleum geochemistry of the North Slope of Alaska: Time and degree of thermal maturity, *in* M. Byorøy, et al., eds., Advances in organic geochemistry 1981: Chichester, United Kingdom, John Wiley and Sons Limited, p. 28–38.

Mayfield, C. F., I. L. Tailleur, and I. Ellersieck, 1988, Stratigraphy, structure, and palinspastic synthesis of the western Brooks Range, northwestern Alaska, chapter 7, *in* G. Gryc, ed., Geology and exploration of the National Petroleum Reserve in Alaska, 1974 to 1982: U.S. Geological Survey Professional Paper 1399, p. 143–186.

McCrea, J. M., 1950, The isotope geochemistry of carbonates and a paleotemperature scale: Journal of Chemistry and Physics, v. 18, p. 849–857.

Miller, E. L., and T. L. Hudson, 1991, Mid-Cretaceous extensional fragmentation of a Jurassic–Early Cretaceous compressional orogen, Alaska: Tectonics, v. 10, p. 781–796.

Miller, J. J., W. F. Agena, W. L. Lee, F. N. Zihlman, J. A. Grow, D. J. Taylor, M. Killgore, and H. L. Oliver, 2000, Regional seismic lines reprocessed using post-stack processing techniques: National Petroleum Reserve-Alaska (NPRA): U.S. Geological Survey Open-file Report 00-286 (CD-ROM).

Miller, J. J., W. F. Agena, W. L. Lee, F. N. Zihlman, J. A. Grow, D. J. Taylor, M. Killgore, and H. L. Oliver, 2001, Four regional seismic lines: National Petroleum Reserve-Alaska (supplement to U.S. Geological Survey Open-file Report 00-286): U.S. Geological Survey Open-file Report 01-0337, http://geology.cr.usgs.gov/pub/open-file-reports/ofr-01-0337/ (accessed October 29, 2001).

Molenaar, C. M., 1988, Depositional history and seismic stratigraphy of Lower Cretaceous rocks in the National Petroleum Reserve in Alaska and adjacent areas, chapter 25, *in* G. Gryc, ed., Geology and exploration of the National Petroleum Reserve in Alaska, 1974 to 1982: U.S. Geological Survey Professional Paper 1399, p. 593–621.

Moore, T. E., 1999, Balanced cross section, Bathtub syncline to Beaufort Sea through Niguanak structural high, Arctic National Wildlife Refuge (ANWR), northeastern Alaska, *in* ANWR Assessment Team, The oil and gas resource potential of the 1002 area, Arctic National Wildlife Refuge, Alaska (chapter BC): U.S. Geological Survey Open-file Report 98-34, p. BC1-BC61. (CD-ROM publication).

Moore, T. E., and T. H. Nilsen, 1984, Regional sedimentological variations in the Upper Devonian and lower Mississippian(?) Kanayut Conglomerate, Brooks Range, Alaska: Sedimentary Geology, v. 38, p. 464–498.

Moore, T. E., and C. J. Potter, 2003, Structural plays in Ellesmerian sequence and correlative strata of the National Petroleum Reserve, Alaska: U.S. Geological Survey Open-file Report OF 03-253, 82 p., http://geopubs.wr.usgs.gov/open-file/of03-253/ (accessed October 11, 2004).

Moore, T. E., W. K. Wallace, K. J. Bird, S. M. Karl, C. G. Mull, and J. T Dillon, 1994, Geology of northern Alaska, *in* G. Plafker and H. C. Berg, eds., The geology of Alaska: Boulder, Colorado, Geological Society of America, The geology of North America, v. G-1, p. 49–140.

Muchez, P., J. D. Marshall, J. L. R. Touret, and W. A. Viaene, 1994, Origin and migration of palaeofluids in the Upper Visean of the Campine Basin, northern Belgium: Sedimentology, v. 41, p. 133–145.

Muchez, P., M. Sintubin, and R. Swennen, 2000, Origin and migration of palaeofluids during orogeny: Discussion on the Variscides of Belgium and northern France: Journal of Geochemical Exploration, v. 69–70, p. 47–51.

Mull, C. G., 1985, Cretaceous tectonics, depositional cycles, and the Nanushuk Group, Brooks Range and Arctic Slope, Alaska, *in* A. C. Huffman, Jr., ed., Geology of the Nanushuk Group and related rocks, North Slope, Alaska: U.S. Geological Survey Bulletin, v. 1614, p. 7–36.

Mull, C. G., and M. B. Werdon, 1994, Generalized geologic map of the western Endicott Mountains, central Brooks Range, Alaska: Alaska Division of Geological and Geophysical Surveys Public Data File map 94-55, scale 1:250,000, 1 sheet.

Mull, C. G., T. E. Moore, E. E. Harris, and I. L. Tailleur, 1994, Geologic map of the Killik River quadrangle: U.S. Geological Survey Open-file Report 94-679, 1 sheet, scale 1:125,000.

Mull, C. G., R. K. Glenn, and K. E. Adams, 1997, Tectonic evolution of the central Brooks Range mountain front: Evidence from the Atigun Gorge region: Journal of Geophysical Research, v. 102, no. B9, p. 20,749–20,773.

Mull, C. G., D. W. Houseknecht, and K. J. Bird, 2003, Revised Cretaceous and Tertiary stratigraphic nomenclature in the central and western Colville basin, northern Alaska: U.S. Geological Survey Professional Paper 1673, http://pubs.usgs.gov/pp/p1673/ (accessed August 15, 2003).

Nilsen, T. H., 1981, Upper Devonian and lower Mississippian redbeds, Brooks Range, Alaska, *in* A. D. Miall, ed, Sedimentation and tectonics in alluvial basins: Geological Association of Canada Special Paper, v. 23, p. 187–219.

Oldow, J. S., C. M. Seidensticker, J. C. Phelps, F. E. Julian, R. R. Gottschalk, K. W. Boler, J. W. Handschy, and H. G. Ave Lallemant, 1987, Balanced cross sections through the central Brooks Range and North Slope, Arctic Alaska: AAPG Publication, 19 p., 8 plates, scale 1:200,000.

Oliver, J., 1986, Fluids expelled tectonically from orogenic belts: Their role in hydrocarbon migration and other geologic phenomena: Geology, v. 14, p. 99–102.

O'Neil, J. R., R. N. Clayton, and T. K. Mayeda, 1969, Oxygen isotope fractionation in divalent metal carbonates: Journal of Chemistry and Physics, v. 51, p. 5547–5558.

O'Sullivan, P. B., 1996, Late Mesozoic and Cenozoic thermal-tectonic evolution of the North Slope foreland basin, Alaska, *in* M. J. Johnsson and D. G. Howell, eds., Thermal evolution of sedimentary basins in Alaska: U.S. Geological Survey Bulletin, v. 2142, p. 45–79.

O'Sullivan, P. B., and W. K. Wallace, 2002, Evidence and implications for out-of-sequence structures in the Sadlerochit Mountains region of the Arctic National Wildlife Refuge, Alaska: Geological Society of America Bulletin, v. 114, no. 11, p. 1356–1378.

O'Sullivan, P. B., P. F. Green, S. C. Bergman, J. Decker, I. R. Duddy, A. J. W. Gleadow, and D. L. Turner, 1993, Multiple phases of Tertiary uplift and erosion in the Arctic National Wildlife Refuge, Alaska, revealed by apatite fission track analysis: AAPG Bulletin, v. 77, no. 3, p. 359–385.

O'Sullivan, P. B., J. M. Murphy, and A. E. Blythe, 1997, Late Mesozoic and Cenozoic thermotectonic evolution of the central Brooks Range and adjacent North Slope foreland basin, Alaska: Including fission-track results from the Trans-Alaska crustal transect (TACT): Journal of Geophysical Research, v. 102, no. B9, p. 20,821–20,845.

O'Sullivan, P. B., T. E. Moore, and J. M. Murphy, 1998, Tertiary uplift of the Mt. Doonerak antiform, central Brooks Range, Alaska: Apatite fission track evidence from the Trans-Alaska crustal transect, *in* J. S. Oldow and H. G. Avé Lallemant, eds., Architecture of a fold and thrust belt: Central Brooks Range, Arctic Alaska: Geological Society of America Special Paper, v. 324, p. 179–194.

O'Sullivan, P. B., T. E. Moore, C. J. Potter, and W. K. Wallace, 2002, Constraints on the Mesozoic and Cenozoic thermotectonic evolution of the northern Brooks Range and the North Slope foreland basin from fission track thermochronology (abs.): Pacific Section, AAPG Bulletin, v. 86, no. 6, p. 1156.

Pagel, M., J. J. Braun, J. R. Disnar, L. Martinez, C. Renac, and G. Vasseur, 1996, Thermal history constraints from studies of organic matter, clay minerals, fluid inclusions and apatite fission tracks at the Ardeche paleo-margin (BA1 drill hole, GPF Program), France: Journal of Sedimentary Research, v. 67, p. 235–245.

Parris, T. M., and R. C. Burruss, 2001, Tectonism, fracturing, and gas generation, northeast Brooks Range, Alaska (abs.): AAPG Annual Meeting Program, v. 10, p. A151.

Potter, C. J., and T. E. Moore, 2003, Brookian structural plays in the National Petroleum Reserve, Alaska: U.S. Geological Survey Open-file Report OF 03-266, http://geopubs.wr.usgs.gov/open-file/of03-266/ (accessed July 16, 2003).

Potter, C. J., J. A. Grow, W. J. Perry, T. E. Moore, P. B. O'Sullivan, J. D. Phillips, and R. W. Saltus, 2004, Tertiary thrust systems and fluid flow beneath the Beaufort coastal plain (1002 area), Arctic National Wildlife Refuge, Alaska, U.S.A., *in* R. Swennen, F. Roure, and J. W. Granath, eds., Deformation, fluid flow, and reservoir appraisal in foreland fold and thrust belts: AAPG Hedberg Series, no. 1, p.187–214.

Roeder, D., and C. G. Mull, 1978, Tectonics of Brooks Range ophiolites, Alaska: AAPG Bulletin, v. 62, p. 1696–1702.

Shelton, K. L., M. B. Underwood, I. B. Burstein, G. T. Haeussler, and D. G. Howell, 1996, Stable-isotope and fluid-inclusion studies of hydrothermal quartz and calcite veins from the Kandik thrust belt of east-central Alaska— Implications for thermotectonic history and terrane analysis, *in* M. J. Johnsson and D. G. Howell, eds., Thermal evolution of sedimentary basins in Alaska: U.S. Geological Survey Professional Paper 2142, p. 111–131.

Travé, A., F. Calvet, A. Soler, and P. Labaume, 1998, Fracturing and fluid migration during Palaeogene compression and Neogene extension in the Catalan coastal ranges, Spain: Sedimentology, v. 45, p. 1063–1082.

Underwood, M. B., D. G. Howell, M. J. Johnsson, and M. J. Pawlewicz, 1996, Thermo-tectonic evolution of suspect terranes in the Kandik region of east-central Alaska, *in* M. J. Johnsson and D. G. Howell, eds., Thermal evolution of sedimentary basins in Alaska: U.S. Geological Survey Professional Paper 2142, p. 81–110.

Underwood, M. B., K. L. Shelton, R. J. McLaughlin, M. M. Laughland, and R. M. Solomon, 1999, Middle Miocene paleotemperature anomalies within the Franciscan complex of northern California: Thermotectonic responses near the Mendocino triple junction: Geological Society of America Bulletin, v. 111, p. 1448–1467.

van Geet, M., R. Swennen, C. Durmishi, F. Roure, and P. Muchez, 2002, Paragenesis of Cretaceous to Eocene carbonate reservoirs in the Ionian fold and thrust belt (Albania): Relation between tectonism and fluid flow: Sedimentology, v. 49, p. 697–718.

Vogl, J. J., A. T. Calvert, and P. B. Gans, 2002, Mechanisms and timing of exhumation of collision-related metamorphic rocks, southern Brooks Range, Alaska: Insights from $^{40}Ar/^{39}Ar$ thermochronology: Tectonics, v. 21, no. 3, doi: 10.1029/2000TC001270, 18 p.

Wahrhaftig, C., 1965, Physiographic divisions of Alaska: U.S. Geological Survey Professional Paper 482, 52 p.

Wallace, W. K., T. E. Moore, and G. Plafker, 1997, Multistory duplexes with forward dipping roofs, north central Brooks Range, Alaska: Journal of Geophysical Research, v. 102, p. 20,773–20,796.

9

Potter, C. J., J. A. Grow, W. J. Perry, T. E. Moore, P. B. O'Sullivan, J. D. Phillips, and R. W. Saltus, 2004, Tertiary thrust systems and fluid flow beneath the Beaufort coastal plain (1002 Area), Arctic National Wildlife Refuge, Alaska, U.S.A., *in* R. Swennen, F. Roure, and J. W. Granath, eds., Deformation, fluid flow, and reservoir appraisal in foreland fold and thrust belts: AAPG Hedberg Series, no. 1, p. 187–214.

Tertiary Thrust Systems and Fluid Flow beneath the Beaufort Coastal Plain (1002 Area), Arctic National Wildlife Refuge, Alaska, U.S.A.

Christopher J. Potter
U.S. Geological Survey, Denver, Colorado, U.S.A.

John A. Grow
U.S. Geological Survey, Denver, Colorado, U.S.A.

William J. Perry
U.S. Geological Survey, Denver, Colorado, U.S.A.

Thomas E. Moore
U.S. Geological Survey, Menlo Park, California, U.S.A.

Paul B. O'Sullivan
Apatite to Zircon, Inc., Viola, Idaho, U.S.A.

Jeffrey D. Phillips
U.S. Geological Survey, Denver, Colorado, U.S.A.

Richard W. Saltus
U.S. Geological Survey, Denver, Colorado, U.S.A.

ABSTRACT

Beneath the Arctic coastal plain (commonly referred to as "the 1002 area") in the Arctic National Wildlife Refuge, northeastern Alaska, United States, seismic reflection data show that the northernmost and youngest part of the Brookian orogen is preserved as a Paleogene to Neogene system of blind and buried thrust-related structures. These structures involve Proterozoic to Miocene (and younger?) rocks that contain several potential petroleum reservoir facies. Thermal maturity data indicate that the deformed rocks are mature to overmature with respect to hydrocarbon generation. Oil seeps and stains in outcrops and shows in nearby wells indicate that oil has migrated through the region; geochemical studies have identified three potential petroleum systems. Hydrocarbons that were generated from Mesozoic source rocks in the deformed belt were apparently expelled and migrated northward in the Paleogene, before

DOI:10.1306/1025691H13117

much of the deformation in this part of the orogen. It is also possible that Neogene petroleum, which was generated in Tertiary rocks offshore in the Arctic Ocean, migrated southward into Neogene structural traps at the thrust front. However, the hydrocarbon resource potential of this largely unexplored region of Alaska's North Slope remains poorly known.

In the western part of the 1002 area, the dominant style of thin-skinned thrusting is that of a passive-roof duplex, bounded below by a detachment (floor thrust) near the base of Lower Cretaceous and younger foreland basin deposits and bounded above by a north-dipping roof thrust near the base of the Eocene. East–west-trending, basement-involved thrusts produced the Sadlerochit Mountains to the south, and buried, basement-involved thrusts are also present north of the Sadlerochit Mountains, where they appear to feed displacement into the thin-skinned system. Locally, late basement-involved thrusts postdate the thin-skinned thrusting. Both the basement-involved thrusts and the thin-skinned passive-roof duplex were principally active in the Miocene.

In the eastern part of the 1002 area, a northward-younging pattern of thin-skinned deformation is apparent. Converging patterns of Paleocene reflectors on the north flank of the Sabbath syncline indicate that the Aichilik high and the Sabbath syncline formed as a passive-roof duplex and piggyback basin, respectively, just behind the Paleocene deformation front. During the Eocene and possibly the Oligocene, thin-skinned thrusting advanced northward over the present location of the Niguanak high. A passive-roof duplex occupied the frontal part of this system. The Kingak and Hue shales exposed above the Niguanak high were transported into their present structural position during the Eocene to Oligocene motion on the long thrust ramps above the present south flank of the Niguanak high. Broad, basement-cored subsurface domes (Niguanak high and Aurora dome) formed near the deformation front in the Oligocene, deforming the overlying thin-skinned structures and feeding a new increment of displacement into thin-skinned structures directly to the north. Deformation continued through the Miocene above a detachment in the basement. Offshore seismicity and Holocene shortening documented by previous workers may indicate that contractional deformation continues to the present day.

INTRODUCTION

Since the 1968 discovery of the giant (13-billion-bbl) Prudhoe Bay oil field, the North Slope of Alaska, United States, located between the Brooks Range and the Arctic Ocean, has attracted considerable interest from the oil industry. This has led to several additional discoveries (Figure 1) and widespread speculation as to the possibilities for additional reserves beneath the large, sparsely drilled tracts of federal lands on the North Slope. All of the producing oil fields (Prudhoe Bay, Alpine, Endicott, Kuparuk River, Lisburne, Tarn, Tabasco, etc.; Figure 1) on the North Slope are situated within relatively undeformed strata that lie to the north of the Jurassic to Tertiary Brookian orogenic belt (Figure 1). Modern exploration strategies for these deposits have involved detailed sequence-stratigraphic analysis of three-dimensional seismic data to define subtle stratigraphic traps. In addition, from the 1940s to the 1970s, several relatively small oil and gas accumulations were discovered (but not developed) in anticlinal and fault-bounded traps related to the Tertiary frontal part of the Brookian orogenic belt beneath the Arctic foothills across much of the North Slope. These hydrocarbon accumulations are shown south of the Tertiary deformation front in Figure 1. Sparse offshore drilling of fold-thrust structures has not yielded any discoveries to date.

Within the Arctic National Wildlife Refuge (ANWR) (Figure 1) in northeasternmost Alaska, the coastal plain and northernmost foothills in the 1002 area were designated in 1980 for possible future oil exploration and development through the Alaska National Interest Lands Conservation Act (United States Congress), Section 1002. This chapter deals with the evolution of thrust systems beneath the coastal plain and foothills of the ANWR and the possible relations among thrusting, fluid flow, and the presence of oil and gas resources. Interpretations are based principally on regional, two-dimensional seismic reflection profiles and potential-field data across the coastal plain and foothills to the south (Figures 2, 3), augmented by field observations in sparse coastal plain and foothills exposures and by fission-track data. The 1002 area is largely underlain by an extensive array of blind thrust faults and folds that constitute the frontal part of the northeastern Brooks Range thrust system. Two large basement-cored domes beneath the northeastern part of the 1002 area have no analogs elsewhere on the North Slope; their resource potential has been the source of considerable speculation in the oil industry, and one of them has been the target of two deep wildcat wells (Kaktovik Inupiat Corporation [KIC]—Jago River—1 and Aurora 1; Figure 2).

REGIONAL GEOLOGIC SETTING

Stratigraphy

The stratigraphy of the eastern North Slope can be summarized in terms of four tectonostratigraphic megasequences (Figure 4). Mildly metamorphosed pre-Mississippian rocks make up the Franklinian megasequence, which forms the stratigraphic basement of the Arctic Alaska plate. Franklinian basement is directly overlain by the Mississippian to Triassic Ellesmerian megasequence that developed on a south-facing passive margin. Subsequent incipient rifting, coeval with the opening of the Canada Basin, is recorded by the Jurassic and Lower Cretaceous Beaufortian megasequence. The Lower Cretaceous to Tertiary Brookian megasequence overlies the Lower Cretaceous unconformity and consists of several northeastward-prograding continental margin sequences derived from the uplifted Brookian orogen; these sequences fill the foredeep and step out across the north-facing Tertiary passive margin.

Petroleum source rocks in the Triassic Shublik Formation (Ellesmerian), the Jurassic to Early Cretaceous Kingak Shale (Beaufortian), and the Cretaceous Hue Shale (Brookian) (Figure 4) compose a composite source for Prudhoe Bay oils (Seifert et al., 1980; Magoon et al., 1987). As discussed in a subsequent section of this chapter, the Shublik is considered as the primary source rock at Prudhoe Bay, whereas the Hue Shale and offshore Tertiary marine shales are probably the most significant source rocks in the 1002 area.

Reservoir units in the North Slope oil fields also include rocks from all of the Mississippian and younger megasequences (Figure 4). Within the Ellesmerian, clastic strata of the Endicott Group (Figure 4) compose the reservoirs in the Endicott field (Figure 1), whereas Lisburne Group carbonates, the Ledge Sandstone of the Sadlerochit Group, and the Sag River Sandstone (Figure 4) are the Prudhoe Bay reservoirs. Within the Beaufortian, the Kuparuk River Sandstone is a major reservoir in the Kuparuk River and other oil fields, and upper Jurassic sandstones are the reservoirs for the Alpine field (Figures 1, 4). Brookian sandstones provide reservoirs for the Point Thomson, Tarn Hammerhead, and Kuvlum fields (Figure 1).

Thrust belt structure

The Brooks Range records two distinct episodes of crustal shortening: (1) Middle Jurassic to Lower Cretaceous deformation that affected a region roughly centered on the main axis of the Brooks Range and involved hundreds of kilometers of north–south shortening (Moore et al., 1994b); and (2) Tertiary shortening that is principally expressed in the northern part of the mountain belt, in the northern foothills belt, and as blind structures beneath large parts of the Arctic

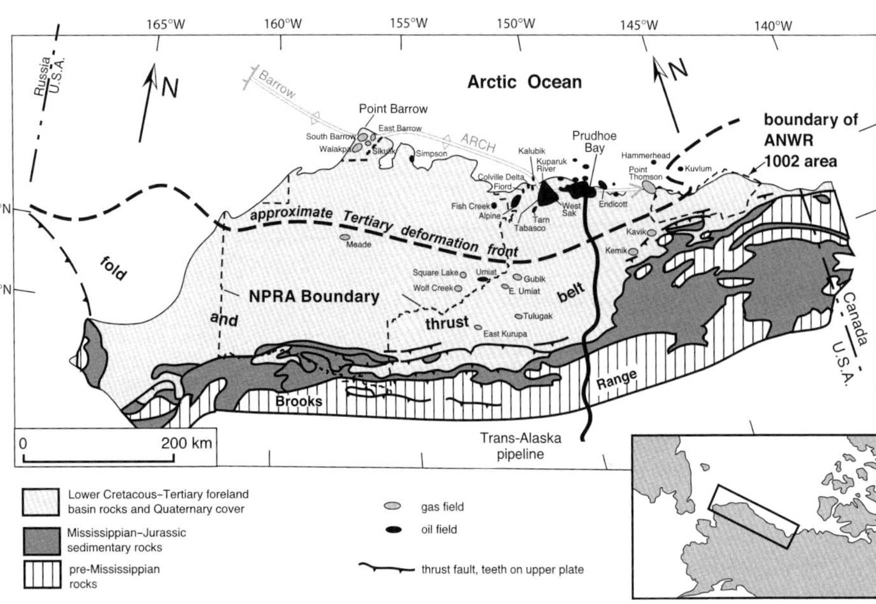

Figure 1. Index map showing generalized geology and location of oil and gas fields on Alaska's North Slope.

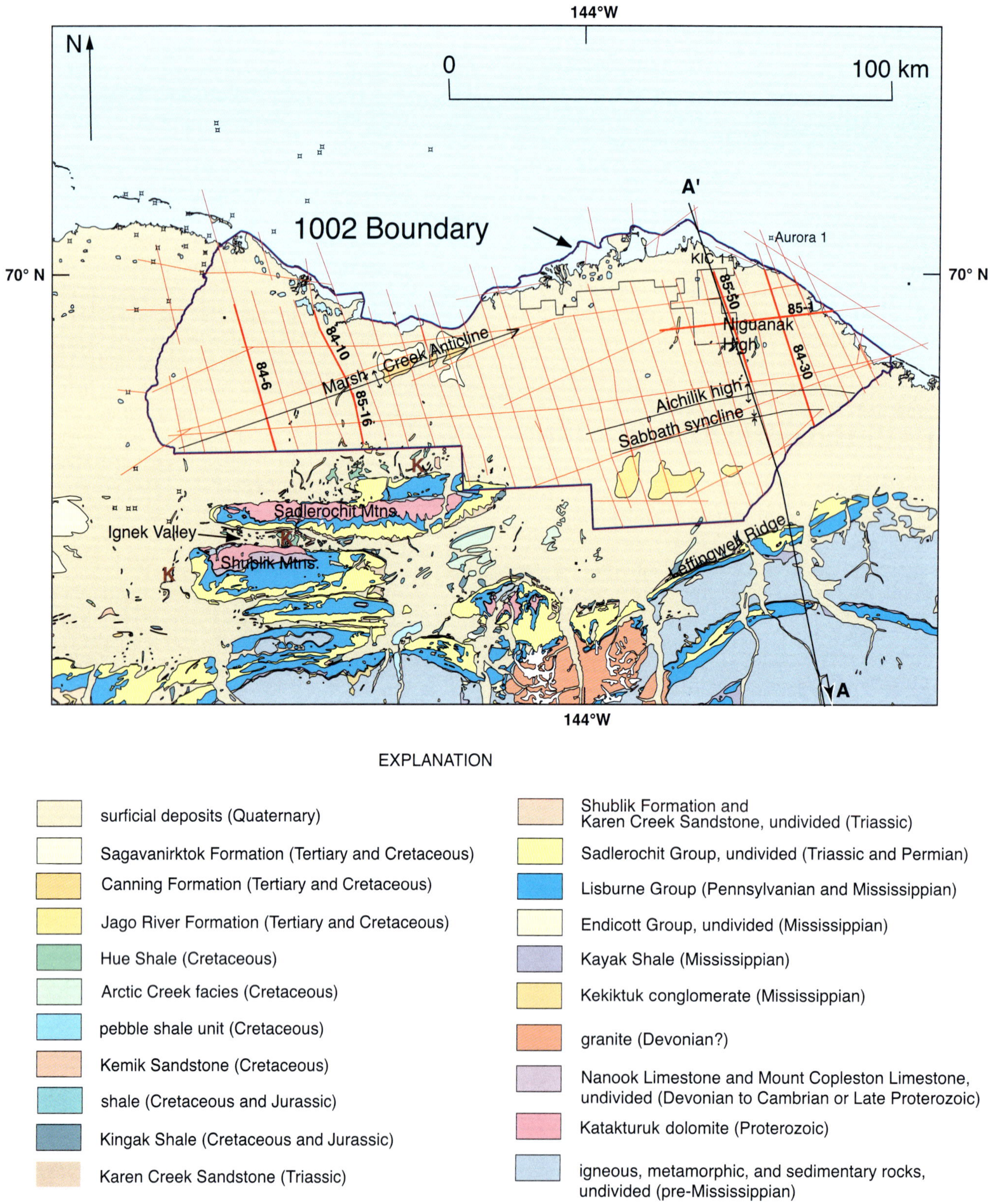

FIGURE 2. Geologic map of the ANWR 1002 area and the surrounding region, based on Bader and Bird (1986). K = location of exposures of Kemik duplex. Labels 84-6, 84-10, 84-16, 84-20, 85-50, and 85-1 refer to locations of seismic lines interpreted in other Figures in this chapter. AA′ refers to location of cross section by Moore (1999) shown in Figure 5.

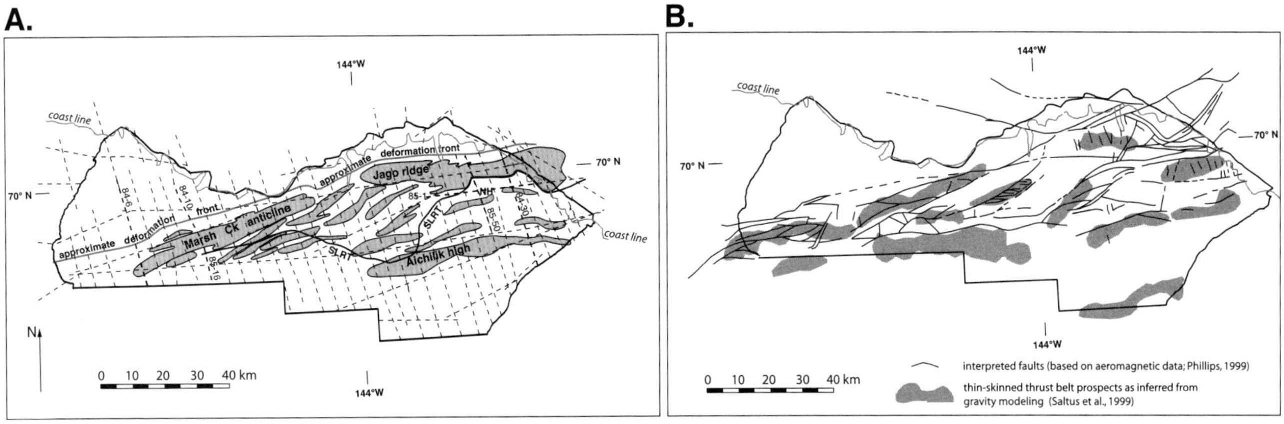

FIGURE 3. Map of the ANWR 1002 area showing geophysical interpretations of thin-skinned structural culminations and faults in the subsurface. (A) Thin-skinned structural culminations inferred from regional network of seismic reflection profiles. The large Marsh Creek and Jago ridge culminations are composite structures that contain many smaller culminations. Dashed lines = locations of seismic reflection profiles. Labels 84-6, 84-10, 84-30, 85-1, and 85-50 refer to locations of seismic lines shown and interpreted in this chapter. NH = Niguanak high. SLRT = southern limit of the (erosionally breached) roof thrust on the passive-roof duplex in the eastern part of the area. (B) Structural highs interpreted from gravity modeling and near-surface faults interpreted from highpass aeromagnetic anomalies. See text for details.

coastal plain (Moore et al., 2004). The Jurassic–Cretaceous phase formed in response to the collision of a south-facing Carboniferous to Triassic passive margin (Arctic Alaska plate) with an arc-related terrane to the south. The Tertiary phase (late Brookian orogenic phase of Moore et al., 1994b) formed as a retroarc thrust belt, hundreds of kilometers behind the arc, in response to plate convergence in southern Alaska. Both episodes of crustal shortening were dominantly north vergent. Foothills and coastal-plain thrust structures in the central and western parts of the North Slope are dominantly Paleogene, whereas the structures beneath the 1002 area are Paleocene to Miocene, based on apatite fission-track analyses (O'Sullivan et al., 1993) and seismic reflection interpretations (Potter et al., 1999). Thus, the coastal plain of the 1002 area overlies the youngest thrust structures in the Brooks Range orogenic belt. They lie directly to the north of the northeastern Brooks Range, a Tertiary orogenic element that is offset to the north relative to the main part of the range and has undergone less than 100 km (60 mi) of principally north-vergent shortening (Wallace and Hanks, 1990). With respect to petroleum systems, potential structural traps beneath the 1002 area formed mainly in the Neogene (Potter et al., 1999), well after petroleum generation and migration from potential source rocks in the mountain belt to the south but perhaps synchronously with southward migration of oil sourced in the Tertiary marine section to the north (Hayba et al., 1999).

Cross sections constructed by Hanks and Wallace (1990), Hanks (1990, 1993), Wallace (1993), Cole et al. (1999), and Moore (1999) (Figures 2, 5) show the broad context of coastal-plain thrust structures in the northeastern Brooks Range thrust system. Beneath the 1002 area, the presence of a relatively shallow detachment near the base of the Brookian section, as well as deeper detachment levels in the Beaufortian, Mississippian, Ellesmerian, and Franklinian sequences (Figure 4), yields a pattern of thin-skinned thrusting with basement-involved thrusting superimposed on it. Structures of both types are evident on geologic maps and seismic data in the region (Bader and Bird, 1986; Bruns et al., 1987; Grow et al., 1999d; Potter et al., 1999).

OVERVIEW OF GEOLOGIC AND GEOPHYSICAL DATA AND INTERPRETATIONS

Seismic reflection, gravity, and aeromagnetic data, combined with outcrop geology and structural modeling, allow detailed definition of the style and distribution of thrust structures. We have reinterpreted a regional network of proprietary seismic reflection data (Figures 2, 3) that were acquired across the 1002 coastal plain during 1984 and 1985 (Foland and Lalla, 1987) and reprocessed by Lee et al. (1999), and we have integrated these seismic interpretations with results of new potential-field studies (Phillips, 1999; Saltus et al., 1999). Regional correlations to the seismic-stratigraphic interpretations of Houseknecht and Schenk (1999) allow us to constrain the ages of several key structural elements. New interpretations of the style, timing, and areal distribution of thrust faulting build on

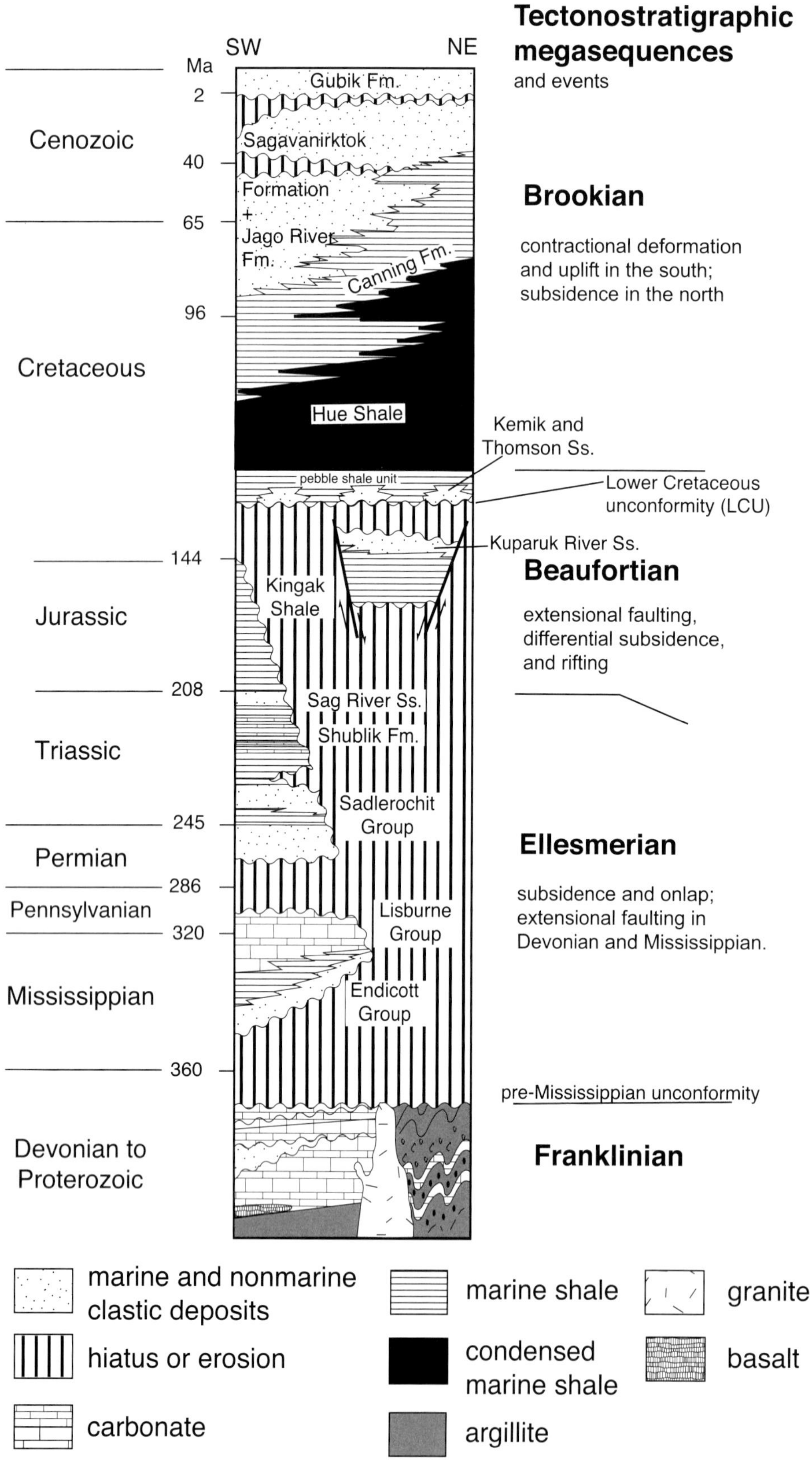

FIGURE 4. Stratigraphic column for northeastern Alaska showing the tectonostratigraphic subdivisions (modified after Hubbard et al., 1987; Bird, 1999).

some of the themes introduced by Bruns et al. (1987) and Kelley and Foland (1987) and provide a comprehensive framework for understanding the structural geology of the 1002 area.

Exposures in the foothills and mountains south of the 1002 area and in sparse bedrock areas in the 1002 area (Figure 2) materially aid structural interpretations of the geophysical data. Just south of the western part of the 1002 area, the Sadlerochit and Shublik Mountains contain extensive exposures of pre-Mississippian basement rocks. These two ranges, bounded on their north sides by major, basement-involved thrusts, contain subsidiary east–west-striking thrust and reverse faults that also involve basement. Geologic map patterns (Robinson et al., 1989) strongly imply that the Sadlerochit uplift breached a formerly continuous east-northeast-trending, thin-skinned duplex or zone of duplexes (Figure 2), in which the Lower Cretaceous Kemik Sandstone was repeatedly thrust-imbricated above a detachment in the Jurassic Kingak Shale and beneath a roof thrust in the Lower Cretaceous pebble shale (Kelley and Foland, 1987; Meigs, 1989). The east-northeast-rending and complexly faulted Marsh Creek anticline (Figure 2) is the only major bedrock structure exposed on the coastal plain in the western half of the 1002 area.

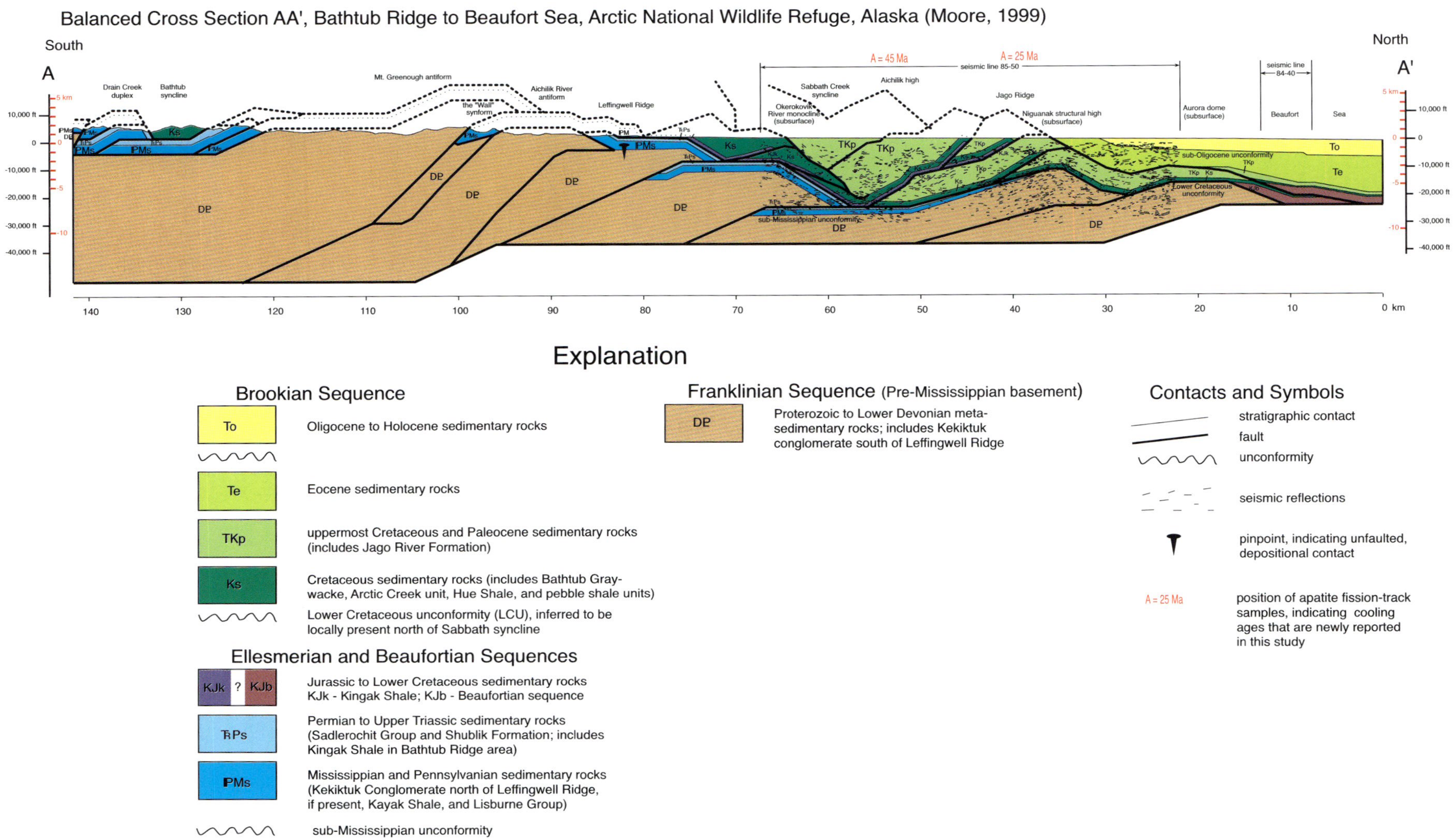

FIGURE 5. Balanced cross section across the 1002 area, modified from Moore (1999), along section line AA′ (Figure 2).

Seismic reflection data (Bruns et al., 1987; Foland and Lalla, 1987; Potter et al., 1999) demonstrate that the Marsh Creek anticline is the approximate northern limit of pervasive thin-skinned deformation (deformation not involving basement rocks) in the western part of the 1002 area.

Bedrock geology exposed in the eastern half of the 1002 area and in the foothills south of it provides further evidence for thrust faulting and folding. Leffingwell Ridge, a 60-km (40-mi)-long, east-northeast-trending ridge along the mountain front just southeast of the 1002 area (Figure 2), is composed of steeply dipping Ellesmerian strata (Bader and Bird, 1986). The position of these strata along the northern flank of a broad tract of pre-Mississippian rocks of the northeastern Brooks Range indicates the presence of a basement-involved thrust at depth beneath Leffingwell Ridge. Pervasively deformed Jurassic and Cretaceous shales are exposed in stream and river valleys north of Leffingwell Ridge (Bader and Bird, 1986; Hanks, 1987). Extensive exposures of north-dipping Cretaceous and lower Tertiary strata of the Jago River Formation (Buckingham, 1985) occupy the foothills in the southeastern part of the 1002 area. Seismic reflection data show that these strata occupy the southern flank of the broad Sabbath syncline (Potter et al., 1999) In the northeastern part of the 1002 area, small outcrops of gently to steeply dipping Jurassic and Cretaceous strata in the Niguanak Ridge area attest to complex deformation beneath the coastal plain. Seismic reflection data (Bruns et al., 1987; Foland and Lalla, 1987; Grow et al., 1999d; Potter et al., 1999) show that these exposures lie above a broad dome in the basement at depth, known as the Niguanak high (Figures 2, 6).

Figure 7 illustrates the distribution of the principal, basement-involved thrusts and folds beneath the coastal plain and foothills in the 1002 area, based on seismic reflection interpretations of the top of pre-Mississippian basement integrated with surface geologic constraints (Grow et al., 1999d). The dominant features of this map include (1) the foothills fault system (south 1002 fault system of Grow et al., 1999d; Potter et al., 1999), a major thrust zone subparallel to the southern boundary of the 1002 area; (2) the large northwest-trending Hulahula structural low beneath the east-central part of the 1002 area; and (3) the Niguanak high and Aurora dome beneath the northeastern part of the 1002 area, with crests at 2680 and 4510 m (8800 and 14,800 ft) below sea level, respectively, bounded by thrust faults on their northern margins. These broad, basement-cored domes are by far the largest individual positive structures in the 1002 area and have no analogs on Alaska's North Slope west of ANWR. As such, these structures may represent significant exploration targets, provided that favorable reservoir and hydrocarbon-charge scenarios can be supported by subsurface interpretations and fluid-flow modeling.

Figure 3 illustrates the distribution of shallower, thin-skinned Brookian structures, as interpreted principally from seismic reflection data, augmented by potential-field data (Phillips, 1999; Potter et al., 1999; Saltus et al., 1999). These shallow highs form elongate structural closures that trend dominantly northeast (in the west) to east–west (in the east). Two regionally extensive composite shallow structural highs, the Marsh Creek anticline and Jago ridge, define the deformation front beneath the coastal plain.

INTERPRETATION OF THRUST STRUCTURES IN THE WESTERN PART OF THE 1002 AREA

Complex structure in southwestern part of 1002

The seismic reflection data show a discrete zone of basement-involved structures (foothills fault system) in the southwestern part of 1002 parallel to the Sadlerochit Mountain front and a broad, wedge-shaped array of thin-skinned structures mainly south of, and subparallel to, the east-northeast-trending Marsh Creek anticline (Figure 3). In the southern parts of the seismic lines in Figures 8 and 9, the Franklinian basement, characterized by a relatively transparent to chaotic seismic character, is overlain by a coherent reflective sequence that we correlate with the lower part of the Brookian section. In these locations, the Mississippian to Triassic Ellesmerian megasequence (Figure 4) has been erosionally removed, so the Lower Cretaceous unconformity lies directly on the basement. The Lower Cretaceous unconformity is cut by north-vergent thrust faults on the southern part of these seismic sections, as are the Franklinian and Ellesmerian rocks in the Sadlerochit Mountains exposed immediately to the south. In the southern part of line 85-16, we interpret a basement duplex consisting of four stacked thrust plates (label a,

FIGURE 6. Seismic reflection, magnetic, and gravity interpretations of the north part of seismic line 85-50 across the crest of the Niguanak high and the south flank of the Aurora dome. Note that seismic profile is displayed at a larger scale than the gravity and magnetic profiles. TPM and BTU = top of pre-Mississippian and basal Tertiary unconformities (coincident), represented by a blue line. Red lines on all panels represent faults interpreted from seismic reflection data. See text for details.

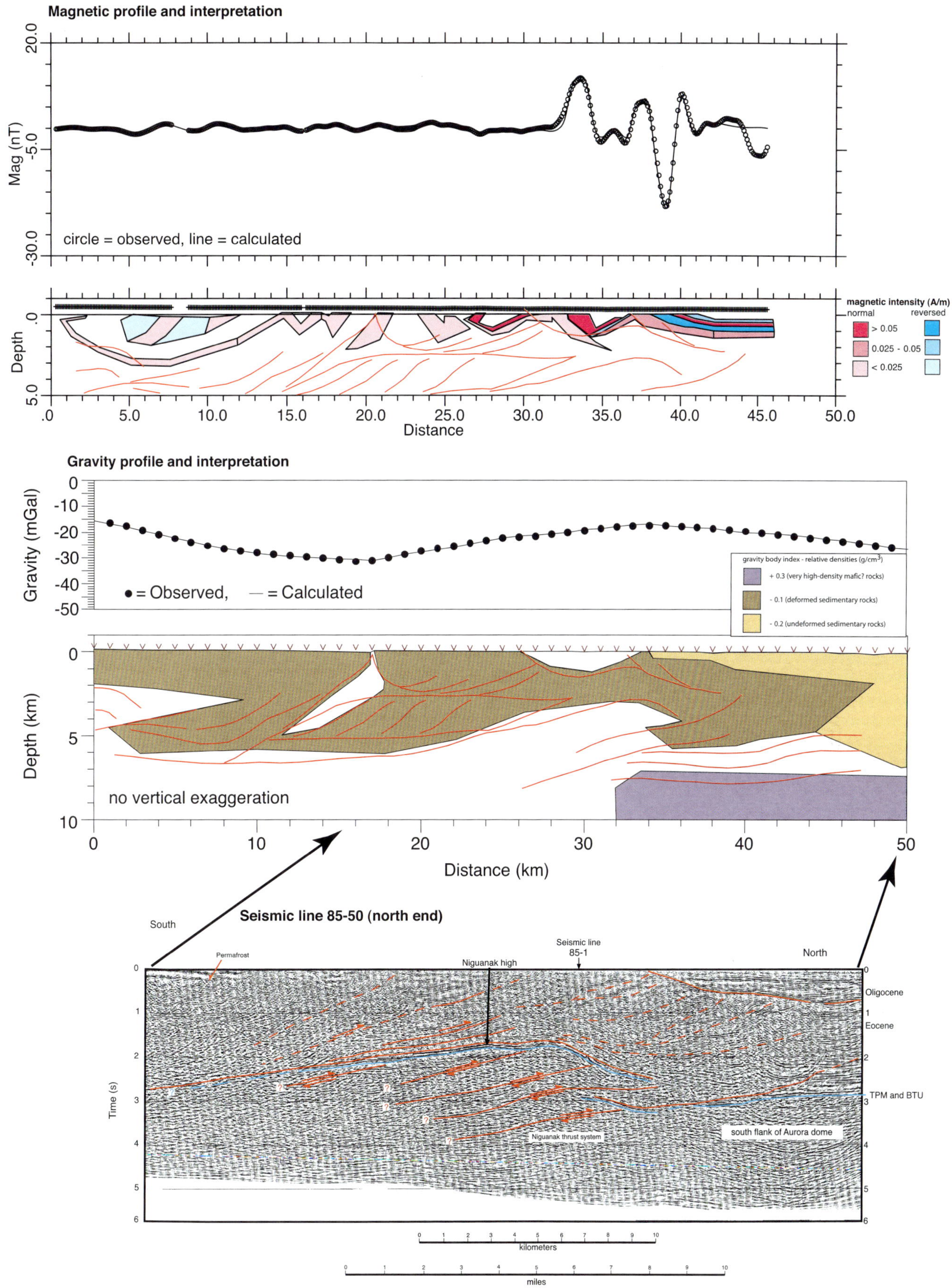

Magnetic profile and interpretation
Mag (nT)
20.0
-5.0
-30.0
circle = observed, line = calculated
Depth
.0
5.0
magnetic intensity (A/m)
normal
reversed
> 0.05
0.025 - 0.05
< 0.025
Distance
Gravity profile and interpretation
Gravity (mGal)
• = Observed, — = Calculated
gravity body index - relative densities (g/cm³)
+ 0.3 (very high-density mafic? rocks)
- 0.1 (deformed sedimentary rocks)
- 0.2 (undeformed sedimentary rocks)
Depth (km)
no vertical exaggeration
Distance (km)
Seismic line 85-50 (north end)
South
North
Permafrost
Niguanak high
Seismic line
85-1
Oligocene
Eocene
TPM and BTU
Time (s)
Niguanak thrust system
south flank of Aurora dome
kilometers
miles

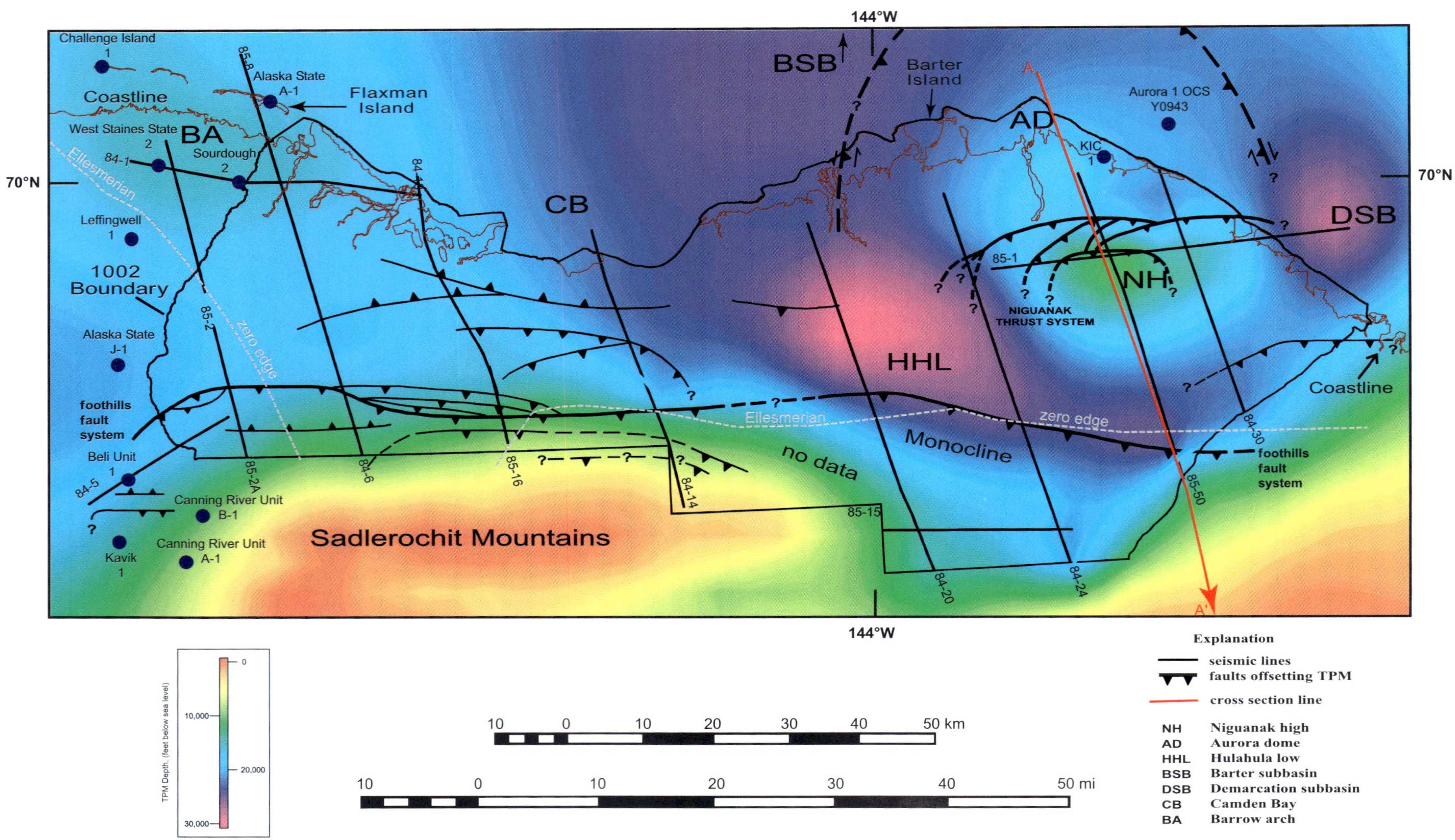

FIGURE 7. Principal faults offsetting the top of the pre-Mississippian (TPM) (Franklinian sequence) in the 1002 area of ANWR, superimposed on a smoothed structure-contour map of the TPM. The structure-contour map was prepared with a grid spacing of 4000 m (1300 ft) using only the seismic reflection profiles shown in black (those published in Bird and Magoon, 1987). Additional outcrop and well data for the TPM were also used. The resulting map has smoothed all structural features less than 8 km (5 mi) in diameter, but the larger structural highs and lows within and surrounding the 1002 area are still clearly represented. The interpretation of fault locations is based on all ANWR seismic lines, which had an average line spacing of 5 × 10 km (3 × 6 mi) (north–south and east–west lines, respectively). Kelley (1999) shows an alternative interpretation of the fault trends. Ellesmerian "zero edge" represents the approximate position of the northern and western limit of the Ellesmerian sequence in the subsurface, based on seismic reflection patterns. Line AA′ represents northern part of cross section AA′ (Figure 5).

Figure 8) in the Franklinian basement, beneath a roof thrust near the base of the Brookian. There was only minor later disruption of lower Brookian reflectors by thrust reactivation of the faults labeled a (Figure 8). This interpretation implies that the shortening in the basement duplex at the south end of line 85-16 was transferred northward along a thin-skinned detachment near the base of the Brookian section. It is also possible that the basement duplex formed a tectonic wedge (passive-roof duplex) driven to the north beneath a major backthrust, similar to interpretations shown by O'Sullivan and Wallace (2002). The basement duplex appears to have produced more than 1500 m (5000 ft) of total vertical relief on the top of Franklinian basement. A basement duplex is also inferred by Cole et al. (1999) to underlie line 84-6, but it is not obvious on that seismic profile (Figure 9).

In outcrops southeast of line 85-16, a detachment in the Jurassic to Lower Cretaceous Kingak Shale underlies a duplex containing 13 horses of imbricated and folded Lower Cretaceous Kemik Sandstone (Figure 2) (Meigs, 1989); higher detachments are documented as well at the top of the pebble shale (roof thrust for the Kemik duplex) and in the lower part of the Brookian section (Hue Shale). Although these detachments are probably present at the south end of the seismic line, this level of detail is not resolvable in the data. On one nearby proprietary seismic line, the reflection geometry may indicate imbricated Kemik Sandstone.

Passive-roof Duplex and Marsh Creek Anticline

The reflective Brookian section is folded in a broad syncline on the south half of Figure 8. Beneath the south-dipping north limb of this syncline, there is a wedge-shaped stack of discordant packets of discontinuous, locally arcuate, south-dipping reflections. We interpret this as a wedge of imbricated Brookian sedimentary rocks above a detachment that lies at or near the base of the Brookian at about 2.5 s two-way traveltime.

A north-dipping, less deformed set of reflectors representing topset and foreset Brookian facies are present above and north of the imbricated Brookian section (Figure 8). Using stratigraphic correlations established with the regional seismic reflection network tied to wells west of the Canning River, these reflectors correlate with Eocene to Miocene sequences A to C of Houseknecht and Schenk (1999). They are tilted northward above the south-dipping imbricate Brookian stack and flatten to subhorizontal dips in the northern part of the profile (Figure 8), where the underlying imbricate stack diminishes in thickness to a featheredge. The base of the less deformed Brookian strata (approximate base of Eocene) must be a roof-thrust detachment surface, because it is consistently discordant to the underlying imbricate stack. Thus, the imbricate stack is a passive-roof duplex (Banks and Warburton, 1986), essentially identical to an Alberta foothills-style triangle zone (Jones, 1982), with a flat floor thrust near the base of the Brookian section and a north-dipping roof thrust that underlies the passively tilted Eocene section.

Kelley and Foland (1987) interpreted the imbricate stack in the passive-roof duplex to be composed of mud-rich Cretaceous and Paleocene rocks, based on surface geology exposed in the Katakturuk River drainage, north of the Sadlerochit Mountains and west of line 84-10. This largely homogeneous and complexly deformed rock mass does not contain recognizable stratigraphic markers that allow calculation of the amount of thrust-related shortening; however, we believe there are many more thrusts than are depicted in Figure 8.

The passive-roof duplex system seen in Figure 8 dominates the thin-skinned structure in the western part of the 1002 area. The east-northeast-trending Marsh Creek anticline records folding of Eocene strata over a bend in the roof thrust. Although this structure is represented on geologic maps as a simple plunging anticline, it is clear from several seismic lines that the core of the fold is cut by numerous faults and, in detail, is a complex structure. A seismic line across the down-plunge (east-northeast) end of the Marsh Creek structure images a well-defined anticline affecting Eocene strata, demonstrating that the structure becomes simpler to the east-northeast. Proprietary high-resolution aeromagnetic data (Phillips et al., 1998) reveal anomaly patterns that faithfully reflect the structure of the Marsh Creek anticline (Figure 10). The most prominent positive anomaly, produced by a highly magnetic Eocene layer, defines a plunging anticline that is locally disrupted as if cut by faults (west of the sample shown in Figure 10), supporting similar interpretations from the reflection data. The magnetic data show similar east-northeast-trending folded patterns south of the Marsh Creek anticline, which indicate folding above the roof thrust of the passive-roof duplex; such folds are also seen less distinctly in the seismic reflection data. The magnetic patterns were also interpreted by Phillips (1999) to be disrupted by faults, which is consistent with seismic data, indicating that the passive roof is breached by several small-displacement faults.

Because of the inability to restore displacements on specific thrust faults, we used an area-balance technique to calculate the amount of shortening recorded by the passive-roof duplex (Potter et al., 1999). We determined that the passive-roof duplex beneath the Marsh Creek anticline records 19–25 km (12–16 mi) of shortening where it is best developed (east of line 84-10) and less shortening (under 15 km [9 mi]) to the west. Because the southward extent of the roof thrust has been removed by erosion, these are minimum figures. The magnitude of shortening associated with the passive-roof duplex that controls the Marsh Creek anticline attenuates to the southwest. Near the southwest corner of the 1002 area, on line 84-6, the deformed wedge that defines the passive-roof duplex (a, Figure 9) is relatively small, and the Eocene reflectors above the wedge are only moderately warped (beneath Marsh Creek anticline label, Figure 9) in response to the underlying thin-skinned structural thickening. Note that the basement-involved anticline on line 84-6 (Figure 9) is genetically related to the foothills fault system, not the Marsh Creek anticline, in this interpretation. Southwest of this location, the Marsh Creek anticline loses definition.

Seismic reflection patterns record the timing of growth of the Marsh Creek anticline. Uniformly northwest-dipping Eocene reflections, forming inclined railroad-tracklike patterns, characterize the north flank of the Marsh Creek anticline on all seismic lines. The tilting of these strata demonstrates that the Marsh Creek anticline is a post-Eocene structure. Unpublished proprietary seismic reflection data, as well as the uninterpreted version of line 84-14 (Kelley and Foland, 1987, their plate 5), contain fanning (growth) patterns in Oligocene to early Miocene strata on the north flank of the Marsh Creek anticline, demonstrating that this structure grew in the late Oligocene to early Miocene. Although O'Sullivan et al. (1993) had inferred an Eocene age for the Marsh Creek anticline based on apatite fission-track data, O'Sullivan and Wallace (2002) used apatite and zircon fission-track thermochronology to conclude that two phases of basement-involved uplift affected the area immediately north of the Sadlerochit Mountains at 45 and 27 Ma, with the younger phase contributing to the growth of the Marsh Creek anticline.

Late Movement of Basement-involved Faults

The basement-cutting foothills fault system identified near the south ends of lines 84-10 and 84-6 (Figures 8, 9) lies 14–21 km (9–13 mi) north of the Sadlerochit Mountain front. This zone is characterized by a right-stepping pattern (Figure 7). The structural interpretations of seismic lines 84-10 and 84-6 (Figures 8, 9) strongly imply that these basement-involved thrusts transferred displacement northward into the thin-skinned passive-roof duplex that underlies the Marsh Creek anticline. O'Sullivan and Wallace (2002) developed a strong case for the 27-Ma basement-involved thrusting along the Sadlerochit front and to the north; the south 2002 fault system probably developed at this time or slightly after. On some unpublished proprietary seismic reflection lines, there is clear evidence for late south-dipping breakthrough reverse faults that cut through and displace the thin-skinned detachment near the base of the Brookian section and appear to continue up to the base of the permafrost (250–600-m [820–2000-ft] depth; Brewer, 1987; Molenaar, 1987). Additional examples of late basement-cutting thrusts that postdated thin-skinned deformation are apparent in reflection data (Figure 8, labels b and c).

INTERPRETATION OF THRUST STRUCTURES IN THE EASTERN PART OF THE 1002 AREA

In the eastern part of the 1002 area, the principal basement-involved structural elements include the foothills fault system and the broad Niguanak and Aurora domes to the north, underlain by major north-vergent basement thrusts (Figure 7). Principal thin-skinned structural elements include the Sabbath syncline, the Aichilik high, a north-vergent imbricate stack across the south flank and top of the Niguanak high, and Jago ridge (Figure 3). All of these structural elements are present on dip lines oriented north-northwest, such as lines 84-30 and 85-50 (Figures 6, 11). Seismic line 85-1 (Figure 12) is an east-northeast-oriented strike line that crosses the Niguanak high and the flanks of the adjacent structural lows (Hulahula low to the west and Demarcation subbasin to the east; Figure 7). These lines, augmented by nearby unpublished seismic lines and by potential-field data, provide valuable insights on the evolution of these structures.

Thrusting of Pre-Brookian Rocks in the Southeastern Part of the 1002 Area

Moore's (1999) balanced cross section (Figure 5) illustrates the general structure of the eastern part of the 1002 area. (Cross section restoration and details on the

FIGURE 8. Seismic reflection, gravity, and magnetic interpretations of merged seismic line 84-10 and 85-16, showing passive-roof duplex geometry. Note that seismic data are displayed at a larger scale than the gravity and magnetic profiles. Red lines on seismic profile are interpreted faults; key interpreted faults have also been transferred to the gravity and magnetic interpretations as red lines. Black lines on gravity and magnetic interpretations are key seismic reflections. Because the Ellesmerian sequence has been erosionally truncated by the Lower Cretaceous unconformity (LCU) in this part of the 1002 area, the top of the pre-Mississippian (TPM) is coincident with the LCU. The top of the Miocene and the top of the Oligocene at the north end of the seismic line are based on seismic reflection tie to West Staines State 2. Locations a–c refer to thrust faults that displace the top of Franklinian basement. See text for details.

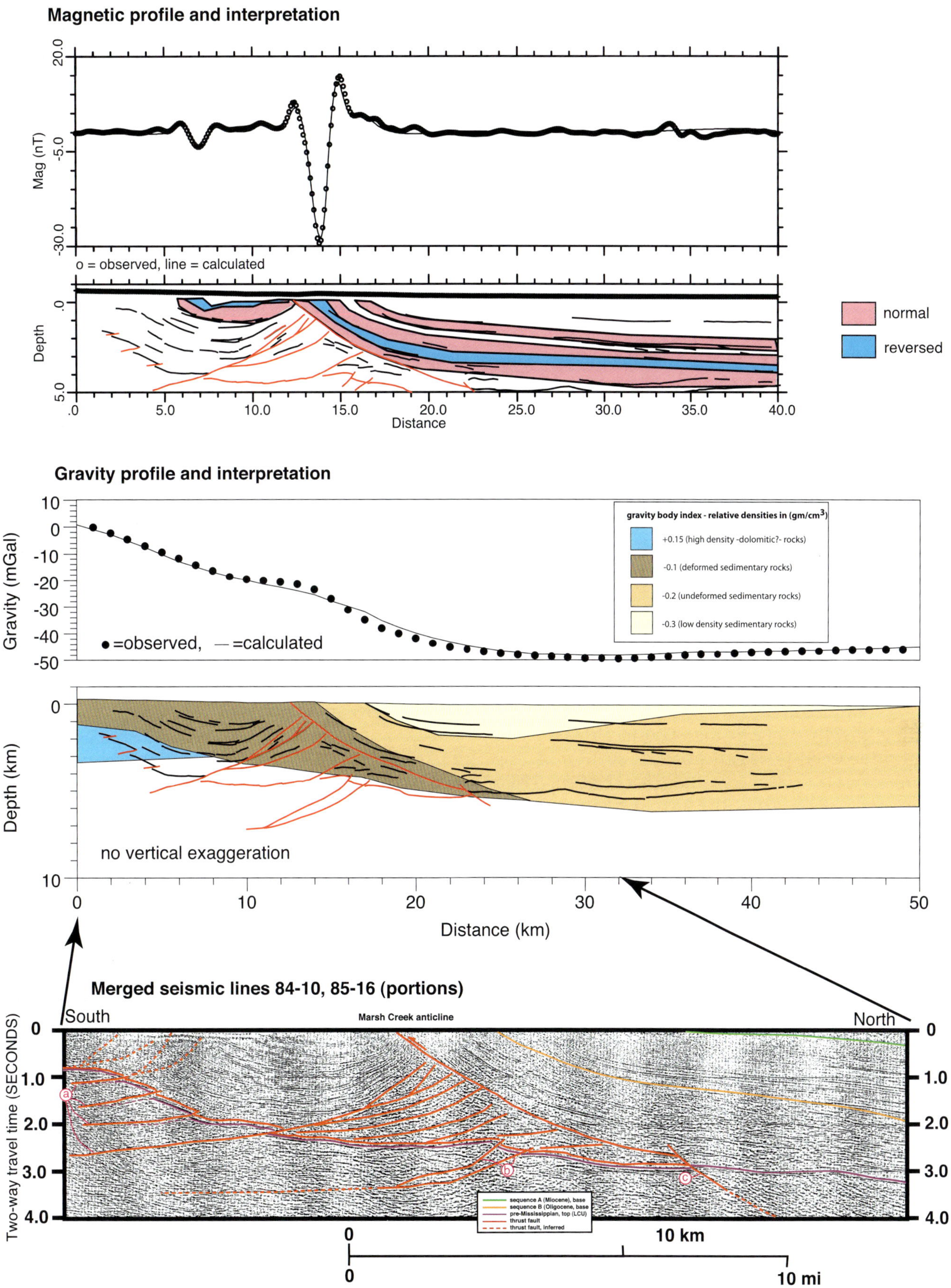
Magnetic profile and interpretation
Mag (nT)
20.0
-5.0
-30.0
o = observed, line = calculated
Depth
.0
5.0
normal
reversed
.0
5.0
10.0
15.0
20.0
25.0
30.0
35.0
40.0
Distance
Gravity profile and interpretation
Gravity (mGal)
10
0
-10
-20
-30
-40
-50
● =observed, — =calculated
gravity body index - relative densities in (gm/cm3)
+0.15 (high density -dolomitic?- rocks)
-0.1 (deformed sedimentary rocks)
-0.2 (undeformed sedimentary rocks)
-0.3 (low density sedimentary rocks)
Depth (km)
0
10
no vertical exaggeration
0
10
20
30
40
50
Distance (km)
Merged seismic lines 84-10, 85-16 (portions)
South
Marsh Creek anticline
North
Two-way travel time (SECONDS)
0
1.0
2.0
3.0
4.0
a
b
c
sequence A (Miocene), base
sequence B (Oligocene), base
pre-Mississippian, top (LCU)
thrust fault
thrust fault, inferred
0
10 km
0
10 mi

balancing of the cross section are presented in Moore, 1999.) This section shows thrust faulting controlled by detachments in the Franklinian basement, in the upper part of the Ellesmerian (Shublik Formation, Sadlerochit Group), the Beaufortian (Kingak Shale), and near the base of the Brookian. Thrust detachments in the Franklinian and Brookian are present through much of the eastern part of the 1002 area. Although the Ellesmerian sequence is absent in the subsurface through much of the western part of the 1002 area, seismic reflection data strongly support its presence beneath the southeastern part, where thrusting above a regional detachment surface in the clastic rocks in the upper part of the Ellesmerian is locally important. As interpreted from a high-quality, proprietary seismic line, the autochthonous Ellesmerian thins depositionally to the north through onlap to a zero-edge well south of the Niguanak high (Grow et al., 1999d). The northern extent of Ellesmerian-bearing thrust sheets is not well constrained. Franklinian, Ellesmerian, and Beaufortian rocks are involved in thrusting at Leffingwell Ridge (Figure 2), as well as along the foothills fault system, which is responsible for 1 km (0.6 mi) or more of throw of these units.

Sabbath Syncline and Aichilik High

The south end of line 84-30 (Figure 11) contains the Sabbath syncline and the Aichilik high, which involve Brookian rocks. On the north flank of the Sabbath syncline (south flank of Aichilik high; Figures 11, 13), there is a pronounced convergence of reflections, indicating growth strata in the 3000 m (10,000 ft) of largely Paleocene [+Maastrichtian(?)] shallow-marine to nonmarine Jago River Formation (Buckingham, 1985; Potter et al., 1999). Thus, the Aichilik high developed during the Paleocene, and the Sabbath syncline was a largely Paleocene piggyback basin behind and above the growing Aichilik high.

The internal structure of the Aichilik high, in which south-dipping reflections have progressively steeper dips in the structurally higher parts of the uplift, is consistent with the geometry of a small passive-roof duplex (triangle zone), bounded above by a smooth, north-dipping roof thrust (Figures 11, 13). As passive-roof duplexes are typical of thrust fronts, it appears likely that the Aichilik high marks the Paleocene front of the Brookian thrust belt. The imbricate stack in the Aichilik duplex is inferred to be shale rich on the basis of reflection character, deformation style, and especially because the Aichilik high (Figure 3) corresponds to a negative gravity anomaly (Figure 11) (Saltus et al., 1999). Figures 11 and 13 show the imbricate stack to be composed of Cretaceous Hue Shale and pebble shale and the underlying detachment to lie in the pebble shale. Moore's (1999) portrayal of an Aichilik duplex consisting of imbricated Paleocene mudstones above a detachment in the Hue Shale (Figure 5) is problematic, because imbricated Paleocene mudstones of the Aichilik high would serve as the basement to syntectonic Paleocene and Maastrichtian(?) clastic strata of the Jago River Formation in the Sabbath syncline.

Thin-skinned Thrust Faults above the Niguanak High

Across the Aichilik high and north to the Beaufort Sea, outcrops are scarce. Near the label NH on Figure 2, small outcrops provide windows through the tundra near Niguanak Ridge and on the nearby Jago and Niguanak rivers; these are valuable clues to the subsurface structure of this part of the region, when placed in the context of the seismic reflection and potential field data. One of these outcrops contains a highly deformed Jurassic shale equivalent to the Kingak Formation, and surrounding patchy outcrops are variably deformed Cretaceous Hue Shale, pebble shale, and Paleocene mudstones (Bader and Bird, 1986).

These outcrops lie directly above the Niguanak high, a large basement dome expressed in the seismic profile as a broad arch in a prominent reflector interpreted as the top of Franklinian basement, cresting at 2680 m (8800 ft) below sea level (Figures 6, 11, 12) (Grow et al., 1999d). Above the south flank and crest of the Niguanak high, there is a uniformly south-dipping seismic reflection fabric, here interpreted as long, north-vergent thrust ramps reaching from a deep detachment at 2.0–2.7 s (near the base of the Brookian section) to the Earth's surface above the Niguanak high (Figure 11). The south-dipping fabric projects to the surface positions of the outcrops described above. The presence of Jurassic outcrops in this area requires that some of the thrust ramps root beneath the base of the Brookian

FIGURE 9. Seismic reflection, gravity, and magnetic interpretations of part of seismic line 84-6, showing superposition of the thin-skinned Marsh Creek anticline above the principal basement-involved reverse fault, at the point where these two trends (Marsh Creek anticline and basement-involved thrusting) intersect. Note that the seismic profile is displayed at a larger scale than the gravity and magnetic profiles. Red lines on seismic profile are interpreted faults; key interpreted faults have also been transferred to the gravity and magnetic interpretations as red lines. Black lines on gravity and magnetic interpretations are key seismic reflections. a = deformed wedge defining passive-roof duplex in seismic data.

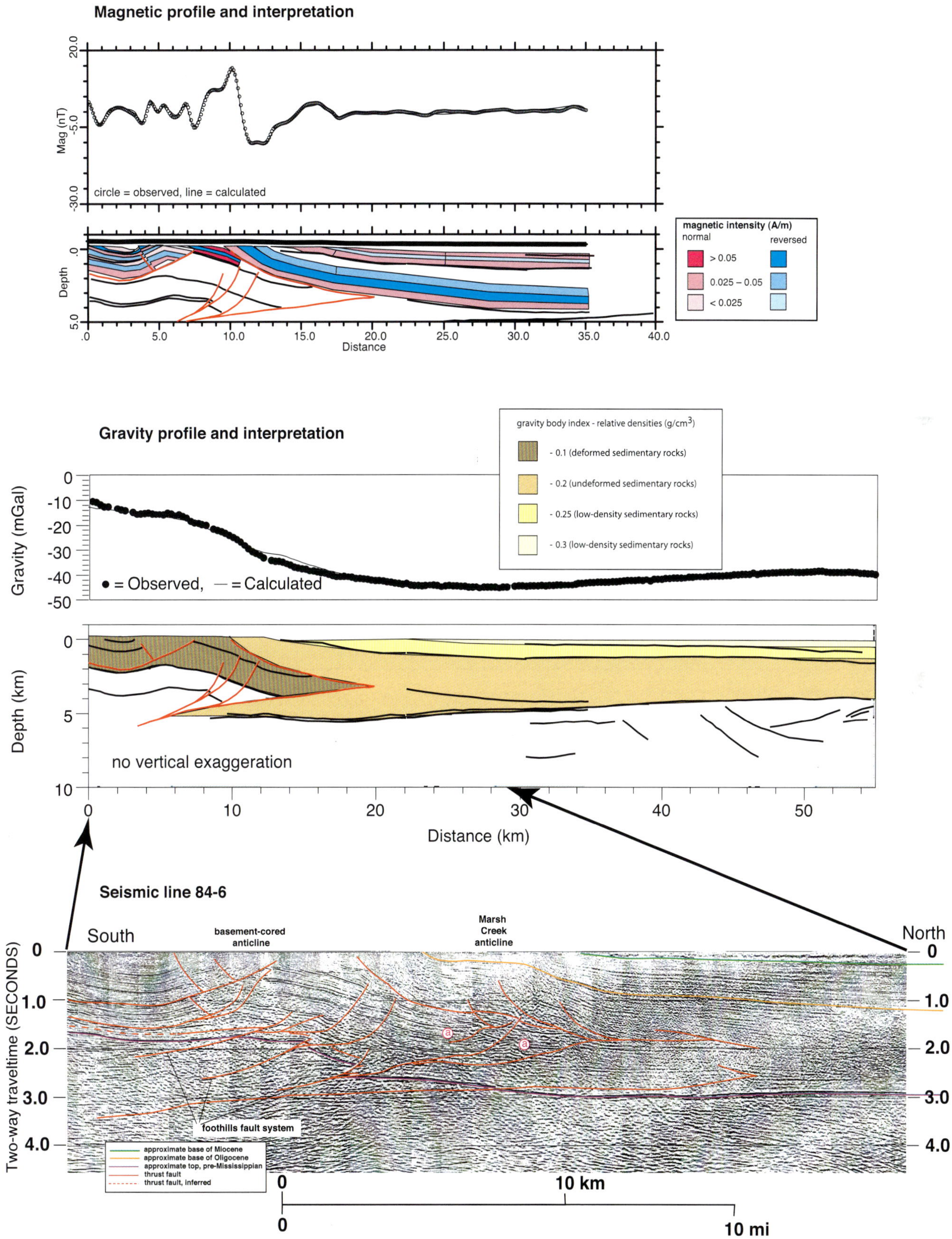
Magnetic profile and interpretation
Mag (nT)
20.0
-5.0
-30.0
circle = observed, line = calculated
Depth
.0
5.0
Distance
10.0
15.0
20.0
25.0
30.0
35.0
40.0
magnetic intensity (A/m)
normal
reversed
> 0.05
0.025 – 0.05
< 0.025
Gravity profile and interpretation
gravity body index - relative densities (g/cm3)
- 0.1 (deformed sedimentary rocks)
- 0.2 (undeformed sedimentary rocks)
- 0.25 (low-density sedimentary rocks)
- 0.3 (low-density sedimentary rocks)
Gravity (mGal)
0
-10
-20
-30
-40
-50
● = Observed, — = Calculated
Depth (km)
5
10
no vertical exaggeration
Distance (km)
Seismic line 84-6
South
North
basement-cored anticline
Marsh Creek anticline
Two-way traveltime (SECONDS)
1.0
2.0
3.0
4.0
foothills fault system
approximate base of Miocene
approximate base of Oligocene
approximate top, pre-Mississippian
thrust fault
thrust fault, inferred
10 km
10 mi

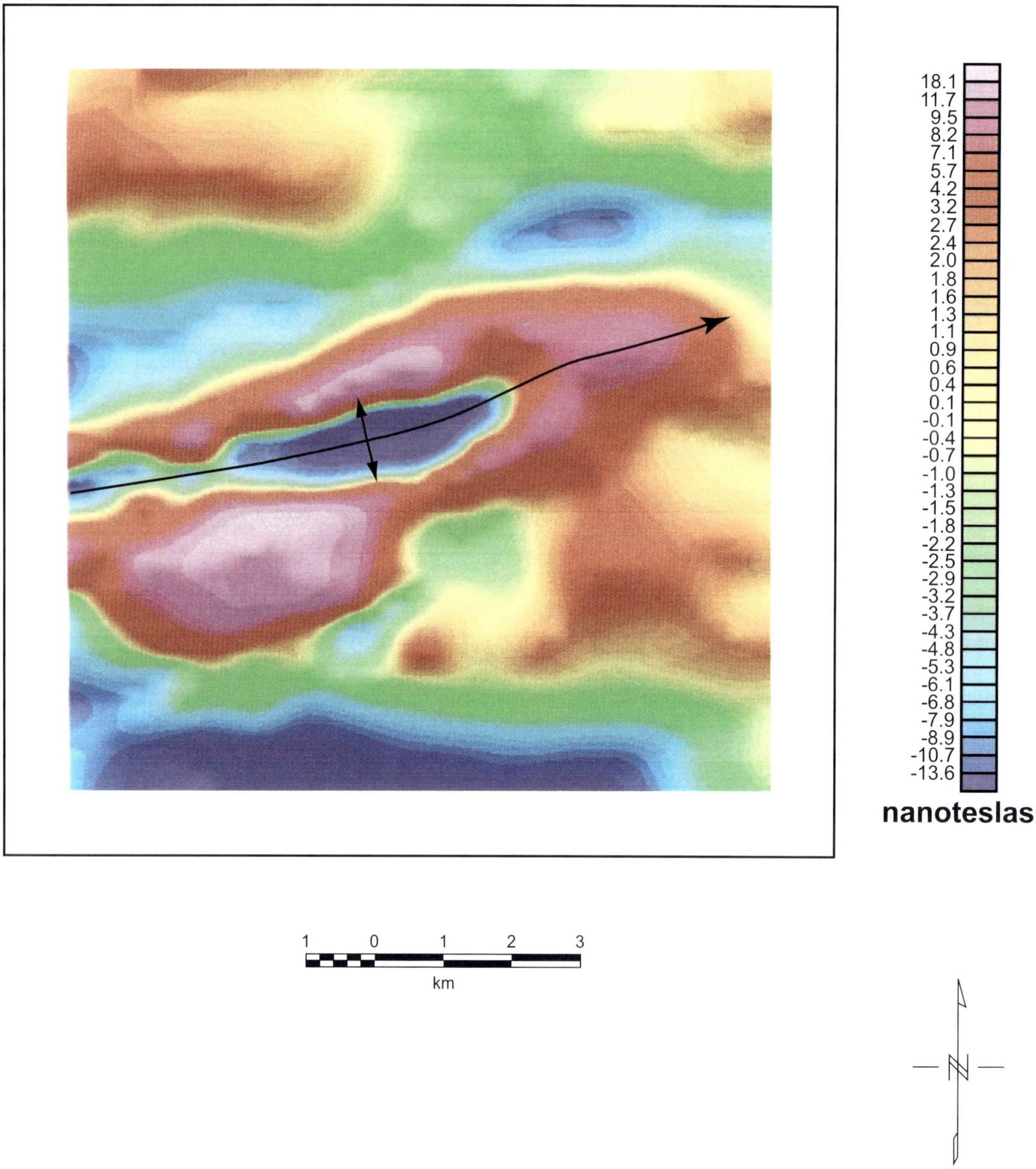

FIGURE 10. Sample of proprietary high-pass aeromagnetic data showing a 10 × 10-km (6 × 6-mi) area over the nose of the Marsh Creek anticline from World Geoscience, Inc., used in this study. Stratigraphically controlled source of magnetic anomalies is apparent around east-northeast-plunging nose of Marsh Creek anticline.

and have transported Jurassic shale into a high structural position, above Brookian strata.

As with the imbricated Brookian section in the western part of the 1002 area, there are no definitive marker horizons to allow rigorous geometric or kinematic interpretation of the faulting in the south-dipping thrust domain. A relatively small proportion of Ellesmerian or Beaufortian clastic rocks may be imbricated with Brookian strata in the south-dipping thrust slices (Potter et al., 1999).

To the north, the south-dipping, ramp-dominated domain gives way to subhorizontal reflection fabrics across the crest and the north flank of the Niguanak high. Individual thrust slices are identified on the basis of discordant reflection patterns, indicating the presence of ramp-and-flat fault geometry (Figure 11).

The regional doming of the thin-skinned detachment surface (which overlies and mimics the broadly arched top of the Franklinian rocks) requires that the broad basement doming (Niguanak high) postdated the shallower, thin-skinned thrusting. An erosionally breached, Brookian, thin-skinned passive-roof duplex is interpreted from seismic lines across the Niguanak high, with an erosional window through the shallow roof thrust evident on lines 84-30, 85-50, and other nearby seismic lines. The erosional breaching of the roof thrust across the top of the Niguanak high might also reflect the late basement-involved doming at depth.

Jago Ridge

A prominent north-vergent thin-skinned imbricate wedge, directly north of and above the Niguanak high on lines 84-30 and 85-50 (Figures 6, 11), corresponds to Jago ridge (Figure 3). This long, east–west-trending culmination forms the thin-skinned thrust front in the northeastern part of the 1002 area. On line 84-30 (Figure 11), the seismic data north of the Niguanak high are represented only as an interpreted line drawing, because data in this area are proprietary. Approximate stratigraphic ties to the Aurora 1 well (Grow et al., 1999d) are shown. The north margin of the Jago ridge imbricate wedge, on this and other seismic lines, includes an abrupt, north-dipping boundary between a deformed sedimentary wedge and relatively undeformed, lower Tertiary strata to the north. Upper Eocene(?) and lower Oligocene(?) reflectors dip to the north above the highly deformed imbricate wedge (point a, Figure 11). At the north end of the seismic line, there is evidence for a thin, thrust-bounded wedge inserted northward in the Paleocene section (point b, Figure 11), characteristic of the tip of a passive-roof duplex. On line 85-50 (Figure 6), the northern limit of thin-skinned imbricate thrusting appears as a complex array of blind thrusts instead of a well-defined north-driven wedge.

Potter et al. (1999) interpreted the Jago ridge passive-roof duplex as a composite structure, recording two different periods of motion, based on relations shown in Figure 11 and on correlation of the detailed sequence-stratigraphic interpretations of Brookian strata (Houseknecht and Schenk, 1999) onto the flanks of Jago ridge (Potter et al., 1999, their figure 13). The thinly tapered imbricate wedge at the north end of the cross section (Figure 11) appears to have developed in the Eocene, and the prominent bulge in the roof thrust, overlain by dipping reflections at location a, appears to record the development of a younger, north-directed triangle zone. This latter deformation is constrained as being Oligocene in age, because it involves probable lower Oligocene reflectors and is overlain by nearly horizontal Miocene and upper Oligocene reflectors. The Oligocene deformation may be related to motion on the basement-cutting Niguanak thrust, as discussed below. Seismic lines show a northward tilting of Miocene strata north of Jago ridge, indicating that less intense regional deformation continued in the Miocene.

Niguanak High and Aurora Dome

The Niguanak high and the Aurora dome are elongate east–west basement-involved structures that merge at depth beneath the northeastern part of the 1002 area (Figure 7) and the immediate offshore area to the north. They define two discrete structural highs, at elevations of −2680 m (−8800 ft) (Niguanak) and −4510 m (−14,800 ft) (Aurora), and have attracted considerable exploration interest because of their uniquely large size and heights of closure. The area of closure for each of these structures is about 600 km^2 (230 mi^2), defined at a structural contour elevation of −5030 m (−16,500 ft). Seismic reflection interpretations indicate that both of these domes are composed of thrust-imbricated Franklinian rocks (Grow et al., 1999d). The composite Niguanak–Aurora high forms the southwestern boundary of the Demarcation subbasin (a compressionally formed offshore basin that straddles the Canada–United States boundary) and effectively separates it from the Barter subbasin to the northwest (Figure 7) (Grantz et al., 1987, 1990).

Line 85-50 (Figure 6) traverses the top of the broad Niguanak antiform. The higher quality line 84-30 passes over the eastern flank of this structure and can be correlated to a deep well (Aurora 1) located 6 km (4 mi) north-northeast of the north end of the seismic line. Both of these lines end on the south flank of the Aurora dome, so the crest of the Aurora dome is not imaged. Line 85-1 (Figure 12), a strike line oriented east-northeast, crosses the Niguanak high as well.

Line 84-30 (Figure 11) clearly images the basement-cutting Niguanak thrust beneath the broad basement dome that defines the Niguanak high. This thrust (imaged less clearly on line 85-50; Figure 6) offsets the top of Franklinian basement and the thin-skinned detachment near the base of the Brookian section. Movement on the Niguanak thrust was probably coeval with late stages of thin-skinned imbrication of Jago ridge (and the thin-skinned deformation across the top of the Niguanak high). As the Niguanak thrust propagated up into the previously deformed Brookian section, its displacement appears to have fed into the thin-skinned passive-roof duplex. It does not cut Eocene reflectors that overlie the roof thrust, so it may have driven a wedge of Paleocene (and possibly lower Eocene) strata forward, producing the prominent bulge in the roof

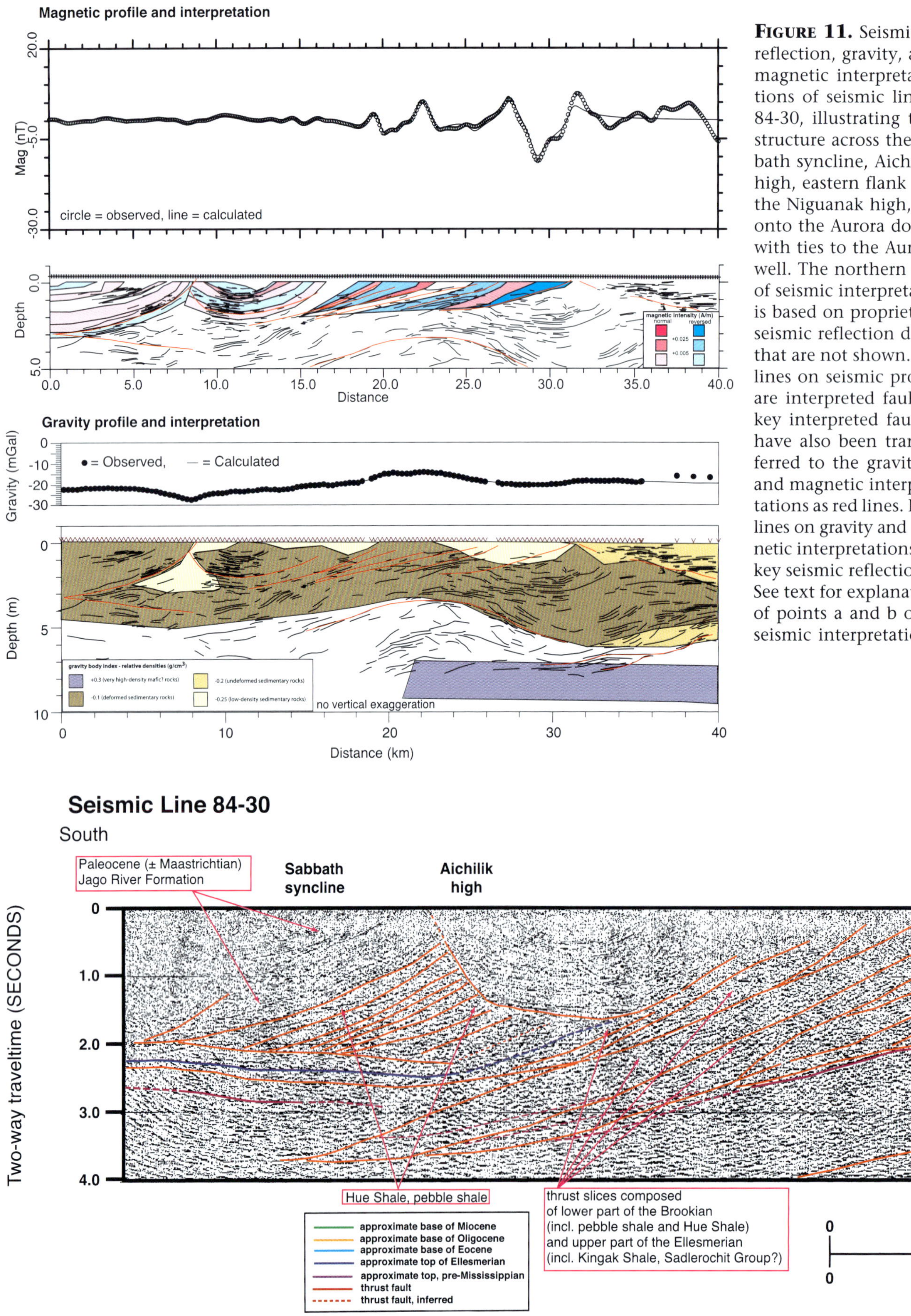

Figure 11. Seismic reflection, gravity, and magnetic interpretations of seismic line 84-30, illustrating the structure across the Sabbath syncline, Aichilik high, eastern flank of the Niguanak high, and onto the Aurora dome, with ties to the Aurora well. The northern part of seismic interpretation is based on proprietary seismic reflection data that are not shown. Red lines on seismic profile are interpreted faults; key interpreted faults have also been transferred to the gravity and magnetic interpretations as red lines. Black lines on gravity and magnetic interpretations are key seismic reflections. See text for explanations of points a and b on seismic interpretation.

thrust (near point a, Figure 11). This strongly implies that the Niguanak thrust is Oligocene in age.

Two exploration wells have been drilled on the Aurora dome, and publicly available stratigraphic information from one of these wells helps to constrain interpretations of seismic lines in the area. The KIC 1 exploration well (Figure 2) was completed to a total depth of 4326 m (15,200 ft) by Chevron USA, Inc., in Kaktovik-Inupiat Corporation lands on the southeast flank of the Aurora dome in 1986, but all of the information from that well remains proprietary. The Aurora 1 well was drilled by the Tenneco Oil Company in the winter of 1987–1988 on the east flank of the Aurora dome in United States federal waters, approximately 10 km (6 mi) northeast of the KIC 1 well. It was drilled to a depth of 5585 m (18,300 ft), and well information has been released (Banet, 1994; Paul et al., 1994). Although there were shows of oil and gas, the well was plugged and abandoned.

The stratigraphic section encountered in the Aurora well includes 4718 m (15,500 ft) of Tertiary (Paleocene to Miocene) sedimentary rocks overlying 867 m (2800 ft) of Middle Jurassic to Lower Cretaceous clastic sedimentary rocks (Figure 11). The latter sequence is interpreted as Beaufortian rift deposits (Paul et al., 1994; Bird, 1999; Poag, 1999). The tops of the Paleocene and Eocene can be correlated to line 84-30 (6 km [4 mi] to the south-southwest) using proprietary seismic data.

Although subhorizontal reflections on proprietary seismic data correlate well with the Beaufortian Jurassic and Lower Cretaceous clastic sedimentary rocks drilled at the Aurora well, they cannot be confidently correlated onto the seismic data shown in Figure 11. Proprietary profiles on the east flank of the Aurora dome indicate that Beaufortian rocks thin westward and may pinch out before reaching the crest of the dome, and that this Beaufortian sequence dips south and does not appreciably change thickness before being overridden by the Niguanak thrust. Because the Aurora well did not penetrate the base of the Beaufortian rift sequence, and the seismic profiles do not delineate a base for this sequence, its thickness and the identity of its basement are unknown.

Gravity models over the Aurora dome and Niguanak high indicate that these two broad domes are composed of rocks having densities averaging more than 2.7 g/cm^3 (Saltus et al., 1999). Seismic velocities derived during the reprocessing of the seismic lines over these two highs indicate that the average velocities of the rock units that core the highs are approximately 5030 m/s (16,500 ft/s) (Grow et al., 1999d; Lee et al., 1999). The velocity and density values are compatible with the Franklinian or Ellesmerian sequences as dominant lithologies in these domes and are much higher than those measured by well logs for the Beaufortian rocks in the Aurora well. Because the Ellesmerian sequence appears to pinch out south of the Niguanak high (Grow et al., 1999d), it is most likely that the Niguanak high and Aurora dome consist of Franklinian rocks.

Seismic reflection patterns, particularly on lines 85-1 and 85-50 (Figures 6, 12), as well as proprietary data, indicate that the Niguanak high and Aurora dome overlie basement duplexes, composed of multiple, north-verging thrust sheets bounded to the west by east-dipping lateral ramps. Low on the west flank of the Aurora dome (north edge of the Hulahula low), the floor

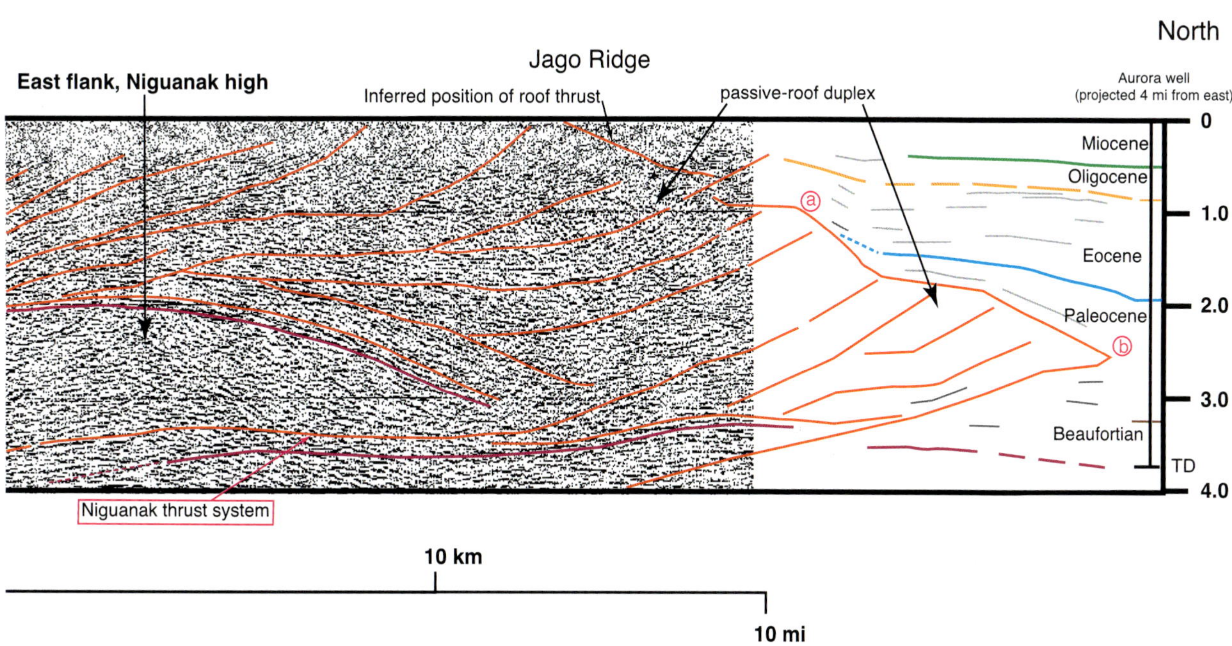

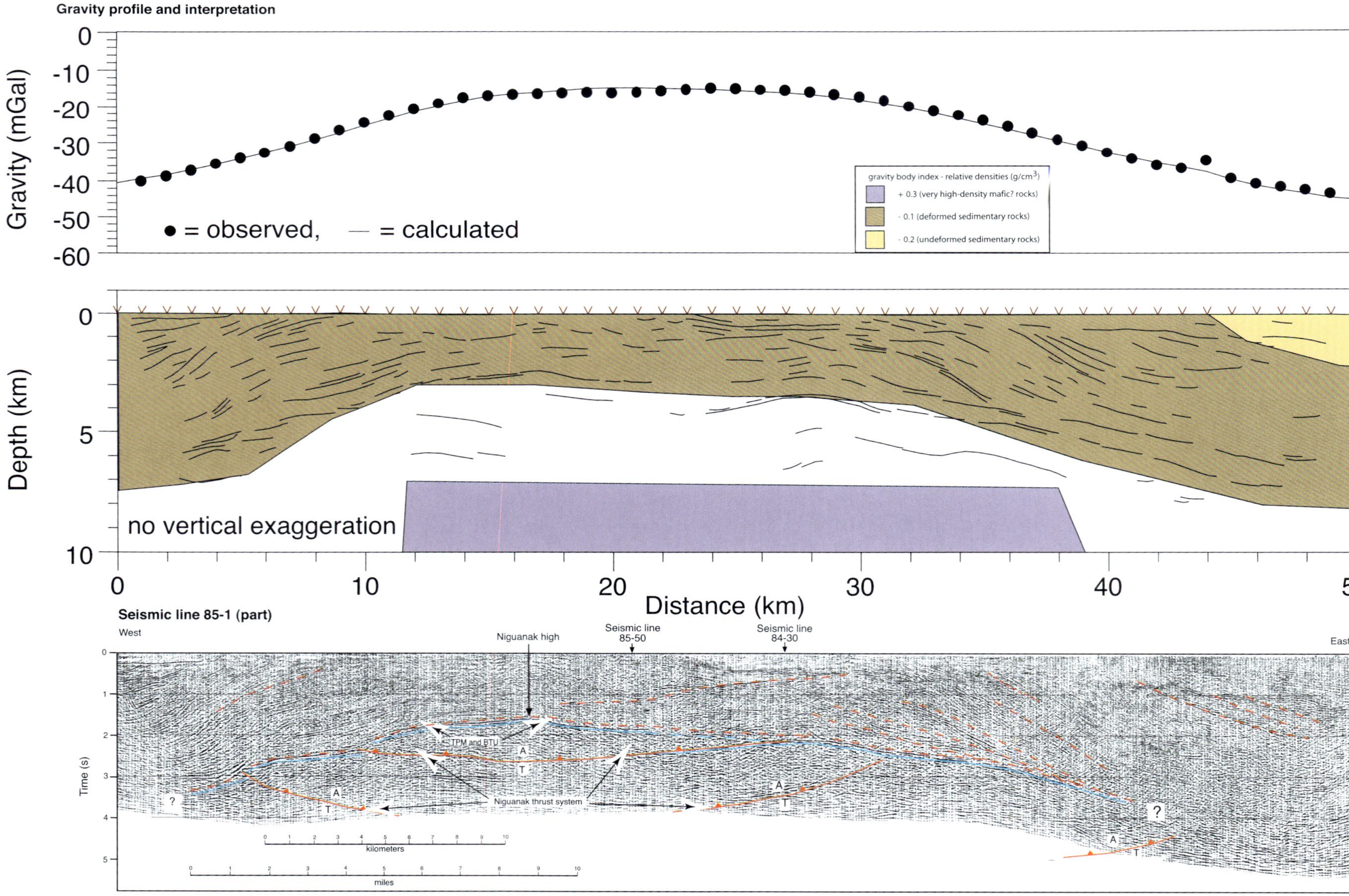

FIGURE 12. Seismic reflection and gravity interpretations of the part of seismic line 85-1 that traverses the crest of the Niguanak high. Red lines on seismic data are interpreted faults. Black lines on gravity model represent key seismic reflections. The interpretation of multiple internal thrust sheets in the Niguanak high for this profile was originally suggested by Bruns et al. (1987, their plate 4). A and T denote relative motion along thrust faults, away (A) and toward (T) the veiwer. TPM and BTU = top of pre-Mississippian rocks, corresponding here to the basal Tertiary unconformity. See text for details.

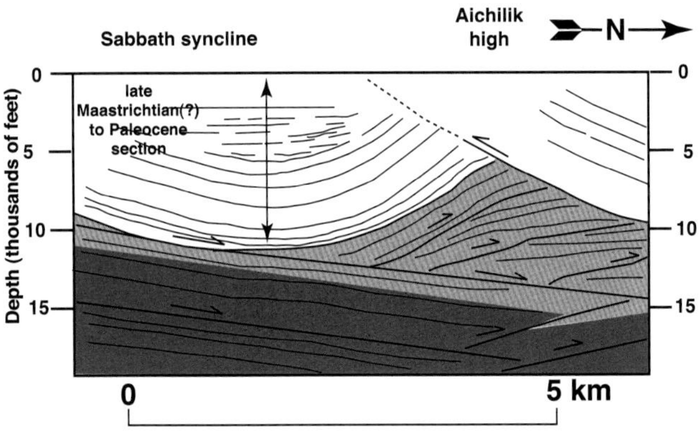

FIGURE 13. Line drawing, based on seismic reflection data, showing detail of the structural and stratigraphic relations beneath the Aichilik high and the northern flank of the Sabbath syncline, demonstrating Paleocene growth of the Aichilik high. White = Maastrichtian(?) to Paleocene section; light gray = Hue Shale and pebble shale; dark gray = Hue Shale, Kingak Shale, and the upper part of Ellesmerian.

thrust beneath the Aurora duplex merges with the top of Franklinian basement at a depth of about 7 km (4 mi), based on our interpretation of proprietary seismic data. From this point, the deep detachment surface has an apparent dip to the east and may be greater than 10 km (6 mi) deep beneath the Aurora dome, defining a large-scale lateral ramp (Grow et al., 1999d). North-directed displacement above the deep basement detachment may feed into deformation offshore in the vicinity of the Camden anticline. Grantz et al. (1987) suggested that offshore domes are active structures, based on present-day seismicity.

DISCUSSION

Evolution of Brookian Thrusting across the Niguanak High

Jurassic and Cretaceous shales exposed in the Niguanak Ridge area of the coastal plain (Figure 2) are key to structural interpretations for two reasons: (1) they provide geologic control on the lithologies involved in the imbricate structure above the Niguanak high; and (2) thermal maturity and apatite fission-track data for these rocks provide constraints on thermal history. The Jurassic and Cretaceous thrust slices are thermally immature ($R_o = 0.6$; Magoon et al., 1987). Their apatite fission-track ages reflect provenance ages, indicating that these rocks have not been subjected to temperatures higher than 60°C for more than 1 m.y., and that their maximum depth of tectonic or sedimentary burial was 1–2 km (0.6–1.2 mi) (O'Sullivan et al., 1993). These observations, and the deformational timing deduced from line 84-30 (Figure 11), are placed in the context of a generalized interpretation across the Niguanak area in a series of schematic cross sections in Figure 14.

Figure 14A illustrates the concept that the Aichilik high formed as a passive-roof duplex at the Paleocene thrust front. The basal detachment is interpreted to lie near the base of the Brookian section. As discussed above, gravity data imply that the volume of rock in the Aichilik high consists dominantly of shale (of probable Cretaceous age), and seismic reflection patterns in the Sabbath syncline demonstrate that it was a nonmarine to shallow-marine piggyback basin behind the thrust front.

In the Eocene and possibly in the early Oligocene, represented by Figure 14B, an imbricate stack, involving the Brookian section as well as Kingak-equivalent shales such as those exposed in the Niguanak Ridge area, propagated to the north, across the current Niguanak high, as far north as the current position of the Jago ridge (although Jago ridge attained most of its current structural relief in the Oligocene; Potter et al., 1999). The basal detachment resided in the Jurassic and Cretaceous Kingak Shale in the southern part of the section, although north of the pinch-out of the Ellesmerian and Beaufortian sequences (point P in Figure 14), the detachment passes up into the basal Brookian strata.

Figure 14C illustrates Oligocene development of basement-cored anticlines through development of basement duplexes beneath the Niguanak and Aurora domes. Because the Niguanak thrust is inferred to have transferred its displacement northward into the thin-skinned passive-roof duplex, this period also corresponds to renewed development of the Jago ridge passive-roof duplex directly north of the Niguanak high.

Previously unpublished apatite fission-track analyses indicate 45-Ma uplift and cooling of the Aichilik high; this uplift is represented in Figure 14B as an Eocene duplex beneath the earlier formed Aichilik high. New apatite fission-track analyses indicate a 25-Ma (latest Oligocene) uplift of the Niguanak high, corroborating the seismic-based interpretations presented here. (Approximate sample positions for these new apatite fission-track analyses are plotted on cross section, Figure 5.)

The Kingak and Hue shales, presently exposed at Niguanak Ridge, were transported into their present structural position by Eocene [to Oligocene(?)] motion on the long thrust ramps above the south flank of the Niguanak high (Figure 14B, C). The Paleocene position of these rocks (Figure 14A), prior to their Eocene to Oligocene ramp-dominated upward transport, represents the deepest structural level attained by these rocks.

A present-day northward continuation of this thrust belt (not depicted in Figure 14) may be manifested by Holocene contractional structures and by upper crustal earthquakes, with epicenters offshore to the north of ANWR (Grantz et al., 1987, 1990; Potter et al., 1999).

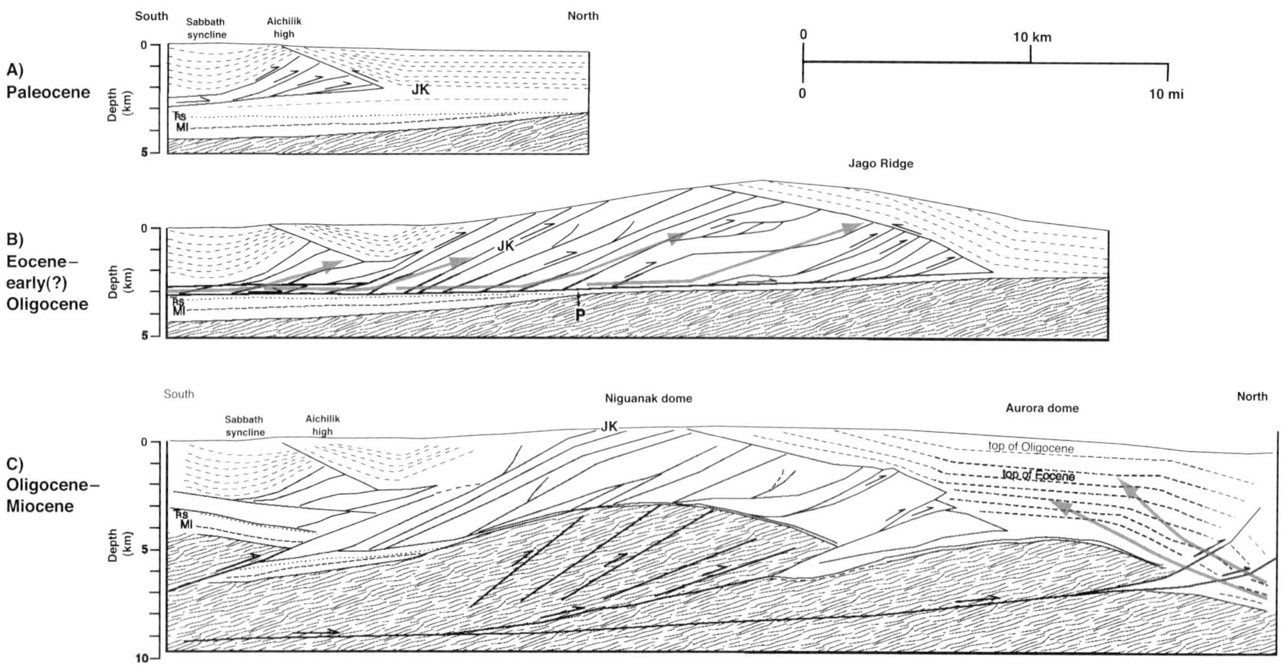

FIGURE 14. Schematic illustrations of the evolution of the structures along a north–south transect across the eastern part of 1002. Bold lines are faults; fault-displacement arrows indicate faults that were active during the indicated time interval. TRs (finely dotted line) = top of Triassic Shublik Formation; Ml (dashed line) = top of Mississippian Lisburne Group; wavy pattern = pre-Mississippian basement rocks. (A) Paleocene frontal triangle zone (Aichilik high), with Sabbath syncline as a piggyback basin to the south; JK indicates Paleocene position of the Jurassic and Cretaceous shale units that are presently exposed near Niguanak Ridge. Basal detachment is near the base of Brookian sequence. (B) Eocene to Oligocene development of ramp-dominated thrusts north of the Aichilik high, transporting strata equivalent to the Cretaceous Hue and Jurassic Kingak shales (JK, presently exposed on the Niguanak high) to shallow structural level, and development of Jago Ridge frontal triangle zone. Basal detachment has stepped down into the Jurassic Kingak Shale in the southern part of the section. In central and northern part of the section, north of the pinch-out of the Ellesmerian sequence (P), basal detachment is near the base of Brookian sequence. Broad light gray arrows indicate northward migration of Hue-sourced oils during the Eocene. (C) Oligocene to Miocene basement-involved thrusting and folding, forming the Aurora dome, Niguanak dome, and the foothills fault system. Three large basement-involved thrust ramps sole into a detachment near 9 km (5 mi) in depth (Grow et al., 1999d). JK indicates position of the Jurassic and Cretaceous shale units that are presently exposed near Niguanak Ridge. Broad light gray arrows indicate migration of Canning-sourced oils during the Neogene.

Summary of the Timing of Deformation in the 1002 Area

In the eastern part of the 1002 area, we interpret a Paleocene to present(?) pattern of progressively younger, thin-skinned structures toward the foreland (Figure 14A, B). Younger basement-involved thrusts were superimposed on this thin-skinned pattern. In the southeastern part, the basement-involved thrusting is post-Paleocene (basement-involved faulting cuts Paleocene structures), but its exact age is indeterminate. In the Niguanak area, basement-involved thrusts folded and rotated early Eocene thin-skinned structures associated with Jago ridge, and we have presented an interpretation that these basement-involved thrusts are Oligocene in age.

In the western part of the 1002 area, east–west-striking, basement-involved faulting parallel to (and north of) the Sadlerochit Mountain front appears to have been synchronous with, and partly later than, the late Oligocene to early Miocene development of the east-northeast-trending Marsh Creek anticline. The northeast-trending, "Kemik duplex" thin-skinned structures that are mapped north of the Sadlerochit Mountains and in Ignek Valley vicinity may be older, but this is not certain (O'Sullivan and Wallace, 2002).

Based on regional analyses of apatite fission-track data, O'Sullivan et al. (1993) identified rapid cooling events at about 62, 45, and 23 Ma in the northeastern Brooks Range thrust belt. These dates presumably represent times of uplift that punctuated the Tertiary tectonic shortening. O'Sullivan and Wallace (2002) documented two stages of basement-involved uplift (45 and 27 Ma) at the northern front of the Sadlerochit Mountains and suggested that the thin-skinned Marsh Creek anticline developed during or shortly after the 27-Ma stage, compatible with the structural evolution presented in this chapter. The Paleocene development of the Aichilik high may be one expression of the 62-Ma deformation of O'Sullivan et al. (1993) (also expressed by deformation in the Bathtub Ridge area in the Brooks Range, 60 km [37 mi] to the south; Figure 5). Using seismic-stratigraphic correlations, we were unable to document structures that may correlate with the 45-Ma event, although an evolving thrust system existed in the eastern part of the 1002 area throughout the Eocene. The growth of Jago ridge was primarily an Oligocene event, as interpreted using seismic-stratigraphic and apatite fission-track techniques. It is possible that all of the young, basement-involved thrusts throughout the 1002 area evolved around 27–23 Ma, a conclusion that is consistent with both seismic reflection and fission-track interpretations.

Relation to Regional Tectonic Framework

Grantz et al. (1990) noted that compressional tectonics have affected the Beaufort continental margin in northeasternmost Alaska throughout Tertiary time and may remain active today. Offshore seismic data provide evidence for the involvement of Holocene sediments in thrusts and folds, and offshore seismicity may be recording continued activity on these structures (Grantz et al., 1990). Although thrust mechanisms have not been reported for these small earthquakes, published focal-plane mechanisms are mainly consistent with north–south compression (Fujita et al., 1990). The young offshore contractional structures are considered to root into the regional detachment of the northeastern Brooks Range orogen and to record the distal effects of the Pacific–North American plate convergence at the Aleutian subduction zone (Grantz et al., 1990). Moore et al. (1994a) suggested that the north-directed Neogene convergence of the Yakatat block against southern Alaska has had tectonic effects across the width of Alaska, including the Neogene shortening recorded both within and to the north of the northeastern Brooks Range. As noted by Wallace and Hanks (1990), the position of the northeastern Brooks Range due north of the anomalously high-standing Alaska Range and actively convergent Yakatat block of southern Alaska gives credence to the link between these youngest "Brookian" structures and plate convergence to the south.

The Niguanak and Aurora highs, unique structures on the North Slope as broad, thrust-driven uplifts of Franklinian rocks beneath the coastal plain, are present because of the northward propagation of a detachment in Franklinian rocks. Along the main Brooks Range front to the west, the basal detachment steps up out of the Franklinian basement much farther south. The anomalous, northerly position of major basement-involved thrusts is an additional manifestation of a unique, young component of thrusting in this region.

Hydrocarbon Generation, Fluid Flow, and Petroleum Occurrence

In a benchmark study of North Slope petroleum geochemistry, Seifert et al. (1980) determined that oil in the Prudhoe Bay field (Figure 1) has a composite source, mainly from the Triassic Shublik Formation, with components from the Kingak Shale and Hue Shale. This corresponds essentially to the Ellesmerian petroleum system of Magoon et al. (1987, 1999), which is also present in the 1002 area but is far less significant there than it is at Prudhoe Bay, as discussed below.

Analyses of seeps and oil-stained rocks in outcrops in the deformed part of the coastal plain and foothills, and hydrocarbon shows in nearby wells, demonstrate the presence of at least three petroleum systems: the Hue-Thomson and Canning-Sagavanirktok systems manifested in the 1002 area (Magoon et al., 1999) and the Ellesmerian petroleum system, which is manifested directly west of ANWR (Anders and Magoon, 1985; Lillis et al., 1999; Magoon et al., 1999). Lillis et al. (1999) determined that oil staining of Brookian sandstones along the Jago River (Figure 2) and in the Katakturuk River drainage 9 km (6 mi) north of the Sadlerochit Mountains has geochemical affinities to the Cretaceous Hue Shale, in the Hue-Thomson petroleum system. Oil shows in the Aurora 1 well on the Aurora dome in the northeastern part of the 1002 area have the geochemical characteristics of a Tertiary marine deltaic source (Lillis et al., 1999) in the Canning-Sagavanirktok petroleum system. Mixed geochemical affinities characterize oil staining of sandstones along the axis of the Marsh Creek anticline, with contributions from both Hue-Thomson and Canning-Sagavanirktok systems and possibly the Ellesmerian petroleum system. The Ellesmerian petroleum system may be present in the westernmost part of, and directly west of, the 1002 area (Lillis et al., 1999) and is apparently the source for the undeveloped Kavik and Kemik gas fields to the southwest (Figure 1) (Magoon et al., 1999) but is at best a minor contributor to hydrocarbons in the 1002 area.

Considering the great significance of the Ellesmerian petroleum system for the greater Prudhoe Bay complex, its relative lack of importance in the 1002 area (Anders et al., 1987; Lillis et al., 1999) is notable. The primary reason for this is the Early Cretaceous erosional truncation of Ellesmerian rocks beneath most of the 1002 area (Figure 7). Additional factors include the following: (1) Shublik and Kingak oils would have been generated and migrated through the region in the Paleocene and earliest Eocene (Hayba et al., 1999) during the onset of Tertiary thrusting, and any early structural traps that captured these oils were almost certainly broken during younger protracted Tertiary deformation; (2) the Kingak Shale is gas prone in the 1002 area, whereas it is oil prone beneath Prudhoe Bay (Magoon et al., 1987); and (3) the Shublik Formation near the 1002 area has a lower total organic content, relative to the Shublik at Prudhoe Bay (Magoon et al., 1987).

Basin modeling by Hayba et al. (1999) indicated that hydrocarbon generation from the Hue Shale began at about 50 Ma south of the 1002 area, and that oil migrated northward from the thrust belt toward the Mikkelsen high in the foreland basin prior to 24 Ma. After 24 Ma, fluid migration beneath the coastal plain was dominantly south directed, driven by compaction offshore. Thus, oils sourced in Tertiary marine deltaic rocks (Canning-Sagavanirktok system) north of the deformation front migrated south toward the Neogene frontal structures during the Neogene (Figure 14C). In contrast, the Hue oils appear to have migrated to the north during the Paleogene (Figure 14B), well before the youngest thrust structures formed, so early formed accumulations of Hue oil were likely disrupted by Neogene thrusting and associated fracturing.

Parris et al. (2003) used fluid-inclusion studies and zircon fission-track dating to conclude that dry gas was a mobile fluid phase in Triassic and Jurassic rocks in the southernmost part of the 1002 area (near the mountain front) at 64 ± 3 Ma, and that these rocks were uplifted from depths of 7–10 km (4–6 mi) at about that time. From this, they concluded that oil had been generated and migrated from the Triassic Shublik Formation and the Jurassic–Cretaceous Kingak Shale (Ellesmerian petroleum system of Magoon et al., 1999) prior to 64 Ma. Thus, Ellesmerian oil would have migrated through the region prior to formation of structural traps.

Parris et al. (2003) also found systematic differences in the salinity of fluid inclusions (from fracture-filling veins) at different Mesozoic stratigraphic levels, and they concluded that fracture pore-fluid compositions were stratigraphically zoned. If fluid flow was stratigraphically controlled on a regional scale, the fluid flow (within which dry gas was the only hydrocarbon phase) must have largely predated thrust faulting, because the thrust faulting thoroughly disrupted the regional stratigraphy, thus preventing regional flow in specific Mesozoic stratigraphic intervals.

Thermal maturity data and modeling for the 1002 area indicate that much of the Brookian section underlying the undeformed area of 1002 is thermally immature, with only the basal part of the Brookian section in the oil window (Houseknecht and Hayba, 1999). Oil generation from Brookian rocks occurred at about 40 Ma in the vicinity of the Marsh Creek anticline and at successively younger ages to the north (Houseknecht and Hayba, 1999). Within the deformed area, vitrinite reflectance values from Ellesmerian, Beaufortian, and Brookian surface samples increase markedly to the south, such that the zone of oil generation (0.6–1.3% R_o) defines a 5–10-km (3–6-mi)-wide band across the southern part of the 1002 area, and the southeastern part of the 1002 area and Leffingwell Ridge to the south are overmature (Bird et al., 1999). Vitrinite reflectance patterns strongly suggest that maximum heating and oil generation occurred in response to sedimentary burial prior to thrusting (Bird et al., 1999).

For the U.S. Geological Survey assessment of oil and gas resources in the 1002 area (ANWR Assessment Team, 1999), we assessed several structural plays, evaluating hydrocarbon charge and timing of deformation, along with detailed considerations of reservoir quality, seal risk, and numbers of prospects (Grow et al., 1999a, b, c; Perry et al., 1999; Potter et al., 1999). Although a south-to-north Paleogene to Neogene progression of thin-skinned thrusting has been demonstrated for much of the 1002 area, there is a strong Neogene deformational overprint on the entire area. Thus, any hydrocarbon accumulation that formed in Paleogene or earlier time, from Shublik-sourced and Hue-sourced oils, would likely have been disrupted during Neogene deformation. The highly disrupted nature of seismic reflections at all stratigraphic levels indicates that there may be considerable seal risk for hypothetical remigrated hydrocarbon accumulations residing in Neogene structural traps. The nearby presence of the Kavik gas field in Triassic rocks in an anticline above a Neogene thrust southwest of ANWR (Magoon et al., 1999, their plate PS4) demonstrates potential for trapping of northward-driven hydrocarbons in Neogene structures. The Ellesmerian rocks in the southern part of the 1002 area are overmature (Bird et al., 1999; Magoon et al., 1999), so any hydrocarbons preserved in similar structural traps in the 1002 area would be expected to be gas.

Oligocene to Miocene structures near the deformation front are favorably located for the trapping of the southward-driven Tertiary marine deltaic oils. The large basement-involved Niguanak and Aurora domes are attractive exploration targets based on their size and their favorable position and timing for trapping Tertiary hydrocarbons. However, there is a large risk associated with the potential for a viable seal in the highly disrupted, thin-skinned structures that overlie the large domes. In addition, the potential reservoirs in the areas of large structural closure in these domes are likely to be mildly metamorphosed pre-Mississippian rocks that probably have extremely low porosities (Grow et al., 1999b, c, d). The Niguanak and Aurora structures appear to have extensive areas of four-way closure above basement-involved thrust faults. However, these structures appear to be internally imbricated, which may have produced many smaller traps in these larger structures. The Aurora 1 well tested Jurassic to Tertiary rocks above the domed basement reflectors but did not reach pre-Mississippian rocks. It is not known what the KIC (Jago River) 1 well encountered, because the data are proprietary.

We determined that the reservoir quality of the thrust-related prospects in the 1002 area is generally poor. Seismic facies analysis, in conjunction with regional seismic-stratigraphic analysis, indicates that most of the thin-skinned structures involve shale-rich facies in Brookian turbidites, particularly in the northeastern part of the 1002 area. Although Ellesmerian rocks elsewhere on the North Slope make excellent reservoirs, Ellesmerian rocks exposed in the southern part of the 1002 area have poor reservoir quality; it is assumed that those involved in thrusting in the subsurface share these characteristics. In the area where Brookian rocks were deformed by thin-skinned thrusting, regional gravity and seismic interpretations imply the presence of large structural traps defined by large, composite, structural highs (Figure 3). However, detailed analyses of seismic and aeromagnetic data indicate that these structures are broken by intense faulting, and that large, effective, structural traps are probably uncommon. Apart from the large composite structures, individual closures may generally be too small to be of economic interest.

CONCLUSIONS

Comprehensive analysis of the northeastern Brooks Range frontal thrust structures reveals a complex array of thin-skinned and basement-involved structures concealed beneath the Arctic coastal plain and foothills (ANWR 1002 area), dominated by a passive-roof duplex structural style. In the western part of the 1002 area, the deformation front is defined by a simple Miocene passive-roof duplex (triangle zone), the tip of which is just north of the Marsh Creek anticline; earlier Tertiary thin-skinned deformation is difficult to resolve. Basement-involved thrusting in the western part also overprinted thin-skinned structures during the Miocene. In the eastern part, there was a regular south-to-north Paleocene to Miocene progression of the thin-skinned structures; the thin-skinned structures were deformed above prominent broad basement domes produced by Oligocene to Miocene basement-involved thrusts that propagated northward to a position near the thrust front. Offshore, contractional deformation may continue to the present.

Three petroleum systems were operative in and near the 1002 area. Hydrocarbons in the Ellesmerian system apparently were generated, expelled, and migrated prior to the Paleocene onset of thrusting in this region. Hydrocarbons sourced in the Lower Cretaceous Hue Shale were generated, expelled, and migrated during the Paleogene, when early thrust structures were forming in the 1002 area but well before the strong Neogene tectonic overprint that affected the region; Neogene faulting and fracturing probably compromised most preexisting oil accumulations. Tertiary marine (Canning-Sagavanirktok) oils were generated in the Neogene north of the thrust front. These Tertiary marine oils could have migrated south into Neogene frontal thrust structures, although the presence of suitable reservoirs and the quality of seals are questionable. In summary, the thrust-deformed area is less favorable for economic accumulations of hydrocarbons relative to the undeformed area in the northeastern part of the 1002 area because of less favorable reservoir characteristics, trapping mechanisms, thermal history, and timing of trap formation relative to hydrocarbon generation and migration. Any economic accumulation of hydrocarbons in the deformed area is likely to be gas instead of oil, because the principal source rocks are overmature.

REFERENCES CITED

Anders, D. E., and L. B. Magoon, 1985, Oil-source correlation study in northeastern Alaska: Advances in Organic Geochemistry, v. 10, p. 407–515.

Anders, D. E., L. B. Magoon, and S. C. Lubeck, 1987, Geochemistry of surface oil shows and potential source rocks, *in* K. J. Bird and L. B. Magoon, eds., Petroleum geology of the northern part of the Arctic National Wildlife Refuge, northeastern Alaska: U.S. Geological Survey Bulletin, v. 1778, p. 181–198.

ANWR Assessment Team, 1999, The oil and gas resource potential of the 1002 area, Arctic National Wildlife Refuge, Alaska: U.S. Geological Survey Open-file Report 98-34 (CD-ROM), 2 disks.

Bader, J. W., and K. J. Bird, 1986, Geologic map of the Demarcation point, Mt. Michelson, Flaxman Island, and Barter Island quadrangles, northeastern Alaska: U.S. Geological Survey Miscellaneous Investigations Series Map I-1791, scale 1:250,000, 1 sheet.

Banet, A. C., 1994, Geological and geochemical analysis of the Aurora well, offshore of the Arctic National Wildlife Refuge 1002 Area, *in* D. K. Thurston and K. Fujita, eds., 1992 Proceedings International Conference on Arctic Margins. Minerals Management Service, United States Department of the Interior, Outer Continental Shelf Report MMS 94-0040, p. 95–100.

Banks, C. J., and J. Warburton, 1986, "Passive-roof" duplex geometry in the frontal structures of the Kirthar and Sulaiman mountain belts, Pakistan: Journal of Structural Geology, v. 8, p. 229–237.

Bird, K. J., 1999, Geographic and geologic setting—Chapter GG, *in* ANWR Assessment Team, The oil and gas resource potential of the 1002 area, Arctic National Wildlife Refuge, Alaska: U.S. Geological Survey Open-file Report 98-34 (CD-ROM), p. GG-1–GG-51.

Bird, K. J., and L. B. Magoon, eds., 1987, Petroleum geology of the northern part of the Arctic National Wildlife Refuge, northeastern Alaska: U.S. Geological Survey Bulletin, v. 1778, 329 p., 5 plates.

Bird, K. J., R. C. Burruss, and M. J. Pawlewicz, 1999, Thermal maturity— Chapter VR, *in* ANWR Assessment Team, The oil and gas resource potential of the 1002 area, Arctic National Wildlife Refuge, Alaska: U.S. Geological Survey Open-file Report 98-34 (CD-ROM), p. VR-1–VR-64.

Brewer, M. C., 1987, Surficial geology, permafrost, and physical processes: *in* K. J. Bird and L. B. Magoon, eds., Petroleum geology of the northern part of the Arctic National Wildlife Refuge, northeastern Alaska: U.S. Geological Survey Bulletin, v. 1778, p. 27–36.

Bruns, T. R., M. A. Fisher, W. J. Leinbach Jr., and J. J. Miller, 1987, Regional structure of rocks beneath the coastal plain, *in* K. J. Bird and L. B. Magoon, eds., Petroleum geology of the northern part of the Arctic National Wildlife Refuge, northeastern Alaska: U.S. Geological Survey Bulletin, v. 1778, p. 249–254.

Buckingham, M. L., 1985, Stratigraphy, petrology, and depositional environments of the Upper Cretaceous to lower Tertiary Sabbath Creek section, Arctic National Wildlife Refuge (ANWR), Alaska: M.S. thesis, University of Alaska, Fairbanks, 120 p.

Cole, F., K. J. Bird, C. G. Mull, W. K. Wallace, W. Sassi, J. M. Murphy, and M. W. Lee, 1999, A balanced cross section and kinematic and thermal model across the northeastern Brooks Range mountain front, Arctic National Wildlife Refuge, Alaska— Chapter SM (structural modeling), *in* ANWR Assessment Team, The oil and gas resource potential of the 1002 area, Arctic National Wildlife Refuge, Alaska: U.S. Geological Survey Open-file Report 98-34 (CD-ROM), p. SM-1–SM-60.

Foland, R. L., and D. J. Lalla, 1987, Seismic reflection data acquisition, processing, and interpretation, *in* K. J. Bird and L. B. Magoon, eds., Petroleum geology of the northern part of the Arctic National Wildlife Refuge, northeastern Alaska: U.S. Geological Survey Bulletin, v. 1778, p. 235–244.

Fujita, K., D. B. Cook., H. Hasegawa, D. Forsyth, and R. Wetmiller, 1990, Seismicity and focal mechanisms of the Arctic region and the North American plate boundary in Asia, *in* A. Grantz, L. Johnson, and J. F. Sweeney, eds., The Arctic Ocean region: The geology of North America, v. L: Geological Society of America, p. 79–100.

Grantz, A., D. A. Dinter, and R. C. Culotta, 1987, Structure of the continental shelf north of the Arctic National Wildlife Refuge, *in* K. J. Bird and L. B. Magoon, eds., Petroleum geology of the northern part of the Arctic National Wildlife Refuge, northeastern Alaska: U.S. Geological Survey Bulletin, v. 1778, p. 271–276.

Grantz, A., S. D. May, P. T. Taylor, and L. A. Lawver, 1990, Canada Basin, *in* A. Grantz, L. Johnson, and J. F. Sweeney, eds., The Arctic Ocean region: The geology of North America, v. L: Geological Society of America, p. 379–402.

Grow, J. A., C. J. Potter, P. H. Nelson, and W. J. Perry, 1999a, Ellesmerian thrust-belt play— Chapter P9, *in* ANWR Assessment Team, The oil and gas resource potential of the 1002 area, Arctic National Wildlife Refuge, Alaska: U.S. Geological Survey Open-file Report 98-34 (CD-ROM), p. P-41–P-44.

Grow, J. A., C. J. Potter, P. H. Nelson, and W. J. Perry, 1999b, Niguanak–Aurora play— Chapter P10, *in* ANWR Assessment Team, The oil and gas resource potential of the 1002 area, Arctic National Wildlife Refuge, Alaska: U.S. Geological Survey Open-file Report 98-34 (CD-ROM), p. P-45–P-50.

Grow, J. A., C. J. Potter, P. H. Nelson, W. J. Perry, and J. S. Kelley, 1999c, Deformed Franklinian play— Chapter P7, *in* ANWR Assessment Team, The oil and gas resource potential of the 1002 area, Arctic National Wildlife Refuge, Alaska: U.S. Geological Survey Open-file Report 98-34 (CD-ROM), p. P-29–P-33.

Grow, J. A., C. J. Potter, W. J. Perry, M. W. Lee, and M. Killgore, 1999d, Structural framework and lithologic composition of the Niguanak high and the Aurora dome— Chapter NA (Niguanak and Aurora structures), *in* ANWR Assessment Team, The oil and gas resource potential of the 1002 area, Arctic National Wildlife Refuge, Alaska: U.S. Geological Survey Open-file Report 98-34 (CD-ROM), p. NA-1–NA-46.

Hanks, C. L., 1987, Preliminary geologic map of central Leffingwell Ridge, Arctic National Wildlife Refuge, northeastern Brooks Range, Alaska: Alaska Division of Geological and Geophysical Surveys Public-data File 86-86i, 12 p., 2 plates.

Hanks, C. L., 1990, Balanced cross sections of the Aichilik River and Okpilak batholith regions, northeastern Brooks Range, Alaska: Alaska Division of Geological and Geophysical Surveys Public-data File 90-2a, 12 p., 3 figures, 2 plates.

Hanks, C. L., 1993, The Cenozoic structural evolution of a fold and thrust belt, northeastern Brooks Range, Alaska: Geological Society of America Bulletin, v. 105, no. 3, p. 287–305.

Hanks, C. L., and W. K. Wallace, 1990, Cenozoic thrust emplacement of a Devonian batholith, northeastern Brooks Range— Involvement of crystalline rocks in a foreland fold and thrust belt: Geology, v. 18, p. 395–398.

Hayba, D. O., D. W. Houseknecht, and E. Rowan, 1999, Stratigraphic, hydrogeologic, and thermal evolution of the Canning River region, North Slope, Alaska— Chapter FF (fluid flow modeling), *in* ANWR Assessment Team, The oil and gas resource potential of the 1002 area, Arctic National Wildlife Refuge, Alaska: U.S. Geological Survey Open-file Report 98-34 (CD-ROM), p. FF-1–FF-80.

Houseknecht, D. W., and D. O. Hayba, 1999, Modeling oil generation in the undeformed part of the Arctic National Wildlife Refuge 1002 area— Chapter HG (hydrocarbon potential of Brookian strata), *in* ANWR Assessment Team, The oil and gas resource potential of the 1002 area, Arctic National Wildlife Refuge, Alaska: U.S. Geological Survey Open-file Report 98-34 (CD-ROM), p. HG-1–HG-45.

Houseknecht, D. W., and C. J. Schenk, 1999, Seismic facies analysis and hydrocarbon potential of Brookian strata— Chapter BS (Brookian sequences), *in* ANWR Assessment Team, The oil and gas resource potential of the 1002 area, Arctic National Wildlife Refuge, Alaska: U.S. Geological Survey Open-file Report 98-34 (CD–ROM), p. BS-1–BS-60.

Hubbard, R. J., S. P. Edrich, and R. P. Rattey, 1987, Geologic evolution and hydrocarbon habitat of the "Arctic Alaska Microplate" *in* I. Tailleur and P. Weimer, eds, Alaskan North Slope geology, v. 2: Bakersfield, California and Anchorage, Alaska, Pacific Section, SEPM and Alaska Geological Society, p. 797–830.

Jones, P. B., 1982, Oil and gas beneath east-dipping underthrust faults in the Alberta foothills, *in* R. B. Powers, ed., Geologic studies of the Cordilleran thrust belt— 1982: Rocky Mountain Association of Geologists, v. 1, p. 61–74.

Kelley, J. S., 1999, Pre-Mississippian rocks beneath the Arctic coastal plain of the Arctic National Wildlife Refuge, Alaska— Chapter BR (basement rocks) *in* ANWR Assessment Team, The oil and gas resource potential of the 1002 area, Arctic National Wildlife Refuge, Alaska: U.S. Geological Survey Open-file Report 98-34 (CD-ROM), p. BR-1–BR-28.

Kelley, J. S., and R. L. Foland, 1987, Structural geology and framework geology of the coastal plain and adjacent Brooks Range, *in* K. J. Bird and L. B. Magoon, eds., Petroleum geology of the northern part of the Arctic National Wildlife Refuge, northeastern Alaska: U.S. Geological Survey Bulletin, v. 1778, p. 255–270.

Lee, M. W., W. F. Agena, J. A. Grow, and J. J. Miller, 1999, Seismic processing and velocity assessments— Chapter SP (seismic processing), *in* ANWR Assessment Team, The oil and gas resource potential of the 1002 area, Arctic National Wildlife Refuge, Alaska: U.S. Geological Survey Open-file Report 98-34 (CD-ROM), p. SP-1–SP-32.

Lillis, P. G., M. D. Lewan, A. Warden, S. M. Monk, and J. D. King, 1999, Identification and characterization of oil types and their source rocks— Chapter OA (oil analyses), *in* ANWR Assessment Team, The oil and gas resource potential of the 1002 area, Arctic National Wildlife Refuge, Alaska: U.S. Geological Survey Open-file Report 98-34 (CD-ROM), p. OA-1–OA-101.

Magoon, L. B., P. V. Woodward, A. C. Banet Jr., S. B. Griscom, and T. A. Daws, 1987, Thermal maturity, richness, and type of organic matter of source-rock units, *in* K. J. Bird and L. B. Magoon, eds., Petroleum geology of the northern part of the Arctic National Wildlife Refuge, northeastern Alaska: U.S. Geological Survey Bulletin, v. 1778, p. 127–180.

Magoon, L. B., K. J. Bird, R. C. Burruss, D. Hayba, D. W. Houseknecht, M. A. Keller, P. G. Lillis, and E. L. Rowan, 1999, Evaluation of hydrocarbon charge and timing using the petroleum system— Chapter PS (petroleum systems), *in* ANWR Assessment Team, The oil and gas resource potential of the 1002 area, Arctic National Wildlife Refuge, Alaska: U.S. Geological Survey Open-file Report 98-34 (CD-ROM), p. PS-1–PS-66.

Meigs, A. J., 1989, Structural geometry and sequence in the eastern Sadlerochit Mountains, northeastern Brooks Range, Alaska: M.S. thesis, University of Alaska, Fairbanks, 220 p.

Molenaar, C. M., 1987, Introduction, *in* K. J. Bird and L. B. Magoon, eds., Petroleum geology of the northern part of the Arctic National Wildlife Refuge, northeastern Alaska: U.S. Geological Survey Bulletin, v. 1778, p. 3–6.

Moore, T. E., 1999, Balanced cross section, Bathtub syncline to Beaufort Sea through Niguanak structural high, Arctic National Wildlife Refuge (ANWR), northeastern Alaska— Chapter BC (balanced cross section), *in* ANWR Assessment Team, The oil and gas resource potential of the 1002 area, Arctic National Wildlife Refuge, Alaska: U.S. Geological Survey Open-file Report 98-34 (CD-ROM), p. BC-1–BC-61.

Moore, T. E., A. Grantz, and S. M. Roeske, 1994a, Continent-ocean transition in Alaska: The tectonic assembly of eastern Denalia, *in* R. C. Speed, ed., Phanerozoic evolution of North American continent–ocean transitions: Geological Society of America, p. 399–441.

Moore, T. E., W. K. Wallace, K. J. Bird, S. M. Karl, C. G. Mull, and J. T. Dillon, 1994b, Geology of northern Alaska, *in* G. Plafker and H. C. Berg, eds., The geology of Alaska: Geological Society of America, p. 49–140.

Moore, T. E., C. J. Potter, P. B. O'Sullivan, K. L. Shelton, and M. B. Underwood, 2004, Two stages of deformation and fluid generation for the central Brooks Range fold and thrust belt, northern Alaska, *in* R. Swennen, F. Roure, and J. W. Granath, eds., Deformation, fluid flow, and reservoir appraisal in foreland fold and thrust belts: AAPG Hedberg Series, no. 1, p. 157–186.

O'Sullivan, P. B., and W. K. Wallace, 2002, Out-of-sequence, basement-involved structures in the Sadlerochit Mountains region of the Arctic National Wildlife Refuge, Alaska— Evidence and implications from fission-track thermochronology: Geological Society of America Bulletin, v. 114, p. 1356–1378.

O'Sullivan, P. B., P. F. Green, S. C. Bergman, J. Decker, I. R. Ruddy, A. J. W. Gleadow, and D. L. Turner, 1993, Multiple phases of Tertiary uplift and erosion in the Arctic National Wildlife Refuge, Alaska, revealed by apatite fission-track analysis: AAPG Bulletin, v. 77, p. 359–385.

Parris, T. M., R. C. Burruss, and P. B. O'Sullivan, 2003, Deformation and the timing of gas generation and migration in the eastern Brooks Range foothills, Arctic National Wildlife Refuge, Alaska: AAPG Bulletin, v. 87, p. 1823–1846.

Paul, L. E., D. R. Choromanski, R. F. Turner, and T. O. Flett, 1994, Geological, geochemical, and operational summary, Aurora Well, OCS Y-0943-1, Beaufort Sea, Alaska: Outer Continental Shelf Report, Minerals Management Service 94-0001, 71 p., 1 plate.

Perry, W. J., C. J. Potter, and P. H. Nelson, 1999, Thin-skinned thrust-belt play— Chapter P8, *in* ANWR Assessment Team, The oil and gas resource potential of the 1002 area, Arctic National Wildlife Refuge, Alaska: U.S. Geological Survey Open-file Report 98-34 (CD-ROM), p. P-34–P-40.

Phillips, J. D., 1999, An interpretation of proprietary aeromagnetic data over the northern Arctic National Wildlife Refuge and adjacent areas, northeastern Alaska— Chapter AM (aeromagnetic modeling), *in* ANWR Assessment Team, The oil and gas resource potential of the 1002 area, Arctic National Wildlife Refuge, Alaska: U.S. Geological Survey Open-file Report 98-34 (CD-ROM), p. AM-1–AM-19.

Phillips, J. D., R. W. Saltus, and R. L. Reynolds, 1998, Sources of magnetic anomalies over a sedimentary basin— Preliminary results from the coastal plain of the Arctic National Wildlife Refuge, Alaska, *in* R. I. Gibson and P. S. Milligan, eds., Geologic applications of gravity and magnetics— Case histories: Society of Exploration Geophysicists and American Association of Petroleum Geologists, p. 130–134.

Poag, C. W., 1999, Biostratigraphic framework— Chapter BI, *in* ANWR Assessment Team, The oil and gas resource potential of the 1002 area, Arctic National Wildlife Refuge, Alaska: U.S. Geological Survey Open-file Report 98-34 (CD-ROM), p. BI-1–BI-9.

Potter, C. J., W. J. Perry, J. A. Grow, and M. W. Lee, 1999, Style and timing of thrust-faulting in the 1002 Area, Arctic National Wildlife Refuge— Chapter BD (Brookian deformation), *in* ANWR Assessment Team, The oil and gas resource potential of the 1002 area, Arctic National Wildlife Refuge, Alaska: U.S. Geological Survey Open-file Report 98-34 (CD-ROM), p. BD-1–BD-32.

Robinson, M. S., J. Decker, J. G. Clough, R. R. Reifenstuhl, A. Bakke, J. T. Dillon, R. A. Combellick, and S. A. Rawlinson, 1989, Geology of the Sadlerochit and Shublik mountains, Arctic National Wildlife Refuge, northeastern Alaska: Alaska Division of Geological and Geophysical Surveys Professional Report 100, scale 1:63,360, 1 sheet.

Saltus, R. W., R. L. Morin, J. D. Phillips, and M. W. Webring, 1999, Gravity data processing, modeling, and interpretation in support of structural studies in the northern part of the Arctic National Wildlife Refuge, Alaska— Chapter GR (gravity modeling), *in* ANWR Assessment Team, The oil and gas resource potential of the 1002 area, Arctic National Wildlife Refuge, Alaska: U.S. Geological Survey Open-file Report 98-34 (CD-ROM), p. GR-1–GR-36.

Seifert, W. K., J. M. Moldowan, and J. W. Jones, 1980, Application of biological marker chemistry to petroleum exploration, *in* Proceedings of the 10th World Petroleum Congress, Bucharest, Romania, 1979: London, Heyden and Son, Ltd., p. 425–438.

Wallace, W. K., 1993, Detachment folds and a passive-roof duplex— Examples from the northeastern Brooks Range, Alaska, *in* D. N. Solie and F. Tannian, eds., Short notes on Alaskan geology 1993: Alaska Division of Geological and Geophysical Surveys Geologic Report 113, p. 81–99.

Wallace, W. K., and C. L. Hanks, 1990, Structural provinces of the northeastern Alaska Brooks Range, Arctic National Wildlife Refuge, Alaska: AAPG Bulletin, v. 74, p. 1100–1118.

10

Travé, A., F. Calvet, R. Salas, E. Playà, and E. Roca, 2004, Fluid flow during Paleogene compression in the Linking Zone fold and thrust belt (northeast Spain), *in* R. Swennen, F. Roure, and J. W. Granath, eds., Deformation, fluid flow, and reservoir appraisal in foreland fold and thrust belts: AAPG Hedberg Series, no. 1, p. 215–243.

Fluid Flow during Paleogene Compression in the Linking Zone Fold and Thrust Belt (Northeast Spain)

Anna Travé, Francesc Calvet,* Ramon Salas, Elisabet Playà

Departament de Geoquímica, Petrologia i Prospecció Geològica, Facultat de Geologia, Universitat de Barcelona, Barcelona, Spain

Eduard Roca

Departament de Geodinàmica i Geofísica, Facultat de Geologia, Universitat de Barcelona, Barcelona, Spain

ABSTRACT

The Linking Zone (northeast Spain) is a fold and thrust belt that, together with the Catalan Coastal Range and the Iberian Chain, constitutes the southern flank of the Ebro Basin. Compressive deformation occurred during the late Eocene–early Miocene, affecting a Hercynian basement, a Mesozoic cover, and syntectonic Tertiary conglomerates.

The integration of the structural study of the fractures with the petrologic and geochemical study of the cements filling the fractures reveals two episodes of fluid circulation during compressive deformation. The first episode occurred during the early stages of the fold and thrust belt development and includes two different fluids responsible for precipitation of two types of calcite cement. The first type of calcite cement is characterized by $\delta^{18}O$ ranging from −9 to −3.8‰ Peedee belemnite (PDB), $\delta^{13}C$ ranging from −5.6 to −2.6‰ PDB, a $^{87}Sr/^{86}Sr$ ratio of 0.70768, between 440 and 3565 ppm of Mg, between 275 and 720 ppm of Fe, between 180 and 410 ppm of Mn, and Sr content always below 275 ppm (below the detection limit). The fluids precipitating these cements were meteoric fluids that were derived directly from the surface and circulating through the undeformed and highly porous Tertiary syntectonic conglomerates in an open paleohydrogeological system. The second type of calcite cement is characterized by $\delta^{18}O$ ranging from −10.1 to −5.4‰ PDB, $\delta^{13}C$ ranging from −7.5 to −3.5‰ PDB, a $^{87}Sr/^{86}Sr$ ratio ranging between 0.70759 and 0.70778, between 350 and 5150 ppm of Mg, Fe content as much as 11,490 ppm, Mn content below 180 ppm (below

*Deceased, January 2002.

DOI:10.1306/1025692H13118

the detection limit), and Sr content as much as 795 ppm. The fluid that was precipitating these cements was originally a meteoric fluid that evolved to a formation water composition probably because of a high interaction with Mesozoic rocks. The Triassic shales and evaporites of the detachment levels served as barriers, forcing the fluids to move laterally above them.

The second episode of fluid circulation occurred during the last stages of fold and thrust belt development, after the major uplift of the interior belt and formation of the relief. Two different fluids are recognized during this episode as being responsible for precipitation of two types of calcite cements. The first type of calcite cement is characterized by $\delta^{18}O$ ranging from −3.4 to −3.3‰ PDB, $\delta^{13}C$ ranging from −4.0 to −3.8‰ PDB, $^{87}Sr/^{86}Sr$ ratio of 0.70845, between 475 and 4200 ppm of Mg, Fe always below 275 ppm (below the detection limit), between 180 and 655 ppm of Mn, and between 280 and 850 ppm of Sr. The second type of calcite cement is characterized by $\delta^{18}O$ ranging from −12.7 to −8.1‰ PDB, $\delta^{13}C$ ranging from −6.4 to −3.9‰ PDB, a $^{87}Sr/^{86}Sr$ ratio ranging between 0.70792 and 0.70795, between 260 and 5240 ppm of Mg, Fe content as much as 1155 ppm, Mn content as much as 925 ppm, and Sr content as much as 925 ppm. These two groups of cements that precipitated from meteoric waters evolved to a formation water composition probably because of interaction with Mesozoic and Paleozoic rocks, being the low $\delta^{18}O$ of the second probably because of high temperatures of precipitation.

A late episode of meteoric fluid circulation, occurring during the Tertiary extension, is recorded by the calcite cements filling normal faults, which are characterized by $\delta^{18}O$ ranging from −9.2 to −7.6‰ PDB, $\delta^{13}C$ ranging from −2.6 to +1.1‰ PDB, $^{87}Sr/^{86}Sr$ ratio of 0.70788, between 870 and 2695 ppm of Mg, between 470 and 1420 ppm of Mn, and Fe and Sr always below the detection limit, which is 275 ppm for both.

The study of the fluid-flow evolution during the Linking Zone fold and thrust belt development shows that the main factors controlling the fluid-flow dynamics are the existence of a high relief, whereas the host rocks and the type of structure are not important factors.

INTRODUCTION

The reconstruction of the relationship between fractures and migration of fluids in fold and thrust belts and their forelands has economic motives, because commonly, they are sites of hydrocarbon accumulations and ore deposits formation (Bethke and Marshak, 1990; Qing and Mountjoy, 1992; Garden et al., 2001). The aim of this reconstruction is to determine the type and origin of the fluids, the fluid pathways, the intensity of fluid-rock interaction, and the paleohydrogeological regime during thrusting along the different zones of the fold and thrust belt. The study of the evolution of the fluid flow during deformation involves the structural and microstructural analysis of the fractures and the petrology and geochemistry analyses of the cements that fill them (e.g., Budai, 1985; Burrus et al., 1985; Marquer and Burkhard, 1992; Cosgrove, 1993; Muchez et al., 1995; Calvet et al., 1996; Travé et al., 1997, 1998). Several studies exist related to fold and thrust belts, such as those in the Albanides (e.g., Swennen et al., 2000; Van Geet et al., 2002), the Andes (e.g., Labaume and Moretti, 2001; Labaume et al., 2001), the Appalachians (e.g., Evans and Battles, 1999), the Canadian Rockies (e.g., Machel et al., 2000), the Helvetic nappes (e.g., Burkhard and Kerrich, 1988), and the Pyrenees (e.g., McCaig et al., 1995; Travé et al., 1997, 1998, 2000).

The study of fluid-flow evolution during fold and thrust belt development is complex because normally, the rocks record the entire history of the fluids, not only during fold and thrust belt development, but also prior to it and subsequent to it. Therefore, the first problem is to discern which fluid events occurred during fold and thrust belt development. Second, the spatial distribution of the belt commonly implies a great variability of lithologies and the coeval presence of fluids from different origins, either in different parts of the belt or mixed in a specific area. Furthermore, the temporal evolution of the belt commonly implies successive fluid events in the same area. This complexity is reflected in most fold and thrust belts by a complex network of fractures filled by one or several generation of cements that vary not only in the different outcrops but also in the different fractures in the same outcrop.

To learn more about the fluid-flow regime in fold and thrust belts, some of these problems need to be minimized. To that end, the present study focuses on a fold and thrust belt (Linking Zone) that has the peculiarities of (1) not being, or being only slightly, deformed after fold and thrust belt formation; (2) having developed always in subaerial conditions; (3) having a rather homogeneous preorogenic substratum in relation to its lithology (Mesozoic carbonates and Paleozoic siliciclastics), uniform depositional environment (deposited in marine platforms), as well as a similar tectonic regime that affected it prior to the fold and thrust belt development (extension); and (4) having an extensive record of synorogenic rocks that allows us to discern, by comparison with the preorogenic rocks, which fractures and cements are synorogenic and which are preorogenic.

GEOLOGICAL SETTING

The Linking Zone (Guimerà, 1984) is a belt of folded and thrusted Mesozoic cover and Hercynian basement that, together with the Catalan Coastal Range and the Iberian Chain, constitutes the southern flank of the Ebro Basin (Figure 1). It is made up of a complex system of folds and north- to northwest-directed thrusts (Figure 2A), which, from a structural point of view, can be divided into three longitudinal belts: interior, central, and frontal (Figures 2B, 3) (Lawton et al., 1999).

The interior belt is the southernmost one and comprises the higher parts. It consists of an antiformal culmination of thick Mesozoic strata cored by Hercynian metamorphic rocks that thrusts north to northwestward along a major thrust.

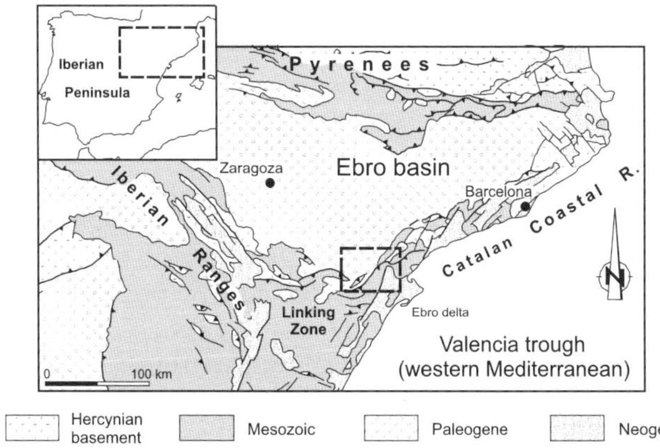

FIGURE 1. A simplified geological map of the northwestern Iberian Peninsula. The location of the studied area in the Linking Zone is indicated by the inset rectangle (enlarged in Figure 2).

The central belt consists of a complex system of folds cut by imbricate thrust sheets of Mesozoic strata detached in Triassic shale and evaporite. The thrust plates are dominated by Upper Triassic to Upper Cretaceous limestone and dolomite that thin steadily toward the north-northwest.

Finally, the frontal belt develops over a thin Mesozoic cover and mainly consists of northeast-trending folds also detached in the Triassic shale and evaporite. These major folds involve not only the Mesozoic cover but also most of the Tertiary sequences deposited along the southern margin of the Ebro foreland basin. The Tertiary sequences are made up of alluvial sediments, mainly conglomerates, and were deposited synchronously to the development of the belt, as the presence of internal angular and progressive unconformities attests. From the syntectonic feature of these deposits, compressive deformation of the Linking Zone has been dated as late Eocene–early Miocene (Villena et al., 1995).

The structure of the Linking Zone records the tectonic inversion of the northern margin of the Mesozoic Maestrat basin, which was filled by uppermost Permian to Lower Cretaceous strata (Salas and Casas, 1993). This basin was affected by two rifting stages (Salas et al., 2001): the first occurred in the latest Permian–earliest Jurassic and created a system of high-angle faults that deformed the Hercynian basement and controlled the sedimentation of the Lower–Middle Triassic and lowermost Jurassic rocks; the second occurred in the late Jurassic–early Cretaceous and created an extensive system of listric faults detached at the Triassic shale and evaporite units or at deeper Paleozoic horizons (Roca et al., 1994).

METHOD AND ANALYTICAL TECHNIQUES

Four structural cross sections of the Linking Zone were studied (from northeast to southwest): the Gandesa section (AA′ in Figures 2B, 3); the Orta de Sant Joan section (BB′ in Figures 2B, 3); the Beseit section (CC′ in Figures 2B, 3); and the Herbers section (DD′ in Figures 2B, 3). These sections comprise several outcrops that are located in the three differentiated belts of the Linking Zone and are made up of Jurassic to Cretaceous limestones, dolostones, or syntectonic Tertiary conglomerates. The outcrops are located on major thrust surfaces or in areas where minor-scale, reverse, normal, or strike-slip faults and bed-to-bed slidings are present. To characterize the evolution of the fluid flow in these outcrops, we combined a structural analysis with a petrological and mineralogical study of standard and polished thin sections of approximately 200 samples that were collected both in the veins-filling fractures and in the host rocks.

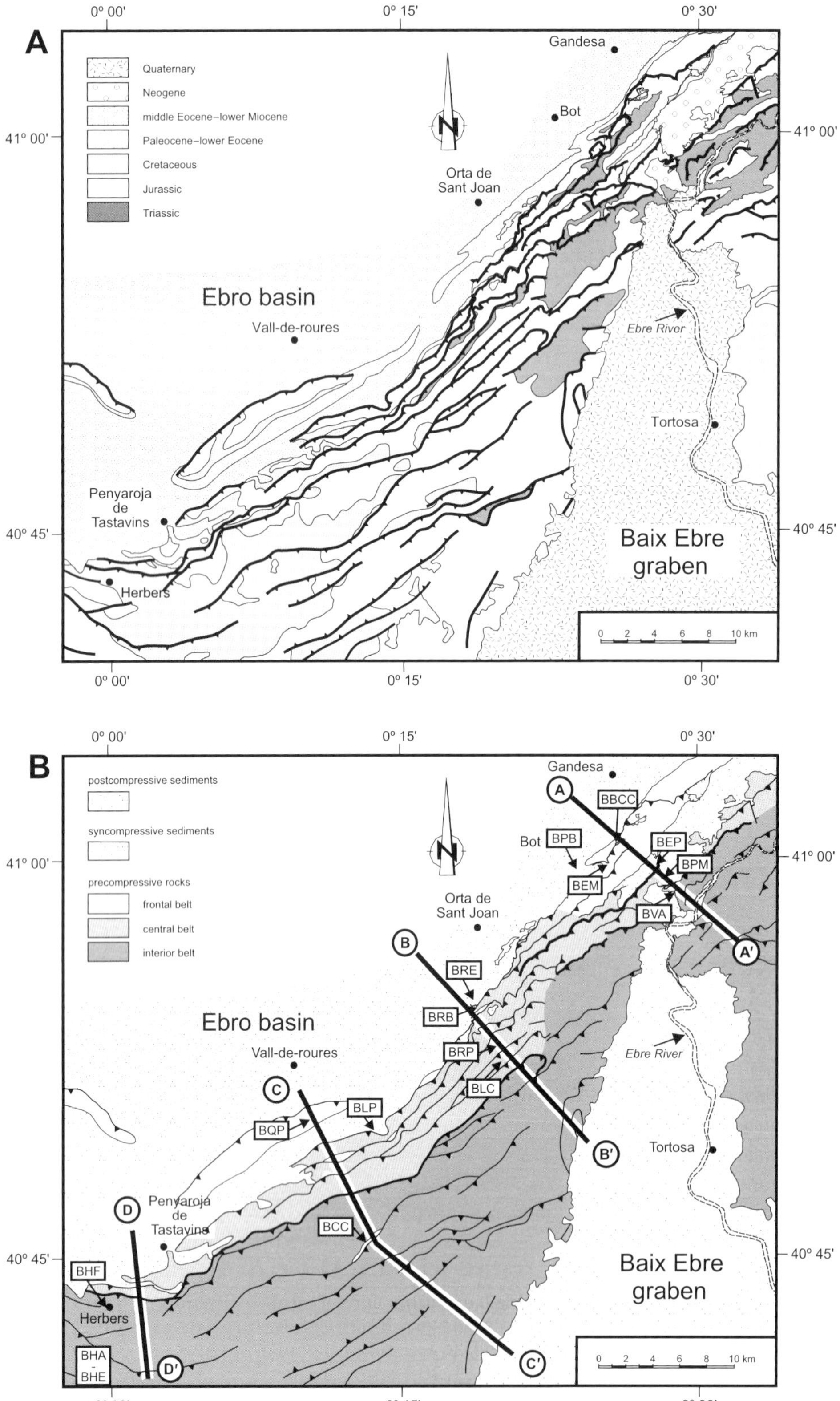

FIGURE 2. (A) The geological map of the studied area. (B) The structural map of the studied area in which the interior, central, and frontal belts of the Linking Zone are differentiated. Thick lines show the location of the four studied structural cross sections given in Figure 3: AA′ = Gandesa section; BB′ = Orta de Sant Joan section; CC′ = Beseit section; and DD′ = Herbers section.

The geochemical analysis of the calcite cements (elemental geochemistry, stable isotopes, and $^{87}Sr/^{86}Sr$ ratios) allowed us to characterize the different fluids. The comparison of the calcite cement geochemistry with the Mesozoic host carbonate geochemistry gave us clues regarding rock-fluid interaction. The host conglomerates were not analyzed because of the different nature of the carbonate clasts. C-metallized polished thin sections were analyzed for Ca, Mg, Fe, Mn, Sr, and Na content, using a Cameca model SX-50 microprobe equipped with four vertically displayed wavelength-dispersive x-ray spectrometers. The microprobe was operated using 20 kV excitation potential, 15 nA current intensity, and a 10-μm beam diameter. The detection limits were 395 ppm for Mg, 275 ppm for Fe, 180 ppm for Mn, 275 ppm for Sr, and 195 ppm for Na. Precision on major-element analyses averaged 6.32% standard error at 3σ confidence level. The Mg/Ca, Mn/Ca, Sr/Ca, and Ca/Fe molar ratios of the fluid responsible for calcite cement precipitation were calculated, applying the distribution coefficient equation of McIntire (1963).

Following the petrographic study, 120 microsamples (39 from Gandesa, 24 from Orta de Sant Joan, 7 from Beseit, and 50 from Herbers section) were carefully prepared to determine the stable isotope composition. For carbon and oxygen stable isotope analyses, CO_2 was extracted from 10 mg

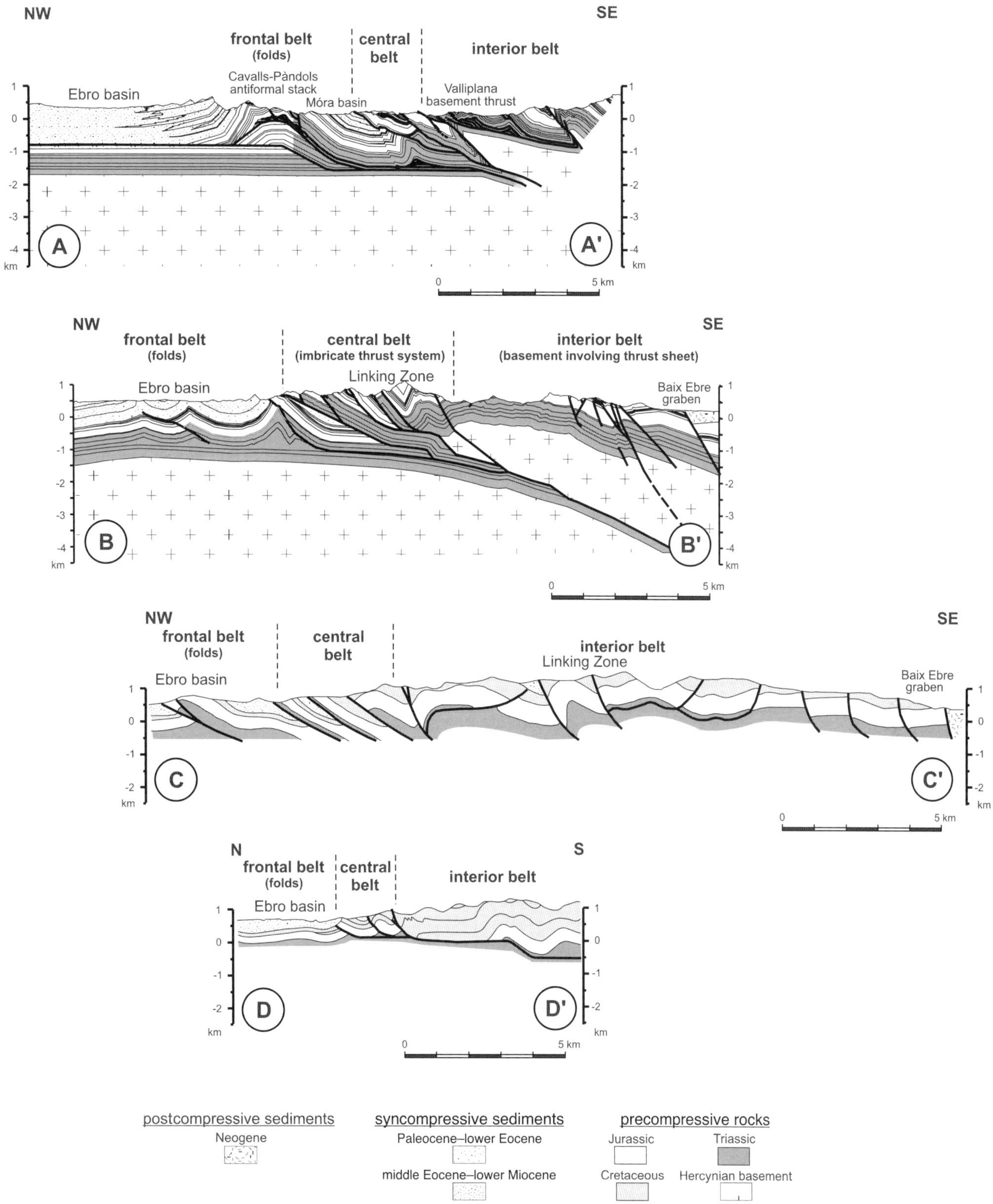

FIGURE 3. Studied geological cross sections of the eastern part of the Linking Zone: AA′ = Gandesa section; BB′ = Orta de Sant Joan section (from Lawton et al., 1999); CC′ = Beseit section (modified from Salas et al., 1997); and DD′ = Herbers section (from J. Guimerà, 1996, personal communication).

Table 1. $^{87}Sr/^{86}Sr$ of all the host rocks and calcite cements.

Sample	*Section*	*Description*	$^{87}Sr/^{86}Sr$	*Instrumental Error*
BVA 2B	Gandesa	calcite cement	0.70772	6
BVA 4A	Gandesa	host dolomite	0.70789	6
BVA 4B	Gandesa	calcite cement	0.70759	6
BPM 2A	Gandesa	host limestone	0.70743	5
BPM 2B	Gandesa	calcite cement	0.70778	5
BEP 3	Gandesa	calcite cement	0.70795	6
BEM 6A	Gandesa	host dolomite	0.70806	6
BEM 6B	Gandesa	calcite cement	0.70845	4
BEM 7B	Gandesa	calcite cement	0.70772	6
BLC 3A	Orta	host limestone	0.70744	6
BLC 3B	Orta	calcite cement	0.70788	5
BRB 2A	Orta	host dolomite	0.70788	6
BRB 2B	Orta	calcite cement	0.70773	5
BRE 2	Orta	calcite cement	0.70768	6
BHA 5A	Herbers	host limestone	0.70750	5
BHA 5B	Herbers	calcite cement	0.70766	6
BHD 1C	Herbers	calcite cement	0.70768	6
BHF 6A	Herbers	host limestone	0.70751	5
BHF 6B	Herbers	calcite cement	0.70792	5

of powdered carbonate sample that was reacted with 103% phosphoric acid for 60 min in a vacuum at 25°C. The CO_2 was analyzed under a Finnigan MAT delta S mass spectrometer. The results were precise to ±0.02‰ for $\delta^{13}C$ and ±0.12‰ for $\delta^{18}O$. The results were corrected using standard procedures (Craig and Gordon, 1965) and are expressed in per mil with respect to the Peedee belemnite (PDB) standard. For the $^{87}Sr/^{86}Sr$ analyses, 19 powdered microsamples (9 from Gandesa, 5 from Orta de Sant Joan, and 5 from Herbers section) of 100% calcite cement from the veins and 100% calcite or dolomite from the host rocks (average 10^{-5} g) were fully dissolved in HCl. During the analysis of these microsamples on a VG Sector 54 mass spectrometer, the mean value of the measured NBS-987 (n = 10) was $^{87}Sr/^{86}Sr = 0.710258 \pm 0.00002$, which is within the error range of the laboratory ($^{87}Sr/^{86}Sr = 0.710260 \pm 0.00002$ [2σ, n = 179]). The error noted in Table 1 is instrumental, estimating a total error of 0.01% in the $^{87}Sr/^{86}Sr$ ratio.

THE GANDESA SECTION

Structural Description

The Gandesa structural section (AA′ in Figures 2B, 3) crosses the three structural belts of the Linking Zone and includes large outcrops of synorogenic conglomerates located both in the central and frontal belts, which allows us to recognize the pattern of the fluid regime during the building of these two belts by comparing the calcite cements that developed in the precompressive and syncompressive rocks. This is the only section in which the calcite cements filling the major thrust (Valliplana thrust) separating the interior and central belts were studied. Six outcrops were analyzed (Benifallet railway [BVA], Mas de Riberola [BPM], Pinell railway station [BEP], Estudi mossen [BEM], Coll d'en Canar [BBCC], and Carretera Prat-Bot [BPB] in Figure 2B and in Table 2).

The syncompressive fluid-flow regime in the central belt was studied in BPM and BEP outcrops, which are located near the frontal thrust of the imbricated thrust system: BPM in the Lower Jurassic limestones outcropping in their hanging wall, and BEP in the overthrusted upper Eocene conglomerates, which were deposited during the emplacement of this thrust. In the two outcrops, the calcite cements form veins located in strike-slip fault planes. In the frontal belt, the synorogenic fluid-flow regime was inferred from the analysis of BEM, BBCC, and BPB outcrops. BEM and BBCC outcrops are made up of Lower Jurassic limestones and dolomitic breccias outcropping in the core of the frontal fold, whereas the BPB outcrop corresponds to the Tertiary conglomerates cropping out in the hinge of this fold. In these host rocks, the studied calcite cements are located in (1) the fracture plane of gash fractures (BEM), (2) minor thrusts with hectometric slip (BEM and BBCC), (3) minor normal faults (BEM and BPB), and (4) bed-to-bed slidings (BPB) formed during the flexural-slip folding of the frontal folds.

Finally, calcite cements filling the major thrust plane separating the interior and central belts were studied in the southernmost outcrop (BVA) of the section. In this outcrop, overturned Triassic limestones of the interior belt overthrust Lower Jurassic dolomitic breccia with a stratified, nondolomitized limestone interval in the uppermost part of the central belt.

Petrology of the Calcite Cements

Significant petrological differences occur between the calcite cements filling the fracture planes that developed in the Tertiary conglomerates and those that developed in the Mesozoic cover. The fracture planes that developed in the Tertiary conglomerates (BEP and BPB outcrops) are filled by euhedral to subhedral calcite crystals from 0.5 mm to 1 cm (0.02 to 0.4 in.) in size, with abundant, nondeformed, mechanical twin planes and uniform extinction (Figure 4a). Fracture porosity is not totally occluded by this generation of cement.

In contrast, the fracture planes developed in the Lower Jurassic rocks (BVA, BPM, BBCC, and BEM outcrops) are filled by euhedral to anhedral calcite crystals from 100 μm to 1 cm in size, and the intercrystalline boundaries are affected by tectonic stylolite surfaces (Figure 4b). In this group of cements, the cements filling thrust planes in the frontal fold (BEM and BBCC outcrops) are generally smaller in size (from 100 μm to 3 mm), showing subgrain formation (Figure 4c); abundant, slightly deformed mechanical twin planes (Figure 4d); and subangular extinction.

Stable Isotopes

The host rock shows a wide range of δ^{18}O and δ^{13}C values (Figure 5), depending on the outcrop, with δ^{18}O values ranging from −7.2 to −1.3‰ and δ^{13}C values ranging from −8.1 to +2.9‰ PDB. The δ^{18}O value of the host Lower Jurassic limestone (BVA and BPM outcrops) range from −3.7 to −2.2‰ PDB, and the δ^{13}C values range from −1.0 to +1.7‰ PDB (Figure 5; Table 3), indicating a marine origin. The δ^{18}O and δ^{13}C values of the host Lower Jurassic dolomitic breccia vary according to the outcrop. In BEM outcrop, the δ^{18}O values range between −1.9 and −1.3‰ PDB, and the δ^{13}C values range between −3.7 and −2.7‰ PDB. In contrast, in BBCC outcrop, the δ^{18}O values range between −7.2 and −6.6‰ PDB, and the δ^{13}C values range between −8.1 and −8.0‰ PDB. The δ^{18}O and δ^{13}C values of the host Upper Triassic dolomite are −1.5‰ PDB and +2.9‰ PDB, respectively. This wide range of δ^{18}O and δ^{13}C values of the host dolomites indicates different dolomitization processes. Dolomites with less negative δ^{18}O and δ^{13}C values precipitated from marine waters, whereas dolomites with more depleted δ^{18}O and δ^{13}C values point to recrystallization by meteoric waters.

In relation to the calcite cement, three major groups with different stable isotopic composition can be distinguished. The first group includes the calcite cements sampled in the interior and central belts (BVA, BPM, and most of the samples in BEP), together with those that do not coincide with major thrust planes in the frontal belt (BPB and most of the samples in BEM). The group is characterized (box a, Figure 5) by δ^{18}O values ranging from −8.8 to −5.3 and δ^{13}C ranging from −6.9 to −3.1‰ PDB. The samples from each individual fracture show a relatively wide range of δ^{18}O values (varying as much as 3.7‰) and a relatively narrow range of δ^{13}C values (varying as much as 1.3‰ PDB). The similarity of the δ^{18}O range of values in these outcrops indicates that although the fractures affected different structural positions and host lithologies (limestones, dolomites, or conglomerates), the δ^{18}O isotopic composition of the fluids circulating through the different fractures was similar, which suggests that all these calcite cements might have been precipitated from meteoric water derived directly from the surface. The δ^{13}C isotopic composition, relatively homogeneous in each outcrop, may indicate a single source of C in each outcrop. The heavier δ^{13}C values of the cements in the fractures affecting the Tertiary conglomerates (BEP and BPB outcrops), compared with those in the fractures affecting the Lower Jurassic rocks (BVA and BPM outcrops), suggest a very low interaction of the meteoric water with the conglomerates (surface runoff) and a slightly higher interaction of the meteoric water with the host Lower Jurassic rocks (subsurface infiltration). In both cases, however, the great differences between the isotopic values of the host rock and the calcite cements from each outcrop indicate low interaction between the fluid circulating through the fractures and the host rock, constituting an open paleohydrogeological system along which meteoric waters that were derived directly from the surface circulate.

The second group, which includes the calcite cements present along the BEM thrusts formed in the frontal fold, shows δ^{18}O values ranging from −3.4 to −3.3‰ and δ^{13}C values ranging from −4.0 to −3.8‰ PDB (box b, Figure 5). Finally, the third group is integrated by the calcite cements present along the BBCC thrusts formed in the frontal fold and by one sample of calcite cement present in the fractures affecting the Tertiary conglomerates of the central belt (BEP outcrop) and show δ^{18}O values ranging from −10.5 to −8.1‰ PDB and δ^{13}C ranging from −6.4 to −3.9‰ PDB (box c, Figure 5).

Assuming that the studied samples are representative, the fact that two of the three groups defined by the stable isotope values consist mainly of calcite cements from two outcrops located in the frontal belt

Table 2. Synthesis for each outcrop of the main structural elements, stratigraphic elements, geometry, and kinematics of the veins.

	Outcrop	*Structural Belt Location*	*Stratigraphy*	*Main Structural Element*	*Sampled Structures*	*Location**
Gandesa	BVA	interior and central belt boundary	Upper Triassic mudstone over Lower Jurassic limestone	major thrust fault	thrust fault planes	footwall peripheral damage zone (<2 m; <6 ft)
	BPM	central belt	Lower Jurassic limestone	backlimb of a fold-bend anticline	minor strike-slip faults	background fracturing
	BEP	frontal belt	upper Eocene conglomerate	major thrust fault	minor left-handed strike-slip faults	footwall peripheral damage zone of the central and frontal bounding thrust (150–200 m; 500–660 ft)
	BEM	frontal belt	Lower Jurassic limestone over Lower Jurassic dolomitic breccia	thrust fault	minor faults, tension gashes, and bed-to-bed slidings	hanging-wall and footwall peripheral damage zone (<3 m; <10 ft)
	BBCC	frontal belt	Lower Jurassic dolomitic breccia over uppermost Triassic dolomite	thrust fault	thrust fault planes, tension gashes, and minor strike-slip faults	slip plane
	BPB	frontal belt	upper Eocene conglomerates	backlimb of a detachment anticline	bed-to-bed slidings and minor reverse and normal faults	background fracturing
Orta de Sant Joan	BLC	central belt	Upper Jurassic limestone	backlimb of a fold-bend anticline	minor faults with strike-slip and normal motions	background fracturing
	BRP	central belt	Lower Jurassic dolomitic breccia	backlimb of the frontal fold	minor strike-slip faults	background fracturing and hanging-wall peripheral damage zone of a major thrust (300 m; 1000 ft)
	BRE	frontal belt	upper Eocene conglomerate	major thrust fault	minor faults	footwall peripheral damage zone of the central and frontal bounding thrust (50 m; 160 ft)
	BRB	central belt	Lower Jurassic dolomitic breccia	major thrust fault	minor strike-slip faults	hanging-wall peripheral damage zone of the central and frontal bounding thrust (<100 m; <330 ft)
Beseit	BCC	interior belt	Tertiary conglomerate	thrust fault	minor strike-slip faults	footwall peripheral damage zone
	BLP	central/frontal belt boundary	Lower Jurassic dolomite breccia over upper Eocene conglomerate	major thrust fault	thrust fault plane and tension gashes	fault plane and footwall peripheral damage zone (<15 m; <50 ft)
	BQP	frontal belt	Lower Jurassic dolomite	backlimb of the frontal fold	minor reverse and strike-slip faults	background fracturing

TABLE 2. Synthesis for each outcrop of the main structural elements, stratigraphic elements, geometry, and kinematics of the veins (cont.).

	Outcrop	*Structural Belt Location*	*Stratigraphy*	*Main Structural Element*	*Sampled Structures*	*Location**
Herbers	BHA	interior belt	Lower Cretaceous limestone	fold	minor strike-slip faults and bed-to-bed slidings	background fracturing
	BHB	interior belt	Lower Cretaceous limestone	fold	minor strike-slip faults and bed-to-bed slidings	background fracturing
	BHC	interior belt	Lower Cretaceous limestone	fold	minor strike-slip faults and bed-to-bed slidings	background fracturing
	BHD	interior belt	Lower Cretaceous limestone	fold	minor strike-slip faults and bed-to-bed slidings	background fracturing
	BHE	interior belt	Lower Cretaceous limestone	fold	minor strike-slip faults and bed-to-bed slidings	background fracturing
	BHF	interior belt	Lower Cretaceous limestone	detachment fold	minor strike-slip faults and bed-to-bed slidings	background fracturing

*The value inside the parenthesis indicates the distance from the main structure.

may be interpreted as a more complex fluid-flow system in this part of the belt (an antiformal stack in this case) than in the other parts, at least in the Gandesa section. Moreover, the $\delta^{13}C$ values of the calcite cements in the major thrust of BEM outcrop (box b, Figure 5) are quite similar to their adjacent dolomite host, indicating a high interaction between the fluid circulating through the fractures and the host rock, making it a relatively closed paleohydrogeological system. However, the $\delta^{18}O$ values of BBCC thrust fault (box c, Figure 5) suggest circulation of hotter meteoric fluids.

$^{87}Sr/^{86}Sr$

The host Lower Jurassic limestones (BPM outcrop) show a $^{87}Sr/^{86}Sr$ ratio of 0.70743 (Figure 6; Table 1), consistent with the $^{87}Sr/^{86}Sr$ ratio of Jurassic seawater (Burke et al., 1982; Koepnick et al., 1990; Jones et al., 1994a, b). The host Lower Jurassic dolomite (BVA and BEM outcrops) shows a $^{87}Sr/^{86}Sr$ ratio ranging from 0.70789 to 0.70806, indicating that dolomitization occurred from radiogenic fluids. Interaction of the dolomitizing fluids with marine Carboniferous, Permian, or Triassic rocks (Burke et al., 1982; Popp et al., 1986; Koepnick et al., 1990) could explain this radiogenic signal.

The first group of calcite cements defined from their stable isotopic composition (BVA, BPM, and one sample from BEM) shows $^{87}Sr/^{86}Sr$ ratios ranging from 0.70759 to 0.70778 (Figure 6; Table 1). These values could have been acquired by interaction of the fluid with marine Triassic or Lower Jurassic rocks (Burke et al., 1982; Koepnick et al., 1990; Jones et al., 1994a, b). The second group of calcite cements, which were sampled in the major thrust located in the BEM outcrop, shows a very high $^{87}Sr/^{86}Sr$ ratio of 0.70845, which indicates precipitation from high radiogenic fluids probably because of interaction with the Hercynian basement (Ordovician, Silurian, or Carboniferous rocks; Burke et al., 1982; Popp et al., 1986). Finally, one sample of the third group of calcite cements (BEP outcrop) shows a $^{87}Sr/^{86}Sr$ ratio of 0.70795, indicating precipitation from radiogenic fluids. Interaction of the fluids with marine Carboniferous, Permian, or Triassic rocks or with Lower Jurassic dolomites will explain this signal.

Elemental Geochemistry

The Lower Jurassic dolomitic breccia host contains between 21.6 and 22.4% of Ca, between 12.5 and 13.8 % of Mg, and between 275 and 1205 ppm of Fe, with Mn and Sr always below the detection limit (Tables 4, 5).

Within the calcite cements, several differences are observed between the three groups differentiated by their stable isotopic and $^{87}Sr/^{86}Sr$ composition. The first one (BVA, BPM, and one sample in BEM) contains between 350 and 3360 ppm of Mg, with Fe, Mn, and Sr mostly below the detection limit. The calculated Mg/Ca molar ratio of the fluid responsible of calcite cement precipitation ranges between 0.04 and 0.71. The second (calcite cements filling the BEM major thrust) contains slightly higher values of Mg, Mn, and Sr: between 475 and 4200 ppm of Mg, between 180 and 655 ppm of Mn, and between 280 and 850 ppm of Sr, with Fe always below the detection limit. The calculated Mg/Ca, Mn/Ca,

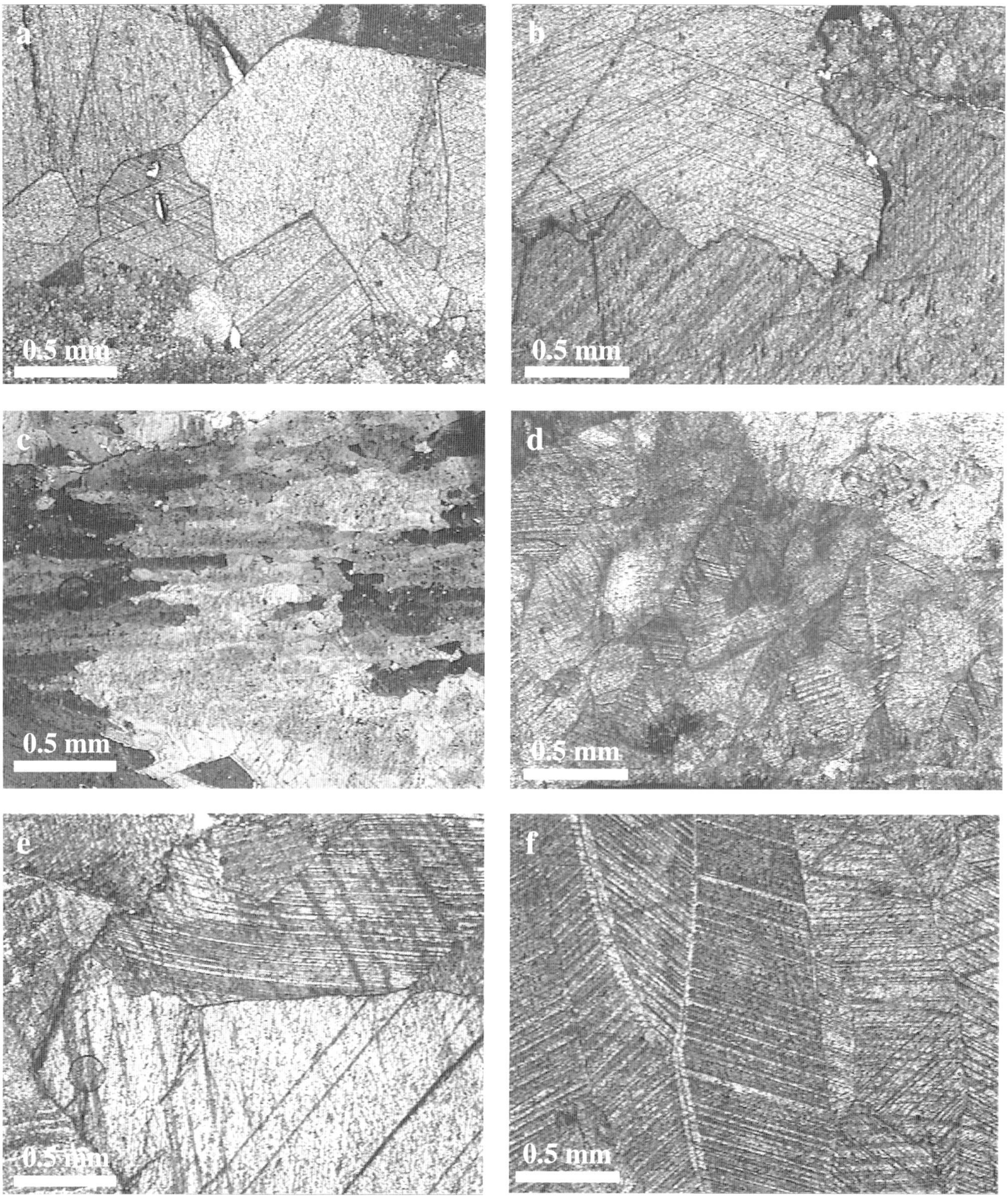

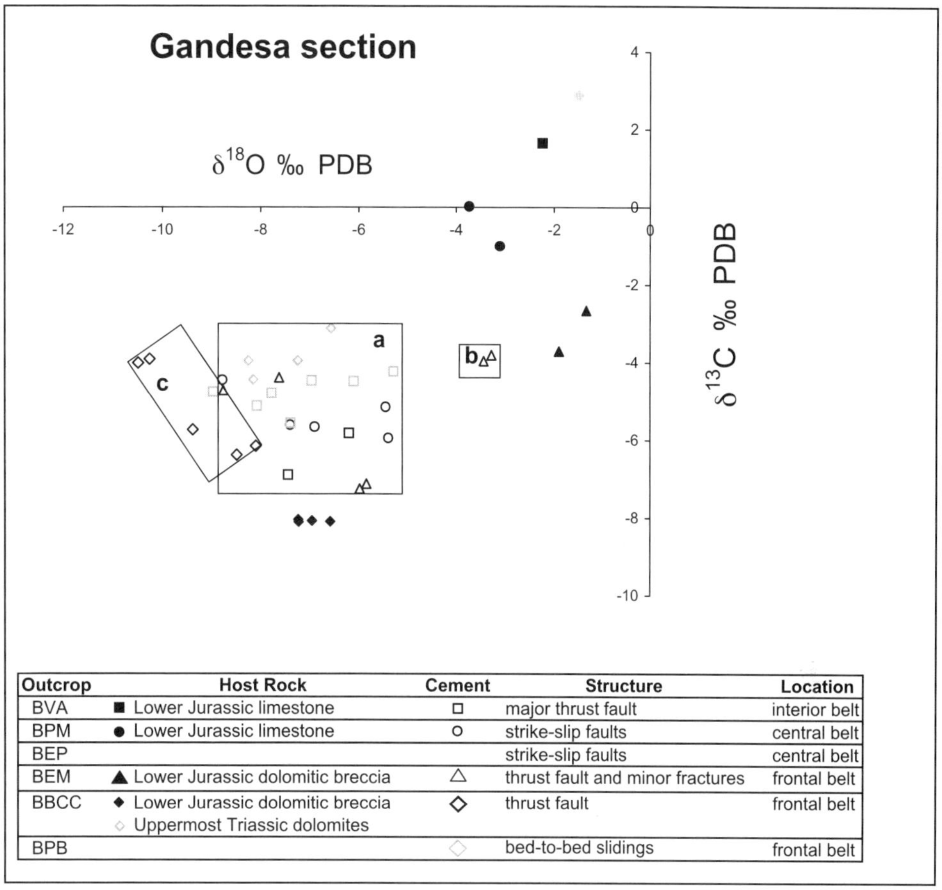

FIGURE 5. Crossplot of the $\delta^{18}O$ and $\delta^{13}C$ values of the samples from the Gandesa section.

and Sr/Ca molar ratios of the fluid responsible of calcite cement precipitation range between 0.16 and 1.46, between 4.2×10^{-5} and 1.5×10^{-4}, and between 0.005 and 0.016, respectively. The third (BEP) contains between 260 and 655 ppm of Mg and Fe, with Mn and Sr mostly below the detection limit. The calculated Mg/Ca molar ratio of the fluid ranges between 0.01 and 0.41.

All these values are consistent with meteoric water evolved toward a formation water composition (Skougstad and Horr, 1963; Kinsman, 1969; Howson et al., 1987; Tucker and Wright, 1990). The higher values of Mg and Sr of some of the cements are probably caused by interaction with marine carbonates.

THE ORTA DE SANT JOAN SECTION

Structural Description

The Orta de Sant Joan section (BB′ in Figures 2B, 3) shows the greatest structural complexity of the central belt. Because of that, here, the focus was on recognizing the pattern of the fluid-flow regime in the imbricated thrust system of the central belt of the Linking Zone. Four outcrops were analyzed in different structural position of this belt: (1) BLC, located in the innermost parts of the belt and made up of Upper Jurassic limestone; (2) BRP, integrated by Lower Jurassic dolomitic breccia outcropping in the middle parts of the belt; (3) BRE, located near the frontal thrust in the Tertiary conglomerates of the footwall of the thrust; and (4) BRB, in the Lower Jurassic dolomitic breccia outcropping in

FIGURE 4. Calcite cements in Gandesa and Orta de Sant Joan sections. Gandesa section: (a) Euhedral to subhedral calcite crystals with nondeformed, mechanical twin planes and uniform extinction filling strike-slip faults and bed-to-bed sliding planes affecting the Tertiary conglomerates (sample BEP 1). (b) Anhedral to subhedral calcite crystals with abundant, slightly deformed, mechanical twin planes and subangular extinction, affected by stylolite surfaces filling compressive fractures that were developed in the Lower Jurassic dolomites (sample BPM 5). (c) Anhedral to subhedral calcite crystals with subgrain formation filling a thrust fault in the frontal belt. The subgrains are elongated parallel to the fracture wall (sample BEM 6). (d) Fracture consisting of different shear surfaces with anhedral to subhedral calcite crystals with abundant, slightly deformed, mechanical twin planes, subangular extinction, and subgrain formation filling a thrust fault in the frontal belt (sample BBCC 8). Orta de Sant Joan section: (e) Calcite cement filling strike-slip faults, affecting the Lower Jurassic dolomitic breccia that was formed by euhedral to subhedral calcite crystals with abundant, mechanical twin planes, locally deformed, and uniform extinction (sample BRP 2). (f) Second-generation calcite cements filling normal faults made up of euhedral crystals 1 mm (0.04 in.) thick and 1 cm (0.4 in.) long, forming a palisade with the *c*-axis normal to the fracture wall, with abundant, nondeformed, mechanical twin plane and uniform extinction (sample BLC 1).

TABLE 3. $\delta^{18}O$ **and** $\delta^{13}C$ **of the host rocks and calcite cements from (a) Gandesa, (b) Orta de Sant Joan, (c) Beseit, and (d) Herbers sections.**

Outcrop	*Structure*	*Sample*	*Type*	$\delta^{18}O$ (‰ PDB)	$\delta^{13}C$ (‰ PDB)
(a)					
BVA (Benifallet railway)	Valliplana thrust: Major thrust separating the interior zone and the central zone, placing Triassic limestones over Lower Jurassic dolomitic breccia	BVA 2B	calcite cement	−6.2	−5.8
		BVA 4A	host limestone	−2.2	+1.7
		BVA 4B	calcite cement	−7.4	−6.9
BPM (Mas de Riberola)	El Empalme thrust: Imbricated thrust system of the central zone; Lower Jurassic limestones outcropping in the hanging wall	BPM 2A	host limestone	−3.7	0.0
		BPM 2B	calcite cement	−7.4	−5.6
		BPM 5	calcite cement	−6.9	−5.7
		BPM 7A	host limestone	−3.1	−1.0
		BPM 7B	calcite cement	−8.8	−4.4
		BPM 8	calcite cement	−5.5	−5.1
		BPM 9	calcite cement	−5.4	−5.9
BEP (Pinell railway station)	El Empalme thrust: Imbricated thrust system of the Central zone; upper Eocene conglomerates of the footwall	BEP 1A	calcite cement	−6.1	−4.5
		BEP 1B	calcite cement	−7.0	−4.5
		BEP 2B	calcite cement	−7.8	−4.8
		BEP 3	calcite cement	−9.0	−4.8
		BEP 5	calcite cement	−7.4	−5.6
		BEP 6A	calcite cement	−8.1	−5.1
		BEP 6B	calcite cement	−5.3	−4.2
BEM (Estudi mossen)	frontal zone (antiformal stack): Thrust that places Lower Jurassic dolomites over the Lower Jurassic dolomitic breccia	BEM 1A	host dolomite	−1.3	−2.7
		BEM 1B	calcite cement	−5.8	−7.1
		BEM 2	calcite cement	−3.3	−3.8
		BEM 3	calcite cement	−7.6	−4.4
		BEM 6A	host dolomite	−1.9	−3.7
		BEM 6B	calcite cement	−3.4	−4.0
		BEM 7A	calcite cement	−6.0	−7.3
		BEM 7B	calcite cement	−8.8	−4.7
BBCC (Coll d'en Canar)	frontal zone (antiformal stack): Thrust that places Lower Jurassic dolomites of the Imón Formation over the Lower Jurassic dolomitic breccia	BBCC 1	host dolomite	−6.6	−8.1
		BBCC 4	calcite cement	−10.5	−4.0
		BBCC 7	calcite cement	−9.4	−5.7
		BBCC 8	calcite cement	−10.3	−3.9
		BBCC 9	calcite cement	−8.1	−6.2
		BBCC 15A	host dolomite	−7.2	−8.0
		BBCC 15B	calcite cement	−8.5	−6.4

Table 3. $\delta^{18}O$ and $\delta^{13}C$ of the host rocks and calcite cements from (a) Gandesa, (b) Orta de Sant Joan, (c) Beseit, and (d) Herbers sections (cont.).

Outcrop	*Structure*	*Sample*	*Type*	$\delta^{18}O$ (‰ PDB)	$\delta^{13}C$ (‰ PDB)
		BBCC 16A	host dolomite	−7.0	−8.1
		BBCC 16B	calcite cement	−7.2	−8.1
		BBCC 19	host dolomite	−1.5	2.9
BPB (Carretera Prat-Bot)	frontal zone: Conglomerates footwall	BPB 3	calcite cement	−8.2	−4.4
		BPB 7	calcite cement	−8.3	−4.0
		BPB 8	calcite cement	−7.2	−4.0
		BPB 10	calcite cement	−6.6	−3.1
(b)					
BRE (Estret River I)	strike-slip faults in Tertiary conglomerates	BRE 1	calcite cement	−8.5	−3.1
		BRE 2	calcite cement	−3.8	−2.6
BRB (Estret River II)	strike-slip faults in Mesozoic breccias	BRB 1	calcite cement	−6.9	−5.2
		BRB 2A	host dolomite	−6.3	−7.2
		BRB 2B	calcite cement	−6.2	−5.9
		BRB 5A	host dolomite	−6.1	−6.6
		BRB 5B	calcite cement	−5.4	−6.6
		BRB 7A	host dolomite	−5.9	−5.9
		BRB 7B	calcite cement	−7.8	−4.7
BRP (quarry)	strike-slip faults in breccias	BRP 1	calcite cement	−6.8	−7.1
		BRP 2A	host dolomite	−6.2	−4.2
		BRP 2B	calcite cement	−8.7	−3.5
		BRP 3	calcite cement	−6.0	−6.1
		BRP 5A	host dolomite	−6.9	−5.3
		BRP 5B	calcite cement	−6.3	−6.7
BLC (Les Clotes)	normal faults in Mesozoic (Formation Iàtova)	BLC 1A	calcite cement	−8.4	+1.1
		BLC 1B	calcite cement	−8.1	−0.8
		BLC 1C	calcite cement	−8.4	−1.9
		BLC 3A	host limestone	−3.7	+2.1
		BLC 3B	calcite cement	−9.1	+0.7
		BLC 4A	calcite cement	−7.6	+0.2
		BLC 4B	calcite cement	−8.1	−2.6
		BLC 7A	host limestone	−3.6	+1.8
		BLC 7B	calcite cement	−9.2	+0.2
(c)					
BQP (Querol-Periganyol)	strike-slip faults, bed-to-bed slidings, and reverse faults	BQP 1A	host dolomite	−3.6	−0.5
		BQP 1B	calcite cement	−11.5	−2.1
		BQP 3A	host dolomite	−3.8	−0.8
		BQP 3B	calcite cement	−11.7	−2.9
BLP (La Pesquera)	thrust in Tertiary conglomerates	BLP 1	calcite cement	−6.4	−6.8
		BLP 3	calcite cement	−10.1	−4.1
BCC (La Coscollosa)	strike-slip faults in Tertiary conglomerates	BCC 2	calcite cement	−10.0	−7.5

TABLE 3. $\delta^{18}O$ and $\delta^{13}C$ of the host rocks and calcite cements from (a) Gandesa, (b) Orta de Sant Joan, (c) Beseit, and (d) Herbers sections (cont.).

Outcrop	*Structure*	*Sample*	*Type*	*$\delta^{18}O$ (‰ PDB)*	*$\delta^{13}C$ (‰ PDB)*
(d)					
Hervers	strike-slip faults and bed-to-bed slidings	BHA 02 AA	first calcite cement	–8.8	–6.9
		BHA 02 AB	second calcite cement	–10.0	–2.5
		BHA 03A	first calcite cement	–10.5	–4.5
		BHA 03B	second calcite cement	–11.4	–1.8
		BHA 05A	host limestone	–5.2	–5.6
		BHA 05B	calcite cement	–11.3	–4.2
		BHA 06A	host limestone	–5.4	–5.9
		BHA 06B	calcite cement	–10.2	–4.4
		BHB 2A	host limestone	–7.4	–5.5
		BHB 2B	calcite cement	–10.1	–2.1
		BHC 2A	host limestone	–5.9	–4.8
		BHC 2B	first calcite cement	–10.2	–5.3
		BHC 2C	second calcite cement	–11.8	–3.2
		BHC 4A	host limestone	–8.5	–5.2
		BHC 4B	first calcite cement	–11.8	–4.0
		BHC 4C	second calcite cement	–12.7	–3.1
		BHC 5A	first calcite cement	–11.5	–3.5
		BHC 5B	second calcite cement	–12.0	–2.1
		BHD 1A	host limestone	–4.7	–6.3
		BHD 1B	first calcite cement	–10.1	–6.2
		BHD 1C	second calcite cement	–10.0	–6.2
		BHD 3	calcite cement	–10.2	–6.3
		BHD 4A	host limestone	–5.0	–5.8
		BHD 4B	calcite cement	–10.4	–6.2
		BHD 5BA	host limestone	–4.9	–6.7
		BHD 5BB	calcite cement	–10.3	–6.4
		BHD 6	calcite cement	–10.2	–6.1
		BHE 1A	host limestone	–4.7	–3.4
		BHE 1B	calcite cement	–9.6	–3.6
		BHE 2	calcite cement	–9.1	–4.0
		BHE 3	calcite cement	–9.7	–3.2
		BHF 01A	host limestone	–6.0	–3.1
		BHF 01B	calcite cement	–8.9	–3.0
		BHF 03A	first calcite cement	–11.0	–3.2
		BHF 03B	second calcite cement	–12.0	–3.4
		BHF 05A	host limestone	–4.2	–3.4
		BHF 05B	calcite cement	–11.9	–2.7
		BHF 06A	host limestone	–4.6	–2.5
		BHF 06B	calcite cement	–10.8	–3.0
		BHF 07	calcite cement	–10.5	–4.3

Table 3. $\delta^{18}O$ and $\delta^{13}C$ of the host rocks and calcite cements from (a) Gandesa, (b) Orta de Sant Joan, (c) Beseit, and (d) Herbers sections (cont.).

Outcrop	*Structure*	*Sample*	*Type*	$\delta^{18}O$ (‰ PDB)	$\delta^{13}C$ (‰ PDB)
		BHF 08	calcite cement	−9.5	−2.4
		BHF 09	calcite cement	−8.9	−2.3
		BHF 10A	host limestone	−3.0	−1.3
		BHF 10B	calcite cement	−7.3	−2.1
		BHF 11A	host limestone	−2.9	−1.2
		BHF 11B	calcite cement	−8.3	−1.8
		BHF 12A	first calcite cement	−9.2	−2.8
		BHF 12B	second calcite cement	−11.7	−4.0
		BHF 14	calcite cement	−9.0	−2.7
		BHF 15	calcite cement	−9.5	−2.6

the hanging wall near the frontal thrust. In these host rocks, the studied calcite cements are located in the fracture planes of the minor normal faults affecting the Upper Jurassic limestone (BLC), strike-slip faults affecting the Lower Jurassic dolomitic breccia (BRP and BRB), and in millimetric fractures affecting the Tertiary conglomerates (BRE) (Table 2).

Petrology of the Calcite Cements

The petrology of the calcite cements in the compressive fractures affecting the Tertiary conglomerates is different from that of the strike-slip faults affecting the Lower Jurassic rocks and from that of the normal faults affecting the Upper Jurassic limestones. The millimetric fractures affecting the Tertiary conglomerates (BRE outcrop) are filled by euhedral to subhedral calcite crystals, as much as 2 mm (0.08 in.) in size, with nondeformed mechanical twin planes and uniform extinction. The strike-slip faults affecting the Lower Jurassic rocks (BRP and BRB outcrops) are filled by euhedral to subhedral calcite crystals, from 0.5 mm to as much as 1 cm (0.02 to as much as 0.4 in.) in size, with abundant, locally deformed, mechanical twin planes and uniform extinction, affected by stylolite surfaces (Figure 4e).

Finally, the faults of the BLC outcrop are filled by two generations of calcite cements. The first generation shows horizontal slickensides, indicating strike-slip movement, and consists of elongated calcite crystals 40 μm thick and as much as 5 mm (0.2 in.) long, with the *c*-axis oblique to the fracture border. The second generation of cement shows vertical slickensides, indicating normal movement, and consists of euhedral calcite crystals 1 mm (0.04 in.) thick and 1 cm (0.4 in.) long, forming a palisade with the *c*-axis normal to the fracture wall and with abundant, nondeformed, mechanical twin planes and uniform extinction (Figure 4f).

Stable Isotopes

In the studied outcrops of the Orta de Sant Joan section, two different Mesozoic host rocks are present: Lower Jurassic dolomites (BRB and BRP outcrops) and

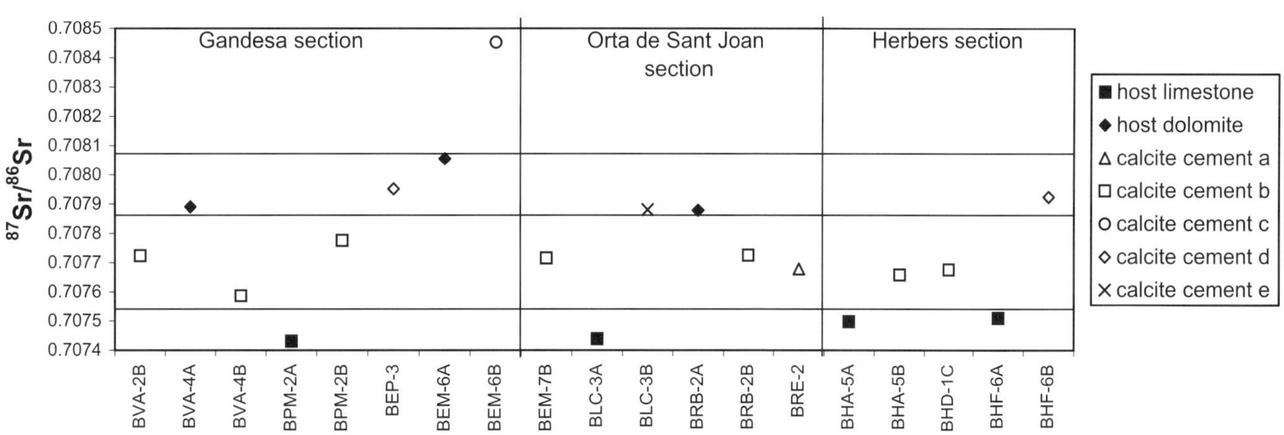

Figure 6. $^{87}Sr/^{86}Sr$ of the calcite cements and host limestones and dolomites.

TABLE 4. Synthesis of the elemental geochemistry results and the calculated Mg/Ca, Mn/Ca, Sr/Ca, and Ca/Fe molar ratios in the precipitating fluid. The values inside the parenthesis are the average values.

	Sample	*Fracture Type*	*Analyses*	*Ca*	*Mg*	*Fe*
Gandesa	BPM 2	strike slip	24	393,445–406,180	415–2485	<d.l.–2605
	BEM 7	minor thrust	29	392,790–409,285	350–3345	<d.l.–1300
	BVA 2	thrust fault	26	401,895–407,770	440–2425	<d.l.
	BVA 4	thrust fault	19	390,045–408,705	905–3360	<d.l.
	BEM 6	main thrust	54	392,070–410,705	475–4200	<d.l.
	BEP 3	strike slip	32	386,000–407,820	260–3485	<d.l.–3485
Orta de Sant Joan	BLC 3	normal faults	18	389,335–404,665	870–2475	<d.l.
	BRB 2	strike slip	24	397,265–408,715	420–1690	<d.l.–750
	BRE 2	strike slip	20	393,290–403,785	440–3565	<d.l.–720
Herbers	BHD 1	strike slip	24	376,150–393,545	730–5150	3055–11,490
	BHA 5	bed to bed	25	391,450–405,945	635–4375	310–1835
	BHF 6	bed to bed	26	393,915–406,940	480–5240	295–1155

<d.l. = below detection limit.

Upper Jurassic limestones (BLC outcrop). The $\delta^{18}O$ and $\delta^{13}C$ values of the Upper Jurassic limestone host range from −3.7 to −3.6‰ PDB and from +1.8 to +2.1‰ PDB, respectively (Figure 7; Table 3). In contrast, the $\delta^{18}O$ and $\delta^{13}C$ values of the Lower Jurassic dolomites host range from −6.9 to −5.9‰ PDB and from −7.2 to −4.2‰ PDB, respectively (Figure 7; Table 3). These values reflect a marine origin for the Upper Jurassic limestone and a recrystallization process by meteoric waters.

In the calcite cements filling the fractures, major differences exist between those that developed in the compressive faults and those that developed in the normal faults. The calcite cements filling the compressive faults affecting the Lower Jurassic dolomitic breccia (BRB and BRP outcrops) and the Tertiary conglomerates (BRE outcrop) show $\delta^{18}O$ values ranging from −8.7 to −3.8‰ PDB and $\delta^{13}C$ ranging from −7.1 to −2.6‰ PDB. The $\delta^{18}O$ and $\delta^{13}C$ values of these calcite cements indicate precipitation from meteoric water.

The similar carbon isotopic composition of these cements and the host Lower Jurassic dolomite can be caused by soil-derived CO_2 during dolomite recrystallization and/or by the buffering of the pore-water carbon isotopic composition by the extensive dissolution of the host rock (Marshall, 1992), resulting in a geochemical composition of fluids similar to the values of the host rock, thus indicating high interaction between the fluid circulating through the fractures and the host rock and a relatively closed paleohydrogeological system.

The less depleted $\delta^{13}C$ values of the calcite cements present in the Tertiary conglomerates (between −3.1 and −2.6‰ PDB) with respect to the values observed in the Lower Jurassic dolomites (between −7.1 and −3.5‰ PDB) suggest again a lower interaction of the meteoric water with the conglomerates (surface runoff) and a slightly higher interaction of the meteoric water with the host Lower Jurassic dolomites (subsurface infiltration).

The calcite cements filling normal faults affecting the Upper Jurassic limestone (BLC outcrop) show $\delta^{18}O$ values ranging from −9.2 to −7.6‰ PDB and $\delta^{13}C$ values ranging from −2.6 to +1.1‰ PDB (Figure 7; Table 3), indicating that the fluids precipitating the calcite cements in these fractures were meteoric, circulating through a relatively open paleohydrogeological system. The small $\delta^{13}C$ depletion of these cements with respect to the host rock can be caused by minor contribution of soil-derived CO_2 in the meteoric water.

$^{87}Sr/^{86}Sr$

The 0.70744 $^{87}Sr/^{86}Sr$ ratio of the Upper Jurassic host limestone (BLC outcrop) (Figure 6; Table 1) is consistent with the $^{87}Sr/^{86}Sr$ ratio of the Jurassic seawater (Koepnick et al., 1990). The 0.70788 $^{87}Sr/^{86}Sr$ ratio of the Lower Jurassic dolomitic breccia (BRB outcrop) indicates, as in the Gandesa section, that dolomitization occurred from radiogenic fluids. Interaction of the dolomitizing fluids with marine Carboniferous, Permian, or Triassic rocks could explain this radiogenic signal.

The calcite cements filling the fractures developed in the Lower Jurassic dolomitic breccia (BRB outcrop) and the upper Eocene–lower Oligocene conglomerates

Table 4. Synthesis of the elemental geochemistry results and the calculated Mg/Ca, Mn/Ca, Sr/Ca, and Ca/Fe molar ratios in the precipitating fluid. The values inside the parenthesis are the average values (cont.).

Mn	*Sr*	*Mg/Ca*	*Mn/Ca*	*Sr/Ca*	*Ca/Fe*	*Fluid Type*
<d.l.–400	<d.l.–455	0.07–0.52 (0.24)			>1000	evolved meteoric
<d.l.	<d.l.	0.05–0.71 (0.31)				evolved meteoric
<d.l.	<d.l.	0.04–0.5 (0.18)				evolved meteoric
<d.l.	<d.l.–600	0.19–0.70 (0.35)				evolved meteoric
<d.l.–655	<d.l.–850	0.16–1.46 (0.58)	4.2×10^{-5}– 1.5×10^{-4}	0.005–0.016		evolved meteoric
<d.l.	<d.l.–390	0.01–0.41 (0.21)				evolved meteoric
470–1420	<d.l.	0.18–0.53 (0.39)	1.07×10^{-4}– 3.26×10^{-4}			meteoric
<d.l.	<d.l.	0.04–0.35 (0.19)				evolved meteoric
180–410	<d.l.	0.04–0.75 (0.35)	4.2×10^{-5}– 9.5×10^{-5}		>3860	meteoric
<d.l.	<d.l.–795	0.26–1.85 (0.97)			230–770	evolved meteoric
<d.l.	<d.l.–310	0.13–0.93 (0.47)			>1500	evolved meteoric
<d.l.–345	<d.l.–925	0.03–1.11 (0.36)	4.5×10^{-5}– 7.9×10^{-5}	0.0–0.018	>2370	evolved meteoric

(BRE outcrop) show a $^{87}Sr/^{86}Sr$ ratio ranging from 0.70768 to 0.70773 (Figure 6; Table 1). These values could have been acquired by interaction of the precipitating fluid with marine Triassic or Lower Jurassic rocks (Burke et al., 1982; Koepnick et al., 1990; Jones et al., 1994a, b).

The $^{87}Sr/^{86}Sr$ ratio (0.70788) of the calcite cements filling normal faults affecting the Upper Jurassic limestone (BLC outcrop) indicates that the fluid precipitating the calcite cement in these normal faults was a radiogenic fluid. This signal could have been acquired by the interaction of the fluid with siliciclastic rocks or with marine Carboniferous, Permian, or Triassic rocks or with Lower Jurassic dolomites. The nature of the fluid precipitating the synkinematic cement in these distensive fractures suggests that the normal faults were formed during the Tertiary extension, which occurred in a meteoric setting, and not during the Mesozoic extension, which occurred in a marine setting.

Elemental Geochemistry

From the elemental geochemistry of the calcite cements in this cross section (Tables 4, 5), different fluids were attested to be responsible for calcite precipitation in the compressive structures developed in the Mesozoic, in the compressive structures developed in the Tertiary, and in the distensive structures developed in the Mesozoic. Thus, the calcite cements filling the compressive fractures developed in the Lower Jurassic dolomitic breccia (BRB and BRP outcrops) contain low values of Mg (between 420 and 1690 ppm), with Fe, Mn, and Sr mostly below the detection limit. They also show a calculated Mg/Ca molar ratio of the precipitating fluid ranging from 0.04 to 0.35. The low values of Mg, Mn, Fe, and Sr suggest a meteoric water composition that slightly evolved toward a formation water composition.

The calcite cements found in the compressive fractures developed in the Tertiary conglomerates (BRE outcrop) have similar or higher contents of Mg (between 440 and 3565 ppm) and higher contents of Fe and Mn (between 275 and 720 ppm and between 180 and 410 ppm, respectively). In these cements, Sr is mostly below the detection limit, and the calculated Mg/Ca, Ca/Fe, and Mn/Ca molar ratios of the precipitating fluid range between 0.04 and 0.75, between 3860 and 10155, and between 4.2×10^{-5} and 9.5×10^{-5}, respectively. The presence of Fe and Mn and the absence of Sr suggest a meteoric water composition that poorly evolved toward a formation water composition.

The calcite cements filling the normal faults affecting the Upper Jurassic (BLC outcrop) are also different. They contain between 870 and 2695 ppm of Mg and between 335 and 1420 ppm of Mn, with Fe and Sr always below the detection limit. The calculated Mg/Ca and Mn/Ca molar ratios of the precipitating fluid range between 0.18 and 0.58 and between 1.07×10^{-4} and 3.26×10^{-4}, respectively. These relatively high values of Mn and Mg and low values of Fe and Sr suggest a meteoric water composition probably in suboxic conditions.

Table 5. Distribution of the calcite cements in the different outcrops and sections showing the range of $\delta^{18}O$, $\delta^{13}C$, $^{87}Sr/^{86}Sr$, Mg, Fe, Mn, and Sr and interpretation of the hydrogeological system, fluid type, and geodynamics for each type of calcite cement.

Calcite Cement	*Outcrops*	*Structural Section*	$\delta^{18}O$ (‰ PDB)	$\delta^{13}C$ (‰ PDB)	$^{87}Sr/^{86}Sr$	*Mg*	*Fe*
a	BEP, BPB, BRE	Gandesa, Orta de Sant Joan	−9 to −3.8	−5.6 to −2.6	0.70768	440–3565	275–720
b	BVA, BPM, BEM, BRB, BRP, BLP, BCC, BHA to BHF	Gandesa, Orta de Sant Joan, Beseit, Herbers	−10.1 to −5.4	−7.5 to −3.5	0.70759–0.70778	350–5150	<d.l.–11,490
c	BEM	Gandesa	−3.4 to −3.3	−4.0 to −3.8	0.70845	475–4200	<d.l.
d	BEP, BBCC, BLP, BQP, BHA, BHC, BHF	Gandesa, Beseit, Herbers	−12.7 to −8.1	−6.4 to −3.9	0.70792–0.70795	260–5240	<d.l.–1155
e	BLC	Orta de Sant Joan	−9.2 to −7.6	−2.6 to + 1.1	0.70788	870–2695	<d.l.

<d.l. = below detection limit.

THE BESEIT SECTION

Structural Description

The Beseit structural section (CC′ in Figures 2B, 3) crosses the three structural belts of the Linking Zone in an area where most of the thrusts, detached in the Middle Muschelkalk evaporites, clearly resulted from the inversion of the extensional faults that bounded the Late Jurassic–Early Cretaceous Maestrat basin to the north (Salas et al., 1997).

Three outcrops (BCC, BLP, and BQP), located in different structural settings along this section, were analyzed. The southernmost outcrop (BCC), located in the interior belt and therefore originally at the highest parts of the belt, corresponds to an outcrop of Tertiary conglomerates overthrusted by Lower Cretaceous limestones. In this outcrop, the calcite cements are found in veins filling minor strike-slip faults. The second outcrop (BLP) corresponds to the major thrust plane that separates the central and the frontal belts and places the Lower Jurassic dolomitic breccia over Tertiary conglomerates. The northernmost outcrop (BQP) is located in the Lower Jurassic dolomites outcropping in the core of the frontal fold, where studied calcite cements are found in veins filling minor reverse faults (Table 2).

Petrology of the Calcite Cements

In the Tertiary conglomerates of BCC outcrop, the calcite cements filling minor strike-slip faults are characterized by anhedral to subhedral crystals from 100 μm to as much as 1 mm in size, with abundant, slightly deformed, mechanical twin planes and subangular extinction and are affected by stylolite surfaces (Figure 8a). In the Tertiary conglomerates of the BLP outcrop, the calcite cements filling fractures are formed by euhedral to subhedral crystals from 400 μm to as much as 1 cm in size, with abundant, nondeformed, mechanical twin planes and with uniform extinction (Figure 8b).

Finally, the reverse faults developed in the Lower Jurassic dolomites of BQP outcrop are filled with calcite cement, forming veins consisting of shear bands filled by anhedral to subhedral crystals, from equant to elongated, parallel to the shear band boundary 100 μm to 3 mm in size, with abundant, deformed, mechanical twin planes, subangular extinction, and subgrain formation. The boundary of the shear bands is shear surfaces that, in some cases, coincide with stylolite surfaces.

Stable Isotopes

Stable isotopic analysis of the host rocks was carried out only in the Lower Jurassic dolomites of BQP outcrop. It shows $\delta^{18}O$ and $\delta^{13}C$ values ranging from −3.8 to −3.6‰ PDB and from −0.8 to −0.5‰ PDB, respectively (Figure 9; Table 3c). These values are different from the values obtained in the Lower Jurassic dolomites of the former sections, which may indicate less recrystallized dolomites under meteoric waters.

In the Beseit section, samples of calcite cements analyzed for their stable isotopic composition show a

TABLE 5. Distribution of the calcite cements in the different outcrops and sections showing the range of $\delta^{18}O$, $\delta^{13}C$, $^{87}Sr/^{86}Sr$, Mg, Fe, Mn, and Sr and interpretation of the hydrogeological system, fluid type, and geodynamics for each type of calcite cement (cont.).

Mn	*Sr*	*Hydrogeological System*	*Fluid Type*		*Geodynamic Stage*	
180–410	<d.l.	open	meteoric	surface runoff	comprehensive deformation; fold and thrust belt development	first stages
<d.l.	<d.l.–795	relatively closed	meteoric evolved to formation water	subsurface infiltration		
180–655	280–850	relatively closed	meteoric evolved to formation water	high fluid-rock interaction with Ordovician, Silurian, or Carboniferous rocks		Last stages
<d.l.–925	<d.l.–925	relatively open	meteoric evolved to formation water	hot meteoric fluid, interaction with Carboniferous, Permian, or Triassic rocks		
470–1420	<d.l.	relatively open	meteoric		Extension	

great dispersion of $\delta^{18}O$ and $\delta^{13}C$ values. The calcite cement sampled in the BCC outcrop presents a $\delta^{18}O$ value of −10.0‰ PDB and a $\delta^{13}C$ value of −7.5‰ PDB, which are consistent with meteoric waters. In contrast, the two samples analyzed in the thrust plane of BLP outcrop are so different that it is difficult to interpret them as derived from the same fluid: One shows $\delta^{18}O$ and $\delta^{13}C$ values of −6.4 and −6.8‰ PDB, respectively, and the other shows $\delta^{18}O$ and $\delta^{13}C$ values of −10.1 and −4.1‰ PDB, respectively, both consistent with meteoric waters. Although it is not proved, the former could have precipitated from meteoric water derived directly from the surface and influenced by C-depleted, soil-derived CO_2, and the second could have precipitated from hotter meteoric waters migrating upward.

Finally, the two cements sampled in the Lower Jurassic dolomites of the frontal fold (BQP outcrop) show $\delta^{18}O$ values ranging from −11.7 to −11.5‰ PDB and $\delta^{13}C$ values ranging from −2.9 to −2.1‰ PDB

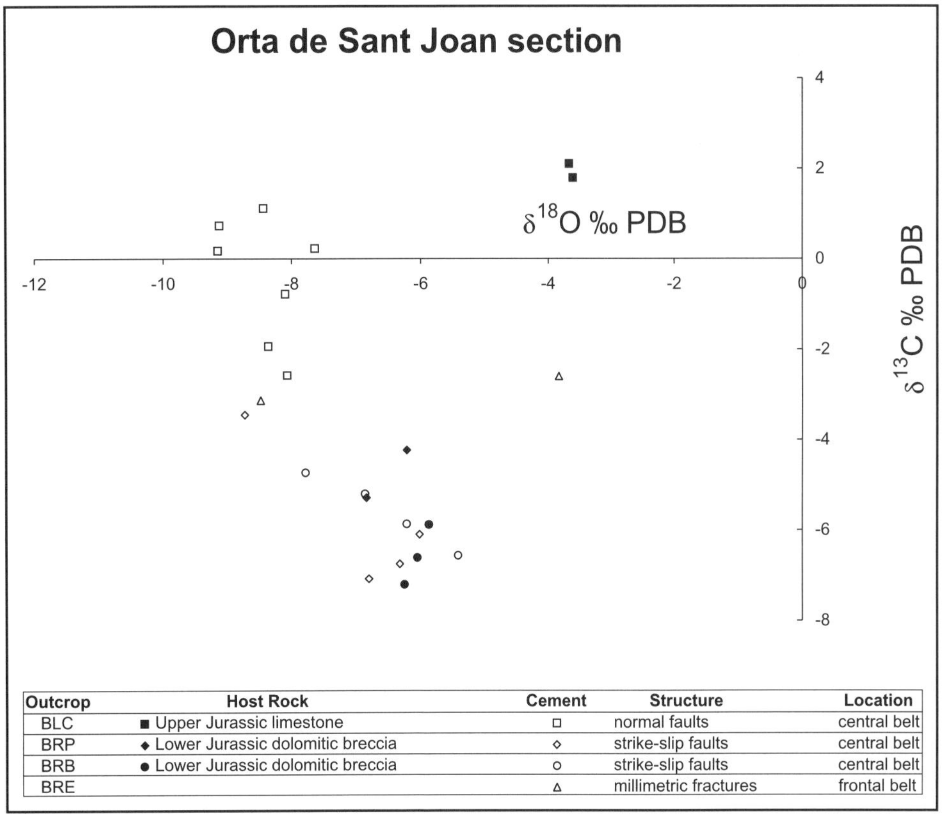

FIGURE 7. $\delta^{18}O$ and $\delta^{13}C$ crossplot of the samples from the Orta de Sant Joan section.

(Figure 9; Table 3c). These $\delta^{18}O$ values are consistent with hot meteoric waters. The relatively similar $\delta^{13}C$ values between the host dolomite and these calcite cements might indicate soil-derived CO_2 during dolomite recrystallization and/or might indicate that the main source of CO_3^{2-}, for calcite precipitation in the veins, was the host dolomites and therefore constituted a relatively closed paleohydrogeological system.

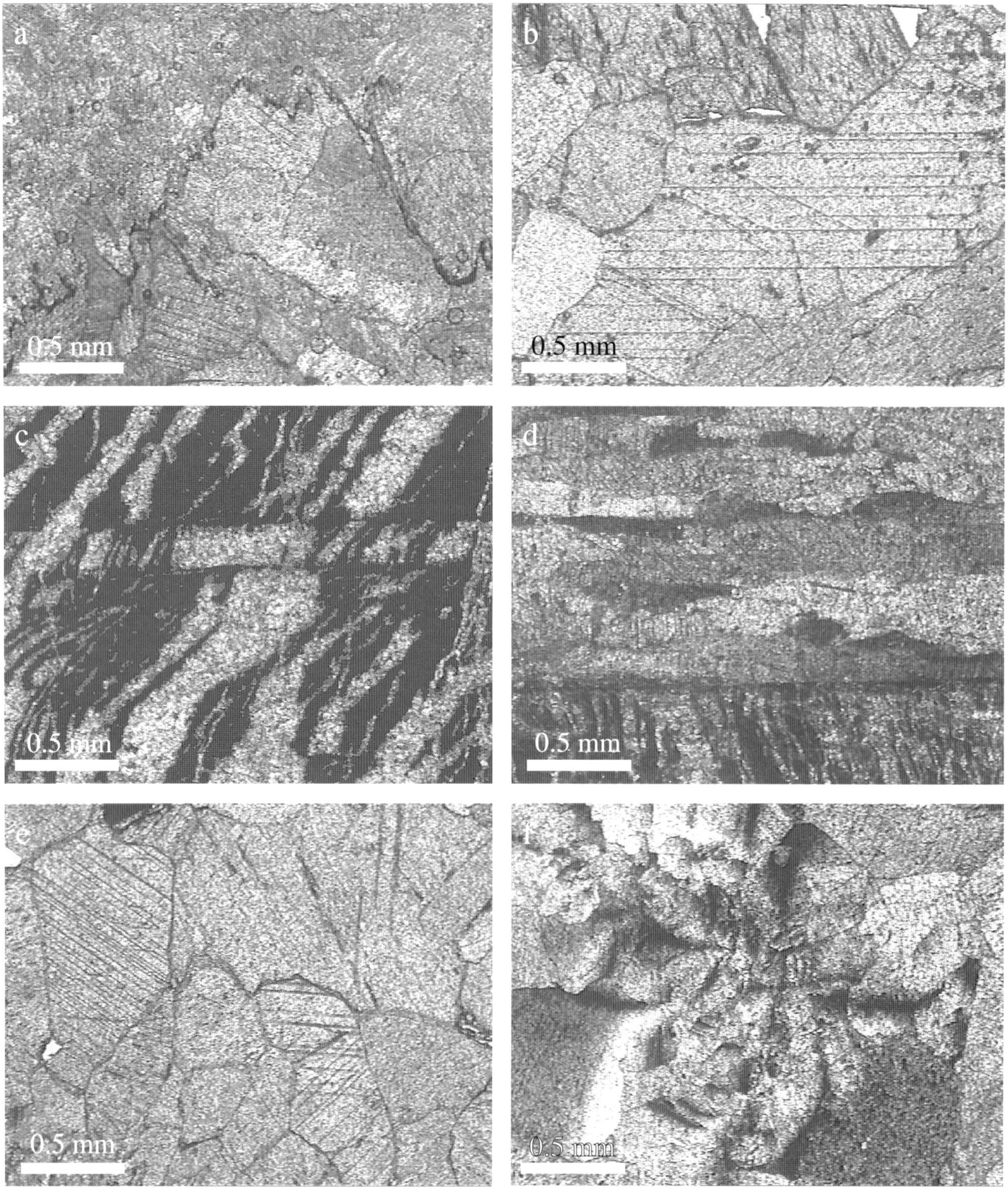

FIGURE 9. $\delta^{18}O$ and $\delta^{13}C$ crossplot of the samples from the Beseit section.

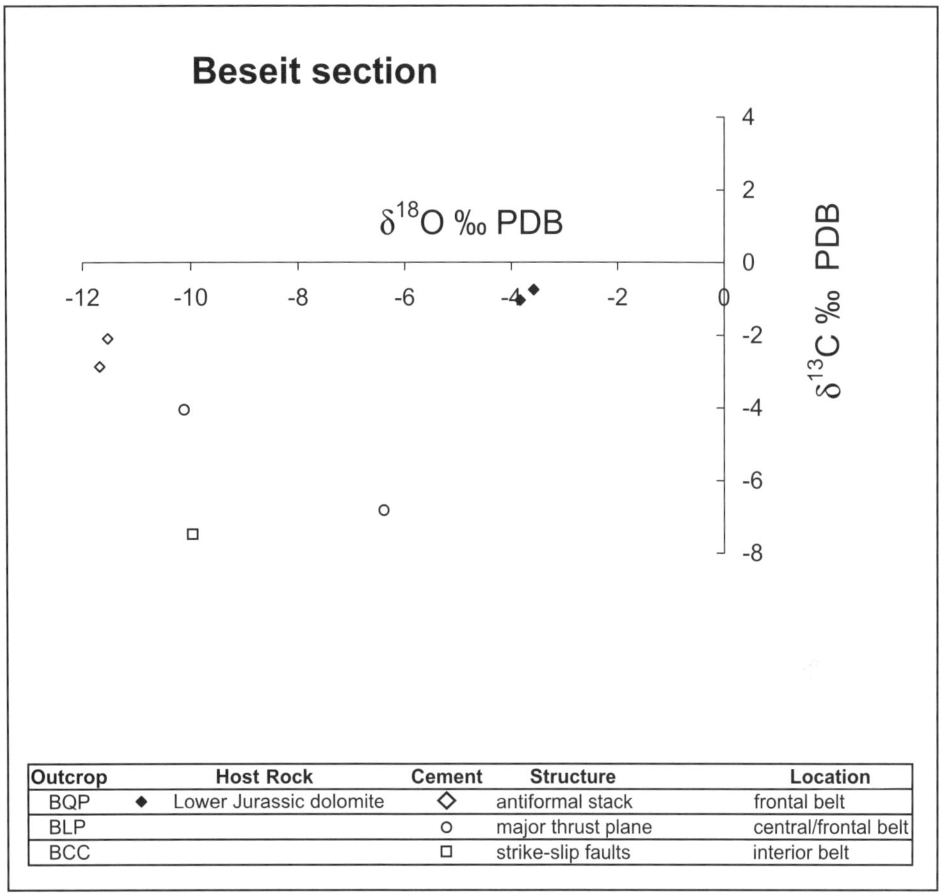

THE HERBERS SECTION

Structural Description

The Herbers section (DD′ in Figures 2B, 3) includes six studied outcrops (BHA, BHB, BHC, BHD, BHE, and BHF) that are located in the interior belt of the Linking Zone and, therefore, in the originally highest and least compressively deformed part of the belt (Table 2). In all these outcrops, the host rocks are Barremian lacustrine limestones and marls deposited in the marginal areas of the Mesozoic Maestrat basin. Minor strike-slip faults and bed-to-bed slidings affecting these host rocks are filled by calcite cements that have been sampled and analyzed for their petrology and geochemistry.

Petrology of the Calcite Cements

Two generations of calcite cement are observed in the strike-slip faults and bed-to-bed slidings planes. Both generations are found in both types of fractures, either isolated as individual fracture fill or together with the other generation. First-generation calcite cement is found in all the outcrops, filling millimeter- to centimeter-thick shear veins, bounded by striated shear surfaces, and containing different shear bands separated by internal shear surfaces. The microstructure of the shear bands, with intervals of sediment separating intervals of calcite cement (Figure 8c), indicates that the fracture was formed by a great number of millimetric openings developed successively by a crack-seal mechanism (Labaume et al., 1991; Travé et al., 1997). The cement consists of subhedral to anhedral elongated crystals, 40 μm thick and as much as 5 mm (0.2 in.) long, containing

FIGURE 8. Calcite cements in Beseit and Herbers sections. Beseit section: (a) Anhedral to subhedral calcite crystals, with abundant, slightly deformed, mechanical twin planes and subangular extinction, affected by stylolite surfaces filling the strike-slip faults developed in the Tertiary conglomerates (sample BCC 2). (b) Euhedral to subhedral calcite crystals, with nondeformed, mechanical twin planes and uniform extinction, filling a major thrust fault separating Lower Jurassic dolomitic breccia and Tertiary conglomerates (sample BLP 4). Herbers section: (c) Veins filling strike-slip faults and bed-to-bed sliding planes developed in Barremian lacustrine limestones and marls. The veins consist of different shear bands, separated by shear surfaces. The microstructure of the shear bands consists of intervals of sediment separating intervals of calcite cement interpreted as being formed by the crack-seal mechanism (first generation of calcite cements, sample BHF 9). (d) The calcite cement of the shear bands consists of anhedral to subhedral calcite crystals, elongated parallel to the shear band boundary, with abundant, deformed, mechanical twin planes, subangular extinction, and subgrain formation (first generation of calcite cements, sample BHF 15). (e) Second-generation calcite cements filling strike-slip faults and bed-to-bed sliding planes developed in Barremian lacustrine limestones and marls. The cement is made up of subhedral to anhedral calcite crystals with nondeformed, mechanical twin planes (sample BHF 12). (f) Baroque dolomite constituting the last generation of cement (sample BHF 11).

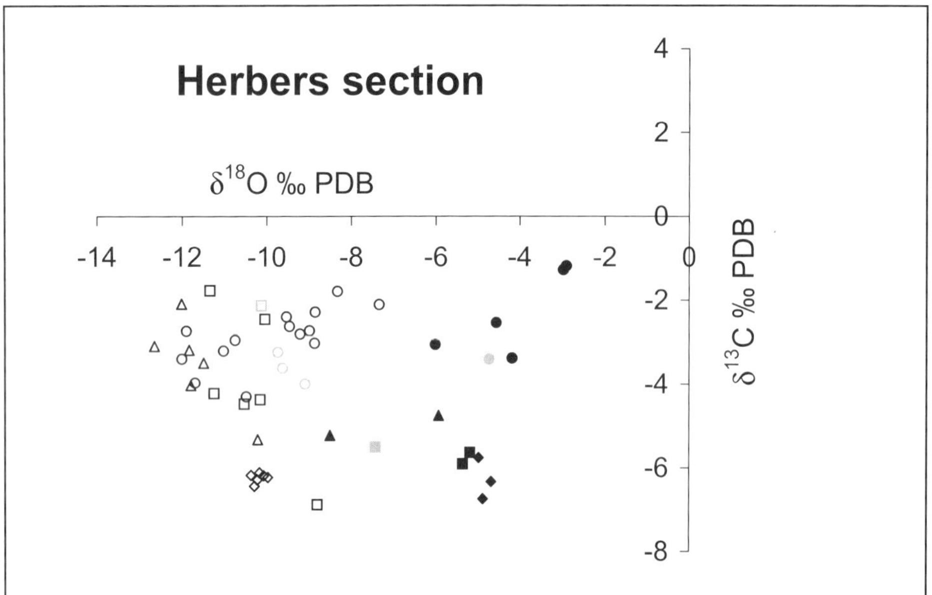

Outcrop		Host Rock	Cement	Structure	Location
BHA	■	Barremian limestones and marls	□	strike-slip / bed-to-bed slidings	interior belt
BHB	■ (grey)	Barremian limestones and marls	□ (grey)	strike-slip / bed-to-bed slidings	interior belt
BHC	▲	Barremian limestones and marls	△	strike-slip / bed-to-bed slidings	interior belt
BHD	◆	Barremian limestones and marls	◇	strike-slip / bed-to-bed slidings	interior belt
BHE	● (grey)	Barremian limestones and marls	○ (grey)	strike-slip / bed-to-bed slidings	interior belt
BHF	●	Barremian limestones and marls	○	strike-slip / bed-to-bed slidings	interior belt

FIGURE 10. $\delta^{18}O$ and $\delta^{13}C$ crossplot of the samples from the Herbers section.

abundant, deformed, mechanical twin planes and elongated parallel to the shear band boundary (Figure 8d).

The second generation, only found in BHA, BHB, BHC, and BHF outcrops, is made up of equant, subhedral to anhedral calcite crystals, from 40 μm to as much as 1 cm in size, with nondeformed, mechanical twin planes (Figure 8e). The calcite cement of these two generations does not occlude the total fracture porosity. A third generation of cement that is very low widespread is made up of baroque dolomite (Figure 8f), but its small size did not allow it to be analyzed.

Stable Isotopes

The $\delta^{18}O$ and $\delta^{13}C$ values of the host limestones and marls, ranging from −8.5 to −2.9‰ PDB and from −6.7 to −1.2‰ PDB, respectively (Figure 10; Table 3d), indicate that these rocks were mainly deposited in continental waters, which is consistent with the interpretation of freshwater carbonates deposited in an estuarine gulf based on the charophyte content (Salas et al., 1995). The two host limestones and marls of the BHF outcrop, with $\delta^{18}O$ values ranging from −3.0 to −2.9‰ PDB and $\delta^{13}C$ values ranging from −1.3 to −1.2‰ PDB, may indicate more important marine influence during their deposition in this estuarine context. The $\delta^{13}C$ trend from less negative values in the stratigraphically lower outcrops (BHE and BHF) to more negative values in the stratigraphically higher outcrops (BHA to BHD) can be explained by a decrease though time, during deposition, of the input of light organogenic carbon to the system (Irwin et al., 1977) and/or a change to more marine conditions.

The $\delta^{18}O$ values of the calcite cements are always more negative, and the $\delta^{13}C$ values are commonly similar or more positive than their adjacent host rock (Figure 10). Calcite cements sampled in this section show $\delta^{18}O$ values ranging between −12.7 and −7.3‰ PDB and $\delta^{13}C$ values ranging between −6.9 and −1.8‰ PDB (Figure 10; Table 3d). These values indicate that the calcite cements might have precipitated from meteoric waters under near-surface temperature conditions.

In relation to the two generations of calcite cements petrologically distinguished, some significant differences are observed with respect to their $\delta^{18}O$ and $\delta^{13}C$ composition. The first generation shows $\delta^{18}O$ values more negative and $\delta^{13}C$ values similar to those of their adjacent host rock (Figure 10; Table 3d), whereas the second generation shows $\delta^{18}O$ values more negative than the $\delta^{18}O$ values of the first cement generation and a narrow range of the $\delta^{13}C$ values (Figure 11), with no similarity to their adjacent host rock.

From these data, it can be inferred that the first generation of calcite cements precipitated from meteoric water derived directly from the surface. The similar carbon isotopic composition of these cements and the host rock in each outcrop is probably caused by the buffering of the pore-water carbon isotopic composition by the extensive dissolution of the host limestone and marl (Marshall, 1992), resulting in a geochemical composition of the fluids similar to the values of the host rock and indicating a high rock-fluid interaction and, therefore, a relatively closed paleohydrogeological system during this early stage. The difference between the $\delta^{18}O$ values of the cements and of those of the host rock may reflect differences in the meteoric water

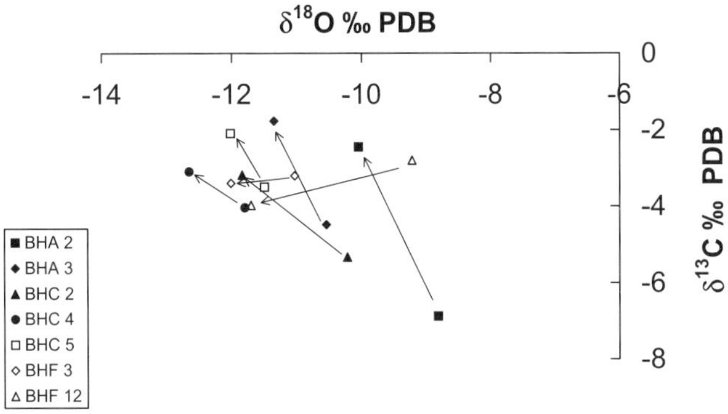

FIGURE 11. $\delta^{18}O$ and $\delta^{13}C$ crossplot of seven samples from the Herbers section showing the isotopic evolution of the two generations of calcite cements developed in the compressive structures of the interior belt. The arrows go from the first generation to the second generation of calcite cement in each sample.

composition or in precipitation temperature. The burial depth of this section during formation of the fractures could account for the increase (around 15°C) in precipitation temperature of the calcite cements with respect to the fluid in equilibrium with the host rock.

In contrast, the more negative $\delta^{18}O$ values of the second generation of calcite cements suggest that these cements might have precipitated from hotter meteoric water, probably ascending fluids, and the difference of $\delta^{13}C$ values between these cements and the host rock, together with the more radiogenic $^{87}Sr/^{86}Sr$ values (see the next section), indicates the input of an external fluid that did not interact with the host limestone and therefore constituted a relatively open paleohydrogeological system during this later stage.

$^{87}Sr/^{86}Sr$

The host Barremian lacustrine limestones show a $^{87}Sr/^{86}Sr$ ratio ranging from 0.70750 to 0.70751 (Figure 6; Table 1), which is consistent with meteoric fluids that interacted with marine Triassic or Lower Jurassic rocks (Koepnick et al., 1990; Jones et al., 1994a, b).

The two generations of calcite cements filling the compressive structures also show a different $^{87}Sr/^{86}Sr$ signature. The $^{87}Sr/^{86}Sr$ ratio of the first generation of calcite cements ranges from 0.70766 to 0.70768 (Figure 6; Table 1), which may be interpreted as a precipitation from meteoric fluids having interacted with marine Triassic or Lower Jurassic rocks (Burke et al., 1982; Koepnick et al., 1990; Jones et al., 1994a, b).

The only analyzed sample of the second-generation calcite cement shows a higher $^{87}Sr/^{86}Sr$ value (0.70792; Figure 6; Table 1), which indicates precipitation from radiogenic fluids. Interaction of the fluids with marine Carboniferous, Permian, or Triassic rocks or with Lower Jurassic dolomites can explain this signal.

Elemental Geochemistry

In the Herbers section, elemental geochemistry analysis was done on both the host rocks and cements. The host Barremian lacustrine limestone contains between 2480 and 5910 ppm of Mg and between 765 and 4835 ppm of Fe, with Mn and Sr mostly below the detection limit. The two samples of the first group of calcite cement (BHA 5 and BHD 1) contain between 635 and 5150 ppm of Mg, between 310 and 11490 ppm of Fe, and between a value below the detection limit and 795 ppm of Sr, with Mn always below the detection limit. The calculated Mg/Ca molar ratio of the precipitating fluid ranges between 0.13 and 1.85. The second group of calcite cement (sample BHF 6) contains similar quantities of Mg (between 480 and 5240 ppm), lower values of Fe (between 295 and 1155 ppm), Sr values ranging from below the detection limit to 925 ppm, and Mn values from below the detection limit to 345 ppm. The calculated Mg/Ca molar ratio of the precipitating fluid for this second generation of calcite cements (0.03–1.11) is in the same range as that calculated for the first generation. These values indicate that both generations of cement that precipitated from meteoric waters evolved toward a formation water composition.

SYNTHESIS OF RESULTS

The integration of the structural study of the fractures (Figure 12) with the petrologic and geochemical study of the cements filling the fractures in the Linking Zone allows to differentiate five groups of calcite cements (Table 5), four associated with compressive structures and one with extensional faults.

The first group of calcite cements (a, Table 5) is formed by 20 of the studied samples and fill the compressive structures in the Tertiary conglomerates of the frontal and central belts (BEP, BPB, and BRE outcrops). They are characterized by $\delta^{18}O$ ranging from −9 to −3.8‰ PDB, $\delta^{13}C$ ranging from −5.6 to −2.6‰ PDB, a low $^{87}Sr/^{86}Sr$ ratio of 0.70768, between 440 and 3565 ppm of Mg, between 275 and 720 ppm of Fe, between 180 and 410 ppm of Mn, and Sr content always below the detection limit. From these data, it has been inferred that they formed in an open paleohydrogeological system by a meteoric water derived directly from the surface, and that they had a low interaction with the conglomerates (surface runoff).

The second group of calcite cements (b, Table 5) is formed by 72 of the studied samples and fills the compressive structures (thrust faults, strike-slip faults, and bed-to-bed slidings) that were developed mostly in the Mesozoic cover and locally in the Tertiary

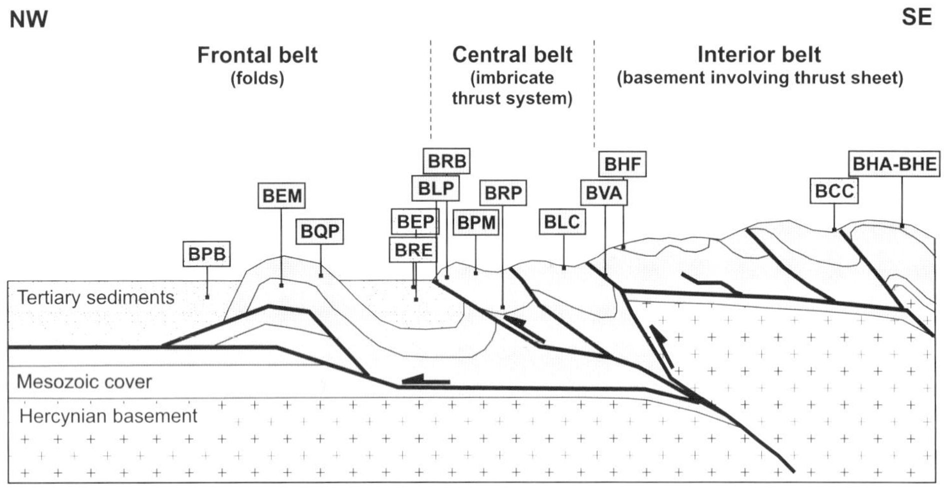

FIGURE 12. Schematic cross section of the Linking Zone showing the structural location of the studied outcrops. White rectangle: outcrops with a Tertiary conglomerate host rock; gray rectangle: outcrop with a Mesozoic host rock. Arrows indicate the thrust planes.

conglomerates. They were found in the four studied cross sections in the interior belt (BHA-BHF, BVA, and BCC outcrops), in the central belt (BLP, BPM, BRP, and BRB outcrops), and in the frontal belt (BEM outcrop). The cements of this group are rather similar to those of the first group. They are characterized by $\delta^{18}O$ ranging from −10.1 to −5.4‰ PDB, $\delta^{13}C$ ranging from −7.5 to −3.5‰ PDB, a low $^{87}Sr/^{86}Sr$ ratio ranging between 0.70759 and 0.70778, between 350 and 5150 ppm of Mg, variable content of Fe (from below the detection limit in the cements affecting marine limestones to as much as 11,490 ppm in the cements affecting lacustrine limestones), Mn content below the detection limit, and variable content of Sr (from below the detection limit to as much as 795 ppm). From these data and by comparison with the host rock, it has been deduced that they were also formed from a meteoric water derived directly from the surface. However, in this case, a fluid that interacted more with the host rock (subsurface infiltration), which therefore is a fluid that circulated in a more closed hydrological system, is considered. The negative $\delta^{13}C$ values suggest soil-derived CO_2.

The third group of calcite cements was only found in two samples from the antiformal stack of the frontal belt in the Gandesa section (outcrop BEM), filling a thrust fault developed in Lower Jurassic rocks. They are characterized by $\delta^{18}O$ ranging from −3.4 to −3.3‰ PDB, $\delta^{13}C$ ranging from −4.0 to −3.8‰ PDB, a very high $^{87}Sr/^{86}Sr$ ratio of 0.70845, between 475 and 4200 ppm of Mg, Fe content always below the detection limit, between 180 and 655 ppm of Mn, and between 280 and 850 ppm of Sr. The fluids that generate these cements were high radiogenic and suboxic meteoric fluids in origin that evolved to a formation water composition because of interaction with the adjacent Lower Jurassic dolomite. Topographic relief was probably the main driving force for meteoric water migration.

The fourth group of calcite cements associated with compressive structures was recognized in 58 samples from the frontal belt of the Gandesa and Beseit sections (BBCC and BQP outcrops), in the central belt of the Gandesa and Beseit sections (BEP and BLP outcrops), and in the interior belt of the Herbers section (BHA, BHC, and BHF outcrops). These cements are mainly found not only in the structures developed in the Mesozoic cover but also locally in the Tertiary conglomerates. This fourth group is characterized by $\delta^{18}O$ ranging from −12.7 to −8.1‰ PDB, $\delta^{13}C$ ranging from −6.4 to −3.9‰ PDB, a $^{87}Sr/^{86}Sr$ ratio ranging between 0.70792 and 0.70795, between 260 and 5240 ppm of Mg, variable content of Fe (from below the detection limit to as much as 1155 ppm), variable content of Mn (from below the detection limit to as much as 925 ppm), and variable content of Sr (from below the detection limit to as much as 925 ppm). These values and their comparison with those of the host rocks indicate that this group of calcite cements derived from hot meteoric fluids that evolved to a formation water composition because of interaction with either the Hercynian rocks or the Lower Jurassic dolomites. This fluid, according to these data, circulated in a relatively open paleohydrogeological system. Topographic relief was probably the main driving force for meteoric water migration.

The fifth group of calcite cements is related to extensional structures and was only found in seven samples from the Orta de Sant Joan section, filling the normal faults affecting Upper Jurassic limestones in the central belt (BLC outcrop). They are characterized by $\delta^{18}O$ ranging from −9.2 to −7.6‰ PDB, $\delta^{13}C$ ranging from −2.6 to +1.1‰ PDB, a high $^{87}Sr/^{86}Sr$ ratio of 0.70788, between 870 and 2695 ppm of Mg, between 470 and 1420 ppm of Mn, and Fe and Sr content always below the detection limit. These values indicate that the cements precipitated from suboxic meteoric waters circulating in a relatively open paleohydrogeological system. The $\delta^{13}C$ values indicate that the C for calcite precipitation in the fractures could have arrived from the adjacent limestones.

DISCUSSION

Episodes of Fluid Circulation

Considering the structural evolution of the Linking Zone, the five differentiated groups of calcite cements can be attributed to three main episodes of fluid circulation (Figure 13), two related to the development of the fold and thrust belt, and the other related to a later extensional phase that affected very locally the innermost part of the Orta de Sant Joan section.

The first episode includes the first and second groups of calcite cements (a and b, Table 5). It corresponds to the early stages of the fold and thrust development that are characterized by (1) the emplacement of thrust sheets of the interior and the central belts that are detached at the Triassic shales and evaporites; and (2) the sedimentation of most of the present-day outcropping Tertiary conglomerates. It is also a period during which the compressive belt widened quickly, creating a wide but relatively low relief formed by Mesozoic rocks (M. Jones, P. Haller, E. Roca, M. Garcés, L. Cabrera, 2004, personal communication).

Accordingly, the geochemical features of the first and second groups of calcite cements indicate that they precipitated from meteoric waters that were derived directly from the surface that did not interact with the Paleozoic basement. The geochemical values of the calcite cements of the first group indicate that this meteoric water circulated through the undeformed and highly porous Tertiary conglomerates in an open paleohydrogeological system with very low interaction with the host rock (surface runoff). In contrast, the geochemical values of the calcite cements of the second group indicate that this meteoric water evolved to a formation water composition when it circulated through the Mesozoic rocks because of a higher interaction with the adjacent rock in a more closed paleohydrogeological system (subsurface infiltration). This second group of calcite cements occurred in the central and interior belts that are formed by several folded limestone and dolomite bodies (thrust sheets) that were isolated one from the other by Triassic shales and evaporites (detachment level). The low $^{87}Sr/^{86}Sr$ ratio of these cements (between 0.70759 and 0.70778) indicates that the thrust faults detached at the Keuper and Middle Muschelkalk shales and evaporites ($^{87}Sr/^{86}Sr$ ratio between 0.7075 and 0.7078) acted as barriers for these fluids, forcing an upward migration parallel and above them.

The second episode corresponds to the latest stages of the compressive development of the Linking Zone. During these stages, compression led to (1) the uplift of the interior belt by the motion of the sole basement thrust; (2) the development of the frontal belt, where the Mesozoic cover and the overlying Tertiary conglomerates were folded in complex anticlines (e.g., antiformal stack of the Gandesa frontal fold); and (3) the development of some out-of-sequence thrusts in the central belt. These structures denote that compressive deformation did not propagate toward the foreland as fast as during the previous stages, and it was concentrated longer in a given area. As a result, during these last stages, there was a major uplift in the Linking Zone and, thus, an increase in relief. This relatively high relief was mainly concentrated in the interior belt, where it was cored by Hercynian rocks and in the frontal folds.

The geochemical features of the third and fourth groups (c and d, Table 5) of cements are in accordance with this tectonic scenario. The elemental geochemistry and the high $^{87}Sr/^{86}Sr$ ratio of both cements indicate that they precipitated from meteoric fluids that evolved toward a formation water composition either arriving from the interior belt and having interacted with the siliciclastic Hercynian basement or arriving directly from the surface and having interacted with the Tertiary conglomerates. The $\delta^{18}O$ and $\delta^{13}C$ values indicate for the third group a very high interaction of the fluid with the adjacent Lower Jurassic rocks. Precipitation of this group of cements in the frontal folds (c, Table 5) could have occurred before folding, when the Mesozoic was flat and covered by the impermeable mudstones of the lower part of the Tertiary sequence, or after folding, when the Mesozoic rocks cropped out along the hinge of the fold. In the first case, the fluids probably derived from the interior belt, with a relatively low relief, whereas in the second case, meteoric fluids could arrive directly from the outcropping Mesozoic rocks. For the fourth group of cements, the stable isotope values indicate relative high temperatures of precipitation that fit very well with the last stages of the building of the Linking Zone, when a high topographic gradient appeared between the interior belt and the foreland basin.

The third episode, postcompressive and represented by the fifth group of calcite cements (e, Table 5), was recognized very locally in the southernmost part of the Orta de Sant Joan section. It developed during the Neogene extension that created the western Mediterranean, and it is characterized by the formation of normal faults. The geochemical features of the cements filling those faults indicate that the precipitating fluid was a meteoric water that interacted with the host Upper Jurassic limestones and with the underlying Lower Jurassic dolomites.

New Clues on Fluid Migration during Fold and Thrust Belt Development

The study of the fluid-flow evolution during the Linking Zone fold and thrust belt development shows that the main factors controlling the fluid-flow dynamics are the existence of a high relief, whereas the host rocks and the type of structure are not important factors.

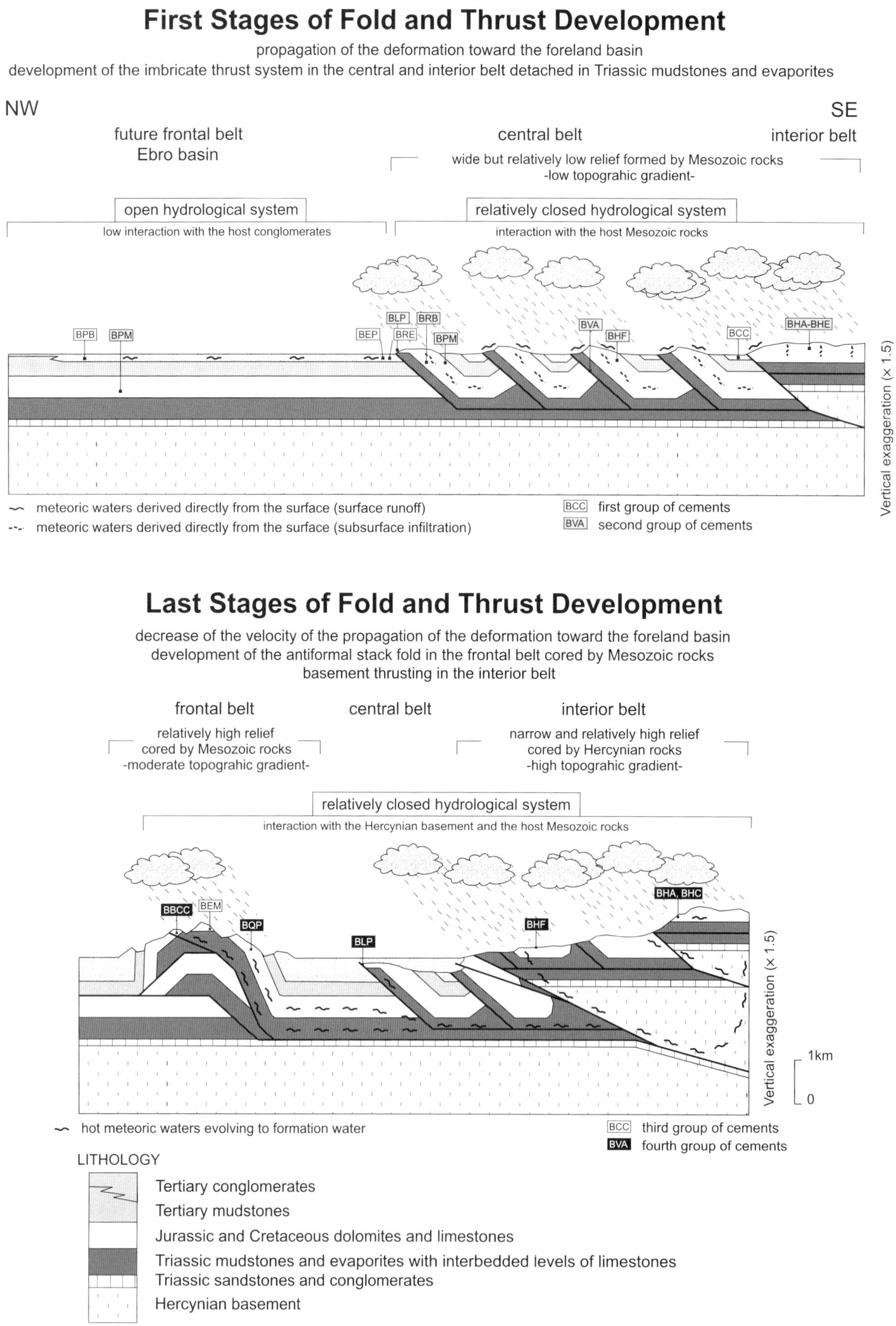

FIGURE 13. Model of the fluid evolution during compressive deformation in the studied area.

Thus, before the development of the main relief, when thrusts were emplaced in the interior and central belts, only meteoric waters without interaction with Paleozoic rocks have been detected. These waters circulated through the Tertiary conglomerates, in an open paleohydrogeological system, and through the Mesozoic rocks, in a more closed paleohydrogeological system, probably being the Triassic shales and evaporites of the detachment levels, the barrier against which fluids were forced to flow.

In contrast, when a high relief was created during the major uplift of the interior belt, synchronously to development of the frontal belt and to the emplacement of some out-of-sequence thrust in the central belt, the detected fluids are meteoric in origin that changed to a formation water composition because of interaction with the Paleozoic rocks. These fluids circulated through the Mesozoic carbonates and through the Tertiary conglomerates, having been interacted with Hercynian siliciclastic rocks and Mesozoic carbonates and sometimes at high temperatures.

This fluid-flow evolution observed during the Linking Zone fold and thrust belt development shows a parallelism with the fluid-flow evolution during Eocene thrust emplacement in the south Pyrenean foreland basin (Travé et al., 1998). In both cases, two main episodes of fluid circulation are observed: an earlier episode dominated by local fluids and a later episode where the local fluids were mixed with external fluids. The local fluids in the south Pyrenean foreland basin were cognate seawaters, and the local fluids in the Linking Zone were meteoric waters. The external fluids were, in both cases, relatively hot fluids arriving from the underthrusted parts of the basin.

CONCLUSIONS

The results of this work allowed us to establish the following conclusions on fluid-flow migration during foreland and fold and thrust belt development of the Linking Zone.

1) Two episodes of fluid circulation, with circulation of four different fluids, have been recognized during the Paleogene compression in the Linking Zone.
2) The fluids circulating through the studied structure during the whole compressive deformation were meteoric fluids in origin. During the different episodes of fluid circulation, these meteoric fluids were more or less modified to a formation water composition.
3) Surface runoff, subsurface infiltration, and interaction with different rocks have been recognized. Interaction between the fluids and the Triassic shales and evaporites are only recorded during the early stages of the fold and thrust belt development. At that time, the Triassic rocks were probably used as detachment horizons for the thrust faults and acted as barriers, forcing the lateral migration of meteoric fluids. Interaction between the fluids and Hercynian rocks are only recorded during the last stages of fold and thrust belt development. Interaction of the fluids with the host Lower Jurassic dolomites is recorded during both stages of compressive deformation.
4) In the Linking Zone, the fluid-flow dynamics was controlled by the existence of a high relief, whereas the host rock and the type of structure were not important factors.

ACKNOWLEDGMENTS

The $^{87}Sr/^{86}Sr$ analyses were carried out at the C.A.I. de Geocronología y Geoquímica Isotópica of the Universidad Complutense of Madrid, and the $\delta^{18}O$ and $\delta^{13}C$ isotope and electron microprobe analyses were carried out at Serveis cientificotècnics of the Universitat de Barcelona. This research was realized in the framework of Dirección General de Investigación del Ministerio de Ciencia y Tecnología Project BTE2002 4453-CO2-01 and Grup Consolidat de Recerca Geologia Sedimentaria 2001/SGR/75. We are indebted to R. Swennen, P. Labaume, and M. V. Murgia for their valuable criticism in reviewing the manuscript and their useful comments, which improved the manuscript. We also wish to thank F. Luttikhuizen for the revision of the English version.

REFERENCES CITED

Bethke, C. M., and S. Marshak, 1990, Brine migrations across North America— The plate tectonics of groundwater: Annual Review of Earth and Planetary Sciences, v. 18, p. 287–315.

Budai, J. M., 1985, Evidence for rapid fluid migration during deformation, Madison Group, Wyoming and Utah overthrust belt, *in* M. W. Longman, K. W. Shanley, R. F. Lindsay, and D. E. Eby, eds., Rocky Mountain carbonate reservoirs— A core workshop: SEPM Core Workshop, v. 7, p. 377–407.

Burke, W. H., R. E. Denison, E. A. Hetherington, R. B. Koepnick, H. F. Nelson, and J. B. Otto, 1982, Variation of seawater $^{87}Sr/^{86}Sr$ throughout Phanerozoic time: Geology, v. 10, p. 516–519.

Burkhard, M., and R. Kerrich, 1988, Fluid regimes in the deformation of the Helvetic nappes, Switzerland, as inferred from stable isotope data: Contributions to Mineralogy and Petrology, v. 99, p. 416–429.

Burrus, R. C., K. R. Cercone, and P. M. Harris, 1985, Timing of hydrocarbon migration: Evidenced from fluid inclusions in calcite cements, tectonics and burial history, *in* N. Schneidermann and P. M. Harris, eds., Carbonate cements: SEPM Special Publication 36, p. 277–289.

Calvet, F., A. Travé, E. Roca, A. Soler, and P. Labaume, 1996, Fracturación y migración de fluidos durante la evolución tectónica Neogena en el Sector Central de las Cadenas Costero Catalanas: Geogaceta, v. 20, p. 1715–1718.

Cosgrove, J. W., 1993, The interplay between fluids, folds and thrusts during deformation of a sedimentary succession: Journal of Structural Geology, v. 15, p. 491–500.

Craig, H., and I. Gordon, 1965, Deuterium and oxygen-18 variations in the ocean and marine atmosphere, *in* E. Tongiorgi, ed., Stable isotopes in oceanographic studies and paleotemperatures: Pisa, Italy, Consiglio Nazionale delle Richerche, Laboratorio di Geologia Nucleare, p. 9–130.

Evans, M. A., and D. A. Battles, 1999, Fluid inclusion and stable isotope analyses of veins from the central Appalachian Valley and Ridge province: Implications for regional synorogenic hydrologic structure and fluid migration: Geological Society of American Bulletin, v. 111, p. 1841–1860.

Garden, I. R., S. C. Guscott, S. D. Burdley, K. A. Foxford, J. J. Walsh, and J. D. Marshall, 2001, An exhumed palaeo-hydrocarbon migration fairway in a faulted carrier system, Entrada Sandstone of SE Utah, U.S.A.: Geofluids, v. 1, p. 195–213.

Guimerà, J., 1984, Palaeogene evolution of deformation in the northeastern Iberian Peninsula: Geological Magazine, v. 121, p. 413–420.

Howson, M. R., A. D. Pethybridge, and W. A. House, 1987, Synthesis and distribution coefficients of low-magnesium calcites: Chemical Geology, v. 64, p. 79–87.

Irwin, H., C. Curtis, and M. Coleman, 1977, Isotopic evidence for source of diagenetic carbonates formed during burial of organic-rich sediments: Nature, v. 269, p. 209–213.

Jones, C. E., H. C. Jenkyns, and S. P. Hesselbo, 1994a, Strontium isotopes in Early Jurassic seawater: Geochimica et Cosmochimica Acta, v. 58, p. 1285–1301.

Jones, C. E., H. C. Jenkyns, and S. P. Hesselbo, 1994b, Strontium isotopic variations in Jurassic and Cretaceous seawater: Geochimica et Cosmochimica Acta, v. 58, p. 3061–3074.

Kinsman, D. J. J., 1969, Interpretation of Sr^{2+} concentration in carbonate minerals and rocks: Journal of Sedimentary Petrology, v. 39, p. 486–508.

Koepnick, R. B., R. E. Denison, W. H. Burke, E. A. Hetheringtonn, and D. A. Dahl, 1990, Construction of the Triassic and Jurassic portion of the Phanerozoic curve of seawater $^{87}Sr/^{86}Sr$: Chemical Geology (Isotope Geoscience Section), v. 80, p. 327–349.

Labaume, P., and I. Moretti, 2001, Diagenesis— Dependence of cataclastic thrust fault zone sealing in sandstones. Example from the Bolivian sub-Andean zone: Journal of Structural Geology, v. 23, p. 1659–1675.

Labaume, P., C. Berty, and P. Laurent, 1991, Syn-diagenetic evolution of shear structures in superficial nappes: an example from the northern Apennines (NW Italy): Journal of Structural Geology, v. 13, p. 385–398.

Labaume, P., S. M. F. Sheppard, and I. Moretti, 2001, Fluid flow in cataclastic thrust fault zones in sandstones, sub-Andean zone, southern Bolivia: Tectonophysics, v. 340, p. 141–172.

Lawton, T. F., E. Roca, and J. Guimerà, 1999, Kinematic-stratigraphic evolution of a growth syncline and its implications for tectonic development of the proximal foreland basin, southeastern Ebro Basin, Catalunya, Spain: Geological Society of America Bulletin, v. 111, p. 412–431.

Machel, H. G., P. A. Cavell, B. E. Buschkuehle, and K. Michael, 2000, Tectonically induced fluid flow in Devonian carbonate aquifers of the Western Canada sedimentary basin: Journal of Geochemical Exploration, v. 69–70, p. 213–217.

Marquer, D., and M. Burkhard, 1992, Fluid circulation, progressive deformation and mass-transfer processes in the upper crust: The example of basement-cover relationships in the external crystalline massifs, Switzerland: Journal of Structural Geology, v. 14, p. 1047–1057.

Marshall, J. D., 1992, Climatic and oceanographic isotopic signals from the carbonate rock record and their preservation: Geological Magazine, v. 129, p. 143–160.

McCaig, A. M., D. M. Wayne, J. D. Marshall, D. Banks, and I. Henderson, 1995, Isotopic and fluid inclusion studies of fluid movement along the Gavernie thrust, central Pyrenees: Reaction fronts in carbonate mylonites: American Journal of Science, v. 295, p. 309–343.

McIntire, W. L., 1963, Trace element partition coefficients— A review of theory and applications to geology: Geochimica et Cosmochimica Acta, v. 27, p. 1209–1264.

Muchez, P., M. Slobodnik, W. A. Viaene, and E. Keppens, 1995, Geochemical constraints on the origin and migration of palaeofluids at the northern margin of the Variscan foreland, southern Belgium: Sedimentary Geology, v. 96, p. 191–200.

Popp, B. N., F. A. Podosek, J. C. Brannon, T. F. Anderson, and J. Pier, 1986, $^{87}Sr/^{86}Sr$ ratios in Permo-Carboniferous sea water from the analyses of well-preserved brachiopod shells: Geochimica et Cosmochimica Acta, v. 50, p. 1321–1328.

Qing, H., and E. Mountjoy, 1992, Large-scale fluid flow in the Middle Devonian Presqu'ile barrier, Western Canada sedimentary basin: Geology, v. 20, p. 903–906.

Roca, E., J. Guimerà, and R. Salas, 1994, Mesozoic extensional tectonics in the southeast Iberian Chain: Geological Magazine, v. 131, p. 155–168.

Salas R., and A. Casas, 1993, Mesozoic extensional tectonics, stratigraphy, and crustal evolution during the Alpine cycle of the eastern Iberian Basin: Tectonophysics, v. 228, p. 33–55.

Salas, R., C. Martín-Closas, X. Querol, J. Guimerà, and E. Roca, 1995, Evolución tectonosedimentaria de las cuencas del Maestrazgo y Aliaga-Penyagolosa durante el Cretácico inferior, *in* R. Salas and C. Martín-Closas, eds., El Cretácico inferior del Nordeste de Iberia:

Barcelona, Publicaciones Universitat de Barcelona, p. 13–94.
Salas, R., J. Guimerà, J. Gimenez-Montsant, C. Martín-Closas, and E. Roca, 1997, Mesozoic rift structure, stratigraphy and Palaeogene inversion of the Maestrat basin, Iberian Range (NE Spain): International Geological Correlation Programme 369 Field Trip Guide, 1–3 September, Barcelona: Barcelona, Publicaciones Universitat de Barcelona, 37p.
Salas, R., J. Guimerà, R. Mas, C. Martín-Closas, A. Meléndez, and A. Alonso, 2001, Evolution of the Mesozoic central Iberian rift system and its Cainozoic inversion (Iberian Chain), *in* P. A. Ziegler, W. Cavazza, A. F. H. Robertson, and S. Crasquin-Soleau, eds., Peri-Tethys Memoir 6: Peri-Tethyan rift/wrench basins and passive margins: Mémoires du Muséum National d'Histoire Naturelle, v. 186, p. 145–185.
Skougstad, M. W., and C. A. Horr, 1963, Occurrence and distribution of strontium in natural waters: U.S. Geological Survey Water Supply Paper 1496D, p. 55–97.
Swennen, R., K. Muskha, and F. Roure, 2000, Fluid circulation in the Ionian fold and thrust belt (Albania): Implications for hydrocarbon prospectivity: Journal of Geochemical Exploration, v. 69–70, p. 629–634.
Travé, A., P. Labaume, F. Calvet, and A. Soler, 1997, Sediment dewatering and pore fluid migration along thrust faults in a foreland basin inferred from isotopic and elemental geochemical analyses (Eocene south Pyrenees, Spain): Tectonophysics, v. 282, p. 375–398.
Travé, A., P. Labaume, F. Calvet, A. Soler, J. Tritlla, M. Buatier, J. L. Potdevin, M. Séguret, S. Raynaud, and L. Briqueu, 1998, Fluid migration during Eocene thrust emplacement in the south Pyrenean foreland basin (Spain): An integrated structural, mineralogical and geochemical approach, *in* A. Mascle, C. Puigdefàbregas, H. P. Luterbacher, and M. Fernàndez, eds., Cenozoic foreland basins of western Europe: Geological Society (London) Special Publication 134, p. 163–188.
Travé, A., F. Calvet, M. Sans, J. Vergés, and M. Thirlwall, 2000, Fluid history related to the Alpine compression at the margin of the south Pyrenean foreland basin: The El Guix anticline: Tectonophysics, v. 321, p. 73–102.
Tucker, M. E., and V. P. Wright, 1990, Carbonate sedimentology: Oxford, Blackwell Scientific Publications, 482 p.
Van Geet, M., R. Swennen, C. Durmishi, F. Roure, and P. Muchez, 2002, Paragenesis of Cretaceous to Eocene carbonate reservoirs in the Ionian fold and thrust belt (Albania); relation between tectonism and fluid flow: Sedimentology, v. 49, p. 697–718.
Villena, J., G. Pardo, A. Muñoz, and A. González, 1995, The Tertiary of the Iberian margin of the Ebro Basin: Sequence stratigraphy, *in* P. F. Friend and C. J. Dabrio, eds., Tertiary basins of Spain, the stratigraphic record of crustal kinematics: New York, Cambridge University Press, p. 77–82.

11

Orlov, A., V. Omelchenko, A. Trubenko, and F. Schneider, 2004, Incidence of tectonic stress on overpressure development in Ukrainian basins, *in* R. Swennen, F. Roure, and J. W. Granath, eds., Deformation, fluid flow, and reservoir appraisal in foreland fold and thrust belts: AAPG Hedberg Series, no. 1, p. 245–255.

Incidence of Tectonic Stress on Overpressure Development in Ukrainian Basins

Alexander Orlov

National Technical University of Oil and Gas, Ivano-Frankovsk, Ukraine

Alexander Trubenko

National Technical University of Oil and Gas, Ivano-Frankovsk, Ukraine

Valery Omelchenko

National Technical University of Oil and Gas, Ivano-Frankovsk, Ukraine

Frédéric Schneider

Institut Français du Pétrole, Rueil Malmaison Cedex, France

ABSTRACT

The data from oil-bearing Ukrainian basins outline the tectonic control of overpressure development in areas characterized by recent episodes of folding and active compression (dominantly horizontal stress regime). Based on the data presented, vertical compaction alone is not sufficient to account for the observed overpressure development.

Tentatively, tectonic deformation is proposed as the dominant factor for overpressure development in sedimentary layers in recently folded regions. During folding, the opening of fractures results in local density decrease and helps in driving the migrating fluids, i.e., both hydrocarbon and formation waters, in the carrier beds toward the top of the structures. Ductile flow of shaly interbeds toward anticlinal closures can also enhance the sealing capacity of these layers. In the Ukrainian Carpathians and salt domes of the Dniepr-Donetz Basin, direct relationships have been evidenced to link and predict the amount of overpressure as a function of the amplitude of the folds.

The use of measurements and of the monitoring of overpressures for evaluating the horizontal stress intensity and for the forecasting of earthquakes is also suggested.

INTRODUCTION

Many parameters control fluid migration in the crust, apart from vertical fluid migration during compaction and dewatering of sedimentary strata, which is classified here as primary migration. Lateral, secondary fluid escape might also contribute to oil and gas accumulation and is believed to be of major importance.

This chapter focuses on how folding stimulates secondary fluid migration in sedimentary strata. For

DOI:10.1306/1025693H13119

instance, secondary migration of fluids is thought to be responsible for generating overpressure in sedimentary layers near the crest of anticlinal closures. During this process, a portion of fluids is squeezed and expelled from ductile layers into adjacent permeable rocks. Fracturing and hydrofracturing are associated with that fluid escape, and overpressuring also causes a decrease in bulk density (γ_b), inducing zones of fluid invasion above the overpressured layers.

This study focuses on folded structures of the Borislav-Pokut unit of the Ukrainian Carpathians, as well as other oil-bearing Ukrainian basins.

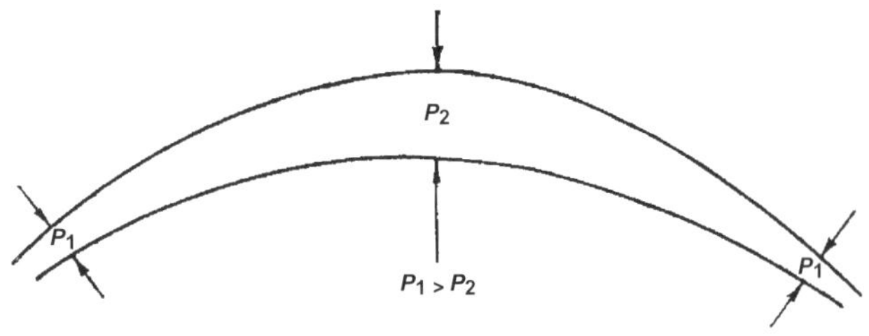

FIGURE 2. Sketch showing correlation between tensions caused by folding (Azhgirey, 1956); P_1 = distorting force along the flanks of the fold; P_2 = distorting force at the crest of the fold.

DISHARMONIC FOLDING AND SECONDARY FLUID MIGRATION IN PRODUCTIVE RESERVOIRS

When interbedded competent and incompetent sedimentary layers are folded, more plastic rocks are squeezed from the flanks toward the fold closures, as shown in Figure 1. As a result, similar or disharmonic folding takes place. The magnitude of distortion is noticeably higher along the flanks than at the crest, as shown in Figure 2. More porous and permeable layers (reservoir rocks) that are interbedded in compacted shales act as fluid collectors, through which fluids are also forced to migrate from the flanks toward the crest of the fold, in analogy with the fold-forming process itself.

Fluid migration in the reservoirs is considerably increased during the fracturing of the rocks. Open fractures act as migration pathways for channeling the fluids. Therefore, the types of fractures developed or reopened during folding is indicative of the process (Belousov, 1962; Hottman and Johnson, 1965; Bulin, 1972). The concentration of fissures and joints in the limbs of folds, as illustrated in Figures 3–6, facilitates the migration of fluids at the onset of the deformation. With progressive strain, however, such collectors become progressively damaged in the limbs of growth anticlines, where most fissures are closed, whereas fractures remain open near the crestal closures of the folds.

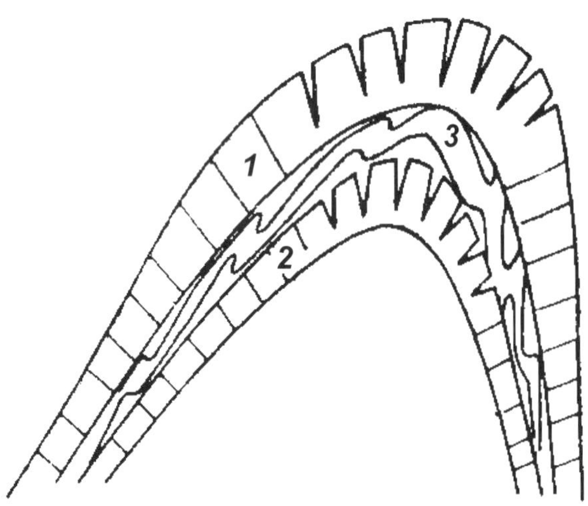

FIGURE 1. Similar or disharmonic folding. 1 and 2 = competent layers; 3 = incompetent rocks, squeezed into the crest of the fold, forming a disharmonic fold (Azhgirey, 1956; Belousov, 1962).

DEVELOPMENT OF FLUID OVERPRESSURE DURING FOLD DEVELOPMENT

High pore-fluid pressures have been encountered in anticlinal structures and mud or salt diapirs in some parts of the Carpathian Neogene orogenic system. For example, Novosiletsky (1961) investigated several Carpathian hydrocarbon-productive fields and observed that the maximum formation pressure in the different reservoir layers always occurred at the crest of the reservoir. On average, overpressure gradients near the anticline crest ranged from 0.135 to 0.157 MPa per 10 m (33 ft) in oil pools to 0.207 MPa per 10 m (33 ft) in aquifer sections. These observations suggest that fluid differentiation based on density occurs in reservoir units as fluid migration progresses. Gas will be forced to the highest part of the reservoir, and liquid hydrocarbons will accumulate at a lower level, below which water will occur. Overpressure will occur in this gas-oil-water system. In areas where folding will be most intense, the highest overpressure values will be encountered. Intra-reservoir fluid migration during folding occurs in a variety of structural environments because the mechanics of fold growth and crack formation are similar. Folds in orogenic belts, in platform domains (where they relate to differential vertical motion of basement blocks associated with draping or folding of overlying sedimentary strata), and in salt domes are all equally potential settings for this process, as illustrated in Figures 7–9.

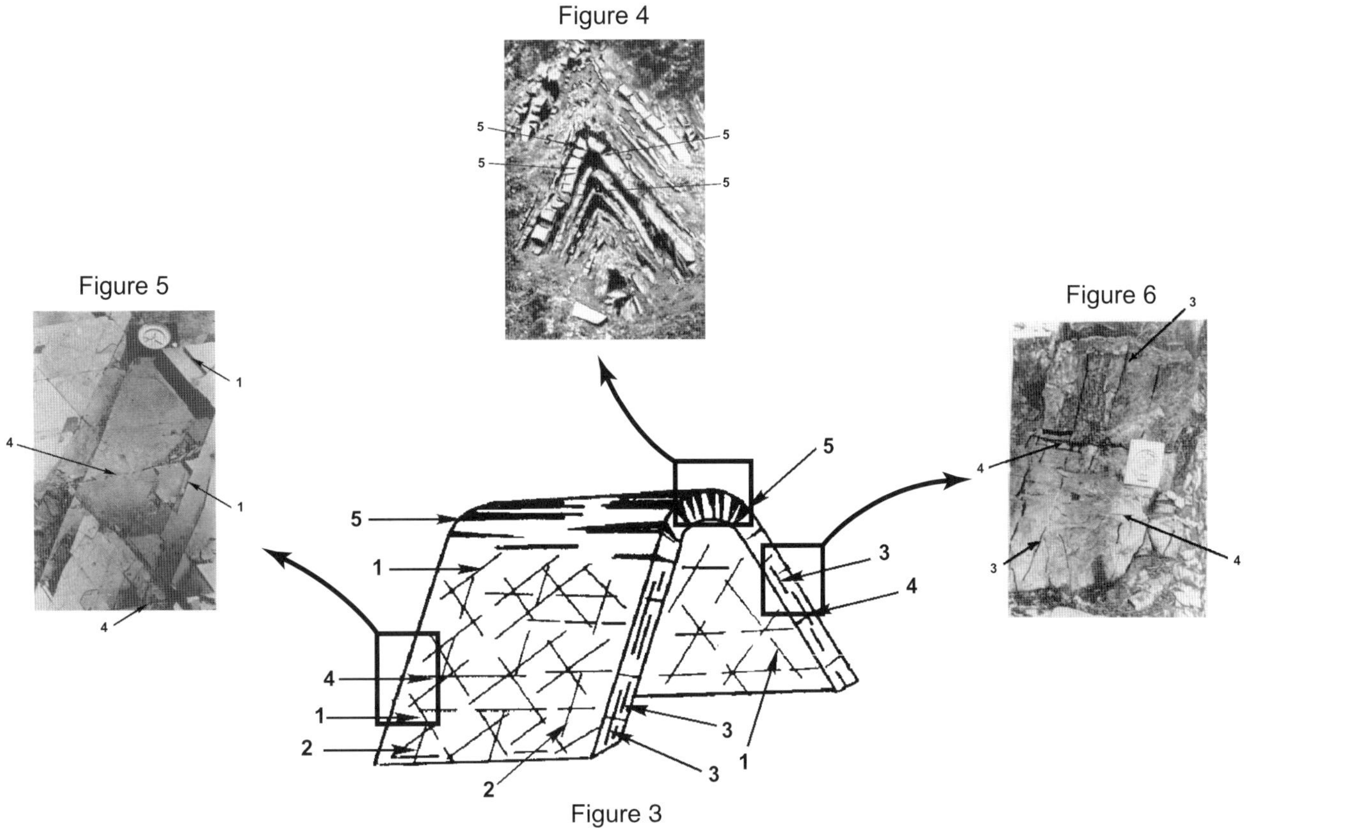

FIGURE 3. Fold-scale distribution of fractures and joints in an anticlinal structure of the Borislav-Pokutt zone in the Ukrainian Carpathians (Orlov, 1980). 1 and 2 = primary conjugate shear joints developing prior to folding; 3 and 4 = shear joints developing during folding; 5 = open joints developing along the outer arc of the fold.

FIGURE 4. Anticlinal fold exposed in the flysch outcrops of the Rybnitsa River (sandstone and argilite deposits of Stryi Svita in the Borislav-Pokutt zone of the Ukrainian Carpathians) (photo by A. Orlov). Outer arc joints with maximum opening in each of the competent layers are visible near the crestal closure of the fold. The joints are opening parallel to the direction of the fold axis; 5 = outer arc joints formed at the crest of the fold.

FIGURE 5. Joints on the bedding surface of a Neogene marl layer on a fold flank ("Brusnyi," inner Carpathian Foredeep, Polianitsa Svita) (photo by A. Orlov). Labeling of joints is the same as in Figure 3.

FIGURE 6. Outcrop section of an Upper Cretaceous sandstone layer exposed at Stryi Svita along the flank of a fold (Pistynka River bank in the Borislav-Pokutt zone of the Ukrainian Carpathians) (photo by A. Orlov). Notice that younger deformation features are parallel and reactivate former shear joints. Labeling of joints is the same as in Figure 3.

Because of the large overpressure in the crest of folds, fluids will tend to migrate progressively from reservoir lithologies toward adjacent strata. This will result in a decrease in formation density, whereby the petrophysical characteristics of the lithologies in the zones of fluid invasion will be affected, as shown in Figures 7–9.

This concept can explain why most overpressures occur in young folded regions and why the thickness of the invasion zones where overpressures develop increases laterally from the flanks toward the crest of an anticline. Thus, the maximum thickness of the invasion zone occurs above the closure of the anticlinal structures.

OVERPRESSURE FORECASTING IN FOLDED REGIONS

We have developed empirical formulas to compute pore-fluid pressure in natural basins over a range of burial depths and distinct values of the fold intensity parameter (i) using geological data derived from oil and gas fields in the inner and outer zones of the Carpathian Foredeep, and from the Dnipro-Donets'k depression. The fold intensity parameter (i) is defined as the ratio of the structure amplitude (h, in meters) to its total area

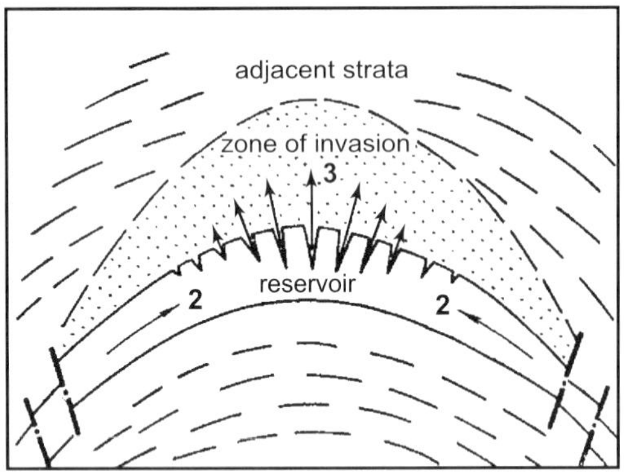

FIGURE 7. Sketch diagram showing the fluid migration occurring in the reservoir during fold development under tangentially directed tectonic stress (Orlov, 1980). 1 = fluid flow inside a reservoir layer; 2 = expulsion of fluids squeezed from a reservoir into adjacent strata; 3 = under conditions of increased high pressure in the crest of fold, fluids migrate progressively from reservoir to the adjacent adjoining rock.

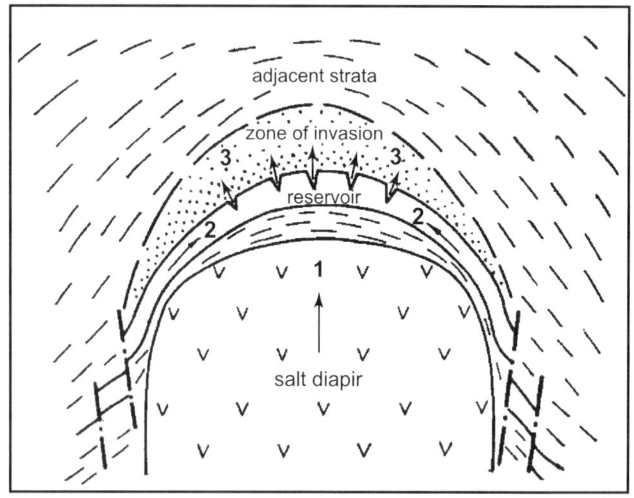

FIGURE 9. Sketch diagram showing the fluid migration occurring during the halokinetic rise of a salt diapir. 1 = direction of salt movement; 2 = fluid flow in a reservoir; 3 = expulsion of fluids squeezed from a reservoir into adjacent strata.

(S, in square meters) inside the last closed isodepth (isohypse), whereby the extrapolation of isodepth contour lines is admitted. Therefore, $i = h/S$ (Orlov, 1980).

With increasing folding, the layer pressures will also increase. The latter is confirmed by the correlation of overpressure factor (k_a) to the value of structure intensity (i). The overpressure factor of a sedimentary layer (k_a) is defined as the ratio of the pore-fluid pressure (P_f) to the conventional hydrostatic pressure, $P_h = P_{atm} + \gamma g z$, expressed in pascals, where P_{atm} is the atmospheric pressure (0.1 MPa); z is the burial depth of the layer (in meters); γ is the density of formation water (1020 kg/m^3); and g is the acceleration caused by gravity (9.81 m/s^2). It should be noted that this formula assumes that the water table is at ground surface.

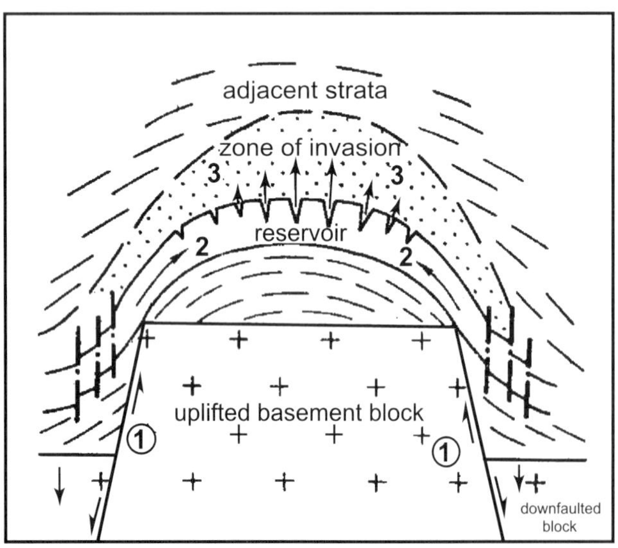

FIGURE 8. Sketch diagram showing the fluid migration occurring in the reservoir during fold development (draping) associated with differential vertical motion of basement blocks. 1 = fault motion between the downfaulted and uplifted basement blocks. 2 = fluid flow inside a reservoir. 3 = expulsion of fluids squeezed from a reservoir into adjacent strata.

The location of petroleum fields from the Carpathians used in this study is shown in Figure 10, whereas the location of petroleum fields from the Dnipro-Donets'k is shown in Figure 11.

Figures 12–14 display empirical linear relations for the function $k_a = f(i)$ from hydrocarbon reservoirs of different petroleum provinces in Ukraine, particularly from the inner and outer Carpathian Foredeep (Figure 10) and the Dnipro-Donets'k depression (Figure 11).

The overpressure empirical correlation law $k_a = f(i)$, obtained by computer calculations, can be defined as: $k_a = Ai + B \pm C$, where A and B refer to constant factors specific for a given geological setting, C constitutes an error factor (referring to the middle quadratic error at a significance level of 0.95, according to Student criterion) and i is the fold intensity parameter.

A, B, and C are calculated by computing the data for specific geological settings ($\pm C$, the mean quadratic error by Student criterion). By substituting empirically derived values for A, B, and C, we used the measured overpressure ratios to compute the initial (i.e., prefolding formation) pressure in natural basins for a given burial depth.

1) For the inner Carpathian Foredeep and the adjacent Skyba foothills zone (Figures 10, 12),

$$k_a^{ICF} = 0.0058i + 0.965 \pm 0.210$$

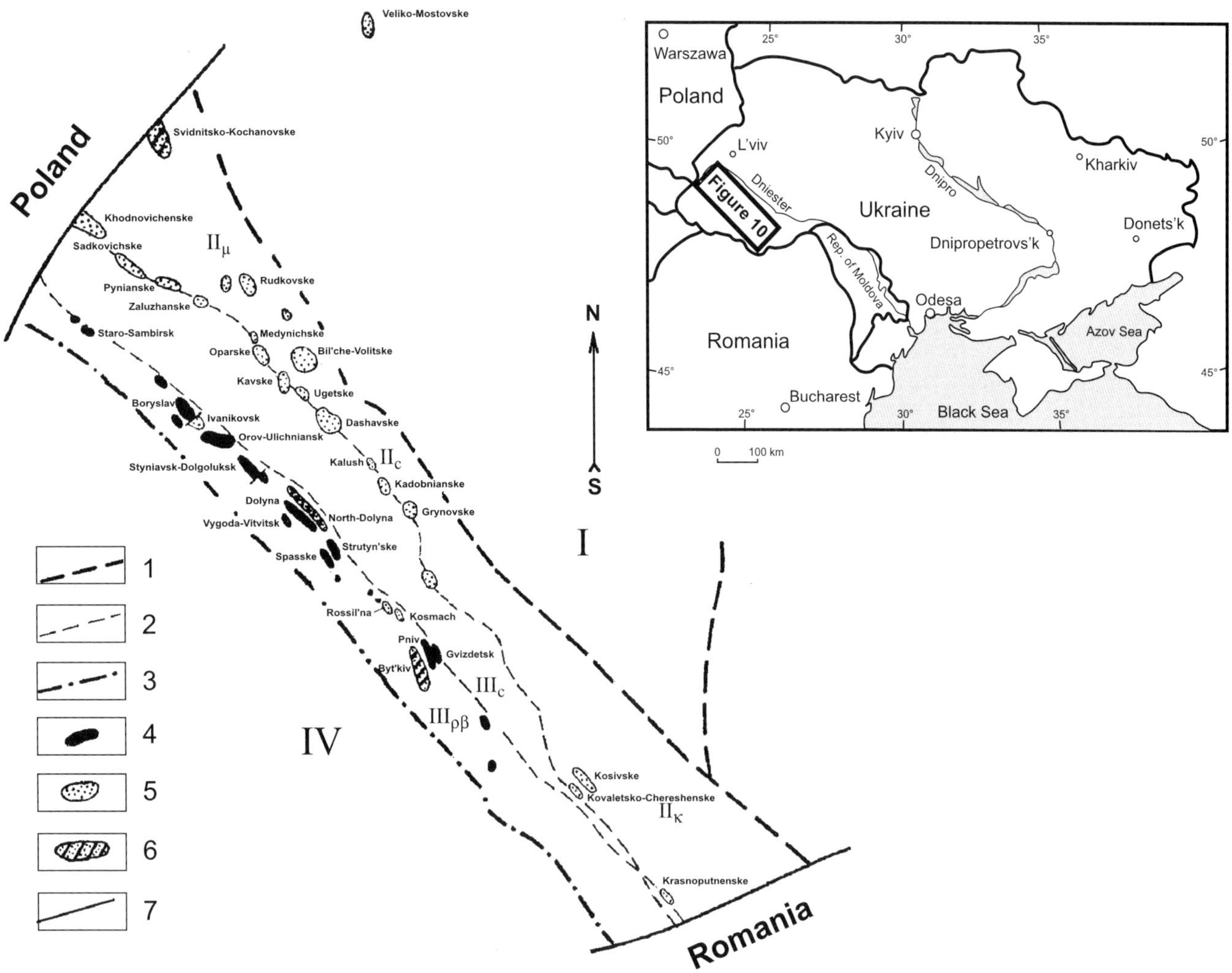

FIGURE 10. Schematic tectonic map of southwestern Ukraine outlining the Skiba zone (allochthon), the Carpathian Foredeep, and the eastern European foreland platform: I = Volyn-Podolsk province of the eastern European foreland platform; II = outer Carpathian Foredeep; II_u = Ugersk-Krukepech depression; II_c = Stanislav uplift; II_k = Kosov depression; III = inner Carpathian Foredeep: III_s = Sambor subzone; III_{pb} = Boryslav-Pokutt subzone; IV = Skiba zone of the Carpathians. 1 = basement faults; 2 = surface thrust fronts; 3 = speculative location of the buried thrust front of III_{pb} Boryslav-Pokutt subzone beneath the Skiba nappe of the Carpathians; 4 = oil fields; 5 = gas fields; 6 = oil and gas fields; 7 = political boundaries.

2) For the outer Carpathian Foredeep (Figures 10, 13),

$$k_a^{OCF} = 0.0180i + 0.914 \pm 0.240$$

3) For the Dnipro-Donets'k depression (Figure 11),

$$k_a^{DDD} = 0.0065i + 1.003 \pm 0.110$$

The fact that B coefficients are very close to 1 indicates that there is no overpressure in nonfolded area (e.g., when i equals 0, the pore pressure equals the hydrostatic pressure). Values greater than 1 could be expected in areas where overpressures have been generated previously to the folding. Thus, a first approximation of the overpressure in a given area could be provided by the following formula:

$$\delta P = P_f - P_h = P_h A i$$

One interesting point for the future will be the study of coefficient A and what it represents. Actually, fold intensity parameter (i) could be seen as a macroscopical strain parameter. In this case, the previous formula could be considered as part of a rheology for porous medium at macroscopical scale, which relates the volumetric strain (ε) to the pore pressure evolution (δP). Thus, A can be seen as a rheological parameter.

The previous relationships, once calibrated in a given area, can be used to predict the amount of overpressure in a folded structure. As a consequence, this method may be applied for forecasting of outburst hazards

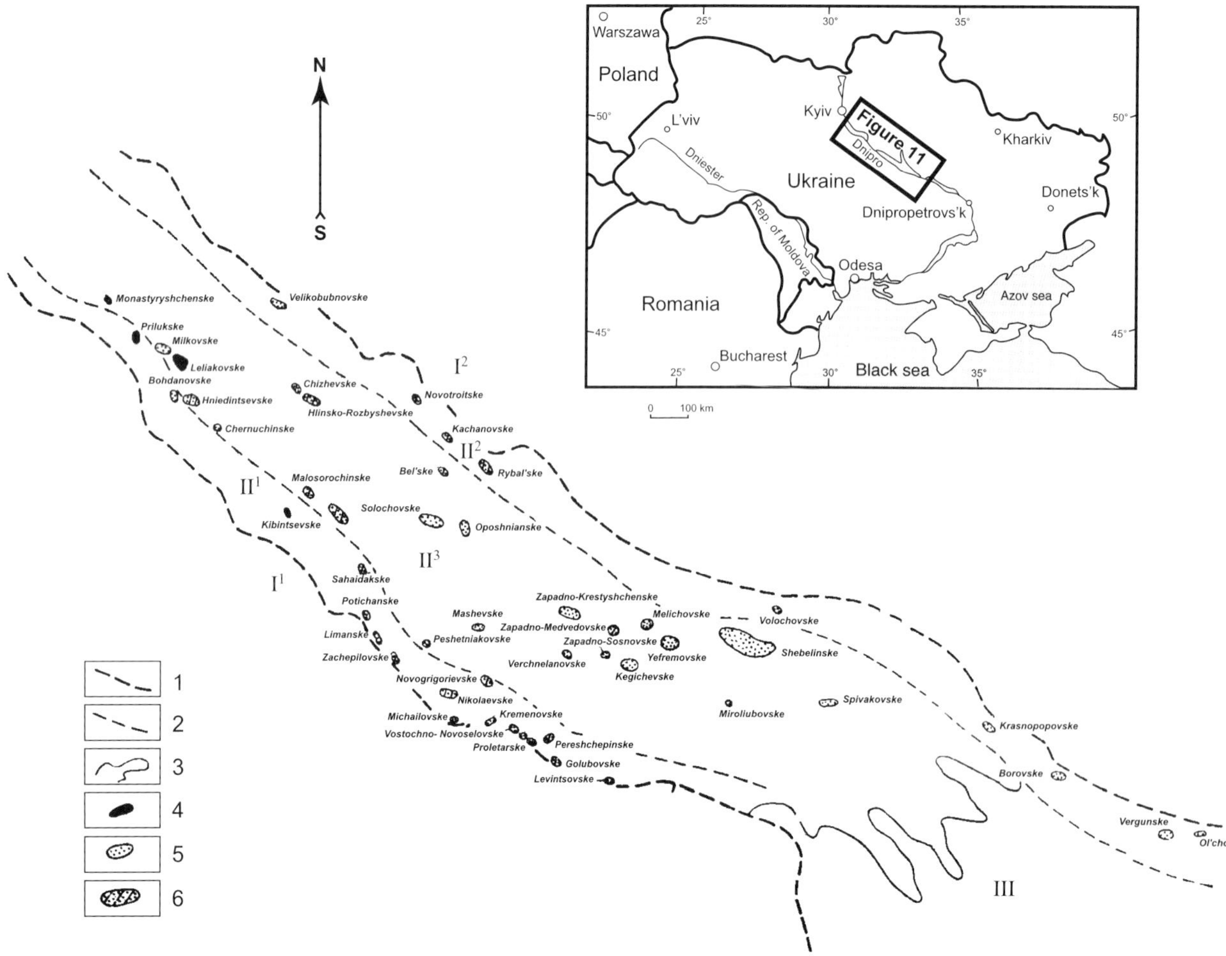

FIGURE 11. Schematic tectonic map of the Dnipro-Donets'k depression. I^1, I^2 = southwestern and northeastern sides of the depression; II^1, II^2 = lateral subzones of the central graben of the Dnipro-Donets'k depression; II^3 = axial zone of the central graben; III = folded region of Donbass. 1 = borders of the central graben; 2 = borders of the lateral subzones of the central graben; 3 = border of the folded region of Donbass; 4 = oil fields; 5 = gas fields; 6 = oil and gas fields.

when exploring and prospecting for hard mineral deposits (oil shale, coal, and others), because sudden outbursts of gas and rocks commonly occur when drilling in compressional areas.

DELINEATION OF CREST ANTICLINE

In Figure 15, an example from the Gvizd oil pool is given, where the distribution of factor k_a is typical. This is a representative example where a gradual increase of the factor k_a occurs from the flanks toward the crest of the anticlines. In the upper Oligocene part of the Gvizd oil field, the following distribution of layer pressures k_a is present:

- in well 45, at isohypse −2000 m (6600 ft), $k_a = 1.65$ (eastern flank)
- in well 204, at isohypse −1300 m (4300 ft), $k_a = 1.75$ (southeastern periclinal)
- in well 203, at isohypse −1200 m (3900 ft), $k_a = 1.85$ (at the crest of the fold)
- in wells 218 and 212, at the crest of the structure, abnormal values reaching a maximum of 2.03 are calculated
- on the western flank of the structure in well 200, at the depth (isohypse) −1100 m (3600 ft), the value of factor k_a decreases to 1.89
- in well 105, at the depth (isohypse) −1800 m (5900 ft) (northern periclinal), k_a decreases to 1.70

The pressure-vs.-depth plot shown in Figure 16 indicates that this reservoir is overpressured. The value of the pressure at the top of the structure is very close to the lithostatic pressure (σ_v). That may prove that the pressure regime is controlled by the fracturation of the seal at the top of the structure. Indeed, in a compressive area, one can expect that the minimum principal stress

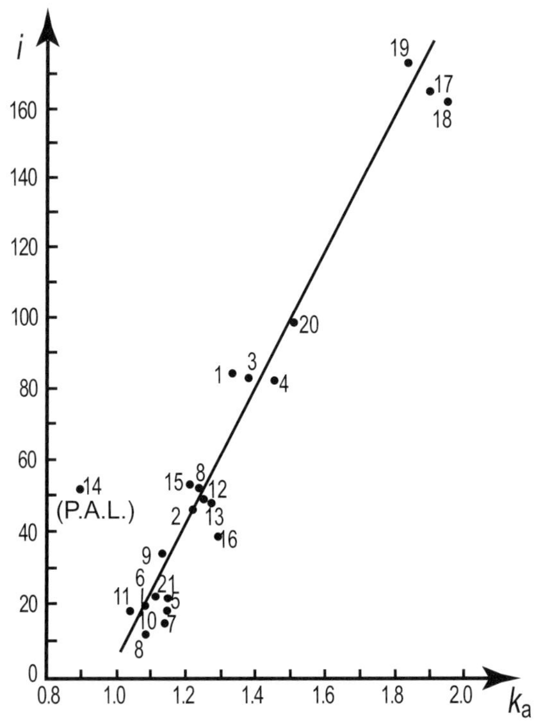

Figure 12. Empirical linear law $k_a = f(i)$ for individual reservoirs of the inner Carpathian Foredeep (current depth of the reservoirs studied is also indicated below; see location of the fields in Figure 10). Fields: 1 = Staro-Sambirsk (3460 m; 11,400 ft); 2 = Boryslav, upper unit (1800 m; 5900 ft); 3 = Boryslav, Popels'k area, second duplex (2350 m; 7700 ft); 4 = Ivanikovsk-Pomyarkovsk and Ivanikovsk units (2748 m; 9000 ft); 5 = Orov-Ulichniansk (3134 m; 10,300 ft); 6 = Styniavsk-Dolgolutsk unit, Oligocene (3785 m; 12,400 ft); 7 = Styniavsk-Dolgolutsk unit, Eocene (3805.5 m; 12,500 ft); 8 = Strutyn'sk (1900 m; 6200 ft); 9 = Vygoda-Vitvitsk (3350 m; 11,000 ft); 10 = North-Dolyna, Oligocene (3017.5 m; 9900 ft); 11 = North-Dolyna, Eocene (3255 m; 10,700 ft); 12 = Dolyna, Oligocene (2303 m; 7600 ft); 13 = Dolyna, Eocene (2500 m; 8200 ft); 14 = Spassk (1870 m; 6100 ft) (the pressure is abnormally low [P.A.L.] here, which is an exception for this region); 15 = Kosmach (3100 m; 10,200 ft); 16 = Rossil'na (3002 m; 9800 ft); 17 = Gvizdetsk, Oligocene (1985.5 m; 6500 ft); 18 = Gvizdetsk Oligocene (1985.5 m; 6500 ft); 19 = Gvizdetsk, Eocene (2219 m; 7300 ft); 20 = Pniv (2110 m; 6900 ft); 21 = Bytkiv, second duplex: Bytkiv-Pasichna and Babche-Starunia units (2475 m; 8100 ft).

is vertical, with an intensity equal to the lithostatic pressure. The maximum principal stress is expected to be in the horizontal plane. The intensity of the maximum principal stress could be related to the lithostatic pressure through the following formula (Schneider et al., 1999):

$$\sigma_h = Ko\sigma_v \text{ where } Ko > 1$$

In such compressional area, one should expect that the lithostatic pressure is an upper boundary for the pore pressure. Thus, an upper boundary for k_a is $k_a^* = \frac{\sigma_v}{P_h} \approx \frac{\gamma_b}{\gamma}$, where γ_b is the bulk density of the surrounding sediments. For example, assuming an average bulk density of 2100 kg/m^3 and a fluid density of 1020 kg/m^3, the maximum value for k_a is 2.06, which should be compared to the 2.03 measured in the upper Oligocene of the Gvizd field. Furthermore, the analysis

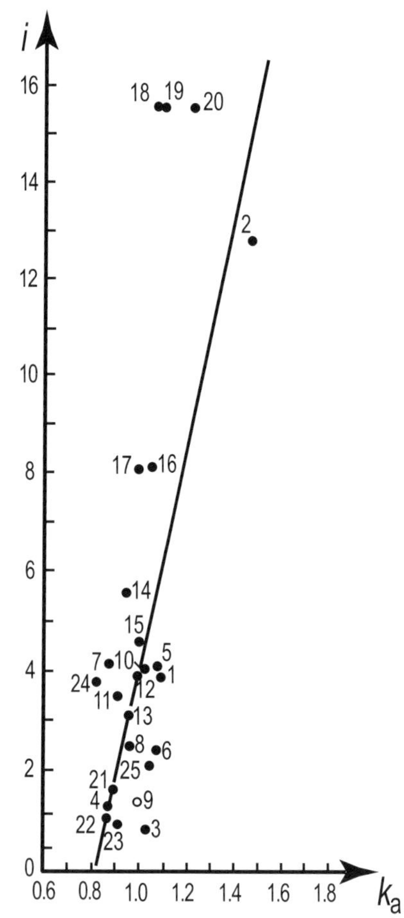

Figure 13. Empirical linear law $k_a = f(i)$ for individual reservoirs of the outer Carpathian Foredeep, including Veliko-Mastova area of Volyn-Podolia (current depth of the reservoirs studied is also indicated below; see location of the fields in Figure 10). Fields: 1 = Zaluzhansk (2075.5 m; 6800 ft); 2 = Zaluzhansk (2910 m; 9500 ft); 3 = Svidnitsk (700 m; 2300 ft); 4 = Kochanovsk (1275 m; 4200 ft); 5 = Khodnovichsk (1170 m; 3800 ft); 6 = Pyniansk (2058 m; 6800 ft); 7 = Sadkovich (1345 m; 4400 ft); 8 = Rudkovsk, lower Sarmatian (1056 m; 3500 ft); 9 =B-Rudkovsk (1505 m; 4900 ft); 10 = Oparsk (774 m; 2500 ft); 11 = Kavsk (851 m; 2800 ft); 12 = Medynichsk (1393 m; 4600 ft); 13 = Bil'che-Volitsk (1030 m; 3400 ft); 14 = Ugersk (1091 m; 3600 ft); 15 = Dashavsk (720.5 m; 2400 ft); 16 = Kadobniansk (687 m; 2300 ft); 17 = Kadobniansk (734 m; 2400 ft); 18 = Kalush (350 m; 1100 ft); 19 = Kalush (550 m; 1800 ft); 20 = Kalush (600 m; 2000 ft); 21 = Grynov (1155 m; 3800 ft); 22 = Kosiv (76 m; 250 ft); 23 = Kovaletsk-Chereshensk (2025 m; 6600 ft); 24 = Krasnoputnensk (826 m; 2700 ft); 25 = Veliko-Mostovsk (2394 m; 7900 ft).

of the pressure distribution in the Gvizd oil pool shows that there is an overpressure gradient from the flank toward the crest equal to 0.036 MPa per 10 m (33 ft). The overpressure at the top of the structure (at 1000-m [3300-ft] depth) is about 10 MPa, whereas the overpressure at 2000-m (6600-ft) depth is about 13 MPa. This indicates that fluids are flowing toward the crest of the structure at present day.

One should remark that assuming a hydrostatic overpressured regime as represented by the dashed line in Figure 16, the factor k_a naturally decreases from the top toward the flank following the formula

$$k_a(z) = 1 + \frac{\delta P(z^*)}{P_h(z)} \approx 1 + \frac{\delta P(z^*)}{\gamma g z}$$

where z^* is the depth of the crest of the structure. As P_h ($P_h \approx \gamma g z$) increases with depth, then k_a decreases with increasing depth.

The same pattern occurs in other basins, such as in the pre-Caucasus (e.g., Skiv province of Ukraine and Russia). This information has also some practical consequences, because it can be used in exploration to locate new wells and delineate the crests of anticline structures, where the best conditions for oil and gas accumulation commonly occur (Orlov et al., 2002).

Concerning the Gvizd oil field, the overpressure (δP = 20 MPa) measured in the Eocene–Oligocene layer is greater than the one measured in the upper Oligocene layer (δP = 10–13 MPa). This probably indicates that lower layers are disconnected from the upper one.

HOW TO USE OVERPRESSURE MEASUREMENTS FOR ESTIMATING THE HORIZONTAL STRESS

In the Earth's crust and in sedimentary basins, rock pressure consists of two components, the lithostatic pressure and the tectonic stress. The lithostatic

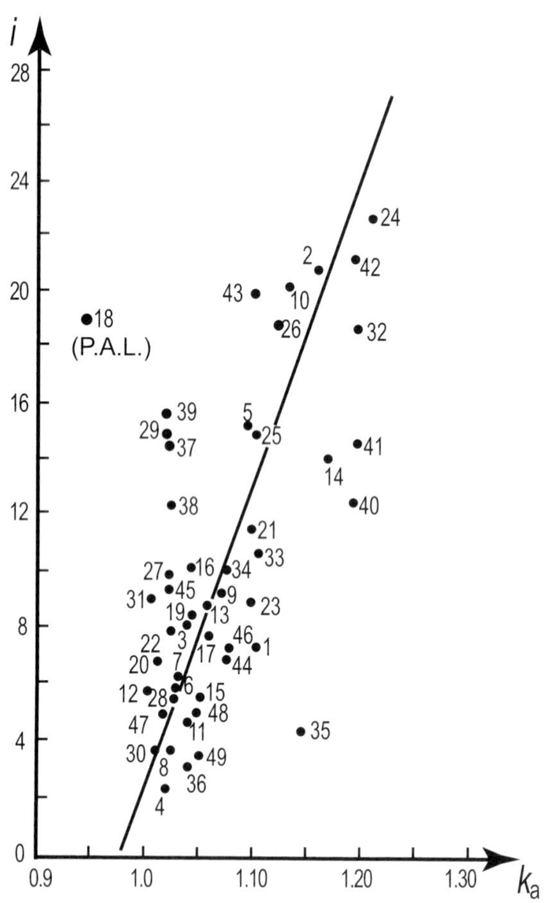

FIGURE 14. Empirical linear law $k_a = f(i)$ for individual reservoirs of Dnipro-Donets'k depression (current depth of the reservoirs studied is also indicated below; see location of the fields on Figure 11). Fields: 1 = Monastyryshchensk (3355 m; 11,000 ft); 2 = Priluky (1602 m; 5300 ft); 3 = Milkovsk (3000 m; 9800 ft); 4 = Leliakovsk (1850 m; 6100 ft); 5 = Bohdanovsk (2495 m; 8200 ft); 6 = Hniedintsevsk (1730 m; 5700 ft); 7 = Hniedintsevsk (3350 m; 11,000 ft); 8 = Velikobubnovsk (3105 m; 10,200 ft); 9 = Chizhevsk (3750 m; 12,300 ft); 10 = Hlinsko-Rozbyshevsk (3748 m; 11,400 ft); 11 = Novotroitsk (3400 m; 11,200 ft); 12 = Chernuchinsk (2957 m; 9700 ft); 13 = Kachanovsk (3085 m; 10,100 ft); 14 = Rybal'sk (3384 m; 11,100 ft); 15 = Bel'sk (3950 m; 13,000 ft); 16 = Kibintsevsk (1590 m; 5200 ft); 17 = Malosorochinsk (2257 m; 7400 ft); 18 = Sahaidaksk (1698 m; 5600 ft) (the pressure is abnormally low [P.A.L.], which is an exception for this region); 19 = Potichansk (1720 m; 5600 ft); 20 = Limansk (1700 m; 5600 ft); 21 = Zachepilovsk (1350 m; 4400 ft); 22 = Peshetniakovsk (3244 m; 10,600 ft); 23 = Mashevsk (4060 m; 13,300 ft); 24 = Oposhniansk (3695 m; 12,100 ft); 25 = Solochovsk (3450 m; 11,300 ft); 26 = Novogrigorievsk (2000 m; 6600 ft); 27 = Pereshchepinsk (2640 m; 8700 ft); 28 = Proletarsk (1837 m; 6000 ft); 29 = Vostochno-Novoselovsk (1922 m; 6300 ft); 30 = Nikolaevsk (2693 m; 8800 ft); 31 = Michailovsk (782 m; 2600 ft); 32 = Yefremovsk (3000 m; 9800 ft); 33 = Zapadno-Sosnovsk (4000 m; 13,100 ft); 34 = Kegichevsk (2882 m; 9500 ft); 35 = Shebelinsk (2450 m; 8000 ft); 36 = Miroliubovsk (2420 m; 8000 ft); 37 = Golubovsk (1112 m; 3600 ft); 38 = Levintsovsk (750 m; 2500 ft); 39 = Zapadno-Krestyshchensk (3400 m; 11,200 ft); 40 = Melichovsk (3100 m; 10,200 ft); 41 = Verchnelanovsk (3400 m; 11,200 ft); 42 = Zapadno-Medvedovsk (3190 m; 10,500 ft); 43 = Kremenovsk (2215 m; 7300 ft); 44 = Spivakovsk (457 m; 1500 ft); 45 = Volochovsk (3060 m; 10,000 ft); 46 = Krasnopopovsk (489 m; 1600 ft); 47 = Borovsk (1510 m; 500 ft); 48 = Vergunsk (1165 m; 3800 ft); 49 = Ol'chovsk (1169 m; 3800 ft). For numbers in parentheses, see Figure 10.

pressure at any point in the Earth's crust is equal to the load (gravitational pressure) of the overburden. Vertically directed stress (σ_v) is the reaction to this pressure:

$$\sigma_v = gz\gamma_b$$

where z refers to the burial depth in meters, γ_b is the average density of the overburden expressed in kilograms per cubic meter, and g is the acceleration caused by gravity in meters per square second.

In the elastic domain, horizontal pressures in rocks (σ_h) relate to Poisson's ratio for the immediate surrounding rock, modified by the tangential component of the tectonic (or geodynamic) stresses (Markov, 1977). It has been established that for a compressive area, the relation σ_v/σ_h in sedimentary layers is in the range 0.25–0.75 (Bulin, 1972). In practice, it is rather difficult to directly measure the tectonic stress. It would require good knowledge of the Young and Poisson factors and strain in deformed rocks that were derived from real geological settings. It would also require good control on the lithologies and anisotropies of sedimentary layers, as well as the determination of the precise direction of tectonic forces. Nevertheless, tectonic stress has been well characterized in compressional settings where instrumental measurements have been possible.

We have shown previously that in the considered area, the overpressure intensity is related to the intensity of folding by the empirical linear law $k_a = f(i)$. Using the porothermoelastic theory, the increase of horizontal stress could be related to the increase of pore pressure by the following formula: $\delta\sigma_h = 3(\frac{K}{bM} + b)\delta P_f$, where b is the Biot coefficient, K is the elastic modulus, and M is the Biot modulus (Schneider et al., 1999). It should be noted that the previous formula assumes a closed system and planar deformation (only σ_h is modified during the deformation; the two other principal stresses are assumed

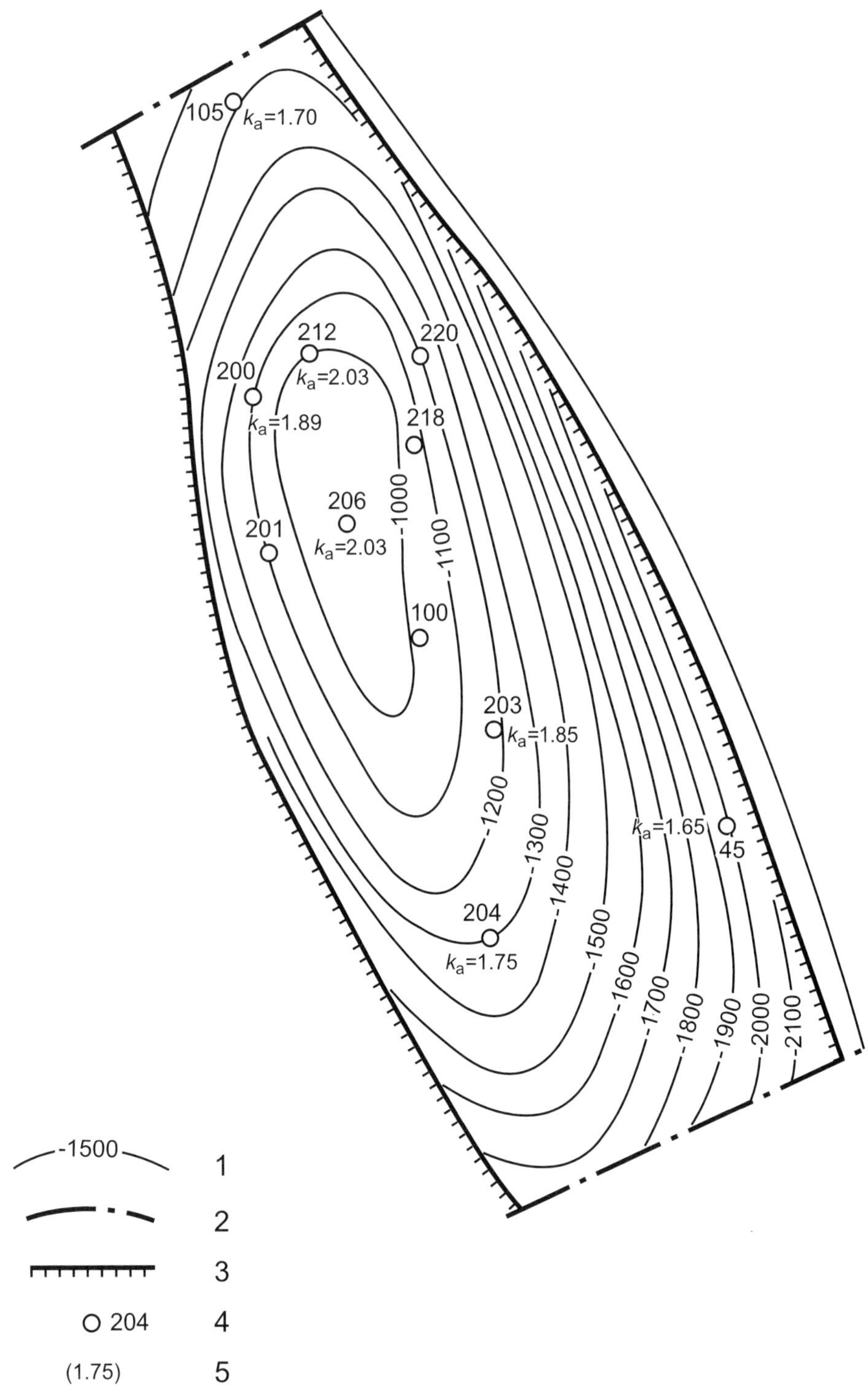

FIGURE 15. Distribution of overpressures in a productive horizon of the upper Oligocene of the Gvizd anticlinal field. 1 = isohypses of reservoir top in meters; 2 = tear faults; 3 = reverse faults; 4 = wells; 5 = values of k_a.

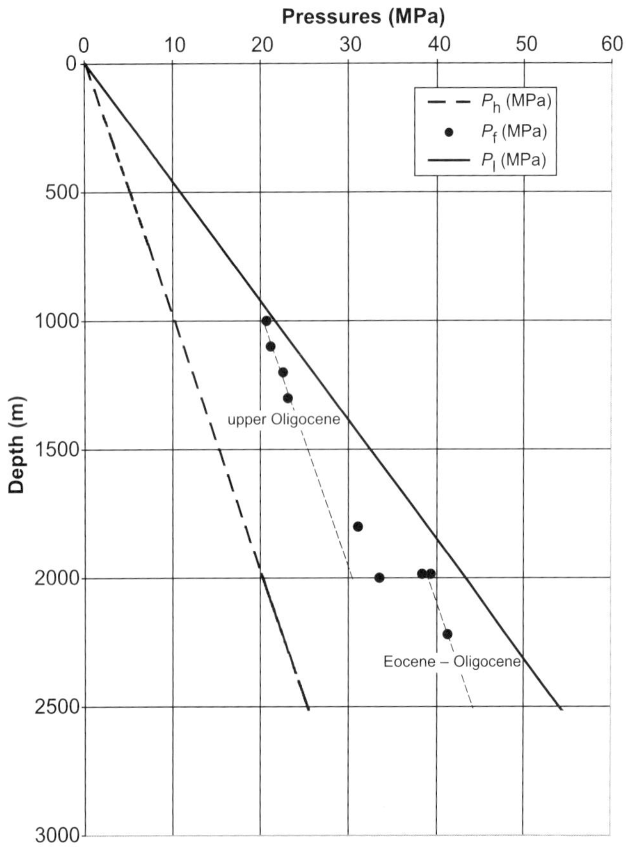

FIGURE 16. Pressure vs. depth for the Gvizd anticlinal field.

to be constant). In a closed system with axisymmetric condition, we have $\delta\sigma_h = \frac{3}{2}(\frac{K}{bM} + b)\delta P_f$ (Schneider et al., 1999).

As an example, $\delta\sigma_h$ has been calculated for the Gvizd field in the Carpathian Foredeep (Figure 15). At a depth of about 2000 m (6600 ft), the Eocene–Oligocene series in the basin display a hydrostatic value of P_h = 20 MPa, whereas the formation pressure (P_f) measured in the reservoir is 40 MPa; the compressibility of the reservoir matrix is $\beta_r = 7.5 \times 10^{-10}$ Pa^{-1}, the pore-fluid compressibility is $\beta_f = 4.5 \times 10^{-10}$ Pa^{-1}, and the reservoir porosity is $\phi = 0.20$.

Let us assume that before folding, the pore pressure was hydrostatic, and the horizontal stress was 70% of the vertical stress. Thus, at 2000-m (6600-ft) depth, we have σ_v = 42 MPa, σ_h = 30 MPa, and P_f = 20 MPa. Now, let us assume that the deformation was performed in the elastic domain, and that the present-day overpressure is the result of this deformation. Thus, as the observed overpressure is 20 MPa, then δP_f = 20 MPa. Neglecting the rock compressibility, we have $b = 1$ and $M = 1/(\phi\beta_f)$, and from the Gassmann formula, we derive $K = \beta/(\phi\beta_f\beta_r)$, where $\beta = \phi\beta_f + \beta_r$ is the layer elastic-capacity factor as defined by Shchelkachev (1959). The numerical application gives K = 12.4 GPa and M = 11.1 GPa. The increase of horizontal stress that corresponds to the increase of pore pressure is then equal to $\delta\sigma_h = 6.35\delta P_f$ = 127 MPa. Thus, the calculated maximum horizontal stress at present day is 157 MPa. This value is very close to the in-situ value measured in the reservoir.

HOW TO USE OVERPRESSURE MONITORING TO FORECAST SEISMIC HAZARD

In these specific areas (e.g., folded area), the horizontal stress variations are correlated with the variations of the overpressure. Thus, using the abovementioned formula, the monitoring of the overpressures in the reservoirs may be used to control the horizontal stress variations. Consequently, any increase of tectonic stress value $\delta\sigma_h$ derived from these calculations will be indicative of a seismic hazard in the given region.

CONCLUSIONS

Overpressure in natural basins is of major importance and locally develops simultaneously with tectonic folding. Fold amplitude can be used for forecasting overpressure when drilling, whereas overpressure values may help to find the exact location of anticlinal closures. Studying the overpressure distribution and evolution is important both for petroleum exploration and for stress estimates in seismically active domains. Furthermore, the intensity of the overpressure in a given reservoir may be related to the increase of the intensity of the horizontal maximum stress during folding. Moreover, the monitoring of the overpressure in a specific reservoir may be useful to predict any modification of the horizontal stress as an indicator of seismic hazard.

REFERENCES CITED

Azhgirey, G. D., 1956, Structural geology: Moscow, Moscow State University, 193 p.

Belousov, V. V., 1962, Basic questions on geotectonics, *in* Elements of mechanism crumpling of the strata on the folds: Moscow, Gosgeoltechizdat, p. 238–244.

Bulin, N. K., 1972, Modern tensions in rocks on measurement data of under-earth developments in the U.S.S.R., Novisibirsk: Journal of Geology and Geophysics, v. 8, p. 56–66.

Hottman, C. E., and R. K. Johnson, 1965, Estimation of formation pressures from log-derived shale properties: Journal of Petroleum Technology, v. 17, no. 6, p. 717–722.

Markov, G. A., 1977, Tectonic pressures and rock tension in Khibinskiy massif mines: Leningrad, Nauka, 213 p.

Novosiletsky, R. M., 1961, Abnormally high layer pressures in the Carpathian Foredeep reservoir: Kiev. Journal of Oil and Gas Industry, v. 1, p. 5–7.

Orlov, A. A., 1980, The influence of tectonic dislocations on overpressure layer forming of porous fluid in sedimentary strata, Moscow: Journal of Geotectonics, v. 1, p. 106–114.

Orlov, A. A., A. N. Trubenko, A. V. Loktev, M. M. Chornyi, and V. G. Omelchenko, 2002, Method of anticlinal structure crests definition/exposure during drilling process: Bulletin of the State Department of Intellectual Property of Ministry of Education and Science of Ukraine, Kiev, v. 12, 5 p.

Schneider, F., M. Boutéca, and J. P. Sarda, 1999, Hydraulic fracturing at sedimentary basin scale: Oil and Gas Science Technology— Revue Institut Français du Pétrole, v. 54, no. 6, p. 797–806.

Shchelkachev, B. N., 1959, Investigation of petroleum layers at elastic conditions: Moscow, Gostoptechizdat, 467 p.

Deformation, Overall Reservoir Characteristics, and Diagenesis

12

Tavani, S., L. Louis, C. Souque, P. Robion, F. Salvini, and D. Frizon de Lamotte, 2004, Folding-related fracture pattern and physical properties of rocks in the Chaudrons ramp-related anticline (Corbières, France), *in* R. Swennen, F. Roure, and J. W. Granath, eds., Deformation, fluid flow, and reservoir appraisal in foreland fold and thrust belts: AAPG Hedberg Series, no. 1, p. 257–275.

Folding-related Fracture Pattern and Physical Properties of Rocks in the Chaudrons Ramp-related Anticline (Corbières, France)

Stefano Tavani
Dipartimento di Scienze Geologiche, Università degli Studi Roma 3, Rome, Italy

Laurent Louis
Département des Sciences de la Terre et de l'Environnement, Université de Cergy-Pontoise, France

Christine Souque
Département des Sciences de la Terre et de l'Environnement, Université de Cergy-Pontoise, France

Philippe Robion
Département des Sciences de la Terre et de l'Environnement, Université de Cergy-Pontoise, France

Francesco Salvini
Dipartimento di Scienze Geologiche, Università degli Studi Roma 3, Rome, Italy

Dominique Frizon de Lamotte
Département des Sciences de la Terre et de l'Environnement, Université de Cergy-Pontoise, France

ABSTRACT

Reservoir appraisal is commonly a difficult task in fold and thrust belts. Thrust-related folding leads to the development of meso- to microscopic brittle structures that can significantly alter the porosity and permeability properties of reservoir rocks, thus influencing fluid migration and accumulation. The aim of this chapter is to describe at different scales the deformation associated to the development of the Chaudrons thrust-related anticline (Corbières, France) and to discuss its influence on reservoir quality. Pervasive solution cleavage sets at high angle to bedding (ATB) were found and measured along the fold. In addition, collected core samples were used to measure the rock physical properties (anisotropy of magnetic susceptibility and anisotropy of the P-wave velocity). The distribution of deformation in the anticline was used to identify three different deformational panels: crest, rounded forelimb, and constantly steeping forelimb. The crest is characterized by the lowest cleavage intensity. Nonpenetrative solution cleavages

DOI:10.1306/1025694H13120

and the magnetic foliation are orthogonal to bedding. In the rounded forelimb, bedding dip progressively rotates from 0 to 60°. Cleavage intensity progressively increases, and cleavage and magnetic foliation ATBs progressively increase from 80 to 120° and then remain constant in the steep forelimb. The timing of the development of mesoscale structures, as well as changes of physical properties occurring before and during folding, is also discussed.

INTRODUCTION

The possibility to predict the spatial outline and evolution of permeability in reservoirs has a primary importance in oil and gas research and development. In fold and thrust belts, fluid flow is strongly controlled by folding-related deformational features (e.g., Srivastava and Engelder, 1990; Cooper, 1992; Storti and Salvini, 1996; Jamison, 1997), whose origin (i.e., contractional, extensional, and shear), frequency, and geometry can significantly alter the porosity and permeability of the reservoir rocks (e.g., Nelson, 1985; Mitra, 1988; Cooke, 1997). In particular, development of pressure-solution cleavage can enhance the reservoir permeability (e.g., Leythaeuser et al., 1995; Van Geet et al., 2002). Pressure-solution cleavage networks can drive hydrocarbon flow when subjected to fluid overpressure (e.g., Aydin, 2000), to compression parallel to the cleavage planes, or to their repeated opening and closure, as what commonly occurs during folding (e.g., Srivastava and Engelder, 1990). The evidence that in many cases, pressure-solution cleavages are among the most commonly developed deformational feature associated to folding (e.g., Marshak and Engelder, 1985; Mitra and Yonkee, 1985; Mitra, 1988; Holl and Anastasio, 1995; Dunne and Caldanaro, 1997; Davidson et al., 1998) implies that understanding the time-space evolution of syntectonic solution cleavage networks can significantly impact the reservoir modeling and performance.

The aim of this chapter is to describe the formational pattern at the different scales associated with the development of the Chaudrons ramp-related anticline (Corbières, France; Figure 1). In particular, we analyzed the red silty sandstone of the Paleogene Vitrollian formation, where deformation is particularly present. The spatial distribution and major characteristics of brittle deformational features are compared to the rock physical properties at the sample scale, including anisotropy of magnetic susceptibility (AMS) and ultrasonic measurements, which are directly linked to the rock structure and composition. The time-space evolution of deformational features and their influence onto the fluid flow are discussed.

GEOLOGICAL SETTING

The Chaudrons anticline belongs to the Corbières transfer zone (i.e., the arc joining the north Pyrenean zone and the Languedoc-Provence thrust belt; Figure 1). This domain is characterized by a thin sedimentary pile formed by limestone, sandstone, and silt of Maastrichtian to Lutetian age supporting Bartonian molasses (Ellenberger, 1967; Plaziat, 1984). This cover and its Paleozoic substratum are involved in large ramp-related folds (Averbuch et al., 1992; Frizon de Lamotte et al., 1997; Grelaud et al., 2000). At the regional scale, the folds exhibit an en echelon pattern relative to the North Pyrenean front, locally called "nappe des Corbières orientales" (Figure 1).

The Chaudrons anticline is an east–west-trending, 3-km (2-mi)-wide, and 9-km (6-mi)-long anticline where exposed rocks range in age from Upper Cretaceous to upper Paleocene. Three distinct formations are exposed in the area of interest. Older exposed rocks are lacustrine limestones of the Rognacian Formation (Maastrichtian). They are conformably covered by clastics (conglomerates and siliciclastic continental deposits) of the Vitrollian formation (lower Paleocene). Upper Paleocene (Thanetian) is again represented by lacustrine limestones and siliciclastic deposits. Folding occurred in the upper Eocene (Ellenberger, 1967). Geometrically, the Chaudrons anticline is characterized by a steep forelimb between flat foreland and crest (Figure 2). The transition between the foreland and the forelimb is exposed in the Thanetian limestone but complicated by secondary collapse structures. On the contrary, the transition between the forelimb and the crest can be continuously followed in the Vitrollian formation (Figure 2). The back of the anticline is offset by normal faults (Figure 1) formed during a late Oligocene–Miocene event, and the backlimb is not outcropping. The deformation features associated to this extensional event are restricted to the normal fault damage zones and do not involve the area of study.

SAMPLING, MEASUREMENTS

It has been demonstrated that stratigraphy is a primary factor in controlling folding-related deformation patterns (e.g., Corbett et al., 1987; Protzman and Mitra, 1990; Gross, 1995; Fischer and Jackson, 1999). To rule out, or at least significantly reduce, the influence of lithology, we collected mesostructural data and samples in the Vitrollian formation of the Chaudrons anticline. This formation is characterized by an alternation of sandstones and silts and clays, having an average bed thickness of 30 cm (12 in.). Conglomeratic levels are also present, but data were collected only in sandy and clayish beds.

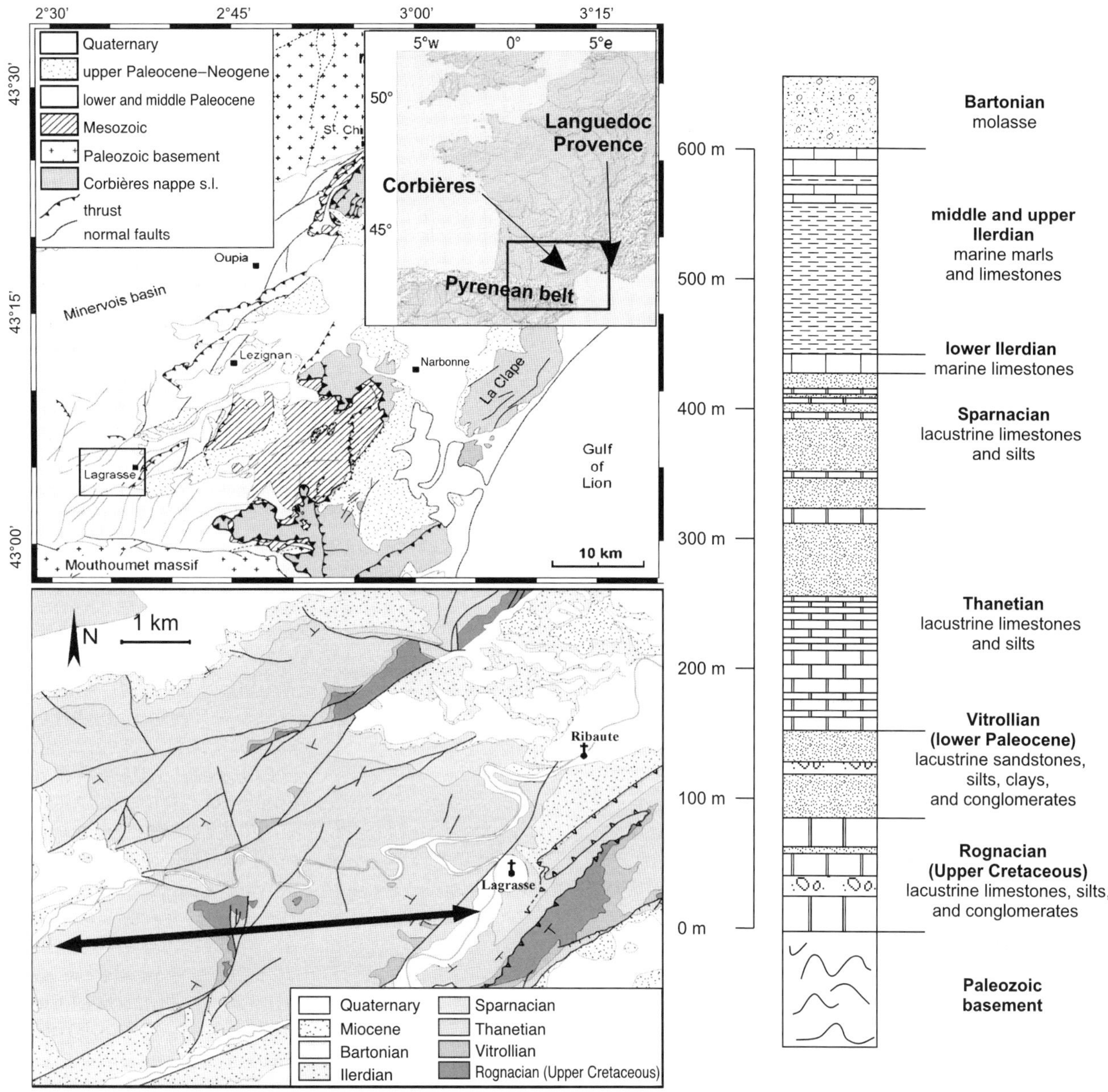

FIGURE 1. Geological map of the Chaudrons fold.

Data were collected through two properly oriented, along-dip transects (Figure 3), with the maximum across a distance of 20 m (66 ft) in the first, shorter (151 m; 500 ft) transect and 40 m (130 ft) in the second, longer (633 m; 2100 ft) transect.

Data location was accomplished by global positioning system (GPS) measurements that provided 20-m (66-ft)-accuracy relative location. This approximation induces a negligible error in the transect Ch-2 (the error with respect to the transect length is 3%). In the transect Ch-1, the error is higher (13%); in this case, data location along each path was ensured by a local coordinate system, consisting of the rectification of the path into a series of oriented (azimuth and strike) segments of a given length with an accuracy of about 0.1 m (0.3 ft). Each measurement or sample has then a local coordinate reference consisting of its distance along the path of the transect from the initial edge. This local reference system was then automatically converted into geographic coordinates (Salvini, 2002), using the absolute coordinate (from GPS) of the leading edge.

Structural Data

At the outcrop scale, the collected structural data include bedding, cleavage, and mesofaults orientation and dimensions of lithons isolated between adjacent cleavages ("cleavage lithons;" e.g., Durney and Kisch, 1994).

FIGURE 2. The forelimb and frontal crest of Chaudrons fold, as seen toward the east.

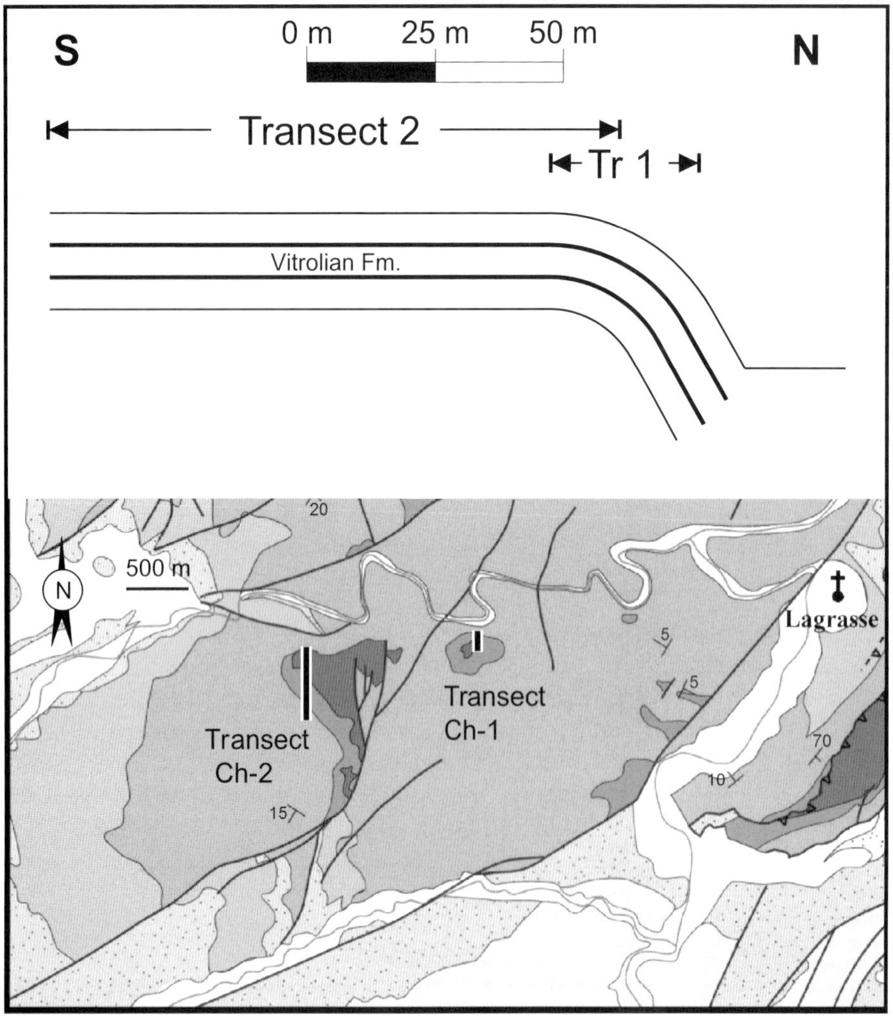

FIGURE 3. Transect locations.

Cleavage consists of a generally well-organized set of rather regular pressure-solution surfaces (disjunctive, nonsutured intrabed surfaces) striking parallel to the anticline axis (Figures 4, 5). Cleavage is isolated in sandy beds, whereas in clayish beds, it is anastomosed. Cleavage angles with respect to the bedding surfaces range from 80 to 130° (value greater than 90° indicates forelandward-dipping surfaces once the bedding has been rotated to the horizontal). In sandy beds, this angle ranges from 80 to 110°, whereas in clayish beds, it ranges from 80 to 130°. Values close to 90° are typical of the crestal sector, whereas larger values characterize the forelimb.

Cleavage spacing is the lithon dimension orthogonal to the cleavage (i.e., the distance between adjacent cleavage planes). Cleavage height is the bed thickness or, in the case of anastomosed cleavage, the dimension of the lithon along the direction normal to spacing and normal to the anticline axis (Figure 4). Cleavage width is the distance, parallel to bedding, between two contiguous intersections of adjacent cleavage surfaces. In this way, a parallelepiped having spacing, height, and width as dimensions would represent the average lithon produced by the cleavage, and its shape would be related to the intensity of deformation.

In the studied area, pressure-solution cleavage surfaces, whose spacing ranges from 10^{-1} to 10^{-2} m (0.3 to 0.03 ft), divide the rock into fairly regular cleavage lithons in a section parallel to the tectonic transport direction (and orthogonal to the bedding surfaces); in sandy beds, cleavage lithon boundaries are defined by two solution cleavages and by the top and the bottom of the beds (Figures 4, 6) or by discontinuities parallel to the bedding surfaces. With increasing clay content, cleavage becomes anastomosed, and lithons form between two adjacent cleavage surfaces (Figure 6). Cleavage surfaces are well organized in the forelimb, and in the crest, they are less penetrative.

Several derived factors were then used to describe the deformation state at the bed scale. These factors include the cleavage angle to bedding (ATB), the

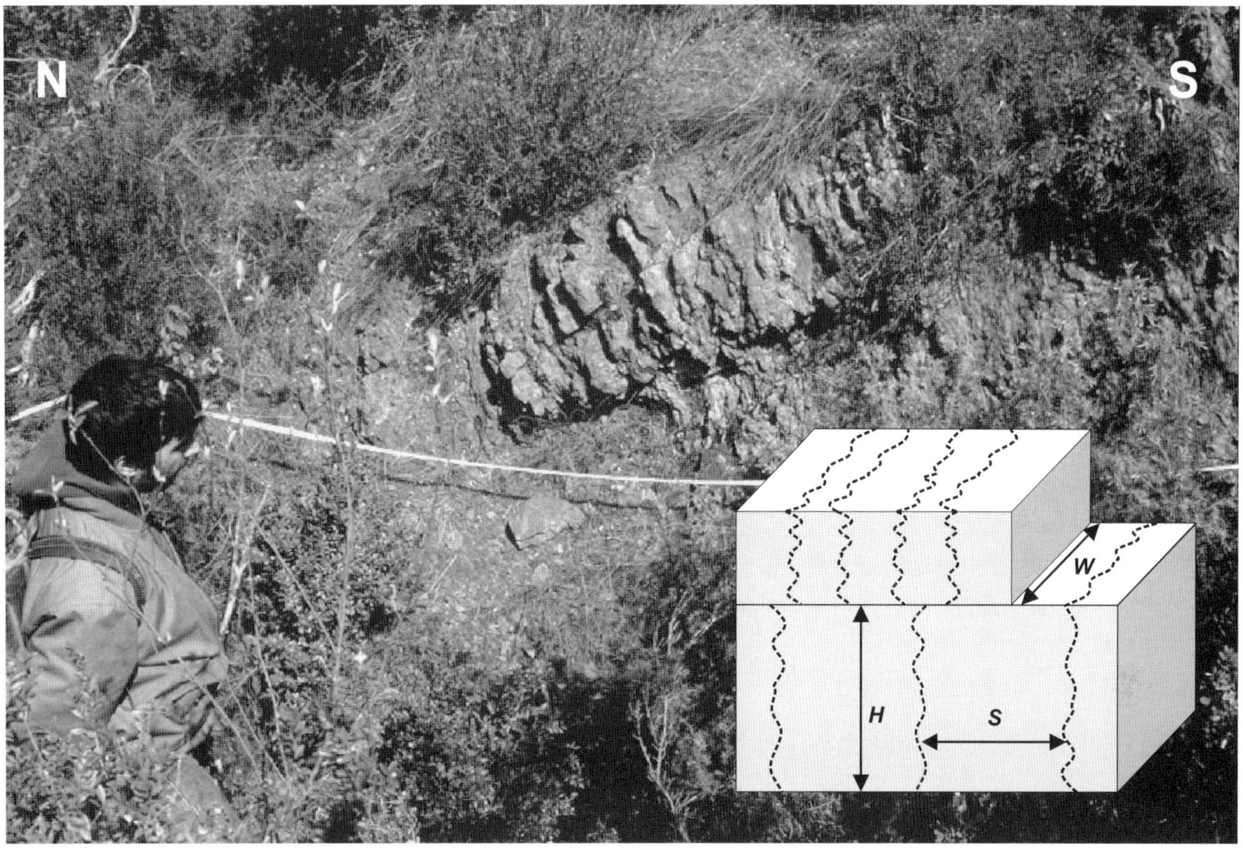

FIGURE 4. Cleavage-lithon dimensions.

cleavage height/spacing ratio (*H*/*S* or cleavage/bedding dimension ratio by Durney and Kisch, 1994), and the lithon shape factor (Sf). The *H*/*S* ratio represents a normalization of the spacing by bed thickness. The shape factor (Sf) is the ratio between the difference of height minus width and width minus spacing.

$$Sf = (H - W)/(W - S)$$

This value describes the shape of the lithons with respect to bedding and cleavage orientation. We investigate the possibility of using spacing, *H*/*S*, and Sf to characterize the deformation intensity. In the Chaudrons anticline, *H*/*S* and Sf range from 1 to 9.2 and from −1.5 to 3.1, respectively; average values are 3.8 and 0.9, respectively.

A conjugate set of reverse faults is also present in the more competent layers (Figure 5) and almost always cuts through the cleavage, thus suggesting a relatively younger age as already observed (Ellenberger, 1967; Frizon de Lamotte et al., 2002). This fault system clearly cuts across beds and shows an intrabed character. Northwest-dipping faults are more abundant than their conjugates (Figure 5), suggesting a component of top-to-the-southeast bed-parallel shear. Slip indicators are also locally present on bedding plane, proving that flexural slip concurs to the overall deformational picture.

Laboratory Measurements

In addition to structural data measured directly in the field, we have collected 100 core samples in the same layer to perform measurements of the physical properties of rocks. It is worth noting that the representative elementary volumes (REV) are different and complementary for these two kinds of measurements: For fractures, the REV is about 1 m^3 (35 ft^3), whereas for physical properties, it is about 10^{-5} m^3 (3.5×10^{-4} ft). Thus, fracture analysis gives information at outcrop scale, whereas physical properties give access to the matrix properties.

Anisotropy of Magnetic Susceptibility

One hundred standard cylindrical specimens (2.5 cm diameter × 2.2 cm length), drilled and oriented in the field, have been sampled in the Chaudrons anticline to perform anisotropy of low-field magnetic susceptibility (AMS). Anisotropy of magnetic susceptibility measurements have been made with the KLY3S kappabridge susceptibility Advanced Geoscience Instruments Company (AGICO) instrument. This specimen-spinning equipment measures the susceptibility during three spins of the sample in three perpendicular planes. During each spin, 64 measurements are made and subsequently give access to a very accurate determination of the susceptibility

FIGURE 5. Deformation types found in Chaudrons anticline.

tensor (K). Principal magnitudes and directions of the tensor are thus calculated using a least-square method. The maximum (K_1), intermediate (K_2), and minimum (K_3) principal values are the susceptibility values along the principal directions ($K_1 > K_2 > K_3$). The cluster of K_1 defines the magnetic lineation, and the cluster of K_3 defines the pole to magnetic foliation.

Basically, the magnetic susceptibility is the proportionality coefficient recorded between the induced magnetization and the applied magnetic field (Bhathal, 1971). The anisotropy of magnetic susceptibility tensor K describes the ability of certain rocks to change their induced magnetization as orientation of applied field is modified (Hrouda, 1982). For most of the crystals, K is governed by crystallographic directions. For crystals with high intrinsic susceptibilities such as magnetite or maghemite, K reflects mainly shape anisotropy (Uyeda et al., 1963). Therefore, it is generally assumed that analysis of magnetic tensors is a way to access the preferred crystallographic orientation or shape orientation (or a combination of both) of all minerals in the rocks (Graham, 1954). This bulk magnetic signature includes the matrix contribution, consisting of diamagnetic, paramagnetic, and antiferromagnetic minerals, as well as the ferromagnetic contribution, consisting of ferromagnetic and ferrimagnetic minerals.

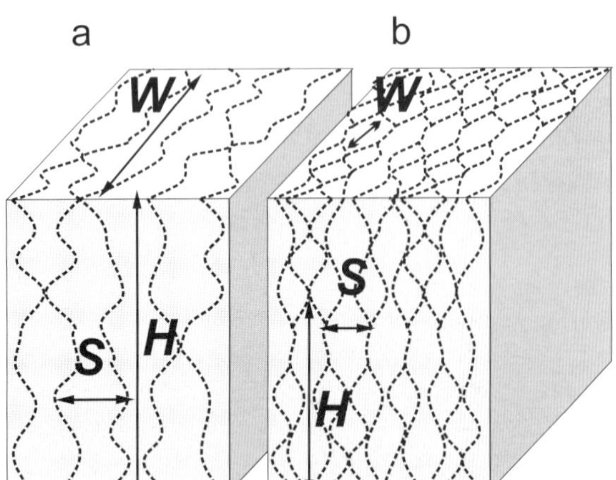

FIGURE 6. (a) Cleavage in sandy bed. (b) Cleavage in clayish bed.

At a sample scale, AMS is thus caused by preferential orientation of grains, which is generated by grains reorientation and chemical and mechanical compactions during diagenetic and tectonic events. Anisotropy of magnetic susceptibility studies have been developed by structural geologists in weakly deformed zones where sedimentary rocks are free from continuous macroscopic deformation and have mainly undergone brittle deformation (Graham, 1966; Kliegfield et al., 1981; Kissel et al., 1986; Lowrie and Hirt, 1987; Averbuch

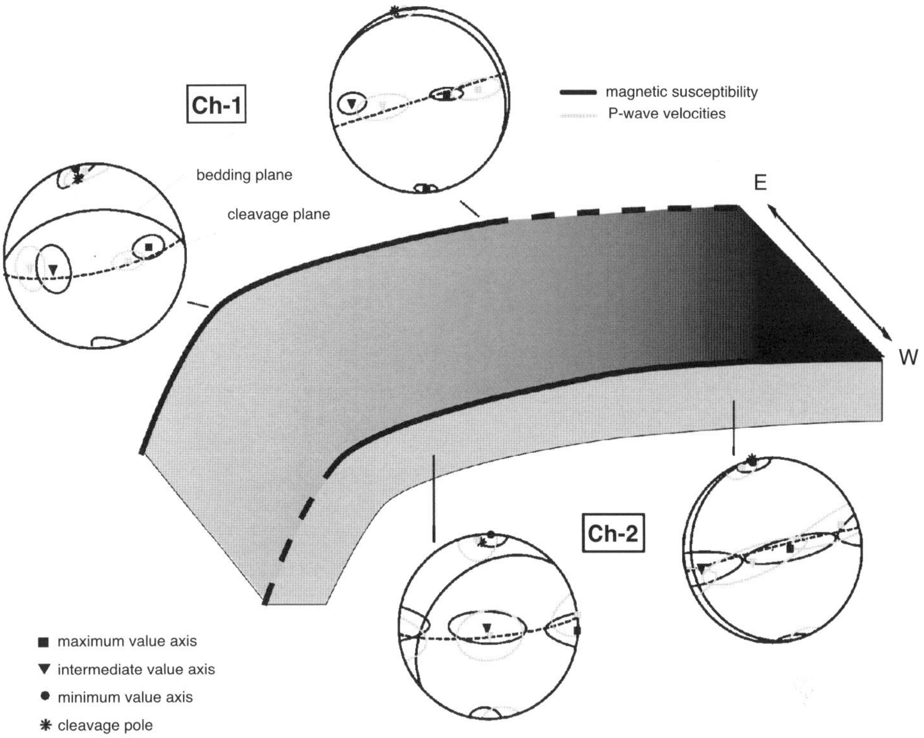

FIGURE 7. Stereoplots obtained from magnetic susceptibility and ultrasonic velocity measurements (geographical reference).

et al., 1992; Sagnotti et al., 1998). The Corbières area has been the site of extensive AMS studies (Averbuch, 1993; Grelaud, 2001; Souque, 2002), which all showed that AMS is of tectonic origin. In the Vitrollian formation, Averbuch et al. (1992) showed that hematite was the main ferromagnetic mineral, and that the matrix was dominated by quartz, calcite, and clay minerals. The anisotropy of magnetic susceptibility was thus interpreted in terms of preferred orientation of clay minerals and hematite. These results are confirmed in the Chaudrons anticline, where we observed tectonic magnetic fabrics (Figure 7) characterized by magnetic foliation oblique (forelimb) or normal to bedding (crest of the anticline). Synthesizing the AMS work conducted in the Corbières, Frizon de Lamotte et al. (2002) assumed that, away from thrust faults, AMS records internal deformation resulting from layer-parallel shortening (LPS), which occurred before folding and fracturing. However, the relationships between AMS and microstructure of a given sample are not straightforward, because it can reflect only a small volumetric fraction of the whole rock, especially when the signal is dominated by ferromagnetic and ferrimagnetic minerals. This is the reason why we have tested the significance of magnetic fabric measured in the Chaudrons anticline with another method: acoustic anisotropy.

Acoustic Anisotropy

It is known that anisotropy of the P-wave velocity is directly related to the microstructures of rocks (nature and shape of the constitutive elements, including intergranular or fractures porosity; Lo et al., 1986). As the sediments remain unstrained, the spatial distribution of P-wave velocities commonly presents a planar shape, with maximum values corresponding to the bedding plane, whereas the bedding pole exhibits the minimum velocity value (Lo et al., 1986; Hrouda et al., 1993; Louis et al., 2003). During deformation, from sedimentary to tectonic compaction, the whole structure of the rock is modified, and the distribution of the acoustic velocities at sample scale can suffer strong modifications. As several processes affecting the microstructures (i.e., grain dissolution and growth, rotation, and fracturing) can work simultaneously, the interpretation is generally not straightforward. Nevertheless, in consolidated clastic rocks, losses of stiffness are generally larger than the gains, and a process that leads to a reduction of the inter- or intragrain cohesion (fractures, solution cleavage) is expected to dominate the acoustic signal.

To estimate the anisotropy of the P-wave velocity, we followed the protocol described by Louis et al. (2003), drilling three perpendicular rock cylinders from a geographically oriented block. Each block was taken out from the outcrop and drilled properly in the laboratory to guarantee as much as possible the orthogonality of each sample with respect to the others. In this method, we assume that P-wave velocities can be correctly spatially represented by a second-order symmetric tensor. On each sample, eight measurements were made every 22.5° across the diameter. Then, the 24 values obtained were used to retrieve three principal values and axes by a least-squares calculation. Following Hext (1963), a scattering angle was computed for each principal axis with respect to the others.

Contrary to AMS, the number of acoustic measurements is not enough to compute trends along the transect (see below). Thus, we will only present a comparison between structural data, AMS, and ultrasonic P-wave velocity of four blocks from different locations (crest and forelimb; Figure 7). The net result is a fairly good correlation between acoustic and magnetic data. Indeed, maximum and intermediate axes are very close (the ellipses of confidence present a common intersection) and are situated in the cleavage. In all locations, minimum axes are very well defined and are parallel to the pole to cleavage. Our results attest that the structural anisotropy, which is observed at outcrop scale, also exists at sample scale.

DATA SPATIAL ANALYSIS

Spatial variations of the studied parameters were analyzed using the transect tool (Wise and McCory, 1982; Salvini et al., 1999). It consists of the preparation of scattering diagrams of a given parameter with the distance along the transect in the *X*-axis and the given values along the *Y*-axis, respectively. These "scatterograms" can be filtered along both axes, and a best fit multiple-Gaussian analysis can be applied to identify and quantify the presence and variation of uni- and polymodal distributions of deformational features along the transect. This methodology has been shown to be very effective for detecting trends from populations of scattered data (e.g., Salvini et al., 1999). Figure 8

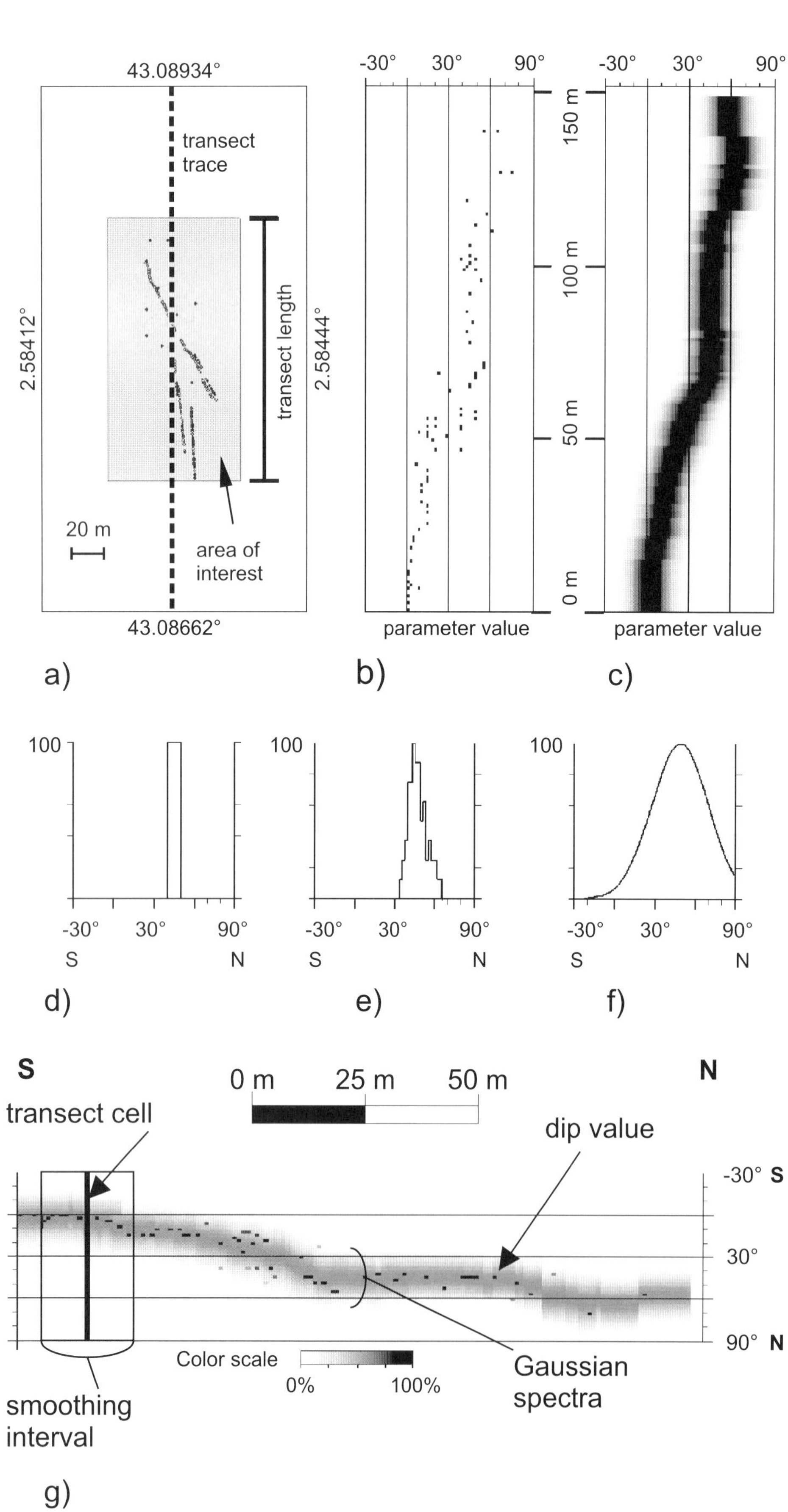

FIGURE 8. Transect methodology (see text for discussion): (a) data location; (b) nonsmoothed histogram transect—black points represent data values; (c) smoothed Gaussian transect—gray scale represents frequency; (d) cell nonsmoothed histogram; (e) cell smoothed histogram; (f) cell smoothed Gaussian fit; (g) final transect output, obtained by overlying nonsmoothed histogram transect and the Gaussian fit transect.

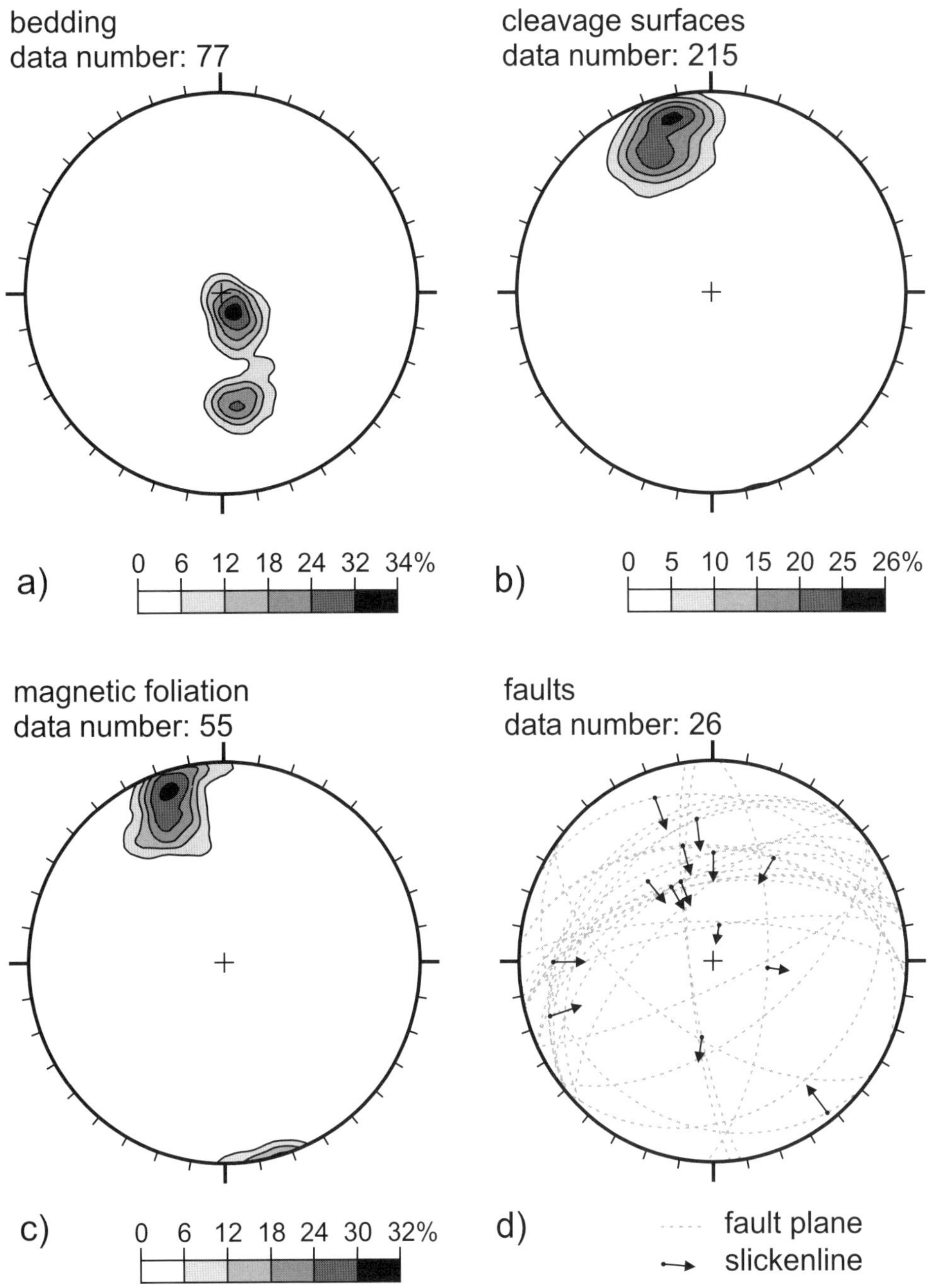

FIGURE 9. Mesostructural analysis of data along transect Ch-1: (a) bedding surface poles, (b) cleavage surfaces poles, (c) AMS K_3 axes, and (d) fault poles and slickenlines.

illustrates the application of this methodology to the analysis of bedding dip along transect Ch-1. The transect is a segment of given direction and length, along which data from the area of interest are projected perpendicularly (Figure 8a). The statistical analysis is performed by dividing the transect length into N cells of a given width (in the example, the cell width is 1 m [3.3 ft]). For each cell, a histogram (normalized to 100) is produced using data whose projection is within the cell (Figure 8d). Data can be smoothed along the transect, and the corresponding histogram represents data whose projections have a distance from the cell lower than the half value of the smoothing interval (Figure 8e). A Gaussian fit is finally applied to the histogram (Figure 8e). Each cell is therefore characterized by its position along the transect, the frequency histogram, its smoothed histogram, and its Gaussian fit curve.

Transect CH-1

The first transect is north–south oriented, and its total length is 151 m (495 ft). It is located in the central part of the Chaudrons anticline (Figure 3). The starting and ending points are located, respectively, in the crest-forelimb transition and in the steep forelimb. A total of 318 mesostructural data (77 bedding surfaces, 215 cleavage surfaces, and 26 mesofaults) and 55 core samples for AMS measurements have been collected along it. Cumulative analysis of mesostructural data is shown in Figure 9. Bedding surface poles (Figure 9a) show two relative maxima, corresponding to the crest and steeper forelimb, respectively. Cleavage poles (Figure 9b) and K_3 axes (Figure 9c) are nearly orthogonal to both the strike of bedding and the anticline axis. Fault planes and fault slickenlines (Figure 9d) show as the movement, which is mainly top to the southeast, is accomplished by fault planes dipping toward the northwest.

Figure 10a shows the bedding dip across the transect. This transect can be divided into two parts. In the left (southern) part, the bedding dip progressively increases, passing from a near-horizontal orientation to a value of 40°. This homogeneous variation testifies to the round shape of the anticline in the upper part of the forelimb. In the second part, the bedding dip increment is slower and rotates from 40 to 60°. A similar distribution

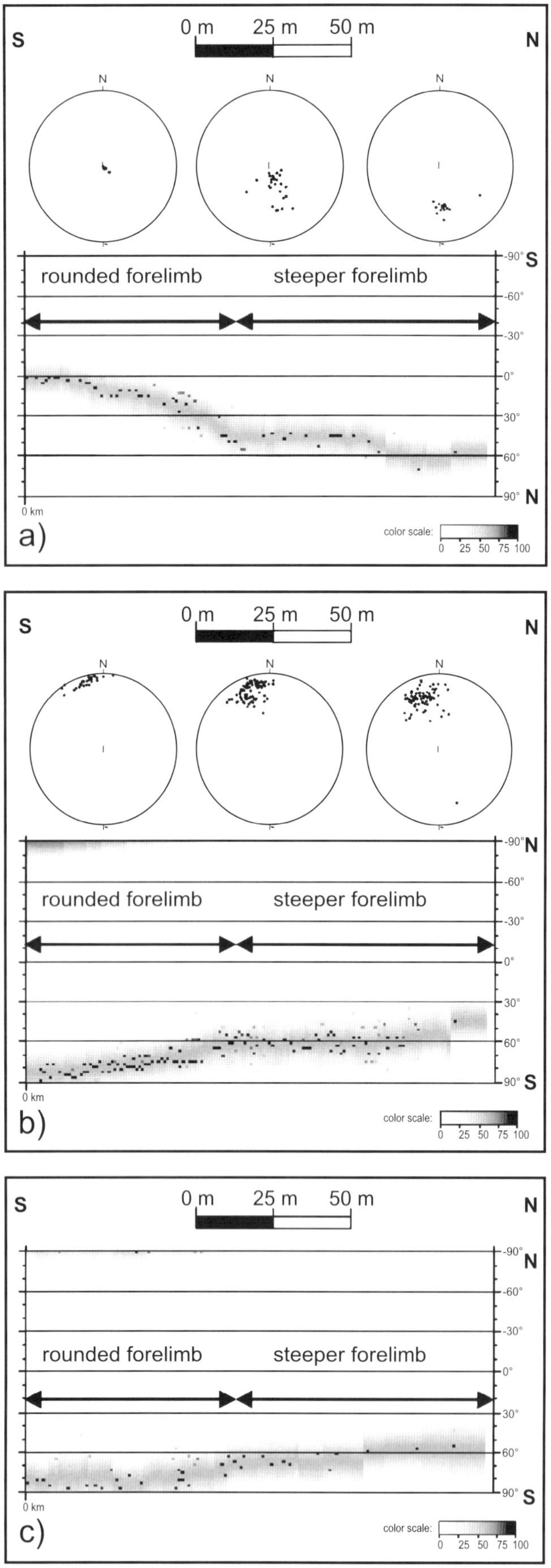

FIGURE 10. Transect Ch-1. (a) Bedding surfaces dip. Bottom: transect analysis; top: stereonet of the left, central, and right part of the transect. (b) Cleavage surfaces dip. (c) AMS magnetic foliation plane dip.

is present in the transect relative to cleavage (Figure 10b) and magnetic foliation (Figure 10c). In the rounded forelimb (southern part of the transect), cleavage and magnetic foliation dip rotates from a near-vertical orientation to an average dip of 60° southward. Moving into the steeper forelimb (northern part of the transect), the decrease of cleavage and magnetic foliation dip becomes slower.

The comparison between cleavage and AMS data is presented in Figure 11, where the values along the transect are shown with respect to the bedding surfaces (i.e., after bedding dip removal). This angle (ATB) is measured as the counterclockwise angles between bedding and cleavage and magnetic foliation looking west. Cleavage (Figure 11a) and magnetic foliation (Figure 11b) ATBs show different behaviors in the southern and northern part of the transect. In the rounded forelimb (south), they undergo a counterclockwise rotation and rotate from 80 to 120°. In the steeper forelimb (north), both the cleavage and the magnetic foliation ATBs remain rather constant until the end of the transect (in the frontal part of the forelimb). Cleavage ATB trends in weak (clayish) and stiff (sandy) beds are also illustrated (Figure 11a). Differences are remarkable only in the steeper part of the forelimb, where cleavages collected in weak beds are characterized by ATBs greater than those of cleavages collected in stiff beds. Differences between cleavages collected in weak and stiff beds are negligible in the rounded forelimb and are not shown.

Cleavage spacing along the transect (Figure 12a) is variable and strongly influenced by bed thickness. Large values characterize the rounded forelimb, indicating that this sector is less deformed than the northern one, as confirmed by field evidence. The H/S ratio (Figure 12b) and the Sf (Figure 12c) reduce the bedding thickness-related bias. H/S shows two different values for the two sectors (3 in the rounded forelimb and 4.5 in the steeper forelimb), divided by a narrow (about 15-m [50-ft]-wide) transition zone. The shape factor shows a progressive increase in the rounded forelimb (from 0 to as much as 1.5). Constant values (about 1.5) characterize the steeper forelimb. The H/S (Figure 12b) analysis is also characterized by a general increase in the scattering of the data (gray-shaded zone) northward.

Transect Ch-2

Figures 13–16 present the data relative to the transect Ch-2, which is north–south oriented. It is located in the western part of the anticline, and its length is 633 m (2100 ft). The transect starts in the middle of the

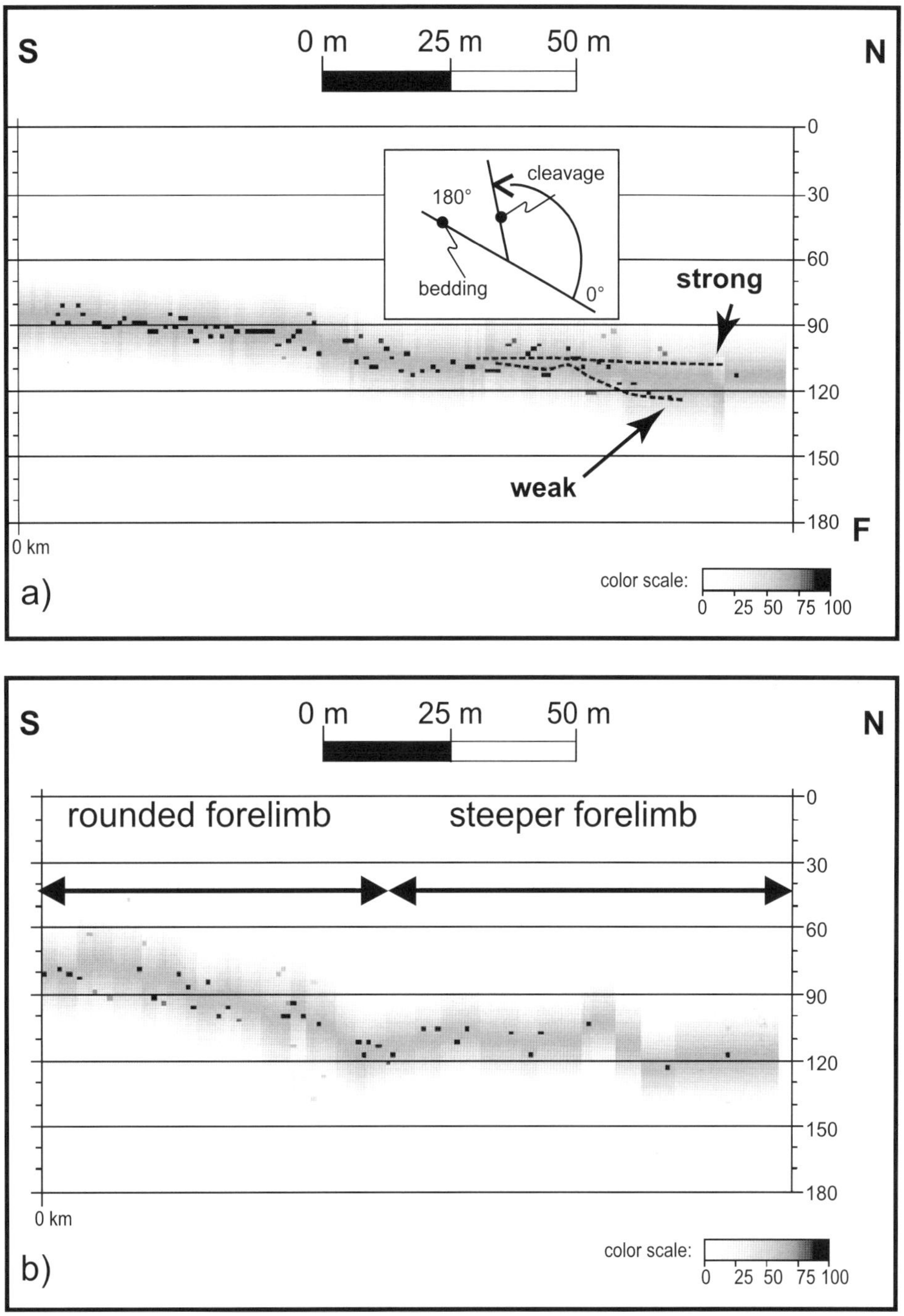

FIGURE 11. Transect Ch-1. (a) Cleavage ATB, trends of data collected in weak and strong beds are illustrated. (b) AMS magnetic foliation plane ATB.

crestal sector and reaches the forelimb to the north. A total of 406 mesostructural data (67 bedding surfaces, 321 cleavage surfaces, and 18 mesofaults) and 45 core samples for AMS measurements have been collected. Cumulative analyses are shown in Figure 13. Bedding surface poles (Figure 13a) show a relative maximum, corresponding to the crest. Poles to cleavage (Figure 13b) and magnetic foliation (Figure 13c) are rather orthogonal and subhorizontal to the bedding surfaces strike. Collected faults (Figure 13d) are organized in two sets: a first one with a north-northwest–south-southeast movement, including both faults parallel and at high ATB, and a second one at high ATB, with a east-northeast–west-southwest movement.

Figure 14a shows the bedding dip variation along the transect. Most part of the transect (southern and central part) presents near-horizontal bedding orientations (crest sector), whereas in the northern part, bedding dip progressively increases to as much as 40° (rounded forelimb). Solution cleavage (Figure 14b) and magnetic foliation (Figure 14c) are near vertical in the southern and central part of the transect (crest sector) while they progressively rotate in the rounded forelimb with a minimum dip of 60° at the edge of the transect. Cleavage (Figure 15a) and magnetic foliation (Figure 15b) ATBs along the transect individuate two sectors. The first one corresponded to the crest, where cleavage and magnetic foliation are near orthogonal to the bedding. The second sector corresponded to the rounded forelimb, where both cleavage and magnetic foliation ATBs progressively rotate from near orthogonal to a value of 120°.

Cleavage spacing along the transect (Figure 16a) is characterized by strongly scattered data. However, the *H*/*S* ratio (Figure 16b) and the Sf (Figure 16c) are more homogeneously distributed along the transect. In particular, the *H*/*S* ratio is rather constant at about 3, with the exception of a narrow sector just south of the forelimb where values are about 1.5. This narrow zone (about 30 m [100 ft] wide) is also characterized by the emergence of a minor thrust. Notice that the same value of

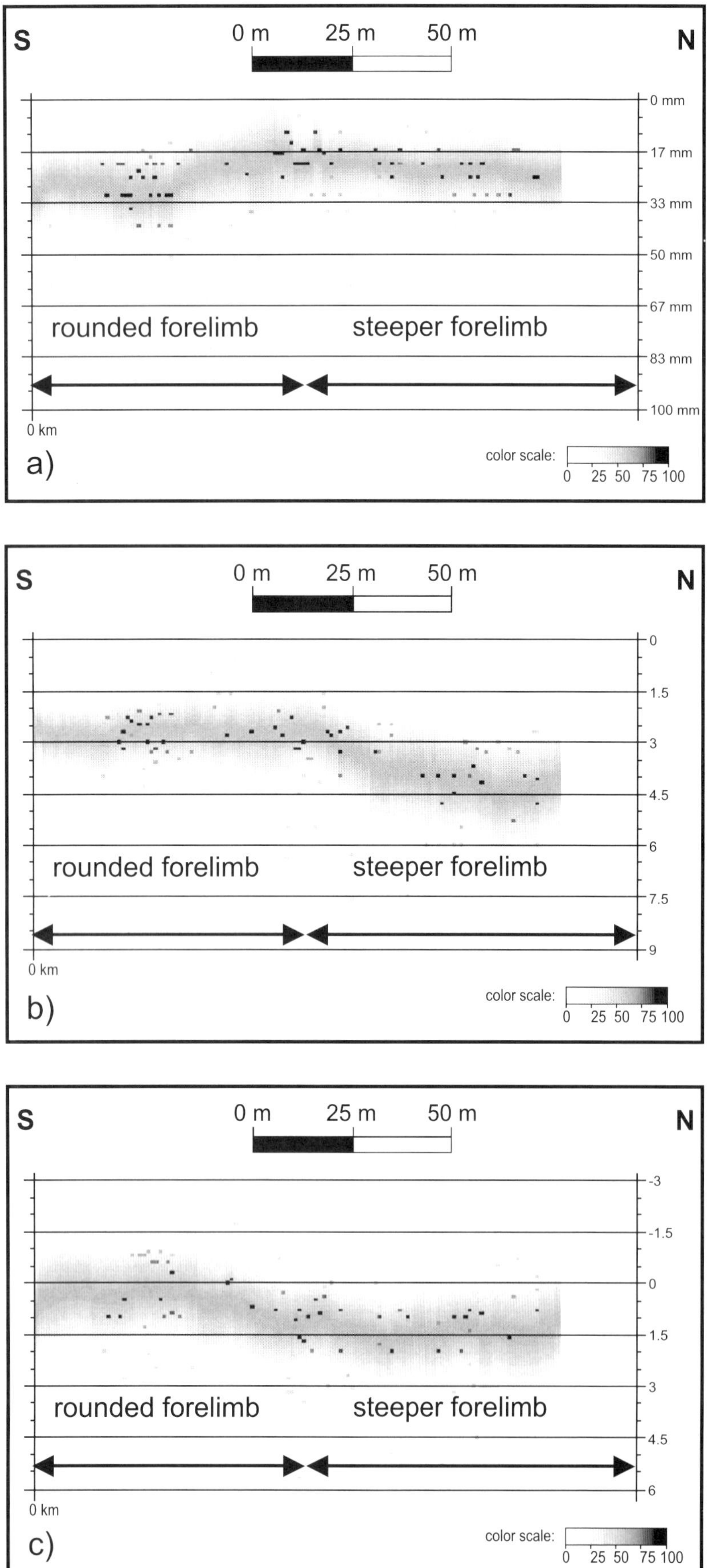

Figure 12. Transect Ch-1. (a) Cleavage spacing. (b) Cleavage *H*/*S*. (c) Cleavage shape factor.

3 also characterizes the southern part of transect Ch-1. This overlapping in the *H*/*S* ratio is in agreement with the overlapping in the geometry (Figures 10a, 14a), thus confirming the possibility of integrating the results of the two transects. Shape factor variations along the transect are similar to those observed for the *H*/*S* ratio. In particular, the southern and central parts of the transect are characterized by undulations in the shape factor between 0 and 1.5, whereas in the forelimb, it is around 1.5. The narrow and abrupt passage between the crest and the forelimb, which is observed in the *H*/*S* ratio transect, is still present.

DISCUSSION

Deformational Features Evolution

Bedding dip and cleavage and magnetic foliation ATBs spatial distributions define three deformational panels (Storti and Salvini, 1996) of comparable size in the Chaudrons anticline. In the first one that corresponds to the crest (southern and central part of transect Ch-2), the bedding orientation is near horizontal, and the cleavage and magnetic foliation ATBs are near 90°. The second panel corresponds to the rounded forelimb sector (northern part of transect Ch-2 and southern part of transect Ch-1). In this panel, the bedding dip progressively rotates from 0 to 40° northward. In addition, cleavage and magnetic foliation ATBs undergo rotations; they rotate from near 90 to 120°. The third panel corresponds to the steeper forelimb (northern part of transect Ch-1), where the bedding dip rotates from 40 to 60°, with an increment lower than in the previous panel. Cleavage and magnetic foliation ATBs in this panel are constant or present small variations.

These three panels are evident in the *H*/*S* and Sf distribution, whereas the spacing distribution is poorly significant. In the crest panel, *H*/*S* and Sf show

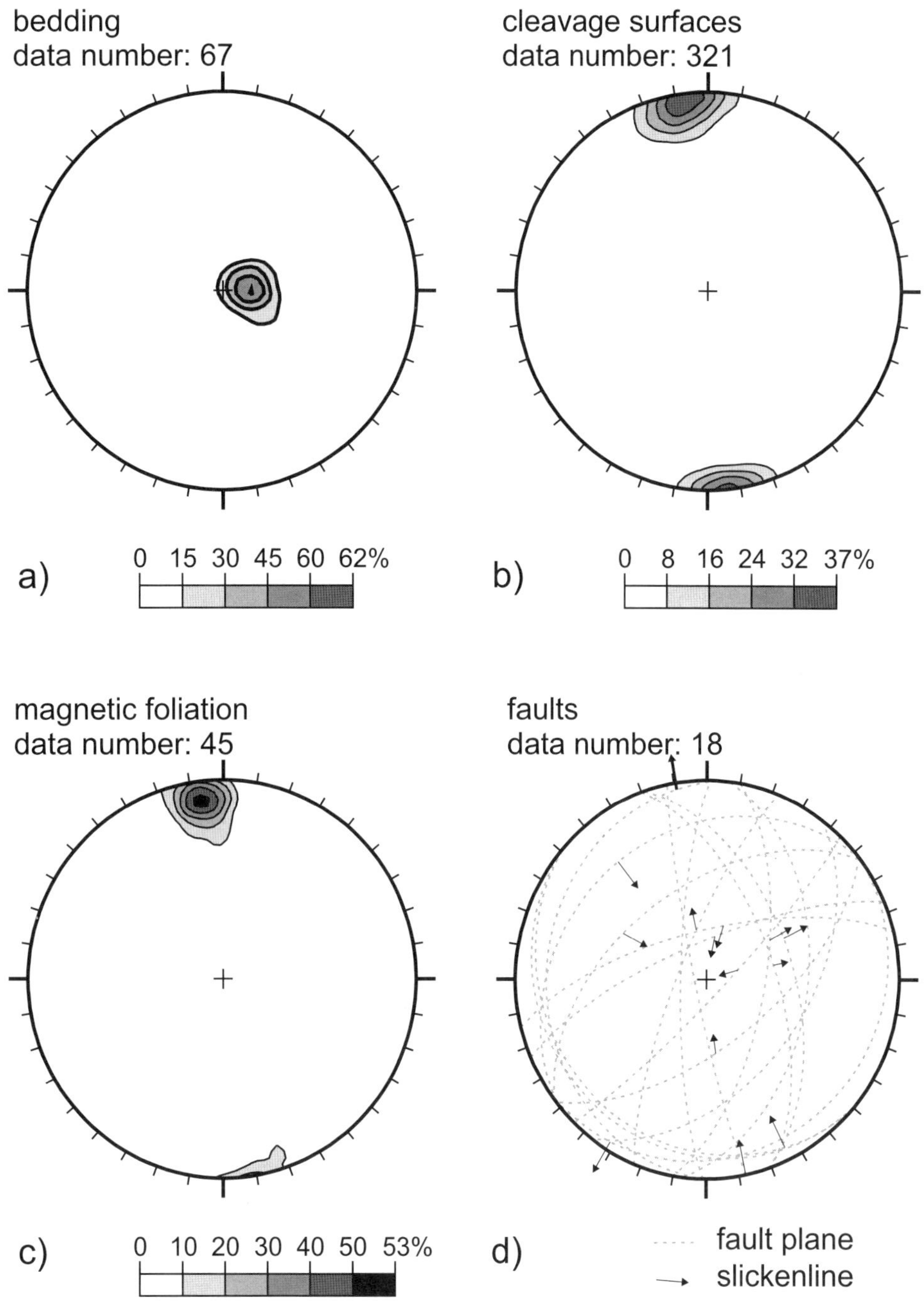

FIGURE 13. Mesostructural analysis of data along transect Ch-2. (a) Bedding surface poles, (b) cleavage surface poles, (c) AMS K_3 axes, and (d) fault poles and slickenlines.

fluctuations around 3 and 1, respectively. The second panel of the anticline (rounded forelimb; corresponding to the northern part of transect Ch-2 and the southern part of transect Ch-1) matches within the two transects for *H*/*S* distribution, with values about 3. Sf ranges from 0.0 to 1.5 when passing from the crest to the forelimb. However, behavior in this panel is different between the two transects; in transect Ch-1, Sf progressively increases toward the forelimb, whereas in transect Ch-2, the passage is abrupt. The third panel is characterized by constant values for both *H*/*S* ratio and Sf that are 4.5 and 1.5, respectively.

Cleavage and magnetic foliation ATBs in the crest (panel 1) measure about 90° and indicate a shortening axis parallel to bedding surfaces and to the regional shortening direction. This angle progressively rotates in the rounded forelimb (panel 2) and stabilizes to a constant value in the steeper forelimb (panel 3). The intrabed feature of cleavage surfaces and the different ATBs characterizing weak and stiff beds allow the correlation of these rotations with a top-to-the-hinterland bed-parallel shearing (e.g., Marshak and Engelder, 1985), which is likely related to the flexural slip (e.g., Tanner, 1989) that is expected in the forelimb of fault-bend and fault-propagation anticlines (e.g., Suppe, 1983; Suppe and Medwedeff, 1990). Cleavage and magnetic foliation ATB is assumed to be dependent on the relative importance of the flexural slip vs. LPS. Pure LPS would result in ATBs close to 90°. By increasing the relative amount of top-to-hinterland (or top-to-foreland) flexural slip, ATBs increase (or decrease). In the Chaudrons anticline, the intensity of flexural slip progressively increases from the crest toward the steeper forelimb. In the crest (panel 1), it is negligible (cleavage and magnetic foliation ATBs are close to 90°); in the rounded forelimb (panel 2), it progressively increases with increasing the bedding dip (that rotates from 0 to as much as 40°), and in the steeper forelimb (panel 3), it stabilizes at high values (cleavage and magnetic foliation ATBs are 120°).

The relationships between orientation of bedding, cleavage, magnetic foliation, and acoustic minimum velocity axes can be used to investigate the relative timing of meso- and microscale deformation during the

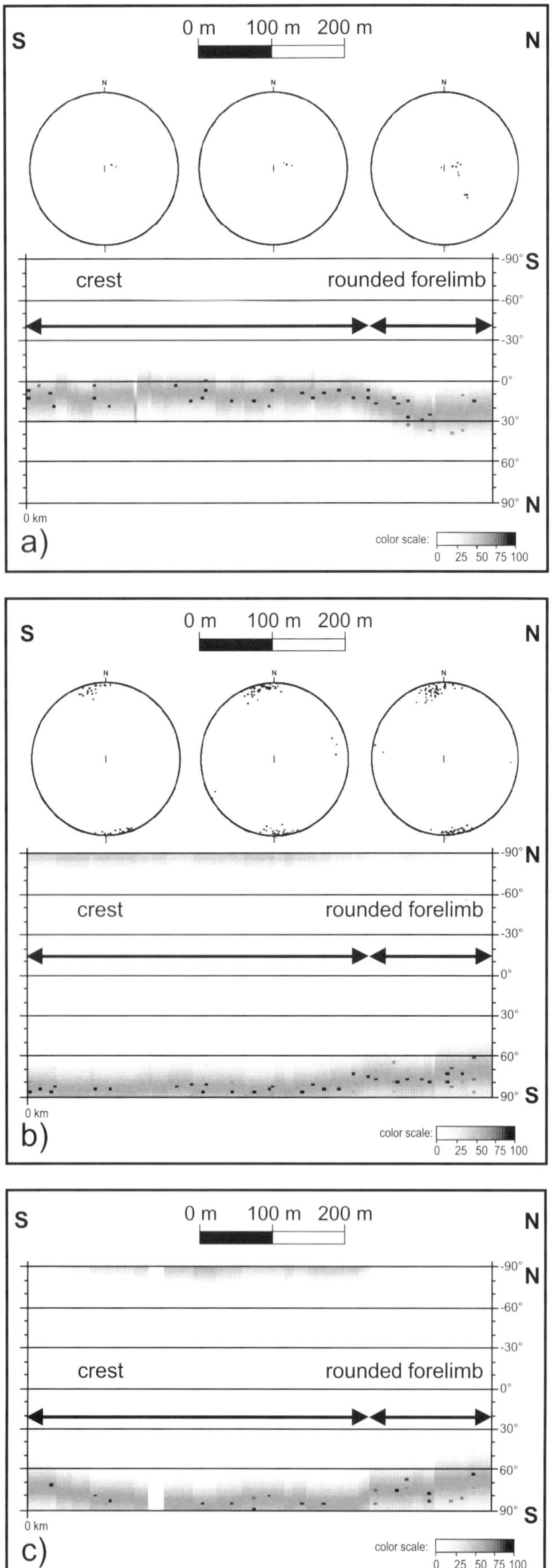

FIGURE 14. Transect Ch-2. (a) Bedding surfaces dip. Bottom: transect analysis; top: stereonet of the left, central, and right part of the transect. (b) Cleavage surfaces dip. (c) AMS magnetic foliation plane dip.

folding process (e.g., Mitra et al., 1984; Averbuch et al., 1992; Harris and Van Der Pluijm, 1998; Grelaud et al., 2000; Silliphant et al., 2002). In the case of prefolding deformation development, angular relationships to the bedding (ATB) would be constant all around the fold; in the case of synfolding deformation, it would be dependent on the structural position; and finally, in the case of postfolding deformation, its features would be independent on the bedding dip and constant all along the folds (Figure 17a–c).

Our data do not really match with these end-member conceptual models. Cleavage and magnetic foliation ATBs are constantly orthogonal to bedding in the crestal sector, suggesting that they result from prefolding LPS. By contrast, in the forelimb, the same elements are oblique to bedding, suggesting a synfolding development (Figure 17d). This apparent discrepancy is consistent with a two-step scenario. Cleavage, related magnetic foliation, and conjugate reverse faults started to develop first before folding and result from a pure-shear LPS as elsewhere in the Corbières (see Cluzel, 1977; Averbuch et al., 1992; Grelaud et al., 2000; Frizon de Lamotte et al., 2002). The tilt of these different elements in the forelimb suggests that during folding, the function of the flexural slip becomes prevalent with respect to the LPS. The increasing of deformation intensity, rather independently from the cleavage ATB (i.e., from the flexural slip intensity), strongly suggests that in the forelimb, LPS partially continued during folding, contemporary to the flexural slip. The mesofaults observed in the stiff layer also show a different geometry depending on their location in the anticline. Out of the forelimb, the conjugate pairs have similar displacements in agreement with LPS pure shear. In the forelimb, by contrast, the southeast-verging faults are more abundant (see above) and accommodate greater displacement in agreement with a global top-to-the-southeast bed-parallel shearing.

Hydrocarbon Research and Development Implications

The development of pressure-solution cleavage can be responsible for strong modifications of the pretectonic rock permeability and porosity. In particular, three events affect rocks when pressure solution occurs: (1) porosity reduction by tectonic compaction, induced by grain-to-grain solution (e.g., Beutner, 1978; Engelder and Marshak, 1985; Holl and Anastasio, 1995); (2) porosity

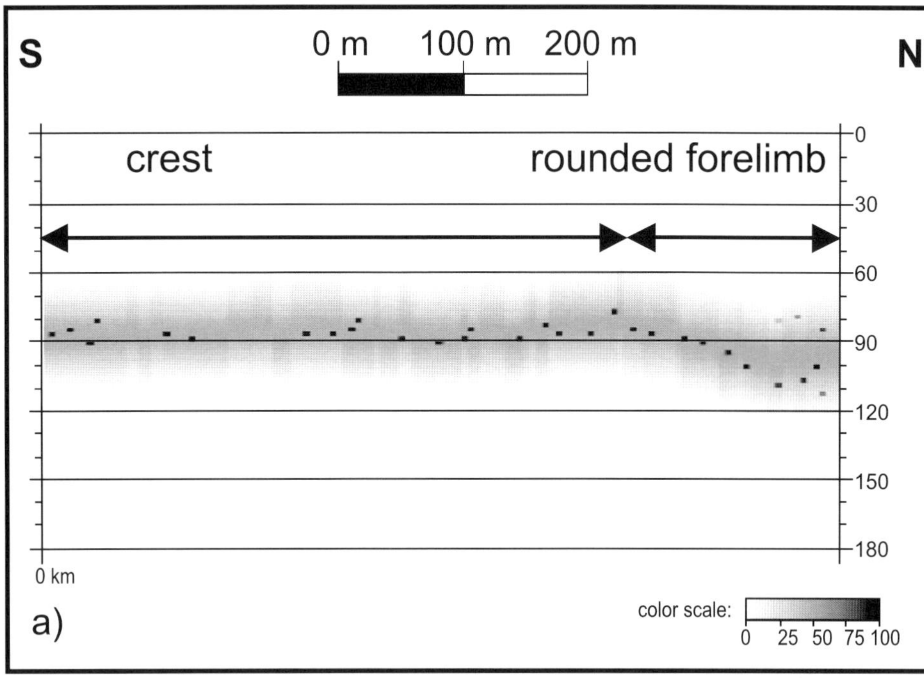

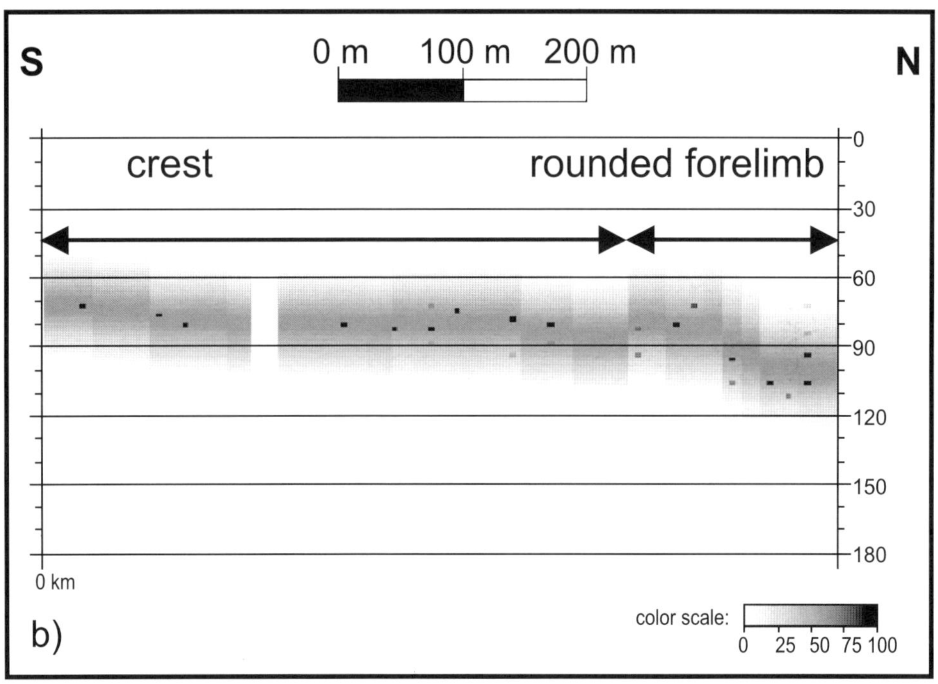

FIGURE 15. Transect Ch-2. (a) Cleavage ATB. (b) AMS magnetic foliation plane ATB.

reduction induced by mass transfer from the solution surfaces to both the pores and the extensional features in the neighboring areas of the solution surfaces (e.g., Elliott, 1973; Fletcher and Pollard, 1981; Mitra et al., 1984); (3) a permeability increase that can be induced by the development of mesoscale solution cleavage networks that can increase rock connectivity (e.g., Alsharhan and Whittle, 1994; Davidson et al., 1998; Van Geet et al., 2002).

Tectonic compaction induced by LPS mainly occurs before folding, whereas pressure-solution cleavage development and mass transfer start prior to folding and continue during folding (e.g., Mitra, 1988; Dunne and Caldanaro, 1997). It follows that the prefolding permeability and porosity distributions in folded layers can be considered roughly homogeneous. The development of pressure-solution cleavage sets during folding can thus be referred to as the main mechanism that modifies permeability and porosity distribution in cleaved folds.

The ratio between the porosity reduction induced by mass transfer and the permeability increase cause by the pressure-solution networks development is strongly controlled by the rock-water system (e.g., Geiser and Sansone, 1981; Engelder, 1984; Davidson et al., 1998). In a closed system where water circulation is restricted, mass dissolved along pressure-solution surfaces is precipitated in the surrounding zones (e.g., Mitra et al., 1984). In this case, pressure solution produces an increase of permeability and a decrease of porosity. However, in an open system, saturated water is removed (e.g., Engelder, 1984; Engelder and Marshak, 1985), pressure solution increases permeability, and porosity may be preserved, or at least, its reduction is much lower than in a closed system. In the Vitrollian formation of the Chaudrons anticline, veins are sparse, supporting the possibility of fluid removal both before (LPS) and during folding (i.e., open system). Preliminary data (Louis et al., 2002) do not show significant porosity reduction in the anticline. It follows that the Chaudrons anticline provides a representative field analog for studying the evolution of a cleaved open-system reservoir. In particular, the permeability of the Vitrollian formation in the Chaudrons anticline increases

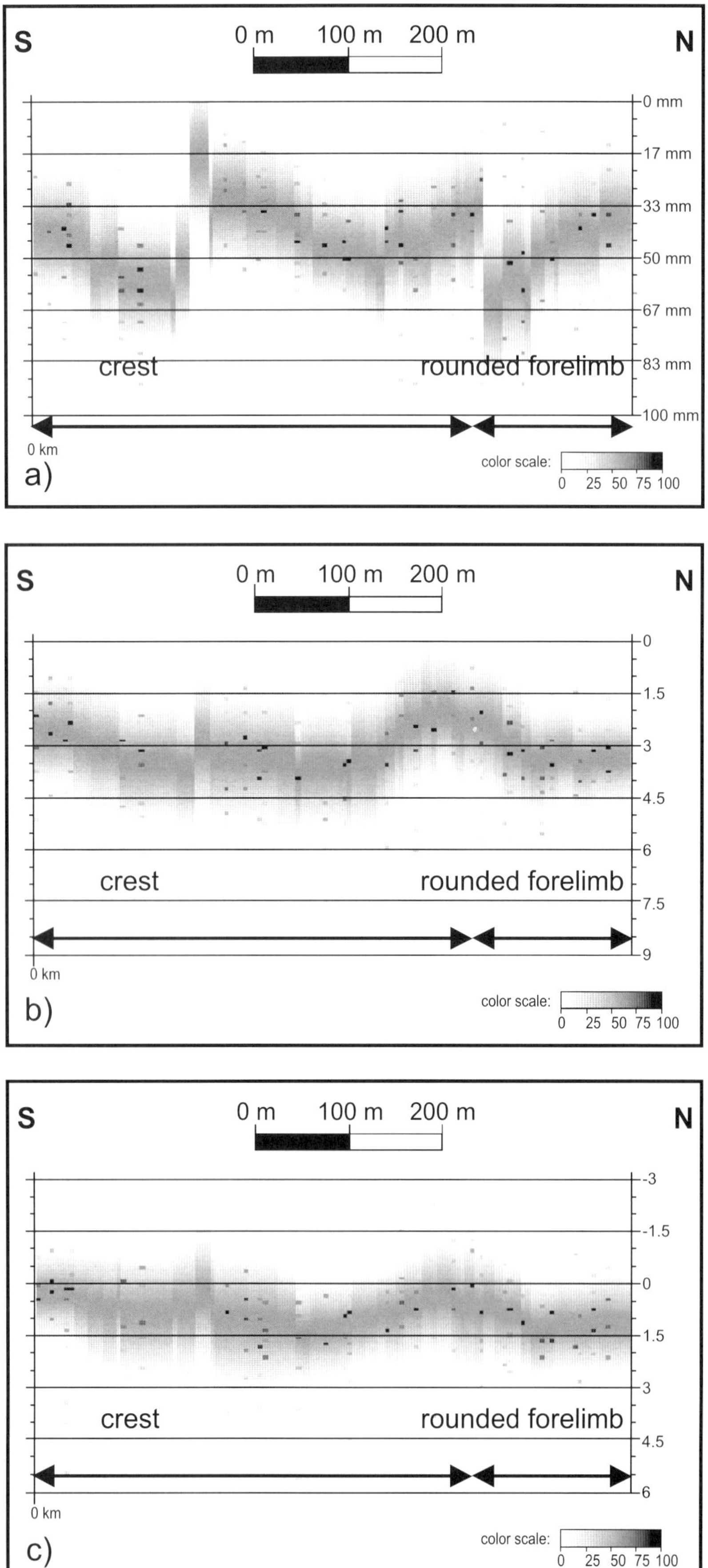

FIGURE 16. Transect Ch-2. (a) Cleavage spacing. (b) Cleavage *H*/*S*. (c) Cleavage shape factor.

from the crestal sector toward the forelimb. In the latter, longitudinal (parallel to the cleavage strike) and vertical (roughly parallel to the cleavage dip) connectivities are enhanced, whereas transversal connectivity (orthogonal to the cleavage surfaces) is reduced.

CONCLUSIONS

The distribution of the deformational features at the various scales in the Chaudrons anticline shows the following basic features:

1) Deformation patterns at the bedding (solution cleavage) and matrix (AMS and acoustic velocity anisotropy) scales show strong analogies and are likely related to similar tectonic processes.
2) Deformation intensity can be conveniently described by the *H*/*S* ratio and, in particular, by the shape factor (Sf), which is a three-dimensional parameter.
3) Spatial distribution of deformational features in the Chaudrons anticline individuates three deformation panels corresponding to the crest, the rounded forelimb, and the steeper forelimb, respectively.
4) Deformation intensity in the Chaudrons anticline is rather low and constant in the crestal sector and progressively increases toward the forelimb as the bedding dip increases. As the forelimb dip assumes constant values (i.e., in the steeper forelimb sector), the deformation intensity stabilizes as well.
5) Deformational features in the Chaudrons anticline started developing before folding by LPS. Fold development caused the selective infilling of deformational features in the forelimb by the combination of bed-parallel simple shear (flexural slip) and bed-parallel pure shear (LPS). The LPS fabric

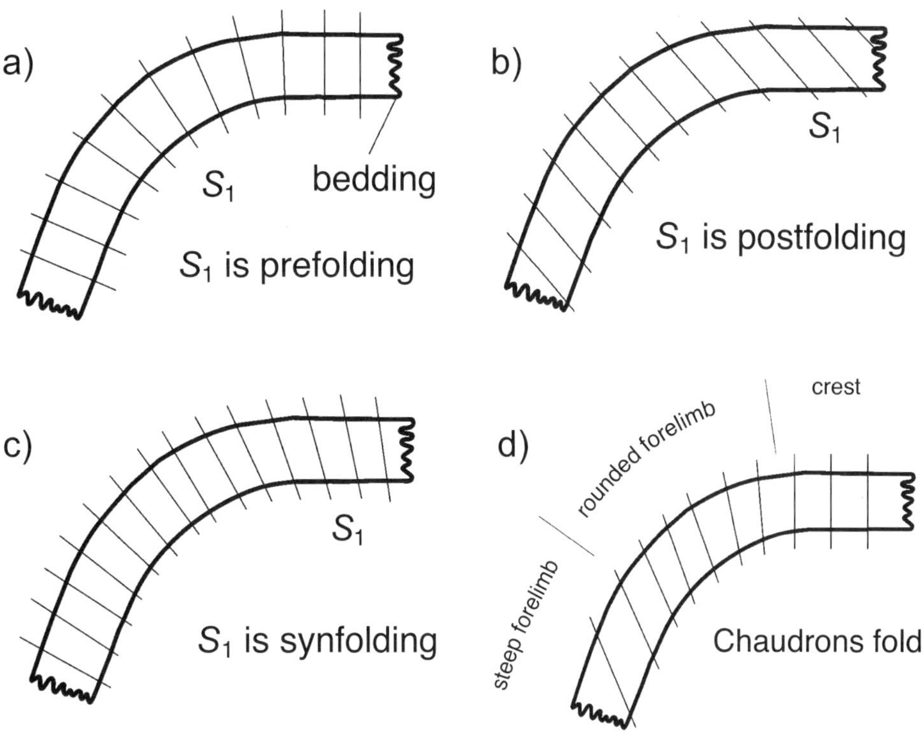

FIGURE 17. Angular relationships between cleavage and magnetic foliation surfaces (S_1) and bedding surfaces in the (a) prefolding, (b) postfolding, and (c) synfolding hypotheses. (d) S_1 distribution in the Chaudrons anticline.

is preserved in anticline crest, whereas the cumulative (LPS + flexural slip) deformation pattern characterizes the rounded and steeper part of the forelimb.

6) At the proper environmental conditions for solution cleavages to become preferential conduits for fluid flow, their anisotropic distribution can significantly influence reservoir characterization and modeling. In particular, in the Chaudrons anticline, secondary porosity and permeability related to solution surface sets increase from the crest to the forelimb.

ACKNOWLEDGMENTS

This study results from an agreement between Roma 3 (Italy) and Cergy-Pontoise (France) universities. Research councils of both institutions are acknowledged for their support. This work has been partly elaborated during a stay of F. Salvini at Cergy-Pontoise University as an invited professor. The chapter benefited from discussion with Christian David, Pascale Leturmy, Bertrand Maillot, François Roure, and Fabrizio Storti. The authors also acknowledge Rudy Swennen and Jim Granath for their helpful review of the chapter.

REFERENCES CITED

Alsharhan, A. S., and G. L. Whittle, 1994, Sedimentary-diagenetic interpretation and reservoir characteristics of the Middle Jurassic (Araej Formation) in the southern Arabian Gulf: Marine and Petroleum Geology, v. 12, p. 615–628.

Averbuch, O., 1993, Caractérisation de la déformation dans les structures de chevauchement-plissement: Utilisation couplée du magnétisme des roches et de l'analyse structurale. Exemples dans les Corbières, l'Apennin central et le Taurus occidental: Ph.D. thesis, Université de Paris-Sud, Orsay, 278 p.

Averbuch, O., D. Frizon de Lamotte, and C. Kissel, 1992, Magnetic fabric as a structural indicator of the deformation path within a fold-thrust structure: A test case from the Corbère (NE Pyrenees, France): Journal of Structural Geology, v. 14, p. 461–474.

Aydin, A., 2000, Fractures, faults, and hydrocarbon entrapment, migration, and flow: Marine and Petroleum Geology, v. 17, p. 797–814.

Beutner, E. C., 1978, Slaty cleavage and related strain in Martinsburg slate, Delaware Water Gap, New Jersey: American Journal of Science, v. 278, p. 1–23.

Bhathal, R. S., 1971, Magnetic anisotropy in rocks: Earth Science Review, v. 7, p. 227–253.

Cluzel, D., 1977, Etude microtectonique de l'avant-pays de la nappe des Corbières orientales (Aude-France): Ph.D. thesis, Université de Paris-Sud, Orsay, 292 p.

Cooke, M. L., 1997, Predicting fracture localization in folded strata from mechanical stratigraphy and fold shape: Case study of east Kaibab monocline, Utah: International Journal of Rock Mechanics and Mineral Sciences, v. 34, p. 351.

Cooper, M., 1992, The analysis of fractures system in subsurfaces thrust structures from the foothills of the Canadian Rockies, *in* K. McClay, ed., Thrust tectonics: London, Chapman & Hall, p. 391–405.

Corbett, K., M. Friedman, and J. Spang, 1987, Fracture development and mechanical stratigraphy of Austin chalk, Texas: AAPG Bulletin, v. 71, p. 17–28.

Davidson, S. G., D. J. Anastasio, G. E. Bebout, J. E. Holl, and C. A. Hedlund, 1998, Volume loss and metasomatism during cleavage formation in carbonate rocks: Journal of Structural Geology, v. 20, p. 707–726.

Dunne, W. M., and A. J. Caldanaro, 1997, Evolution of solution structures in deformed quartz arenite: Geometric change related to permeability changes: Journal of Structural Geology, v. 19, p. 663–672.

Durney, D. W., and H. J. Kisch, 1994, A field classification and intensity scale for first-generation cleavages: Journal of Australian Geology and Geophysics, v. 15, p. 257–295.

Ellenberger, F., 1967, Les interferences de l'érosion et de la tectonique tangentielle tertiatie dans le Bas-Languedoc; note sur les charriages cisaillantes: Revue de Géographie Physique et de Géologie Dynamique, v. 9, p. 87–142.

Elliott, D., 1973, Diffusion flow laws in metamorphic rock: Geological Society of America Bulletin, v. 84, p. 2645–2664.

Engelder, T., 1984, The role of pore water circulation during the deformation of foreland fold and thrust belts: Journal of Geophysical Research, v. 89, p. 4319–4325.

Engelder, T., and S. Marshak, 1985, Disjunctive cleavage formed at shallow depths in sedimentary rocks: Journal of Structural Geology, v. 7, p. 327–343.

Fischer, M. B., and P. B. Jackson, 1999, Stratigraphic controls on deformation patterns in fault-related folds: A detachment fold example from the Sierra Madre Oriental, northeast Mexico: Journal of Structural Geology, v. 21, p. 613–633.

Fletcher, R. C., and D. D. Pollard, 1981, Anticrack model for pressure solution surfaces: Geology, v. 9, p. 419–424.

Frizon de Lamotte, D., E. Mercier, A. Dupré la Tour, P. Robion, and O. Averbuch, 1997, Cinématique du plissement et déformation interne des roches: l'Exemple du pli de Lagrasse (Aude, France): Compte Rendus de l'Academie des Sciences, Paris, v. 324, p. 591–598.

Frizon de Lamotte, D., C. Souque, S. Grelaud, and P. Robion, 2002, Early record of tectonic magnetic fabric during inversion of a sedimentary basin: Short review and examples from the Corbières transfer zone: Bulletin de la Société Géologique de France, v. 173, no. 5, p. 461–469.

Geiser, P. A., and S. Sansone, 1981, Joint, microfractures, and the formation of solution cleavage in limestone: Geology, v. 9, p. 280–285.

Graham, J. W., 1954, Magnetic susceptibility anisotropy, an unexploited petrofabric element: Geological Society of America Bulletin, v. 65, p. 1257–1258.

Graham, J. W., 1966, Significance of magnetic anisotropy in Appalachian sedimentary rocks, *in* J. S. Steinhart and T. J. Smith, eds., The Earth beneath the continents: American Geophysical Union Geophysical Monograph Series 10, p. 627–648.

Grelaud, S., 2001, Couplage socle-couverture et style de la déformation au front des orogènes: Comparaison entre le Minervois (front des Pyrénées, France) et le Potwar (front de l'Himalaya, Pakistan): Ph.D. thesis, Université de Cegy-Pontoise, France, 212 p.

Grelaud, S., D. Buil, S. Hardy, and D. Frizon de Lamotte, 2000, Trishear kinematic model of fault-propagation folding and sequential development of minor structures: The Oupia anticline (NE Pyrenees, France): Bulletin de la Société Géologique de France, v. 171, p. 441–449.

Gross, M. R., 1995, Fracture partitioning: Failure mode as a function of lithology in the Monterey Formation of coastal California: Geological Society of America Bulletin, v. 107, p. 779–792.

Harris, J. H., and B. A. Van Der Pluijm, 1998, Relative timing of calcite twinning strain and fold-thrust belt development; Hudson Valley fold-thrust belt, New York, U.S.A.: Journal of Structural Geology, v. 20, p. 21–31.

Hext, G. R., 1963, The estimation of second-order tensors, with related tests and designs: Biometrika, v. 50, nos. 3–4, p. 353–373.

Holl, J. E., and D. J. Anastasio, 1995, Cleavage development within a foreland fold and thrust belt, southern Pyrenees, Spain: Journal of Structural Geology, v. 17, p. 357–369.

Hrouda, F., 1982, Magnetic anisotropy of rocks and its application in geology and geophysics: Geophysical Surveys, v. 5, p. 37–82.

Hrouda, F., P. Zdenek, and J. Wohlgemuth, 1993, Development of magnetic and elastic anisotropies in slates during progressive deformation: Physics of the Earth and Planetary Interiors, v. 77, p. 251–265.

Jamison, W. R., 1997, Quantitative evaluation of fractures on Monkshood anticline, a detachment fold in the foothills of Western Canada: AAPG Bulletin, v. 81, p. 1110–1132.

Kissel, C., E. Barrier, C. Laj, and T.-Q. Lee, 1986, Magnetic fabric in "undeformed" marine clays from compressional zones: Tectonics, v. 5, p. 769–781.

Kliegfield, R., H. Owens, and W. Lowrie, 1981, Magnetic susceptibility anisotropy, strain and progressive deformation in Permian sediments from the Maritime Alps (France): Earth and Planetary Science Letters, v. 55, p. 181–189.

Leythaeuser, D., O. Borromeo, F. Mosca, R. Di Primo, M. Randke, and R. G. Shaefer, 1995, Pressure solution in carbonate source rocks and its control on petroleum generation and migration: Marine and Petroleum Geology, v. 12, p. 717–733.

Lo, T.-W., K. B. Coyner, and M. N. Toksöz, 1986, Experimental determination of elastic anisotropy of Berea Sandstone, Chicopee Shale, and Chelmsford Granite: Geophysics, v. 51, p. 164–171.

Louis, L., P. Robion, C. David, and D. Frizon de Lamotte, 2002, Anisotropic physical properties at sample scale and their evolution within fold-thrust structures (abs.): AAPG Annual Meeting Program, v. 90011, p. 25.

Louis, L., C. David, and P. Robion, 2003, Comparison of the anisotropic behaviour of undeformed sandstones under dry and saturated conditions: Tectonophysics, v. 370, p. 193–212.

Lowrie, W., and A. M. Hirt, 1987, Anisotropy of magnetic susceptibility in the Scaglia Rossa pelagic limestone: Earth and Planetary Science Letters, v. 82, p. 349–356.

Marshak, S., and T. Engelder, 1985, Development of cleavage in limestones of a fold-thrust belt in eastern New York: Journal of Structural Geology, v. 7, p. 345–359.

Mitra, S., 1988, Effects of deformation mechanisms on

reservoir potential in central Appalachian overthrust belt: AAPG Bulletin, v. 72, p. 536–554.

Mitra, G., and W. A. Yonkee, 1985, Relationship of spaced cleavage to fold and thrust in the Idaho–Utah–Wyoming thrust belt: Journal of Structural Geology, v. 7, p. 361–373.

Mitra, G., W. A. Yonkee, and D. J. Gentry, 1984, Solution cleavages and its relationship to major structures in the Idaho–Utah–Wyoming thrust belt: Geology, v. 12, p. 354–358.

Nelson, R. A., 1985, Geological analysis of naturally fractured reservoir: Houston, Texas, Gulf Publishing, 320 p.

Plaziat, J. C., 1984, Le domaine pyréneen de la fin du Crétacé à la fin de l'Eocène. Stratigraphie, paléoenvironnements et évoluzion paléo-géographique: Ph.D. thesis, University Paris-Sud, 1362 p.

Protzman, G. M., and G. Mitra, 1990, Strain fabric associated with the Meade thrust sheet: Implications for cross-section balancing: Journal of Structural Geology, v. 12, p. 403–417.

Sagnotti, L., F. Speranza, A. Winkler, M. Mattei, and R. Funiciello, 1998, Magnetic fabric of clay sediments from the external northern Apennines (Italy): Physics of the Earth and Planetary Interiors, v. 105, p. 73–93.

Salvini, F., 2002, Daisy 3 software: Rome.

Salvini, F., A. Billi, and D. U. Wise, 1999, Strike-slip fault-propagation cleavage in carbonate rocks: The Mattinata fault zone, southern Apennines, Italy: Journal of Structural Geology, v. 21, p. 1731–1749.

Silliphant, L. J., T. Engelder, and M. R. Gross, 2002, The state of stress in the limb of the Split Mountain anticline, Utah: Constraints placed by transacted joints: Journal of Structural Geology, v. 24, p. 155–172.

Souque, C., 2002, Magnétisme structural dans les chaînes de chevauchement-plissement: Développements analytiques et exemples d'utilisation dans les Corbières: Thèse Sciences, Université de Cergy-Pontoise, Cergy-Pontoise, France, 231 p.

Srivastava, D. C., and T. Engelder, 1990, Crack-propagation sequence and pore-fluid circulations during fault-bend folding in the Appalachian Valley and Ridge, central Pennsylvania: Geological Society of America Bulletin, v. 102, p. 116–128.

Storti, F., and F. Salvini, 1996, Progressive rollover fault-propagation folding: A possible kinematic mechanism to generate regional-scale recumbent folds in shallow foreland belts: AAPG Bulletin, v. 80, p. 174–193.

Suppe, J., 1983, Geometry and kinematics of fault-bend folding: American Journal of Sciences, v. 283, p. 684–721.

Suppe, J., and D. A. Medwedeff, 1990, Geometry and kinematics of fault-propagation folding: Ecoglae Geologicae Helvetiae, v. 83, p. 409–454.

Tanner, P. W. G., 1989, The flexural slip mechanism: Journal of Structural Geology, v. 11, p. 635–655.

Uyeda, S., M. D. Fuller, J. C. Belshe, and R. W. Girdler, 1963, Anisotropy of magnetic susceptibility of rocks and minerals: Journal of Geophysical Research, v. 68, p. 279–291.

Van Geet, M., R. Swennen, C. Durmishi, F. Roure, and P. H. Muchez, 2002, Paragenesis of Cretaceous to Eocene carbonate reservoirs in the Ionian fold and thrust belt (Albania): Relation between tectonism and fluid flow: Sedimentology, v. 49, p. 697–718.

Wise, D. U., and T. A. McCory, 1982, A new method of fracture analysis: Azimuth versus traverse distance plot: Geological Society of America Bulletin, v. 96, p. 952–967.

13

Murgia, M. V., P. Ronchi, and A. Ceriani, 2004, Dolomitization processes and their relationships with the evolution of an orogenic belt (central Apennines and peri-Adriatic foreland, Italy), *in* R. Swennen, F. Roure, and J. W. Granath, eds., Deformation, fluid flow, and reservoir appraisal in foreland fold and thrust belts: AAPG Hedberg Series, no. 1, p. 277–294.

Dolomitization Processes and Their Relationships with the Evolution of an Orogenic Belt (Central Apennines and Peri-Adriatic Foreland, Italy)

M. V. Murgia
Eni-E&P Division, Milan, Italy

P. Ronchi
Eni-E&P Division, Milan, Italy

A. Ceriani
University of Pavia, Pavia, Italy

ABSTRACT

In the Jurassic–lower Cretaceous sequence of central Apennines and their foreland (onshore and offshore Marche–Abruzzi regions), the dolomitization processes enhanced the petrophysical properties of the carbonate platform and slope series.

The area experienced a first phase of passive-margin regime, from the Lower Jurassic to the Miocene: the Liassic carbonate platform underwent extensional tectonics that established the southern Apulian–Apennines persisting platforms and the northern Umbria-Marche basin; within the basin, differential subsidence rates created faulted horsts characterized by condensed series. From upper Miocene, the area entered the collision-margin phase when its eastern part was involved in the Apennines orogeny. The aim of this study was to analyze the dolomitization process in relation with the fluid-flow changes caused by the evolution of the geological framework.

Dolomitized bodies are mainly located at the Jurassic–Early Cretaceous platform edges and in the paleohigh areas, particularly in the platform Calcare Massiccio and basinal Corniola formations, with minor extension to younger slope successions (up to Maiolica Formation). The petrographic observations evidenced a multiphase dolomitization of alternated dolomite replacement, dissolution, and recrystallization. The carbon and oxygen isotopic analyses suggest a seawater-derived diagenesis in a wide temperature range; this is confirmed by the fluid-inclusion analyses that detected a few stages of dolomitization

DOI:10.1306/1025695H13121

events during a progressive heating of the carbonate series. The reconstructed paragenetic sequence is almost the same in all the studied successions, both in outcrop and subsurface.

The data collected show that the dolomitization processes changed during the evolution of the area from a passive-margin domain to the collision-margin regime. In both phases, the interaction between the increasing burial temperature and the fluid migration paths, which is driven by the approaching orogenic wave, is suggested.

The first dolomitization event (dolomite 1) is characterized by low homogenization temperature (T_h) and is interpreted as a replacement of calcite precursor; its formation occurred during the passive-margin domain and is strongly dependent on the rate of subsidence of the different areas. The dolomite 1 time of generation varies from Cretaceous, in the highly subsiding areas of the thrust zone, to the Miocene in the paleohighs of the foreland. Dolomite overgrowths (dolomite 2) precipitated at higher temperatures. In the depocenters of thrust zone, those temperatures (as much as 130°C) were reached during maximum overburden during the upper Miocene, whereas in the foredeep and foreland areas, the high-temperature dolomitizing fluids seem to flow through faults during the orogenic phases (Pliocene). In the whole studied area, pore-filling dolomite cements (dolomite 4 and saddle dolomite) are supposed to be precipitated from heated fluids coming from deep strata along fault planes. Regional considerations and salinity data of the fluid inclusions support the hypothesis that the dolomitizing fluids of the last phases could come from Triassic evaporites that are present in the area and represent the detachment surface of the thrusts.

INTRODUCTION

For a long time, in the study area, the uppermost part of carbonate upthrown fault blocks, sealed by clastic sequences, was considered the only drillable oil trap (Pieri and Mattavelli, 1986). However, if the structural setting of a region is fundamental in defining the trap geometry (Gibbs, 1984; Mitra, 1986; Harding and Tuminas, 1989), the petrophysical characteristics are also very important, and particularly in carbonate rocks, they are strongly influenced by diagenesis. For example, several studies point out that the dolomitization process commonly improves the petrophysical properties of carbonate rocks (Amthor and Friedman, 1991), and sometimes, only the dolomitized section of a carbonate sequence constitutes a reservoir facies. This fact is evident in the slope facies, where the porous dolomitized layers, draping paleohighs, may form a structural-stratigraphic trap that is sealed by less porous muddy sedimentary strata, which resulted in a successful play concept in recent hydrocarbon exploration in the studied area. The comprehension of the dolomitization process for the prediction of the spatial distribution of porous dolomite bodies is, in these cases, of paramount importance both in exploration and production. Even the origin of dolomitization is still debated and yet to be solved (Hardie, 1987). The number of genetic models is reduced if the massive replacement is regarded; many authors pointed to the burial dolomitization as one of the most likely processes for it in different geological settings (Braithwaite and Rizzi, 1997; Drivet and Mountjoy, 1997; Swennen et al., 2002). The integration of petrographic, geochemical, and fluid-inclusion analyses commonly allowed the reconstruction of the timing of late dolomitization related to the main burial and structural phases (Amthor et al., 1993; Zempolich and Hardie, 1997; Lonnee and Al-Aasm, 2000). This approach was applied for the first time to the dolomitized carbonate facies of Early Jurassic to Lower Cretaceous in the central Apennines and peri-Adriatic foreland. In particular, the carbonate platform of Calcare Massiccio (lower Lias), the basinal facies of Corniola Formation (middle Lias), and the slope calcarenitic succession of equivalent Maiolica Formation (Lower Cretaceous) show good reservoir characteristics when dolomitized. The proposed model of burial dolomitization and fluid-flow pattern represents a tool in the prediction of the distribution of the dolomitized potential hydrocarbon reservoirs in the region.

GEOLOGICAL FRAMEWORK

Structural Setting

The study area is located in central Italy, facing the Adriatic Sea (Figure 1). It includes a portion of the Apennines chain and its foreland. The present geological framework is the result of a complex tectonic evolution: extension during the Mesozoic (Neo-Tethys passive margin;

Decourt et al., 1993; Marchegiani et al., 1999) and compression starting from the Miocene (Apennines orogeny) (Bigi et al., 1992; Cipollari et al., 1999). According to the present structural frame (Figures 1, 2), three zones can be distinguished: the thrust zone (A zone), where Mesozoic to Tertiary sequences are involved in fold and thrust sheets from late Miocene to Pliocene; the intermediate zone (B zone), where Miocene–Pliocene foredeep sediments are deformed during middle to late Pliocene (Patacca et al., 1992); and the foreland (C zone).

Particularly from the Serravallian to middle Pleistocene, the tectonic compression wave migrated northeastward, progressively involving, in overthrusts and folds, the outer zone of the Apennines chain (Centamore et al., 1992). In the studied area, the most important tectonic phase, whose main effect is the overthrusts of A zone, is dated lower Pliocene by means of regional, geological studies based on seismic interpretation, surface, and subsurface data (unpublished Eni data): In the lower Pliocene, a tensional pattern starts to act in the inner part of the chain (Brunamonte et al., 1992), coaxial with the former one, and lasting until the Pleistocene (Dramis et al., 1992), when part of the foreland is involved mainly with transpressive features.

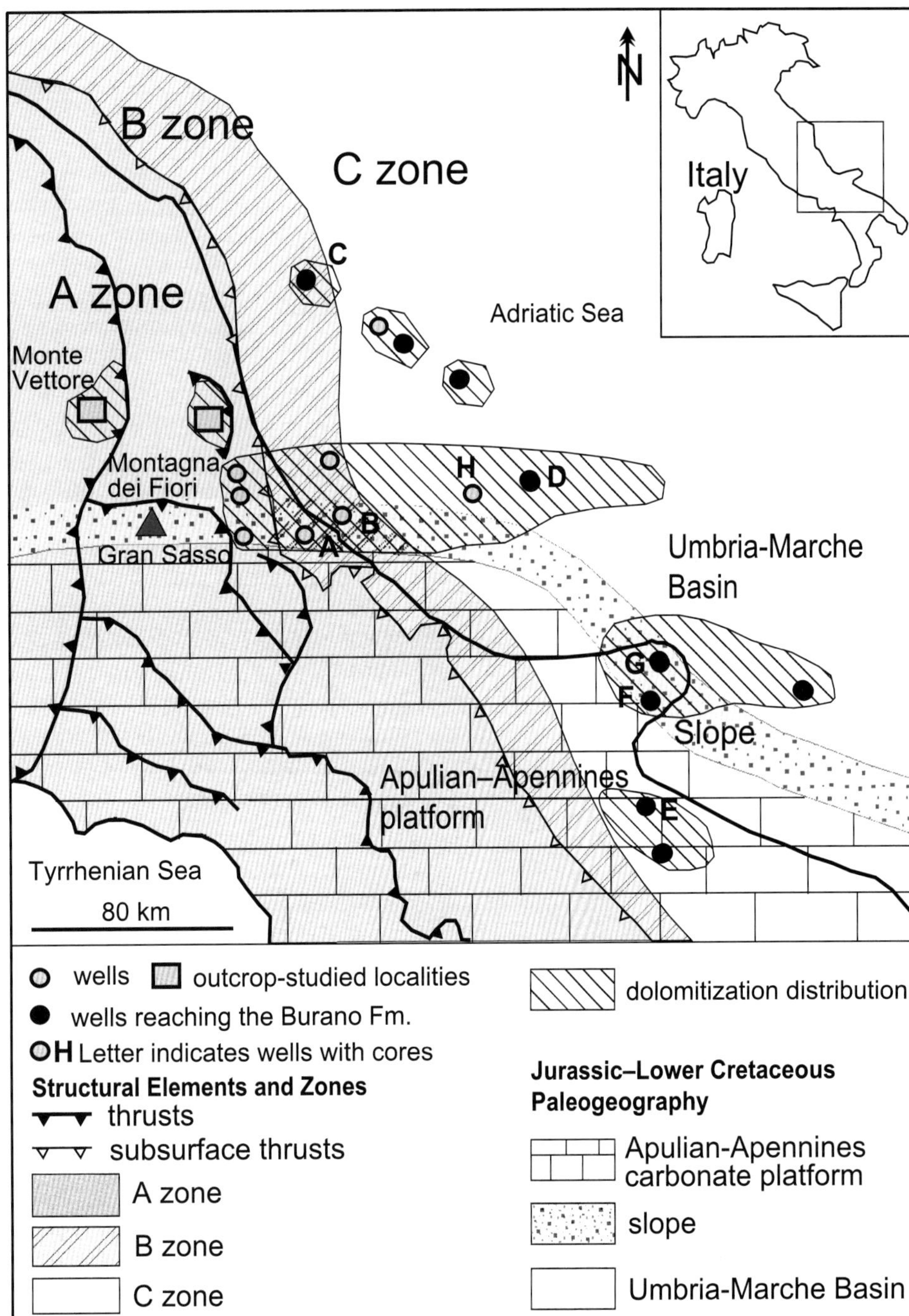

Figure 1. Schematic map showing present structural elements and Jurassic–Cretaceous paleogeographic domains. A zone = Apenninic thrust belt involving Mesozoic units, both outcropping and subsurface; B zone = Apenninic foredeep with thrusting in the Tertiary successions; C zone = foreland with transcurrent faults. During Jurassic–Cretaceous in the southern area, a persisting carbonate platform grew (Apulian–Apennines Platform, AAP), facing northward to the pelagic basin (Umbria-Marche basin, UMB); in the transition area, resedimented facies developed.

Sedimentary Evolution

The surface (Damiani et al., 1992; Ciarapica and Passeri, 1998) and subsurface data allow the reconstruction of the sedimentary evolution of the area (Figure 3).

The Neo-Tethys passive-margin evolution begins in the Late Triassic. The whole area is characterized by peritidal dolomite platform growth, within which sabkha and anoxic intraplatform basins develop (Ciarapica et al., 1987; Bigozzi et al., 1992). The Norian–Rhaetian peritidal dolomites with thick evaporative layers (Burano Formation) are widespread over most of the area. The early Lias is represented by the extensive carbonate platform of the Calcare Massiccio Formation, spotted by small, intraplatform anoxic basins. The Jurassic deposits record the rifting and drowning of this vast carbonate platform, coupled with rapid subsidence (Adamoli et al., 1982; Santantonio, 1993). The middle Lias is marked by platform tearing and the establishment of the northern Umbria-Marche basin (UMB, with Corniola Formation) (Santantonio, 1993; Santantonio et al., 1996), whereas a southern carbonate platform persists (Apulian–Apennines Platform realm, AAP). Both the Calcare Massiccio and Corniola formations are locally pervasively dolomitized in the central and northern Apennines (Crescenti, 1969). From the late Lias to Early Cretaceous in the UMB, the different thickness in basinal areas and paleohighs (characterized by condensation and hiatuses) of the carbonate and marly facies deposits causes strong variation in the overburden of the Lower Jurassic carbonates (Adamoli et al., 1978, 1990). The carbonate platform growth continues in the southern area; in the AAP-UMB transition area, bioclastic resedimented facies develop. Breccias and calcarenites characterize the escarpment and slope of the Tithonian–early Neocomian Maiolica Formation; this facies here is called the equivalent Maiolica and, in the study area, is locally pervasively dolomitized. Similar lithotypes outcrop in the Gargano area (Pavan and Pirini, 1966; Luperto Sinni, 1996).

Late Cretaceous–middle Miocene sediments consist of limestones and marls in the UMB (Scaglia and Bisciaro-Schlier formations), whereas in the southern platform (Accarie et al., 1986; Eberli et al., 1993), the shallow carbonate deposition is locally interrupted by subaerial exposure phases during the Paleocene–Oligocene (Ciarapica and Passeri, 1998). In the AAP-UMB transition zone, slope deposition continues. In upper Miocene age, the eastern migration of the Apennines orogeny involves the above-mentioned A zone, causing the formation of troughs where the turbiditic sequences of the Laga formation deposit. During the lower Pliocene, acme of the orogenic phase, silicoclastic sedimentation characterized the whole area; overpressure and reverse faults can be observed in the turbiditic sequences of the B zone (unpublished Eni data). During the middle–upper Pliocene, thrusting and folding activity continues in the northern part of the studied area (zones A and B).

METHODOLOGY

The Calcare Massiccio, Corniola, and Maiolica formations' facies have been examined in a large group of wells, among which seven were chosen for this study after careful analysis of cutting and core thin sections; the wells represent sequences belonging to all of the three structural zones (Figure 1). At the Montagna dei Fiori and Monte Vettore, located in the central Apennines and representing the A zone, the dolomitized facies of Calcare Massiccio and Corniola formations' outcrops have also been studied.

All the samples, about 50, were observed both through transmitted light optical microscopy and cathodoluminescence (CL, using a Technosyn, model 8200 MK 2; 15 kV and 300–400 μA gun current); then the polished thin sections were examined using a scanning electron microscope (SEM) in backscattered electron (BSE) mode.

Geochemical studies include stable isotope, trace-element, and fluid-inclusion analyses.

The carbon and oxygen stable isotopes have been analyzed, when possible, on dolomite samples showing a homogeneous crystalline size; microsamples of about 3 mm (0.1 in) were obtained using a dental drill. Carbonate (4–6 mg) was allowed to react with concentrated orthophosphoric acid at 25°C for calcites and cements and at 60°C for dolomite. The CO_2 content was measured in a Finnigan MAT 252 mass spectrometer (Peedee belemnite formation international standard, PDB), and

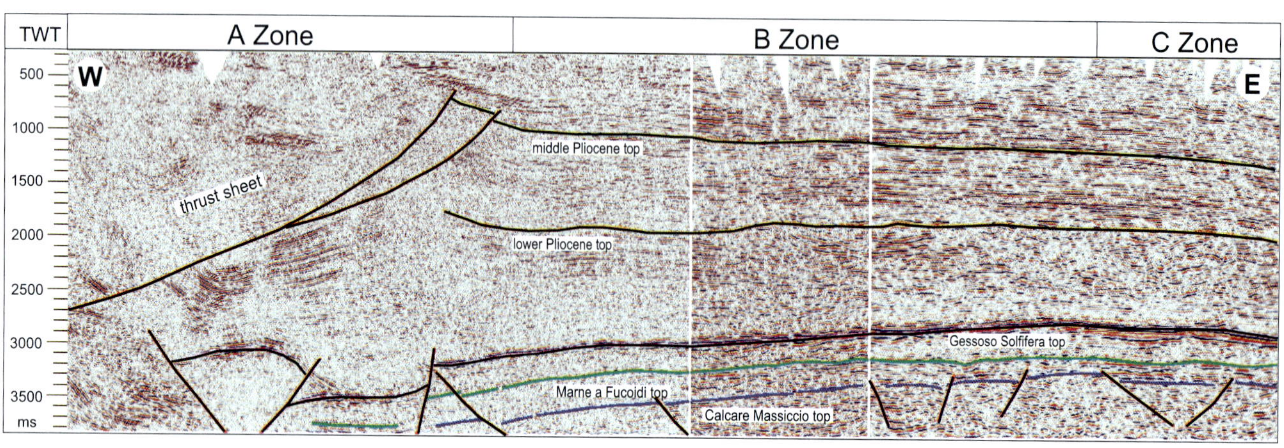

FIGURE 2. Seismic profile crosscutting the structural zones. The length is about 60 km (10 mi). TWT = two-way traveltime.

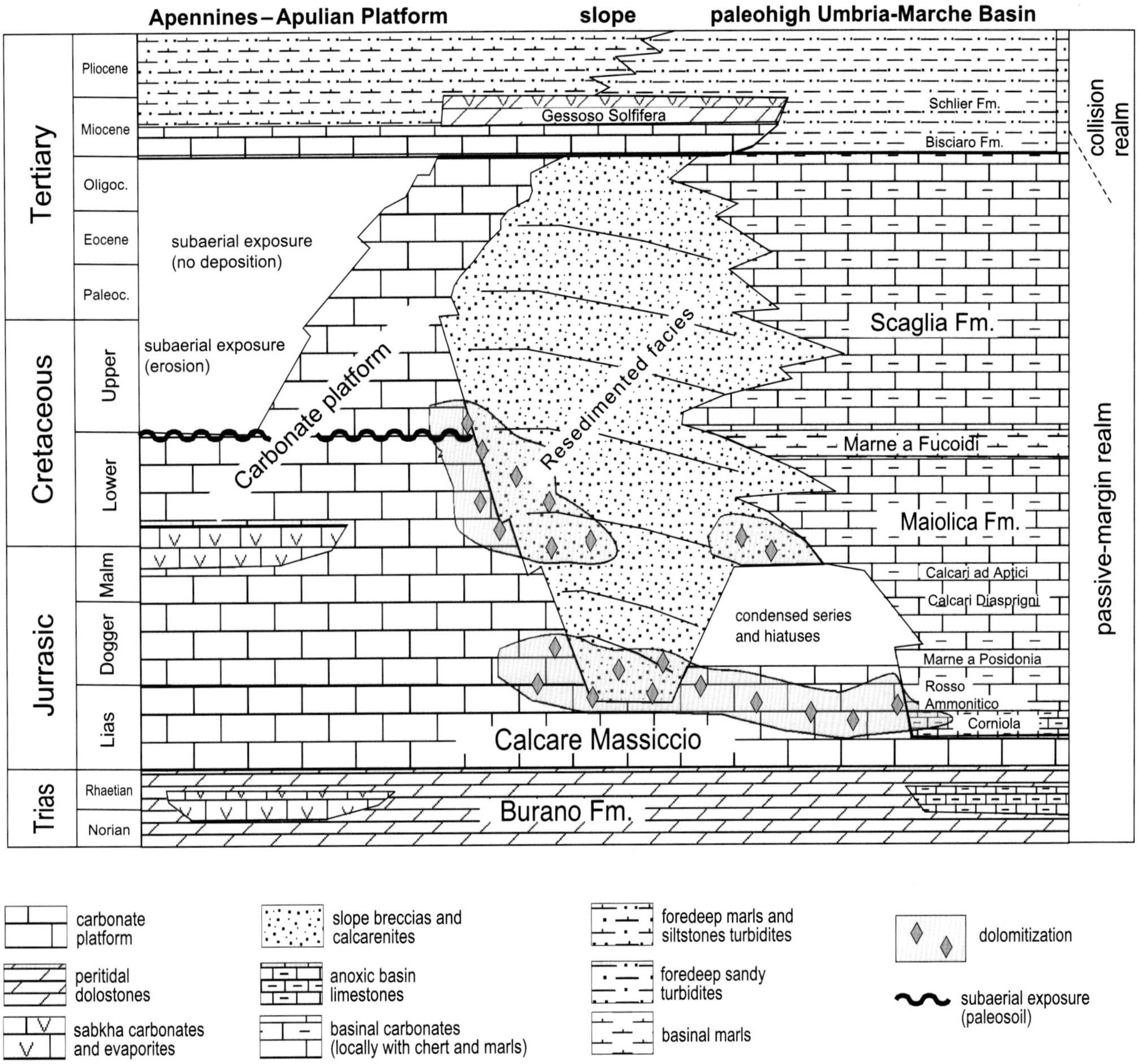

FIGURE 3. Stratigraphic scheme of the different paleogeographic domains from Triassic to Pliocene in the area, including the Apulian–Apennines Platform and the Umbria-Marche basin. The dolomitization object of the study is found in the Lower Jurassic succession (Calcare Massiccio and Corniola) and in the slope resedimented deposits of Lower Cretaceous (equivalent Maiolica Formation).

results were expressed in parts per thousand (delta units). Trace elements were determined with semiquantitative analysis by x-ray fluorescence using a Philips PW1404 x-ray diffractometer.

Fluid inclusions in dolomite cements were studied on double-polished, thick sections treated with cold preparation techniques (Goldstein and Reynolds, 1994) to avoid reequilibration. The petrography of the fluid inclusions was studied on the selected, unheated, and unstained thick sections, using a petrographic microscope equipped with ultraviolet epi-illumination.

Fluid-inclusion assemblages (FIAs) were selected to discriminate petrographically associated groups of fluid inclusions, probably trapped at about the same time. After the petrographic study, the thick sections were cut into small chips and selected for microthermometry. The measurements were performed using a Linkam THMSG 600 heating-freezing stage, calibrated with synthetic pure water and CO_2 inclusions. Low heating rates were used for measuring homogenization temperatures (T_h) that were measured in the order of increasing T_h to avoid overheating. Low-temperature phase changes were measured after all T_h in a sample had been measured. In all-liquid fluid inclusions, bubbles were generated by cooling the sample at $-100°C$ and holding the temperature for at least 2 min. This procedure stretches the fluid-inclusion cavities and allows the bubbles to nucleate to obtain final ice-melting temperature ($T_{m\ ice}$) measurements (Bodnar, 1992). Most T_h and $T_{m\ ice}$ were measured by cycling (cf. Goldstein and Reynolds, 1994).

RESULTS

Generalities and Dolomite Distribution

In the Jurassic–Lower Cretaceous sequence of central Apennines and their foreland (onshore and offshore Marche–Abruzzi regions), the dolomitization processes enhanced the petrophysical properties of the carbonate platform and related basin series (Figure 4), giving them good hydrocarbon reservoir characteristics (intercrystalline and vuggy porosity).

In both offshore and onshore subsurface successions, the Calcare Massiccio and the lower part of the overlying Corniola Formation are found locally dolomitized (Figures 1, 3); time-equivalent dolomitized sediments crop out in the central Apennines and are known as Castel Manfrino dolostones (Crescenti, 1969). The dolomitization process also involves the bioclastic resedimented facies that characterizes the AAP-UMB transition zone (equivalent Maiolica Formation).

The Calcare Massiccio (as much as 600 m [2000 ft] thick) consists of thick to massive beds of peritidal cycles having subtidal shoal deposit intercalations (Colacicchi et al., 1975). The microfacies, observed on the outcrop samples, cores, and thin cutting sections of selected wells, varies from mudstone, bioclastic wackestone-packstone to oolitic grainstones; fenestral fabric and pisolitic crusts are associated to peritidal facies and tepee horizons. In a cored interval of well H, the dolomitized Calcare Massiccio shows layers of fine crystalline dolomite containing centimetric anhydrite nodules. In outcrop, the dolomitization forms patches along fractures, reactivated Jurassic extensional faults (Ronchi et al., 2003), and locally, it can be very extensive and completely pervasive. In the subsurface, there is a low control on the three-dimensional geometry of dolomitized bodies; commonly, the dolomitization involves the whole carbonate platform and extends for a few tens of meters in the overlying Corniola Formation. Sometimes, the Calcare Massiccio shows only intercalations of dolomitized rocks independent of the sedimentary facies along fractured zones. The study is focused on the mesodiagenetic dolomite, the fine crystalline horizons of dolomite coupled with anhydrite layers interpreted as early diagenetic dolomite (Qing et al., 2001), which have been left out. Dolomite is coarse crystalline, and porosity is of intercrystalline, vuggy, and solution-enlarged type (Figure 5A).

The Corniola Formation consists of pelagic (50–200-m [160–660-ft] water depth), gray micritic limestone that is characterized by regular strata 20–40 cm (8–16 in.) thick, containing chert nodules with sponge spicules, radiolaria, foraminifera, small gastropods, and sparse ammonites; thickness ranges from a few meters to 200 m (660 ft) (Passeri, 1990). Bioclastic resedimented calcarenites, together with breccias and slumps, are present in areas near the persisting carbonate platform. Both in outcrop and in subsurface, the dolomitization involves mainly the lower part of the formation and particularly the resedimented facies (Ronchi et al., 2003) (Figure 5B).

The equivalent Maiolica Formation, representing the slope connecting the arc of the southern Apulian–Apennines Platform to the northern Umbria-Marche pelagic basin, was observed in cores of wells A and B (Figure 1) and in outcrop sections in the Gran Sasso area (Adamoli and Mancinelli, 1978). It is formed by thinning and fining-upward cycles of bioclastic, crinoid-rich sediments; locally, breccia lenses and layers are also present. This facies grades upward and basinward to thin-bedded limestone with chert nodules (Maiolica Formation). The microfacies show mudstone in the fine-grained textures, whereas in the coarse ones, microfacies show packstones and grainstone. The equivalent Maiolica thickness ranges from about 100 m (330 ft) to as much as a few hundred along the slope; in the outcrops of central Apennines, the dolomitization is not common, whereas in the Gargano area, dolostones that are equivalent to the studied succession are described and called Calcari e Dolomie di Monte Iacotenente

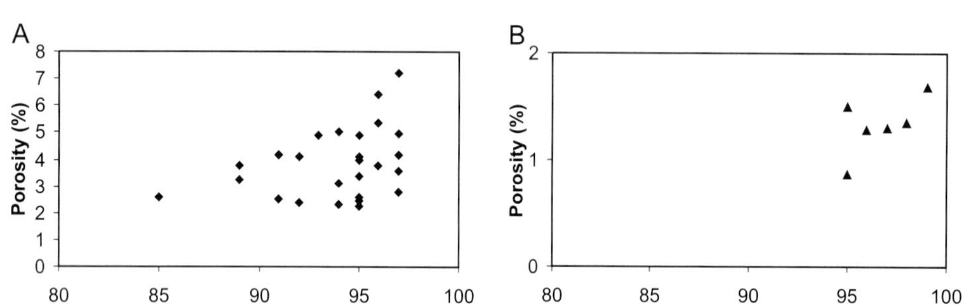

FIGURE 4. Standard core analysis porosity values of dolostones (A) compared with nondolomitized samples (B) in Calcare Massiccio. Dolomitization improves the porosity, enlarging the vugs and forming intercrystalline porosity; average porosity is 4% in the dolostone compared to 1.3% in limestones.

Figure 5. Photographs of dolomitized facies. (A) Outcrop example of Calcare Massiccio dolomitized facies at Montagna dei Fiori; note the large vugs caused by solution enlargement of bird's eye view; (B) Montagna dei Fiori outcrop showing a dolomitized calcarenitic bed of Corniola; the selective dissolution of clasts provided moldic porosity, and intergranular porosity is preserved. (C) Slabbed core of dolomitized resedimented facies of equivalent Maiolica Formation (well A). The sedimentary lamination is preserved by the dolomitization; note the alternation of porous dolomite (intercrystalline porosity), former coarse-grained calcarenites, and tight dolomite, former finer grained facies.

(Pavan and Pirini, 1966; Luperto Sinni, 1996). In two wells (A and B), the dolomitization is confined to the coarser texture layers in the lower part of the formation (Figure 5C).

Samples Petrography

All the Jurassic–Lower Cretaceous dolostones studied in this work, regardless of the original formation, show almost the same texture types and paragenetic evolution, revealing a multistage dolomitization that results in a generally medium to coarse crystalline dolomite (Figure 6). The fine crystalline, mimetic dolomite locally present in the Calcare Massiccio, as stated before, was not considered.

Two dolomite textures have been recognized:

- Texture A: Tight, massive replacement dolomite, represented by planar-S to nonplanar texture (Sibley and Gregg, 1987), with a medium crystalline

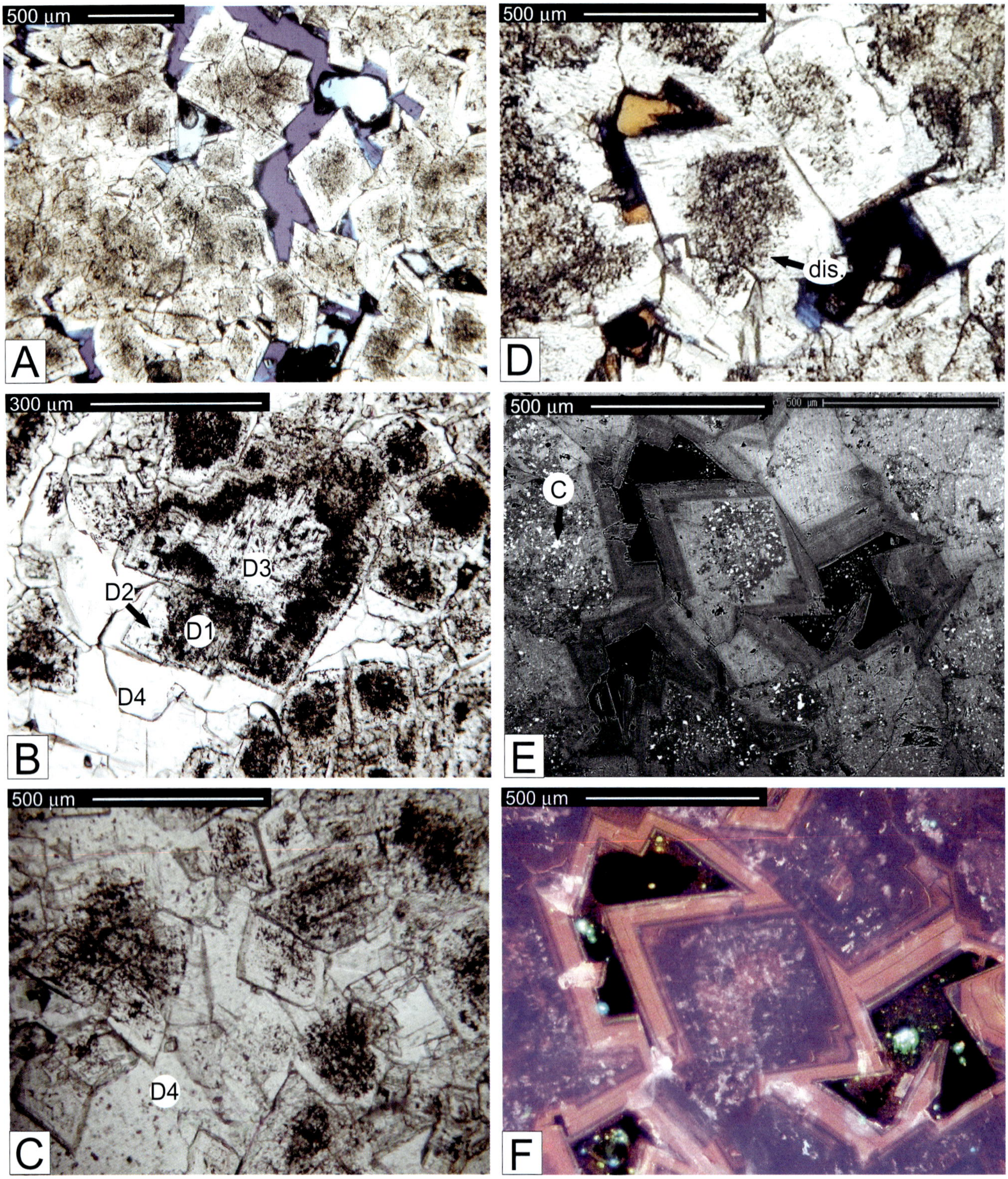

FIGURE 6. Dolomite texture photomicrographs. (A) Planar-E texture of dolomite with intercrystalline porosity, sample from well B, equivalent Maiolica Formation. (B) Planar-E texture of dolomite; note the turbid rhombohedra nuclei (dolomite 1, D1) partly recrystallized (dolomite 3, D3), the clear dolomite overgrowth (dolomite 2, D2), and the dolomite cement (dolomite 4, D4), sample from well G, dolomitized Calcare Massiccio. (C) Planar-E texture dolomite; note the complete occlusion of intercrystalline porosity caused by dolomite cementation (dolomite 4), sample from Montagna dei Fiori outcrop, dolomitized Corniola. (D–F) The same field of planar-E dolomite texture in planar light, SEM, and CL; note that the turbid dolomite 1 was partially dissolved by a dissolution event before the precipitation of dolomite 3 ("dis." in micrograph D). In dolomite 1, relicts of calcite are detectable under SEM ("C" in micrograph E). Dolomite 2 is characterized by zonation visible both in SEM and CL (micrographs E and F, respectively).

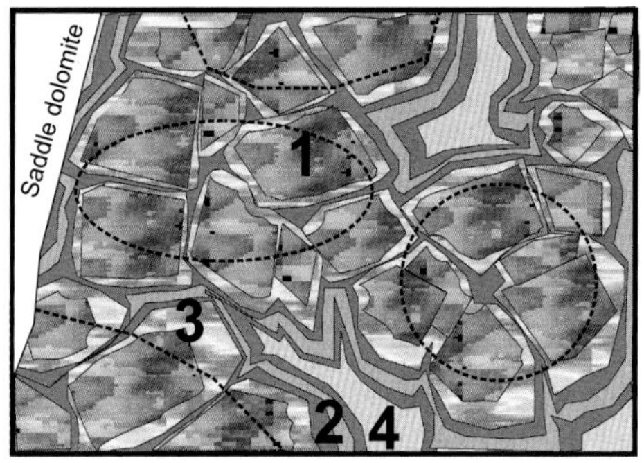

	Petrography	FI T_h (°C)
Dolomite 1	turbid nuclei of rhombohedra; mottled in CL	<50–70
Dolomite 2	dolomite rhomb. overgrowths; zoned orange and dull in CL	60–125
Dolomite 3	recrystallization of the rhomb nuclei and Dolomite 2; dull orange in CL	
Dolomite 4	pore- and fracture-filling dolomite cement	95–110
Saddle Dolomite	pore- and fracture-filling dolomite cement	110–140

FIGURE 7. Schematic paragenesis of dolomite. 1 = dolomite 1 turbid nuclei; 2 = dolomite 2, rim dolomite cement; 3 = dolomite 3, recrystallization of dolomite 1; 4 = dolomite 4, clear blocky pore-filling cement, S = saddle dolomite pore- and fracture-filling cement, FI = fluid inclusion.

size (from 50 to 150 μm). This dolomite is present mostly in originally fine-grained facies.

- Texture B: porous, white, sucrosic dolomite showing a planar-E texture and coarse crystalline size (from 150 to about 500 μm); this dolomite maintains the original grainstone texture, whose allochems are visible as relicts as evidenced by impurities [organic matter inclusions(?)] and calcite remains in the crystal cores.

Several dolomite types, reconstructed through petrographic observations, have been recognized (Figure 7):

- Dolomite 1 is represented by the dolomite turbid crystal cores (Figure 6A–D) and shows a mottled orange aspect in CL (Figure 6F); many small (few micrometers in size) inclusions and some traces of calcite have been detected by SEM analyses (Figure 6E). The edges of the crystals are corroded by a phase of dissolution. In well C, partial dolomitization involved the high-energy facies of the Calcare Massiccio; scattered turbid rhombohedra are present.
- Dolomite 2 consists of clear dolomite rim cement commonly zoned in CL (Figure 6E, F). Dolomite 2 is extremely well developed in the sucrosic dolomite texture (type A), where the original porous grainstone textures allowed the growth of large, clear dolomite void-filling cements.
- Dolomite 3 is a turbid dolomite, dull orange in CL and relatively inclusion rich; it represents the recrystallization of dolomite 1 (Figure 6B).
- Dolomite 4 is made of dolomite cement that grew over dolomite 2. This dolomite is generally void-filling and fracture-filling cement (Figure 6C).
- Saddle dolomite, in crystals as much as 600 mm (24 in.) in size, shows curved crystal faces, cleavage, and sweeping extinction; it is present locally inside large vugs and in fractures (Radke and Mathis, 1980).

In Figure 7, the schematic paragenesis of all the detected dolomites are sketched out.

Stable Isotopes

Carbon and oxygen stable isotope analyses show a narrow range of $\delta^{13}C$ values (+1.5 to +4.8‰), whereas the oxygen values spread from −2.8 to +2.64‰ (Figure 8). In the relatively wide distribution range of $\delta^{18}O$ values, the Jurassic dolomites, coming from outcrops of A zone, form a group that represents the most depleted oxygen samples; the equivalent Maiolica dolomitized samples from cores of B zone wells show $\delta^{18}O$ values slightly less depleted than those of A zone; the dolomite samples of C zone wells are characterized by a wide range of values reaching 1.5‰ PDB.

Fluid-inclusion Petrography and Microthermometry

The fluid-inclusion analyses were performed on samples of dolostones representing the three structural areas (Figure 1): five samples of dolomitized Calcare Massiccio and Corniola formations of Montagna dei Fiori area, representing the A zone; two samples of dolomitized equivalent Maiolica of wells A–C cores (B zone); and nine samples of dolomitized Calcare Massiccio belonging to cores of D–H wells (C zone).

Fluid-inclusion petrography was aimed at determining

- fluid-inclusion origin (primary, secondary, pseudosecondary, uncertain)
- the most finely discriminated, petrographically distinguishable groups of fluid inclusions (this defines FIAs)
- the inclusions' composition (aqueous or oil filled)
- the relative consistency of the liquid-to-vapor volumetric ratio in any inclusion among a single FIA

Among the mineral phases described above, the most appropriate FIAs were selected in the following dolomite types:

- Dolomite 1 (inclusion-rich turbid crystal cores): Commonly, in this mineral phase, the dense distribution of inclusions in the crystals reduces the optical characteristic of dolomite 1, making the analyses difficult. In wells B and D, all liquid fluid inclusions of variable size (from 4 to 50 μm) are present.

- Dolomites 2–4 (clear dolomite rims): In these dolomites, primary, two-phase (liquid + vapor), aqueous fluid inclusions are very common. The size of the inclusions is variable from a few to several micrometers (from generally 4 and 5 to as much as 40–50 μm).
- Saddle dolomite: Primary, two-phase (liquid + vapor), aqueous fluid inclusions (average size, 10–12 μm) are present.

No oil-filled fluid inclusions are present in any FIAs. After petrography was completed and the most appropriate FIAs were selected, we performed microthermometric measurements to determine both temperatures (T_h) and salinity ($T_{m\ ice}$) of fluids responsible for mineral precipitation.

The dolomite 1 turbid cores commonly contain small fluid inclusions that do not give microthermometric information because of their average small size and poor optical characteristics of this mineral phase, except in wells A, B, and D (Figure 9B, C), where all-liquid (well B) and two-phase (wells A and D) fluid inclusions are present. All-liquid inclusions are in the same size range as two-phase inclusions (see dolomite 2) but not petrographically paired with them and do not nucleate bubbles during cooling unless stretched. Therefore, all-liquid inclusions do not result from necking down of two-phase fluid inclusions or significant metastability. Freezing-point depression ($T_{m\ ice}$) for all-liquid fluid inclusions is −7.5 to −3.3°C. If we assume an H_2O-NaCl model (Bodnar, 1992), these would translate to salinity of 11–5.3% weight NaCl equivalent. We can presume that dolomite 1, at least in well B, precipitated at a low temperature (below about 50°C) from hypersaline fluids comparable to those fluids found in inclusions in the Burano Formation dolomites (15% weight NaCl equivalent; P. Ronchi, 2001, personal communication).

In well A (Figure 9B), temperature data from two-phase inclusions in dolomite 1 are confined in a very narrow range (except for one point), giving an initial precipitation temperature between 71 and 76°C and salinity of 3–4% weight NaCl equivalent.

In well D, C zone, we do not have any data on temperature of homogenization, but from petrographic observations, data suggest that dolomite 1 was formed during the first phase of dolomitization. Because of this petrography, it is possible to ascribe dolomite 1 to the low-temperature phase with respect to those postdating it.

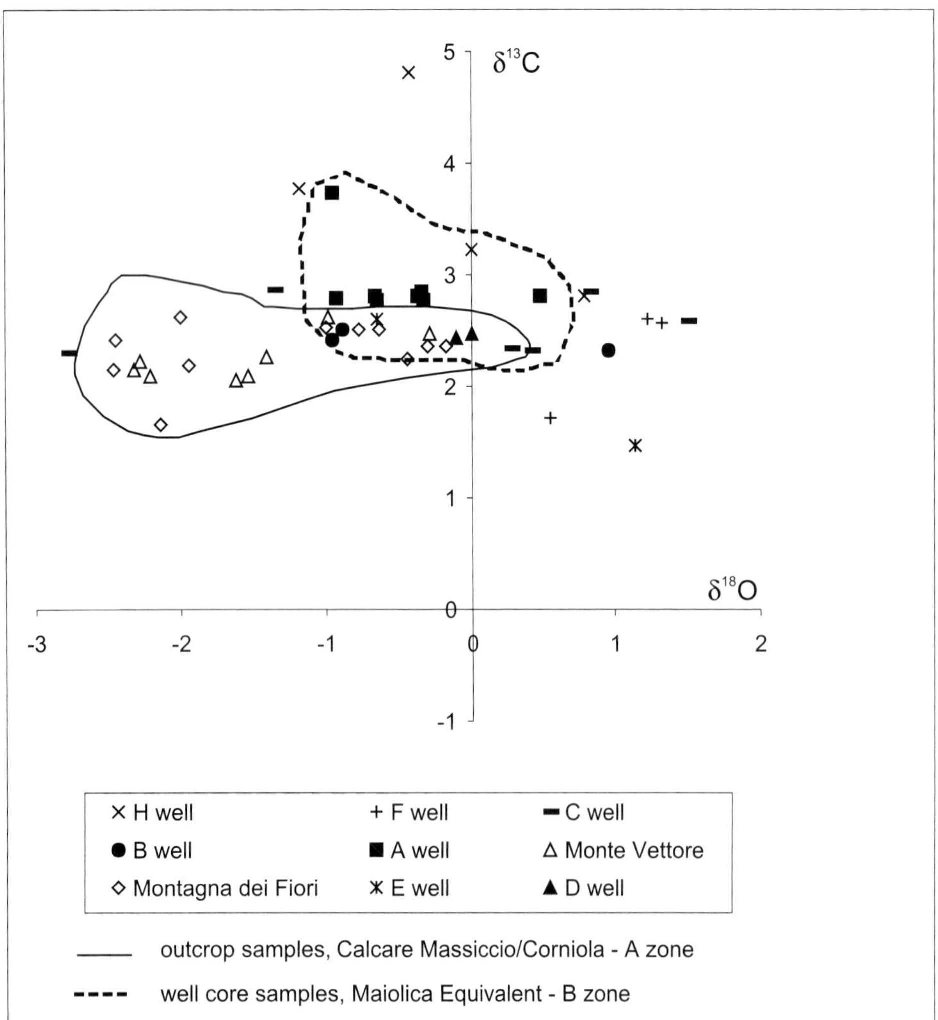

FIGURE 8. Stable isotopes diagram of the dolomite bulk samples coming from wells and outcrops. The $\delta^{13}C$ positive values are interpreted as caused by the presence of seawater-derived dolomitizing fluids; the large range of $\delta^{18}O$ values may be explained as the result of admixtures of signals coming from different dolomite phases; the most coherent data are those belonging to the A zone that, with negative values, are consistent with high diagenetic temperatures proven by the fluid-inclusion analyses.

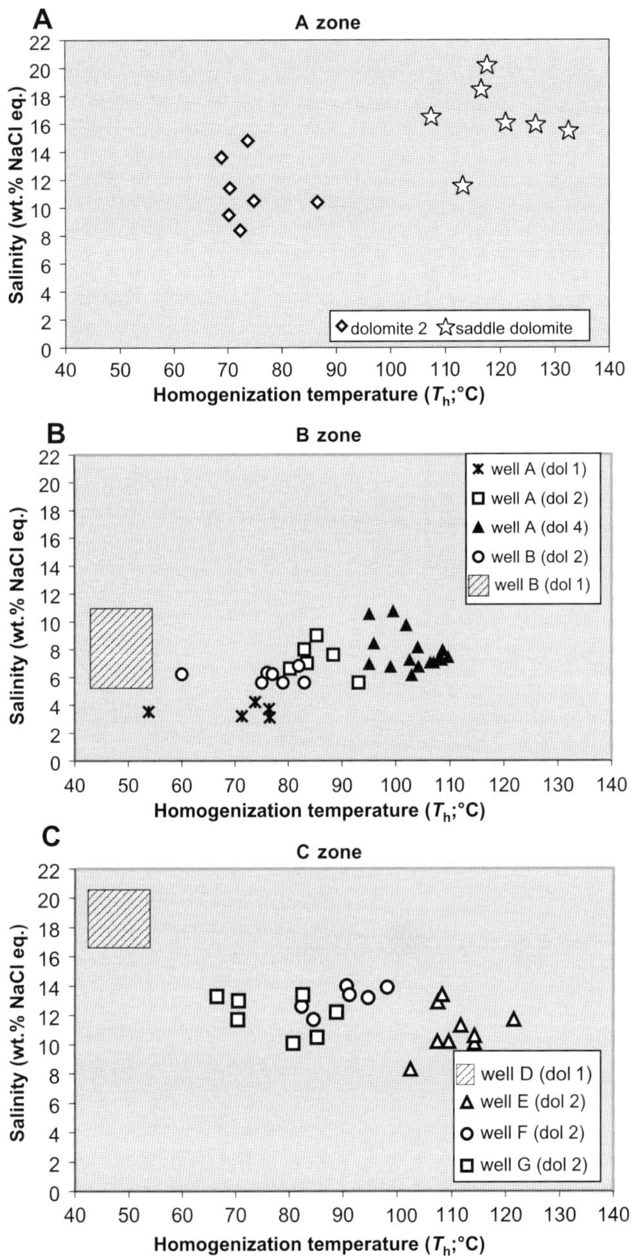

FIGURE 9. FI diagrams showing the T_h vs. the calculated NaCl equivalent salinity.

Dolomite 2 contains two-phase inclusions and gives good information both on temperature of homogenization and salinity. Microthermometric analyses show T_h values spread in different groups of samples, ranging from 60 to 125°C (Figure 9A–C), indicating that this phase of dolomite cement grew at different temperatures in different samples. It is important to note that the FIAs in each sample yield consistent evidence that they were not reequilibrated after entrapment.

The fluid-inclusion salinity in dolomite 2 shows values ranging from 5 to 17% weight NaCl equivalent. Salinity data are commonly well constrained (Figure 9B, C) in each sample, except for salinities recorded in dolomite 2 of A zone (Figure 9A). The wide salinity range can probably be explained through leakage and refilling of inclusions during the presence of later higher salinity fluids. Dolomite 4 is recorded only in well A (B zone), where the data show high temperature (from 95 to 110°C) and salinity (between 6 and 11% weight NaCl equivalent) values. This is further proof of increasing temperature and salinity during progressive burial (in fact, dolomite 4 is a void- and fracture-filling cement that grows, postdating dolomite 2).

The saddle dolomite has been analyzed in Montagna dei Fiori, where it shows the highest T_h values (average value 119°C) of the dolomite paragenetic sequence. The salinity for this dolomite phase ranges between 15.5 and 20% weight NaCl equivalent (Figure 9A).

In general, two trends can be noticed in the analyzed samples:

1) In each succession, temperature and salinity increase in the paragenetic sequence (from dolomite 1 to saddle dolomite) (cf. Montagna dei Fiori, Figure 9A, and well A, Figure 9B).
2) In the same dolomite phase, salinity increases moving from one structural zone to another. For example, NaCl content in dolomite 2 increases moving toward the foreland (from B zone to C zone), with average values of 7.3–8.0% weight NaCl equivalent (wells A and B) in the B zone and from 10.5 to 13% weight NaCl equivalent in wells E–G in the C zone. An exception is dolomite 2 of A zone probably because of the effects of leakage and refilling. The same trend of increasing salinity can be seen in dolomite 1 from wells B and D (B and C zones, respectively).

BURIAL HISTORY AND STRUCTURAL EVOLUTION

The burial-history diagrams have been performed using the Software GET® (Integrated Basin Modeling System for Unix) adopted by Eni. The stratigraphic input data come from literature (Passeri, 1990) and well data, according to the studied successions. It is worthwhile to note that the many uncertainties in the input variables required for this simulation (such as paleodepth, paleotemperatures, sedimentation rate, and heat paleoflux) influence the isotherm reconstruction reliability.

The studied Lias and Lower Cretaceous successions underwent different burial evolution, depending on their Mesozoic–Paleogene paleogeographic deposition areas and Neogene structural evolution. In the A zone, the Montagna dei Fiori area (Figure 10A) shows a UMB sedimentary evolution characterized by progressive overburden and relevant temperature increase to as much as 80°C until the late Miocene, when it was involved in the Apennines belt. In the B zone, represented by A and B wells (Figure 10B), the paleohigh Mesozoic succession, followed by uncompleted Paleogenic series, causes a low overburden; from Pliocene, the thick clastic sedimentary wedge deposited with consequent heating. The compressive folds and thrusts that involved the B zone are not considered in the model, because their influence on the total amount of overburden is small. The C zone is characterized by the low thickness of Neogenic sediments, caused by external position regarding the Apennines chain; the overburden and relevant heating of the Jurassic–Cretaceous succession depend mainly on the Mesozoic–Paleogenic sequence thickness. The highest overburden and temperatures of the Jurassic–Early Cretaceous series are reached in the UMB sequence (about 80°C), whereas in the AAP areas, low overburden is recorded (with 60–70°C).

DISCUSSION

Geometry of the Dolomite

The analysis of the spatial distribution of dolomitized bodies is one of the criteria to define a model of dolomitization (Hardie, 1987; Amthor et al., 1993). Although it is possible to achieve only

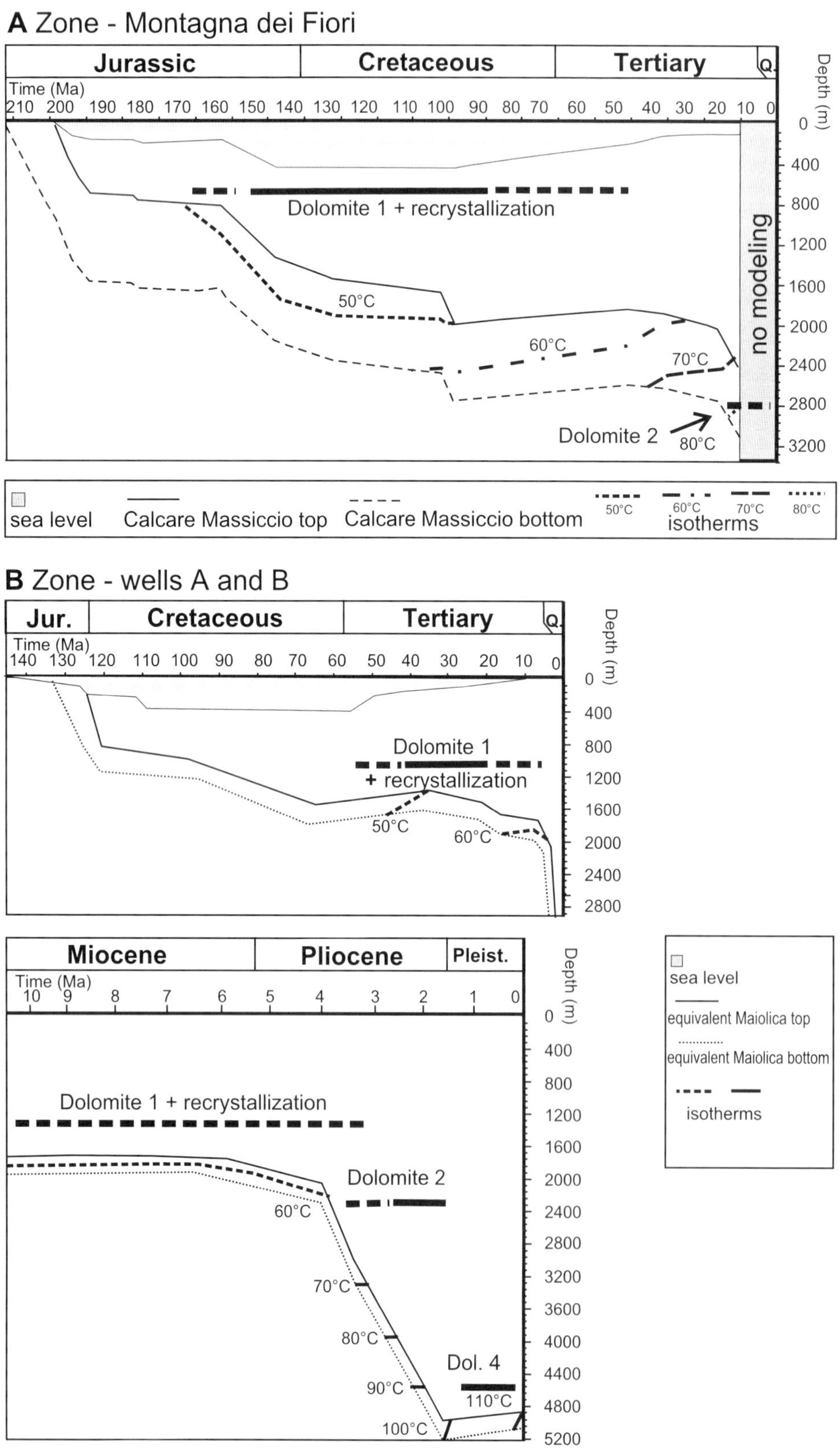

FIGURE 10. Burial-history diagrams with isotherms and the main dolomite phases plotted. (A) A zone, the evolution of the Calcare Massiccio at Campo dei Fiori; (B) B zone, the evolution of the equivalent Maiolica Formation succession in wells A and B.

incomplete information on the dolostones geometry in drilled successions, a scattered distribution of dolomitization with a preferential development around fractured and porous horizons can be observed both in drilled successions and in the Montagna dei Fiori and Monte Vettore areas. Similar settings have been recently interpreted as burial dolomites (Braithwaite and Rizzi, 1997; Drivet and Mountjoy, 1997). This pattern suggests that the dolomitizing fluids rose along faulted and fractured zones and extended laterally inside porous bodies of the Calcare Massiccio and widely in the breccia and calcarenites porous layers of Corniola and equivalent Maiolica formations; the dolomitization front, in fact, stops upward in the tight muddy facies of Corniola and Maiolica Formations.

Petrography and Geochemistry

The petrographic observations evidenced that the dolomites experienced a multiphase dolomitization (Qing and Mountjoy, 1989; Braithwaite and Rizzi, 1997); it consists of an early stage of dolomite replacement (dolomite 1), several phases of dolomite precipitation (dolomites 2 and 4 and saddle dolomite), and phases of dissolution (between dolomites 1 and 2) and dolomite recrystallization (dolomite 3).

The stable isotope values, coupled with the fluid-inclusion data, suggest that this process occurred in the presence of heated fluids. This is particularly remarkable for the samples coming from A zone (Figure 8), which show negative $\delta^{18}O$ values lying in the range of dolomites attributed to heated waters (Mattes and Mountjoy, 1980; Sibley and Gregg, 1987), whereas the positive $\delta^{18}O$ values of samples of B and C zones may be explained as the result of admixtures of signals coming from different dolomites (Frisia and Wenk, 1993). In fact, the stable isotope analyses were performed on bulk samples, and therefore, each value represents an average value resulting from the contribution of all the dolomite phases. Therefore, the contribution of an earliest dolomite can be assumed (Qing et al., 2001). This early dolomite may be represented by the fine crystalline dolomite that has been found in well H and not analyzed in the present study. The fluid-inclusion analyses allowed the constraint of the T_h range of the different dolomite types, from lower than 50 to as much as 130°C, confirming the presence of heated fluids. Moreover, in the foreland (C zone), the T_h values of dolomite 2 are higher than those measured at well total depth, accounting for the presence of hydrothermal fluid during that dolomite cementation phase. The saddle dolomite commonly suggests high temperatures (Radke and Mathis, 1980).

$\delta^{13}C$ values (+1.5 to +4.8‰) indicate a seawater-derived dolomitizing fluid (Allan and Wiggins, 1993), whereas the fluid-inclusion analyses evidence a salinity greater than that typical of mean seawater; all these information indicate that the fluids may derive from basinal modified seawater.

The Dolomitization in the Three Bends

The above-discussed data allow the reconstruction of the evolution of dolomitization in the different zones, as exemplified in Figure 11.

The A Zone

In Montagna dei Fiori and Monte Vettore, representative areas of the thrust zone (A), the dolomitization process involved the Calcare Massiccio and part of the Corniola Formation (Lower Jurassic). The monophase fluid inclusions found in dolomite 1, representing the first dolomitization stage, do not have reliable T_h data; by comparison with values obtained for other dolomite 1 samples, it is likely to have started to form in a shallow burial environment (50–60°C). Two-phase fluid-inclusion measurements established that dolomite 2 occurred under temperatures of about 70°C, whereas the saddle dolomite points to temperatures above 110°C (as much as 132°C).

An evaluation of salinity in the fluid inclusions shows values for dolomite 2 (average 12%) lower than saddle dolomite (average 16% NaCl equivalent), suggesting a change in fluid geochemistry through time.

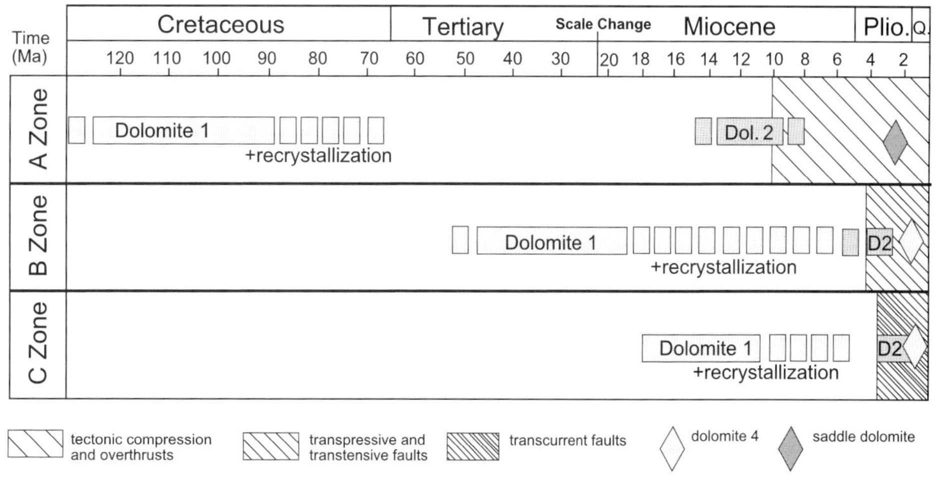

FIGURE 11. Dolomitization phases in the different zones.

The burial history reconstructed from regional and outcrop data enables us to match the diagenetic phases with the geological evolution of the A zone (Figures 10A, 11). The shallow burial phase of dolomitization (dolomite 1) could have occurred over a large time span, ranging from the Upper Jurassic to the lower Tertiary. The dolomite cementation phase (dolomite 2) occurred during middle Miocene, when the Calcare Massiccio and the Corniola formation units may have reached temperatures of about 70°C.

After this period, the area was involved in Apennines orogeny, and hydrothermal fluids (with temperatures of 110–130°C as recorded by the fluid inclusion of saddle dolomite), arriving from deep-seated faults, are thought to be the dolomitizing fluids of the last event.

The B Zone

In this area, the dolomitization process involved both the Calcare Massiccio Formation and the resedimented facies of Lower Cretaceous (equivalent Maiolica Formation). The modeling has been performed using the more reliable data of the Cretaceous succession of well A (Figure 9B).

The monophase fluid inclusion found in dolomite 1 indicates 55–78°C of homogenization temperature in a shallow burial environment; biphase fluid-inclusion measurements established that dolomite 2 occurred at temperatures between 75 and 92°C, whereas dolomite 4 points to temperatures between 95 and 110°C. An evaluation of fluid-inclusion salinity shows increasing salinity of fluid inclusion with proceeding burial: dolomite 1, from 2 to 5.5% weight; dolomite 2, from 2 to 9% weight; dolomite 4, as much as 11% weight of NaCl equivalent. This again suggests a change in the fluid geochemistry through time. The burial history reconstructed from well data enables us to match the diagenetic phases with the geological evolution of the B zone (Figure 10B). The temperatures of dolomite 1 were reached starting from the Eocene. The dolomite cementation phase (dolomite 2) started from the middle Pliocene, whereas part of the intermediate zone was involved in the peripheral bulge, and the Lower Cretaceous sediments may have reached temperatures of about 75°C. Dolomite 4 and the saddle dolomite should have formed more recently because of influx of heated fluids along open faults. This could be possible at that time, when the present stress field (σ_3 south–north oriented) was established in the area.

The C Zone

In this area, the dolomitization process involved the Calcare Massiccio Formation and the resedimented facies of the Lower Cretaceous; the modeling has been performed only on the available Calcare Massiccio cores. No successful measurements were performed on dolomite 1; biphase fluid-inclusion analyses established that dolomite 2 developed at temperatures of about 65–100°C, whereas dolomite 4 points to temperatures between 100 and 125°C.

Fluid-inclusion salinity data show similar values for dolomites 2 and 4, with a wide range of values from 9 to as much as 14% weight NaCl equivalent.

The burial-history diagrams of the C well zone show that only dolomite 1, whose temperatures of crystallization are assumed to be below 60–70°C, could have occurred in the burial regime; the temperatures obtained by fluid-inclusion analyses of the late dolomitization phases (dolomite 2 and further phases) were never reached in the Calcare Massiccio Formation during burial. This demonstrates that they occurred in the presence of hydrothermal fluids.

Possible Fluid Source and Flow Mechanisms

The above-discussed data allow us to interpret the studied dolomites as caused by late diagenetic processes, which can be compared to analog cases of massive burial dolomitization. This phenomenon has been related to basinal formation fluid flow in both passive- and collisional-margin realms (Mountjoy et al., 1997; Zempolich and Hardie, 1997; Swennen et al., 2002). The elevated temperature present in the burial environment favors dolomitization both by removing kinetic barriers and by raising the Ca/Mg ratio at which dolomitization can occur (Hardie, 1987; Allan and Wiggins, 1993). Nevertheless, a source of Mg and favorable flow mechanism are required.

The area experienced the evolution from passive margin (Mesozoic to Paleogene) to collisional margin (Neogene Apennines orogeny), with consequent changes in the diagenetic realm and fluid-flow patterns (Figure 12). In the tensional setting (Figure 12A), the burial fluid-flow pattern is dominated by compactional-driven movements, with lateral components (from the basin toward the platform), upward component, and thermohaline circulation (Heidary, 1997). In this phase, it is likely that the dolomitizing fluids were the Mg-rich residual evaporative brines (Broomhall and Allan, 1987) coming from strata lying in the Burano Formation. In the compressional tectonic phases (Figure 12B), the fluid flow is episodic and dominated by lateral movements from the collision zone to the foreland through dipping porous and permeable beds and vertically through faults and fractures (Oliver, 1986; Montanez, 1994; Heidary, 1997). The lateral flow can be of prime importance in the underthrust area because of fluid squeezing. In the studied area, only A and B zones were directly involved by compressional tectonics. Nevertheless, the heated

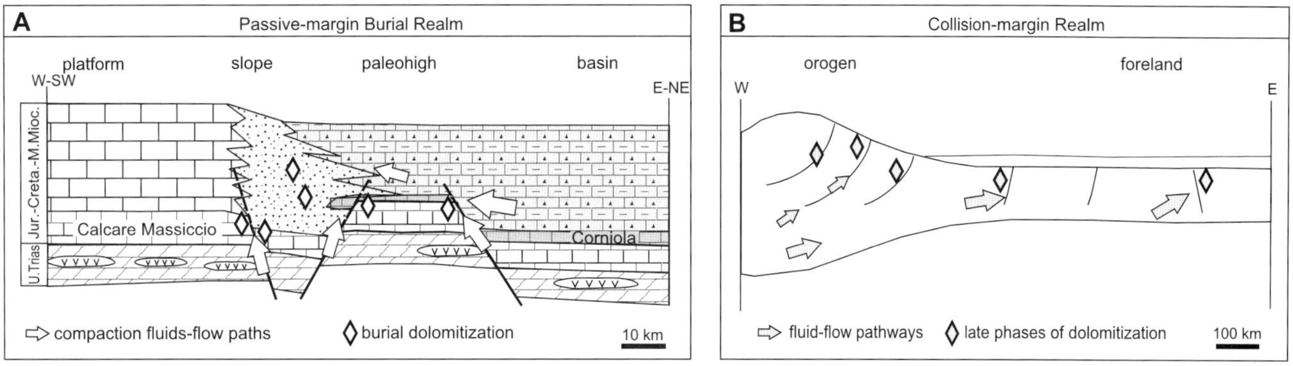

FIGURE 12. (A) Fluid-flow paths during the passive-margin burial realm; (B) fluid-flow paths during the collision-margin realm.

fluid flows in these settings may reach the foreland (Heidary, 1997). In A zone, it is suggested by the modeled hydrocarbon flow path (Eni unpublished data). Again, the source of Mg may be provided by Mg-rich residual evaporative brines coming from the Burano Formation or by shale formations present in the sedimentary succession (cf. Mattes and Mountjoy, 1980).

CONCLUSIONS

The dolomitization affecting the Lower Jurassic carbonate platform Calcare Massiccio, its related basinal limestone Corniola Formation, and the resedimented facies of the Lower Cretaceous (equivalent Maiolica Formation) improved the porosity and permeability of the succession and may indicate new exploration hydrocarbon plays. The petrographic and geochemical study of these dolostones, integrated with the reconstructed geological and tectonic evolution, provides a predictive model for the dolomitization process, which could support hydrocarbon explorative projects.

The paragenesis of the dolomites consists of a first phase of dolomite replacement (dolomite 1) and a few phases of dolomite cementation (dolomite 2 overgrowths, dolomite 4, and saddle dolomite) that is intercalated to dolomite dissolution and recrystallization (dolomite 3).

The collected data demonstrated that the dolomitization was a mesodiagenetic multistage process influenced by several factors: the burial temperature linked to the subsidence rate, the presence of porous horizons, and the permeable ways for fluid flow. All these factors varied through time, when the area changed from the passive-margin realm to collision-margin realm, and in the function of the structural position relative to the advancing Apennines thrust.

Three main domains have been distinguished in the studied area. In the thrust zone (A), characterized by differentiated subsidence during the Mesozoic and involved in thrusting from the lower Pliocene, the dolomitization process started early (Cretaceous–Eocene) and was over when the Laga formation basin was established (Miocene). Only the saddle dolomite precipitated during the main Apennines phase (lower Pliocene). In the foredeep zone (B), the timing of the first dolomitization event varies according to the subsidence rates. In the examined paleohigh succession, the process started during the Tertiary, but the most pervasive dolomitization occurred immediately after the lower Pliocene tectonic phase, when the area became a foredeep. In the foreland zone (C), characterized by low subsidence in the passive-margin realm, the dolomitization process started (dolomite 1) in the late Tertiary, when the siliciclastic sediments provided the overburden, whereas in the far foreland, the whole process happened in recent times (from the Pliocene) because hydrothermal fluids flowed along transpression faults.

On the whole, a migration of the dolomitization process through time, from the chain toward the foreland, passing from a passive margin to a collisional realm, is observed. Meanwhile, the fluid-flow pathways changed from reactivated tensional Jurassic faults to compressional and transpressional Tertiary faults. The source of Mg in the dolomitizing fluids is thought to be found in the Burano Formation evaporates that may have been squeezed during compaction, and later, they represent the detachment horizons for the Apennines thrusts. Another source of Mg may be found in shaly intervals present in the sedimentary succession that have been compacted during burial.

ACKNOWLEDGMENTS

The authors thank all Eni-Agip Division colleagues who contributed to the accomplishment of this work, particularly V. Scola, for the constant help; S. Merlini, for the discussion; T. Ricchiuto, for isotopic analyses; A. De Poli, for burial histories reconstruction; and M. Garioni, for the remarkable contribution to a previous study.

REFERENCES CITED

Accarie, H., B. Berard, R. Cussey, P. Joseph, and S. Triboulet, 1986, Dynamique sédimentarire et structurale au passage plateforme/bassin. Les faciès carbonatés crétacées du massif de la Majella: Memorie della Societa Geologica Italiana, v. 36, p. 217–231.

Adamoli, L., and A. Mancinelli, 1978, Ricerche geologiche sul Gran Sasso d'Italia (Abruzzo): III. Correlazioni microbiostratigrafiche tra facies di margine della piattaforma carbonmatica e facies pelagiche del Giurassico e Cretacico inferiore: Studi Geologici Camerti, v. IV, p. 19–36.

Adamoli, L., T. Bertini, M. Chiocchini, G. Deiana, A. Mancinelli, U. Pieruccini, and A. Romano, 1978, Ricerche geologiche sul Mesozoico del Gran Sasso d'Italia (Abruzzo): II. Evoluzione tettonico sedimentaria dal Trias superiore al Cretoacico inferiore dell'area compresa tra il Corno Grande e S.Stefano di Sessano (F-140 Teramo): Studi Geologici Camerti, v. IV, p. 7–18.

Adamoli, L., T. Bertini, M. Chiocchini, G. Deiana, A. Mancinelli, U. Pieruccini, and A. Romano, 1982, Ricerche geologiche sul Gran Sasso d'Italia (Abruzzo): V. Evoluzione tettonico sedimentaria dal Trias superiore al Cretacico inferiore dell'area compresa tra M. Camicia ed Ofena (F-140 Teramo): Studi Geologici Camerti, v. VII, p. 89–95.

Adamoli, L., A. Bigozzi, G. Ciarapica, S. Cirilli, L. Passeri, A. Romano, F. Duranti, and V. Venturi, 1990, Upper Triassic bituminous facies and Hettangian pelagic facies in the Gran Sasso Range: Bollettino della Societa Geologica Italiana, v. 109, p. 219–230.

Allan, J. R., and W. D. Wiggins, 1993, Dolomite reservoirs: Geochemical techniques for evaluation origin and distribution: AAPG Continuing Education Course Note Series 36, 170 p.

Amthor, J. E., and G. M. Friedman, 1991, Dolomite-rock textures and secondary porosity development in Ellemburger Group carbonates (Lower Ordovician), west Texas and southeastern New Mexico: Sedimentology, v. 38, p. 343–362.

Amthor, J. E., E. W. Mountjoy, and H. G. Machel, 1993, Subsurface dolomites in Upper Devonian Leduc Formation buildups, central part of Rimbey–Meadowbrook reef trend, Alberta, Canada: Bulletin of Canadian Petroleum Geology, v. 41, no. 2, p. 164–185.

Bigi, S., F. Calamita, E. Centamore, G. Deiana, M. Ridolfi, and R. Salvucci, 1992, Assetto strutturale e cronologica della deformazione della "zona d'incontro" tra le aree Umbro-Marchigiana e Laziale Abruzzese (Marche meridionali e Lazio-Abruzzo settentrionali): Studi Geologici Camerti, Volume Speciale (1991–1992), Deep Crust Project, Crosta Profonda 11, p. 21–26.

Bigozzi, A., G. Ciarrapica, S. Cirilli, and L. Passeri, 1992, Eteropie di facies nel trias superiore e nel Lias inferiore del Gran Sasso: Studi Geologici Camerti, Volume Speciale (1991–1992), Deep Crust Project, Crosta Profonda 11, p. 115–118.

Bodnar, R. J., 1992, Revised equation and table for freezing point depressions of H_2O-NaCl solutions: Geochimica et Cosmochimica Acta, v. 57, p. 683–686.

Braithwaite, C. J. R., and G. Rizzi, 1997, The geometry and petrogenesis of hydrothermal dolomites at Navan, Ireland: Sedimentology, v. 44, p. 421–440.

Broomhall, R. W., and J. R. Allan, 1987, Regional cap-rock destroying dolomite on the Middle Jurassic to Early Cretaceous Arabian Shelf: Society of Petroleum Engineers Formation Evaluation, v. 2, p. 435–441.

Brunamonte, F., A. M. Michetti, L. Serva, and E. Vittori, 1992, Evidenze paleosismologiche nell'Appennino centrale ed implicazioni neotettoniche: Studi Geologici Camerti, Volume Speciale (1991–1992), Deep Crust Project, Crosta Profonda 11, p. 265–270.

Centamore, E., G. Cantalamessa, A. Micarelli, M. Potetti, D. Berti, S. Bigi, C. Morelli, and M. Ridolfi, 1992, Stratigrafia ed analisi di facies dei depositi del Miocene e del Pliocene inferiore dell'avanfossa marchigiano-abruzzese e delle zone limitrofe: Studi Geologici Camerti, Volume Speciale (1991–1992), Deep Crust Project, Crosta Profonda 11, p. 125–132.

Ciarapica, G., and L. Passeri, 1998, Evoluzione paelogeografica degli Appennini: Atti Ticinesi Scienze della Terra, v. 40, p. 233–290.

Ciarapica, G., S. Cirilli, B. D'Argenio, E. Marsella, L. Passeri, and L. Zaninetti, 1987, Late Triassic opening and euxinic basins in Italy: Rendiconti della Societa Geologica Italiana, v. 9, p. 10–21.

Cipollari, P., D. Cosentino, and E. Ghiozzi, 1999, Extension and compression-related basins in central Italy during the Messinian Lago-Mare event: Tectonophysics, v. 315, p. 163–185.

Colacicchi, R., L. Passeri, and G. Pialli, 1975, Evidence of tidal environment deposition in the Calcare Massiccio Formation (central Apennines–lower Lias), *in* R. N. Ginsburg, ed., Tidal deposits: A casebook of recent examples and ancient counterparts: Berlin, Springer-Verlag, p. 345–353.

Crescenti, U., 1969, Serie stratigrafiche della serie calcarea dal Lias al Miocene nella regione Marchigiano Abruzzese: parte I e II: Memorie della Societa Geologica Italiana, v. 8, p. 155–204 and 343–420.

Damiani, A. V., M. Chiocchini, R. Colacicchi, G. Mariotti, M. Parotto, L. Passeri, and A. Praturlon, 1992, Elementi litostratigrafici per una sintesi delle facies carbonatiche meso-cenozoiche dell'Appennino centrale: Studi Geologici Camerti, Volume Speciale (1991–1992): Deep Crust Project, Crosta Profonda 11, p. 187–213.

Decourt, J., L. E. Ricuo, and B. Vrielynck, eds., 1993, Atlas Tethys paleoenvironmental maps and explanatory notes: Paris, Gauthier-Villars, 307 p., 14 maps.

Dramis, F., G. Panbianchi, O. Nesci, and M. Consoli, 1992, Il ruolo di elementi strutturali trasversali nell'evoluzione tettonico-sedimentaria e geomorfologia della regione marchigiana: Studi Geologici Camerti, Volume Speciale (1991–1992), Deep Crust Project, Crosta Profonda 11, p. 287–293.

Drivet, E., and E. W. Mountjoy, 1997, Dolomitization of the Leduc Formation (Upper Devonian), southern Rimbey–Meadowbrook reef trend, Alberta: Journal of Sedimentary Research, v. 67, p. 411–423.

Eberli, G. P., D. Bernoulli, D. Sanders, and A. Vecsei, 1993, From aggradation to progradation: The Macella Platform (Abruzzi, Italy): AAPG Memoir 56, p. 213–232.

Frisia, S., and H. R. Wenk, 1993, TEM and EAM study of pervasive, multi-step dolomitization of the Upper Triassic Dolomia Principale (northern Italy): Journal of Sedimentary Petrology, v. 63, no. 6, p. 1049–1058.

Gibbs, A. D., 1984, Structural evolution of extensional basin margin: Journal of the Geological Society (London), v. 141, p. 609–620.

Goldstein, R. H., and T. J. Reynolds, 1994, Systematics of fluid inclusions in diagenetic minerals: SEPM Short Course 31, 199 p.

Hardie, L. A., 1987, Dolomitization: A critical review of some current views: Journal of Sedimentary Petrology, v. 57, p. 166–183.

Harding, T. P., and A. C. Tuminas, 1989, Structural interpretation of hydrocarbon traps sealed by basement normal block faults at stable flank of foredeep basins and at rift basins: AAPG Bulletin, v. 73, p. 812–840.

Heidary, E., 1997, Hydrotectonic models of burial diagenesis in platform carbonates based on formation water geochemistry in North American sedimentary Basins, *in* I. P. Montanez, J. M. Gregg, and K. Shelton, eds., Basin-wide diagenetic patterns: Integrated petrologic, geochemical, and hydrologic considerations: SEPM Special Publication 57, p. 54–79.

Lonnee J., and I. S. Al-Aasm, 2000, Dolomitization and fluid evolution in the Middle Devonian Sulphur Point Formation, Rainbow South field, Alberta: Petrographic and geochemical evidence: Bulletin of Canadian Petroleum Geology, v. 48, p. 262–282.

Luperto Sinni, E., 1996, Schema stratigrafico del Cretaceo del Gargano basato su risultati di recenti ricerche: Memorie della Societa Geologica Italiana, v. 51, p. 1019–1036.

Marchegiani, L., G. Bertotti, A. Cello, G. Deiana, S. Mazzoli, and E. Tondi, 1999, Pre-orogenic tectonics in the Umbria-Marche sector of the Afro-Adriatic continental margin: Tectonophysics, v. 315, p. 123–143.

Mattes, B. W., and E. W. Mountjoy, 1980, Burial dolomitization in the Upper Devonian Miette buildup, Jasper National Park, Alberta, *in* D. H. Zenger, J. B. Dunham, and R. L. Ethington, eds., Concepts and models of dolomitization: SEPM Special Publication 28, p. 259–297.

Mitra, S., 1986, Duplex structures and imbricate thrust systems: Geometry, structural position, and hydrocarbon potential: AAPG Bulletin, v. 70, no. 9, p. 1087–1112.

Montanez, I. P., 1994, Late diagenetic dolomitization of Lower Ordovician, upper Knox carbonates: A record of the hydrodynamic evolution of the southern Appalachian basin: AAPG Bulletin, v. 78, p. 1210–1239.

Mountjoy, E., S. Whittaker, A. Williams-Jones, H. Qing, E. Drivet, and X. Marquez, 1997, Variable fluids and heat flow regimes in the three Devonian dolomite conduit systems, Western Canada sedimentary basin: Isotopic and fluid inclusion evidences/constraints, *in* I. P. Montanez, J. M. Gregg, and K. Shelton, eds., Basin-wide diagenetic patterns: Integrated petrologic, geochemical, and hydrologic considerations: SEPM Special Publication 57, p. 119–137.

Oliver, J., 1986, Fluids expelled tectonically from orogenic belts: Their role in hydrocarbon migration and other geologic phenomena: Geology, v. 14, p. 99–102.

Passeri, L., ed., 1990, Guide geologiche regionali, Appennino Umbro Marchigiano: Società Geologica Italiana, 300 p.

Patacca, E., P. Scandone, M. Bellatalla, N. Perilli, and U. Santini, 1992, La zona di giunzione tra l'arco appenninico settentrionale e l'arco appenninico meridionale nellìAbruzzo e nel Molise: Studi Geologici Camerti, Volume Speciale (1991–1992), Deep Crust Project, Crosta Profonda 11, 417–441.

Pavan, G., and C. Pirini, 1966, Stratigrafia del Foglio 157 "Monte S. Angelo": Bollettino del Servizio Geologico d'Italia, v. 86, p. 123–189.

Pieri, M., and L. Mattavelli, 1986, Geologic framework of Italian petroleum resources: AAPG Bulletin, v. 70, no. 2, p. 103–130.

Qing, H., and E. W. Mountjoy, 1989, Multistage dolomitization in Rainbow buildups, Middle Devonian Keg River Formation, Alberta, Canada: Journal of Sedimentary Petrology, v. 59, p. 114–126.

Qing, H., D. W. J. Bosence, and E. P. F. Rose, 2001, Dolomitization by penesaline sea water in Early Jurassic peritidal carbonates, Gibraltar, western Mediterranean: Sedimentology, v. 48, p. 153–163.

Radke, B. M., and R. L. Mathis, 1980, On the formation and occurrence of saddle dolomite: Journal of Sedimentary Petrology, v. 50, p. 1149–1168.

Ronchi, P., F. Casaglia, and A. Ceriani, 2003, The multiphase dolomitization of the Liassic Calcare Massiccio and Corniola formations (Montagna dei Fiori, northern Apennines, Italy): Bollettino Società Geologica Italiana, v. 122, p. 157–172.

Santantonio, M., 1993, Facies associations and evolution of pelagic carbonate platform/basin systems: Examples from the Italian Jurassic: Sedimentology, v. 40, p. 1039–1067.

Santantonio, M., F. Galluzzo, and G. Gill, 1996, Anatomy and palaebathimetry of a Jurassic pelagic carbonate platform/basin system. Rossa Mts, central Apennines (Italy): Geological implications. Palaeopelagos, v. 6, p. 123–169.

Sibley, D. F., and J. M. Gregg, 1987, Classification of dolomite rock texture: Journal Sedimentary Petrology, v. 57, p. 967–975.

Swennen, R., J. L. Faure, K. Osadetz, P. Robion, K. Herbos, and F. Roure, 2002, Mesodolomite formation as a result of layer parallel shortening and secondary porosity development by cooling of formation waters or cooling due to thrust emplacements: An example from the Canadian foreland fold and thrust belt (abs.): Presented at the American Association of Petroleum Geologists-Institut Française du Pétrole Hedberg Research Conference, May 2002, Palermo, Italy, http://www.searchanddiscovery.com/documents/abstracts/hedberg2002sicily/index.htm (accessed October 25, 2004).

Zempolich, W. G., and L. A. Hardie, 1997, Geometry of dolomite bodies within deep-water resedimented oolite of Middle Jurassic Vajont Limestone, Venetian Alps, Italy: Analogs for hydrocarbon reservoirs created through fault-related burial dolomitization, *in* J. A. Kupecz, J. Gluyas, and S. Bloch, eds., Reservoir quality prediction in sandstones and carbonates: AAPG Memoir 69, p. 127–162.

Regional Fluid Flow and Diagenesis

Al-Aasm, I. S., and J. D. Clarke, 2004, The effect of hydrothermal fluid flow on early diagenetic dolomitization: An example from the Devonian Slave Point Formation, northwest Alberta, Canada, *in* R. Swennen, F. Roure, and J. W. Granath, eds., Deformation, fluid flow, and reservoir appraisal in foreland fold and thrust belts: AAPG Hedberg Series, no. 1, p. 297–316.

The Effect of Hydrothermal Fluid Flow on Early Diagenetic Dolomitization: An Example from the Devonian Slave Point Formation, Northwest Alberta, Canada

Ihsan S. Al-Aasm
Department of Earth Sciences, University of Windsor, Windsor, Ontario, Canada

Julie D. Clarke[1]
Department of Earth Sciences, University of Windsor, Windsor, Ontario, Canada

ABSTRACT

The Middle Devonian carbonates of the Slave Point Formation, Hamburg field, northwestern Alberta, are composed mainly of stromatoporoid and *Amphipora* floatstones and rudstones, with interbedded mudstone and grainstone facies characteristic of deposition in open to slightly restricted marine platform environments. These carbonates have undergone a complex diagenetic history, from shallow to deep burial, as represented by fracturing, calcite cementation, silicification, and dolomitization.

Petrographically, four different types of dolomite have been identified (from early to late): (1) fine-crystalline matrix dolomite; (2) pseudomorphic dolomite; (3) medium-crystalline pervasive dolomite; and (4) saddle dolomite. Fine-crystalline dolomite (5–50 μm) replaces the mud matrix and slightly penetrates the edges of allochems. It occurred in mud-supported facies and was precipitated by marine fluids. Oxygen isotope values range from −11.62 to −9.34‰ (Peedee belemnite), lower than postulated values for Devonian carbonates. The enriched $^{87}Sr/^{86}Sr$ isotope value from this phase (0.71002) suggests that later diagenetic fluids may have recrystallized this dolomite. Pseudomorphic dolomite (50–100 μm) replaces crinoids and occurs as single, large dolomite crystals. Its oxygen and carbon isotopic values range from −10.58 to −9.65 and

[1]*Present address:* Duke Energy, Chatham, Ontario, Canada.

DOI:10.1306/1025696H13122

+4.24 to +4.49‰, respectively. Medium-crystalline pervasive dolomite (10–100 μm) occurs along dissolution seams and obliterates all previous fabrics. It is proposed that this medium-crystalline dolomite formed during shallow to intermediate burial because of its association with dissolution seams and high iron content. The range of oxygen isotope values for this dolomite (−11.74 to −9.5‰) suggests precipitation from a warm fluid, possibly in a burial environment, and/or later recrystallization by hydrothermal fluids. The relatively wide range of carbon isotope values (+1.19 to +4.49‰) and enriched strontium isotope ratio (0.710020) suggests recrystallization. Saddle dolomite (250–2000 μm) partially to completely occludes void spaces (both fractures and vugs) and also occurs as a minor replacement mineral. The oxygen isotope values for saddle dolomite (−13.95 to −11.97‰), as well as the nonradiogenic to enriched strontium isotope ratios for saddle dolomite (0.70494 to 0.710351), and the fluid-inclusion data (homogenization temperature, T_h, range between 125 and 161°C and estimated salinity, between 22.2 and 24.7 wt.% NaCl equivalent) indicate precipitation from hot, highly saline, hydrothermal fluids, which were probably expelled tectonically during the Late Devonian–Mississippian Antler thrust belt development.

INTRODUCTION

Paleofluid flow, its mechanisms, chemical attributes, pathways, timing, and rates of migration constitute a fundamental aspect of basin evolution studies and analysis of hydrocarbon reservoir development. The origin and composition of fluids impose a major control on the diagenetic history of sedimentary basins under various tectonic settings.

In many sedimentary basins, including the Western Canada sedimentary basin, regionally extensive fluid flow may have occurred during tectonic thrusting, sedimentary loading, uplift, and compression (e.g., Garven, 1985; Oliver, 1986; Muchez et al., 2000). Such fluids were instrumental in the formation of sediment-hosted ore deposits, petroleum migration, and dolomitization (e.g., Bethke and Marshak, 1990; Bachu, 1999). In the Western Canada sedimentary basin, laterally extensive dolomitization affected the majority of Devonian and Mississippian carbonates to varying degrees. Debate continues with respect to the mechanism and timing of dolomitization and fluid flow. At least three models have been suggested to explain the massive dolomitization in the basin. The "topography or gravity-driven flow" (Garven, 1985) and "tectonically driven squeegee" models (Oliver, 1986) have attracted wide attention of many investigators. However, Machel and Cavell (1999) suggested that large-scale hydrothermal dolomitization is only possible if the fluids are focused into, and funneled through, relatively narrow aquifers. More recently, Morrow (1998) advocated a thermal convection flow model as a viable alternative to the other two models. Wendte et al. (1998) proposed a model that involves both seepage reflux and thermal convection of residual evaporitic brines to explain the dolomitization pattern of the Devonian Swan Hills carbonates from west-central Alberta. The timing of formation of dolomite also remains a controversial issue. Pre-, syn-, and post-Laramide fluid flows have been invoked in the literature based on a multitude of paleomagnetic, geochemical, and other evidences (Nesbitt and Meuhlenbachs, 1994; Mountjoy and Amthor, 1994; Lewchuck et al., 1998; Machel and Cavell, 1999; Symons et al., 1999; Al-Aasm et al., 2002).

The aim of this study is to unravel the influence of hydrothermal fluids associated with tectonics on the formation and alteration of dolomite in the Devonian Slave Point Formation and related reservoir characteristics. To accomplish this, the source(s) and chemistry of the diagenetic fluids that formed dolomite, which are controlled by tectonic history of Western Canada sedimentary basin, are evaluated.

REGIONAL GEOLOGY AND TECTONIC SETTING

The Western Canada sedimentary basin is composed of a wedge of sedimentary rocks thickening from the exposed Canadian Shield in the east to more than 6 km (4 mi) thick in the west. The internal structure of the basin reflects a complex history involving a Late Jurassic to early Tertiary foreland basin that was superimposed on a cratonic platform (Porter et al., 1982). However, Nelson et al. (2002) interpret the Western Canada sedimentary basin at the Devonian–Mississippian boundary to be in a back-arc setting.

Preexisting basement structures had a pronounced effect on sedimentation of the Slave Point Formation.

The Peace River Arch was a major east-northeast-trending basement structure (Cant, 1988) that extended toward the craton from the Cordillera across northeastern British Columbia and northwestern Alberta. The arch was a stable feature prior to the Middle Devonian, including the time of deposition of the Slave Point Formation, but during the Middle Devonian, a period of instability commenced, and the Arch began to collapse, resulting in a complex of small, tilted fault blocks that formed development sites for numerous reef carbonates (O'Connell et al., 1990). Situated about 40 km (25 mi) north of the Hamburg field is the Hay River fault zone (Figure 1). It is a 1300-km (810-mi)-long shear zone that extends from the foothills of northeastern British Columbia to the southern side of Great Slave Lake. A series of northwest-trending wrench faults emanating from the Hay River fault zone cut through the Hamburg field and have been a major factor in diagenesis of the Slave Point Formation (e.g., Davies, 1997). These faults may have served as the principal conduits for hydrothermal fluid flow. By the end of the Mississippian, burial of the Slave Point had reached depths of as much as 1.5 km (1 mi). The time of maximum burial (4–6 km; 2.5–4 mi) of Devonian rocks corresponds to the culmination of the Laramide orogeny in the Late Cretaceous–early Tertiary (Porter et al., 1982; Hitchon, 1984). Since then, erosion has removed between 1.1 and 2.5 km (0.68 and 1.6 mi) of strata in the Peace River Arch area (Hitchon, 1984; Issler et al., 1990).

STRATIGRAPHY AND FACIES

The Slave Point Formation is part of a multicyclic, Middle to Upper Devonian sequence that was confined to the east by the Precambrian shield and to the west by the Peace River Arch (Tooth and Davies, 1989). It is part of the transgressive Beaverhill Lake Group that is composed of carbonate platform deposits and minor evaporites. The Fort Vermilion, Slave Point, and Waterways formations make up the Beaverhill Lake Group (Figure 2). The Fort Vermilion Formation is composed mainly of anhydrite, the Slave Point Formation is a fossiliferous limestone that is variably dolomitized, and the overlying Waterways Formation is made up of shales and argillaceous carbonates.

The Slave Point Formation is composed of fossiliferous, shallow-water limestones that were deposited on a broad, semirestricted carbonate platform that stretched from northeastern British Columbia into eastern Alberta, with pronounced thinning to the east and south (Opalinski, 1984). The main skeletal frame builder of the Slave Point Formation is stromatoporoids. Corals, crinoids, brachiopods, and gastropods are accessory fauna and occur in fine-grained, muddy limestones that were variably dolomitized. The fossils show different degrees of preservation that may reflect their original mineralogy and shell structure. Fabric-selective dolomitization allowed for the retention of many original textures, structures, and fossil components.

The Slave Point Formation is an important producer of hydrocarbons in northeastern British Columbia, northwestern Alberta, and the southern Northwest Territories. The Hamburg, Cranberry, and Chinchaga gas fields are located immediately north of the Peace River Arch in northwestern Alberta (Figure 1). These reservoirs are stratigraphic traps; some primary intra- and interparticle porosities in limestones are retained along with secondary vuggy, fracture, and intercrystalline porosity in dolostones. The southeastern and eastern flanks of the arch yield oil from platform reefs and shelf limestone deposits, whereas higher on the arch, oil is contained in dolostones where the reservoir porosity is the result of dissolution of *Amphipora* (Craig, 1987).

SAMPLING AND ANALYTICAL PROCEDURES

Eleven cored wells from the Hamburg field (Figure 1) were examined and sampled in this study. Petrographic examination of 150 stained thin sections was performed using a standard petrographic microscope and cathodoluminescence microscopy using a Technosyn cold cathodoluminescence stage with a 12–15-kV beam and a current intensity of 0.42–0.43 mA. Fluorescence characteristics were studied with a Nikon Episcopic (EPI) fluorescence connected to a petrographic microscope. Fluid-inclusion analyses were carried out for eight doubly polished thin sections on a Linkam THM600 stage.

Semiquantitative analysis of major, minor, and trace elements were performed on eight highly polished, carbon-coated thin sections with a Cameca Camebax SX50 electron microprobe analyzer equipped with three crystal spectrometers and a backscatter electron detector. The operating conditions were set at an acceleration voltage of 15 kV, a beam current of 6 nA, and a focused beam diameter of 4–10 μm. The standards used were $CaSiO_3$ (Ca), MgO (Mg), Fe_2O_3 (Fe), $MnTiO_3$ (Mn), and $SrCO_3$ (Sr). Detection limits were 0.1 wt.% oxide for Ca and Mg and better than 0.03 wt.% oxide for Sr, Mn, and Fe. Precision was better than 0.5 mol% for all elements. Oxygen and carbon isotopes were extracted from calcite and dolomite samples ($n = 80$) that were obtained using a microscope-mounted drill assembly. The samples were reacted in vacuo with 100% pure phosphoric acid for 4 hr at 25 and 50°C for calcite and dolomite, respectively. CO_2 gas from samples that

A

B

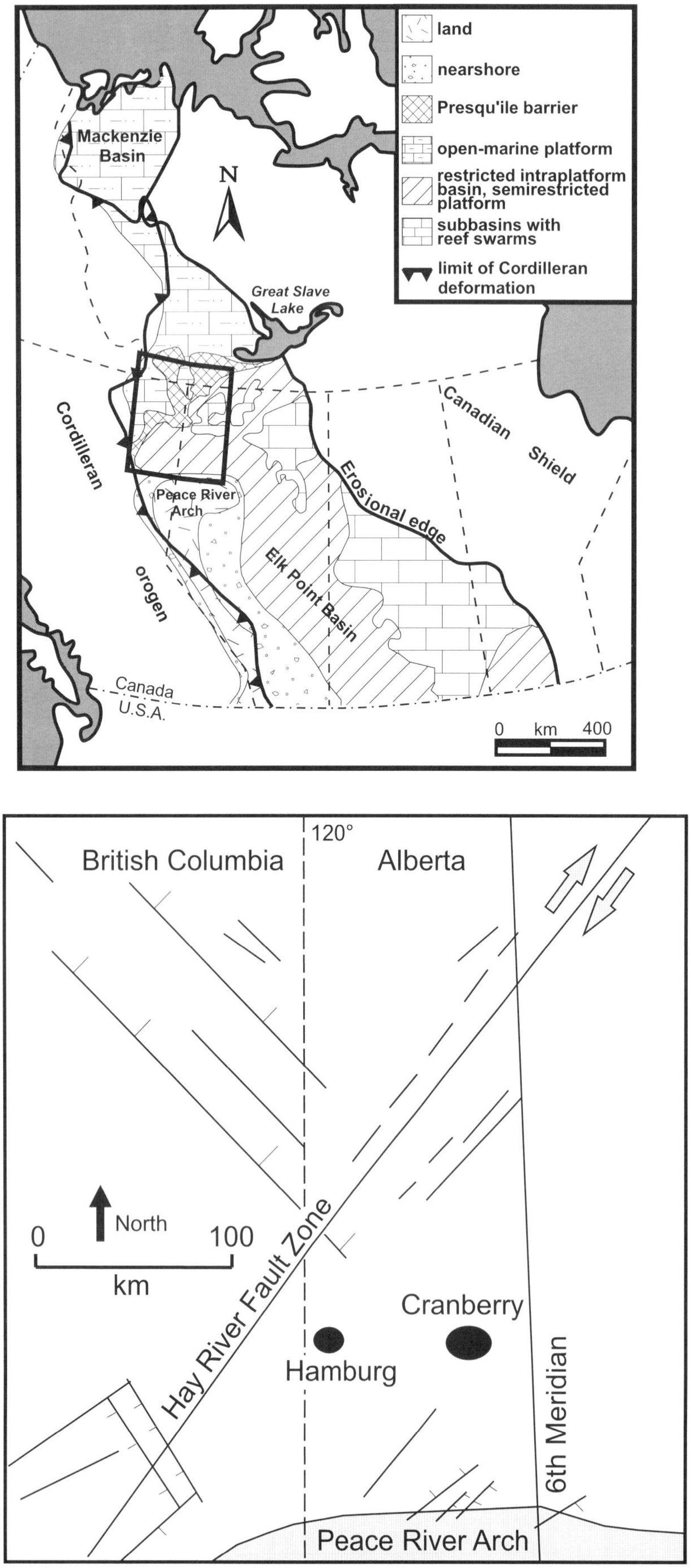

FIGURE 1. (A) Location map of the study area showing Eifelian to Givetian facies and facies successions of the Western Canada sedimentary basin (modified from Moore, 1988). The bottom map (B) shows the Hamburg field in relation to the tectonic elements in the area (modified after Davies, 1997).

contained a mixture of calcite and dolomite was extracted using the chemical separation method of Al-Aasm et al. (1990). The evolved CO_2 gas was analyzed for isotopic ratios on a SIRA-12 mass spectrometer. Delta (δ) values for oxygen and carbon are reported in per mil (‰) relative to the Vienna Peedee belemnite (VPDB) standard. Precision was better than 0.05‰ for both $\delta^{18}O$ and $\delta^{13}C$. Strontium isotopes were analyzed on a Finnigan MAT 262 with five fixed collectors. National Bureau of Standards (NBS)-87 and ocean water were used as standard references, and $^{87}Sr/^{86}Sr$ ratios were normalized to $^{88}Sr/^{86}Sr$ = 8.375209. Precision was better than 0.00010.

DIAGENESIS

Carbonates from the Slave Point Formation have undergone a complex diagenetic history. Diagenesis occurred in a wide range of tectonic settings, each having distinct hydrologic regimes and varying fluid compositions. The most important diagenetic processes that occurred were dolomitization and calcite cementation, although several other minor processes have occurred. The paragenetic sequence presented in Figure 3A is based on petrographic observations, crosscutting relationships, compactional features, and geochemical evidence. Only diagenetic processes relevant to dolomitization will be discussed below.

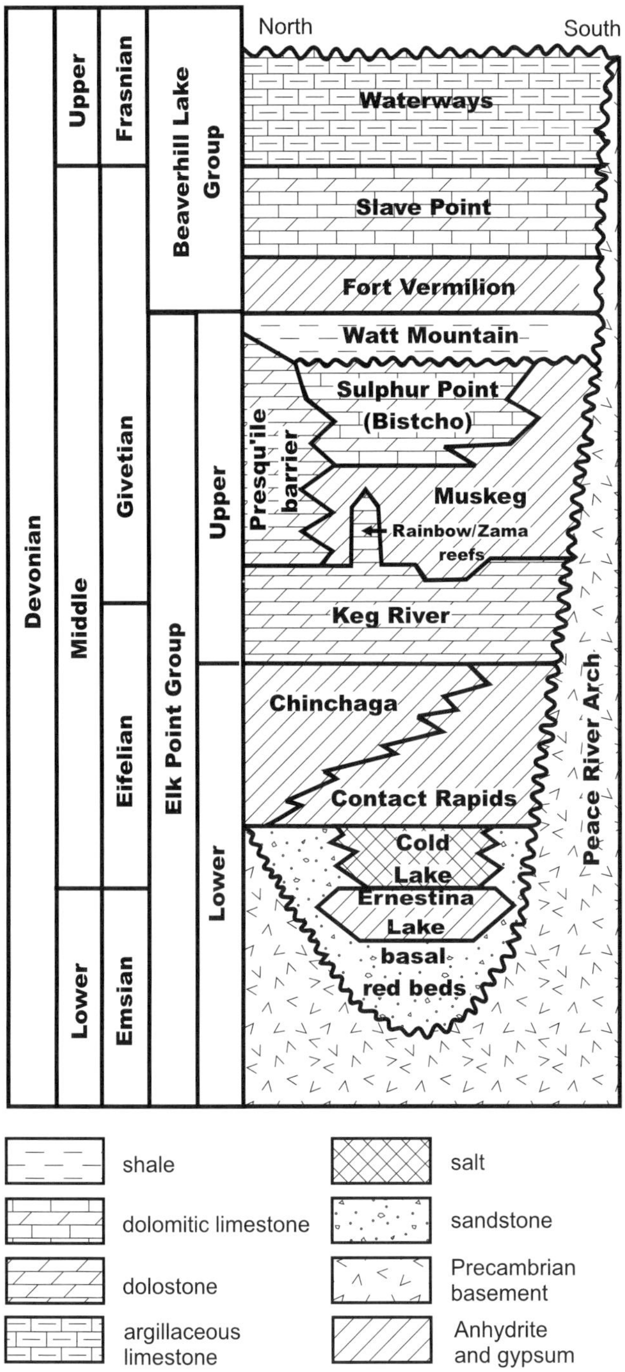

FIGURE 2. Schematic diagram for the stratigraphy of the Elk Point and Beaverhill Groups in northwestern Alberta (modified from Meijer Dress, 1994).

Calcite Cementation

Calcite cementation spans the entire diagenetic history of the Slave Point carbonates, where it occludes early to late porosity (Figure 4A). Several types of calcite cement have been identified petrographically. Although some overlap occurs, their general order of formation is (1) radial fibrous cement, (2) bladed and prismatic cement, (3) fine equant calcite, and (4) blocky calcite.

Radial fibrous calcite is isopachous, inclusion rich, and nonluminescent and occupies interparticle pores (Figure 5A). The bladed and prismatic cement is composed of turbid, nonluminescent crystals with scalenohedral terminations. Lengths range between 100 and 275 μm, and widths range from 20 to 65 μm, resulting in length-to-width ratios between 3:1 and 6:1. Bladed and prismatic cements nucleate on the septa of corals and stromatoporoids and are the earliest cements in these intraparticle pores. Equant calcite spar is the most abundant calcite cement, with crystals that range in size from 30 to 150 μm and have few inclusions. Homogeneous, dull brown and red colors are displayed under cathodoluminescence. This cement type is most commonly found occluding matrix porosity in limestones and intraparticle porosity in stromatoporoids. Blocky calcite cement occurs primarily as a fracture and void-filling cement in both limestones and dolostones but does occur as a replacement mineral for saddle dolomite (Figure 5B, C). Crystals are coarse (as much as 250 μm in size), euhedral, and nonluminescent and become progressively iron rich toward the centers of voids and veins. Blocky calcite contains two-phase fluid inclusions and numerous small hydrocarbon inclusions, which are recognized by their strong yellow-green fluorescence under ultraviolet (UV) illumination.

Dolomitization

Dolomitization is the most significant diagenetic process that has occurred in the Slave Point carbonates, and all facies have been affected to varying degrees. In the Hamburg field, the best reservoir quality is in dolomitized sequences. Four types of dolomite have been recognized: (1) microcrystalline matrix dolomite, (2) pseudomorphic dolomite, (3) medium-crystalline pervasive dolomite, and (4) saddle dolomite. Although each dolomite type is discussed separately, some are genetically related and likely represent a continuous spectrum of dolomitization processes.

Microcrystalline Matrix Dolomite

Matrix dolomite occurs in mud-supported facies as well as in grainstone facies, selectively replaces the mud matrix, and penetrates the outer edges of fossils (Figure 4B, C). It is volumetrically minor and commonly accounts for a relatively small portion of the total dolomite volume (<10%). Dissolution seams deflect around individual dolomite crystals, and matrix dolomite crystals are truncated by low-amplitude stylolites. In most cases, this dolomite does not occur in association with medium-crystalline pervasive dolomite.

Matrix dolomite consists of isolated, euhedral crystals that range in size from 5 to 50 μm (Figure 6A, B).

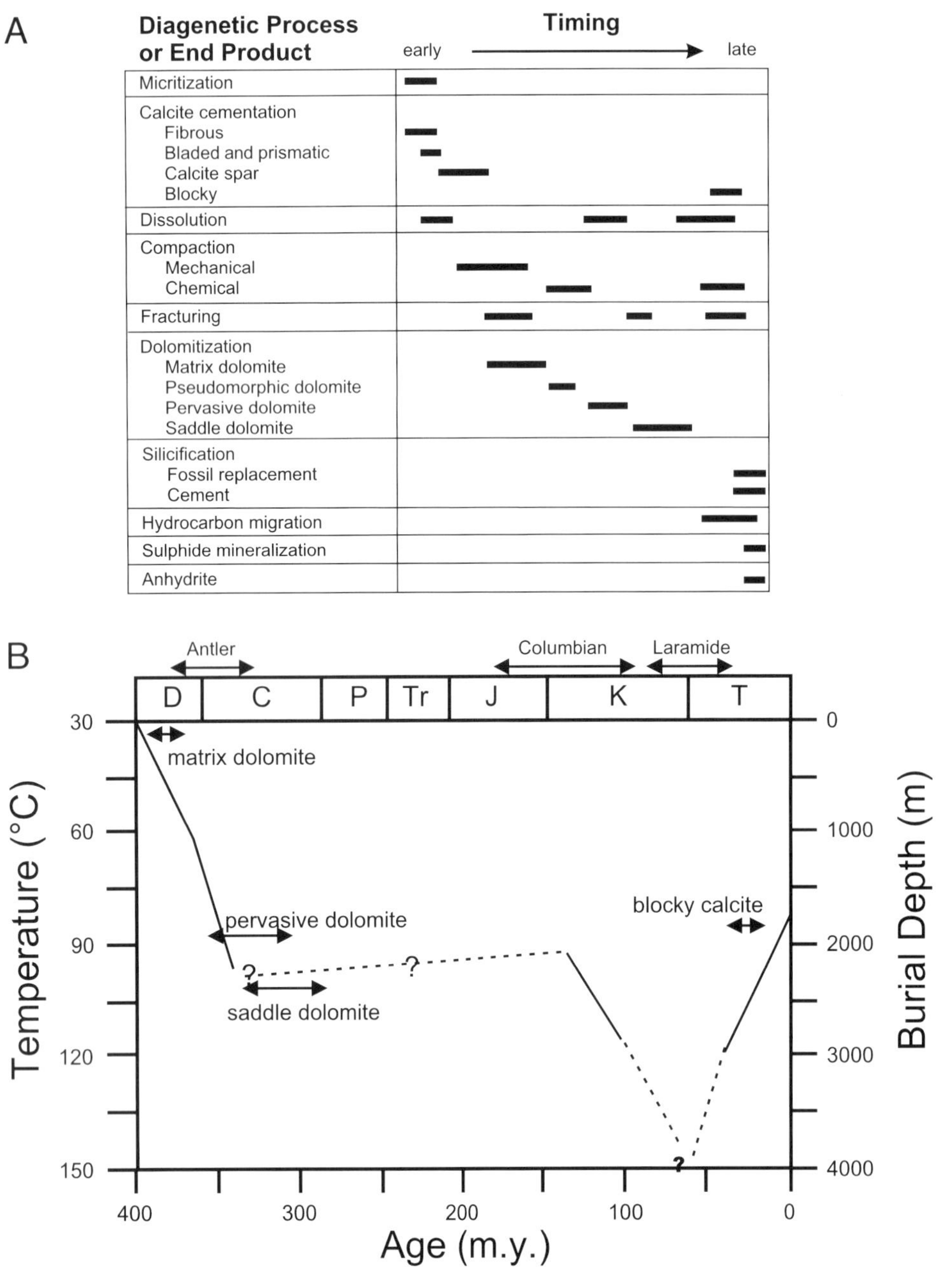

Figure 3. (A) Diagenetic sequence for carbonates from the Slave Point Formation, Hamburg field. (B) Theoretical burial-history curve for the Slave Point Formation showing the approximate timing of dolomite formation with respect to temperature (assuming 30°C/km geothermal gradient) and orogenic events influencing the Western Canada sedimentary basin (see text for details).

Crystals are generally turbid, somewhat with etched corners but at times display small, clear overgrowth rims. Under cathodoluminescence, microcrystalline matrix dolomite is luminescent with dull cores and bright red rims, and UV fluorescence reveals crystal cores that are dull green or orange with lighter green or orange outer rims that correspond to the clear overgrowth rims.

Pseudomorphic Dolomite

Pseudomorphic dolomite mimics the morphology of crinoid fragments and occurs as single, inclusion-rich crystals of dolomite (Figure 7A). Crystals tend to be large (as much as 450 μm) because of relatively few nucleation sites (Sibley and Gregg, 1987), and when viewed under partial extinction, the original crinoids fragment can be observed. Dull red colors are displayed under cathodoluminescence.

Medium-crystalline Pervasive Dolomite

Pervasive dolomite is volumetrically the most prominent form of dolomite in the Hamburg field, accounting for 50% of all dolomite phases. It has affected all facies and is fabric destructive, replacing both the matrix and allochems. This dolomite also occurs within and around bed-parallel dissolution seams and stylolites. Individual crystals range in size from 50 to 150 μm and exhibit planar-E and planar-S fabrics with cloudy (inclusion-rich) cores and clear (inclusion-free) rims; cathodoluminescence colors are dull brownish red with bright red rims (Figure 6D, E). The contacts between the cores and the rims are irregular. Pervasive dolomite locally replaces pseudomorphic dolomite, as indicated by its precipitation across pseudomorphic dolomite crystal boundaries.

Saddle Dolomite

Saddle dolomite is characterized by large crystal size, curved crystal faces, sweeping extinction, and abundant, small, two-phase fluid inclusions that define growth banding (Figure 7C). Saddle dolomite occurs primarily as fracture, occluding bores in undolomitized fossils (Figure 7B) and also as vug-filling cement; however, it also occurs as a replacement mineral after medium-crystalline pervasive dolomite (Figure 4A–C). Its occurrence in subvertical fractures (Figure 4B, C) is crosscut by high-amplitude, bed-parallel stylolites and fluidized internal sedimentation (Figure 4B). It is volumetrically significant, accounting for about 30% of all dolomite phases, and crystals range in size from 250 to 2000 μm, depending on the size of the void in which it was precipitated. Under cathodoluminescence, saddle dolomite displays homogeneous dull red colors, with occasional bright red rims occurring along etched crystal boundaries. This dissolution can also be seen under UV illumination as orange rims on dull green saddle dolomite crystals. Bitumen is present as a coating surrounding saddle dolomite crystals.

GEOCHEMISTRY

Carbon and Oxygen Isotopes

Calcitic Components

The stable carbon and oxygen isotope compositions of skeletal and nonskeletal components are plotted in Figure 8A and B (see also Table 1). Brachiopods, some crinoids, some corals, micrite, and fibrous calcite cement have comparable or slightly depleted isotopic values in comparison to the postulated

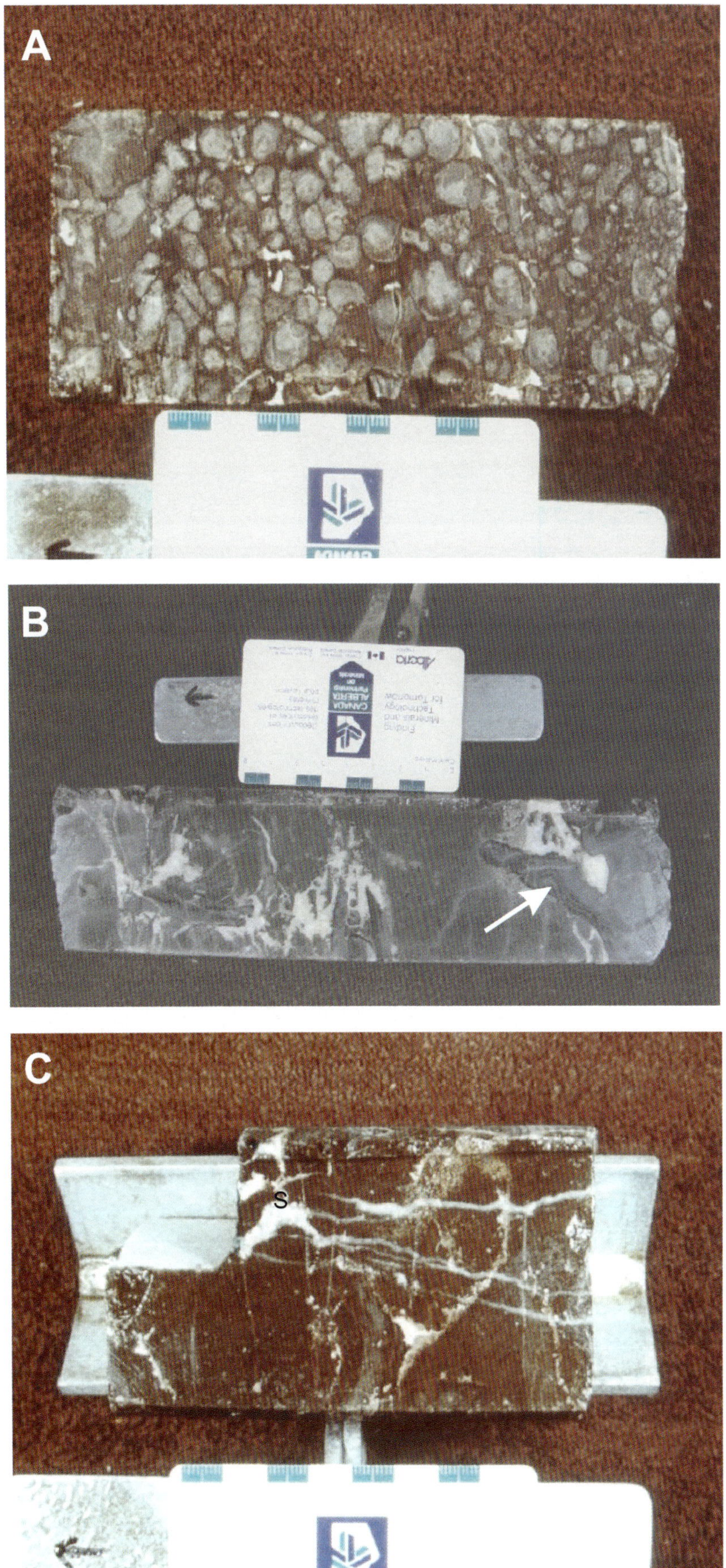

FIGURE 4. (A) Core photograph of *Amphipora* rudstone facies. Vugs were filled with saddle dolomite cement, which is cut by fractures, some of which are filled with blocky calcite. (B) Core photograph showing dolomitized matrix, fluidized channel (arrow) filled by internal sediment crosscut fracture filled by earlier precipitated saddle dolomite cement. (C) Core photograph showing subvertical fractures filled with saddle dolomite (s). These are crosscut by thin fractures filled by blocky calcite.

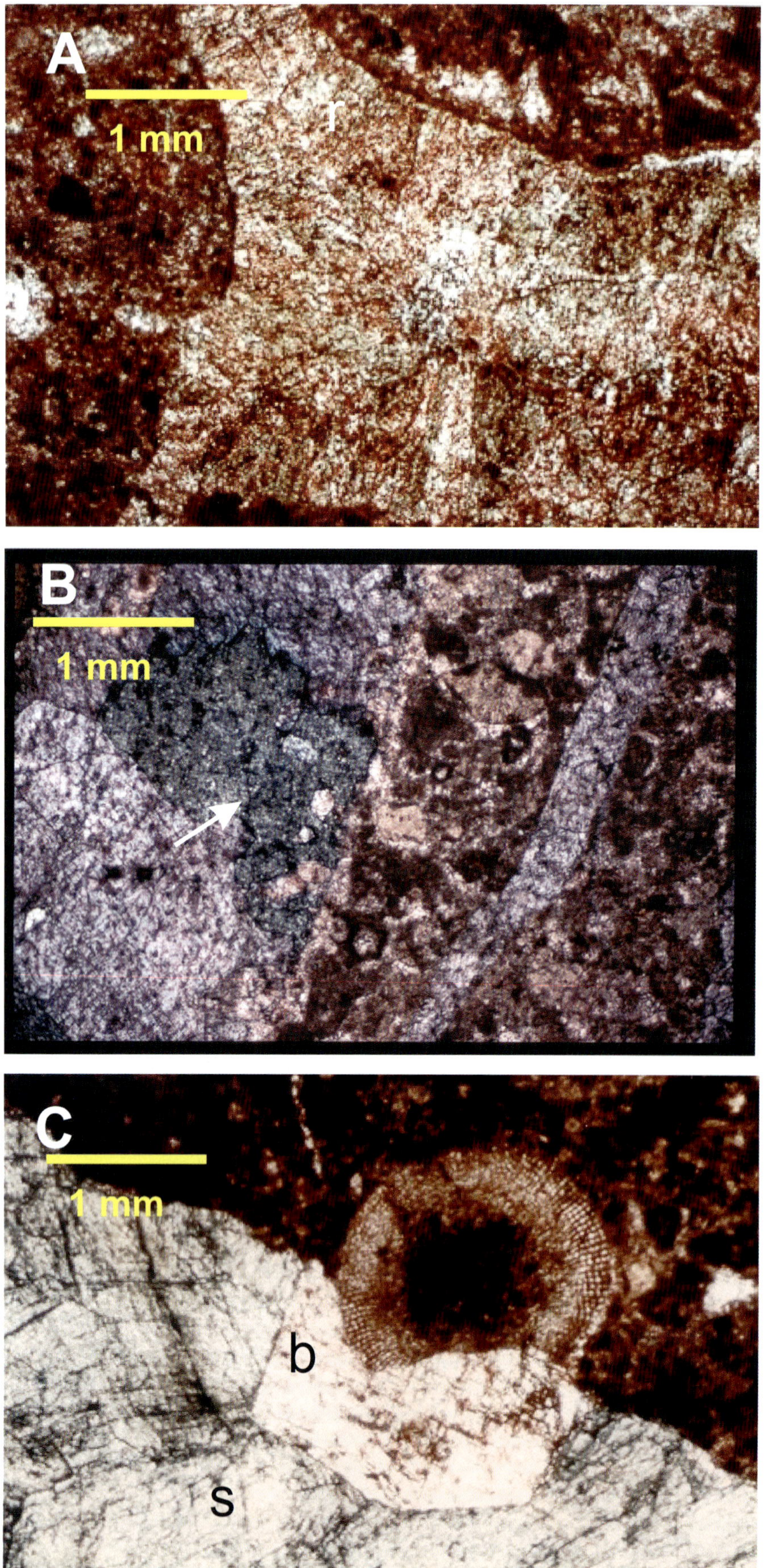

Middle Devonian marine calcite (cf. Hurley and Lohmann, 1989). However, other fossils, such as stromatoporoids, some corals, equant calcite spar, and blocky calcite show more depleted $\delta^{18}O$ values but comparable $\delta^{13}C$ values to Middle Devonian carbonates.

Dolomitic Components

The following is a summary of the carbon and oxygen isotopic composition of the dolomites of the Slave Point carbonates. A complete compilation of the isotope results is found in Table 1.

1) Microcrystalline matrix dolomite: $\delta^{18}O$ values range from −11.62 to −9.34‰ (VPDB) (mean = −10.42, standard deviation (σ) = 0.82), and $\delta^{13}C$ values range from +0.20 to +3.17‰ (VPDB) (mean = +1.61, σ = 0.1); values for oxygen do not fall within the expected range for postulated middle marine Devonian dolomite (Figure 9A)
2) Pseudomorphic dolomite: $\delta^{18}O$ values −10.58 and −9.65‰ (VPDB) and $\delta^{13}C$ values +4.24 and +4.49‰ (VPDB) (Figure 9A)
3) Medium-crystalline pervasive dolomite: $\delta^{18}O$ values ranging from −11.74 to −9.50‰ (VPDB) (mean = −10.57, σ = 0.71) and $\delta^{13}C$ values ranging from +2.01 to +3.50‰ (VPDB) (mean = +2.54, σ = 0.60) (Figure 9A)
4) Saddle dolomite: $\delta^{18}O$ values ranging from −13.95 to −11.97‰ (VPDB) (mean = −12.91, σ = 0.61) and $\delta^{13}C$ values ranging from +1.19 to +3.02‰ VPDB) (mean = +2.09, σ = 0.43) (Figure 9A)

FIGURE 5. (A) Photomicrograph of radial fibrous calcite (r) occluding interparticle porosity between *Amphipora*. (B) Late fracture-fill, ferroan, blocky calcite postdated saddle dolomite (arrow). (C) Late-stage blocky calcite cement (b) replacing saddle dolomite (s) and fossil.

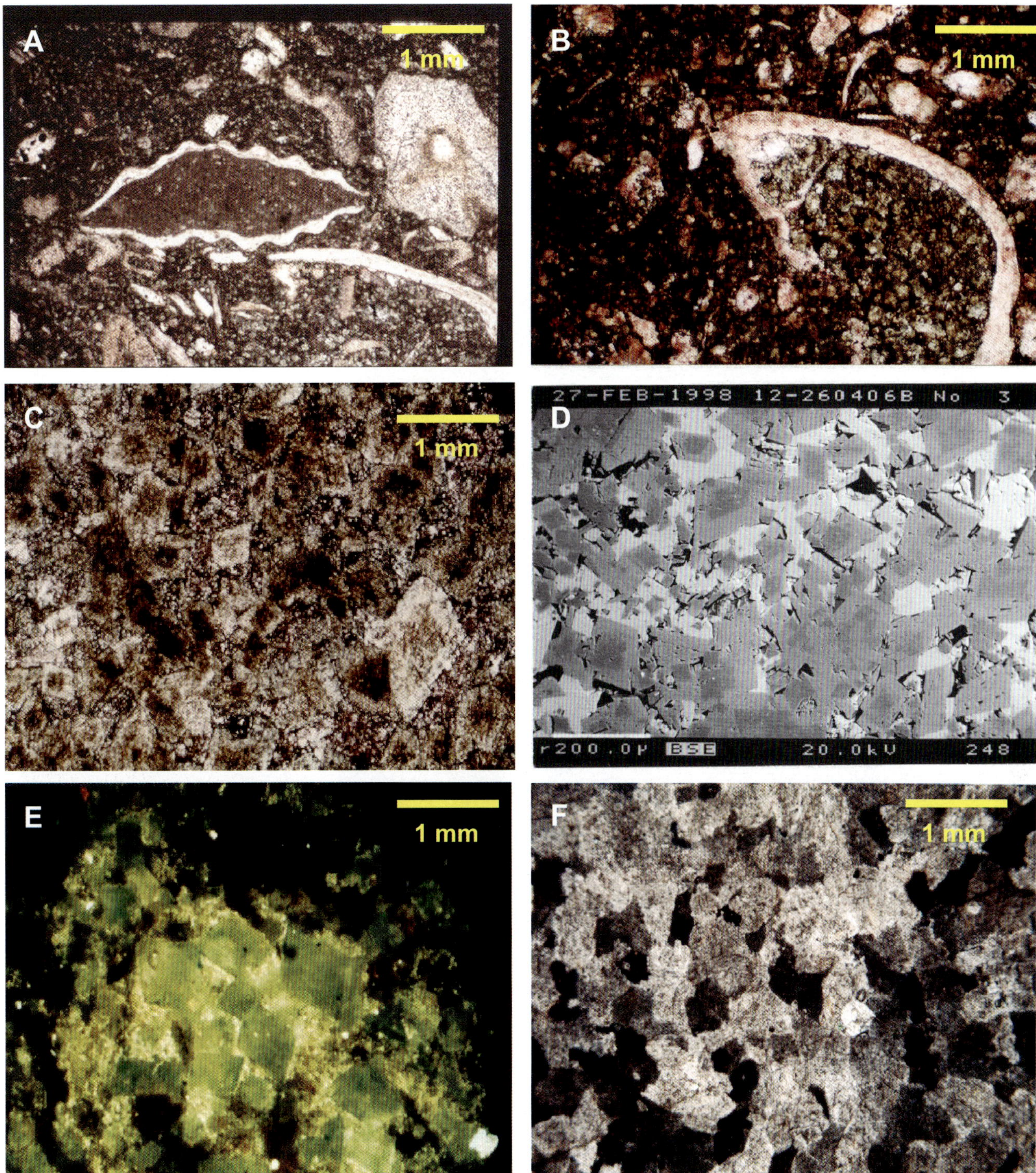

FIGURE 6. (A) Photomicrograph showing scattered crystals of matrix dolomite in a fossiliferous limestone. (B) Photomicrograph showing early microcrystalline dolomite replacing micritic sediment inside a fossil cavity. (C) Photomicrograph of pervasive dolomite replacing fossil grains and earlier cement generation. Note the cloudy cores and clear rims. (D) Backscatter electron detector image of pervasive dolomite. Note the zoned crystals and the porosity. (E) Pervasive dolomite under UV illumination. Crystals have dull green cores followed by bright green rims. Rims may represent overgrowths that developed during later diagenetic stages (see text for details). (F) Photomicrograph of pervasive, anhedral dolomite.

FIGURE 7. (A) Pseudomorphic dolomite with cloudy cores and clear rims replacing crinoids fragments. (B) Photomicrograph showing dissolution seam-related euhedral dolomite (far right corner; also seen is a fossil boring that is occluded by saddle dolomite in a still calcitic fossil). (C) Iron-rich saddle dolomite filling a fracture. Note also the blocky calcite (b).

Strontium Isotopes

Figure 9B illustrates $^{87}Sr/^{86}Sr$ ratios plotted vs. $\delta^{18}O$ for selected calcite and dolomite types from the Slave Point Formation. Most cement types have enriched ratios relative to Middle Devonian seawater. A single sample of fibrous calcite has a value of 0.70799, which lies within the expected range of marine calcite and therefore supports an interpretation of precipitation from marine fluids (consistent with fibrous geometry of crystals), whereas a sample of equant calcite spar is slightly more enriched at 0.70863. The dolomite samples analyzed have a wide range of values. Microcrystalline matrix dolomite has a value of 0.71002, medium-crystalline pervasive dolomite is 0.70943, and pseudomorphic dolomite has a value of 0.70939. Saddle dolomite has values of 0.70494 and 0.71035; this second value is the most enriched sample. The late-stage blocky calcite cement was also analyzed, and it has a value of 0.70963.

Major, Minor, and Trace Elements

All dolomite phases measured in this study are nonstoichiometric, having high calcium and low magnesium ratios; however, the cores of measured phases tend to be more stoichiometric than

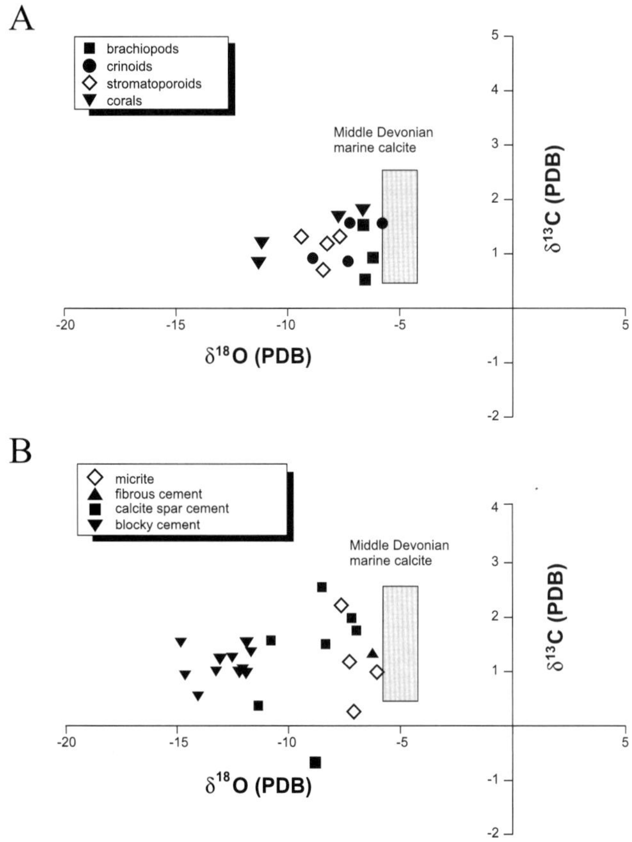

FIGURE 8. (A) Carbon and oxygen isotopic composition of calcitic skeletal components. The box represents Middle Devonian marine calcite (Hurley and Lohmann, 1989). (B) Carbon and oxygen isotopic composition of calcite cements.

the outer rims (Table 1). Microcrystalline matrix dolomite and medium-crystalline pervasive dolomite have iron-rich rims compared to their cores, which could indicate an evolving fluid chemistry during precipitation or later alteration. Pseudomorphic dolomite and saddle dolomite are very iron rich throughout. In the calcite phases, iron concentrations were significantly lower. Manganese concentrations varied from 10 to 2000 ppm, and strontium contents were very minor, ranging from 300 to 700 ppm.

Fluid Inclusions

Measurements of homogenization temperatures and final melting temperatures of ice were made on primary, two-phase, liquid-vapor inclusions from pseudomorphic dolomite, saddle dolomite, and late-stage blocky calcite. Two-phase fluid inclusions were also observed in medium-crystalline pervasive dolomite. However, they were too small to measure. Melting temperature values, T_m, were converted to weight percent NaCl equivalent using the expression of Bodnar (1993).

Pseudomorphic dolomite samples (n = 8) had homogenization temperature values (T_h) ranging from 76 to 92°C and salinities of 22.6–25.3 wt.% NaCl (Figure 10). Inclusions were small, isolated, and concentrated at the perimeter crystals of the clear rims. Saddle dolomite inclusions (n = 12) were very small and were arranged so that they defined growth banding in a single crystal. The inclusions varied in shape from circular and elliptical to irregular. Homogenization temperatures ranged from 125 to 161°C, and salinities ranged from 22.2 to 24.7 wt.% NaCl (Figure 10).

Primary fluid inclusions in the blocky calcite were relatively large compared to other diagenetic phases and varied in size to as much as 25 μm. T_h values ranged from 96 to 127°C, and salinities were 21.8 to 22.5 wt.% NaCl (Figure 10). This late-stage cement also contained numerous small, spherical hydrocarbon inclusions that were identified by strong fluorescence under UV illumination.

DISCUSSION OF DOLOMITE ORIGINS

Microcrystalline Matrix Dolomite

An early, shallow burial origin has been implied from petrographic evidence from the size of the crystals and the presence of these crystals replacing muddy substrate as well as selectively replacing fossils. Their occurrence is always predated solution seams as evidenced from deflection of solution seams and truncation of dolomite rhombs by stylolites limbs; however, original marine oxygen isotopic compositions have not been preserved in the matrix dolomite. The δ^{18}O values (11.62 to −9.34‰ VPDB; Figure 9A) are depleted relative to estimated Middle Devonian dolomite. The radiogenic $^{87}Sr/^{86}Sr$ isotope value (0.71002; Figure 9B) deviates from the estimated Middle Devonian seawater value. One likely explanation for this depletion of oxygen isotopes and enriched Sr isotopic ratios is the resetting of these isotopes during subsequent diagenesis (White and Al-Aasm, 1997; Al-Aasm, 2000). Microprobe analysis reveals that stoichiometry differs slightly between the rims and the cores of matrix dolomite crystals; in addition, the rims are much more iron rich than the cores. All of these analyses suggest that matrix dolomite was altered after precipitation. Geochemical modeling (see below) supports this premise.

Pseudomorphic Dolomite

Pseudomorphic dolomite mimically replaces crinoid fragments and is present only in pervasively dolomitized sections of the Slave Point Formation. Fluid inclusions (T_h = 76–92°C and salinities of

TABLE 1. Isotopic composition and major-, minor-, and trace-element concentrations of Slave Point carbonate components. Br = brachiopod; Cr = crinoid; STR = stromatoporoid; COR = coral; MCT = micrite; FIB = fibrous calcite; SP = calcite spar; LSB = blocky calcite; MD = fine-crystalline matrix dolomite; PD = medium-crystalline pervasive dolomite; PMD = pseudomorphic dolomite; and SD = saddle dolomite; VPDB = Vienna Peedee belemnite.

Sample	*Depth (m)*	*$\delta^{18}O$ (SMOW)*	*$\delta^{18}O$ (VPDB)*	*$\delta^{13}C$ (VPDB)*	*$^{87}Sr/^{86}Sr$*
Oxygen, carbon, and strontium isotope results for calcitic and dolomitic components					
12-26-96-12W6-BR	2551.70	24.14	−6.52	0.52	
13-35-96-12W6-BR	2546.23	24.50	−6.17	0.92	
15-12-96-12W6-BR	2534.00	23.85	−6.80	1.50	
03-06-98-10W6-CR	2555.32	23.29	−7.34	0.82	
10-19-96-12W6-CR	2553.68	23.47	−7.17	1.61	
10-19-96-12W6-CR	2553.68	24.84	−5.84	1.60	
12-26-96-12W6-CR	2552.30	21.87	−8.72	0.92	
03-33-96-11W6-STR	2544.52	21.17	−9.40	1.31	
06-07-96-11W6-STR	2522.96	22.43	−8.18	1.17	
13-35-96-12W6-STR	2549.20	22.14	−8.46	0.70	
16-25-97-11W6-STR	2537.40	22.98	−7.64	1.34	
03-06-98-10W6-COR	2543.35	23.84	−6.81	1.70	
03-33-96-11W6-COR	2516.20	19.36	−11.61	0.82	
03-33-96-11W6-COR	2516.20	19.44	−11.08	1.10	
10-19-96-12W6-COR	2541.13	22.65	−7.96	1.59	
03-33-96-11W6-MCT	2546.13	23.57	−7.07	0.27	
12-26-96-12W6-MCT	2580.75	23.03	−7.60	2.22	
13-35-96-12W6-MCT	2546.23	24.66	−6.01	0.98	
15-12-96-12W6-MCT	2541.10	23.40	−7.24	1.18	
03-06-96-10W6-FIB	2544.60	24.47	−6.20	1.34	0.707988
06-07-96-11W6-SP	2519.90	23.73	−6.92	1.76	
10-19-96-12W6-SP	2546.82	23.51	−7.13	2.00	0.708166
12-26-96-12W6-SP	2550.40	21.83	−8.76	−0.70	
13-35-96-12W6-SP	2546.64	19.21	−11.30	0.37	
13-35-96-12W6-SP	2550.64	19.79	−10.74	1.57	
13-35-96-12W6-SP	2565.30	22.14	−8.46	2.56	
15-12-96-12W6-SP	2535.24	22.28	−8.32	1.51	
06-07-96-11W6-LSB	2523.84	18.61	−11.88	1.00	
12-26-96-12W6-LSB	2523.00	16.40	−14.03	0.58	
13-35-96-12W6-LSB	2552.35	18.38	−12.11	1.02	
13-35-96-12W6-LSB	2565.30	17.25	−13.20	1.03	
13-35-96-12W6-LSB	2567.83	15.60	−14.80	1.53	
15-12-96-12W6 LSB	2531.26	15.82	−14.59	0.96	0.709630
15-12-96-12W6-LSB	2534.00	18.83	−11.67	1.36	
15-12-96-12W6-LSB	2535.76	17.99	−12.48	1.28	
16-25-97-11W6-LSB	2537.80	18.45	−12.04	1.08	
16-25-97-11W6-LSB	2539.64	18.65	−11.84	1.53	
16-25-97-11W6-LSB	2548.58	17.43	−13.03	1.25	

TABLE 1. Isotopic composition and major-, minor-, and trace-element concentrations of Slave Point carbonate components. Br = brachiopod; Cr = crinoid; STR = stromatoporoid; COR = coral; MCT = micrite; FIB = fibrous calcite; SP = calcite spar; LSB = blocky calcite; MD = fine-crystalline matrix dolomite; PD = medium-crystalline pervasive dolomite; PMD = pseudomorphic dolomite; and SD = saddle dolomite; VPDB = Vienna Peedee belemnite (cont.).

Sample	*Depth (m)*	*$\delta^{18}O$ (SMOW)*	*$\delta^{18}O$ (VPDB)*	*$\delta^{13}C$ (VPDB)*	*$^{87}Sr/^{86}Sr$*
03-06-98-10W6-MD	2555.32	18.88	−11.62	1.41	
03-06-98-10W6-MD	2524.56	19.76	−10.77	3.17	
03-06-98-10W6-MD	2538.25	20.41	−10.14	2.18	
10-19-96-12W6-MD	2557.62	20.30	−10.24	0.97	
10-19-96-12W6-MD	2548.23	21.14	−9.43	2.39	
12-26-96-12W6-MD	2550.40	19.97	−10.56	0.67	0.710020
13-35-96-12W6-MD	2546.23	21.23	−9.34	0.20	
13-35-96-12W6-MD	2562.80	20.46	−10.09	2.60	
15-12-96-12W6-MD	2550.65	18.90	−11.60	0.87	
03-06-98-10W6-PD	2552.18	20.24	−10.30	3.08	
03-06-98-10W6-PD	2526.60	20.14	−10.40	3.50	
10-19-96-12W6-PD	2536.00	18.76	−11.74	2.01	
10-19-96-12W6-PD	2541.13	20.32	−10.22	2.32	
12-26-96-12W6-PD	2558.00	19.49	−11.03	2.04	
13-35-96-12W6-PD	2555.54	19.70	−10.83	2.79	0.709426
13-35-96-12W6-PD	2558.61	21.07	−9.50	2.02	
12-26-96-12W6-PMD	2575.60	20.91	−9.65	4.24	
12-26-96-12W6-PMD	2576.70	19.95	−10.58	4.49	0.709386
03-06-98-10W6-SD	2546.65	16.50	−13.93	1.35	
03-06-98-10W6-SD	2526.60	16.65	−13.78	2.26	0.707100
06-07-96-11W6-SD	2521.65	17.76	−12.71	2.00	
06-07-96-11W6-SD	2439.60	18.52	−11.97	2.26	
06-07-96-11W6-SD	2541.60	17.86	−12.61	2.20	0.710351
06-07-96-11W6-SD	2541.60	17.48	−12.98	2.11	0.704940
10-10-96-12W6-SD	2561.63	17.90	−12.57	2.18	
10-10-96-12W6-SD	2569.33	17.86	−12.61	2.14	
12-08-96-11W6-SD	2424.70	18.30	−12.18	2.06	
13-35-96-12W6-SD	2563.17	17.76	−12.71	3.02	0.709140
15-12-96-12W6-SD	2543.07	17.60	−12.86	2.26	0.708626
15-12-96-12W6-SD	2543.07	17.29	−13.16	1.19	
16-25-97-11W6-SD	2533.60	17.73	−12.74	2.17	
16-25-97-11W6-SD	2549.96	16.48	−13.95	2.05	
	Summary of microprobe result for calcitic and dolomitic components				
	$CaCO_3$ (mol%)	*Samples*	*Sr (ppm)*	*Mn (ppm)*	*Fe (ppm)*
Fibrous	99.80	6	783	BDL	BDL
Spar	99.66	8	803	BDL	360
Blocky	99.46	15	684	283	1170
MD (cores)	55.93	2	300	200	5600
MD (rims)	56.43	2	500	BDL	20,500
PD (cores)	51.55	2	400	BDL	1250
PD (rims)	54.27	2	400	600	20,250
PMD (cores)	54.65	2	560	2000	23,000
PMD (rims)	55.90	2	700	1600	36,200
SD (core)	56.65	8	612	588	35,425
SD (rims)	56.65	3	433	567	33,300

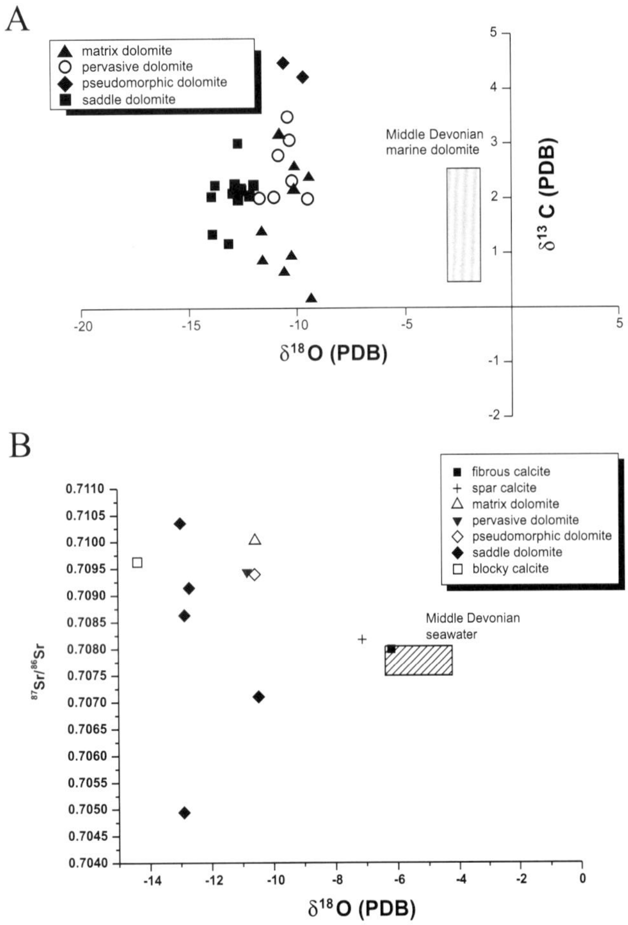

FIGURE 9. (A) Carbon and oxygen isotopic composition of dolomite phases. The box represents the postulated Middle Devonian marine dolomite (Hurley and Lohmann, 1989). (B) $^{87}Sr/^{86}Sr$ ratios and oxygen isotopic composition for calcite and dolomite phases. The box represents Middle Devonian seawater values from Denison et al. (1997).

22.6–25.3 wt.% NaCl equivalent; Figure 10) and depleted $\delta^{18}O$ values (−10.58 and −9.65‰ VPDB; Figure 9A) point to the replacement of the calcitic crinoids by dolomite formed from warm, saline fluids. In addition, high iron concentrations in both rims and cores indicate reducing conditions, which are all consistent with precipitation from basinal brines (e.g., Moore, 1989; Morse et al., 1997) during progressive burial. Precipitation of pseudomorphic dolomite by basinal fluids is also supported by the radiogenic strontium isotopes (0.70939).

Medium-crystalline Pervasive Dolomite

Any mechanism used to explain the origin of pervasive dolomite must be able to account for (1) its position in the paragenetic succession; (2) its presence in and around dissolution seams, replacing both matrix and allochems; (3) depleted $\delta^{18}O$ values (−11.74 to −9.5‰ VPDB; Figure 9A) and radiogenic $^{87}Sr/^{86}Sr$ isotope ratios (Figure 9B); and (4) the source of magnesium.

The formation of medium-crystalline pervasive dolomite from hypersaline, seawater, and meteoric-marine fluids at near-surface conditions can be excluded because of the presence of two-phase fluid inclusions in pervasive dolomite. All of the aforementioned models would precipitate minerals with single-phase inclusions because of low formation temperatures (Allan and Wiggins, 1993). Hypersaline environments are typically associated with evaporite deposits and tend to form thin beds, crusts, and nodules of fine-crystalline dolomite, which are not present in the Slave Point Formation. However, the possibility that evaporitic brines may have emanated from the underlying Muskeg succession during burial cannot be discounted. Seawater dolomitization is also rejected, because this model has been partially invoked to account for the presence of matrix dolomite that is texturally and temporally distinct from pervasive dolomite. Geochemically, depleted oxygen isotopes and trace-element concentrations argue against these types of dolomite formation (see below).

Medium-crystalline dolomite in the Slave Point Formation seems to have formed at shallow to medium burial stages. The occurrence of medium-crystalline, subhedral to euhedral dolomite crystals in pressure dissolution seams gives the best evidence for formation in a diagenetic setting, whereby the effect of compaction becomes important and also suggests that the seams acted as conduits for the dolomitizing fluid. Magnesium sources could include pore water, the incongruent dissolution of high magnesium calcite, and magnesium remobilized by pressure solution of older dolomites (Qing and Mountjoy, 1994); it would then be transported to the dolomitization site along the stylolites. Considerable volumes of fluid are squeezed out of sediments subjected to mechanical compaction. These fluids would be slightly modified seawater having Mg/Ca ratios close to those of seawater.

Geochemical data for pervasive dolomite includes $\delta^{18}O$ values that are more depleted than the expected range for dolomite precipitated from Middle Devonian seawater, data that are consistent with precipitation from fluids with elevated temperatures (or with equilibrium during burial). The majority of $\delta^{13}C$ values for pervasive dolomite falls within the expected range for Middle Devonian carbonates, suggesting that carbon was derived internally from the carbonates. $^{87}Sr/^{86}Sr$ isotope value for this dolomite is radiogenic (0.70943), suggesting precipitation from basinal fluids or reequilibration with later diagenetic fluids. The cores and rims of pervasive dolomite analyzed by microprobe show that the overgrowth rims are enriched in iron and less stoichiometric relative to the cores that may also indicate an evolving fluid chemistry.

Saddle Dolomite

Petrographic and geochemical characteristics of saddle dolomite from the Slave Point Formation include (1) the occurrence in subvertical fractures, (2) relatively late diagenetic timing in the paragenetic succession, (3) depleted $\delta^{18}O$ values (Figure 9A), (4) nonradiogenic to enriched $^{87}Sr/^{86}Sr$ ratios (Figure 9B), and (5) high homogenization temperatures and salinities as indicated by fluid inclusions (Figure 10). The saddle dolomite is turbid, coarsely crystalline with sweeping extinction, Ca^{2+} rich (56–57 mol% $CaCO_3$) with high Fe^{2+} concentrations (Table 1), and is interpreted to have precipitated during later diagenetic stages, because it postdates pervasive dolomite and precipitated as cement in vugs and late-stage fractures.

Several authors of diagenetic carbonate studies have interpreted that saddle dolomite in Devonian reservoirs of Western Canada has been precipitated from fluids of burial origin (Mattes and Mountjoy, 1980; Wong and Oldershaw, 1981; Qing and Mountjoy, 1989). The oxygen isotopic composition of Slave Point saddle dolomite ($\delta^{18}O$ = −13.95 to −11.97‰ VPDB) is consistent with precipitation from warm fluids, and the $\delta^{13}C$ composition (1.19–3.02‰ PDB) reflects an internal source for carbon. $^{87}Sr/^{86}Sr$ isotopic ratios (0.70494–0.71035) are significantly different from Middle Devonian marine values (0.7078–0.70803), indicating that marine carbonates and evaporites are not the source for these ratios. Brines may incorporate radiogenic ^{87}Sr during passage through shales or clastic sequences (Mountjoy et al., 1992); in addition, Precambrian rocks containing feldspars and micas underlying the Western Canada sedimentary basin may be a source of radiogenic ^{87}Sr. The iron, which is incorporated into the saddle dolomite, may have been derived from underlying and adjacent shales (Wong and Oldershaw, 1981). However, the nonradiogeic value may reflect a juvenile source for Sr. The data can indicate a mixing trend between mantlelike source for fluids and radiogenic fluids that originated from basinal brines. Fluid-inclusion data (T_h = 125–161°C and $T_{m\ ice}$ = −19.8 to −23.6°C) indicate that warm, saline brines (22–25 wt.% NaCl equivalent) were responsible for precipitating saddle dolomite in the Slave Point Formation. Using these homogenization temperatures, under burial conditions, saddle dolomite would have formed at depths greater than 3000 m (10,000 ft) (Figure 3B, assuming a geothermal gradient of 30°C/km and surface temperature of 20°C) or by hydrothermal fluids that were derived from deeper within the basin or migrated some distance.

In the Hamburg field, faults associated with tectonic events may have funneled fluids into the carbonate sequence, causing dolomitization. Therefore, the distribution of dolomite may be constrained by fault and fracture systems available at the time of precipitation. The Hay River–Great Slave Lake fault zone lies 30 km (20 mi) northwest of the Hamburg field and may have served as the principal conduit for hydrothermal fluid flow. Several authors have suggested that the upward and lateral flow of hydrothermal fluids may have been responsible for the dolomitization of many different regions: (1) Packard et al. (1990) for the Wabamun formation, southeast of the Peace River Arch; (2) Morrow and Aulstead (1995) for the Devonian Manetoe dolomite in the Northwest Territories; (3) Kaufman et al. (1990) for the Middle Devonian Swan Hills Formation; (4) Lonnee and Al-Aasm (2000) for the Middle Devonian Sulphur Point Formation in northwest Alberta; and (5) Aulstead and Spencer (1985) for the Middle Devonian Zama-Rainbow region in northwestern Alberta.

In the Western Canada sedimentary basin, large-scale fluid movement caused by tectonic compression and sedimentary loading (e.g., Root, 2001; Nelson et al., 2002) may have occurred twice since the deposition of the Slave Point Formation: (1) between the Late Devonian and early Mississippian (Antler orogeny), and (2) between the Late Jurassic and early Tertiary (Columbian and Laramide orogenies). Either event may be sufficient to cause the amount of fluid flow required to precipitate saddle dolomite.

Saddle dolomite precipitated before the development of high-amplitude stylolites, implying that formation preceded the final phase of stylolitization. This would constrain the timing of dolomite precipitation that have occurred before burial associated with the Columbian and Laramide orogenies (4–6 km; 2.5–4 mi; Hitchon, 1984). This is also suggested by the presence of saddle dolomite along the subvertical generation of fractures, which may have developed coincident with the subsidence of the Peace River Arch during Late Devonian to early Mississippian. Vertical fractures may form during extensional tectonic events such as the Antler. The Laramide orogeny was compressional; thus, these fractures probably did not form during this event.

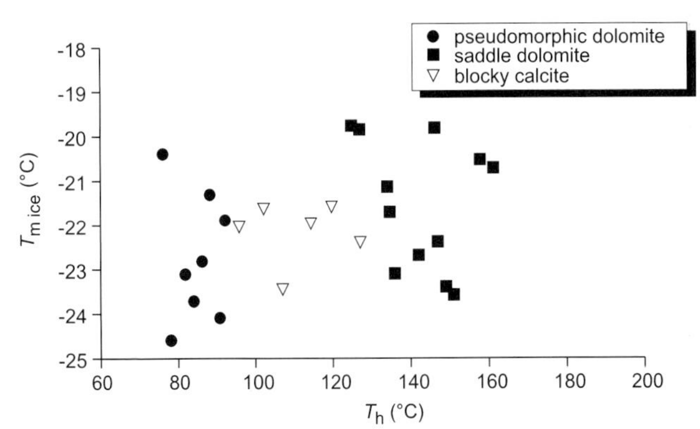

FIGURE 10. Plot of homogenization temperature vs. $T_{m\ ice}$/wt.% NaCl equivalent for fluid inclusions in dolomite and calcite phases.

Root (2001) suggested that Antler fold and foreland basin development might have occurred along the length of the Canadian part of the ancient Cordilleran continental margin during that time. He also suggested that hot fluids that were expelled in response to the Antler thrust belt development might have significantly influenced diagenetic processes in Devonian carbonate reservoirs. Nelson et al. (2002) proposed that carbonate-hosted Zn-Pb deposits and hydrothermal dolomite in the Canadian Cordillera and the Western Canada sedimentary basin were far-field effects of subduction, and the fluids were driven along reactivated back-arc structures and permeable strata.

Geochemical evidence that may help constrain the timing for saddle dolomite precipitation includes both nonradiogenic and enriched $^{87}Sr/^{86}Sr$ ratios and high salinities in the precipitating fluids as indicated by fluid inclusions. As discussed previously, the $^{87}Sr/^{86}Sr$ ratios for saddle dolomite are indicative of an allochthonous source. Values of $T_{m\ ice}$ from fluid-inclusion analyses indicate that the fluids that precipitated saddle dolomite were extremely saline (22–25 wt.% NaCl equivalent). Nesbitt and Meuhlenbachs (1994) suggested that salinities in this range are associated with pre-Laramide fluid flow (i.e., Devonian–Mississippian), whereas fluids associated with the Laramide orogeny tend to be mixed brines and meteoric waters, resulting in a salinity range of 0–10 wt.% NaCl equivalent. The homogenization temperatures and salinity values for the Slave Point saddle dolomite overlap with those reported in the literature for Devonian saddle dolomite from the Wabamun formation southeast of the Peace River Arch area (Packard et al., 1990, 2001) and Manetoe dolomite facies in the Northwest Territories, north of the Presqu'ille barrier (Morrow et al., 1986). Based on the above petrographic and geochemical evidence, it is proposed herein that saddle dolomite precipitation occurred as a result of hydrothermal fluid flow along faults and fractures associated with deformation during the Late Devonian and Mississippian.

Recrystallization of Matrix and Pervasive Dolomite

Recrystallization of the matrix and pervasive dolomite in the studied rocks has been suggested because of petrographic and/or geochemical data that are inconsistent with the proposed genesis of dolomites. Alteration of early dolomite phases is a common process during progressive burial of platform carbonates (Land, 1985; Hardie, 1987; Mazzullo, 1992; Durocher and Al-Aasm, 1997; Al-Aasm, 2000). Chemical evidence that supports recrystallization comes from (1) depleted $\delta^{18}O$ values and radiogenic $^{87}Sr/^{86}Sr$ isotopic compositions inconsistent with precipitation under shallow and intermediate burial conditions; (2) irregular contacts between the dull cores and bright red rims, which may correspond to the distinctly different iron concentrations between the cores and clear rims; and (3) differences in stoichiometry between the cores and rims.

It was stated previously that matrix dolomite has not retained the postulated original isotopic composition that accompanies precipitation in a shallow burial environment. The $\delta^{18}O$ values are depleted relative to the expected values, suggesting that matrix dolomite was altered after precipitation. Matrix dolomite displays small, clear overgrowth rims and recrystallization fronts that are extremely enriched with iron relative to the cores, suggesting that the overgrowths and recrystallization fronts were precipitated by a fluid chemically different from the one that precipitated the cores. Using the range of homogenization temperatures and the range in $\delta^{18}O$ values for all dolomite types, the $\delta^{18}O$ composition of the ambient fluid was calculated (Figure 11). From Figure 11, it can be seen that the fluid composition evolved with time and with the introduction of hot, basinal brines. To confirm the proposed alteration of matrix dolomite in particular, geochemical modeling is discussed below.

Similar trends in cathodoluminescence colors, oxygen and strontium isotopic compositions, trace-element chemistry, and stoichiometry were observed for matrix dolomite and pervasive dolomite, which suggests that pervasive dolomite may have also been altered by later diagenetic fluids.

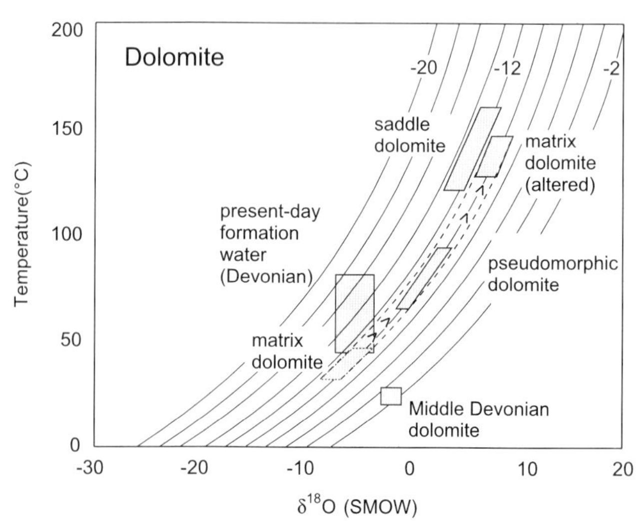

Figure 11. Oxygen isotope composition of fluids vs. crystallization temperature for dolomite phases. Isochore lines show the oxygen isotope composition of the dolomite phases with respect to VPDB. Formation water values are from Connolly et al. (1990). SMOW = standard mean ocean water.

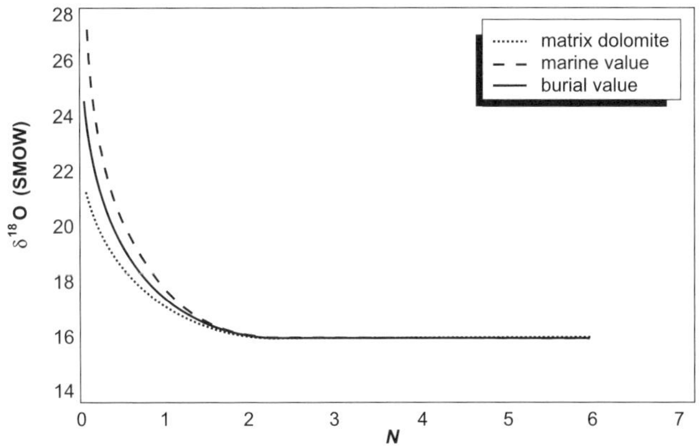

FIGURE 12. Model for recrystallization of matrix dolomite and pervasive dolomite by hydrothermal fluids using the model by Banner and Hanson (1990). *N* is the cumulative water/rock ratio. See text for details. SMOW = standard mean ocean water.

Geochemical Modeling

Geochemical modeling was applied to this study to (1) determine the fluid composition that altered the matrix and pervasive dolomite and (2) compare the results of this modeling with actual geochemical data. The geochemical model used in this study was developed by Banner and Hanson (1990). This model uses mass-balance equations to monitor simultaneous variations in isotopic compositions of the fluid and solid, as a function of water/rock ratios, during carbonate mineral alteration. The mass-balance equations allow comparisons to be made between the proposed diagenetic model and the actual geochemical data.

Banner and Hanson's model has two inherent assumptions that are not necessarily true for diagenetic systems. The first assumption is that the fluid and solid reach isotopic equilibrium; however, Banner and Hanson (1990) indicate that it is sufficient to recognize that the system advances toward equilibrium. The second assumption is that the reactions are considered to be between the fluid and a single mineral phase, when in reality, reactions commonly involve several mineral phases simultaneously.

It was proposed earlier that the oxygen isotopic compositions of matrix and pervasive dolomite were altered by the interaction with later diagenetic fluids. It is proposed herein that the hydrothermal fluids that precipitated saddle dolomite are the altering fluids (based on similar isotopic compositions and elemental chemistry); thus, the following parameters were used for model calculations. (1) The oxygen isotopic composition of matrix dolomite was chosen to be −9.34‰ VPDB (or 21.28‰ standard mean ocean water [SMOW]). This value represents the least altered value of matrix (and pervasive) dolomite. The validity of this model parameter is questionable; therefore, the model was also run using an initial value of −3‰ PDB (or 27‰ SMOW), which represented pristine Middle Devonian marine carbonate values, and a value of −6‰ PDB (or 24.72‰ SMOW), which represented dolomite precipitation under burial conditions. (2) The original oxygen isotopic composition of the diagenetic fluid was −0.35‰ SMOW based on calculations using the equation of Land (1983). (3) The temperature used in the calculations was 142°C, which represented the average homogenization temperature measured from fluid inclusions in saddle dolomite. (4) Porosity was assumed to be 15%, because matrix and pervasive dolomite occur in mud-supported facies that may have had initial porosities to as much as 70%. Final porosities in these facies are less than 10%, so a value of 15% is reasonable. (5) Open-system behavior was assumed based on the presence of fractures and stylolites, which would allow constant flow.

Results from calculations using these parameters are shown in Figure 12. Under these conditions, using the model of Banner and Hanson (1990), matrix and pervasive dolomite would reach complete equilibrium at δ^{18}O values of 16.04‰ SMOW, with a corresponding water/rock ratio of 2. The model nearly predicts the δ^{18}O compositions that are observed from the Slave Point Formation; thus, modeling confirms that the hydrothermal fluids that precipitated saddle dolomite may have altered the matrix and pervasive dolomite. The radiogenic Sr isotopic values (0.71002 and 0.70943) obtained for matrix and pervasive dolomite, which lie between the measured values for saddle dolomite, may confirm this conclusion.

CONCLUSIONS

Petrographic observations, isotopic, and elemental data on dolomite from the Slave Point Formation, Hamburg field, northwestern Alberta, suggest that the four types of dolomite formed from different fluid regimes with distinctive chemical and isotopic composition and were affected by tectonic activity prior to the Laramide deformation event in the Late Cretaceous–early Tertiary. These data also indicate that later diagenetic, warm, diagenetic fluids recrystallized early microcrystalline matrix and medium-crystalline pervasive dolomite. These fluids were also responsible for the formation of saddle dolomite that was the last dolomite phase to form. This fluid flow was cross-formational and focused along faults and fractures. Late blocky calcite cements were precipitated following saddle dolomite and occurred concurrently with hydrocarbon migration.

ACKNOWLEDGMENTS

This research was supported by a Natural Sciences and Engineering Research Council (NSERC) of Canada grant and additional grants from EnCana awarded to I. S. Al-Aasm. The authors thank I. M. Samson for assistance with fluid-inclusion measurements. Earlier versions of this chapter benefited from constructive reviews by J. Lonnee, S. Mazzullo, and J. Gregg. Reviews by R. Swennen and an anonymous reviewer of a later version of this chapter were greatly appreciated.

REFERENCES CITED

Al-Aasm, I. S., 2000, Chemical and isotopic constraints for recrystallization of sedimentary dolomites from the Western Canada sedimentary basin: Aquatic Geochemistry, v. 6, p. 227–248.

Al-Aasm, I. S., B. E. Taylor, and B. South, 1990, Stable isotopes analysis of multiple carbonate samples using selective acid extraction: Chemical Geology (Isotope Geoscience Section), v. 80, p. 119–125.

Al-Aasm, I. S., J. Lonnee, and J. Clarke, 2002, Multiple fluid flow events and the formation of saddle dolomite: Case studies from the Middle Devonian of the Western Canada sedimentary basin: Marine and Petroleum Geology, v. 19, p. 209–218.

Allan, J. R., and W. D. Wiggins, 1993, Dolomite reservoirs: Geochemical techniques for evaluating origin and distribution: AAPG Continuing Education Course Note Series 36, 170 p.

Aulstead, K. L., and R. J. Spencer, 1985, Diagenesis of the Keg River Formation, northwestern Alberta: Fluid inclusion evidence: Bulletin of Canadian Petroleum Geology, v. 33, p. 167–183.

Bachu, S., 1999, Flow systems in the Alberta Basin: Patterns, types and driving mechanisms: Bulletin of Canadian Petroleum Geology, v. 47, p. 455–474.

Banner, J. L., and G. N. Hanson, 1990, Calculation of simultaneous isotopic and trace element variations during water-rock interaction with applications to carbonate diagenesis: Geochimica et Cosmochimica Acta, v. 54, p. 3123–3137.

Bethke, C. M., and S. Marshak, 1990, Brine migrations across North America— The plate tectonics of groundwaters: Annual Review of Earth and Planetary Sciences, v. 18, p. 287–315.

Bodnar, R. J., 1993, Revised equation and table for determining the freezing point depression of H_2O-NaCl solutions: Geochimica et Cosmochimica Acta, v. 57, p. 683–684.

Cant, D. J., 1988, Regional structure and development of the Peace River Arch, Alberta: A Paleozoic failed-rift system?: Bulletin of Canadian Petroleum Geology, v. 36, p. 284–295.

Connolly, C. A., L. M. Walter, H. Baadsgaard, and F. J. Longstaffe, 1990, Origin and evolution of formation waters, Alberta Basin, Western Canada sedimentary basin: I. Chemistry: Applied Geochemistry, v. 5, p. 375–395.

Craig, J. H., 1987, Depositional environments of the Slave Point Formation Beaverhill Lake Group, Peace River Arch, *in* F. F. Krause and O. G. Burrowes, eds., Devonian lithofacies and reservoir styles in Alberta: Canadian Society of Petroleum Geologists International Symposium on the Devonian System, p. 181–199.

Davies, G. R., 1997, Hydrothermal dolomite reservoir facies, *in* Graham Davies Geologic Consultants, compilers, Canadian Society of Petroleum Geologists–Society of Sedimentary Petrologists Joint Convention Short Course Notes: Calgary, Alberta, Figure 54, 167 p.

Denison, R. E., R. B. Koepnick, W. H. Burke, E. A. Hetherington, and A. Fletcher, 1997, Construction of the Silurian and Devonian seawater $^{87}Sr/^{86}Sr$ curve: Chemical Geology, v. 140, p. 109–121.

Durocher, S., and I. S. Al-Aasm, 1997, Dolomitization and neomorphism of Mississippian (Visean) upper Debolt Formation, Blueberry field, northeastern British Columbia: Geologic, petrologic and chemical evidence: AAPG Bulletin, v. 81, p. 954–977.

Garven, G., 1985, The role of regional fluid flow in the genesis of the Pine Point deposits, Western Canada sedimentary basin: Economic Geology, v. 80, p. 307–324.

Hardie, L. A., 1987, Perspectives—Dolomitization: A critical view of some current views: Journal of Sedimentary Petrology, v. 57, p. 166–183.

Hitchon, B., 1984, Geothermal gradients, hydrodynamics, and hydrocarbon occurrences, Alberta, Canada: AAPG Bulletin, v. 68, p. 713–743.

Hurley, N. F., and K. C. Lohmann, 1989, Diagenesis of Devonian reefal carbonates in the Oscar Range, Canning Basin, Western Australia: Journal of Sedimentary Petrology, v. 59, p. 127–145.

Issler, D. R., C. Beaumont, S. D. Willett, R. A. Donelick, J. Mooers, and A. Grist, 1990, Preliminary evidence from apatite fission track data concerning the thermal history of the Peace River Arch region, Western Canada sedimentary basin: Bulletin of Canadian Petroleum Geology, v. 38A, p. 250–269.

Kaufman, J., W. J. Meyers, and G. N. Hanson, 1990, Dolomitization of the Swan Hills Formation (Devonian), Rosevear field, Alberta, Canada: Journal of Sedimentary Petrology, v. 60, p. 918–939.

Land, L. S., 1983, The application of stable isotopes to studies of the origin of dolomite and to problems of diagenesis of clastic sediments, *in* Stable isotopes in sedimentary geology: SEPM Short Course 10, p. 4-1–4-22.

Land, L. S., 1985, The origin of massive dolomite: Journal of Geologic Education, v. 33, p. 112–125.

Lewchuck, M. T., I. S. Al-Aasm, D. T. A. Symons, and K. P. Gillen, 1998, Dolomitization of Mississippian carbonates in the Shell Waterton gas field, southwestern Alberta: Insights from paleomagnetism, petrography and geochemistry: Bulletin of Canadian Society of Petroleum Geology, v. 46, p. 387–410.

Lonnee, J., and I. S. Al-Aasm, 2000, Dolomitization and fluid evolution in the Middle Devonian Sulphur Point Formation, Rainbow South field, Alberta: Petrographic and geochemical evidence: Bulletin of Canadian Petroleum Geology, v. 48, p. 262–283.

Machel, H.-G., and P. A. Cavell, 1999, Low-flux, tectonically-induced squeegee fluid flow ("hot flash") into the Rocky Mountain foreland basin: Bulletin of Canadian Petroleum Geology, v. 47, p. 510–533.

Mattes, B. W., and E. W. Mountjoy, 1980, Burial dolomitization of the Upper Devonian Miette buildup, Jasper National Park, Alberta, *in* D. H. Zenger, J. B. Dunham, and R. L. Etherington, eds., Concepts and models of dolomitization: SEPM Special Publication 28, p. 259–297.

Mazzullo, S. J., 1992, Geochemical and neomorphic alteration of dolomite: A review: Carbonates and Evaporites, v. 7, p. 21–37.

Meijer Dress, N. C., 1994, Devonian Elk Point Group of the Western Canada sedimentary basin, *in* G. D. Mossop and I. Shetsen, compilers, Geological atlas of the Western Canada sedimentary basin: Canadian Society of Petroleum Geology and Alberta Research Council, p. 128–147.

Moore, P. F., 1988, Devonian geohistory of the western interior of Canada, *in* N. J. McMillan, A. F. Embry, and D. J. Glass, eds., Devonian of the world: Canadian Society of Petroleum Geologists Second International Symposium on the Devonian System, v. 1, p. 67–83.

Moore, C. H., 1989, Carbonate diagenesis and porosity: Amsterdam, Elsevier, 338 p.

Morrow, D., 1998, Regional subsurface dolomitization: Models and constraints: Geoscience Canada, v. 25, p. 57–70.

Morrow, D. W., and K. L. Aulstead, 1995, The Manetoe dolomite— a Cretaceous–Tertiary or a Paleozoic event? Fluid inclusion and isotopic evidence: Bulletin of Canadian Petroleum Geology, v. 43, p. 267–280.

Morrow, D. W., G. L. Cumming, and R. B. Koepnick, 1986, Manetoe facies— A gas-bearing megacrystalline, Devonian dolomite, Yukon and Northwest Territories, Canada: AAPG Bulletin, v. 70, p. 702–720.

Morse, J. W., J. S. Hanor, and S. He, 1997, The role of mixing and migration of basinal waters in carbonate mineral mass transport, *in* I. Montanez, I. J. M. Gregg, and K. L. Shelton, eds., Basin-wide diagenetic patterns: SEPM Special Publication 57, p. 41–50.

Mountjoy, E. W., and J. E. Amthor, 1994, Has burial dolomitization come of age? Some answers from the Western Canada sedimentary basin, *in* B. Purser, M. Tucker, and D. Zenger, eds., Dolomites: A volume in honour of Dolomieu: International Association of Sedimentologists Special Publication 21, p. 203–229.

Mountjoy, E. W., H. Qing, and R. H. McNutt, 1992, Strontium isotopic composition of Devonian dolomites, Western Canada sedimentary basin: Significance of sources of dolomitizing fluids: Applied Geochemistry, v. 7, p. 59–75.

Muchez, P., M. Sintubin, and R. Swennen, 2000, Origin and migration pattern of paleofluids during orogeny: Discussion on the Variscides of Belgium and northern France: Journal of Geochemical Exploration, v. 69–70, p. 47–51.

Nelson, J., S. Paradis, J. Christensen, and J. Gabites, 2002, Canadian Cordilleran Mississippi Valley-type deposits: A case for Devonian–Mississippian back-arc hydrothermal origin: Economic Geology, v. 97, p. 1013–1036.

Nesbitt, B. E., and K. Meuhlenbachs, 1994, Paleohydrogeology of the Canadian Rockies and origins of brines, Pb-Zn deposits and dolomitization in the Western Canada sedimentary basin: Geology, v. 22, p. 243–246.

O'Connell, S. C., G. R. Dix, and J. E. Barclay, 1990, The origin, history and regional structural development of the Peace River Arch, Western Canada, *in* S. C. O'Connell and J. S. Bell, eds., Geology of the Peace River Arch: Bulletin of Canadian Petroleum Geology, v. 38A, p. 4–24.

Oliver, J., 1986, Fluids expelled tectonically from orogenic belts, their role in hydrocarbon migration and other geologic phenomena: Geology, v. 14, p. 99–112.

Opalinski, P., 1984, Origin and significance of dolomite reservoirs in the Middle Devonian, Slave Point Formation, northeastern British Columbia, *in* The hydrocarbon occurrence in carbonate rocks: Jakarta, Indonesia, Ascope Secretariat, p. 179–211.

Packard, J. J., G. J. Pellegrin, I. S. Al-Aasm, I. M. Samson, and J. Gagnon, 1990, Diagenesis and dolomitization associated with hydrothermal karst in Famennian upper Wabamun ramp sediments, northwestern Alberta, *in* G. R. Bloy and M. G. Hadley, compilers, The development of porosity in carbonate reservoirs: Canadian Society of Petroleum Geologists Short Course Notes, p. 9.1–9.19.

Packard, J., I. S. Al-Aasm, I. Samson, and Z. Berger, 2001, A Devonian "hydrothermal" chert reservoir: The 225 bcf Parkland field, British Columbia, Canada: AAPG Bulletin, v. 85, p. 51–84.

Porter, J. W., R. A. Price, and R. G. McCrossan, 1982, The Western Canada sedimentary basin: Philosophical Transactions of the Royal Society of London, v. 305, p. 169–192.

Qing, H., and E. W. Mountjoy, 1989, Multistage dolomitization in Rainbow buildups, Middle Devonian Keg River Formation, Alberta, Canada: Journal of Sedimentary Petrology, v. 59, p. 114–126.

Qing, H., and E. W. Mountjoy, 1994, Formation of coarsely crystalline, hydrothermal dolomite reservoirs in the Presqu'ille barrier, Western Canada sedimentary basin: AAPG Bulletin, v. 78, p. 55–77.

Root, K. G., 2001, Devonian Antler fold and thrust belt and foreland basin development in the southern Canadian Cordillera: Implications for the Western Canada sedimentary basin: Bulletin of Canadian Petroleum Geology, v. 49, p. 7–36.

Sibley, D. F., and J. M. Gregg, 1987, Classification of dolomite rock textures: Journal of Sedimentary Petrology, v. 57, p. 967–975.

Symons, D. T. A., R. J. Enkin, and M. T. Cioppa, 1999, Paleomagnetism in the Western Canada sedimentary basin: Dating fluid flow and deformation events:

Bulletin of Canadian Petroleum Geology, v. 47, p. 534–547.

Tooth, J. W., and G. R. Davies, 1989, Gift Lake Slave Point Reef, Middle Devonian, Alberta, *in* H. H. J. Geldsetzer, N. P. James, and G. E. Tebbutt, eds., Reefs, Canada and adjacent areas: Canadian Society of Petroleum Geologists Memoir 13, p. 528–534.

Wendte, J., H. Qing, J. J. Dravis, S. L. O. Moore, L. D. Stasiuk, and G. Ward, 1998, High-temperature saline (thermoflux) dolomitization of Devonian Swan Hills platform and bank carbonates, Wild River area, west-central Alberta: Bulletin of Canadian Petroleum Geology, v. 46, p. 210–265.

White, T., and I. S. Al-Aasm, 1997, Hydrothermal dolomitization of the Mississippian Upper Debolt Formation, Sikanni gas field, northeastern British Columbia, Canada: Bulletin of Canadian Petroleum Geology, v. 45, p. 297–316.

Wong, P. K., and A. Oldershaw, 1981, Burial cementation in the Devonian Kaybob reef complex, Alberta, Canada: Journal of Sedimentary Petrology, v. 51, p. 507–520.

Robion, P., J. L. Faure, and R. Swennen, 2004, Late Cretaceous chemical remagnetization of the Paleozoic carbonates from the undeformed foreland of the western Canadian Cordillera, *in* R. Swennen, F. Roure, and J. W. Granath, eds., Deformation, fluid flow, and reservoir appraisal in foreland fold and thrust belts: AAPG Hedberg Series, no. 1, p. 317–330.

Late Cretaceous Chemical Remagnetization of the Paleozoic Carbonates from the Undeformed Foreland of the Western Canadian Cordillera

Philippe Robion
Département des Sciences de la Terre, Université de Cergy-Pontois, France

Jean Luc Faure
Division Géologie-Géochimie, Institut Français du Pétrole, Rueil Malmaison, France

Rudy Swennen
Laboratorium voor Sediment petrologie en milieu geochemie, Afdeling Fysico-chemische Geologie, Katholieke Universiteit Leuven, Heverlee, Belgium

ABSTRACT

Viscous remanent magnetization (VRM) recently acquired in the Earth's magnetic field provides a compass to recover in-situ orientation of unoriented core material. This method was used to date a late chemical remanent magnetization (CRM) in Paleozoic carbonate rocks (Devonian to Mississippian) from the foreland of the western Canadian Cordillera. The paleomagnetic data showed three distinct components: (1) a low-temperature component, which is commonly removed at temperatures below 180°C and is assumed to be induced by drilling; (2) a medium-temperature component, which is commonly removed below 250–400°C, with a steep normal polarity direction; and (3) a high-temperature component, which is isolated above 250–400°C. In the foothills, high-temperature magnetizations always have steep reverse polarities, whereas in the Interior Plains, both normal and reverse polarities with more scattered inclinations are observed. The agreement between theoretical and laboratory blocking temperatures supports the interpretation that the medium-temperature component has recorded Earth's magnetic field over the normal polarity Brunhes epoch as a VRM. The same approach for

DOI:10.1306/1025697H13123

the high-temperature component led to the interpretation that the high unblocking temperatures indicate a CRM event that affected the Paleozoic carbonates long after deposition. This component is equivalent to the A component observed by Enkin et al. (2000) in exposed strata. By correcting declination values using the medium-temperature component, the direction of the high-temperature component in the Interior Plains is $D = 330.4^\circ$, $I = 74.3$, $k = 84$, $\alpha_{95} = 3.4^\circ$, $N = 22$ specimens, and in the foothills, it is $D = 331.3^\circ$, $I = 82.4^\circ$, $k = 30$, $\alpha_{95} = 8.1^\circ$ $N = 12$ specimens, corresponding to pole position consistent with the Late Cretaceous section of the North American apparent polar wander path.

INTRODUCTION

Paleomagnetism provides a useful tool for dating diagenetic processes and early deformation stages in the foreland fold and thrust belt (McCabe and Elmore, 1989). The model shows that during fluid migration, remanence-carrying minerals, which record the ambient Earth's magnetic field, are formed by precipitation or recrystallization. In a given area, each mineralization event, which ideally can be linked to the geochemical evolution of the host rocks, may be magnetically recorded. Two mechanisms have been proposed to explain the acquisition of such chemical remanent magnetization (CRM) in foreland basins. The first one (Oliver, 1986, 1992; Garven, 1995) relates to a large-scale orogenic fluid flow with patterns reflecting expulsion from the inner part of the belt migrating toward the foreland. A second process relates to the more local circulation of diagenetic fluids expelled during burial (Katz et al., 2000). The age of the induced remagnetization event in relation to these fluid-flow systems can be inferred from the position of the paleomagnetic pole along the craton's apparent polar wander path (APWP).

This approach has been successfully used in oil and gas exploration in the foreland belt of the western Canadian Cordillera (WCC) (Enkin et al., 1997, 2000; Gillen et al., 1999; Symons et al., 1999; Cioppa et al., 2000). These authors were able to constrain the timing of specific diagenetic events. They showed that the remanent magnetization components in Paleozoic rocks are Cretaceous in age. Furthermore, there is no evidence for primary depositional magnetization with the shallow inclinations characteristic of Paleozoic age. This was interpreted in terms of large-scale "magnetic alteration" of rocks coeval to the Laramide orogeny. An asynchronous magnetization signal, related to the west-to-east thrust propagation, was interpreted by Enkin et al. (2000). Normal and reverse polarities appeared in the Front Ranges, whereas in the foothills, only reverse polarities were observed. Furthermore, all these magnetizations were acquired before or during the early stage of the folding. In a recent development, Cioppa and Al-Aasm (2004) measured different steep-inclination CRM in the same formation from foothills, providing evidence that the remagnetization of carbonates records more than one event. Following these previous studies on remagnetizations in the WCC, the present study sets out to investigate how far ahead of the Laramide deformation front the remanence of Paleozoic carbonates has been affected. Paleozoic carbonate rocks were sampled from the undeformed foreland along a transect passing through the outer foothills and Interior Plains. Because outcrops in this area consist of Mesozoic strata, nonoriented boreholes have been sampled. Four exposed field sites were also sampled in the Front Ranges and the inner foothills, adjacent to the end location of the southwest–northeast transect of Enkin et al. (2000) to link our results with reported data.

To deduce the initial drill core orientation, we used the low-temperature viscous remanent magnetization (VRM) component, which records the present-day geomagnetic field. The magnetic declination of this VRM was used as a compass to extract the orientation of the high-temperature magnetic component, which provides an age of remagnetization based on published APWP curves.

METHODS

All samples were thermally cleaned and measured using tools provided by the paleomagnetic laboratory of the Laboratoire des Sciences du Climat et de l'Environement (Gif Sur Yvette). The natural remanence was measured in a 2G Enterprise three-axis cryogenic magnetometer in a magnetically shielded room. As many as 20 measurements were carried out on each sample from room temperature to 550°C by steps of 25 or 50°C. Samples were heated during 40 min and then cooled in the magnetically shielded room of the oven. Bulk susceptibilities were measured after each step to control possible mineralogical changes. The evolution of directions of magnetization was determined using the program Paleomac® developed by J. P. Cogné (2003) and based on Kirschvink's (1980) algorithm. The identification of ferromagnetic minerals was inferred from thermal demagnetization of saturation isothermal remanent magnetization in three perpendicular directions (Lowrie, 1990). A pulse magnetizer (ASC Scientific, IM10) allowed the acquisition of Isothermal Remanent Magnetization (IRM), delivering magnetic fields as much as 1.2 T. We applied successively in three orthogonal directions fields of 1.2, 0.5, and 0.1 T. Subsequently, the samples were thermally demagnetized in a Shaw MMTD18 furnace in temperature steps ranging between 25 and 50°C from 75 to 680°C, and magnetization was measured with a spinner magnetometer (Agico JR5).

SAMPLING AND MAGNETIC MINERALOGY

Samples were collected along a 200-km (120-mi) transect extending through the Front Ranges, foothills, and Interior Plains (Figure 1). This common division in three regions is mainly based on geomorphological and geological arguments. The Front Ranges to the west are bordered by the frontal McConnell thrust, where Devonian and older rocks are exposed. To the east lie the foothills, which can be divided into the inner foothills with Carboniferous outcrops and outer foothills, where younger strata outcrop. East of the foothills lie the Interior Plains, which are bound by the hinge of the Alberta syncline. Surface rocks of the eastern limb of this structure are made up of apparently undeformed Lower Cretaceous strata, overlying conformably older rocks. Our sampling transect forms the continuation of the Trans-Canada Highway transects T1 and T2 of Enkin et al. (2000).

Road and natural outcrop samples were collected in the Front Ranges and inner foothills. They were drilled and oriented in the field in different parts of the belt and in different stratigraphic levels. ALB02 and ALB03 sites were taken in Palisser Formation (Upper Devonian), close to the Rundell thrust at Spray Lake. Site

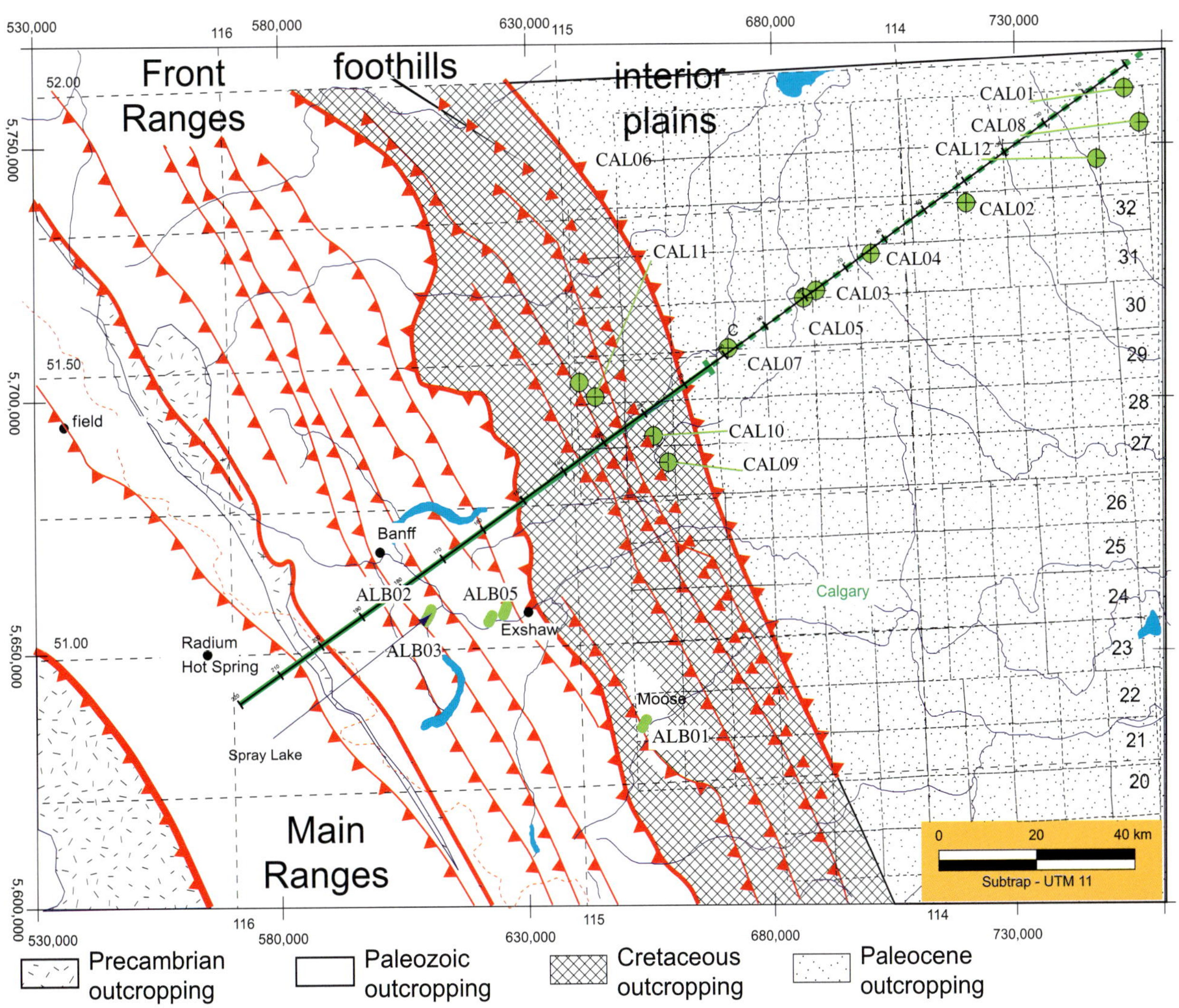

FIGURE 1. Location map of the study area. For the exact location of the wells, see Table 1. The symbols refer to paleomagnetic interpretation of the high-temperature component: filled symbols for normal high-temperature components, open symbol for reverse high-temperature components, mixed open and filed symbols for both reverse and normal components. CAL01 and CAL02 Cretaceous sites were not integrated in the interpretations.

ALB05 is located just above the Exshaw thrust in the Banff–Pekisko Formation (lower Mississippian). Site ALB01 was sampled in the southwestern limb of the Moose Mountain anticline in the Turner Valley and Mount Head Formation (upper Mississippian).

Samples from unoriented wells were taken at the Alberta Energy Utilities Board core facilities in Calgary. Twelve wells were selected along a transect roughly normal to the trend of the regional structures (Figure 1). Both Mississippian and Devonian carbonate rocks were sampled, i.e., one Devonian (CAL06) and three Mississippian sites (CAL09, CAL10, and CAL11) in a foothills setting and two Devonian (CAL03, CAL04, CAL08), two Mississippian (CAL07, 12), and one mixed site with both ages (CAL05) in the plains. In addition, Cretaceous levels were sampled (CAL01, CAL02, CAL03, CAL04, CAL05, and CAL07), but magnetic investigations in these strata failed to record any characteristic magnetization. Consequently, they will not be discussed below.

Stepwise thermal demagnetization of the three-axis IRM was performed on 18 samples. Results are very coherent all along the transect and indicate the occurrence of magnetite and/or titanomagnetite with systematic demagnetization between 450 and 550°C (Figure 2). The soft-coercivity component ($H_c \leq 01$ T) is dominant, and a significant contribution of the medium-coercivity component ($0.1 \leq H_c \leq 0.5$ T) appears systematically. In some case, high-coercivity component ($0.5 \leq H_c \leq 1.4$ T) may have measurable magnetization above 580°C but always with contribution lower than 5% on the whole signal. Such a distribution indicates that magnetite in our samples has a corresponding coercivity spectra and represents a mixture of multidomain and single domain grains. The presence of high-coercivity hematite with unblocking temperature higher than 600°C is also demonstrated, but its contribution is slight. In addition, stepwise three-axis demagnetization cannot give clear evidence for the presence of pyrrhotite as identified in the Moose Mountain reservoir by Cioppa et al. (2000).

RESULTS

Paleomagnetic Directions of Oriented Core Samples from Front Ranges and Inner Foothills

Stepwise thermal demagnetization efficiently distinguished the remanence components. Only 5 samples on a total of 32 were rejected because of spurious remagnetization appearing during heating. To calculate magnetization direction on each single specimen, we used at least four successive points on path demagnetization, and then we get the maximum angular deviation following the Kirschvink's algorithm. For the 27 remaining samples, 96% gave extracted components on orthogonal plots calculated with α_{95} below 15° (sites ALB01, ALB02,

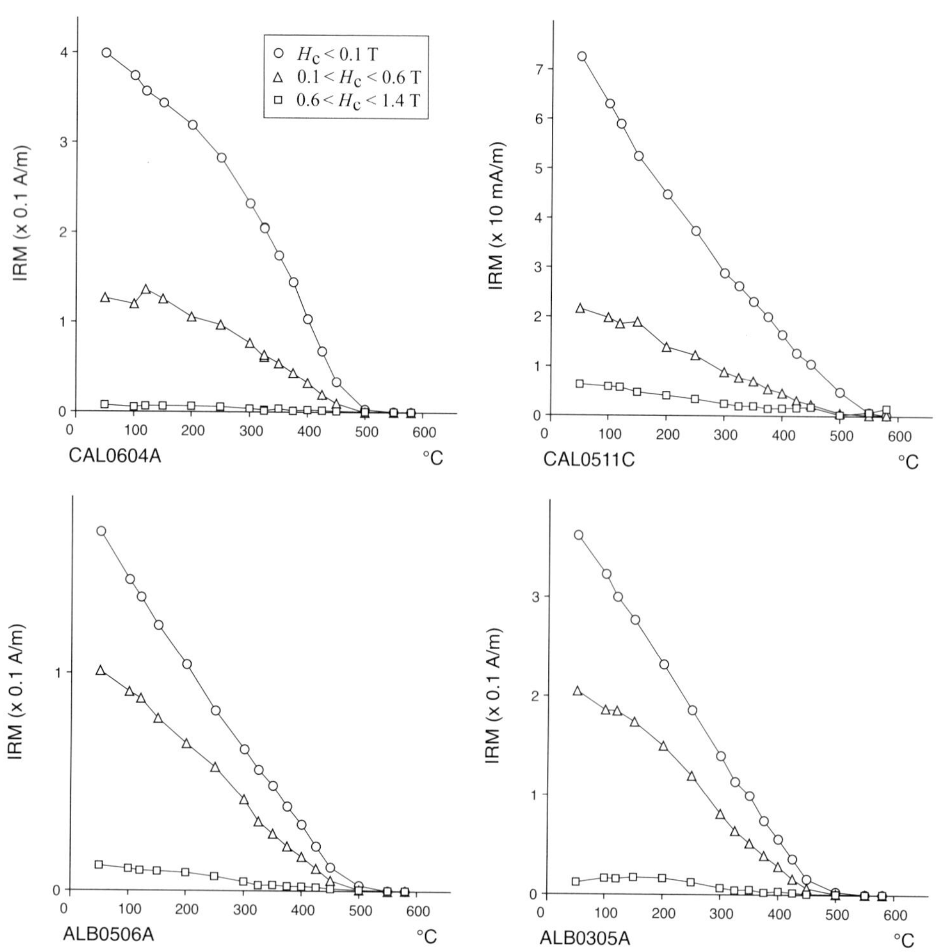

FIGURE 2. Thermal demagnetization of three-axis IRMs [Lowrie's (1990) protocol]. Circles, triangles, and squares are for the soft ($H_c < 0.1$ T), medium ($0.1 < H_c < 0.6$ T), and hard ($0.6 < H_c < 1.4$ T) component of partial IRM ($0.6 < H_c < 1.4$ T), respectively. Because of the device, we cannot perform IRM saturation higher than 1.4 T, and the corresponding coercivity fraction is supposed to be distributed on the three axes.

Table 1. Site information.

Site	*Age*	κ_0 (μ *SI*)	*NRM (m A/m)*	*Strike (°)*	*Dip (°)*	*Longitude/Latitude*
			Front Ranges (Oriented Samples)			
ALB02	Up Miss	53.0	8.5	130	45	115°24′58″/51°3′58″
ALB03	Up Miss	116.6	19.6	115	50	115°24′58″/51°3′58″
ALB05	Up Dev	30.8	8.6	127	45	115°14′19″/51°3′29″
			Foothills (Oriented Samples)			
ALB01	Miss	45.7	3.3	125	30	114°47′55″/50°51′19″
			Foothills (Unoriented Samples)			
CAL06	Dev	6.1	10.7	*	*	114°58′31.56″/51°27′56.88″
CAL11	Miss	29.2	5.2	*	*	114°55′48.49″/51°26′20.99″
CAL10	Miss	9.6	7.9	*	*	114°45′48.53″/51°21′58.95″
CAL09	Miss	7.4	4.2	*	*	114°43′24.76″/51°19′6″
			Plains (Unoriented Samples)			
CAL07	Miss	−2.8	5.8	*	*	114°32′11.16″/51°30′59.70″
CAL05	Miss	0.7	5.9	*	*	114°18′29.84″/51°36′0.78″
CAL05	Dev	16.2	7.4	*	*	114°18′29.84″/51°36′0.78″
CAL03	Dev	3.3	7.9	*	*	114°16′16.49″/51°36′44.37″
CAL04	Dev	0.6	11.0	*	*	114°06′14.57″/51°40′22.9″
CAL08	Miss	31.3	2.0	*	*	113°18′23.33″/51°54′41.69″
CAL12	Miss	3.4	3.4	*	*	113°25′15.02″/51°49′17.10″

*κ_0 = initial mean susceptibility; NRM = mean natural remanent magnetization; Up Miss = Upper Mississippian; Up Dev = Upper Devonian; Miss = Mississippian; Dev = Devonian; *subhorizontal bedding.*

ALB03, and ALB05 in Table 1; Figure 3a, b). A fold test unfortunately could not be performed, because the samples all originate from southwestern-dipping thrust sheets with insufficient variation of bedding strike. We compared our data with those of Enkin et al. (2000) collected in the same area (see samples from the Trans-Canada Highway transects T1 and T2 in table 1 of Enkin et al., 2000) in rocks of similar age and extending from the Front Ranges to the foothills.

On orthogonal plots, three distinct remanence components can be identified, which are classified depending on their declination and polarity (Figure 3a, b). These three components are observed together in the same samples and in each site. The only difference between sites is related to the range of thermal demagnetization needed to remove magnetic components during heating. The first component is a low-temperature normal component with maximum unblocking temperature between 120 and 200°C. Site mean directions are better grouped before bedding dips are removed, indicating that this low-temperature component is acquired after tilting. This component is attributed to a viscous remanent component (VRM) caused by the recent Earth's magnetic field.

The second component is a medium-temperature, reverse-polarity component. The maximum unblocking temperatures are between 260 and 380°C, and in-situ directions are oriented in the southeast direction (Figure 3a, b), which is equivalent to the reverse B component of Enkin et al. (2000) (Figure 4a). A parametric bootstrap fold test (following the method of Tauxe and Watson, 1994) performed on our data, combined with those of Enkin et al., shows that optimal concentration occurs on −20 ± 23%, indicating a negative fold test.

In the four sites studied, the maximum unblocking temperature spectrum of the high-temperature component extends to values between 400 and 500°C. The high-temperature component is a normal-polarity, northeasterly directed component in geographic coordinates and northwesterly in stratigraphic coordinates (Figure 5a). Enkin et al. (2000) found this A component to have normal polarity in the Front Ranges and reverse polarity in the foothills with only one exception. Site ALB01 with normal polarity constitutes a second exception. Maximum concentration occurs on 81 ± 10% untilting (Figure 5b). Thus, following Enkin's interpretation, we interpret that the high-temperature component normal polarity we find both in the foothills and Front Ranges was acquired in the first stage of tectonic tilting.

These results lead us to follow the interpretations of Enkin et al. (2000). All the magnetizations in the foothills and Front Ranges are steep and clearly indicate remagnetizations acquired long after the Paleozoic. The low-temperature component is a VRM recording the present-day field and corresponds to the P component. The medium-temperature component is equivalent to the B component and is interpreted as a thermoviscous remanent magnetization (TVRM) overprint acquired in

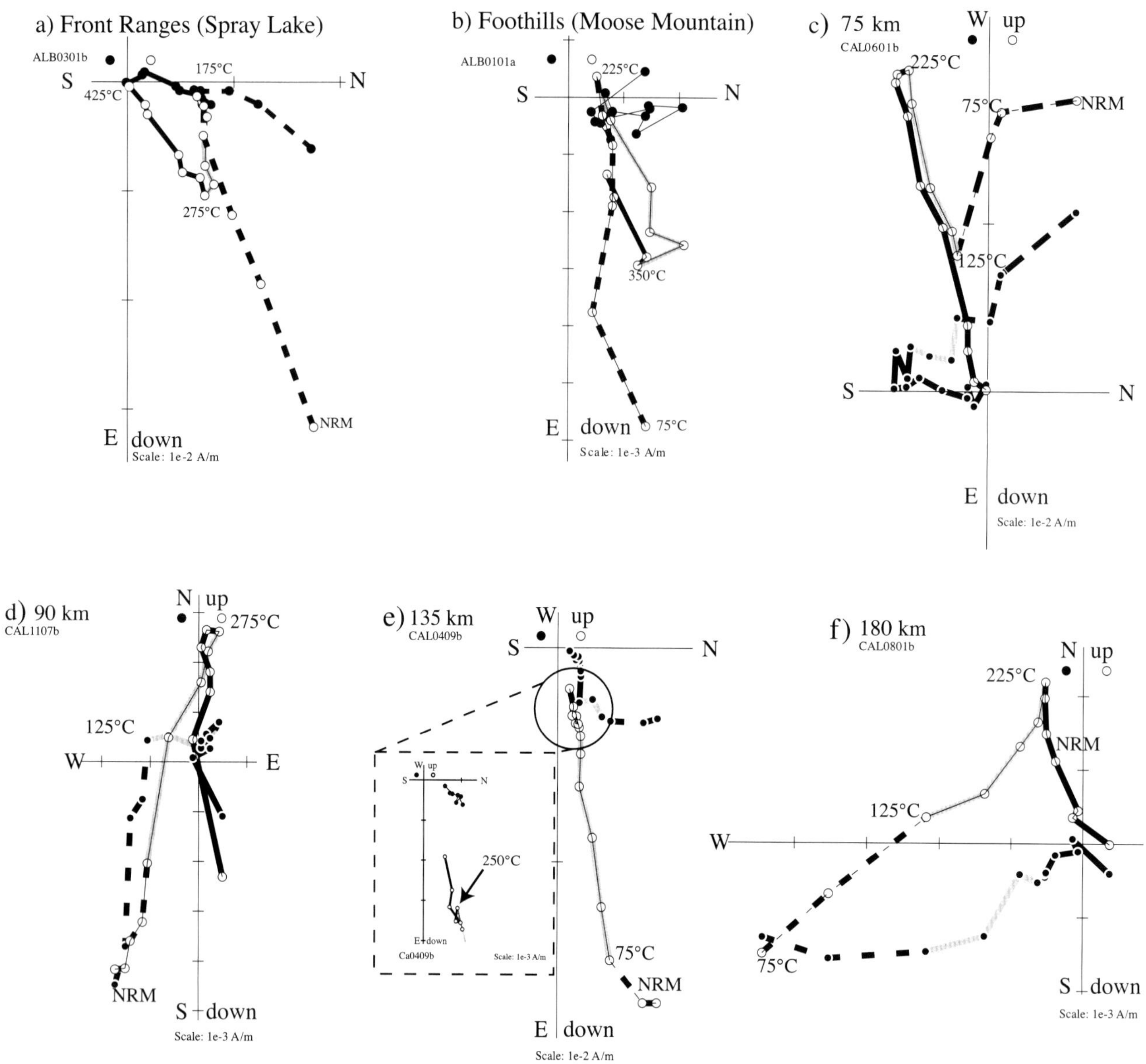

FIGURE 3. Orthogonal plots of stepwise demagnetization in geographic coordinates from Front Ranges (a), foothills (b), and Interior Plains (c–f). Solid (open) symbols represent the horizontal (vertical) plane. Low-temperature overprint is shown as gray solid lines, medium-temperature component is shown as dashed gray lines, and high-temperature component is shown as solid lines. Castle Mountain fault, separating the Main Ranges from the Front Ranges, is the reference for the positions in kilometers.

the late stage of contractional deformation. The high-temperature component equivalent to the A component is a predeformational to early syndeformational CRM.

Unoriented Core Samples: Outer Foothills and Interiors Plains

The unoriented core samples did not perform as well as the oriented samples did. Of the 110 samples we measured, 25% were not suitable for vector analysis. These samples did not show any coherent direction, indicating that signal intensity was close or below the magnetometer noise level. No specimens from site CAL10 were suitable for interpretation. Of the remaining samples, after interpreting components on orthogonal projection plots, about 95% possess directional signals with an α_{95} lower than 25°, with most about 15°. These large values are likely caused by the low amount of ferromagnetic minerals present in the rocks. These samples commonly have low paramagnetic or even diamagnetic susceptibility (Table 1).

Most samples possessed three distinct components of magnetization. Some of them, depending on the temperature at which new ferromagnetic phases were created during laboratory heating (above 350°C), show only

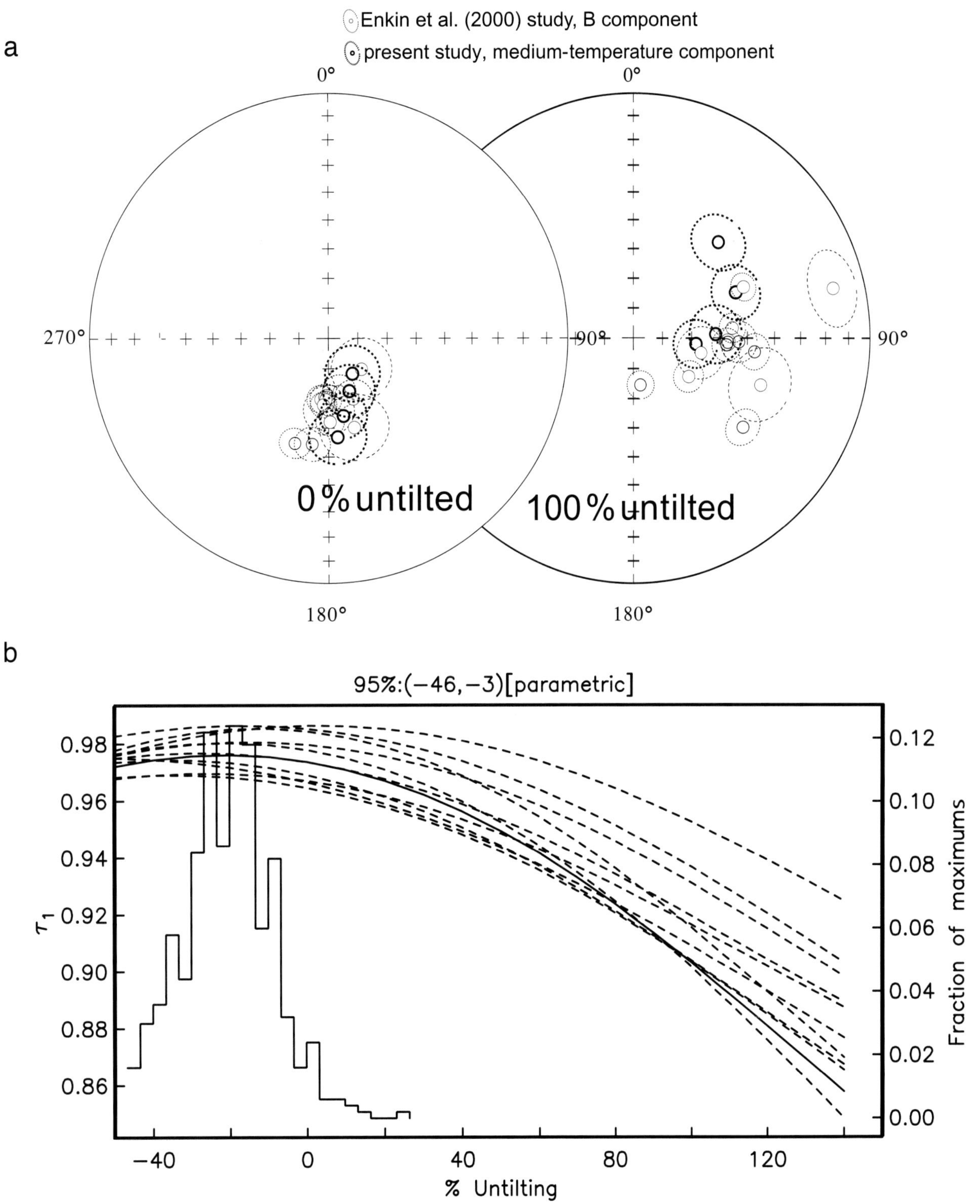

FIGURE 4. (a) Medium-temperature B component site mean directions with 95% confidence circles before and after bedding correction. We added directions of Enkin et al. (2000). The medium-temperature reverse components are better grouped after tectonic correction in good accordance with Enkin's previous study (open symbols represent upper hemisphere projection). (b) Parametric bootstrap fold test performed with the Tauxe's (1998) program. Dashed lines are 20 representative paradata sets that show evolution of maximum tightness (τ_1) for tilt incrementally from −50 to 150%. Histogram localizes the simulations of maximum of τ_1 for 500 paradata sets and gives value of 95% confidence interval. The medium-temperature B components are the best grouping for −20% with a large range for best unfolding (from −46 to −2%). This is interpreted as reflecting different rates of untilting and leads us to associate this component to a late syndeformation magnetization (see text).

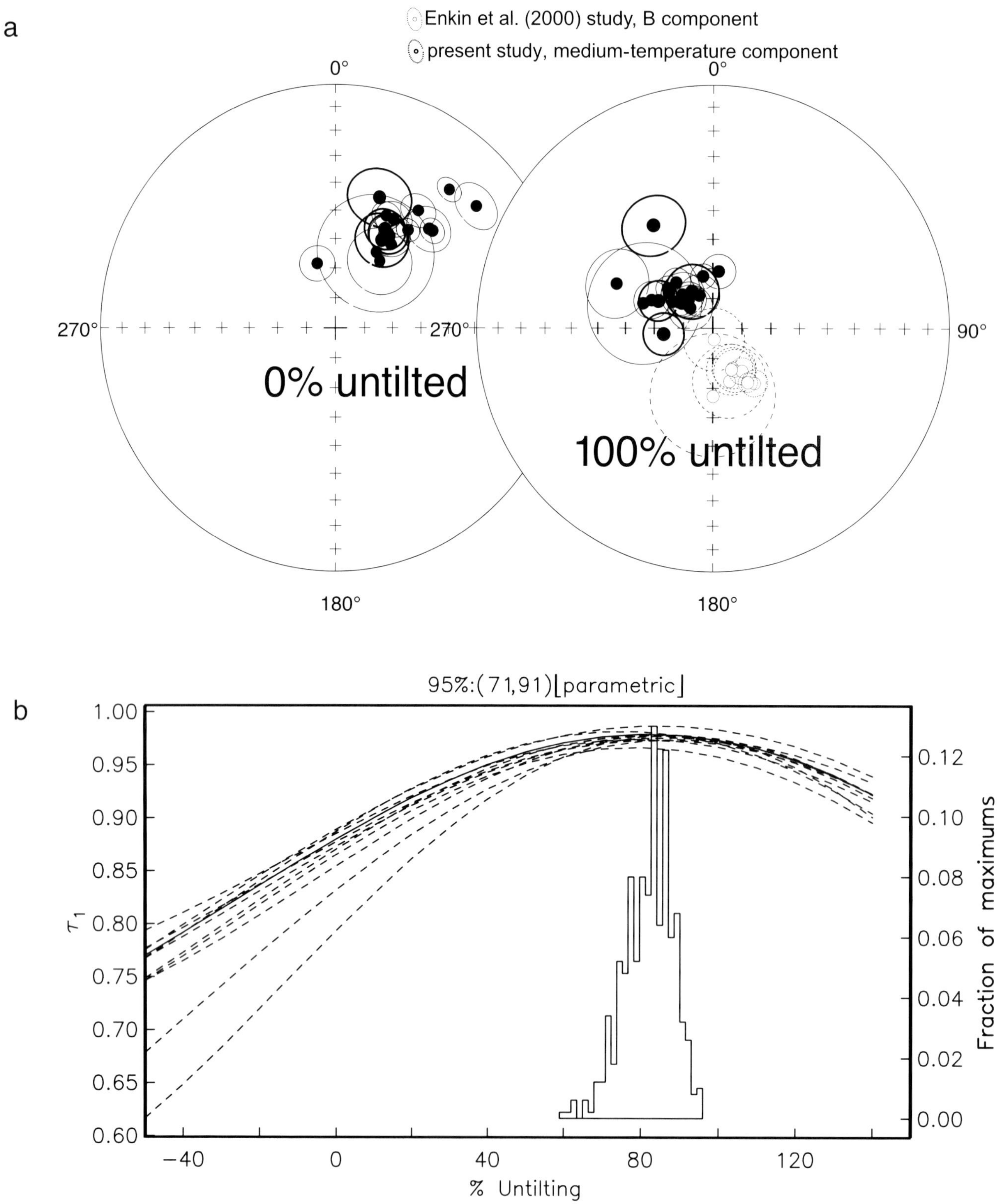

FIGURE 5. (a) The high-temperature A component site mean directions with 95% confidence circles before and after bedding correction measured in field samples. The high-temperature components are better clustered after untilting the bedding (closed and open symbols represent lower and upper hemispheres, respectively). (b) τ_1 reaches its maximum at 81%, with a range from 71 to 91%, indicating early to antetilting magnetization.

two interpretable components. The low-temperature component is removed at temperatures below 50–180°C (Figure 3c–f). Quantification of the orientation of this low-temperature component is commonly difficult, because this signal was commonly represented by no more than three successive points on the demagnetization path to fit a vector direction. This low-stability component shows scattered inclinations. We interpret that the low-temperature component is caused by drilling-induced remanent magnetization (DIRM), as described by Gillen et al. (1999) from the same area.

The second signal corresponds to a medium-temperature component and is commonly removed below 225–350°C. Linear trajectories in orthogonal projection plots with at least five successive points are commonly observed during stepwise thermal demagnetization. All medium-temperature components have normal polarity, with no significant difference between the foothills and plains (Figure 6a). On histogram, the medium-temperature component is generally steep, with a mean inclination of about 70°. A few samples display lower inclination values. Gillen et al. (1999) detected similar low inclinations in the same range of temperature, which they attributed to a possible DIRM effect. Despite the lack of a clear clustering of declination associated with low inclination values, which would rather indicate a DIRM effect, a cutoff of 60° was applied as was used by Gillen et al. (1999). Therefore, all samples that show medium-temperature components with inclinations lower than 60° were excluded.

The high-temperature component was isolated above 275–400°C and is commonly observed to follow a univectorial path to the origin of the orthogonal plot. The high-temperature component is sometimes contaminated above 300–350 °C by mineralogical transformations in the furnace, as commonly observed in remagnetized carbonates. Both normal and reverse polarities have been observed depending on the structural position of sites (Figure 3e, f). In the foothills (CAL06, CAL11, CAL10, and CAL09; see Figure 1 for location), all samples display a reverse polarity with a steep inclination. In the plains, both normal and reverse polarities with scattered inclination have been observed (Figure 6b). Some sites show only reverse polarities (CAL05 and CAL08), normal polarities (CAL07), or both (CAL05, CAL04, CAL03, and CAL12). For most of the samples, steep inclinations were measured, except in two sites (CAL05 and CAL12), where some of the samples displayed low inclinations.

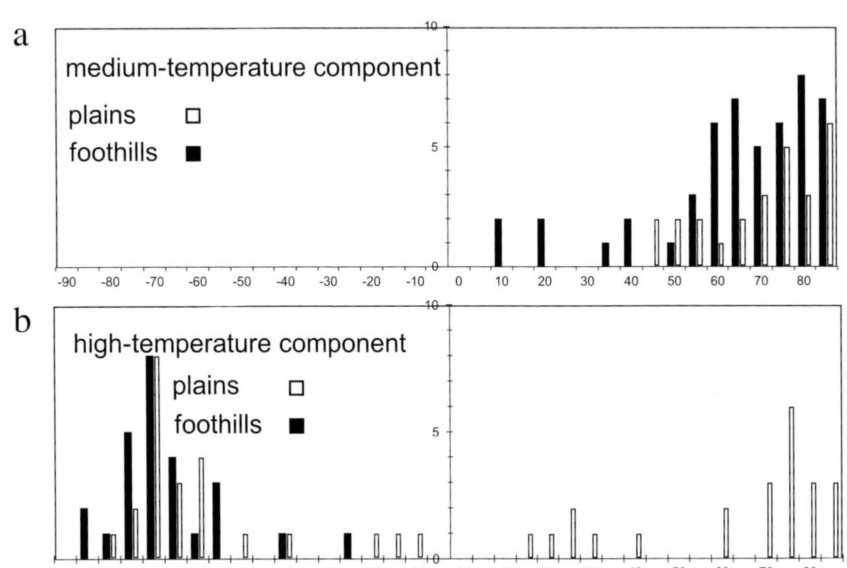

FIGURE 6. Frequency of inclination values for nonoriented foothills and plains samples. (a) Medium-temperature component; (b) high-temperature component.

DISCUSSION

By analyzing thermal demagnetizing paths in non-oriented well plugs, only two natural components were recorded, the medium-temperature and high-temperature components; the low-temperature component is interpreted to relate to a spurious magnetization caused by drilling effects. In this discussion, several questions are addressed. Does the medium-temperature component correspond to a recording of the recent geomagnetic field, and if so, can it be used as a fossil compass needle to recover in-situ orientation of the high-temperature component? What is the nature of the high-temperature component? Does the high-temperature component correspond to a primary magnetization acquired at the time of sedimentation, or is this signal caused by secondary magnetization linked to chemical or thermal effects acting at the stage of tectonic deformation as observed in the Front Ranges? Is it possible to date the characteristic high-temperature magnetization by orientation of their medium-temperature components, i.e., its transformation into the corresponding paleomagnetic pole by positioning the signal on the APWP of the North American craton?

Origin of the Medium-temperature Component

The medium-temperature component always has normal polarity and has inclinations similar to that of the present field. However, its unblocking temperatures as much as 350°C appear too high for it to be a viscous overprint acquired during the recent Brunhes polarity chron (780 ka). Faure et al. (2004) have determined 4–6-km (2.5–4-mi) burial until the Pleistocene, leading to burial temperatures of 95°C in the inner part and 65°C in the outer part. When corrected to laboratory-duration heatings (Pullaiah et al., 1975), these temperatures are perfectly compatible with the unblocking temperatures observed for the medium-temperature component (gray curve on Figure 7).

The gray curve is in good accordance with a mean unblocking temperature of 245°C observed in the well samples at least in western part of the transect. Eastward, unblocking temperature slightly increases. This could have different interpretations: superimposition of unblocking temperature spectrum between medium-temperature and high-temperature components or change in magnetic mineralogy. The consequence is that for two sites, some values are shifted toward the second curves (solid curve) corresponding to the theoretical values expected for maximum burial. For outer foothills samples, the range of 100–120°C (Faure et al., 2004) (unblocking temperature between 250–270°) is a reasonable assumption and in good accordance with unblocking temperature measured. For the field samples, we assume that rocks have resided at surface temperature (15°C) since more than 780 k.y., and we estimate the offset temperature of about 160°C. Expected temperatures about 175°C are in good accordance with mean unblocking temperature values of 165°C measured for Devonian sites (horizontal part of gray curve in Figure 7). Nevertheless, unblocking temperature values measured for Mississippian sites ALB01 and ALB05, whose values varied about 200°C, are slightly higher than the expected theoretical values probably because of magnetic mineralogy effects. Thus, all along the transect (plains, foothills, and Front Ranges), measured unblocking temperature roughly correlates with burial depth reaches of more than 780 k.y.

Origin of the High-temperature Component

The high-temperature component is commonly completely removed above 450°C, and if it was of thermoviscous origin, the rocks would have to have been geologically heated to at least 300°C (for 780 k.y.). Based

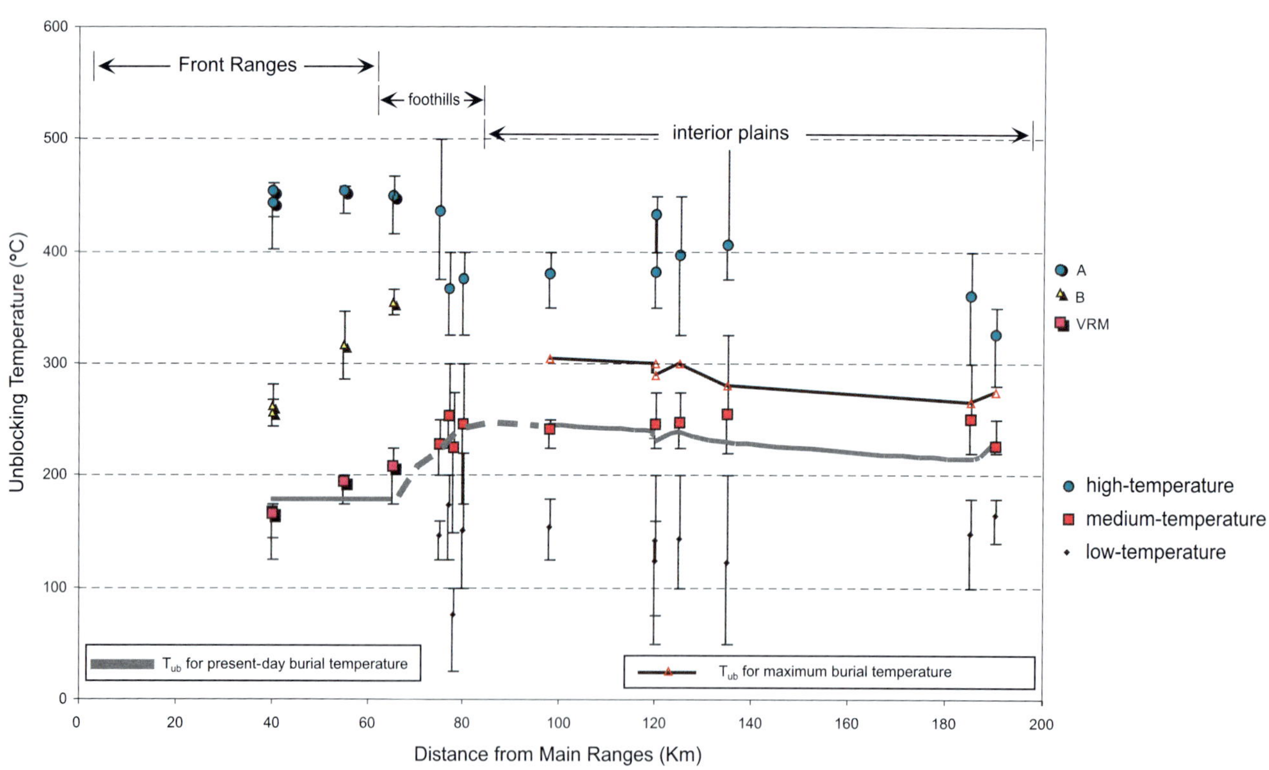

FIGURE 7. Unblocking temperature spectrum for high-temperature, medium-temperature, and low-temperature components measured in nonoriented samples and for A, B, and VRM measured in field samples. Theoretical curves of unblocking temperature (T_{ub}) are calculated for the present-day burial temperature (gray curve) and the maximum burial temperature (solid curve). Theoretical T_{ub} deduced from present-day burial temperature is consistent with laboratory T_{ub} (square). We thus assume consistency between medium-temperature components in plains and foothills and VRM component in Front Ranges. Similarly, high-temperature unblocking values correspond to CRM unblocking temperatures in Front Ranges (see circles).

on thermal modeling of this foreland basin, Faure et al. (2004) came to the conclusion that the maximum temperature that was reached in the plains and outer foothills is in the order of 160°C. This value implies that the theoretical unblocking temperature should be lower than 310°C (solid curve in Figure 7), which is systematically lower than laboratory temperatures measured along the transect for the high-temperature component. Consequently, the high-temperature component cannot be related to the TVRM event equivalent to the Enkin's B component observed in field samples. Because of high unblocking temperature values and steep inclination, the high-temperature component is interpreted here as a CRM similar to Enkin's A component. The polarity of high-temperature magnetization is normal in the Front Ranges and commonly reverse in the inner foothills (Enkin et al., 2000); however, it is always reverse in the outer foothills and mixed in the plains (Figure 8). This implies that the timing of CRM acquisition differs in the function of the position of the sampling site and is thus locality dependent. In the plains, two Devonian sites (CAL03 and CAL04) and one Mississippian site (CAL 05) record both polarities. However, all other Mississippian and Devonian sites are exclusively normal or reverse, respectively. This suggests that remagnetization is controlled by lithology. This also suggests that in the foreland, long-range fluids probably do not play their part in the remagnetizing process and reflect localized fluid circulations. Additionally, in Mississippian site CAL12, more than 80 km (50 mi) from the deformed zone, low normal high-temperature inclinations (<20°) are observed, suggesting that primary depositional magnetization may not be completely overprinted. Therefore, the high inclination, high unblocking temperature, and the evolution of the polarities that were recorded along transect corroborate and complete the interpretation of Enkin et al. (2000) for a diachronous chemical remagnetization.

Core Orientation and Age of CRM

We aligned the medium-temperature components to the dipole of the Earth's magnetic field using the vertical axis of the wells as rotational axis, assuming that all wells are orientated vertically. This assumption seems to be reasonable for the plains samples, where horizontal laminations were observed to be aligned perpendicular to the core axis. For foothills samples, some very steep inclinations close to 85° that are associated with well-defined magnetization directions with high-confidence angles suggest that some difficulties with respect to well orientation might exist. However, this observation is not systematic, and commonly, deviations are less than 10°. To avoid the possible effect of overprinting caused by DIRM on natural VRM, samples with inclinations higher than 85 and lower than 60° were excluded from the process of orientation. Moreover, all single medium-temperature directions that were calculated with an angular deviation higher than 20° were also filtered out. Inspecting maximum unblocking temperatures indicates that they never overstep the limit of 275°C. By doing so, only 33 of 110 samples were used for paleomagnetic reorientation purposes (Figure 8). The small quantity of samples available on each site constrains us to gather samples from different sites in two sets for reconstructing the paleomagnetic pole position. Therefore, a mean latitude and longitude was used for calculating the pole position from the magnetization directions. The boundary was placed between samples coming from the foothills (12 samples) and from the plains (21 samples). This boundary is arbitrary, but from

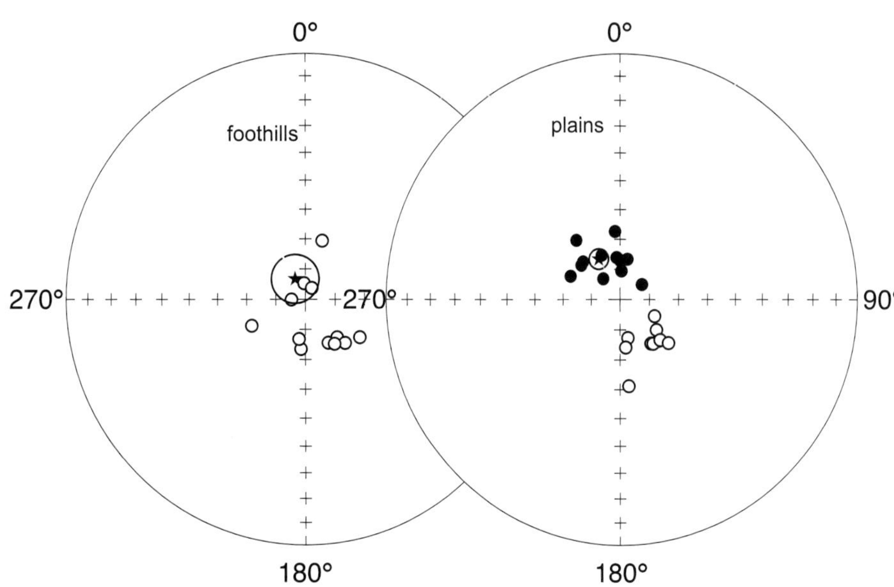

FIGURE 8. Equal-area stereogram for high-temperature components after correction of the VRM. Open circles represent directions from the upper hemisphere (reverse), and closed circles represent directions from the lower hemisphere (normal). Specimens with mean angular deviation higher than 20° have not been considered for the treatment. We have also excluded samples with VRM inclination higher than 85 and lower than 60°C to minimize DIRM and noncorrected well inclination effects.

TABLE 2. Paleomagnetic directions and calculated VGP from outer foothills and Interior Plains.

Site	*Mean Directions*							*VGP*			
	n	*D*	*I*	*k*	α_{95}	*Slat*	*Slon*	*Lat*	*Lon*	*dp*	*dm*
Foothills	12	331.3	82.4	30.0	8.1	51.4	245.2	229.7	63.9	15.3	15.7
Plains	22	330.4	74.3	84.0	3.4	51.7	246.0	198.7	71.7	5.6	6.2
Gillen	17	335.7	73.3	***	7.1	51.0	245.0	194.6	74.0	***	***
14-th-h	13	324.6	71.4	12.2	11.9	50.9	245.2	182.9	68.6	***	***
23-th-h	5	297.0	73.3	24.0	15.9	50.9	245.2	193.3	54.4	***	***

n *is the number of samples;* D, I, k, *and* α_{95}: *Fisher average declination, inclination, precision factor, and 95% confidence interval, respectively; VGP = virtual geomagnetic pole; dp and dm = principal semiaxes of 95% confidence interval about the pole. Sites Gillen, 14-th-h, and 23-th-h correspond respectively to the mean and VGP by Gillen et al. (1999) and Cioppa et al. (2000). ***Not included.*

inclination data, the two areas behave differently; in one area, all magnetizations are reverse, whereas in the other area, both polarities coexist. The high-temperature remanence magnetization in the plains after correction is $D = 330.4°$, $I = 74.3°$, $k = 84$, and $\alpha_{95} = 3.4°$ (Figure 8; Table 2). A parametric bootstrap reversal test (method developed by Tauxe, 1998) was performed, and it suggests that the two means are not significantly different from the antipode (Figure 9). In the foothills, the values are $D = 331.3°$, $I = 82.4°$, $k = 30$,

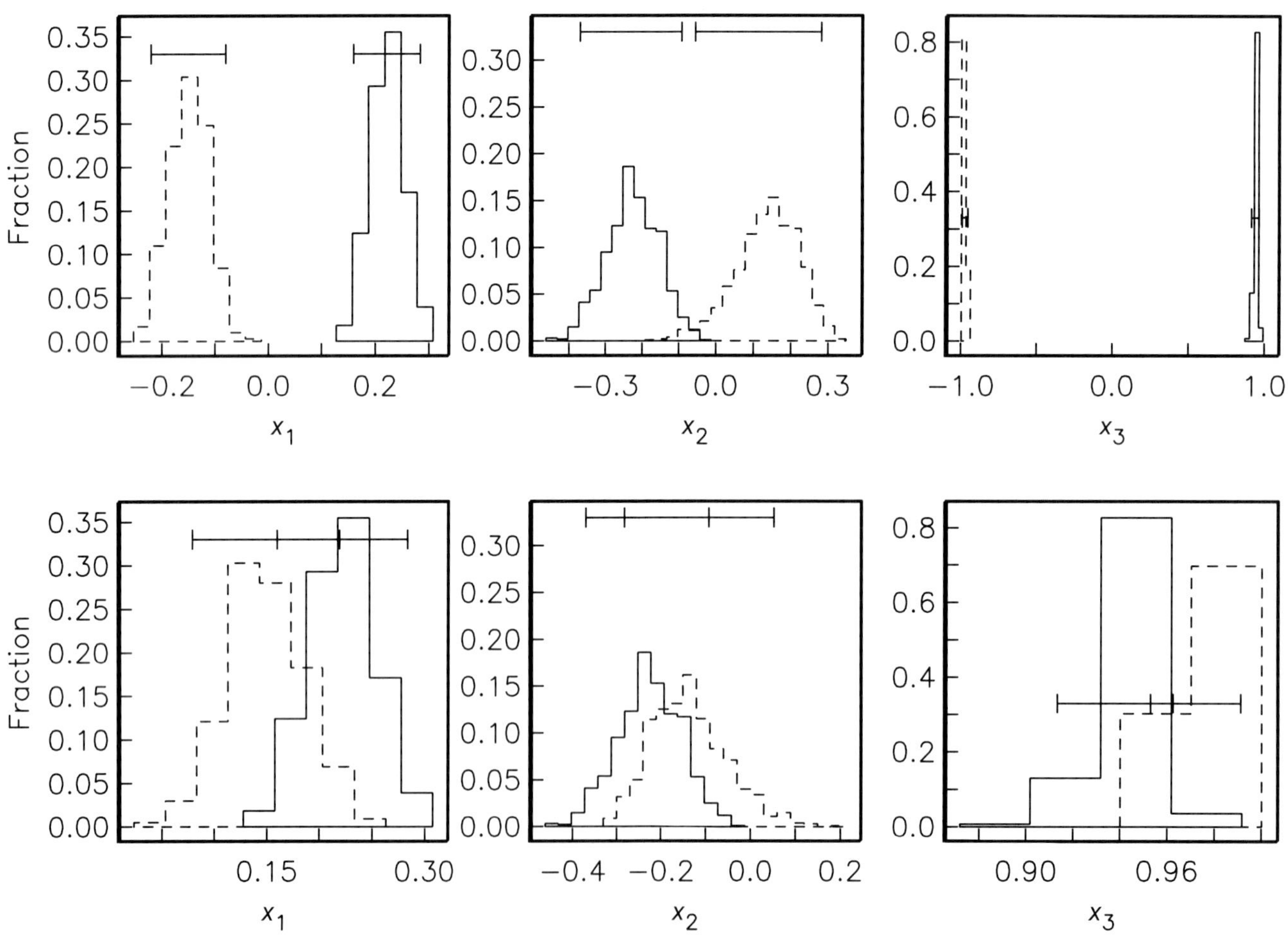

FIGURE 9. Parametric reversal test performed with Tauxe's (1998) algorithm from the plains samples shown in Figure 7 (after correcting the VRM). In the upper part of the diagram, x_1, x_2, and x_3 are the Cartesian coordinates of the input data in both the normal and reverse mode. The lower part of the diagram corresponds to the reversed-polarity mode, which has been flipped to antipode. The intervals containing 95% of the simulated data overlap in all three components, meaning that they pass the reversals test.

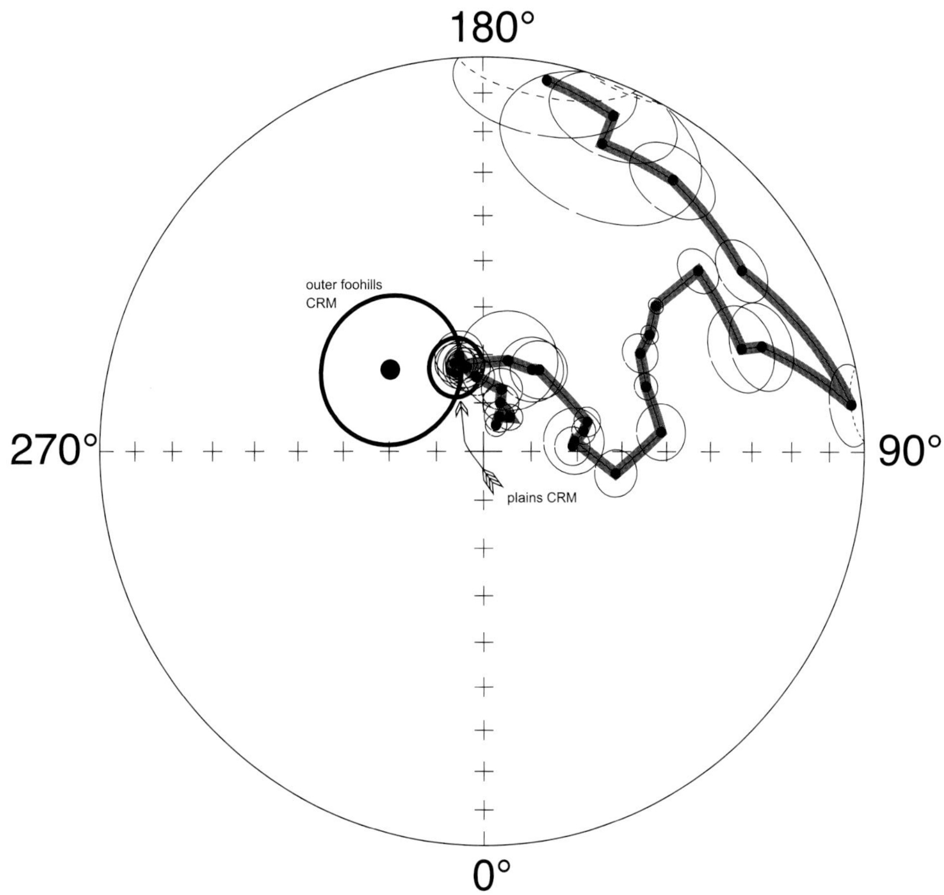

FIGURE 10. Apparent polar wander path for North America craton from Bess and Courtillot (1991). High-temperature components CRM are shown for both plains and foothills. Chemical remanent magnetization component in plains is more clearly defined and falls in the Cretaceous segment of the APWP.

and $\alpha_{95} = 6.8°$, which is somewhat steeper. This is possibly caused by an uncorrected well inclination. The pole corresponding to the plains direction is more precise than that deduced for the foothills. Nevertheless, the two α_{95} overlap the APWP of the North American craton (Figure 10) along the Cretaceous segment of the APWP. Particularly, the CRM pole deduced in the plains is very close to the poles defined by Gillen et al. (1999) and Cioppa et al. (2000) (Table 2). These results show that core orientation is sufficiently reliable to retrieve the mean pole position for a specific area. Unfortunately, having so few samples to work with makes it impossible to establish trends along our transect. Despite some evidence for asynchronous acquisition of the CRM (i.e., change of polarities along the transect), it is not possible to infer whether the CRM in the plains is younger than in the foothills.

CONCLUSION

Viscous remanent magnetization acquired in the present field provides a useful compass to recover core orientation. In the present study, it allowed us to date chemical remagnetization to be Cretaceous in age. However, it was not possible to deduce a precise timing of remagnetization events along the studied transect, because we had to exclude about 70% of all samples studied. Weak magnetization of samples caused by the low amount of ferromagnetic minerals may explain this difficulty. A way to minimize this waste is to carefully check the intensity of magnetic susceptibility during sampling with a hand-held instrument. This should be particularly useful in case of core laboratory sampling, where lowest magnetized samples were encountered. Increasing the amount of samples from the same well is another solution to improve the paleomagnetic investigation (30 samples should be enough to obtain at the end about 10 interpretable samples). Despite these limitations, our paleomagnetic investigations of carbonate rocks from foreland basin of the western Canadian Cordillera are in line with the interpretation of Enkin et al. (2000), namely, the following:

1) A chemical remagnetization, previously well described in the fold and thrust belt of WCC, also affected the Devonian and Mississippian strata in the apparently undeformed Foreland Basin. This CRM is recorded as far as 150 km (90 mi) ahead of the Front Ranges. The CRM is not uniformly recorded in all the wells, and polarities seem to be controlled by the age of rocks in the plains: normal in Mississippian carbonates and reverse in Devonian carbonates. This argues for strata-dependent fluid circulations and their associated magnetite crystallizations in the undeformed basin without excluding the possibility that fluids moved during folding and thrusting in the foothills.
2) Late thermoviscous remagnetization was not recorded in the outer foothills and Interior Plains.

This may be caused by a very short event corresponding to the maximum of burial or an unmeasurable low-magnetization event or simply caused by a slow rate of uplifting event.

ACKNOWLEDGMENTS

The present study has been founded by Subthrust Reservoir Appraisal II Consortium. We thank E. Deville (Institut Français du Pétrole) and B. Gomez (Institut Français du Pétrole), as well as several colleagues at the Geological Survey of Canada, for assistance in the field and at the Alberta Energy and utilities Board. The latter Board is also thanked for permitting sampling of the cores. We also greatly thank C. Kissel and C. Laj (Laboratoire des Sciences du Climat et de l'Environemen, Gif sur Yvette) for providing us their laboratory device measurements and for their constant assistance during paleomagnetic measurements. Special thanks go to F. Roure (IFP) and K. Osadetz (Geological Survey of Canada), who initiated this work. Finally, we acknowledge R. Enkin (Geological Survey of Canada) and D. Elmore (University of Oklahoma) for their reviews and especially R. Enkin and M. Molinaro (University of Cergy-Pontoise) for improving the text.

REFERENCES CITED

Besse, J., and V. Courtillot, 1991, Revised and synthetic apparent polar wander paths of the African, North American, and Indian plates, and true polar wander path since 200 My: Journal of Geophysical Research, v. 96, no. B3, p. 299–316.

Cioppa, M. T., I. S. Al-Aasm, D. T. A. Symons, and K. P. Gillen, 2004, Causes of multiple magnetizations in the Devonian upper member, Wabamun Group, Alberta, Canada: Fluid flow and diagenesis?, *in* R. Swennen, F. Roure, and J. W. Granath, eds., Deformation, fluid flow, and reservoir appraisal in foreland fold and thrust belts: AAPG Hedberg Series, no. 1, p. 331–346.

Cioppa, M. T., I. S. Al-Aasm, D. T. A. Symons, M. T. Lewchuck, and K. P. Gillen, 2000, Correlating paleomagnetic, geochemical and petrographic evidence to date diagenetic and fluid flow events in the Mississippian Turner Valley Formation, Moose field, Alberta, Canada: Sedimentary Geology, v. 131, p. 109–209.

Cogné, J. P., 2003, PaleoMac: A Macintosh(TM) application for treating paleomagnetic data and making plate reconstructions: Geochemistry, Geophysics, and Geosystems, v. 4, no. 1, p. 1007, doi:10.1029/2001GC000227.

Enkin, R. J., K. G. Osadetz, and J. Baker, 1997, Paleomagnetic constrains on the tectonic history of foreland belt, southern Canadian Cordillera: Preliminary results: Canadian Journal of Earth Sciences, v. 34, p. 260–270.

Enkin, R. J., K. G. Osadetz, J. Baker, and D. Kissilevsky, 2000, Orogenic remagnetization in the Front Ranges and inner foothills of the southern Canadian Cordillera: Chemical harbinger and thermal handmaiden of Cordilleran deformation: Geological Society of America Bulletin, v. 112, no. 6, p. 929–942.

Faure, J. L., K. G. Osadetz, Z. N. Benaouli, F. Schneider, and F. Roure, 2004, Kinematic and petroleum modeling of the Alberta foothills and adjacent foreland, west Calgary: Oil & Gas Science and Technology— Revue de'l Institut Francois du Petrole, v. 59, no. 1, p. 81–108.

Garven, G., 1995, Continental-scale groundwater flow and geological processes: Annual Review of Earth and Planetary Sciences, v. 23, p. 89–117.

Gillen, K. P., R. Van der Voo, and J. H. Thiessen, 1999, Late Cretaceous–early Tertiary remagnetization of the Devonian Swan Hills Formation recorded in carbonate cores from the Caroline gas field, Alberta, Canada: AAPG Bulletin, v. 83, no. 8, p. 1223–1235.

Katz, B., R. D. Elmore, M. Cogoini, M. H. Engel, 2000, Association between burial diagenesis of smectite, chemical remagnetization, and magnetite authigenesis in the Vocontian trough, SE France: Journal of Geophysical Research, v. 105, no. B1, p. 851–868.

Kirschvink, J. L., 1980, The least-squares line and plane and the analysis of paleomagnetic data: Geophysical Journal of the Royal Astronomical Society, v. 62, p. 699–718.

Lowrie, W., 1990, Identification of ferromagnetic minerals in rock by coercivity and unblocking temperature properties: Geophysical Research Letters, v. 17, no. 2, p. 159–162.

McCabe, C., and R. D. Elmore, 1989, The occurrence and origin of Late Paleozoic remagnetization in the sedimentary rocks of North America: Review of Geophysics, v. 27, p. 471–494.

Oliver, J., 1986, Fluids expelled tectonically from the orogenic belt: Their role in the hydrocarbon migration and other geologic phenomena: Geology, v. 14, p. 99–102.

Oliver, J., 1992, The spot and stains of plate tectonics: Earth Science Reviews, v. 32, p. 77–106.

Pullaiah, G., E. Irving, K. Buchan, and D. J. Dunlop, 1975, Magnetization changes caused by burial and uplift: Earth and Planetary Science Letters, v. 28, p. 133–143.

Symons, D. T. A., R. J. Enkin, and M. T. Cioppa, 1999, Paleomagnetism in the Western Canada sedimentary basin: Dating fluid flow and deformation: Bulletin of Canadian Petroleum Geology, v. 47, no. 4, p. 534–547.

Tauxe, 1998, Paleomagnetic principles and practice: London, Kluwer Academic Publishers, 299 p.

Tauxe, L., and G. Watson, 1994, The fold test: An eigen analysis approach: Earth and Planetary Science Letters, v. 122, p. 331–341.

16

Cioppa, M. T., I. S. Al-Aasm, D. T. A. Symons, and K. P. Gillen, 2004, Causes of multiple magnetizations in the Devonian upper Member, Wabamun Group, Alberta, Canada: Fluid flow and diagenesis?, *in* R. Swennen, F. Roure, and J. W. Granath, eds., Deformation, fluid flow, and reservoir appraisal in foreland fold and thrust belts: AAPG Hedberg Series, no. 1, p. 331–346.

Causes of Multiple Magnetizations in the Devonian Upper Member, Wabamun Group, Alberta, Canada: Fluid Flow and Diagenesis?

M. T. Cioppa
Department of Earth Sciences, University of Windsor, Windsor, Ontario, Canada

I. S. Al-Aasm
Department of Earth Sciences, University of Windsor, Windsor, Ontario, Canada

D. T. A. Symons
Department of Earth Sciences, University of Windsor, Windsor, Ontario, Canada

K. P. Gillen
Vox Terrae International, Calgary, Alberta, Canada

ABSTRACT

Geochemical and petrologic analysis of three wells from the upper Member of the Wabamun Group in the Limestone gas field indicates a complex diagenetic history, including several episodes of dolomitization, silicification, and calcite cementation. Paleomagnetic analysis showed that several magnetization components are present in the reservoir rocks: (1) a high-temperature magnetization of probable Jurassic age (paleopole long. 81°E, lat. 78°N) carried by magnetite; (2) at least one dominant reversed-polarity characteristic remanence of Late Jurassic or Tertiary age (paleopoles long. 172°E, lat. 78°N; long. 139°E, lat. 79°N) carried by magnetite and pyrrhotite; and (3) a low-temperature and coercivity reversed-polarity magnetization (paleopole long. 235°E, lat. 77°N) that is either an uplift-induced magnetization or a late chemical remanent magnetization. The presence of several remagnetization events implies that the Laramide paleomagnetic history of the Cordillera is incompletely understood.

DOI:10.1306/1025698H13124

INTRODUCTION

In referring to a fluid-flow event to explain a widespread remagnetization, as many authors have for both the Appalachian Alleghanian orogeny (Oliver, 1986, 1992) and the Cordilleran Laramide orogeny (Symons et al., 1999), the tendency is to think of the event as almost instantaneous, geologically speaking. Stamatakos et al. (1996), for example, used the timing of folding with respect to the formation of the Permian Kiaman overprint caused by the Alleghanian orogeny to provide compelling evidence for a very short remagnetization event. They determined that the overprint was postfolding in the east or hinterland of the orogeny, and as one moved west toward the foreland, the overprint became synfolding and then prefolding in origin, suggesting that the Kiaman magnetization was a geologically instantaneous event that trapped a "snapshot" of the orogeny during its formation. Additional evidence for a short remagnetization event is suggested by the single polarity of the Kiaman overprint. In contrast, Pannalal et al. (2004) have shown that Mississippi Valley-type Pb-Zn ore deposits in the mid-continent were formed by Appalachian–Ouachitan orogenic fluid-flow events (Leach et al., 2001) over an about 60-m.y. period from about 303 to about 245 Ma.

The Laramide orogeny presents a more complex situation. Enkin et al. (1997) suggested the presence of several remagnetization events in the orogenic belt, ranging in age from greater than 130 to less than 75 Ma. In a later paper, Enkin et al. (2000) suggested the presence of two magnetizations: a chemical remanent magnetization, which is formed in advance of the Cordilleran tectonic wedge during the Cretaceous, and a thermoremanent magnetization, which is formed in response to tectonic or sedimentary loading. In our past papers, there is also evidence for multiple events in the Western Canada Sedimentary Basin. Although the paleopoles for the major magnetization components in the Waterton (Mount Head Formation; Lewchuk et al., 1998), Moose (Turner Valley Formation; Cioppa et al., 1999), Rainbow (Keg River Formation; Lewchuk et al., 2000), Dunvegan (Debolt Formation; Cioppa et al., 2003), and Rainbow South (Sulphur Point Formation; Lonnee et al., 1998; Cioppa et al., 2001) reservoirs are, in general, Cretaceous and early Tertiary in age, there is some variation observed. Second, this major magnetization component is not always of the same polarity; it is reversed in the Waterton, Moose, and Rainbow reservoirs and normal in the Dunvegan and Rainbow South reservoirs. A third complication is the presence of a low-coercivity and low-temperature component. It is sometimes normal in polarity, can be used for orientation purposes, and is considered to be a viscous remanent magnetization. However, in several instances, this low-coercivity and low-temperature component is reversed and therefore cannot be a viscous remanent magnetization, implying that it must be some sort of chemical remanent magnetization.

Thus, this study of dolomites from the Late Devonian Wabamun Group in the Limestone gas field was formulated to explore for evidence of multiple remagnetization events in one field.

GEOLOGIC AND STRATIGRAPHIC SETTING

The Devonian Wabamun Group is subdivided in the west-central Alberta subsurface into upper, upper middle, lower middle, and lower Members. The upper Wabamun was sampled for this study and is correlated with the Costigan and Big Valley formations of the Wabamun Group elsewhere in the basin (Halbertsma, 1994). Overall, the Wabamun sediments consist of stacked cyclical ramp and shelf carbonates and associated evaporites and are considered to record a regressive sequence that is punctuated by several transgressive pulses (Halbertsma, 1994). The Wabamun Group is overlain by the Exshaw Formation and underlain by the Graminia Formation.

The Limestone natural gas field is located in a structural culmination of the foothills belt, northwest of Calgary (Figure 1). Within the field,

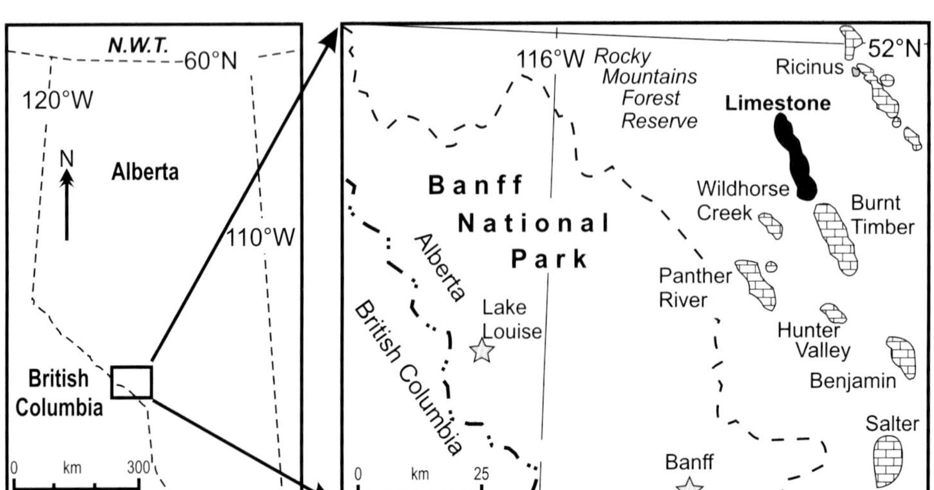

FIGURE 1. Right: well locations in the Limestone gas field sampled for this study. Left: location of the Limestone field in the Western Canada Sedimentary Basin.

Table 1. Wells sampled in the Limestone field.

Well	*Abbreviation*	*Borehole Deviation (°)*	*Interval Sampled (m)*
Shell Canada 10-25-33-10w5	SL13	16–34	3540–3582
Shell Canada 4-12-33-10w5	SL15	3–358	3600–3624
Shell Canada 6-2-32-10w5	SL16	9–70	3360–3395

there are three gas pools with 3.963×10^6 m^3 (1.40×10^8 ft^3) of marketable reserves (Halbertsma, 1994). The Limestone field lies along strike from the northwestern end of the earlier discovered Burnt Timber field (Figure 1). Mississippian, Devonian, and Cambrian imbricate sheets form a structural complex beneath the overlying Marble Mountain or Brazeau thrust sheet, and the Limestone field is found in a structural trap, whereas most Wabamun fields are stratigraphic traps (Halbertsma, 1994). The Limestone field has a very high hydrogen sulfide content of as much as 67%, which has been postulated to be the product of thermochemical sulfate reduction (Eliuk, 1984). A more complete description of the cores sampled is given in the Results.

METHODOLOGY

Three wells were sampled (Table 1): Shell Canada 6-2 (site location [SL] 16), 10-25 (SL15), and 4-12 (SL13). Paleomagnetic, petrologic, and geochemical techniques followed those reported in Lewchuk et al. (1998) and Cioppa et al. (1999). A brief summary is given below.

Petrology and geochemistry

Forty samples were collected for petrographic and geochemical analysis. From these samples, 32 thin sections were examined using standard petrographic microscopy and cathodoluminescence microscopy with a Technosyn cold cathodoluminescence stage with a 12–15-kV beam and a current intensity of 0.42–0.43 mA. Fluorescence characteristics were examined with a Nikon Episcope Fluorescence stage connected to a petrographic microscope.

Oxygen and carbon isotopes were extracted from 36 dolomite and calcite samples using a microscope-mounted drill assembly. The samples were reacted in vacuo with 100% pure phosphoric acid for at least 4 hr at 25°C for calcite and at 50°C for dolomite using the method described by Al-Aasm et al. (1990). The evolved CO_2 gas was analyzed for isotopic ratios on a SIRA-12 mass spectrometer. Values of O and C isotopes are reported in per mil (‰) relative to the Vienna Peedee belemnite (VPDB) standard and were corrected for phosphoric acid fractionation. Precision was better than 0.05‰ for both $\delta^{18}O$ and $\delta^{13}C$. Strontium isotopes were analyzed for selected dolomite, calcite, and anhydrite samples ($n = 12$). Sr isotopic ratios were measured on a Finnigan MAT 261 mass spectrometer equipped with nine collectors. All analyses were performed in the static multicollector mode using Re filaments. Correction for isotope fractionation during the analyses was made by normalization to $^{86}Sr/^{88}Sr = 0.1194$. The mean standard error for mass spectrometer performance was 0.00003 for National Bureau of Standards-987.

Major- and minor-element analyses (Ca, Mg, Fe, Sr, and Mn) were performed on highly polished carbon-coated thin sections with a Cameca Camebax SX50 electron microbeam analyzer equipped with three crystal spectrometers and a backscatter electron detector. The operating conditions were an accelerating voltage of 15 kV, a measured beam current of 6 nA, and a beam diameter of 4–10 μm. The standards used were wollastonite (Ca), MgO (Mg), $MnTiO_3$ (Mn), hematite (Fe), and strontianite (Sr). Precision was better than 0.5 mol% for all elements.

Paleomagnetism

Core collection procedures followed the methods described in Lewchuk et al. (1998) and Cioppa et al. (1999, 2000). The natural remanent magnetization of each specimen was measured using a Canadian Thin Films cryogenic magnetometer (CTF-420), which can detect a magnetization intensity down to about 5×10^{-6} A/m for a standard paleomagnetic specimen. The initial (pilot) specimens were subjected to either alternating field (AF) demagnetization in about 25 steps from 2 to 140 mT using a Sapphire Instruments SI-4 AF demagnetizer or thermal demagnetization in about 30 steps from 80 to 580°C using a Magnetic Measurements MMTD-1 thermal demagnetizer. The remainder of the specimens were demagnetized thermally in about 20 steps from 80 to 500°C. The individual remanence vector components from the specimens were obtained using the principal component method of Kirschvink (1980), and components were defined over at least three demagnetization steps, with maximum angular deviation angles of less than 13°. Because of the depth of sampling of about 3.5 km (2.2 mi) below surface, steps below approximately 150°C or about 15 mT were not used because the formation temperature of about 120°C results in demagnetization steps at lower temperatures, having remanences acquired upon cooling during removal from the reservoir. Dipmeter logs were available for these wells and were the primary means of orientation, together with fracture orientation, as discussed in more detail below. After specimen and segment orientation, the remanence component means for the wells were calculated using Fisher (1953) statistics, followed by correction for borehole deviation.

Rock magnetic experiments were done to help determine the mineralogy and grain size of the dominant magnetic minerals in the samples. Isothermal remanent magnetization (IRM) acquisition paths were measured by progressively magnetizing a specimen in an increasing direct field in 15 steps to a maximum of 900 mT using a Sapphire Instruments SI-6 pulse magnetizer. The specimen was then AF demagnetized in 10 steps to 100 mT to provide information on magnetic grain size.

RESULTS: PETROLOGY AND GEOCHEMISTRY

The Wabamun carbonates have undergone a complex diagenetic history (Packard et al., 1990; Halbertsma, 1994; this study). The most important diagenetic processes that occurred were dolomitization, dolomite recrystallization, calcite cementation, silicification, and anhydritization, as well as some other processes.

Dolomitization: Petrography, Spatial Distribution, and Geochemistry

Early Fine-crystalline Dolomite

The earliest dolomite phase observed is fine-crystalline dolomite (Figure 2B) that replaces original mud matrix and fossils in the investigated carbonates. This dolomite consists of subhedral to euhedral crystals (10–25 μm) that are nonstoichiometric (mean = 55.78 mol% $CaCO_3$). In many samples, this dolomite has been recrystallized to medium-crystalline dolomite.

The $\delta^{18}O$ values of the fine-crystalline dolomite range from −5.53 to −6.07‰ (mean = −5.67‰), and the $\delta^{13}C$ values range from −0.22 to 0.85‰ (mean = 0.27‰; Figure 2A). These values are lower than the postulated values for Late Devonian carbonates (Denison et al., 1997). The $^{87}Sr/^{86}Sr$ values of this dolomite range from 0.708870 to 0.709212 (Figure 3B), which are slightly more enriched than expected for Devonian seawater (e.g., Denison et al., 1997).

Medium-crystalline Dolomite

The fine-crystalline dolomite has commonly been recrystallized to medium-crystalline dolomite, which is the most abundant type of dolomite in the gas field. These dolomite crystals are euhedral to subhedral (40–120 μm) (Figure 2A) and may occur also as vug-rimming zoned cement (Figure 2A, D). This dolomite is also nonstoichiometric (mean = 56.22 mol% $CaCO_3$) and shows evidence of dissolution (Figure 2D), and in some samples, it is associated with stylolites.

Medium-crystalline dolomite in the Wabamun formation has slightly lighter symbol $\delta^{18}O$ values than fine-crystalline dolomite, with a range of −5.45 to −7.36‰ (mean = −6.15‰). The $\delta^{13}C$ values range from −0.84 to 1.17‰ (mean = 0.44‰, Figure 3A). The medium-crystalline dolomite has comparable $^{87}Sr/^{86}Sr$ values to fine-crystalline dolomite, ranging from 0.70843 to 0.70984.

Coarsely Crystalline Dolomite and Saddle Dolomite

Coarse (200–560 μm) anhedral dolomite and patchy dolomite are also observed (Figure 4B). They appear to represent advanced stages of recrystallization of an earlier formed dolomite and are nonstoichiometric (mean = 56.68 mol% $CaCO_3$). These dolomites overlap isotopically with the medium-crystalline dolomite, as their symbol $\delta^{18}O$ values range from −5.83 to −6.81‰ (mean = −6.31‰), and $\delta^{13}C$ values range from −0.29 to 0.83‰ (mean = 0.15‰). In contrast, saddle dolomite crystals with their typical wavy extinction are restricted to fractures and vugs and must occur late in the tectonic history and occur only in a very few samples (Figure 2B). Their $\delta^{18}O$ and $\delta^{13}C$ signatures of −11.3 and −9.07‰, respectively, are very depleted relative to other types of dolomite in this study.

Calcitic Components

Calcite occurs as micritic material and as cement. The cement formed during both early and late diagenetic phases. Early calcite cement occurs as spar infills of vugs and hairline fractures and predates dissolution seams. This cement has been replaced by microdolomite. The second phase of calcite cement is equant and ferroan and occupies larger fractures (Figure 2C) that postdate the formation of dolomite, and that may have been formed in deeper burial settings.

Anhydrite

Anhydrite occurs as disseminated crystals in the matrix and as fracture fill that postdates dolomite. The first form has an early diagenetic origin, whereas the latter precipitated during the burial stage (Figure 4C). This late anhydrite has Sr isotopic values that are more enriched than Devonian seawater composition (Figure 3B). The results suggest that the late anhydrite formed from brines with a different composition than Devonian fluids.

FIGURE 2. Photomicrographs of (A) medium-crystalline dolomite (MD), vug-rimming dolomite (VD), and late calcite cement (CC). Sample 6-2-23 at 3380.69 m (11,091.5 ft). (B) Fine-crystalline dolomite (FD) in the matrix, saddle dolomite (SD), and stylolite (ST) cutting across matrix dolomite and late calcite cement. Sample 6-2-23 at 3388.19 m (11,116.1 ft). (C) Medium-crystalline dolomite (MD) crosscut by stylolite (ST) and succeeded by late calcite cement (CC) in a vein. Sample 6-2-33 at 3379.56 m (11,087.80 ft). (D) Zoned, vug-rimming dolomite under fluorescence light. Sample 6-2-33 at 3379.56 m (11,087.80 ft).

Silica

Silicification occurred both early and late in the diagenetic history of the Wabamun carbonates. The volume and occurrence of silica in the formation is minor. Early silicification occurs as fine-crystalline chert, replacing the original calcitic components (Figure 4A). Evidence of recrystallization of this silica is present in some samples where an increase in the crystal size is apparent. This may have occurred at depth. The early generation of replacement chert predates the first generation of dolomite (Figure 4A). Later fracture-fill silicification occurred in a few samples. This event is late and postdates the last dolomitization stage.

Other Diagenetic Events

Dissolution of the fine-crystalline dolomite, with concurrent formation of microporosity and moldic porosity, was observed in some samples. Dissolution may have occurred during early and later diagenetic events (e.g., associated with the formation of stylolites).

Both physical and chemical compaction phenomena were observed in the formation. Dissolution seams, early small-scale stylolites, and large-scale stylolites formed at variable burial depths. The stylolites formed either before or during the formation of the medium-crystalline and coarsely crystalline dolomite.

The fine-crystalline dolomite in the matrix is crosscut by both early dolomite-filled hairline fractures and by later larger fractures. The latter fractures are infilled by saddle dolomite, calcite, and anhydrite (Figure 2C). In a very few samples, fluorite crystals were present. They postdate the formation of dolomite but may coincide with the precipitation of late calcite cement.

The relationships above were used to construct the paragenetic sequence presented in Figure 5, which is based on the petrographic observations, crosscutting relationships, compactional features, and geochemical evidence.

RESULTS: ROCK MAGNETISM AND PALEOMAGNETISM

Rock Magnetism

Saturation IRM acquisition and demagnetization curves for the three wells are very similar, suggesting little difference in magnetic mineralogy throughout the reservoir (Figure 6). When the acquisition and demagnetization data are plotted on the magnetite and pyrrhotite crossover plots (Symons and Cioppa, 2000), the data follow the pyrrhotite type curves more closely; however, thermal demagnetization to 450°C suggests that magnetite is also present. We point out that the crossover point for almost all of the data is about 0.5, indicating little magnetic interaction between the grains.

Paleomagnetism

Two components of magnetization were isolated in most specimens. A low-temperature and low-coercivity remanence that was removed between about 160 and 300°C is apparent in all three wells (Figures 7–9;

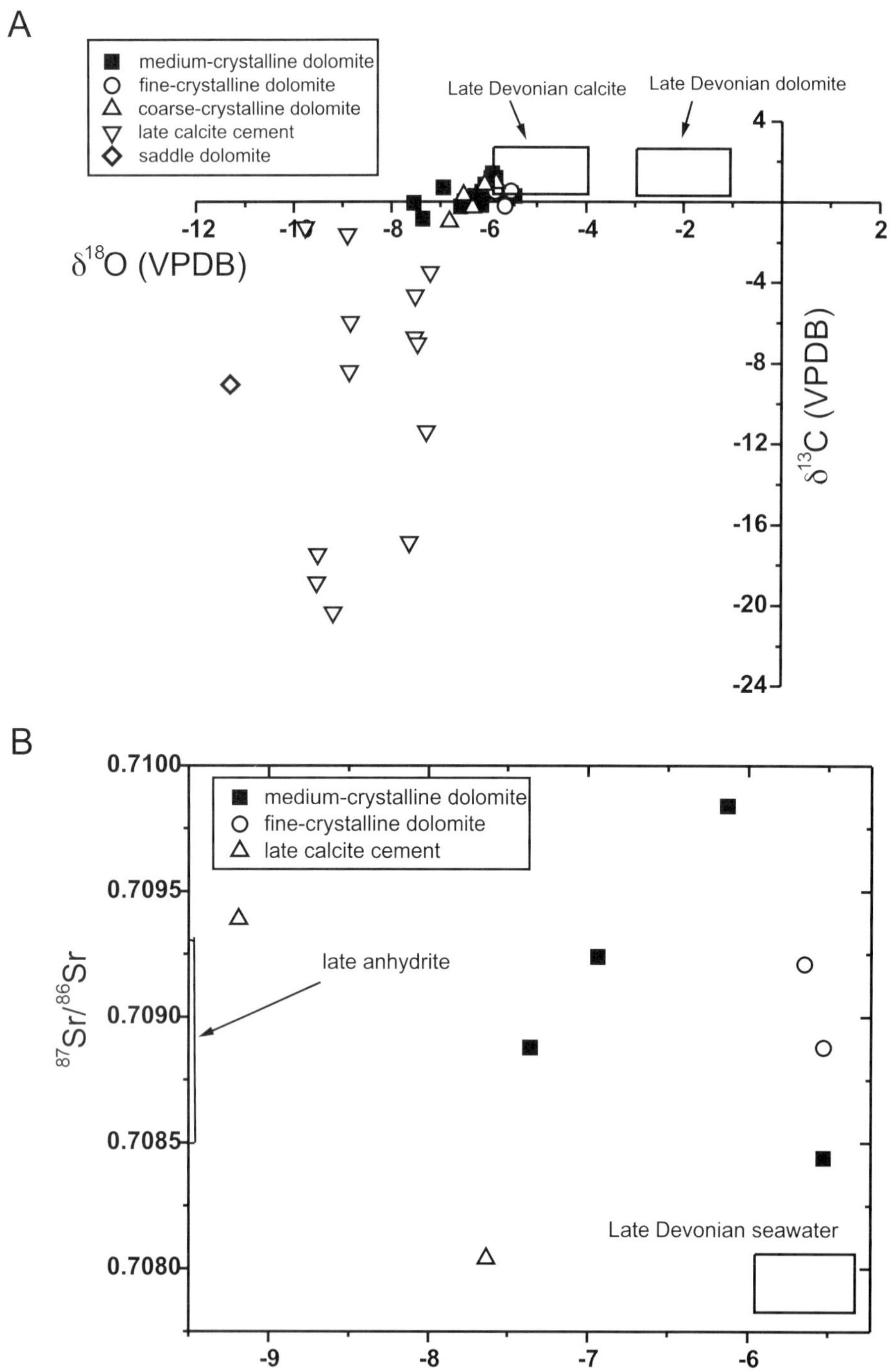

FIGURE 3. (A) Oxygen and carbon isotopic compositions of dolomite and calcite components. The boxes represent Late Devonian calcite and dolomite precipitated from seawater with no further alteration (see text for details). (B) Oxygen isotopes vs. $^{87}Sr/^{86}Sr$ ratios of calcite, dolomite, and anhydrite samples. The box represents Late Devonian seawater (Denison et al., 1997).

Table 2). In SL16, this magnetization has a positive (downward) inclination; in SL13, the magnetization has a negative (upward) inclination; and in SL15, the magnetization has both positive and negative polarities (Figure 10). This magnetization is probably carried by pyrrhotite or a combination of pyrrhotite and magnetite. The characteristic remanent magnetization, which was isolated between 300 and 450°C, is present in all wells and is reversed (Figures 7 –9; Table 2). The thermal demagnetization characteristics suggest that this magnetization is probably carried by magnetite. A small number of samples in well SL15 carry a third, high-temperature component, isolated at temperatures greater than 450°C. This high-temperature component is of normal polarity.

In SL16, both the dipmeter rotation correction and the viscous remanent magnetization method of orientation placed the low-temperature and low-coercivity magnetization subparallel to the present Earth's magnetic field at $D = 5°$, $I = 74°$, $\alpha_{95} = 5°$, and $k = 34$ (where D is declination, I is inclination, α_{95} is the radius of the cone of 95% confidence, and k is the precision parameter, all of which are from Fisher, 1953) (Figure 11, L-16). Use of the same methods places the characteristic remanent magnetization orientation in this well at $D = 162°$, $I = -65°$, $\alpha_{95} = 11°$, and $k = 29$.

In SL13, the reversed low-temperature and low-coercivity magnetization did not permit the viscous remanent magnetization method of orientation to be used. Use of the dipmeter logs and fracture orientation methods placed the low-temperature and low-coercivity component at $D = 175°$, $I = -75°$, $\alpha_{95} = 9°$, and $k = 24$, and its characteristic remanent magnetization direction at $D = 160°$, $I = -69°$, $\alpha_{95} = 7°$, and $k = 26$.

Figure 11 shows the variation in low-temperature and low-coercivity magnetization inclination in SL15.

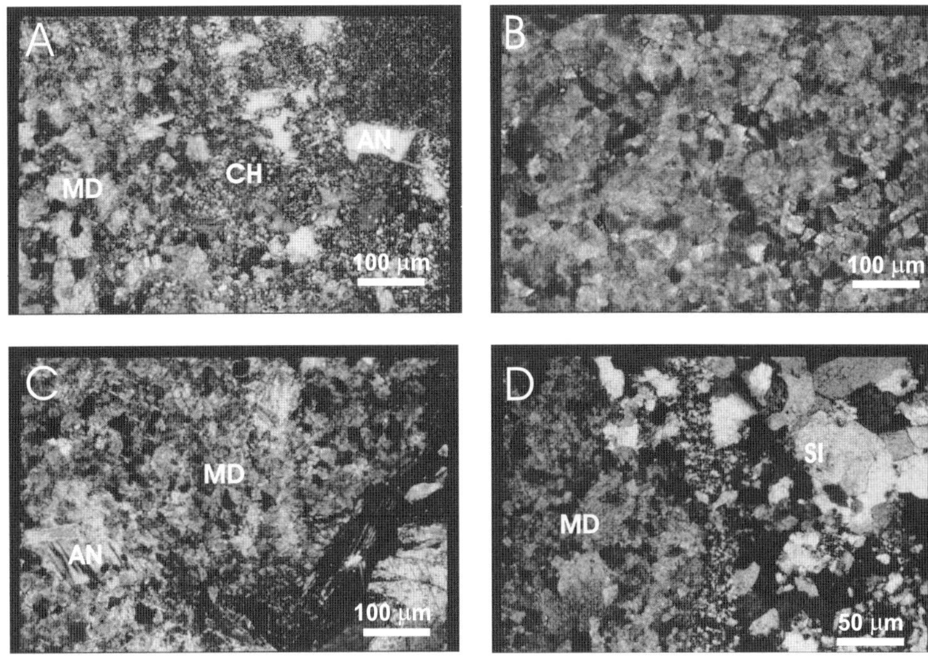

FIGURE 4. Photomicrographs of (A) medium-crystalline dolomite (MD) in the matrix replacing early chert (CH) that was also replaced by anhydrite (AN). Sample 4-12 at 3621.76 m (11,882.4 ft). (B) Coarse, recrystallized mesodolomite, very porous (moldic?), with some euhedral, vug-rimming dolomite. Sample 6-2-33 at 3372.54 m (11,064.8 ft). (C) Fine- to medium-crystalline dolomite (MD) in the matrix. Also shown is a late anhydrite vein (AN). Sample 10-25-33 at 3546.12 m (11,634.25 ft). (D) Medium-crystalline dolomite (MD) and two generations of quartz (SI) filling a fracture. Sample 10-25-33 at 3580.5 m (11,747 ft).

Again, the lack of a viscous remanent magnetization made the use of the dipmeter logs necessary, and the characteristic remanent magnetization direction was $D = 149°$, $I = -58°$, $\alpha_{95} = 9°$, $k = 16$. The high-temperature component had a direction of $D = 356°$, $I = 59°$, $\alpha_{95} = 18°$, and $k = 26$.

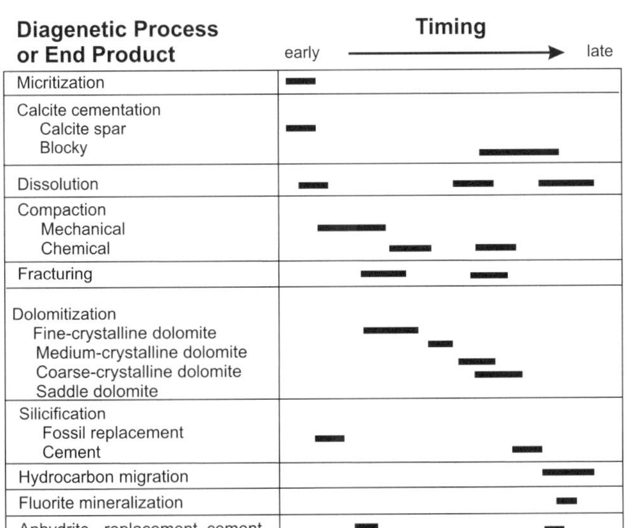

FIGURE 5. Paragenetic sequence and diagenetic history for the Wabamun upper Member, Limestone field, as determined from petrographic observations, crosscutting relationships, compactional features, and geochemical evidence.

DISCUSSION

Dolomitization

The paragenetic sequence (Figure 5), determined through petrological and geochemical analysis, allows constraints to be placed on the timing and origin of dolomitization and other diagenetic events. Petrographic evidence suggests that matrix fine-crystalline dolomite formed early in the diagenetic history during shallow burial of the Wabamun Group. Its micrometer-scale crystal size and nonstoichiometric nature, as well as its habit from replacing original calcitic components, suggest formation in a diagenetic environment, where marine pore fluids likely influenced its formation. This dolomite appears to have formed prior to any appreciable compaction. However, most of the present-day isotopic signatures are inconsistent with those properties characteristic of syndepositional and/or very shallow burial, which may indicate modifications during successive stages of diagenesis. $\delta^{18}O$ isotope values show a negative shift in the magnitude of about 4‰, but the $\delta^{13}C$ values seem to be rock buffered (Figure 3A). The isotopic values of this dolomite also fall very close to, or overlap with, those of the medium-crystalline dolomite.

Recrystallization of fine- to medium-crystalline dolomite is supported by petrographic and geochemical evidence. These include (1) a gradual increase in dolomite crystal size; (2) the presence of zoning in some dolomite; (3) a negative shift in oxygen isotopes; and (4) comparable enrichment of radiogenic Sr ratios (Figures 2 –4).

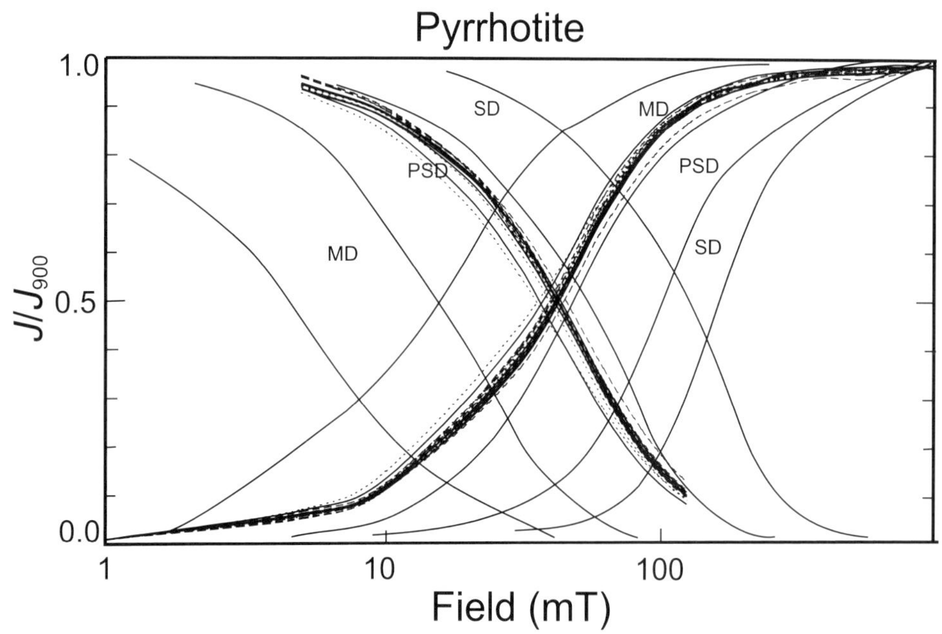

FIGURE 6. Saturation isothermal remanent magnetization acquisition and demagnetization curves for samples from each well plotted on the template curves for pyrrhotite (Symons and Cioppa, 2000).

Recrystallization of early-formed dolomite has been observed previously in other formations in the Western Canada Sedimentary Basin and elsewhere (e.g., Mazzullo, 1992; Smith and Dorobek, 1993; Al-Aasm and Lu, 1994; Kupecz and Land, 1994; Malone et al., 1994; Durocher and Al-Aasm, 1997; Yoo and Lee, 1998; Al-Aasm, 2000). This recrystallization may occur in shallow, low-temperature environments to deep, high-temperature environments, and it could alter both isotopic and geochemical signatures. In the present study, this recrystallization is believed to have occurred during the shallow to intermediate burial stages, after both early silicification and the stabilization of calcitic components but prior to late calcite and late silica cementation. Sr isotopic ratios of both fine- and medium-crystalline dolomite (Figure 3B) show enriched values relative to Devonian seawater composition. This can be interpreted to reflect the nature and source of the dolomitizing and alteration fluids on the formation of these dolomites. Burial brines that could have been affected by interaction with siliciclastics or basement rocks will enrich Sr isotopes in the fluids (e.g., Faure and Powell, 1972; Mountjoy et al., 1992). Packard et al. (1990), in their investigation on the upper Wabamun carbonates in the Peace River Arch area, showed similar isotopic values in matrix dolomite. In contrast, the only saddle dolomite sample analyzed in this study shows much depleted oxygen and carbon isotopic values in comparison to matrix dolomites (Figure 3A). The formation of the saddle dolomite could have occurred in a deep burial environment, where increasing temperature is associated with thermal sulfate reduction (e.g., Machel et al., 1996). Its formation can be also related to the formation of late calcite cement.

Calcite Cementation, Silicification, and Anhydritization

Both early and late calcite cements were observed in this study. Stable oxygen isotopes of the late calcite cement (Figure 3A) show a slight shift with respect to Late Devonian calcite values and a more pronounced shift with respect to carbon isotopes. Both shifts can be explained by increasing temperature during later diagenetic stages. The light carbon could have been produced by thermal abiotic reactions at depth (Machel et al., 1996). However, the $^{87}Sr/^{86}Sr$ ratios of this calcite (Figure 3B) show two end members, with one sample having a nonradiogenic signature and the other samples showing more enriched values. The former could represent an earlier generation of cement, possibly buffered by Late Devonian seawater, and the latter could represent formation from basinal fluids affected by siliciclastics and basement rocks.

Silicification commenced during shallow burial. Petrographic evidence supports this interpretation. The association of silica with microdolomite, the replacement and preservation of internal skeletal structures by the silica (Figure 4A), and the occurrence of silica predating mesodolomite would argue for early replacement of carbonates. The source of this silica could be biogenic (e.g., Al-Aasm and Lu, 1994). The later vein-fill quartz is related to late diagenetic events concurrent with the formation of saddle dolomite and late calcite cement. Packard et al. (2001) observed similar silica formation in the upper Wabamun formation in the Parkland field of British Columbia. However, they relate its formation and the formation of matrix dolomite to hydrothermal fluids during shallow burial that they related to an early Tournaisian tectonic event. In the present study, evidence for hydrothermal alteration is not obvious.

The early disseminated anhydrite probably represents primary deposition in an intertidal setting. In contrast, secondary anhydrite is found in some fractures and can be related to deeper burial brines that originated in the basin. These brines could also have

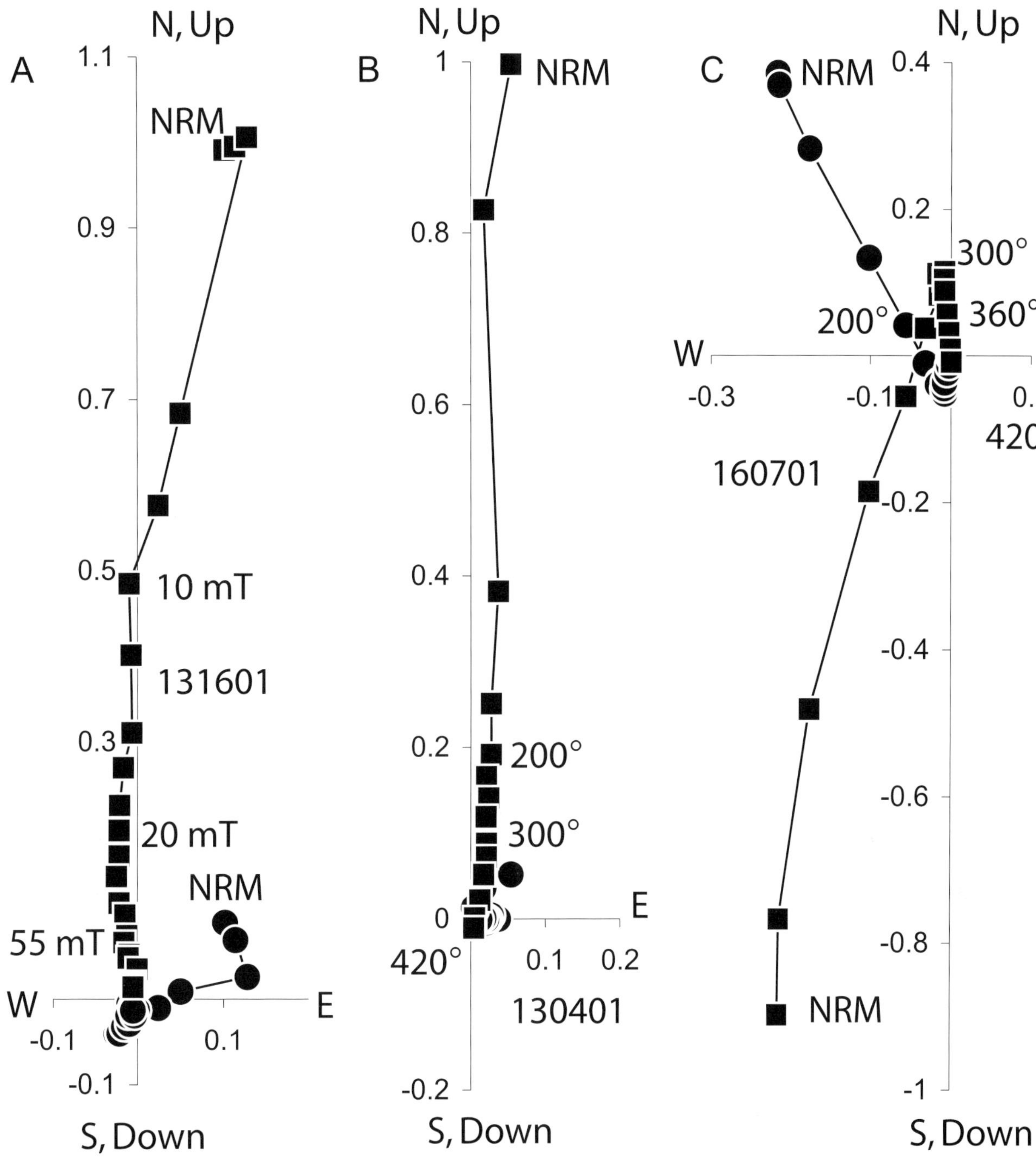

FIGURE 7. Normalized orthogonal demagnetization plots (Zijderveld, 1967) for representative, nonazimuthally oriented specimens from wells SL13 and SL16 in the Limestone field on step-alternating field and thermal demagnetization. (A) Specimen 131601; (B) specimen 130401; (C) specimen 160701. Each point represents a single demagnetization step. The plots overlay two separate graphs: on one, the east and north constituents of the magnetization show the projection of the direction on a horizontal plane (circles), and on the other, the east and horizontal constituents show the projection of the magnetization in the vertical plane (squares). Units on the *X*- and *Y*-axes are normalized magnetization intensity (intensity at step or natural remanent magnetization intensity). Note that for these nonazimuthally oriented specimens, the north and east directions are arbitrary. NRM = natural remanent magnetization.

been responsible for the formation of the minor fluorite that is infilling some late fractures. Sr isotopic signatures of this anhydrite show a departure from Late Devonian seawater composition (Figure 3B).

The high acid gas content of this field led Eliuk (1984) to propose a model for the origin of hydrogen sulfide. In his model, anhydrite, hydrocarbons (or organic matter), and water reacted to form hydrogen sulfide,

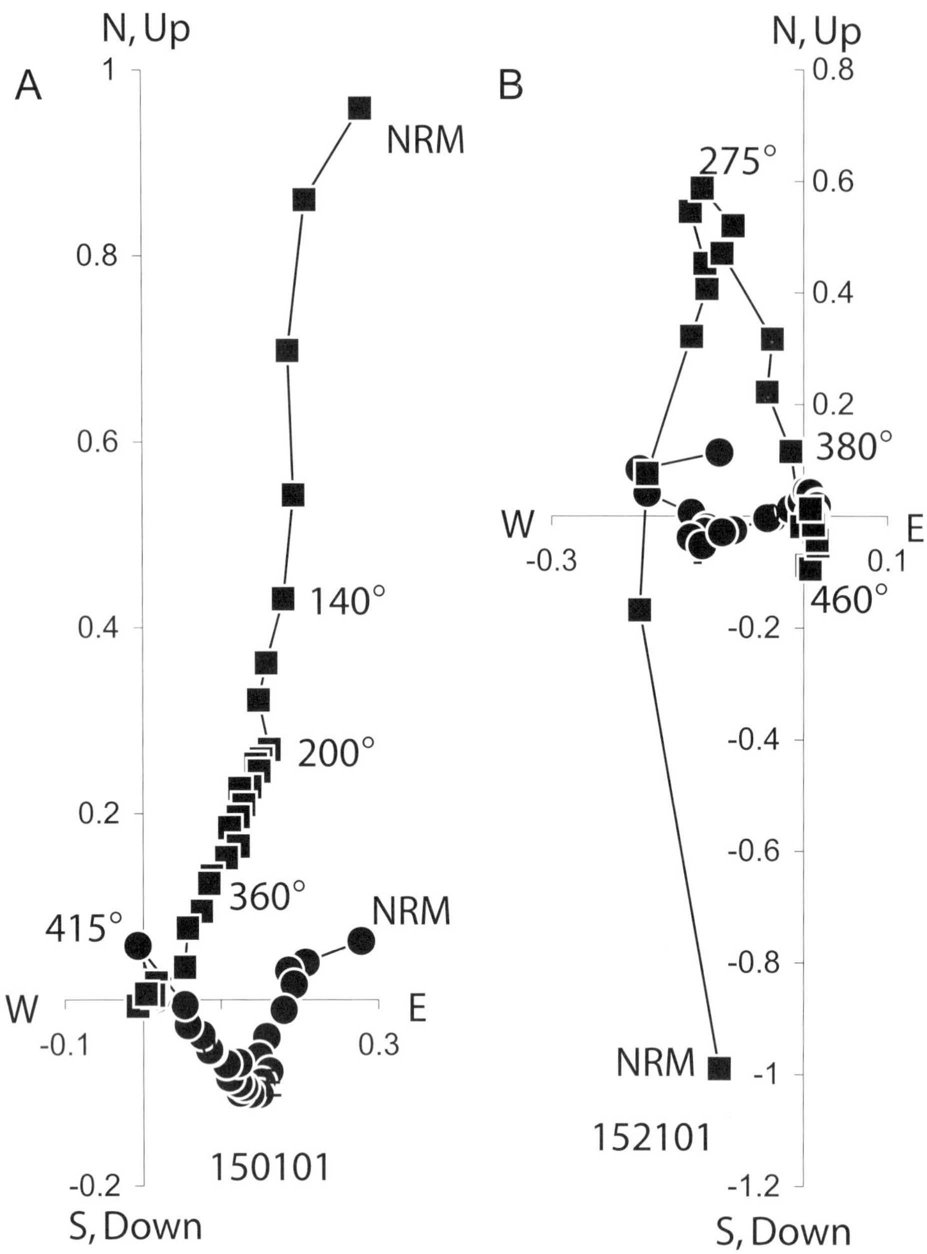

Figure 8. Normalized orthogonal demagnetization plots for representative specimens from the SL15 well of the Limestone field. (A) Grainstone specimen 150101; (B) mudstone specimen 152101. The coarser grained specimen shows a reversed-polarity, low-temperature component, whereas the finer grained specimen displays a normal-polarity, low-temperature component. See text for additional discussion. Symbols are as in Figure 7. NRM = natural remanent magnetization.

carbon dioxide, sulfur, calcite, and water and increased porosity under temperature and pressure conditions equivalent to a depth of about 2.5–3 km (1.5–2 mi). Evidence supporting this model is apparent in the isotopic values for late calcite cement that are depleted in carbon isotopes (Figure 3B), potentially indicating an organic source for carbon resulting from thermochemical sulfate reduction (e.g., Machel et al., 1996).

Paleomagnetic Components

SL16

In SL16, the normal-polarity, low-temperature, and low-coercivity component is clearly a true viscous remanent magnetization. The paleopole of the characteristic remanent magnetization direction determined using the dipmeter, fracture orientation, and viscous remanent magnetization methods is located at long. 139°E, lat. 79°N (9.3°, 11.6°) (radii of the oval of 95% confidence along and perpendicular to the site-pole great circle). This paleopole is located midway between the latest Jurassic and Miocene sections of the apparent polar wander path (APWP) (Figure 11).

SL13

The low-temperature and low-coercivity magnetization from SL13 is reversed (Figures 7, 9) and thus cannot be a viscous remanent magnetization carried by preexisting magnetite or pyrrhotite. Three possible interpretations of this magnetization exist: an uplift magnetization, a chemical remanent magnetization, or a self-reversed magnetization. The magnetization could result from uplift and cooling (see Symons and Cioppa, 2002). Under this hypothesis, the unblocking temperatures of less than 300°C, which is seen for this magnetization in magnetite and in pyrrhotite, would suggest that the rocks should have conodont color alteration indices (CAI) of about 2.5. Observed conodont CAI values are greater than or about 3 immediately to the west of this field in the foreland belt. This suggests that an uplift magnetization in a reversed-polarity Earth's magnetic field is a reasonable explanation. Alternatively, the component's direction is very close to a reversed Tertiary direction, and hence,

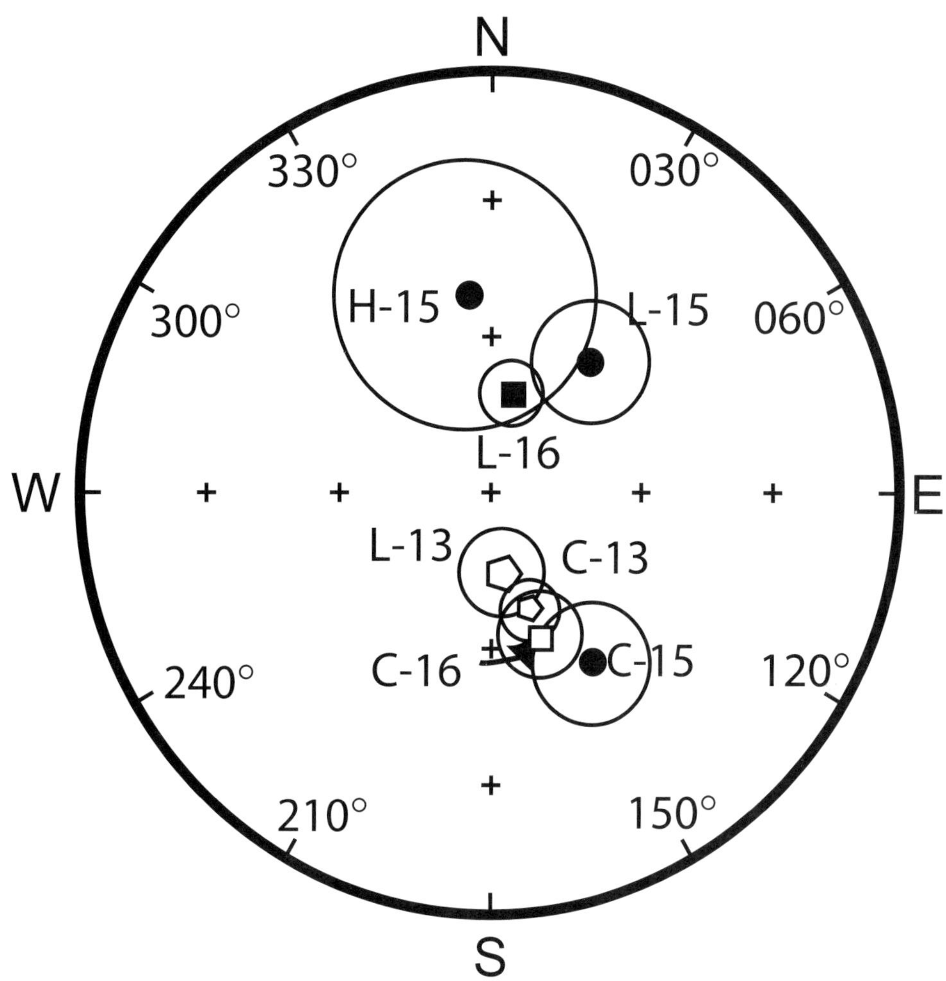

FIGURE 9. Mean magnetization directions for wells plotted on an equal-angle stereonet projection. L indicates the low-temperature and low-coercivity magnetization; C indicates the characteristic magnetization; and H indicates the high-temperature magnetization. Thus, C-13 is the characteristic remanent magnetization from well SL13. The solid circles indicate downward (positive) inclination; the open symbols indicate upward (negative) inclinations.

it could also be a chemical remanent magnetization that formed during a Tertiary event, such as fault reactivation or fluid migration, or the thermochemical sulfate reactions proposed by Eliuk (1984). The thermochemical sulfate reactions would produce sulfide minerals such as pyrrhotite or pyrite that could carry a remanence. The third possibility, that of a self-reversed magnetization, such as that seen in the presence of both magnetite and pyrrhotite (Bina and Daly, 1994), is unlikely, because there is no evidence of the temperatures greater than 500°C required for oxidation of the pyrrhotite to magnetite and the production of a chemical remanence.

The SL13 characteristic remanent magnetization paleopole lies at long. 172°E, lat. 78°N (9.5°, 11.1°) on the Eocene section of the APWP, although the error is such that a latest Cretaceous age cannot be ruled out with 1σ confidence. The characteristic remanent magnetization paleopoles for SL16 and SL13 are similar but not identical; however, a bedding correction has not been applied. Core logging during sampling indicated that bedding was within about 5° of horizontal, and whereas tilt correction is unlikely to affect the results significantly, the potential induced error as well as the similarity in magnetic characteristics suggests that the magnetizations are the same.

SL15

The low-temperature and low-coercivity magnetization in SL15 follows a peculiar pattern (Figure 10). A change in magnetization polarity is seen at a depth of about 3612 m (11,850 ft), corresponding to a downward facies change from grainstone to mudstone. At depths above 3612 m (11,850 ft), the polarity is reversed, resembling that observed in SL13, whereas below 3612 m (11,850 ft), the polarity is normal, similar to SL16, and suggests that the component may be a viscous remanence. Thus, there appears to be some facies or facies-associated control on these magnetization components.

The in-situ paleopole for the SL15 characteristic remanent magnetization (Figure 11, C-15) lies on the Middle to Late Jurassic section of the APWP at long. 137°E, lat. 66°N (10.3°, 13.9°). Although the in-situ pole positions suggest a significant age difference between the SL13 or SL16 and SL15 magnetizations, the characteristic remanent magnetization directions (Figure 9) differ by only about 13° of arc. At face value, the data suggest that the SL15 characteristic remanent magnetization is postdeformational in origin, and that the deformation had to have occurred prior to the Middle Jurassic. This statement has two problems. First, Cordilleran Laramide orogenesis and deformation are known to have occurred after the Middle Jurassic, and second, the rock magnetic results suggest that the

Table 2. Paleomagnetic data from the Limestone field.

Well	n	Magnetization Components								Paleopoles					
		Low T/C				ChRM				Low T/C			ChRM		
		D (°)	I (°)	α_{95} (°)	k	D (°)	I (°)	α_{95} (°)	k	Latitude	Longitude	dp, dm	Latitude	Longitude	dp, dm
SL13	13, 20	175.4	−75.3	8.7	23.8	160.3	−69.3	6.5	25.9	77.3	235.4	14.5, 15.9	77.7	171.7	9.5, 11.1
SL15	5, 16	37.2	64.4	14.2	30.2	149.5	−58.1	9.4	16.2	65.2	331.5	18.2, 22.9	65.5	137.5	10.3, 13.9
H-15	4					355.6	59.2	18.3	26.2				78.6	81.1	18.4, 26.7
SL16	27, 22	4.6	74.4	4.8	33.9	161.5	−65.4	10.9	29.1	78.3	255.8	8.1, 8.8	77.5	146.1	9.3, 11.6

n: (x,y) number of specimens used for low-temperature remanence (x) and characteristic remanence (y) calculations. Magnetization components: Low T/C = low-temperature, low-coercivity component; D = declination; I = Inclination; α_{95} = radius of cone of 95% confidence in degrees; and k = precision parameter of Fisher (1953); ChRM = characteristic remanent magnetization; dp, dm = semi-axes of the oval of 95% confidence along and perpendicular to the site-pole great circle.

magnetic mineralogy in SL15 is identical with that in SL13 and SL16 (Figure 7), implying a single common source for the magnetic minerals.

An alternative explanation of the data is that the SL15 characteristic remanent magnetization is similar in age to that of the characteristic remanent magnetization in SL16 or SL13, and that differential postmagnetization tilting or thrust sheet rotation has occurred in this area. Stockmal et al. (2001) have documented such rotations of thrust sheets in the Cordillera. Therefore, there are two explanations for the difference between the SL15 and SL13 or SL16 characteristic remanent magnetization directions. First, we are indeed dealing with two separate magnetization events. Second, an error has been introduced into the SL15 data by the processes discussed above. Given the similarity in rock magnetic properties and demagnetization characteristics, we prefer the explanation of an induced error and consider the characteristic remanent magnetizations in all three wells to be the same.

High-temperature Component in SL15

The high-temperature component (Figure 9, H-15) is seen only in a few samples in the lower mudstone facies of SL15, again suggesting some facies or facies-associated control of remagnetization. The unblocking temperatures suggest that magnetite carries this magnetization. The bedding correction discussed above, when applied to the high-temperature magnetization, causes little shift in its direction, and its paleopole is located at long. 81°E, lat. 78°N (18°, 27°) to indicate an Early Jurassic age.

Paleomagnetism, Petrology, and Geochemistry: Explanations?

Neither the paleomagnetic data nor the geochemical data show any evidence for primary magnetic or geochemical signatures. The petrologic data suggest a relatively early timing for the formation of the fine-crystalline dolomite. However, this dolomite has been substantially recrystallized. The earliest magnetization is observed in only a few samples and is Early Jurassic and thus may be dating the recrystallization event instead of the dolomitization.

The dominant characteristic remanent magnetization is reversed in polarity; its age could be Late Jurassic, Early Cretaceous, or Tertiary but not middle or Late Cretaceous for two reasons. First, the observed inclinations would need to be significantly steeper to provide a paleopole of the latter age; second, the Earth had a significant episode of normal magnetic polarity, the Cretaceous Normal Superchron, between about 119 and 83 Ma (Opdyke and Channell, 1996). Given

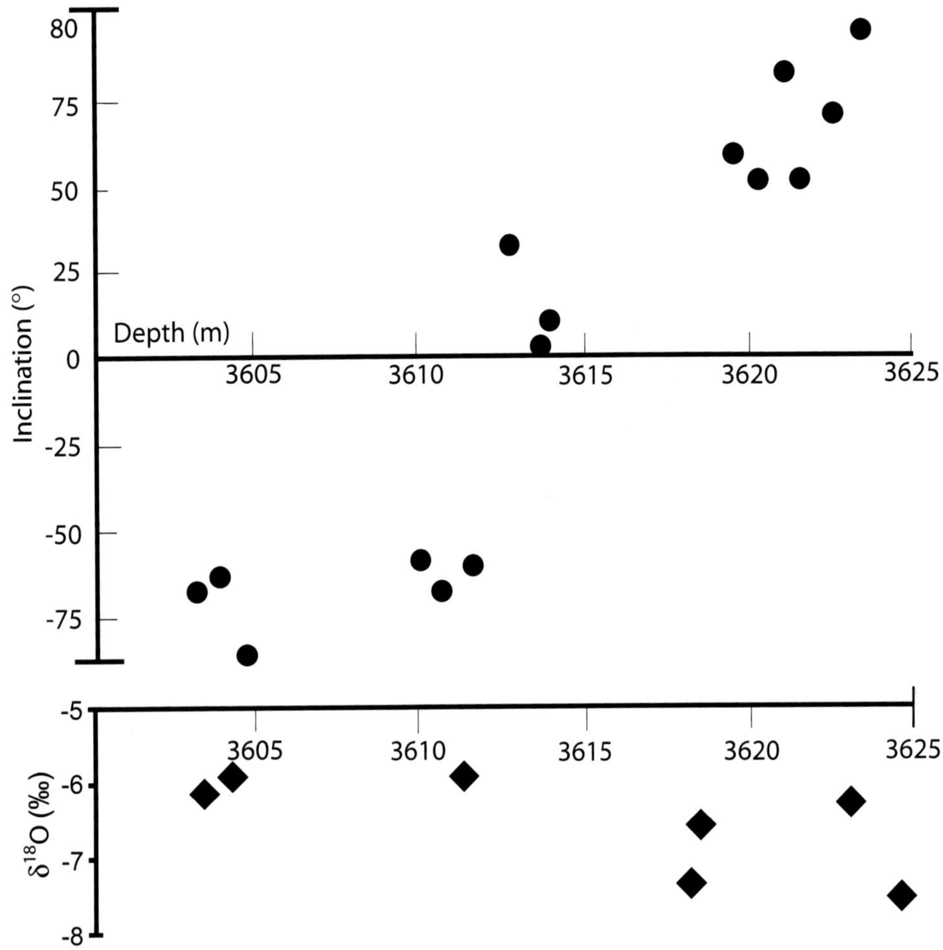

FIGURE 10. Variation in low-temperature and low-coercivity inclination (circles) and oxygen isotopic composition (diamonds) in the medium-crystalline dolomite as a function of depth in core SL15. The uppermost section consists of grainstone facies; the lower section is a mudstone facies. Note the slight change in oxygen isotopic composition that is coincident with the change in magnetization characteristics. This may signify some diagenetic process that affected one facies but not the other, potentially because of variations in porosity and permeability.

the potential errors, as well as the fact that previous paleomagnetic studies in the area south of the Peace River Arch area have yielded Late Cretaceous to Tertiary ages, it is more likely that the characteristic remanent magnetization is Tertiary instead of Jurassic. Thus, the characteristic remanent magnetization was probably formed by a significant diagenetic event that occurred relatively late in the tectonic history of the area. Upon examination of the paragenetic sequence (Figure 5), the most likely diagenetic event corresponding to the remagnetization is hydrocarbon-bearing fluid migration. Alternatively, coarsely crystalline dolomite formation or dolomite recrystallization during deep burial could have produced the magnetization.

The low-temperature, low-coercivity, reversed-polarity magnetization found in SL13 and the lower sections of SL15 has been previously discussed. The Tertiary age of this magnetization suggests a very late cause. The most likely alternatives are uplift or late diagenetic events, such as thermochemical sulfate reduction and precipitation of the coarsely crystalline dolomite, secondary calcite cement, and anhydrite.

Multiple Remagnetization Events and Orogenic Remagnetizations

The observed components can also be examined in the context of Enkin et al.'s (2000) model for the formation of two magnetizations in the Cordillera. Enkin et al. (2000) considered their A component to be a chemical magnetization that is related to diagenesis and their B component to be a partial thermoremanent magnetization, which was created by uplift and tilting resulting from thrusting. Interpretation of a component as a partial thermoremanent uplift remanence implies that any direction removed at temperatures of less than that of the uplift component must also be uplift related. Enkin et al.'s (2000) study indicates that the B component was removed between 150–200 and 200–400°C and the A component between 150–350 and 450–500°C. However, the B component is not always present. The model suggests that differential uplift resulting from thrusting created the uplift remanence closer to the frontal thrust, with the implication that the rocks that were further away from the thrust (back into the Front Ranges) could not have been buried as deeply because no B component is present. However, the conodont CAI values indicate that at least as much burial and heating occurred in the Front Ranges as in the inner foothills (Read et al., 1991). In addition, the age of the B component is equivocal; it could as easily be 120–130 Ma as 60 Ma. Therefore, although there are similarities in our results and interpretations, i.e., we consider the characteristic remanent magnetization in our samples

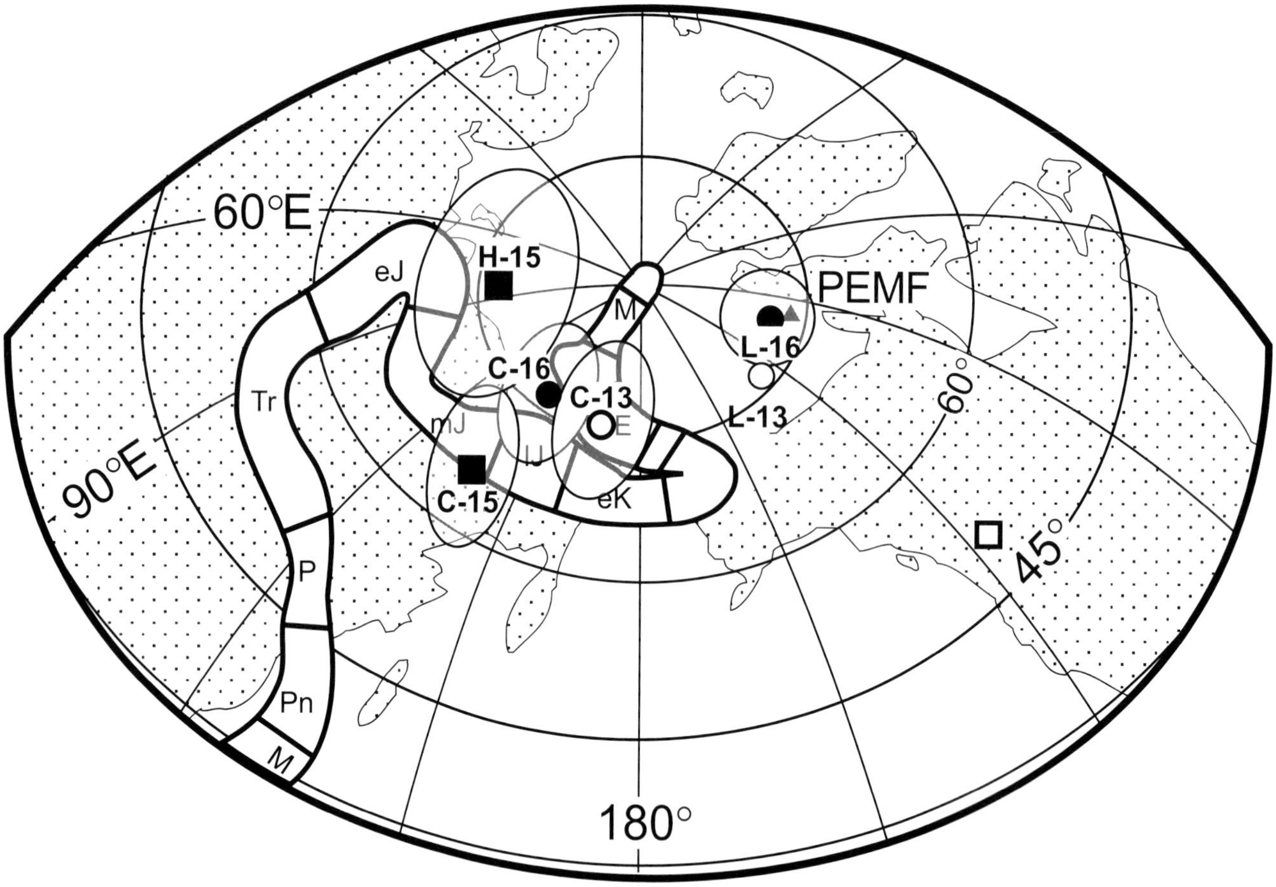

FIGURE 11. Paleopoles and 1σ errors for magnetization components determined for this study plotted on the combined apparent polar wander path for stable North America (Van der Voo, 1993; Besse and Courtillot, 1991). PEMF = present Earth's magnetic field. L indicates the low-temperature and low-coercivity magnetization; C indicates the characteristic magnetization; and H indicates the high-temperature magnetization. Thus, L-16 is the low-temperature component from well 16. Sampling locality is indicated by the hollow square. The solid (open) pole symbols indicate a normal (reversed) polarity for the Earth's magnetic field.

to be a chemical magnetization and the reversed low-temperature and low-coercivity component to be either an uplift remanence or a chemical remanence related to the latest states of diagenesis, we are unable to say with any degree of confidence that the results from this study correspond to Enkin et al.'s (2000) interpretation.

In our data, it is clear that there are at least two and possibly as many as four remagnetization events in the Limestone reservoir rocks. The oldest is the Early Jurassic normal-polarity high-temperature magnetization isolated in SL15, which is probably not related to orogenesis but is instead related to recrystallization of early-formed fine-crystalline dolomite. In the section above, the reversed-polarity characteristic remanent magnetizations in the three wells have been interpreted as the same magnetization, with variations in the directions resulting from either sampling error or rotations on the thrust sheets. We consider this component to have formed as a result of orogenesis, either by hydrocarbon-bearing fluid migration or by deep burial recrystallization. The reversed-polarity, low-temperature, and low-coercivity magnetization in SL13 and SL15 was probably formed as either a partial thermoremanent magnetization caused by late Laramide orogenic uplift or by Tertiary mineral precipitation or alteration caused by processes such as thermochemical sulfate reduction or fluid migration.

CONCLUSIONS

Paleomagnetic, petrologic, and geochemical analysis of the upper section of the Wabamun Group in the Limestone gas field indicates that extensive diagenetic alteration and remagnetization have occurred. No indications of primary Devonian magnetizations or isotopic signatures are present in the data.

Petrologic analysis shows complete dolomitization of the wells sampled, with four generations of dolomite apparent: an early fine-crystalline dolomite phase, which has been extensively recrystallized to the dominant medium-crystalline dolomite, then by the coarsely crystalline dolomite phase, and finally by a minor saddle dolomite phase. The oxygen and carbon stable isotope values of the fine-crystalline, medium-crystalline, and coarsely crystalline dolomites are similar, whereas the saddle dolomite values are considerably more depleted.

Paleomagnetic analysis demonstrates that three magnetization components are present in the Limestone field: (1) a high-temperature magnetization of probable Early Jurassic age carried by magnetite; (2) at least one dominant reversed-polarity characteristic remanent magnetization of Late Jurassic or early Tertiary age; and (3) a low-temperature reversed-polarity magnetization in SL13 and SL15 that is either an uplift-induced magnetization or a late chemical remanent magnetization. These results reinforce the idea that, paleomagnetically speaking, the situation in the Cordillera is significantly more complex than the simple hypothesis commonly cited for the Appalachians (Oliver 1986, 1992), i.e., that a single, fluid-flow event induced remagnetization. Although this single flow event model can be used to explain some of the results, other factors must have produced the disparate remagnetizations in the Limestone gas field.

ACKNOWLEDGMENTS

The authors thank Shell Canada Ltd. for permitting the paleomagnetic and geochemical sampling of their core. Julie Clarke and Jeff Lonnee aided in petrologic and geochemical sampling and analysis. Funding was provided by a Natural Sciences and Engineering Research Council grant to D. T. A. Symons and I. S. Al-Aasm. This chapter was improved greatly by the comments of K. Osadetz and P. Robion.

REFERENCES CITED

Al-Aasm, I. S., 2000, Chemical and isotopic constraints for recrystallization of sedimentary dolomites from the Western Canada sedimentary basin: Aquatic Geochemistry, v. 6, p. 227–248.

Al-Aasm, I. S., and F. Lu, 1994, Multistage dolomitization of the Mississippian Turner Valley Formation, Quirk Creek field, Alberta: Chemical and petrologic evidence, *in* D. J. Glass, B. Beauchamp, and A. F. Embry, eds., Pangea: Global environments and resources: Canadian Society of Petroleum Geologists Memoir 17, p. 657–675.

Al-Aasm, I. S., B. E. Taylor, and B. South, 1990, Stable isotope analysis of multiple carbonate samples using selective acid extraction: Chemical Geology (Isotope Geoscience Section), v. 80, p. 119–125.

Besse, J., and V. E. Courtillot, 1991, Revised and synthetic apparent polar wander paths of the African, Eurasian, North American and Indian plates, and true polar wander since 100 Ma: Journal of Geophysical Research, v. 96, p. 4029–4050.

Bina, M., and L. Daly, 1994, Mineralogical change and self-reversed magnetizations in pyrrhotite resulting from partial oxidation: Geophysical implications: Physics of the Earth and Planetary Interiors, v. 185, p. 83–99.

Cioppa, M. T., I. S. Al-Aasm, D. T. A. Symons, M. T. Lewchuk, and K. P. Gillen, 1999, Correlating paleomagnetic, geochemical, and petrologic evidence to date diagenetic and fluid flow events in the Husky Moose reservoir, Alberta, Canada: Sedimentary Geology, v. 131, p. 109–129.

Cioppa, M. T., K. P. Gillen, M. T. Lewchuk, and D. T. A. Symons, 2000, Orienting drill core: Old and new paleomagnetic methods: Oil & Gas Journal, v. 98.21 (May 22, 2000), p. 46–53.

Cioppa, M. T., J. S. Lonnee, I. S. Al-Aasm, D. T. A. Symons, and K. P. Gillen, 2001, Facies and lithologic controls on paleomagnetism: An example from the Rainbow South field, Alberta, Canada: Bulletin of Canadian Petroleum Geology, v. 49, p. 393–407.

Cioppa, M. T., I. S. Al-Aasm, D. T. A. Symons, and K. P. Gillen, 2003, Dating penecontemporaneous dolomitization: Paleomagnetic and geochemical constraints: AAPG Bulletin, v. 87, p. 71–88.

Denison, R. E., R. B. Koepnick, W. H. Burke, E. A. Heatherington, and A. Fletcher, 1997, Construction of the Mississippian, Pennsylvanian and Permian seawater $^{87}Sr/^{86}Sr$ curve: Chemical Geology (Isotope Geoscience Section), v. 112, p. 145–167.

Durocher, S., and I. S. Al-Aasm, 1997, Dolomitization and neomorphism of Mississippian (Visean) upper Debolt Formation, Blueberry field, NE British Columbia, Canada: Geologic, petrologic and chemical evidence: AAPG Bulletin, v. 81, p. 954–977.

Eliuk, L., 1984, A hypothesis for the origin of hydrogen sulphide in Devonian Crossfield Member dolomite, Wabamun formation, Alberta, Canada, *in* L. Eliuk, ed., Carbonates in subsurface and outcrop: Canadian Society of Petroleum Geologists Core Conference, p. 245–290.

Enkin, R. J., K. G. Osadetz, P. M. Wheadon, and J. Baker, 1997, Paleomagnetic constraints on the tectonic history of the foreland belt, southern Canadian Cordillera: Canadian Journal of Earth Sciences, v. 34, p. 260–270.

Enkin, R. J., K. G. Osadetz, J. Baker, and D. Kisilevsky, 2000, Orogenic remagnetizations in the Front Ranges and inner foothills of the southern Canadian Cordillera; chemical harbinger and thermal handmaiden of Cordilleran deformation: Geological Society of America Bulletin, v. 112, p. 929–942.

Faure, G., and J. L. Powell, 1972, Minerals rocks and inorganic materials: Berlin-Heidelberg, Federal Republic of Germany, Springer-Verlag, v. 5, 171 p.

Fisher, R. A., 1953, Dispersion on a sphere: Proceedings of the Royal Astronomical Society, Series A, v. 217, p. 295–305.

Halbertsma, H. L., 1994, The Devonian Wabamun Group of the Western Canada Sedimentary Basin, *in* G. Mossop and I. Shetsen, eds., Geological atlas of the Western Canada Sedimentary Basin: Canadian Society of Petroleum Geologists and Alberta Research Council, p. 203–220.

Kirschvink, J. L., 1980, The least-squares line and plane and the analysis of paleomagnetic data: Geophysical Journal of the Royal Astronomical Society, v. 62, p. 699–718.

Kupecz, J. A., and L. S. Land, 1994, Progressive recrystallization and stabilization of early-stage dolomite: Lower Ordovician Ellenburger Group, west Texas, *in* B. Purser, M. Tucker, and D. Zenger, eds., Dolomite: A volume in honour of Dolomieu: International Association of Sedimentology Special Publication 21, p. 255–279.

Leach, D. L., D. Bradley, M. T. Lewchuk, D. T. A. Symons, G. deMarsily, and J. Brannon, 2001, Mississippi Valley-type lead-zinc deposits through geological time: Implications from recent age-dating research: Mineralium Deposita, v. 36, p. 711–740.

Lewchuk, M. T., I. S. Al-Aasm, D. T. A. Symons, and K. P. Gillen, 1998, Dolomitization of Mississippian carbonates in the Shell Waterton gas field, southwestern Alberta: Insights from paleomagnetism, petrography and geochemistry: Bulletin of Canadian Petroleum Geology, v. 46, p. 387–410.

Lewchuk, M. T., I. S. Al-Aasm, D. T. A. Symons, and K. P. Gillen, 2000, Late Laramide dolomitization of the Husky Rainbow "A" hydrocarbon reservoir, northwestern Alberta, Canada: Paleomagnetic and geochemical evidence: Canadian Journal of Earth Sciences, v. 37, p. 17–29.

Lonnee, J. S., M. T. Cioppa, I. S. Al-Aasm, D. T. A. Symons, M. T. Lewchuk, and K. P. Gillen, 1998, Correlating paleomagnetic data with the paragenetic sequence in the Middle Devonian Sulphur Point (Bistcho) Formation, northwestern Alberta: Geological Association of Canada Abstracts with Programs, p. A-112.

Machel, H. G., P. A. Cavell, and K. S. Patey, 1996, Isotopic evidence for carbonate cementation and recrystallization, and for tectonic expulsion of fluids in the Western Canada Sedimentary Basin: Geological Society of America Bulletin, v. 108, p. 1108–1119.

Malone, M. J., P. A. Baker, and S. J. Burns, 1994, Recrystallization of dolomite: Evidence from the Monterey Formation (Miocene), California: Sedimentology, v. 41, p. 1223–1239.

Mazzullo, S. J., 1992, Geochemical and neomorphic alteration of dolomite: A review: Carbonates and Evaporites, v. 7, p. 21–37.

Mountjoy, E. W., H. Qing, and R. H. McNutt, 1992, Strontium isotopic composition of Devonian dolomites, Western Canada Sedimentary Basin; significance of sources of dolomitizing fluids: Applied Geochemistry, v. 7, p. 59–75.

Oliver, J., 1986, Fluids expelled tectonically from orogenic belts: Their role in hydrocarbon migration and other geologic phenomena: Geology, v. 14, p. 99–102.

Oliver, J., 1992, The spots and stains of plate tectonics: Earth Science Reviews, v. 32, p. 77–106.

Opdyke, N. D., and J. E. T. Channell, 1996, Magnetic stratigraphy: San Diego, Academic Press, Inc., 346 p.

Packard, J., G. J. Pellegrin, I. S. Al-Aasm, I. Samson, and J. Gagnon, 1990, Diagenesis and dolomitization associated with hydrothermal karst in Famennian upper Wabamun ramp sediments, northwestern Alberta, *in* G. Bloy and M. Hadley, eds., The development of porosity in carbonate reservoirs: Canadian Society of Petroleum Geologists Short Course, p. 9-1–9-27.

Packard, J. J., I. S. Al-Aasm, I. Samson, Z. Berger, and J. Davies, 2001, A Devonian hydrothermal chert reservoir: The 225 bcf Parkland field, British Columbia, Canada: AAPG Bulletin, v. 85, p. 51–84.

Pannalal, S. J., D. T. A. Symons, and D. F. Sangster, 2004, Paleomagnetic dating of upper Mississippi Valley zinc-lead mineralization, Wisconsin, U.S.A.: Journal of Applied Geophysics, v. 56, p. 135–153.

Read, P. B., G. J. Woodsworth, H. J. Greenwood, E. D. Ghent, and C. A. Evenchick, 1991, Metamorphic map of the Canadian Cordillera: Geological Survey of Canada Map 1714A, scale 1:2,000,000, 1 sheet.

Smith, T. M., and S. L. Dorobek, 1993, Alteration of early-formed dolomite during shallow to deep burial: Mississippian Mission Canyon Formation, central to southwestern Montana: Geological Society of America Bulletin, v. 105, p. 1389–1399.

Stamatakos, J., A. Hirt, and W. Lowrie, 1996, The age and timing of folding in the central Appalachians from paleomagnetic results: Geological Society of America Bulletin, v. 108, p. 815–829.

Stockmal, G. S., K. G. Osadetz, D. Lebel, and P. K. Hannigan, 2001, Structure and hydrocarbon occurrence, Rocky Mountain foothills and Front Ranges, *in* Turner Valley to Waterton Lakes field trip guidebook: Geological Survey of Canada Open-file Report 4111, 161 p.

Symons, D. T. A., and M. T. Cioppa, 2000, Crossover plots: A useful method for plotting SIRM data in paleomagnetism: Geophysical Research Letters, v. 27, p. 1779–1782.

Symons, D. T. A., and M. T. Cioppa, 2002, The CAI and magnetic mineral unblocking temperatures: Implications for the Western Canada Sedimentary Basin: Physics and Chemistry of the Earth, v. 27, p. 1189–1193.

Symons, D. T. A., R. E. Enkin, and M. T. Cioppa, 1999, Paleomagnetism in the Western Canada Sedimentary Basin: Dating fluid flow and deformation events: Bulletin of Canadian Petroleum Geology, v. 47, p. 534–547.

Van der Voo, R., 1993, Paleomagnetism of the Atlantis, Tethys and Iapetus Oceans: Cambridge, Cambridge University Press, 411 p.

Yoo, C. M., and Y. Lee, 1998, Origin and modification of early dolomites in cyclic shallow platform carbonates, Yeongheung Formation (Middle Ordovician), Korea: Sedimentary Geology, v. 118, p. 141–157.

Zijderveld, J. D. A., 1967, AC demagnetization of rocks: Analysis of results, *in* D. W. Collinson, K. M. Creer, and S. K. Runcorn, eds., Methods in palaeomagnetism: Amsterdam, Elsevier, p. 254–286.

17

Gliniak, P., R. Laskowicz, A. Urbaniec, G. Lesniak, and P. Such, 2004, The facies development and reservoir properties in Late Jurassic carbonate sediments in the central Carpathian foreland, *in* R. Swennen, F. Roure, and J. W. Granath, eds., Deformation, fluid flow, and reservoir appraisal in foreland fold and thrust belts: AAPG Hedberg Series, no. 1, p. 347–355.

The Facies Development and Reservoir Properties in Late Jurassic Carbonate Sediments in the Central Carpathian Foreland

P. Gliniak
Polish Oil and Gas Company, Krakow, Republic of Poland

R. Laskowicz
Polish Oil and Gas Company, Krakow, Republic of Poland

A. Urbaniec
Polish Oil and Gas Company, Krakow, Republic of Poland

G. Lesniak
Oil and Gas Institute, Krakow, Republic of Poland

P. Such
Oil and Gas Institute, Krakow, Republic of Poland

ABSTRACT

Geological and petrophysical research focused on reservoir properties of Upper Jurassic carbonate deposits was carried out in the central part of the Carpathian foreland, also taking into consideration facies development and diagenetic processes. Petrophysical analyses of porosity, permeability (intergranular and fracture), and capillary pressure measurement were conducted on samples from 28 boreholes. Types and ranges of permeability and porosity were correlated with diagenetic processes (such as karst phenomena) on the basis of extracted lithofacies and regional tectonics. The best petrophysical properties are connected with organogenic buildups of the Upper Jurassic rocks. High values of intergranular permeability only occur on this type of rocks. Dual porosity-fracture permeability and fractures dominate other rocks. The karst phenomena improved storage and filtration properties of rocks.

DOI:10.1306/1025699H13125

INTRODUCTION

The main goals of this chapter are to identify the tectonic, facies, and diagenetic factors affecting the reservoir properties of the Upper Jurassic carbonate series of the Polish Carpathian foreland and to characterize the impact of porosity and permeability distribution on the overall fluid flow in the reservoir intervals.

GEOLOGICAL SETTING AND PETROLEUM SYSTEMS OF THE POLISH CARPATHIAN FORELAND

Location

The area of investigation is located in the central part of the Polish Outer Carpathians between Krakow in the west and Dêbica in the east. (Figure 1). North of the deformation front and below the sole thrust of the Carpathian allochthon, the autochthon is made up of Paleozoic series and Jurassic–Cretaceous platform series, which are overlain by the Neogene flexural sequence of the Carpathian Foredeep basin. The geological setting of this region is complex because of the interplay of sedimentary, tectonic, and erosional processes.

Petroleum Systems

Two oil systems occur in the Polish Carpathian foreland (Karnkowski, 1999). The first one, connected with clastic Miocene sediments, will not be considered in this study. The second petroleum system relates to Mesozoic carbonate rocks. Potential reservoir series occur in the Upper Jurassic and Lower Cretaceous intervals in which gas and oil reservoirs were found.

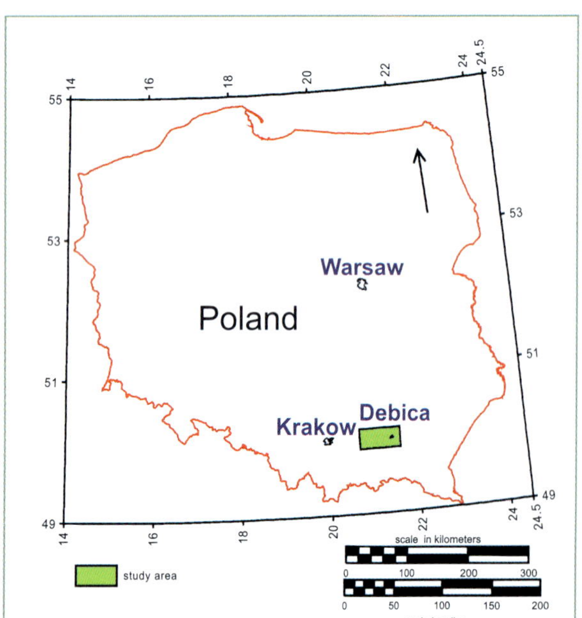

FIGURE 1. Sketch map of the study area.

The basic source rocks of this system comprise coal-bearing Silurian shales that show the capability to generate liquid hydrocarbon and dolomitic Devonian sediments with sapropel-type kerogen. Good source rocks were also found in Lower Carboniferous and Middle Jurassic samples.

Hydrocarbons were generated during two periods (Słupczyński and Biedrzycki, 1993). The first phase lasted from the Namurian to the Triassic (deposition of variegated sandstone, Buntsandstein), when the Devonian and Carboniferous sediments reached a burial depth of 2000 m (6600 ft), resulting in the first oil-generation episode (Lenk et al., 1993). The second and main phase of hydrocarbon generation occurred between the Barremian and the Campanian (generation of thermogenic gases). During this phase, the subsidence of Paleozoic sediments was equal to 3000–4000 m (10,000–13,300 ft). An additional factor affecting hydrocarbon generation is the overlapping of the Carpathian orogen during the Neogene. The gases that accumulated in the Jurassic traps contain as much as 5% of nitrogen and as much as 2% of hydrogen sulfide.

The migration of generated hydrocarbons was determined by dislocation zones and distribution and connectivity of the fracture systems.

Mesozoic Sedimentation and Reservoir Distribution

The main traps comprise fractured Upper Jurassic and Lower Cretaceous carbonate rocks. Fracturing and coeval porosity development result from tectonic uplift during the Kimmeridgian and Paleogene and from subsequent erosion episodes, the first one lasting from the Neocomian to Cenomanian and the second one occurring during the Late Cretaceous. Finally, reservoir burial resumed during the Neogene because of the Carpathian overthrust and coeval deposition of a synflexural sequence in the adjacent foredeep basin, thus inducing new diagenetic processes in relation to the circulation of reservoir waters.

Erosion has changed the thickness and lateral distribution of the Jurassic carbonate succession, which originally covered the whole study area. Such variations mainly relate to two post-Jurassic erosional episodes: one after Hauterivian and the second during the Late Cretaceous. Thick, Upper Jurassic strata occurred in the southeast prolongation of the Mid-Poland Synclinorium, having an average thickness of 1000 m (3300 ft) throughout the study area, but reaching locally as much as 1300 m (4300 ft) in the Outer Carpathians

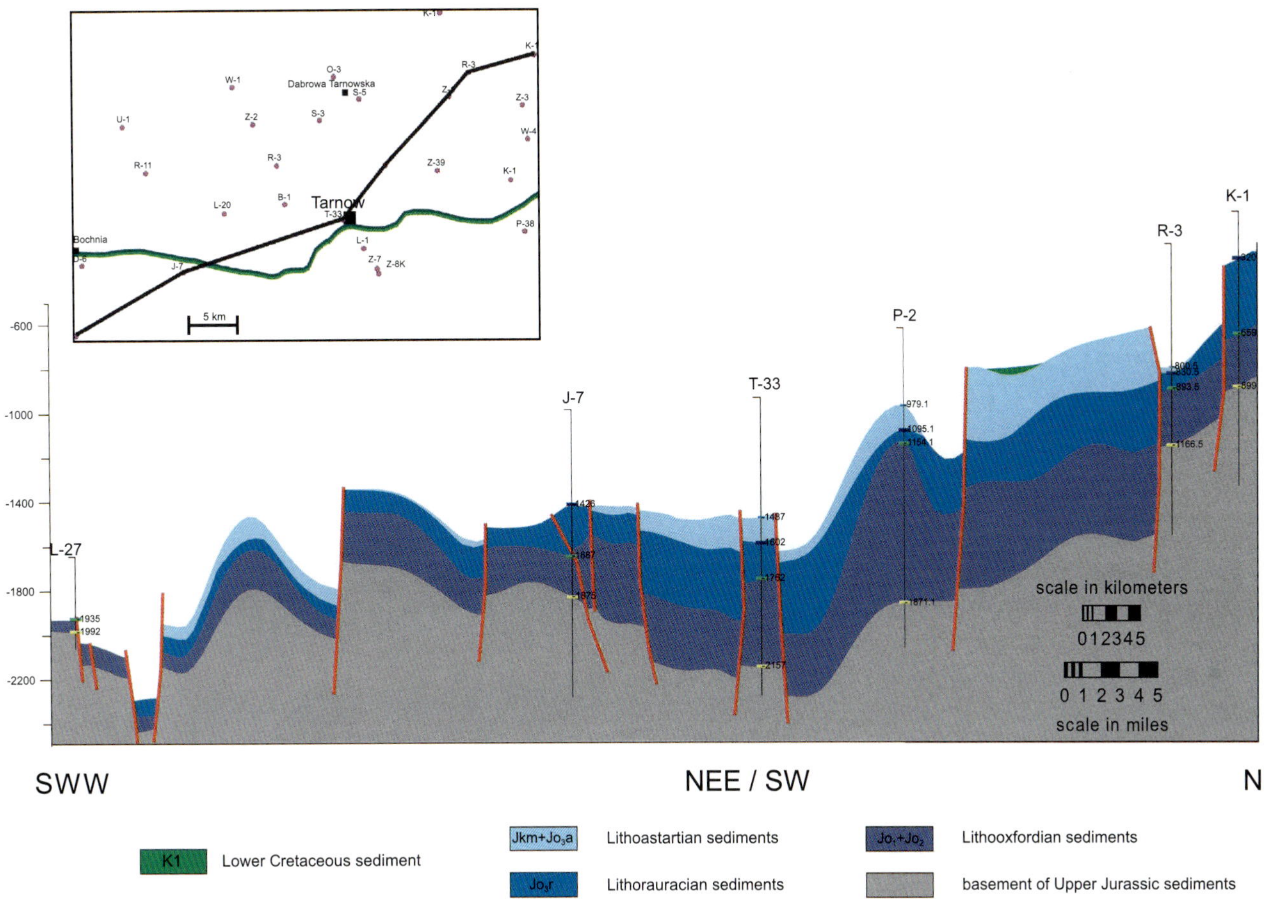

FIGURE 2. Vertical distribution of lithofacies in the Upper Jurassic and Lower Cretaceous series (first transect).

(Figures 2, 3). The Upper Jurassic sediments overlie either Middle Jurassic and Lower Triassic series or even locally Paleozoic and Precambrian rocks. The Upper Jurassic carbonates are overlain by the Lower Cretaceous (Berriasian, locally Valanginian) and Upper Cretaceous sediments of the Carpathian foreland, as well as a complex of Miocene clastic rocks with evaporitic intercalations. Carbonates reflect extensive carbonate platform development. Facies and thickness variations during the Oxfordian indicate that the reservoir strata were affected by several synsedimentary subsidence movements, which were especially pronounced in the southern part of Mid-Poland Synclinorium (Niemczycka, 1997). During the Upper Jurassic, the central part of the Polish Carpathian foreland was characterized by the development of an epicontinental marine setting that forms the reservoir interval. Faunal evidence indicates a connection with the Tethyan Ocean toward the southeast as well as with the Boreal Sea toward the north (Malinowska, 1980, 1986; Niemczycka, 1997). Bathymetry varied during deposition of the reservoir strata, with successive periods of shallowing-upward depositional conditions (Gliniak et al., 2000; Gliniak and Urbaniec, 2001; Gregosiewicz et al., 2001). Low-energy facies reflecting deposition on a deeper shelf are predominant in the lower and middle Oxfordian, whereas shallow, high-energy conditions producing oolithic, oncolithic, and bioclastic carbonates are more characteristic of the upper Oxfordian, the Kimmeridgian, and the Tithonian.

FACIES DEVELOPMENT OF THE UPPER JURASSIC SERIES

Lower Oxfordian

The lower Oxfordian sediments conformably overlie Callovian strata. Generally, the lower Oxfordian is represented by a marly-carbonate complex (Gregosiewicz et al., 2001). Bioturbation, as well as the occurrence of single silica sponges and foraminifera (*Globuligerina* sp.), characterizes this sedimentary series. These rocks are interpreted to represent the deepest Upper Jurassic carbonate sediments in the region and form a characteristic white horizon that is useful for regional correlation (Figure 4).

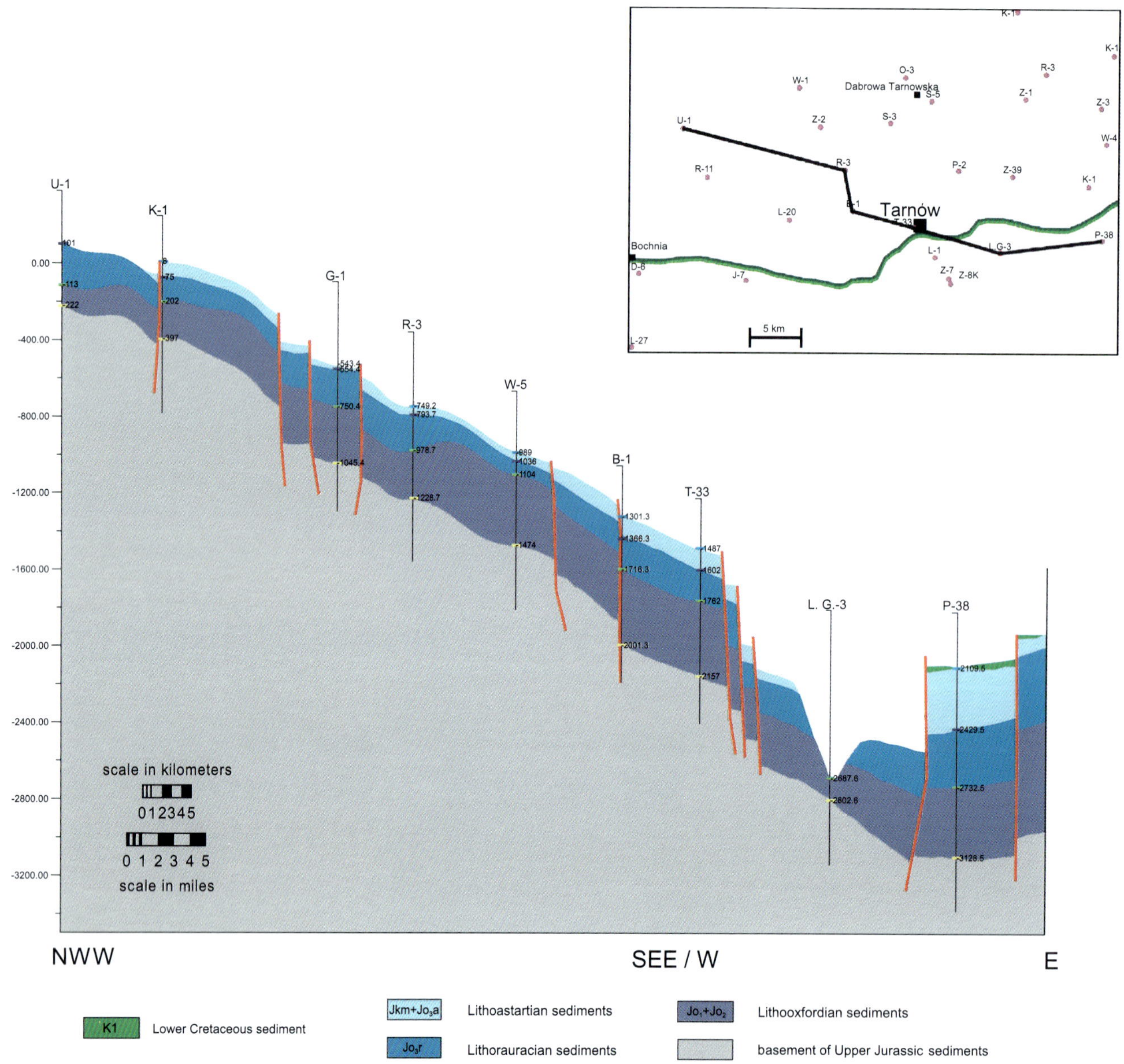

FIGURE 3. Vertical distribution of lithofacies in the Upper Jurassic and Lower Cretaceous series (second transect).

Middle and Upper Oxfordian

The middle Oxfordian succession consists of hundreds of meters of carbonates, with a large variability throughout the strata. It is difficult to refine the age assigned to these rock based on biostratigraphy because of the large variability of microfauna but a lack of major diagnostic taxons. Log interpretation helped to identify three major megafacies. In accordance with traditional, informal lithostratigraphic divisions commonly used, these facies are assigned as Lithooxfordian. The upper part of Oxfordian is represented by Lithorauracian and, partially, the Lithoastartian. Another problem is the presence of organic buildups, which sometimes are of considerable size (few hundred meters). Under favorable circumstances, they can be found all over the Oxfordian succession. Where this is the case, the above-mentioned facial subdivision becomes useful.

The Lithooxfordian

The lower Oxfordian includes biohermal complexes consisting of platy (thin-bedded) and massive dolomites and limestones in the middle part of the Oxfordian. Microbial-sponge bioherms are commonly of considerable size and reflect rapid growth and little differentiation of species. The microbial-sponge bioherms persisted only until the beginning of the upper Oxfordian when, as a result of regression (sea level lowering) and probably the opening of new sediment

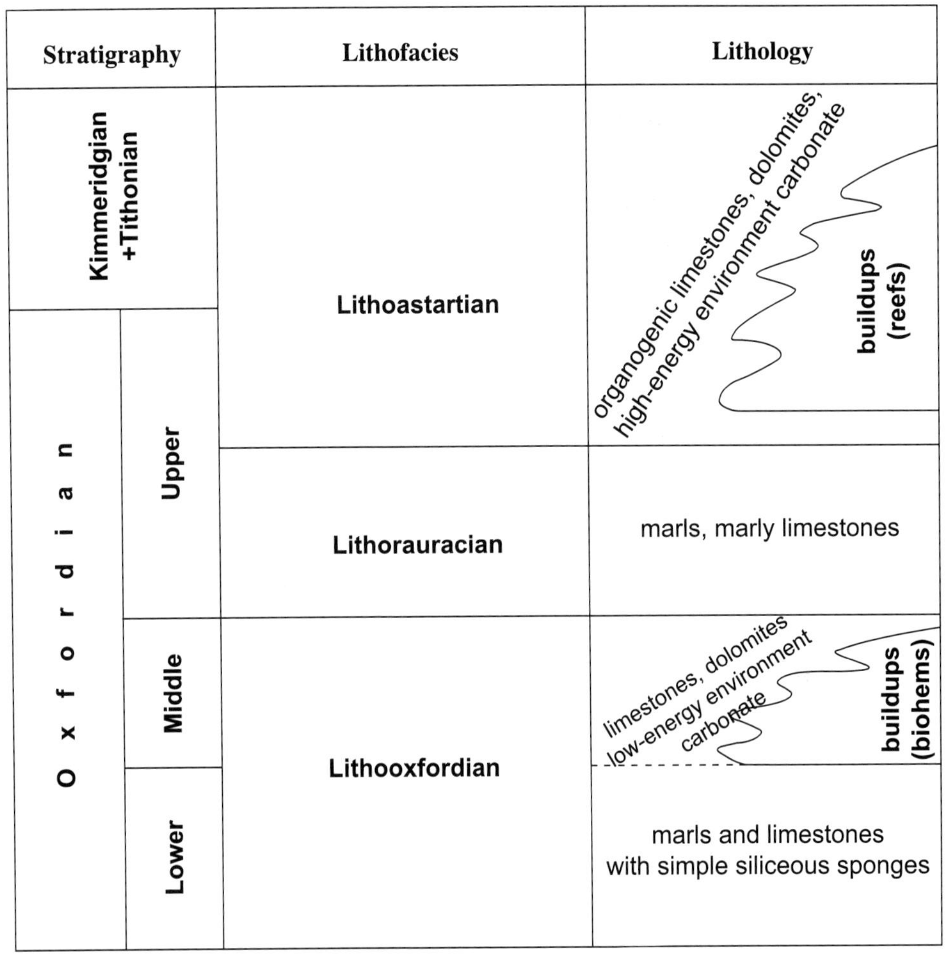

FIGURE 4. Subdivision of the Upper Jurassic based on lithofacies.

source areas, they were covered by silty marls of the Lithorauracian. Only a limited number of bioherms persisted until the end of the Oxfordian.

The platy-type limestone is composed of marly, pelitic limestone having fractures filled with dark-gray clay material. These are basinal sediments deposited in between surrounding biohermal domains. These limestones are characterized by a larger lateral extent and higher content of marly material. They also contain necton remnants, with a practical lack of sessile benthos.

The Lithorauracian

The lower part of the upper Oxfordian is dominated by marly limestones with variable thickness from 200 to 500 m (660 to 1600 ft), deposited in variable conditions. In oxidizing conditions, this series is characterized by layers of bright limestones, whereas dark marls and limestones with only dispersed fauna of crinoids occur in reducing conditions. In some cases, there are intercalations of clastic rocks, such as claystones, siltstones, and fine-grained sandstones. The sequence reflects a gradual shallowing of the depositional environment, which began in the middle Oxfordian with increased detrital input, mainly pelitic. The Lithorauracian sediments are characterized by low poroperm values, with almost no fracturing, thus accounting for good regional seals.

The Lithoastartian

The upper part of the upper Oxfordian comprises dolomitic limestones, sometimes with intercalations of massive marly limestones. The strata dominantly represent high-energy intertidal environments, reflecting a further gradual shallowing of the sedimentary basin. Uniform, depositional conditions blanketed the study area. The most characteristic facies include oolites, oncolites, bioclastic grainstones, packstones, coral meadow, wackestones, and boundstones forming patch and table reefs. Reefs of the Lithoastartian are microbial coral in character. A large variety of reef-building species can be recognized (Bobrek et al., 2000).

Kimmeridgian and Tithonian

Deposition during the uppermost Jurassic is characterized by detrital, oolitic limestones, bioclastic limestones, and marls. Locally, fine-crystalline dolomites and reef algal-coral boundstones are prominent. These sediments are internally affected by paleokarstification below an erosional boundary, above which Cretaceous strata were deposited.

Lithoastartian facies are made up of shallow-water sediments belonging to the Kimmeridgian and the Tithonian. The barrier-type sediments, oolitic limestones and oncolithic-bioclastic limestones, show the largest primary porosity, with dominance of intergranular porosity. Secondary moldic porosity is commonly caused by the dissolution of individual grains and reef components (coated grains or organic fragments) or is related to dolomitization.

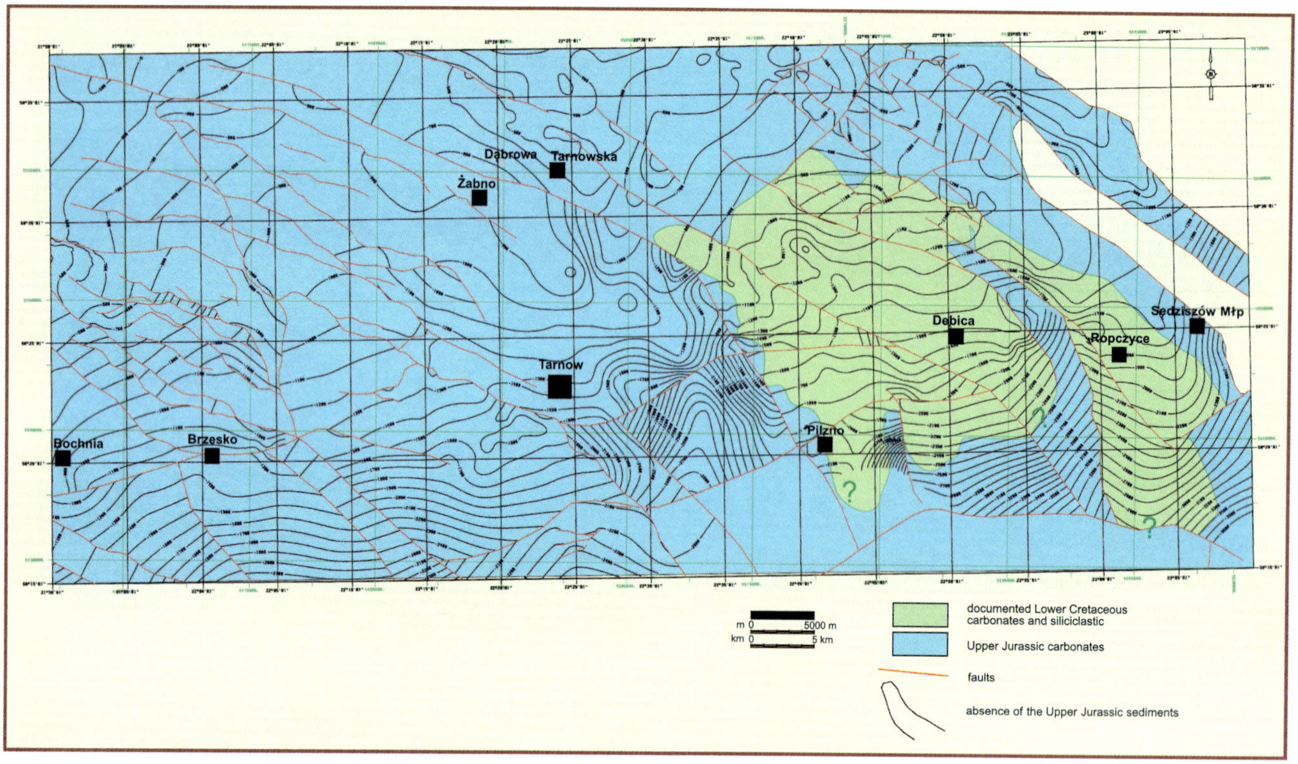

FIGURE 5. Subsurface map of the top of the $J_3 + K_1$ surface in the studied area.

TECTONICS

The main tectonic deformation episodes affected the Mesozoic formations underlying the Outer Carpathians and are only partially linked to the Neogene Carpathian orogeny. Two main directions of faults are distinguished. The earlier one is southwest trending and is controlled by older pre-Mesozoic structural grain. The second direction is southeast-trending and is a result of the Carpathians thrusting. The two systems of faults divide the area into several tectonic blocks. Erosion of variable intensity has led to the removal of a considerable part of the Cretaceous and part of the Jurassic (Figure 5). Fracturing of the Upper Jurassic carbonates of the study area is mainly controlled by the fault system. Fractures occur as two types. The first type of fracture, which is also the largest (biggest), results from the Kimmeridgian phase of the Alpine orogeny. Most fractures are cemented by calcite and anhydrite, which were precipitated from hydrothermal fluids probably during the Cretaceous. The second type of fracture (smallest) is perpendicular to the first type and relates to the Laramian and late Alpine phase of the Alpine orogeny. This type is very commonly uncemented. There is a secondary porosity along the fractures. This porosity is created by aggressive reservoir water dissolution of the rocks. This process started in the Paleocene and persisted until the Pliocene.

PETROPHYSICAL ANALYSES

Petrophysical analyses were performed on Upper Jurassic samples collected from 28 boreholes. A total of 294 cores were analyzed. Porosity, permeability, and capillary-pressure analyses on plug-type samples and laboratory investigations of fracture porosity and permeability with the use of thin sections and polished slides were carried out (Nelson, 1989; G. Leśniak and B. Darłak, 1995, personal communication). Samples were saturated with fluorescent resin.

Porosities were calculated by summarizing all types of free spaces observed in the analyzed thin sections. Porosity was calculated also from mercury injection capillary-pressure measurements. Capillary-pressure analyses were also performed for mixed-type porosity samples. Such procedures give the opportunity to compare values of permeability obtained by various methods and to obtain cumulative pore-size distribution curves and values of threshold diameters (Pittman, 1992). For fractured rocks, permeability was calculated with the use of thin section. A random section procedure was applied (Nelson, 1989; G. Leśniak and B. Darłak, 1995, personal communication). Additionally, the permeability of larger fractures (larger than 0.1 mm) was calculated using the Boussinesqa method (Nelson 1989; G. Leśniak and B. Darłak, 1995, personal communication).

The permeability for rocks that show intergranular porosity was determined in two steps. Samples were divided

TABLE 1. Petrophysical characterization of the lithofacies.

	Lithofacies	*Porosity (%)*	*Permeability (md)*	*Fracture Porosity (%)*	*Fracture Permeability (md)*
Lithooxfordian	average	5.5	24.5	2.37	28.7
	the best	24.5	320	4.97	220
Lithoastartian	average	8.76	11.3	3.78	8.12
	the best	24.82	271	12.1	208
Lithorauracian	average	0.73	0.0	0.9	1.11
	the best	1.42	0.0	1.75	2.43

into two groups on the basis of the values of threshold diameter. Samples with values of threshold diameters lower than 3 μm were grouped as being characterized by fracture-type permeability. The remaining plug samples were cut if possible, and permeability to nitrogen was measured. The grouping was checked by displacement of reservoir water by gas. Correlation between residual water saturation and fractal dimension calculated from cumulative curves was done. The slope of the trendline was used to consider the type of permeability (e.g., matrix dominated, microporous, or mixed) (Such and Leśniak, 2001).

RESULTS

The results allowed quantification of the reservoir properties at bed scale, which suggests the system of transport of reservoir fluids in each of the lithofacies. Reservoir rock types were found to vary between matrix-dominated porous rocks to dual-porosity and dual-permeability rocks and from porous fractured to entirely fractured rocks. The types of reservoir rocks (Table 1) could ultimately be linked to sediment type, diagenetic processes, and position in the basin (Narr and Lerche, 1984; Gliniak et al., 2001).

The Lithooxfordian

Porous rocks with matrix-dominated permeability were found only in organogenic buildups of the Lithooxfordian. These rocks create relatively thin beds in the sequence (as much as 1 m [3 ft]). Porosity ranges from 8 to 20%; threshold diameters of pores cover the range between 10 and 30 μm, sometimes reaching as much as 50 μm. These parameters give intergranular permeability as much as 320 md (Figure 6).

The second type of good reservoir rock in the Lithooxfordian consists of dual-porosity and dual-permeability rocks. Secondary porosity is associated with dolomitization, migration of fluids, and dissolution along the fracture systems. This rock type is shown in Figure 6b. Typically, porosity is greater than 8%, with the threshold pore diameter variation averaging 10 μm. Permeabilities measured on plug samples are lower than 10 md. This type of reservoir rock was recognized in relatively thin intervals (less than 10% of the investigated intervals).

A great part of the investigated sequence is characterized by porosity-fracture permeability. Total porosity

A) Matrix-dominated porous bioclastic packstone (P-2 borehole; 20,18% ; 320 md)

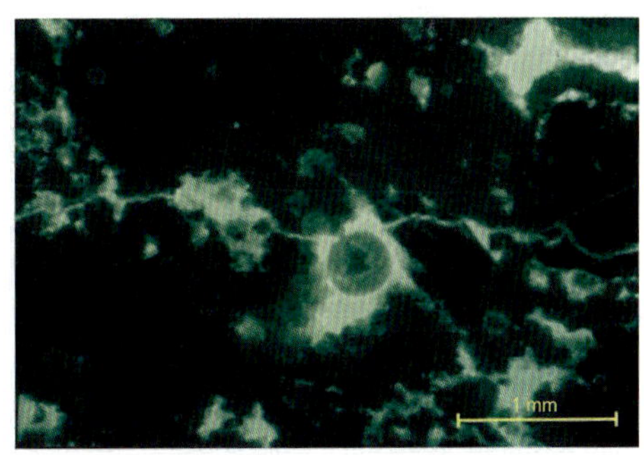

B) Dual porosity and dual permeability (green color indicates free spaces, K-1 borehole; 6.9%; 14.6 md)

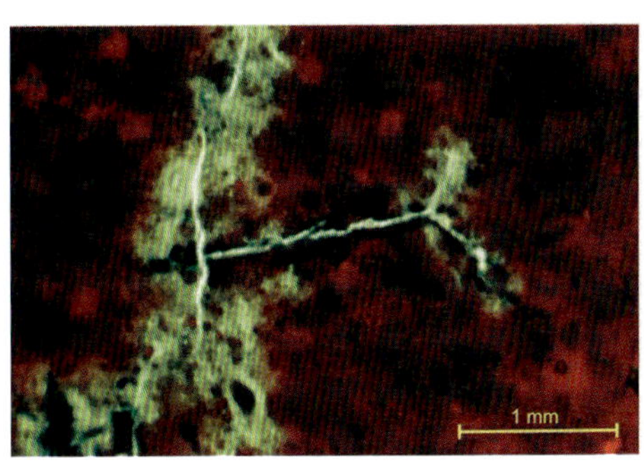

C) Dual porosity and fracture permeability (Z-39 borehole; 4.29%; 5.84 md)

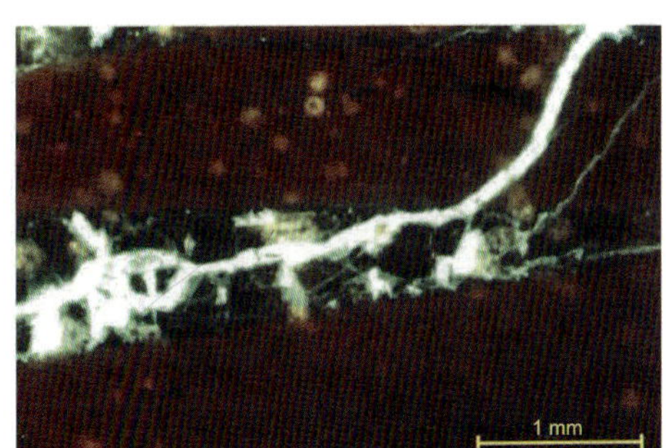

D) Fracture porosity and permeability (P-2 borehole; 11.4% ; 16.1 md)

FIGURE 6. Thin sections outlining various types of pore spaces.

(i.e., the volume of fractures and volume of intergranular porosity along fractures) varies in the range of 4–8%. Threshold pore diameter extracted from capillary-pressure analyses are lower than 1 μm, and fracture permeability that is calculated with the use of thin sections varies in the range of 1–15 md. Average values of porosity and permeability for the Lithooxfordian sediments can be estimated as 4% and 20 md, respectively. Such high values of permeability are caused by the pervasive nature of the fracture system.

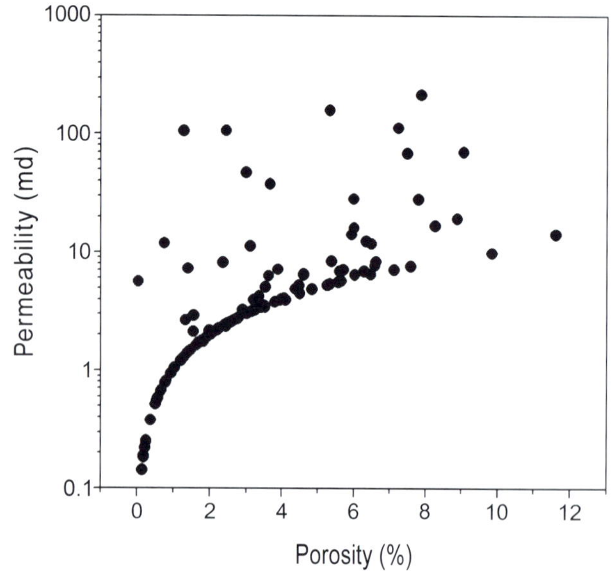

FIGURE 7. Crossplot of porosity vs. permeability for karstified intervals.

The Lithoastartian

Petrophysical results of the Lithoastartian rocks show a large variation in reservoir properties, grading from dominantly fractured rocks to mixed types (dual porosity and fracture) and to dual-porosity and dual-permeability rocks. Porosity and permeability values covered the range 0–24.82% and 0–271 md, respectively. Average values of these parameters cluster around the values of 6% and 10 md.

Reservoir properties of the Lithoastartian rocks were strongly modified by the karst phenomena. Porosity-permeability crossplot for karstified intervals is shown in Figure 7.

The Lithorauracian

The Lithorauracian sediments are practically impermeable rocks, although they are locally cut by a system of fractures. Porosities of these rocks are lower than 1.75%; maximum values of permeability reach as much as 2.43 md, whereas average permeability values remain lower than 1 md.

CONCLUSION

Analysis of petrophysical properties of the Upper Jurassic sediments shows that all intervals are capable of storing and transporting fluids. All types of pore space are involved. Karstification improved hydraulic properties. It is possible to distinguish three geographical zones of interest based on petrophysical properties. Fracture porosity and permeability are predominant in zones I and III, whereas in zone II, various types of pore space occur (Figure 8).

Petrophysical properties depend on the intensity of the Kimmeridgian tectonics. The secondary porosity development is promoted by reservoir water circulation throughout the system of faults in zones I and III.

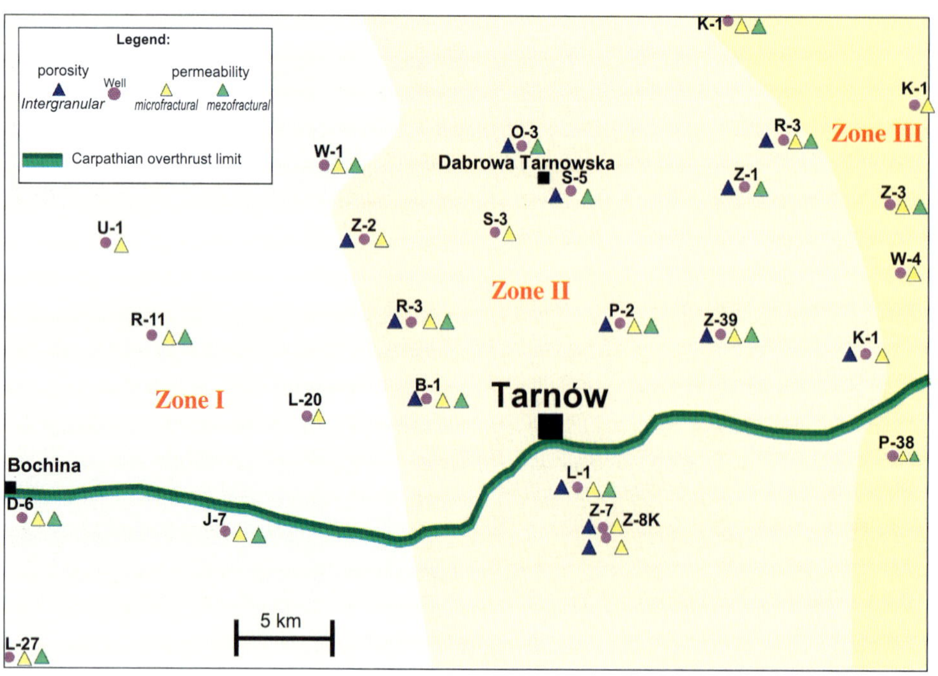

FIGURE 8. Distribution of potential reservoirs based on petrophysical properties distribution.

Circulation of reservoir waters was connected with fault zones and with zones of syngenetic dolomitization. There were three successive episodes of dolomitization: The first is responsible for occurrence of porous horizons; the second episode of dolomitization damaged the pore spaces by cementation processes, whereas the third dolomitization episode restored the pore connectivity in reservoir rocks because of the secondary dissolution phenomena.

The main factor accounting for the development of the pore space and the petrophysical properties relates to the diagenetic processes. The presence of fault zones is only a local factor in the increase of petrophysical properties. The main karst development occurred during the Paleogene erosion episode.

Fluid flow in the studied carbonate rocks is more complicated and quite different than what is known to occur in siliciclastic rocks. Generally, a dominant fracture transport is observed: the fluid flow in the adjacent pore space being limited to short-range circulation because of great space variability in carbonate rock structure.

Fractures, in fact, are the only factor that allows production in such type of reservoirs.

REFERENCES CITED

Bobrek, L., B. Świetlik, and A. Urbaniec, 2000, Faunal assemblages of Upper Jurassic bioherms from the middle part of the Polish Carpathians foreland, *in* XVII Conference of Paleontologists Publication, Kracow, Polish Geological Society, p. 18–20.

Gliniak, P., and A. Urbaniec, 2001, Oxfordian bioherm type of buildups in a 3-D seismic data in Bochnia—Sędziszów area, *in* Nafta-Gaz Press no. 10/2001: Polish Oil and Gas Institute Kracow, p. 545–556.

Gliniak, P., R. Laskowicz, and A. Urbaniec, 2000, Upper Jurassic organogenic buildups in area between Kraków and Dębica. Capabilities of identification from the seismic data and exploratory prospect for hydrocarbons. Institute of Oil and Gas—Kraków Special Paper 110, p. 161–165.

Gliniak, P., R. Laskowicz, A. Urbaniec, P. Such, and G. Leśniak, 2001, Reservoir rocks and facies development of the Upper Jurassic carbonates in Zawada—Lękawica area, *in* Nafta-Gaz Press no. 11/2001: Kracow, Polish Oil and Gas Institute, p. 597–606.

Gregosiewicz, Z., et al., 2001, Microfacial analysis and diagenetic alterations in Upper Jurassic organogenic sediments based on cores from selected wells of the area between Kraków and Dębica: Kracow, Polish Oil and Gas Company Archive, p. 93.

Karnkowski, P., 1999, Oil and gas deposits in Poland: The Geosynoptics Society "GEOS"—University of Mining and Metallurgy, Kracow, p. 380.

Lenk, T., et al., 1993, Complex evaluation of the prospectivity in upper Devonian and Carboniferous based on results of the geochemical, petrophysical, geophysical and geological researches in the area between Tarnow and Kraków: Etap II Kracow, Polish Oil and Gas Company Archive, p. 36.

Malinowska, L., 1980, Regional biostratigraphic subdivision of the Oxfordian of extra-Carpathian Poland: Bulletin de l'Acadamie Polonaise des Sciences. Serie des Sciences de la Terre, v. 28, no. 1., p. 23–29.

Malinowska, L., 1986, Paleozoogeography and paleoecology of the lower Kimmeridgian of Poland: Bulletin de l'Acadamie Polonaise des Sciences. Serie des Sciences de la Terre, v. 34, no. 2, p. 165–173.

Narr, W., and I. Lerche, 1984, A method for estimating subsurface density in cores: AAPG Bulletin, v. 68, no. 5, p. 637–648.

Nelson, A. R., 1989, Geologic analysis of naturally fractured reservoirs: Houston, Gulf Publishing Company, 320 p.

Niemczycka, T., 1997, Upper Jurassic: Sedimentation, paleogeography and paleotectonics, *in* Prace Państwowy Instytut Geologiczny, v. 153: Epikontynentalny perm i mezozoik w Polsce, Warsaw, p. 327–331.

Pittman, E. D., 1992, Relationship of porosity and permeability to various parameters derived from mercury injection—Capillary pressure curves for sandstones, AAPG Bulletin, v. 76, no. 2, p. 191–198.

Słupczyński, K., and W. Biedrzycki, 1993, Evaluation of the prognostic reserves in Devonian and Carboniferous deposits of the Carpathian foreland area, *in* Evaluation of the prognostic reserves of oil and gas in Poland, Kracow, Polish Oil and Gas Company Archive, p. 50.

Such, P., and G. Leśniak, 2001, Application of fractal approach and microscopic images in net model of pore space: 63rd European Association of Geoscientists and Engineers Conference, Amsterdam, 2001, p. 644.

18

Schneider, F., M. Pagel, and E. Hernandez, 2004, Basin modeling in a complex area: Example from the Eastern Venezuelan foothills, *in* R. Swennen, F. Roure, and J. W. Granath, eds., Deformation, fluid flow, and reservoir appraisal in foreland fold and thrust belts: AAPG Hedberg Series, no. 1, p. 357–369.

Basin Modeling in a Complex Area: Example from the Eastern Venezuelan Foothills

Frédéric Schneider
Institut Français du Pétrole, Rueil Malmaison, France

Maurice Pagel
Centre National de la Recherche Scientifique-Université de Paris-Sud, University of Paris XI, Orsay, France

Elizabeth Hernandez
Intevep-Petroleos de Venezuela, S.A. Caracas, Venezuela

ABSTRACT

The main focus of this study is the origin of the paleofluids in the Late Cretaceous–Oligocene sandstone reservoirs of the El Furrial structure (Venezuela). Basin modeling was performed using Thrustpack®, Locace®, and Ceres® tools. The temperature and nature of the fluids obtained by this modeling were compared to fluid inclusions and oxygen isotope data on quartz overgrowth. Four stages should be considered in this area: (1) From 65 to 20 Ma: Fluids were at thermal equilibrium with the sediments. They were continuously expelled vertically toward the surface during compaction-driven dewatering processes. (2) From 20 to 12 Ma: As a result of the regional tilting and the deposition of the synflexural Naricual Formation, the Cretaceous and Oligocene sandstones of the El Furrial structure became efficient conduits for fluids circulating from the north. These fluids were at thermal equilibrium with the Cretaceous and Oligocene sandstones and seem to be correlated with the first generation of quartz overgrowths. This episode is characterized by an increase in the overpressure of the Oligocene and Upper Cretaceous sandstones that is correlated to a hydraulic fracturing of the sealing Carapita black shales. (3) From 12 to 8 Ma: Fluids were expelled laterally from the Cretaceous sediments of the Pirital hanging-wall unit that is located immediately north of the El Furrial structure. These fluids were likely in chemical disequilibrium, and their temperature was higher than the temperature of adjacent sediments, which probably resulted in additional but minor episodes of quartz precipitation. This hypothesis is consistent with the information obtained from oxygen isotope analyses, which suggest

DOI:10.1306/1025700H1504

that subsequent generations of quartz cement probably formed from evolved basinal fluids. (4) A reduction of the intensity of the flow and then an inversion of this flow mark the sealing of the southern structural closure of the structure at about 8 Ma. Then, the closure of the northern flank occured at about 5 Ma, as indicated by a present velocity of the fluids close to zero in the El Furrial reservoirs and the filling of the structure by the hydrocarbons.

INTRODUCTION

The Subthrust Reservoir Appraisal (SUBTRAP) consortium was created in 1996 to document the main geological processes affecting sandstone and carbonate reservoirs in foreland fold and thrust belts to decrease the exploration risk in such tectonically complex areas. The consortium has focused on the study of well-documented structural transects in productive foreland fold and thrust belts around the world (i.e., Albania, Canada, Colombia, Mexico, Pakistan, and Venezuela).

In Venezuela, the area selected for the study is a north–south transect in the Eastern Venezuela Basin (Figure 1). The Eastern Venezuela Basin is bordered to the north by the Serrania del Interior, a fold and thrust belt, and by the El Pilar fault. This basin extends to as far south as the Orinoco River (Hedberg, 1950). This basin is a flexure-related foreland basin that developed in Neogene time on the late Mesozoic passive margin of the South American craton. The two major petroleum provinces of this continent are located in this basin (Gonzalez de Juana et al., 1980). These two petroleum provinces are (1) the fields of the Faja Petrolifera in the southern part of the basin near the Orinoco River (Fiorillo, 1987); and (2) the Quiriquire fields and the Maturin basin, in which the giant El Furrial field has been discovered in the buried overthrusts of the basin (Carnevali, 1988). This area has been described in numerous studies addressing the stratigraphy and the regional tectonic and petroleum systems (e.g., Hedberg, 1950; Sulek, 1961; Arnstein et al., 1985; Chiock, 1985; Rossi, 1985; Rossi et al., 1987; Carnevali, 1988; Potié, 1989; Roure et al., 1994; Parnaud et al., 1995).

The selected transect extends from the Serrania del Interior to the Orinoco River to include the Pirital allochthon, the subthrust plays (El Furrial–Carito–Santa Barbara trends), and the autochthonous foreland (Faja and Oficina fields). The lithostratigraphic column includes rocks from the Jurassic to Quaternary periods, and the nomenclature is summarized in Figure 2.

The main focus of this study is the understanding of the fluid-flow history in the two major reservoir units (Upper Cretaceous and Oligocene formations) of the El Furrial structure. In the first step, mineralogical and geochemical data from the two formations have been collected. Some data have also been collected for the surrounding seal, the Carapita black shales. In the second step, basin modeling techniques have been used to reach some understanding about the fluid, pressure, and temperature evolution in the considered reservoirs. Then, a synthesis proposes a general scheme for the reservoir evolution of the El Furrial structure.

DATA

Two distinct reservoir intervals occur in the western part of the El Furrial–Carito subthrust trend: (1) the Cretaceous San Juan Formation and (2) the Oligocene Merecure Formation. The Eocene and Paleocene series are presently entirely eroded in this part of the basin.

These two formations are the major objectives in term of subthrust prospects in the considered area. They are sealed by the black shale of the Carapita Formation.

Late Cretaceous San Juan Subarkosic Sandstone

The sandstones are moderately to well sorted, with grain size that is generally fine to medium (from 0.125 to 0.5 mm). Monocrystalline quartz grains are the major detrital constituent. Their fraction ranges from 60 to 70%. Feldspars are the second most important detrital grains; their fraction ranges from 4 to 14%. K-feldspar is the most abundant type, with small percentages of microcline. Plagioclase is only present in trace amounts, generally as isolated grains. Polycrystalline quartz and rock fragments are less important constituents; their proportion ranges from 0 to 5%. Chert was also observed, but it represents less than 2% of the grain types. Diagenetic minerals include pyrite, quartz overgrowth (8% on average), clay minerals, and various carbonates (mainly calcite and dolomite).

The sandstones are characterized by longitudinal, concave-convex, and irregular penetration contacts (indentation). Horizontal stylolites are also common features. These characteristics indicate pressure solution as the result of compaction and deformation and, therefore, a possible source of silica for the quartz overgrowths.

Quartz overgrowths around detrital quartz grains are discontinuous and well developed in the coarse-grained samples. Locally, two generations of quartz overgrowths were observed, but the second is very

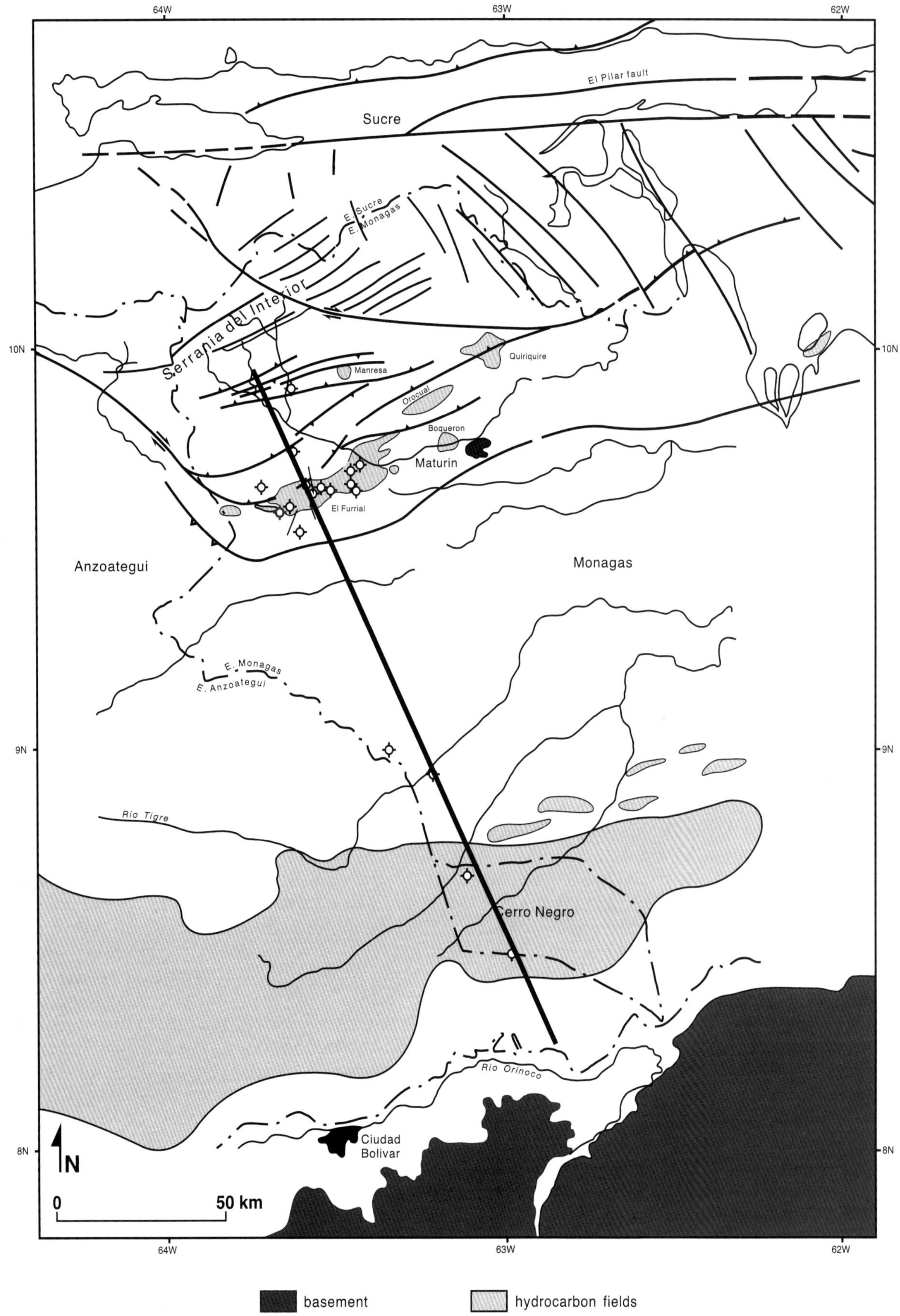

FIGURE 1. Location of the transect selected for the study.

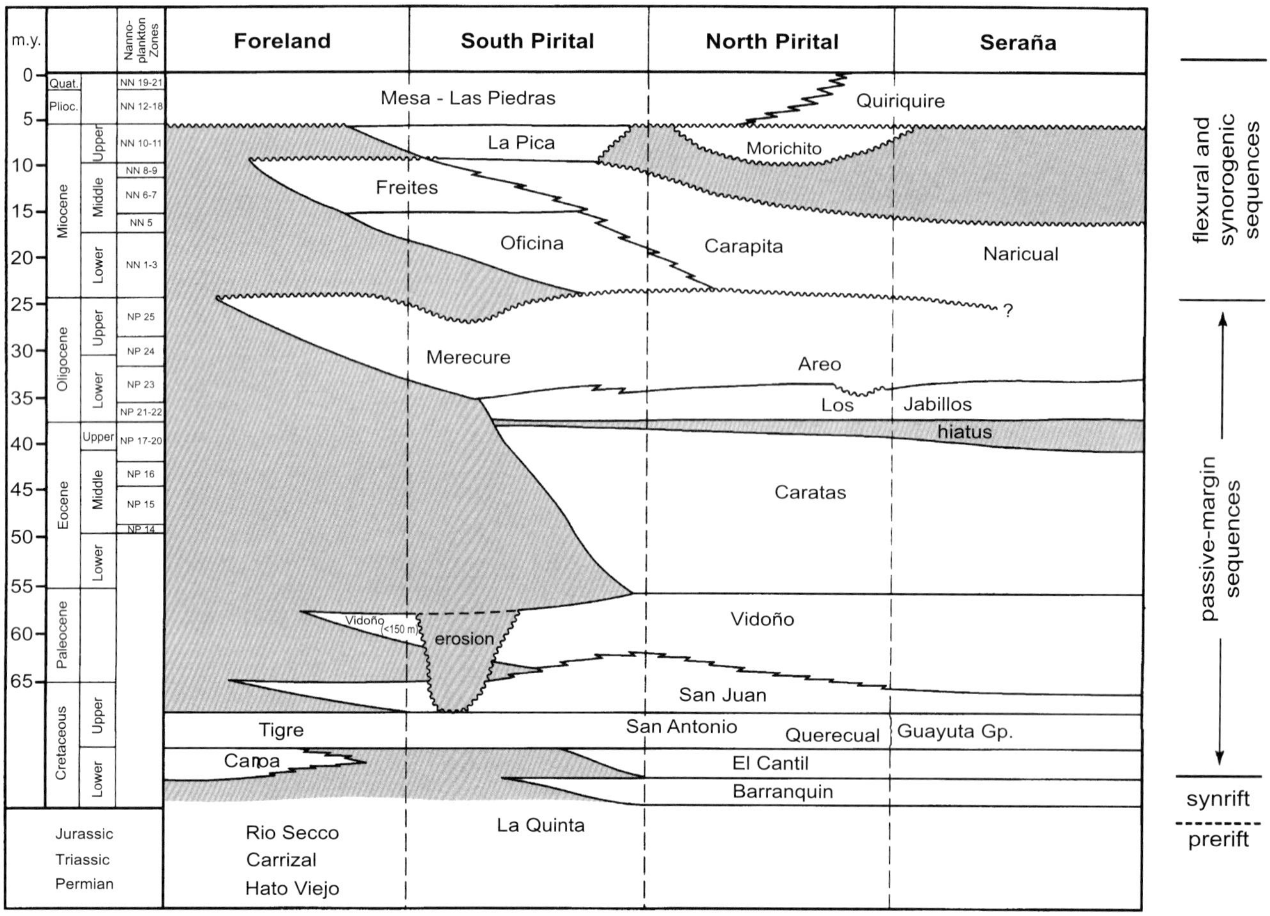

FIGURE 2. Summary of the lithostratigraphic nomenclature used in Eastern Venezuela Basin. The main source rock considered in this study is the Querecual Formation. The two major reservoir intervals are the Upper Cretaceous San Juan Formation and the Oligocene sandstone of the Merecure Formation.

minor. Some fractures in the detrital quartz grains are filled with the same type of luminescent quartz.

From a fluid-inclusion study (M. Pagel, 2002, personal communication), it should be noted that hydrocarbon inclusions have been observed along the detrital quartz grain–quartz overgrowth boundary. This indicates that water and hydrocarbons were both present in the reservoir at the beginning of silicification. Silicification occurs at temperatures ranging from 100 to 125°C, temperatures lower than the present temperatures of sampling (about 140°C).

The measured $\delta^{18}O$ values for the quartz overgrowth range from +18 to +25‰$_{SMOW}$ (SMOW = standard mean ocean water) (M. Pagel, 2002, personal communication). The highest values of about +25‰ correspond to a second generation of quartz overgrowth, which is only present in small volumes. The oxygen isotope values were homogeneous in each sample, a fact that could be associated with the absence of cathodoluminescence zoning. The $\delta^{18}O$ of quartz overgrowths depends on two parameters: (1) its temperature of deposition and (2) the $\delta^{18}O$ of the silicification fluid. If we consider that the deposition temperatures are in the range between 100 and 125°C, the $\delta^{18}O$ of fluid 1 ranges from −1 to +2‰, and the $\delta^{18}O$ of fluid 2 ranges from +2 to +6‰. For fluid 1, the values are closed to marine waters. For fluid 2, it implies formation waters.

Oligocene Merecure Quartz Sandstone

The thick sandstones of the Merecure Formation were deposited during the basal Oligocene transgression. In this area, they correspond to inner platform deposits. However, sedimentation becomes rapidly more distal (outer platform) in the north (Areo formation) and entirely continental farther south in the Cerro Negro.

The detrital components are essentially made up monocrystalline quartz grains (97%), with only a minor contribution of lithic fragments, polycrystalline quartz, mica, feldspar, and heavy minerals. Clay matrix with dominant kaolinite can reach 8% but commonly remains at a much lower level.

Syntaxial quartz overgrowths constitute the most significant diagenetic phase, resulting in an important decrease of the matrix porosity and a strong cementation of the reservoir.

Pressure solution and mechanical interactions between quartz grains account for the indentation and microcracks of the detrital quartz. The results indicate that silicification, generally as syntaxial overgrowths, was the most important diagenetic process in terms of reservoir quality control. Trapping temperatures of aqueous fluid inclusions in the quartz cement range from 80 to 130°C and account for a relatively early diagenetic episode. According to Bordas Le Floch (1999), the lack of hydrocarbon inclusions in the overgrowths suggests that silicification stopped before or during the emplacement of hydrocarbons in the area. This author also proposes that because hydrocarbons reached the area of El Furrial during middle Miocene (Gallango and Parnaud, 1995), and tectonic accretion in El Furrial occurred between 15 and 5 Ma (Roure et al., 1994), then the silicification in the Merecure Formation probably occurred during the lower–middle Miocene, when the reservoir was still part of the foreland.

Carapita Seal

Calcite veins were observed in the black shale layers of the Miocene Carapita Formation. They are typically formed by hydrofracturing, and they are filled by calcite cement. The fluid inclusions do not contain any hydrocarbon except methane that is detected by Raman microprobe. The homogenization temperature measured range from 65 to 95°C, with salinities between 4.2 and 6.6 wt.% equiv. NaCl (Bordas Le Floch, 1999).

BASIN MODELING

Basin modeling aims to reconstruct the time evolution of a sedimentary basin to make quantitative predictions of geological phenomena leading to pressure generation and hydrocarbon accumulations. It accounts for porous medium compaction, heat transfer, hydrocarbon formation, and migration (e.g., Schneider et al., 2000).

Today's state-of-the-art sedimentary basin models are able to handle relatively simple geometries resulting from deposition, erosion, and vertical compaction. Two main areas, however, are insufficiently, if at all, expedited: (1) Basin geometry is commonly not precise enough. Most basins are cut by faults with significant offsets. In addition, basin geometry may be affected by the creeping of salt or mud, which, in turn, may lead to the formation of diapirs. (2) Fluid flow and convective heat transfer do not handle the permeability evolution of faults correctly. A fault can be pervious or act as a seal, and it can change its behavior through time.

The aim of the Ceres project (Schneider et al., 2002a) was to build a prototype that is able to simulate three-phase flow in a two-dimensional section of a basin in which geometry changes because of deposition, compaction, erosion of the sediments, or salt or mud creeping, and blocks displacement along faults.

In addition to the simulation of overpressures, Ceres® modeling was also successful in tracing regional changes in the water flow across the El Furrial reservoirs. The models suggest that Late Cretaceous to Oligocene sandstone reservoirs evolved first in an open system, with episodes when the fluids were expelled toward the south or toward the north, before becoming a dominantly closed system after complete structural closure.

Methodology

The methodology is the result of modeling performed in a Western Canada transect (Schneider et al., 2002b) and the Eastern Venezuelan transect described in this chapter. To balance the section through time and to constrain the eroded parts of the section, we decided to use Thrustpack®, a forward kinematic modeling tool (Deville and Sassi, 1996). Thus, from a section restored to its prethrusting geometry with Locace® (Moretti et al., 1988; Moretti and Larrere, 1989), Thrustpack allows the user to reconstruct any intermediate section during time, which is also consistent from a kinematic point of view. The final section of Thrustpack, which is supposed to be equivalent to the initial section at present day, is then imported in Ceres. The edition of the section, the restoration, and the simulations are then performed with Ceres. The intermediate geometries built with Thrustpack are directly used as templates in Ceres to draw the eroded parts and to perform the displacements along the faults.

Data Preparation

As defined in the methodology, the first work to do at the beginning of a Ceres study is the definition of the initial section. This is generally performed with data coming from the seismic interpretation, well data, field data, and core data.

The available initial section and Thrustpack scenario was used to build the data set for Ceres. The lithology distribution was not studied in detail, and only the main units were considered (Figure 3).

Geometry Evolution

First, one scenario for the geometry evolution was chosen using a limited number of evolutionary stages. Then, additional work was necessary to provide intermediate geometries for the main thrusting period. The resulting scenario is characterized by five major episodes in the geodynamic evolution (see Figure 4).

1) The first episode is represented by a prerift megasequence in the Paleozoic. The corresponding sediments are included in the brittle upper crust of the Ceres section.
2) The second episode corresponds to a rifting period during the Jurassic and earliest Cretaceous. The corresponding sediments are partly included in the brittle upper crust in the Ceres section. Only the upper part of the synrift sediments is considered.
3) The third episode corresponds to a passive-margin period, developing during the Cretaceous–Paleogene. In the Ceres section, this episode begins with the deposition of the Barranquin Formation and ends at 22.5 Ma (see Figure 4).
4) From 22.5 to 15 Ma (see Figure 4), the deposition of the synflexural Naricual–Carapita Formation was characterized by a tilting phase, with deposition of thicker layers in the northern part of the section.
5) From 15 Ma onward, the thrusting period developed. According to the results of Thrustpack modeling, different steps have been differentiated here (Figure 4). This (thrusting) and the previous period (tilting) are considered as one in Parnaud et al. (1995).

Thermal Calibration

Uncorrected temperature data from some wells were available for the thermal calibration. As no data about the thermal conductivity and thermal capacity of the sediment were available, default values have been given to the sediments according to their dominant lithologies. Some simulations with different values of the thermal heat flow and the radiogenic source term have been performed. Then, the computed temperatures are compared to the temperature data to reach a reasonable fit (Figure 5). A good fit is obtained for a thermal heat flow of 40 mW m^{-2} and a radiogenic source term equal to 1.7 μW m^{-2}. The heat flow value is close to the one calibrated previously with Temispack® in the undeformed southern foreland, autochthonous part of the section (Gallango and Parnaud, 1995).

Pressure Calibration

From the data available for this study (Figure 6), it is clear that the El Furrial

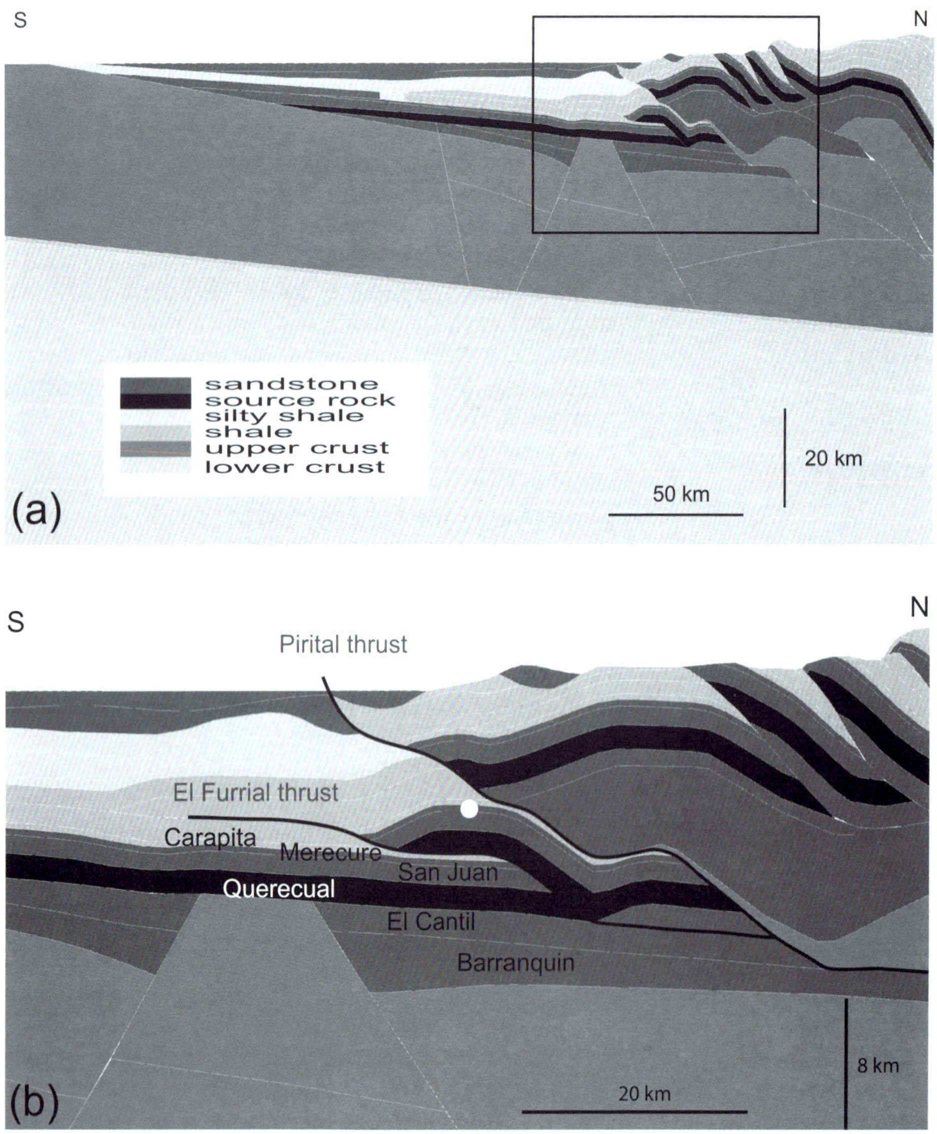

FIGURE 3. (a) Geometry and lithology distribution of the section at present day. (b) Zoom of the section showing the area of main interest. The white dot indicates the El Furrial structure. The main source rock (i.e., the Querecual Formation) is in black. This formation is overlain by the Upper Cretaceous San Juan Formation and the Oligocene Merecure Formation, which constitutes the main reservoir interval.

FIGURE 4. Geometric evolution of the section from 65 Ma to present day. The white dot indicates the El Furrial structure. The passive-margin period developed during the Cretaceous–Paleogene. In the Ceres section, this episode begins with the deposition of the Barranquin Formation and ends at 22.5 Ma. From 22.5 to 15 Ma, the deposition of the synflexural Naricual–Carapita Formation was characterized by a tilting phase with deposition of thicker layers in the northern part of the section. From 15 Ma onward, the thrusting period developed. According to the results of Thrustpack modeling, different steps have been differentiated here.

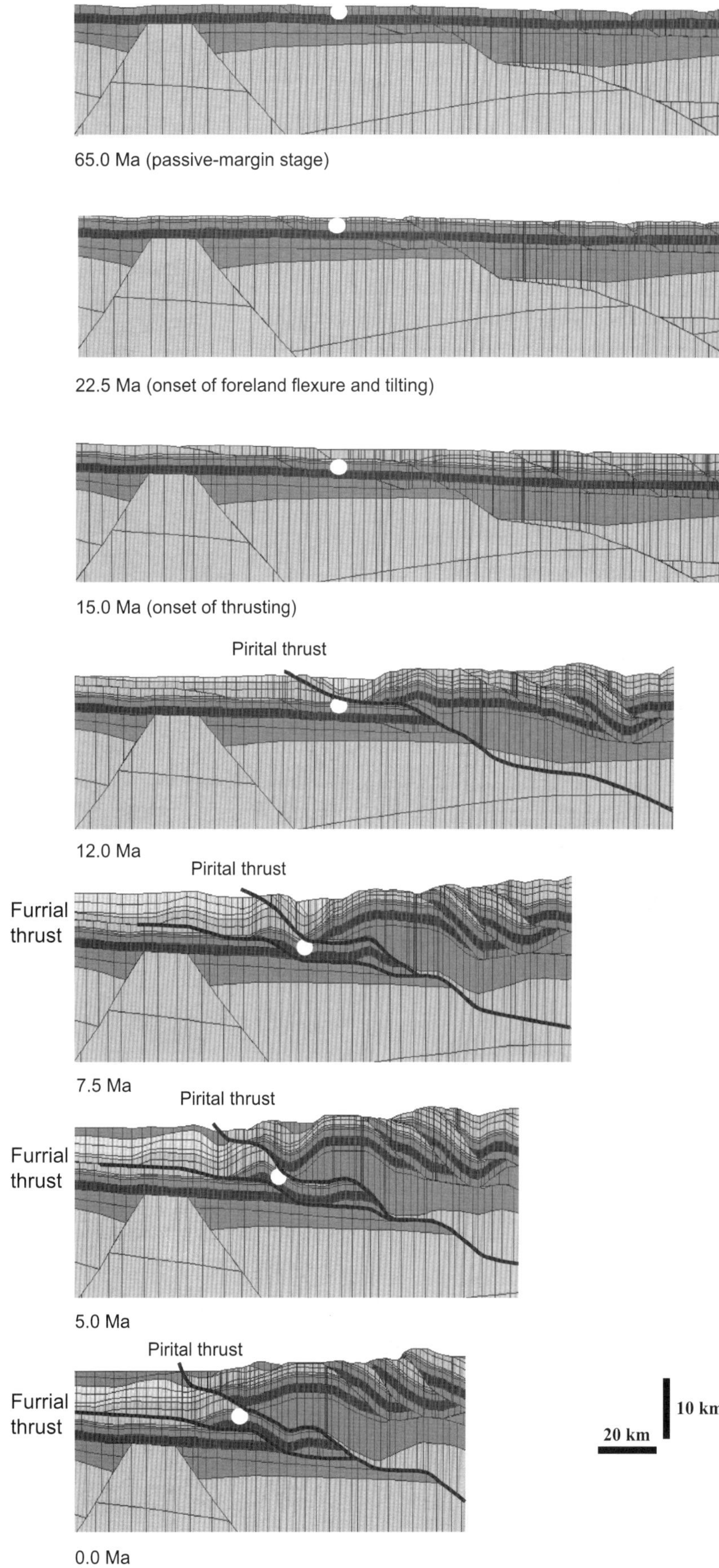

structure, where overpressure greater than 30 MPa are measured, is disconnected at present from the equivalent layers of the southern part of the section, where the pressure is close to the normal hydrostatic pressure. The closure between the two units may be obtained by considering an impermeable fault or by involving impermeable sediments in the El Furrial thrust. Because of geological considerations, the second hypothesis has been chosen. A lithology distribution has been defined by the introduction of a shaly layer at the top of the Oligocene (Figure 3). This shaly layer belongs to the Naricual and lower Carapita formations that overlie the El Furrial thrust in this area.

With realistic values of permeabilities for the different formations, the second lithology distribution reproduces the observed hydrostatic water pressure in the southern part of the section and overpressure in the El Furrial structure. Nevertheless, the pressures computed for the El Furrial structure are lower than 80 MPa, which was the target for the calibration. Pressure close to 80 MPa is calculated in the structure when multiphase flow is considered. Indeed, in this case, the

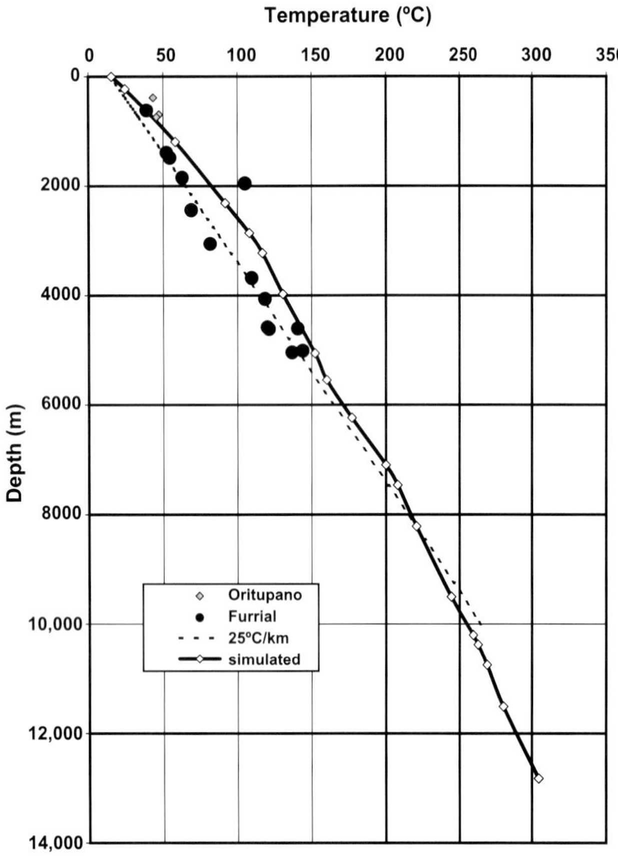

FIGURE 5. Temperature data (dots) and computed temperatures by Ceres at El Furrial location. A good fit is obtained for a thermal heat flow of 40 mW m^{-2} and a radiogenic source term equal to 1.7 $\mu W\ m^{-3}$.

maturation of the organic matter in the source rock provides an additional source term for the pressure equation.

Hydraulic Fracturing

The simulated fracturing ratio of the Carapita seal (Figure 7) shows that these shales might have fractured between 21 and 15 Ma. This fracturing is related to the increase of the fluid pressure as a consequence of the tilting period, where low-permeable sediments are deposited with a quite high sedimentation rate. This increase of the overpressure is correlated with an increase of the fluid flow along the stratification in the Oligocene and Upper Cretaceous sandstones (see below).

Porosity

As pressure solution affecting quartz grains and precipitation of quartz overgrowth have been recognized in the Oligocene and Upper Cretaceous sandstones, the mechanical and chemical compaction model has been activated in Ceres (Schneider et al., 2002a). This model simulates the mechanical compaction with porosity-effective stress relationships. The chemical compaction model is given as a volumetric viscous rheology (Schneider et al., 1993, 1996). Calibration of the viscous parameters has been performed with data from the Upper Cretaceous San Juan Formation. The results (Figure 8) show that a good fit is reached with an activation energy of 17.5 kJ/mol and a macroscopic viscosity of 29.2 GPa Ma. The activation energy is close to the one (e.g., 17 kJ/mol) calibrated for the Garn sandstones of the Haltenbanken area (Schneider and Hay, 2001). For the Garn sandstones, the macroscopic viscosity was equal to 51 GPa Ma. The difference can be explained by the fact that the average grain size of the San Juan Formation is lower than the average grain size of the Garn Formation.

Fluid-flow History

The fluid-flow history has been analyzed for the Merecure Formation of Oligocene age. This formation constitutes the main reservoir level in the El Furrial

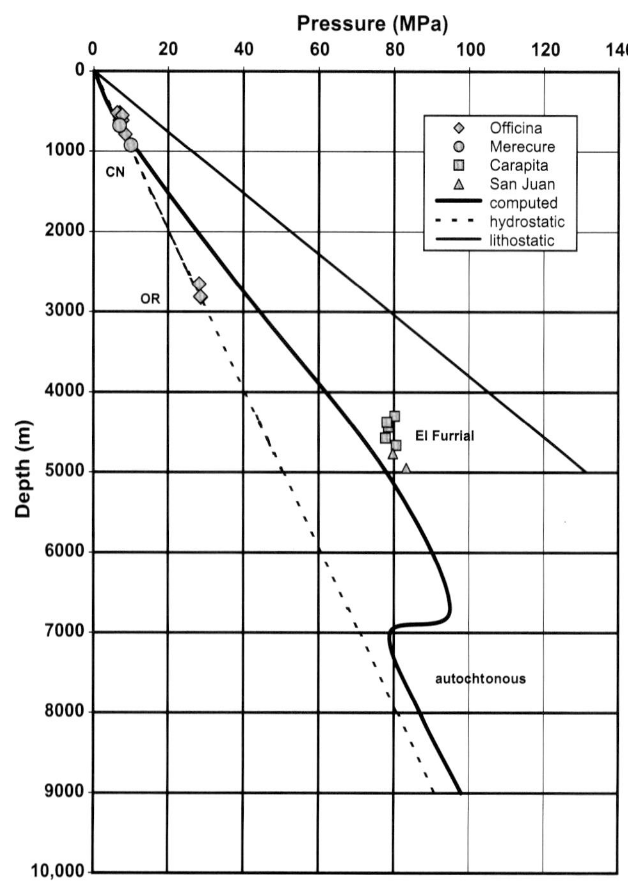

FIGURE 6. Pressure data at different locations along the transect showing close to hydrostatic pressure in the southern part of the transect and overpressures at El Furrial. CN: Cerro Negro field, OR: Oritupano field.

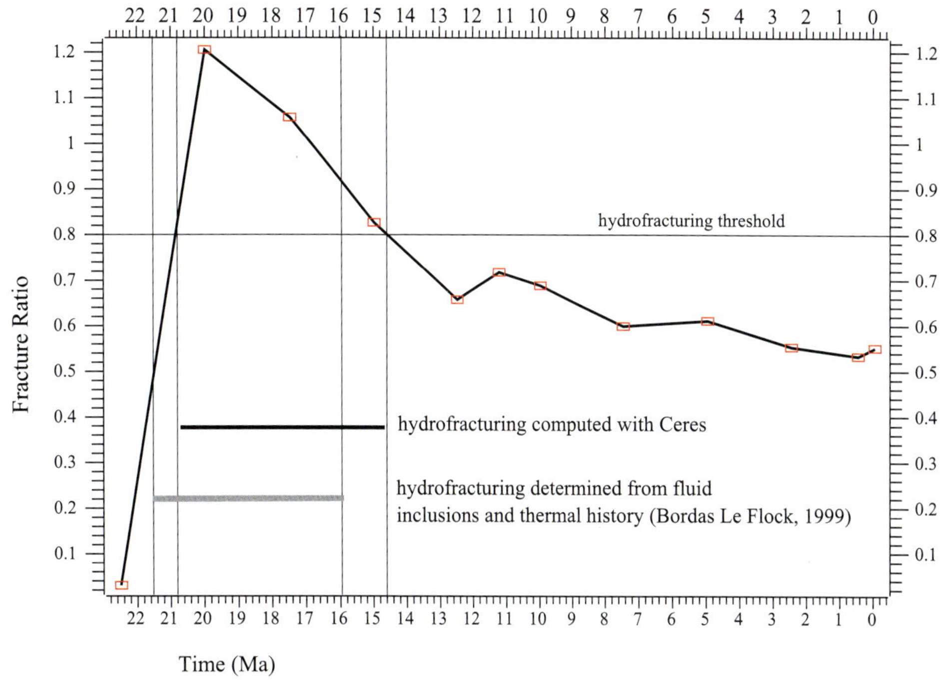

Figure 7. Computed fracture ratio evolution in the Carapita shales of the El Furrial structure. The simulated fracture ratio of the Carapita seal shows that these shales might have fractured between 21 and 15 Ma. These shales are cut by hydraulic fractures sealed with calcite. The fluid-inclusion study associated with the simulated thermal history allowed Bordas Le Floch (1999) to conclude that this fracturing period is dated from 22 to 16 Ma.

structure. For the Oligocene sandstones of the El Furrial structure, the fluid-flow history is a direct consequence of the deformation model and the lithology distribution.

From the deposition of the Late Cretaceous–Oligocene reservoirs until the beginning of the foreland tilting, the fluid flow remained dominantly vertical because of the compaction-driven dewatering of the sediments.

During regional tilting (foreland burial) from 23 to 15 Ma, impervious sediments were deposited with higher sedimentation rates in the north of the transect, and as such, paleofluids escaped southward along the stratification, with an average velocity of 15–20 km/Ma (squeegee 1 in Figure 9).

Between 15.0 and 12.0 Ma, i.e., at the onset of thrusting, the El Furrial structure is partially sealed in its northern part by the Querecual Formation of the Pirital unit (see Figure 4). This is correlated to a decrease of the fluid flow toward the south (Figure 10).

At about 12.0 Ma, as the Pirital thrust was still active, the northern part of the El Furrial structure was then directly in contact with the Cretaceous reservoir formations of the Pirital unit (Figures 4, 10). As the water that was trapped in the Cretaceous aquifers of the Pirital unit could escape farther south, the fluid flow toward the south in the Oligocene of the El Furrial structure increased drastically to reach a velocity of 20 km/Ma (squeegee 2 in Figure 9).

From 10 Ma onward, the El Furrial thrust became active, and as a consequence, the communication of the Oligocene sandstone toward the south was progressively reduced, until it became completely disconnected from the autochthonous foreland at about 7.5 Ma (Figure 4). At this time, the fluid flow toward the south was drastically reduced and ultimately became even

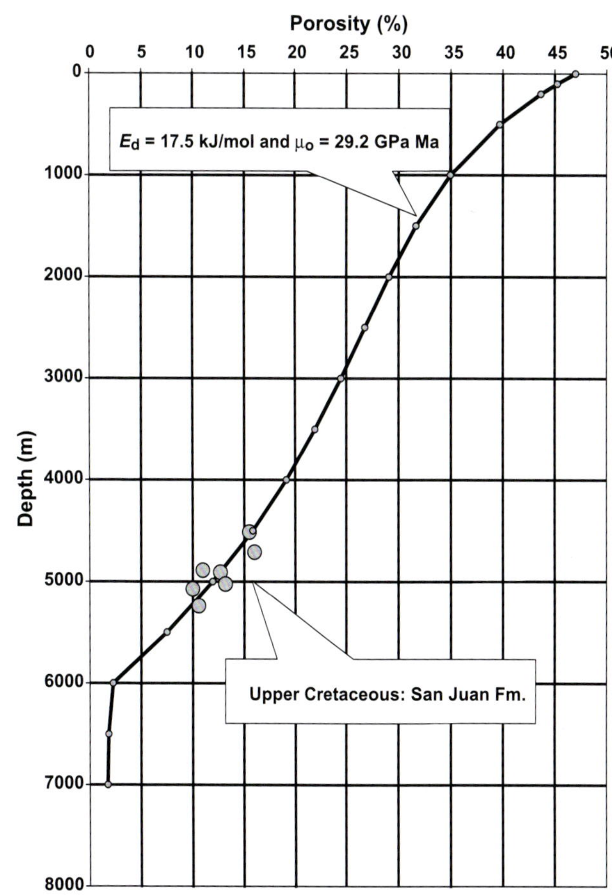

Figure 8. The results of the calibration carried out for the Cretaceous San Juan Formation show that a good fit is reached with an activation energy of 17.5 kJ/mol and a macroscopic viscosity of 29.2 GPa Ma.

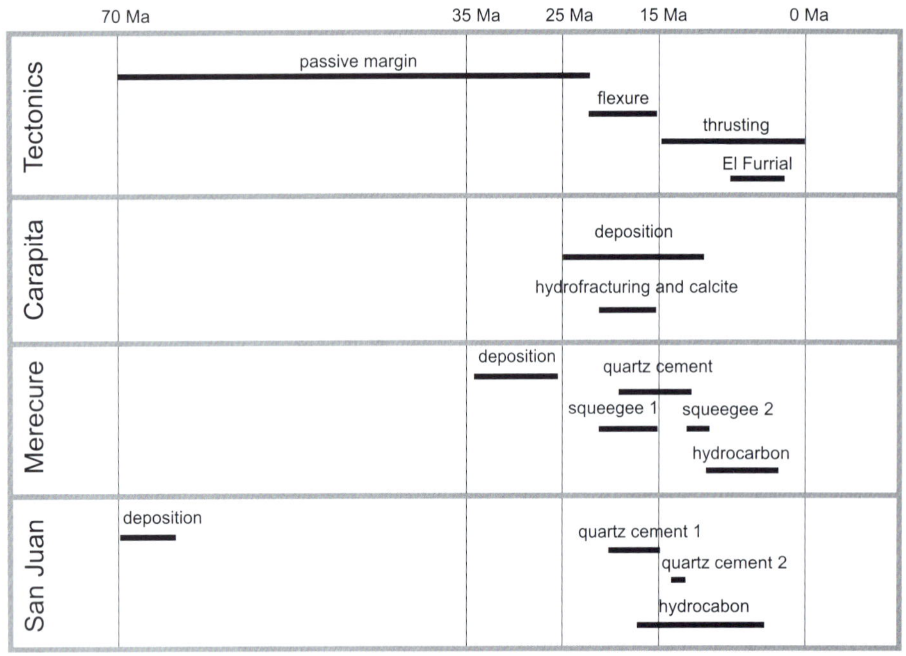

FIGURE 9. Synthesis of the geological events related to the El Furrial structure. The geological history of the structure is characterized by two squeegee episodes, two generations of quartz overgrowth, and a hydrofracturing episode.

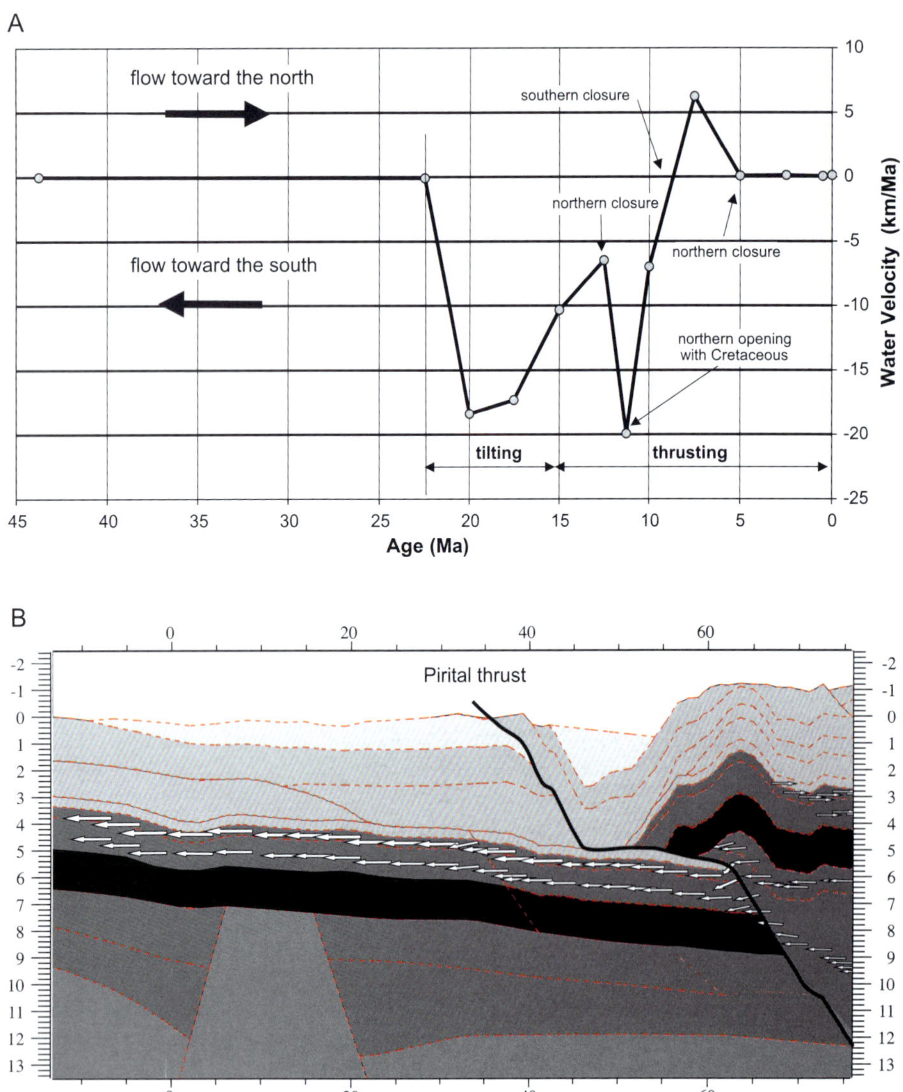

FIGURE 10. Computed fluid velocity evolution along the stratification in the Oligocene sandstones from the El Furrial structure. From the deposition of the late Cretaceous–Oligocene reservoirs until the beginning of the foreland tilting, the fluid flow remained dominantly vertical because of the compaction-driven dewatering of the sediments. During regional tilting, from 23 Ma to 15 Ma, paleofluids (squeegee 1) escaped southward along the stratification with an average velocity of 10–20 km/Ma. Between 15.0 and 12.0 Ma., the fluid flow toward the south decreases. Around 12.0 Ma, as the Pirital thrust was still active, the northern part of the El Furrial structure was then directly in contact with the Cretaceous reservoir formations of the Pirital unit (b), the fluid flow toward the south in the Oligocene of the El Furrial structure increased drastically to reach velocity of 20 km/Ma (squeegee 2). Since 5 Ma, the water flow velocity has been about 0 km/Ma.

reversed, thus inducing a progressive reflux of the fluids toward the north as a direct consequence of the southern closure of the structure (Figure 10).

Since 5 Ma, the El Furrial structure was sealed again at its northern border (Figure 4). From that time to the present, the structure has remained closed. This closure of the El Furrial structure is characterized by a strong and very recent decrease of the water-flow velocity to about 0 km/Ma (Figure 10).

Hydrocarbons Filling

The simulated filling history of the El Furrial structure (Figure 11) shows that the first hydrocarbons generated by the Querecual source rock first reach the sandstones of the San Juan Formation. This occurred at about 17 Ma. At about 12 Ma, the hydrocarbons reached the Oligocene sandstones. The increase of the hydrocarbon saturation in the Oligocene sandstone is correlated with the creation of the El Furrial structure at about 5–7 Ma.

SYNTHESIS

This work is an attempt of basin modeling in a complex area. The aim was to understand the fluid-flow evolution in the Late Cretaceous and Oligocene sandstones of the El Furrial reservoir. However, the results of this work can be extrapolated to other transects to help in the understanding of such tectonically complex areas. The geological history (see Figure 9) of the structure is characterized by two squeegee episodes, two generations of quartz overgrowth, and a hydrofracturing episode.

As it has been discussed, trapping temperatures of fluid inclusions in the quartz overgrowths indicate that precipitation of the main phase began between 100 and 125°C. The average trapping temperature obtained was plotted on the temperature and burial vs. time curves to determine the approximate depth and timing of this diagenetic phase (Bordas Le Floch, 1999). As a result, quartz cementation probably occurred between 19 and 10 Ma, when the depth of the reservoirs was approximately

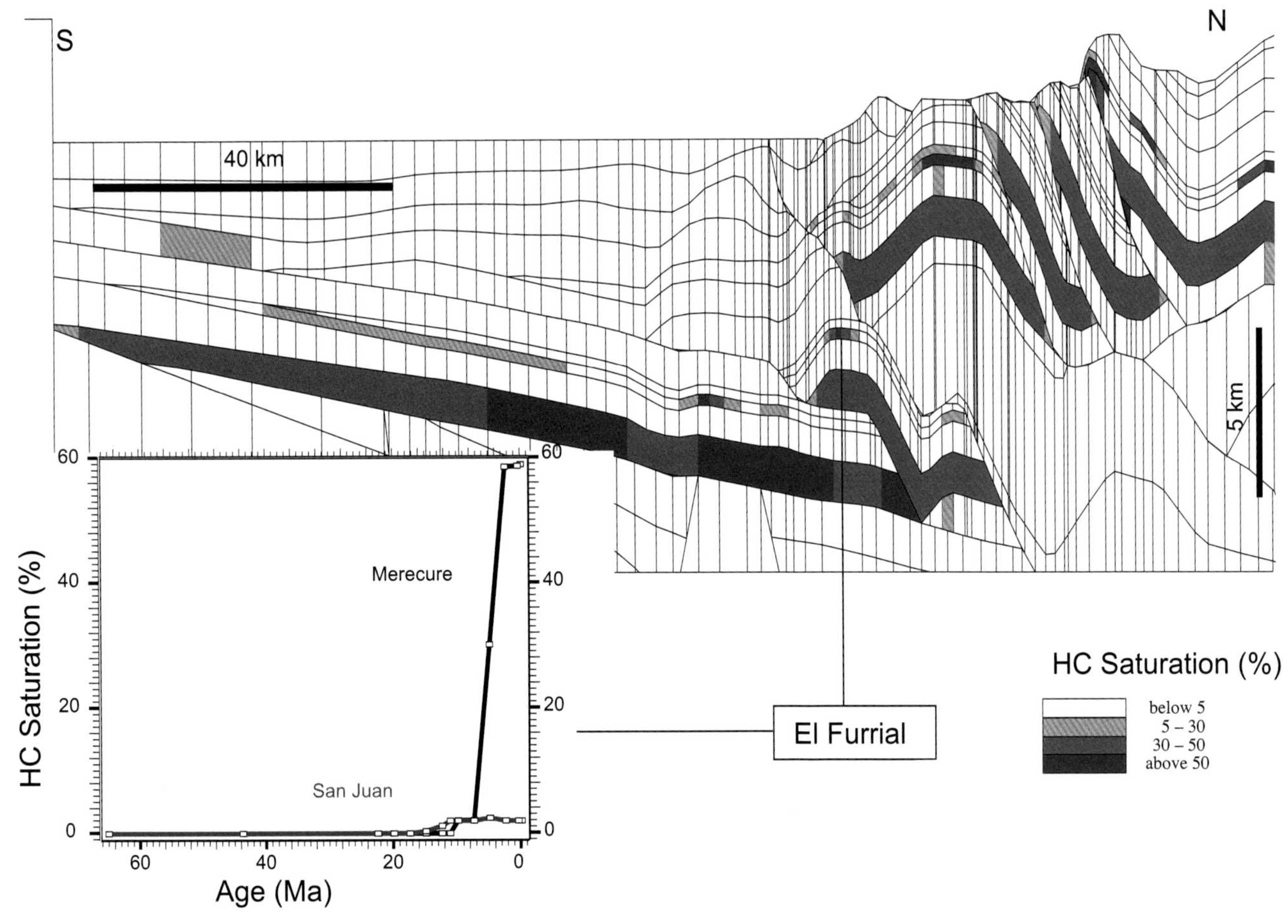

FIGURE 11. The simulated filling history of the El Furrial structure shows that the first hydrocarbons generated by the Querecual source rock reach first the sandstones of the San Juan Formation. This occurred around 17 Ma. At around 12 Ma. the hydrocarbons reached the Oligocene sandstones. The increase of the hydrocarbon saturation in the Oligocene sandstone is correlated with the creation of the El Furrial structure at about 5–7 Ma.

between 2500 and 4000 m (8200 and 13,100 ft). Thus, quartz cementation of the Late Cretaceous series was probably the result of deep-burial and high-temperature conditions, mainly related to the deposition of the thick shales of the Carapita Formation, when the succession was still part of the autochthon.

Quartz overgrowths display oxygen isotope values ($\delta^{18}O$) that suggest that the main phase of quartz cement precipitated from marine waters, whereas the second generation of quartz overgrowth probably formed from evolved basinal fluids. This is consistent with the results from Ceres modeling, which demonstrates that formation (marine) waters were circulating in the system without restraint, until tectonic accretion occurred. Then, because of the deformation, deeper and shallower sedimentary units were in tectonic contact, which probably resulted in the opening of the geochemical system. In other words, fluids with different temperatures and compositions (evolved basinal fluids) entered the system.

The fluids of the first squeegee episode were at thermal equilibrium with the Oligocene sandstones in the north but recorded higher temperatures than adjacent sediments in the El Furrial area. This episode is characterized by an increase in the overpressure in the Oligocene and Upper Cretaceous sandstones correlated to a hydraulic fracturing of the sealing Carapita black shales.

The second squeegee episode gave rise to an exchange and mixture of fluids with different temperatures and compositions, which probably resulted in additional but minor episodes of quartz precipitation. This hypothesis is consistent with the information obtained from oxygen isotope analyses, which suggest that subsequent generations of quartz cement probably formed from evolved basinal fluids.

The fracturing period of the Carapita seal has been recognized by Bordas Le Floch (1999). Indeed, these shales are cut by hydraulic fractures sealed with calcite. The fluid-inclusion study associated with the simulated thermal history allowed Bordas Le Floch to conclude that this fracturing period is dated from 22 to 16 Ma. This interval is very close to the one computed with Ceres.

The hydrocarbon filling history is consistent with the fluid-inclusion studies, which show that hydrocarbons were present during the quartz cementation of the San Juan Formation, but they were not present during the quartz cementation of the Merecure Formation.

REFERENCES CITED

Arnstein, R. E., E. Cabrera, F. Russomanno, and H. Sanchez, 1985, Revisión estratigráfica de la cuenca de Venezuela oriental: VI Conferencia Geológica Venzolana Memoria I, p. 41–69.

Bordas Le Floch, N., 1999, Diagenèse, compaction et déformation des réservoirs gréseux dans les chaînes plissées: Ph.D. thesis, University Paris VI, Paris, 261 p. and annexes.

Carnevali, J. O., 1988, Venezuela nor-oriental: Exploración en frente de montaña, *in* A. G. Belizzaria, A. C. Escoffery, and I. Bass, eds., III Simposio Bolivariano: Sociedad Venezolana de Geología, v. 1, p. 69–89.

Chiock, M., 1985, Cretáceo y Paleógeno en el subsuelo del norte de Monagas, *in* Estratigrafia, paleontologia y sedimentacion, C. A. Espejo, J. H. Rios, and N. P. de Belizzia, eds., Memoria - Congreso Geologico Venezolano, v. 6, no. 1–2, p. 350–383.

Deville, E., and W. Sassi, 1996, Thrustpack; un logiciel de modélisation intégrée (cinématique, thermique et géochimique); application aux Alpes nord-occidentales, *in* L'Exploration pétroliere dans les fronts compressifs: Petrole et Techniques, v. 401, p. 77.

Fiorillo, G., 1987, Exploration and evaluation of the Orinoco oil belt, *in* R. F Meyer, ed., Exploration for heavy crude oil and natural bitumen: AAPG Studies in Geology 25, p. 103–114.

Gallango, O., and F. Parnaud, 1995, Two-dimensional computer modeling of oil generation and migration in a transect of the Eastern Venezuelan Basin, *in* A. J. Tankard, R. Suarez Soruco, and H. J. Welsink, eds., Petroleum basins of South America: AAPG Memoir 62, p. 727–740.

Gonzalez de Juana, C., J. Arozena, and X. Picard-Cadillat, 1980, Geología de Venezuela y de sus cuencas petrolíferas: Caracas, Foninves ediciones, v. 2, 1051 p.

Hedberg, J. D., 1950, Geology of the Eastern Venezuela Basin (Anzoategui–Monagas–Sucre–eastern Guarico portion): Geological Society of America Bulletin, v. 61, p. 1173–1216.

Moretti, I., and M. Larrere, 1989, Locace; computer-aided construction of balanced geological cross sections: Geobyte, v. 4, no. 5, p. 16–24.

Moretti, I., F. Bessis, B. Colletta, T. Cuvillier, M. Larrere, S. Triboulet, J. Letouzey, V. Maisondieu, and P. Werner, 1988, Locace program; a new tool for balancing cross section, *in* Mediterranean basins conference and exhibition: AAPG Bulletin, v. 72, no. 8, p. 1015.

Parnaud, F., Y. Gou, J. C. Pascual, I. Truskowski, O. Gallango, and H. Passalacqua, 1995, Petroleum geology of the central part of the Eastern Venezuelan Basin, *in* A. J. Tankard, R. Suarez Soruco, and H. J. Welsink, eds., Petroleum basins of South America: AAPG Memoir 62, p. 741–756.

Potié, G., 1989, La Serrania del Interior oriental sur le transect Cumana-Urica et la basin de Maturin (Venezuela): Ph.D. dissertation, Université de Bretagne Occidentale, Brest, 240 p.

Rossi, T., 1985, Contribution à l'étude géologique de la frontière sud-est de la plaque caraïbe: La Serrania del Interior oriental (Venezuela) sur le transect Cariaco-Maturin: Ph.D. dissertation, Université de Bretagne Occidentale, Brest, 338 p.

Rossi, T., J. F. Stephan, R. Blanchet, and G. Hernandez, 1987, Etude géologique de le Serrania del Interior oriental sur le transect Cariaco-Maturin: Revue de l'Institut Français du Pétrole, v. 52, no. 1, p. 3–30.

Roure, F., J. Carnevali, Y. Gou, and T. Subieta, 1994, Geometry and kinematics of the North Monagas thrust belt (Venezuela): Marine and Petroleum Geology, v. 11, p. 347–362.

Schneider, F., and S. Hay, 2001, Compaction model for quartzose sandstones. Application to the Garn Formation, Haltenbanken, mid-Norwegian continental shelf: Marine and Petroleum Geology, v. 18, no. 7, p. 833–849.

Schneider, F., J. L. Potdevin, S. Wolf, and I. Faille, 1993, Modèle de compaction élastoplastique et viscoplastique pour simulateur de bassins sédimentaires: Revue de L'institut Français du Pétrole, v. 49, no. 2, p. 141–147.

Schneider, F., J. L. Potdevin, S. Wolf, and I. Faille, 1996, Mechanical and chemical compaction model for sedimentary basin simulators: Tectonophysics, v. 263, p. 307–317.

Schneider, F., S. Wolf, I. Faille, and D. Pot, 2000, A 3-D-basin model for hydrocarbon potential evaluation: Application to Congo offshore: Oil & Gas Science Technology, Revue de l'Institut Français du Pétrole, v. 55, no. 1, p. 3–13.

Schneider, F., H. Devoitine, I. Faille, E. Flauraud, and F. Willien, 2002a, Ceres 2-D: A numerical prototype for HC potential evaluation in complex area: Oil & Gas Science and Technology, Revue de l'Institut Français du Pétrole, v. 54, no. 6, p. 607–619.

Schneider, F., J. L. Faure, and F. Roure, 2002b, How to address petroleum system in complex area: Example from the Canadian Foothills. Emerging concepts in organic petrology and geochemistry (abs.): August 31–September 4, The Banff Center, Banff, Alberta, Canada: Poster abstract no. 25.

Sulek, J. A., 1961, Miocene correlations in the Maturin subbasin: Associación Venezolana de Geólogo Mineros y Petroleros, v. 4, p. 131–139.

Case Studies and Petroleum Systems

19

El Euchi, H., M. Saidi, L. Fourati, and C. El Maherssi, 2004, Northern Tunisia thrust belt: Deformation models and hydrocarbon systems, *in* R. Swennen, F. Roure, and J. W. Granath, eds., Deformation, fluid flow, and reservoir appraisal in foreland fold and thrust belts: AAPG Hedberg Series, no. 1, p. 371–390.

Northern Tunisia Thrust Belt: Deformation Models and Hydrocarbon Systems

Hassen El Euchi, Moncef Saidi, Lotfi Fourati, and Chokri El Maherssi

Entreprise Tunisienne des Activités Pétrolières, Tunisia

ABSTRACT

The integrated use of the geological, geophysical, and geochemical data of northern Tunisia onshore and offshore led to an inversion model from the Tethyan rifting and subsequent evolution of the North African passive margin to the Late Cretaceous–Tertiary orogeny, inducing the fold and thrust belt and associated foreland deformations. Respective deposits characterize each tectonic cycle; Triassic synrift and Jurassic–Cretaceous open-marine series are related to the Mesozoic opening, and Paleogene–Neogene clastic sequences are closely controlled by the Tertiary shortening. For hydrocarbon prospectivity and despite its early stage of exploration, this domain could be considered as an emerging area with encouraging ingredients, particularly three fractured carbonate reservoirs, two sandy reservoirs, five source rocks, and numerous potential structural and stratigraphic traps. The main feature in this area is the close relationship between the Tertiary tectonics and the evolution of the petroleum systems. Hence, the Numidian turbidites identified both as reservoir and source rock were deposited in a foredeep directly generated by the Paleogene thrusting and later displaced as far-traveled nappes above the Ypresian fractured limestones defined also as a source rock and reservoir. The maturity of these source rocks is closely related to the nappes displacement, which assumes the overburden and increases the heat flow. Fracturing is also generated by these events.

INTRODUCTION

Because of the numerous and widespread oil seeps in the area, northern Tunisia has been the focus for oil exploration by pioneering geologists since the beginning of the 20th century. More than 30 wells were drilled, resulting in some encouraging shows and tests.

Since the mid-1950s, the search for hydrocarbons has quickly switched to other more promising and geologically less complex areas, particularly to the Saharan Platform in southern Tunisia. More than 60 discoveries have been made since, either onshore or offshore, which contributed to distract the operators from northern Tunisia.

DOI:10.1306/1025701H13126

Recent worldwide exploration activity, which resulted in large discoveries in comparable thrust belts, has revived interest for northern Tunisia, and exploration slowly resumed in this largely underexplored region.

This study initiated by ETAP aimed to assess the hydrocarbon potential of the area and try to answer basic, but important key questions, relative to the identification of active petroleum systems, which could result in the accumulation and trapping of important volumes of hydrocarbons.

Given the scarcity of the well and seismic data in this area, this evaluation has been mainly performed on outcrop investigations field study (tectonics, sedimentology, and geochemistry) and analyses of collected samples. Some available seismic data in the offshore and surrounding areas were also used.

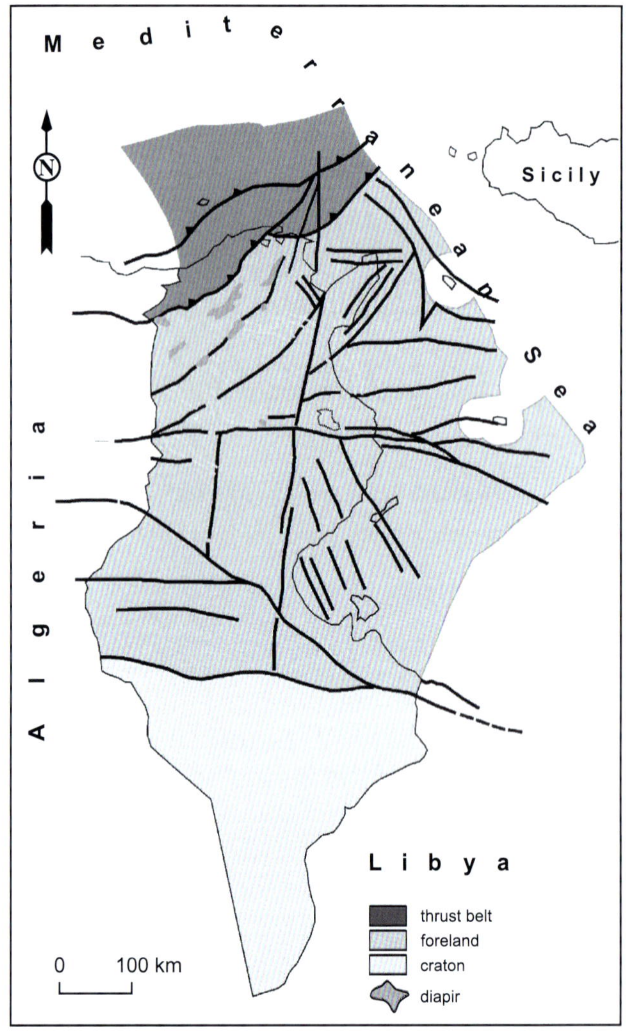

FIGURE 1. Tunisia structural domains.

GEODYNAMICS AND GEOLOGICAL BACKGROUND

Geologically, northern Tunisia (both offshore and onshore) corresponds to a fold and thrust belt (Rouvier, 1977; El Euchi et al., 1997a) linking the North Africa Atlas and the Sicily chains (Figure 1), all belonging to the peri-Mediterranean orogenic arc that occurred during the Cenozoic, following the collision between the African and European plates. Structurally, this northeast–southwest-trending thrust belt is made up of two major units (Figure 2).

The Main Tectonostratigraphic Units and Paleogeographic Domains

The Kabilyan Equivalent Unit

It is situated in the Tunisian offshore, northward the "La Galite" Island (Figure 2) and represents the eastern equivalent of the North African (particularly the Kabylian) crystalline units that were detached and transported during the Late Cretaceous–Oligocene–Eocene thrusting as deep nappes and later deformed during the Miocene shortening.

The Numidian–Tellian Units

It corresponds to vertical and lateral successive allochthonous units (Figure 3) made up of Oligocene–Miocene Numidian flysch and Cretaceous–Eocene shales and limestones. These thrust sheets were mainly displaced during the Langhian–Tortonian compression and folded again during the late Miocene–Pliocene shortening. Detachment levels are represented by either the allochthonous Triassic salt already remobilized during various diapiric episodes or the Cretaceous–Paleogene shales.

These two major domains are separated by the Kabylian domain front fault, which is identified in seismic data as a major thrust front (Figure 2).

In details, these detached and displaced units are mainly represented by the following:

- The Oligocene–early Miocene nappes made up of the so-called Numidian flysch, which is mainly composed of turbidites organized in successive shale-sandstone sequences (Gottis, 1953; El Maherssi, 1992) showing different sedimentary figures, such as ripple marks, slumps, convoluted water expulsion, sills, dykes, flood casts, fining-upward and/or coarsening-upward sandy bodies, etc.
- The Late Cretaceous–Eocene Addissa nappe (Rouvier, 1977) composed of shales interbedded with breccia horizons and crops out only in the northernmost area of the onshore
- The Paleocene–Eocene Kasseb unit (Rouvier, 1977), which is obviously present at the front of the Numidian nappe and is made up of the uppermost

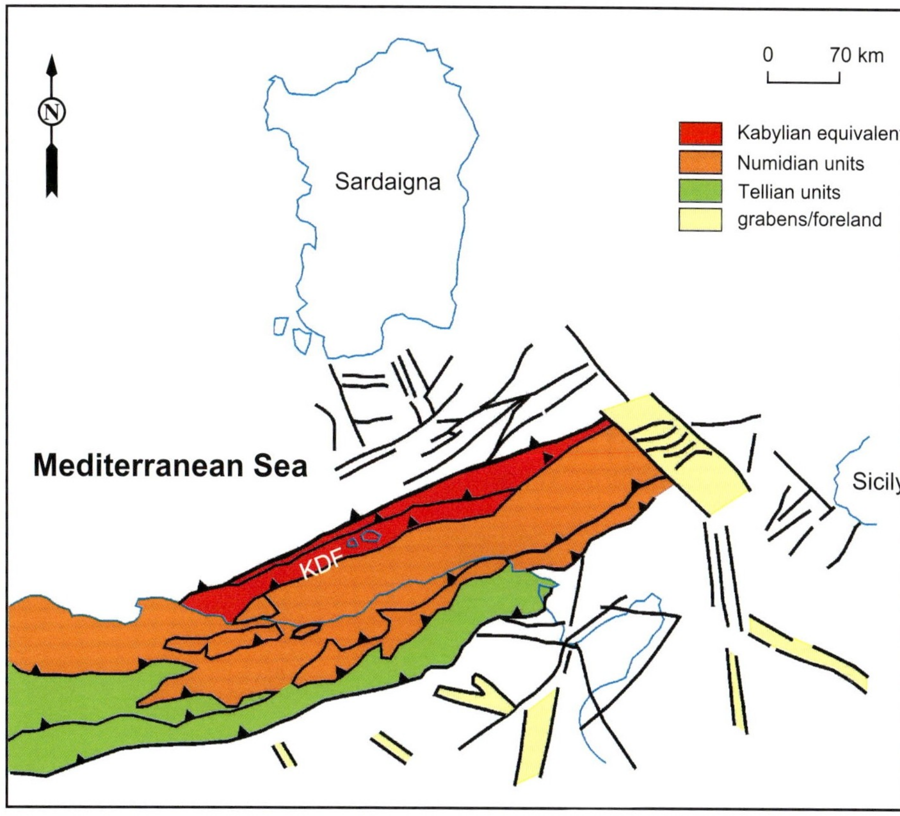

FIGURE 2. Northern Tunisia thrust belt units. KDF = Kabylian domain front.

Paleocene shales, Ypresian open-marine limestones with Globigerinae, and partly eroded Lutetian shales unconformably overlain by autochthonous Oligocene–early Miocene sandstones, shales, and conglomerates (the Numidian equivalent, commonly known as the Bejaoua Group)

- The Cretaceous Ed Diss unit (Rouvier, 1977), which is made of Cretaceous shales andlimestones.

Tethyan Rifting and Subsequent Evolution of the North African Passive Margin

Related to the Tethyan oceanic development along its southern continental margin with a north–south extensional regime, this event starts during the Late Triassic and continued through the Jurassic–Cretaceous time interval. It led to the deposition of the synrift, postrift, and breakup marginal sequences.

The Triassic rifting is characterized by northeast-southwest to north–south spreading, which led to normal faulting and transfer wrench faulting, along which tilted blocks, horsts, and grabens are the most common structures controlling, in turn, the facies and thickness distributions. These movements are evident in the foreland area; however, within the thrust belt zone, the Triassic outcropping sequences are mostly allochthonous and are identified as diapirs and detachment levels. However, the lack of well control and deep seismic data represents a handicap to define this event in this area.

In the foreland, a Middle–Late Triassic to Liassic synrift deposit that is represented by subcontinental evaporites, marls, sandstones, and dolomites marks the Tethyan rifting.

During the Jurassic, the regional facies distribution in Tunisia indicates that the depositional system corresponds to a dominant marine environment, with several lateral variations of thickness and facies (Soussi, 2002). These variations are closely related to a normal and wrench faulting system locally associated with salt doming. The main fault trends are east–west and northeast–southwest, which created grabens,horsts, and pull-apart basins, and led to a local deposition of breccias, debris flows, buildups, and condensed sections (Soussi, 2002).

Particularly, in the "dorsale" trend corresponding to the southern boundary of the northern Tunisia area, Cohen et al. (1980), Turki (1985), and Soussi (2002) demonstrated that the debris flows, oolithic limestones, breccias, and buildups of the Jurassic are related to the development and breakup of the Tunisian margin, driven by the east–west and northeast–southwest normal-wrench faults.

In northernmost Tunisia, the Jurassic sequences crop out only in Jebel El Hairech, Thuburnic, Oued Zarga, Jebel Ichkeul, Tebourba, and Jebel Ammar localities and are obviously metamorphosed and highly fractured and exhibit an intensive schistosity.

It seems to be the same situation for the Cretaceous, which does not exhibit a clear tectonic control of the sedimentation, except for a regional southeast–northwest thickening and local slight facies variations. Elsewhere, in central and southern Tunisia, the Cretaceous time interval gives clear evidence for a dominant extensional regime, ranging from the northwest–southeast to northeast–southwest spreading.

Hence, the Early Cretaceous northeast–southwest extension has several evidences both in seismic data and outcrops. Depositional environments are dominated by the Berriasian–Barremian shallow-marine to fluvial

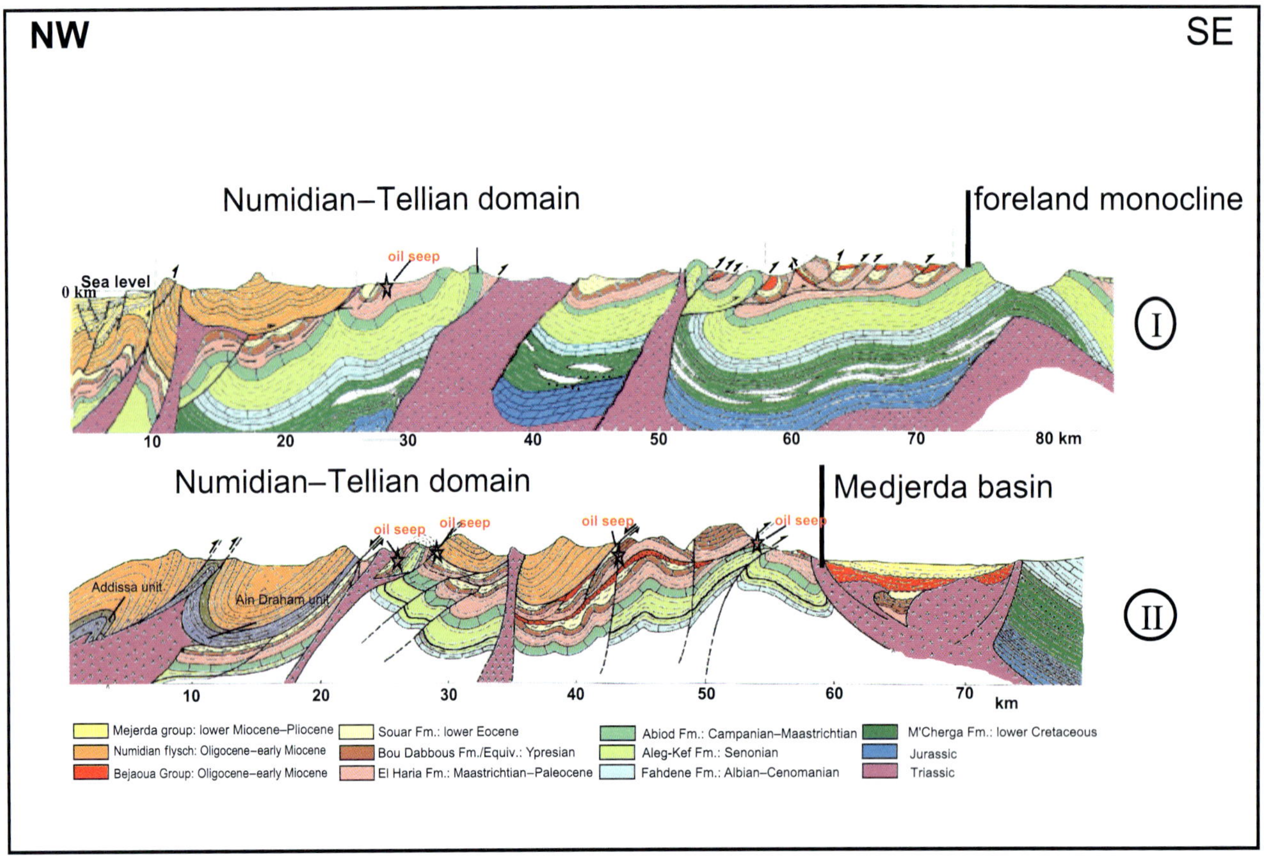

FIGURE 3. Numidian–Tellian domain onshore northern Tunisia.

systems, followed during the Aptian by the development of a widespread carbonate platform in central Tunisia and open-marine turbiditic sequences in northern Tunisia.

The early Albian is marked by an east–west, well-defined, transpressional to compressional regime, which created anticlines, en echelon folds, and reverse faults (El Euchi et al., 2000). Erosion was very active during this phase and is associated with an emersion, which induced early diagenesis phenomena such as karstification, cementation, and dolomitization. This event that assimilated to the Austrian phase is characterized by a regional unconformity separating the partly eroded and early deformed Aptian from its Albian cover.

From Cenomanian to Coniacian, Tunisia was submitted to a mainly northwest–southeast to north–south extensional-transtensional regime that reactivated some of the previous trends and created new basins and uplifts. The salt doming activity was more important during this epoch. Locally, stress field migration and reorientation led to the formation of en echelon folds along the wrench faults, such as in the Gulf of Gabes and Sahel areas.

The Santonian–Campanian is obviously characterized by a sag period and northwest–southeast wrench-normal faulting probably responsible for the northwest-trending grabens in central Tunisia, which are known as Late Cretaceous in age.

Late Cretaceous–Paleogene Orogeny

The earliest pulse of this major Mediterranean event is uppermost Cretaceous in age. The first paroxysm is dated uppermost Eocene–Oligocene and is commonly known as the "Pyrenean compression." It resulted in the convergent plate boundary between Africa and Europe and is characterized by a northwest-trending stress field, which induced a regional detachment and thrusting of the crystalline units and the Mesozoic overburden series. Thrusts, high-angle reverse faults, wrench faults, ramps, overturned anticlines, pop-up anticlines, en echelon folds, foredeep, and piggyback basins are the common associated structures (El Euchi et al., 1997b).

Evidences are frequent either in seismic (Figure 4) or surface geology. In northern Tunisia onshore, immediately southward of the actual Numidian thrust front, particularly in the Bou Gotrane-Kasseb-Sidi Ahmed area, the outcrops show a well-defined angular unconformity between the early folded Paleocene (El Haria Formation) and the later deposited and deformed Oligocene–early Miocene (Figure 4). Westward, in Fernana–Jendouba area, the Triassic overthrust the metamorphosed Jurassic sequences, which are highly deformed and squeezed. Several Tethyan structures, such as wrench and normal faults, were reactivated during this phase (Crampon, 1971). However, some previous

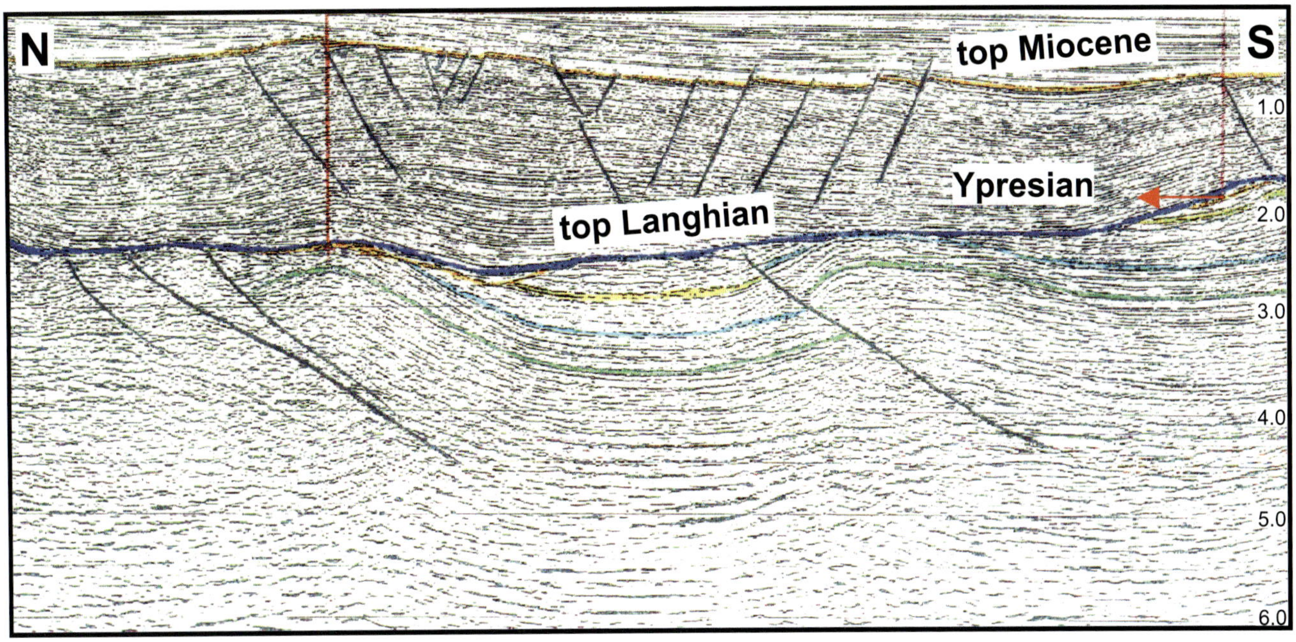

FIGURE 4. Seismic evidence of the Late Cretaceous–Paleogene compression (Gulf of Hammamet). Scale: 1 cm = 2 km.

diapiric salts and shales were reactivated mostly horizontally as local decollement and transport sheets instead of vertical doming. This is the case of Hedhil area, where the Cretaceous–Eocene detached ramp is locally supported by Triassic early diapiric salt (Kujawski, 1969).

This folding-thrusting phase ended by the uppermost Eocene to Oligocene and was followed by a sag period until the early Miocene (Langhian–Serravalian).

The regional deformation model consisting of a northwest-verging fold and thrust belt is composed of far displaced faulted folds and ramps separated by foredeeps and ended to the southeast by a forebulge (platform). The thrust belt and foreland were affected by subsequent normal faults, where Oligocene–Langhian deposits correspond to shallow-water to fluvial detritus sediments.

The foredeeps were filled by thick turbidites as much as 3000 m (10,000 ft). These sequences correspond to syntectonic deposits commonly known in North Africa and Sicily as the Numidian flysch.

Neogene Orogeny

During the Serravalian–Tortonian, the so-called Alpine sensu-stricto compression induced the detachment at different decollement levels and southward displacement of the Numidian sequences and locally the Late Cretaceous–Eocene ramps. The Numidian sequence is duplicated at least in two superposed sheets. Allochthonous Triassic salt, Numidian, Cretaceous, and Eocene shales acted as a detachment horizon.

This thrusting-folding had produced high-amplitude faulted anticlines, ramps, and triangle zones simultaneous to piggyback and foredeep basins.

Later, the Pliocene–Pleistocene north–south shortening induced the deformation of older series. This deformation is characterized by high-angle reverse faults, en echelon folds associated to strike-slip faults, and pull-apart basins.

Hence, the deformation timetable of northern Tunisia is obviously dominated by the Tertiary compressional movements (Figure 5), which contrast with the Mesozoic opening cycle and induced the structural inversion. The latter closely controlled the hydrocarbon prospectivity of the area.

POTENTIAL AND EFFECTIVE SOURCE ROCKS

The Potential Cretaceous Source Rocks

Albian–Vraconian (Lower Fahdene Formation)

The lower Fahdene formation (Albian–Vraconian) gradually overlies the deep-sea clastics of M'Cherga formation and is conformably overlain by the Cenomanian upper Fahdene shales. It covers the northern and eastern area of Tunisia (onshore, Gulf of Hammamet, north of the Gulf of Gabes). Toward the south, this level corresponds to the dominantly dolomitic Zebbag formation (Figure 6a).

The lower Fahdene formation is composed of black platy limestones, black limestones, and black to gray marl and light gray to dark gray, nonlaminated nodular limestone (Figure 6b). Thickness is variable and may

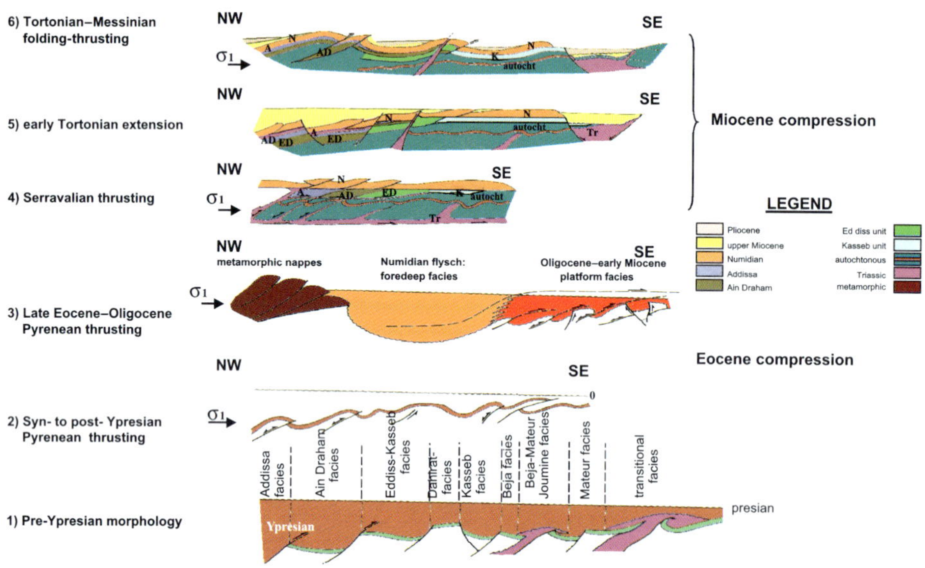

FIGURE 5. Tertiary tectonic evolution of northern Tunisia (El Euchi et al.,1997a).

reach more than 1200 m (4000 ft) in some depocenters or troughs (basinal facies). In several areas located above or close to the Triassic salt domes, the early and middle Albian deposits are commonly missing, and only late Albian and Vraconian series are present (paleohigh-related hiatus).

This unit is known to be a good source rock in northern Tunisia (Talbi, 1991; Saidi, 1993) and is rich in types II and III organic matter, having total organic carbon values as much as 4% (Figure 6c). Average total organic carbon values are about 1% and are generating potential of as much as 15 kg of hydrocarbon/ton of rock. The thermal maturity is relatively low, corresponding to the beginning of the oil window in the paleohigh facies; however, in the basinal facies, it may reach gas window level. A general northeast–southwest-increasing maturity trend is observed in the area.

Petrographically, the organic matter is a mixture of amorphous material, microspores, cuticles, and vitrinite (Figure 6d).

The organic matter associated with this unit displays a Pr/Ph ratio ranging between 1 and 3, a moderate to high diasterane content, a moderate to high C_{28}/C_{29} sterane ratio (0.69–0.76), C_{35}/C_{34} homohopane ratios less than 1, low tricyclic terpane content, and carbon isotope composition ranging between −28.5 and −29‰.

Late Cenomanian–Early Turonian (Bahloul Formation)

The late Cenomanian–early Turonian laminated, black, organic-rich limestones were described in central Tunisia and attributed to the Bahloul Formation. It covers northern Tunisia, the Sahel, and the northern part of the Gulf of Gabes (Figure 7a) and is also encountered in the central west area of the country (Jebel Berda, Bligi-1 well, etc.). The Bahloul Formation is considered as one of the main source rocks in Tunisia and is composed of outer-shelf to slop-laminated and dark-gray to black-colored limestones and marls (Figure 7b). This formation had been deposited in a relatively deep-marine, calm, and anoxic environment (Caron, 1985; Ben Ferjani et al., 1990). Laterally, the Bahloul is equivalent to the Gattar Member toward the south or to the shaly member at the base of the Aleg formation (Gulf of Hammamet).

Geochemical data of the Bahloul Formation in northern Tunisia show this level to be a good potential source rock as indicated by the high content of type II organic matter (Figure 7c, d).

This formation exhibits high total organic carbon contents (as much as 14%), with an average value of 2% and a good to excellent petroleum potential, ranging between 2 and 50 kg of hydrocarbon/ton of rock. The thermal maturity degree is fairly low (immature to mature) except for those intervals outcropping in the western area near the Algerian border (J. Mahsar and Kef Lemsane, outcropping sections), where the organic matter is overmature (spore colorization index, SCI = 8). This high level of maturity is thought to be responsible for the loss of original petroleum potential of the rock. However, the originally good source rock potential of the Bahloul is demonstrated by the high total organic carbon residual values (as much as 4%).

The extracts from the Bahloul Formation are characterized by Pr/Ph ratios ranging between 1.5 and 3, a moderate diasterane content, a C_{27} sterane approximately equal to C_{29} sterane, a high C_{28}/C_{29} ratio (0.82–0.99), a C_{35}/C_{34} homohopane ratio less than 1, a moderate tricyclic terpane, and carbon isotope compositions between −24 and −24.5‰.

The Potential Ypresian Source Rock

The Bou Dabbous formation covers the entire north and northeastern part of Tunisia, including the Pelagian Sea (Figure 8a). This formation is composed

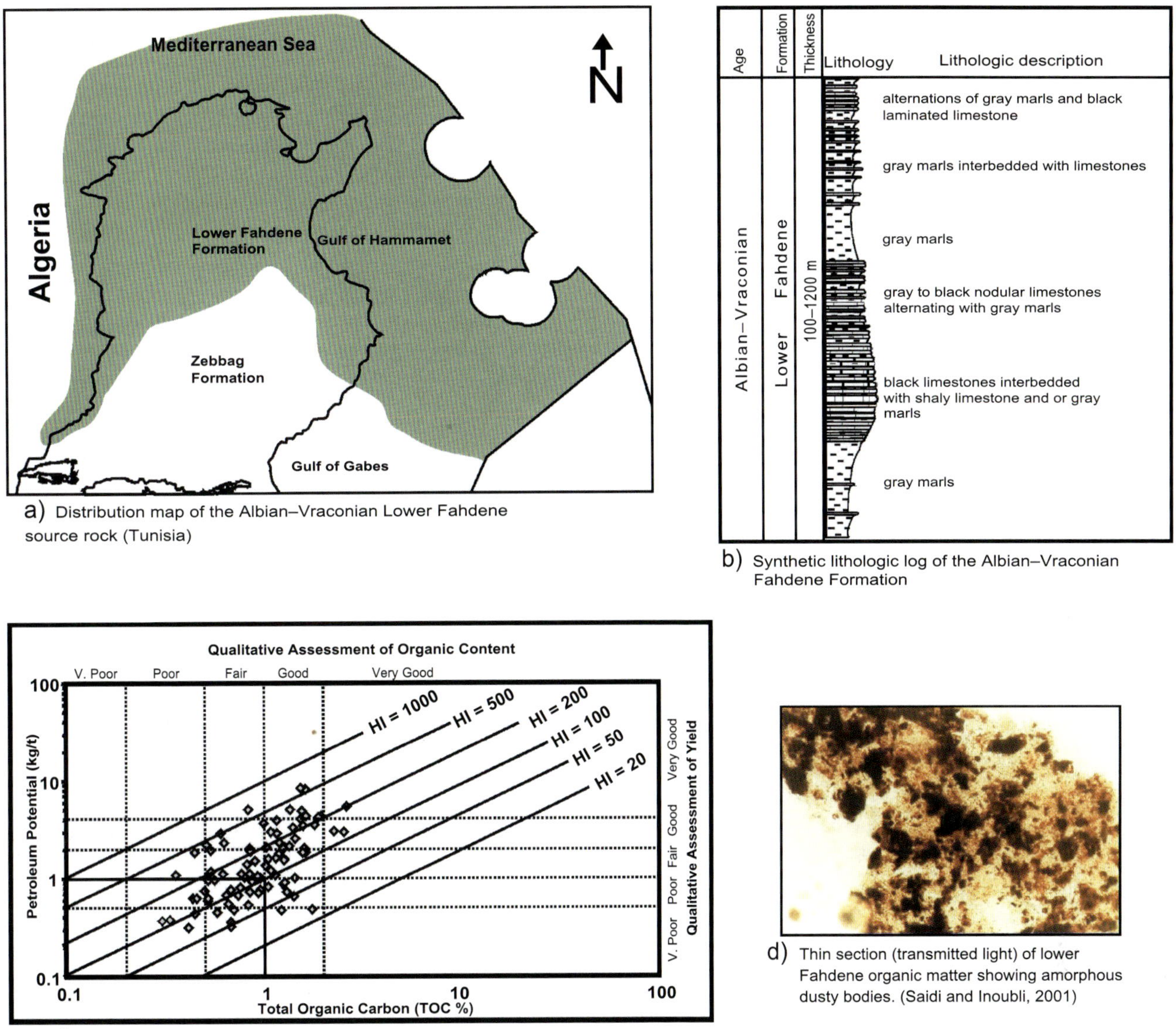

a) Distribution map of the Albian–Vraconian Lower Fahdene source rock (Tunisia)

b) Synthetic lithologic log of the Albian–Vraconian Fahdene Formation

c) Quantitative and qualitative assessment of the Albian–Vraconian source rock

d) Thin section (transmitted light) of lower Fahdene organic matter showing amorphous dusty bodies. (Saidi and Inoubli, 2001)

FIGURE 6. Petroleum potentialities of the Albian–Vraconian lower Fahdene source rock. HI = hydrogen index.

of dark-gray to black, micritic limestone, very rich in Globigerina tests, alternating with marl (Figure 8b).

In northern Tunisia, this formation appears to be a good source rock formation, showing good petroleum potential (as much as 16 kg of hydrocarbon/ton of rock) and high total organic carbon contents (as much as 3.5%) (Figure 8c). The organic matter is particularly well preserved and of type II origin composed of very sapropelic amorphous material (Figure 8d). However, some lateral variations in the organic richness and maturity are observed. Toward the northwest, the Ypresian becomes less organic rich, probably related to increasing maturity level and/or to poor anoxic conditions at the sea bottom and to an imperfect preservation of the organic matter. The maximum maturity level is reached close to the Cape Serrat–Ghardimaou fault, where the source rock is at the end of the oil window or overmature. On both sides of this fault, the maturity decreases toward the northwest and toward the southeast. This high maturity degree along the fault is thought to be related to the circulation of some hot hydrothermal water and/or the presence of some volcanic tuffs that are partly outcropping in the Nefza area.

The extracts are characterized by a low Pr/Ph ratio, a very low diasterane content, C_{27}/C_{29} sterane ratios approximately equal to 1, a high C_{28}/C_{29} ratio (0.81–1.0), a C_{35}/C_{34} homohopane ratio greater than 1, and a carbon isotope composition between −27.5 and −28‰.

Geochemical correlation studies demonstrate the Bou Dabbous formation to be the main source of the oils produced in the Gulf of Gabes (Ashtart, Salambo, Cercina, etc.).

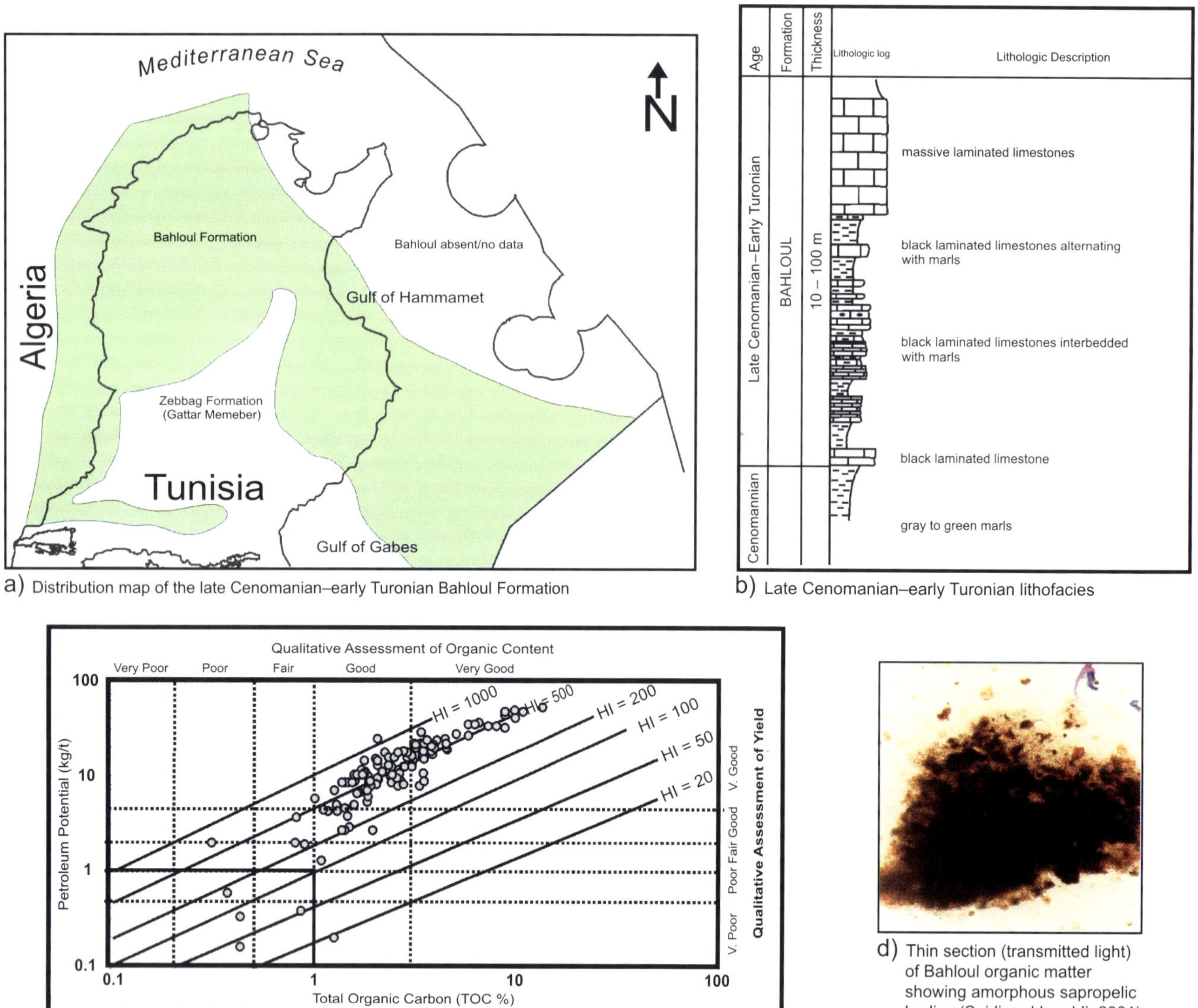

a) Distribution map of the late Cenomanian–early Turonian Bahloul Formation

b) Late Cenomanian–early Turonian lithofacies

c) Quantitative and qualitative assessment of the late Cenomanian–early Turonian Bahloul source rock

d) Thin section (transmitted light) of Bahloul organic matter showing amorphous sapropelic bodies (Saidi and Inoubli, 2001)

FIGURE 7. Petroleum potentialities of the late Cenomanian–early Turonian Bahloul source rock. HI = hydrogen index.

The Potential Numidian Source Rock

Available geochemical data of the shale-interbedded levels of the Numidian flysch series, outcropping in northern Tunisia, show this level to be a fair source rock (Figure 9), as indicated by fair total organic carbon values (as much as 1.1%) and fair petroleum potential (as much as 3.62 kg of hydrocarbon/ton of rock). Thermal maturity level is variable within the area, ranging between an immature and overmature stage. Some shale levels are associated with predominantly type II organic matter, whereas others are rich in type III kerogen. Nevertheless, the remarkable thickness of the Numidian sequence makes these sediments interesting concerning hydrocarbon potential (e.g., offshore Italy and Albania).

Distribution of Seeps and Oil and Gas Shows

In addition to the main source rocks mentioned above, numerous surface seeps have been found throughout northern Tunisia. The seepages (Figure 10) occurred mainly in the nappes zone (Oued Kasseb, Jbel Bou Gotrane, Ain Allega, Babbouch, Jbel Bou Touil and Jbel Ach el Agab, Ouejh El Homs, etc.) and the diapirs zone (J. Ech Cheïd, Oued El Beji, Kef Bou Dabbous, Lakhouat mine, Bougrine mine, Kabbouch mine, Fej el Haddoum mine, Zag Et Tyr, Slouguia). As mentioned earlier, the first exploration wells in Tunisia were drilled in the vicinity of these oil seeps.

Because of their nature, those various seeps could be subdivided into two main types: oil seeps and

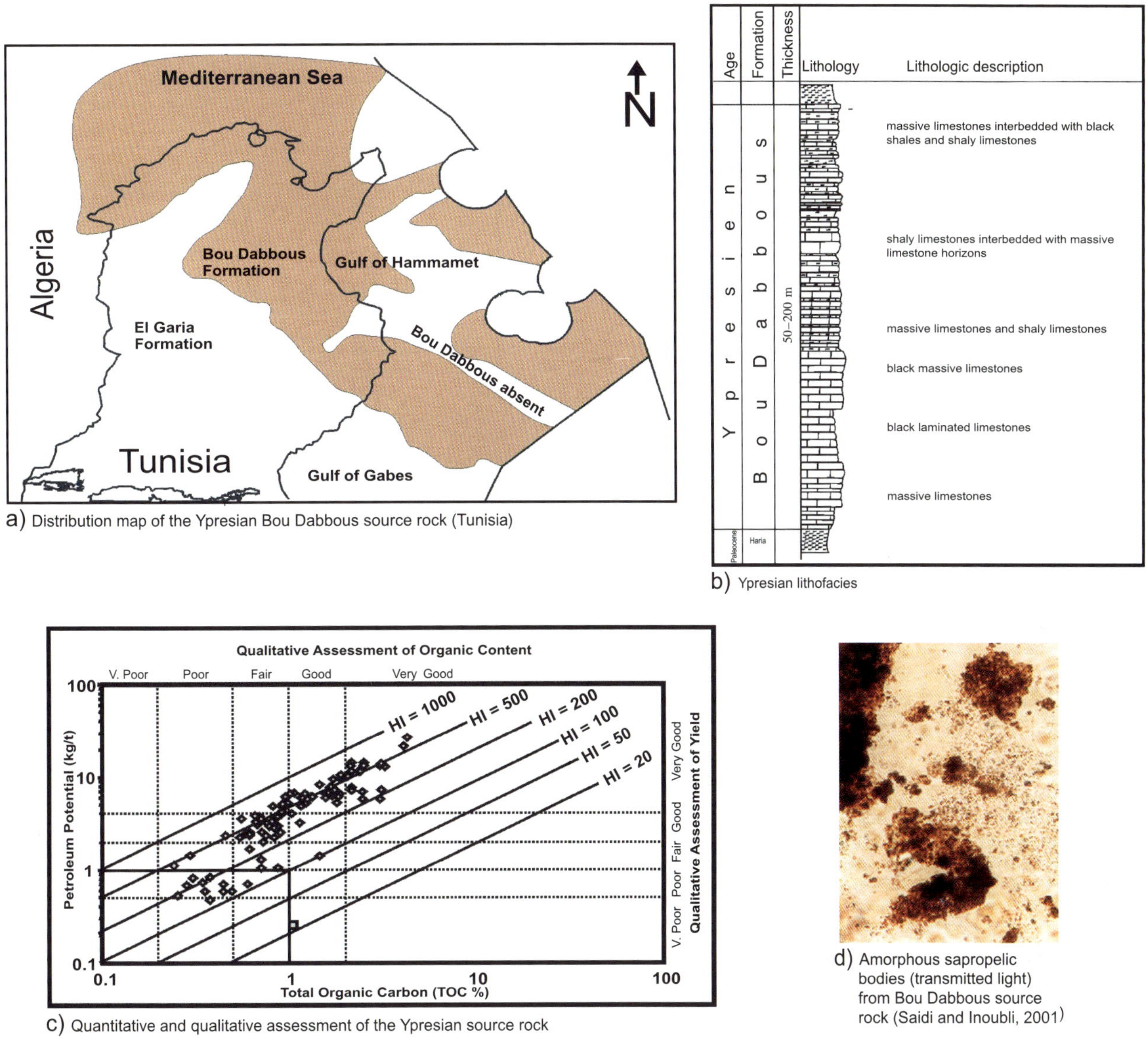

FIGURE 8. Petroleum potentialities of the Ypresian Bou Dabbous source rock. HI = hydrogen index.

asphalt seeps. Asphalt seeps consist of viscous asphalt or bitumen impregnations located mostly in fractures and cap rock that developed in Triassic salt domes or impregnating porous formations.

Effective Source Rocks (Oil Source Correlations)

A detailed organic geochemistry study has been undertaken to provide information on the genetic relationship between oil seeps and the candidate source rocks (Saidi and Philip, 1997, ETAP/Agip project), and was conducted on several oil seeps sampled from Jebel Ech Cheid, Slouguia, Jebel Bou Gotrane, Zag Et Tyr, El Haroun and Kasseb Dam, and Wejh El Homs and on source rock samples, which were extracted from the lower Fahdene, Bahloul, and Bou Dabbous formations and from shaly levels of Numidian series, using gas chromatography, gas chromatography–mass spectrometry, and carbon isotope composition techniques.

The main results obtained are graphically illustrated in Figures 11–13 and are as follows.

The carbon isotope composition of the saturate and the aromatic fractions ranges from −25.48 to −29.64‰ and from −25.12 to −28.61‰, respectively. Based on the Sofer (1984) plot, the analyzed oil seeps and the source rock extracts seem to be nonwaxy and of marine origin. A similar conclusion is also reached from canonical variable values that are less than 1. This carbon isotope composition allowed to make some genetic relationships between the Bou Dabbous formation and the Bou Gotrane and the Kasseb Dam oil seeps; the Slouguia, El Haroun, and Echeid oil seeps are well correlated with the Albian lower Fahdene formation source rock. The Zag Ettir and Jebel Dabadib oil seeps are

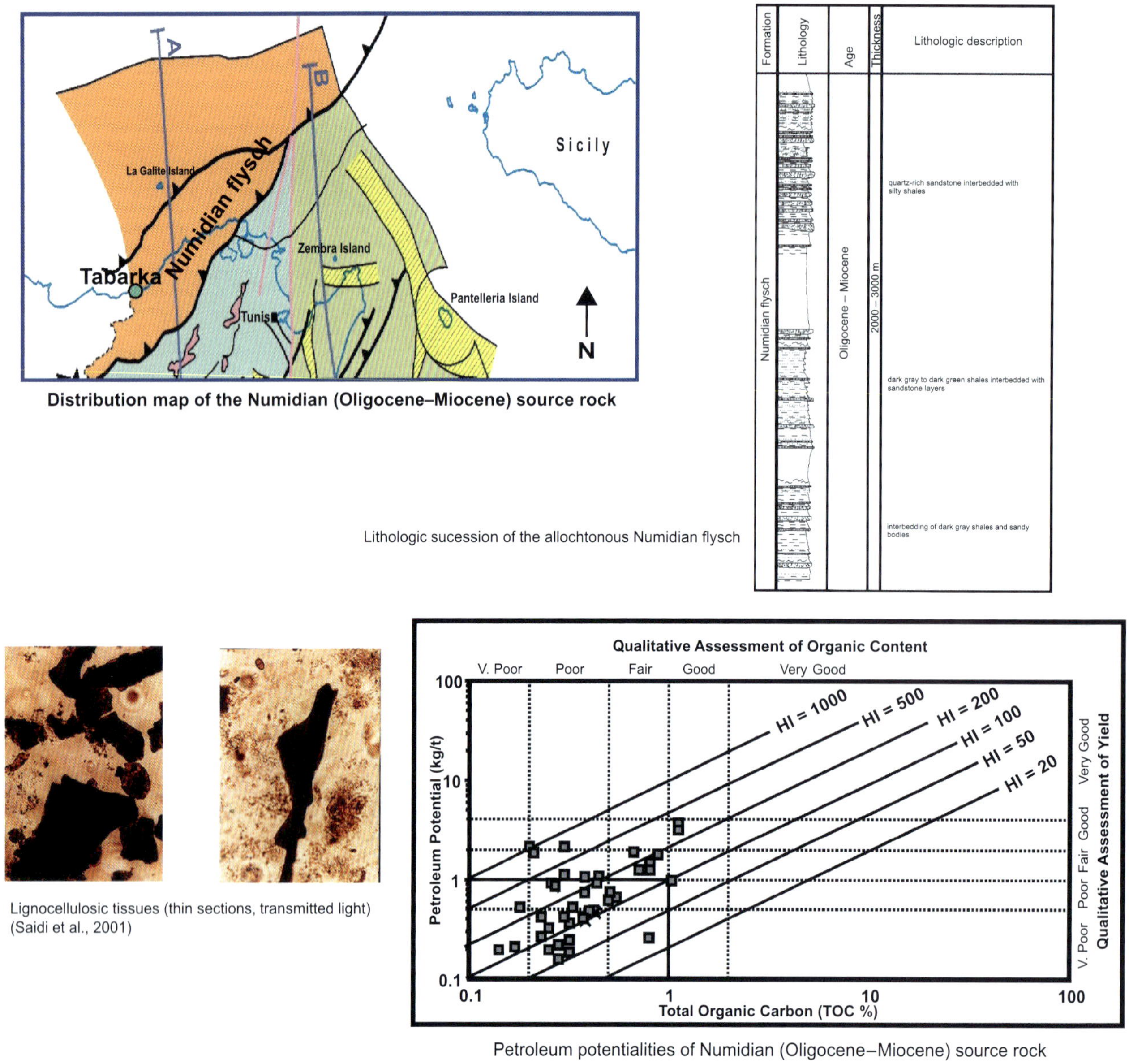

FIGURE 9. Petroleum potentialities of the Numidian source rock. HI = hydrogen index.

sourced by the Turonian Bahloul source rock. Finally, there is an excellent correlation between the Wejh El Homs seepage and the Miocene Numidian shales of the Cap Serrat outcropping source rock.

POTENTIAL RESERVOIRS, SEALS, AND TRAPS

Reservoirs and Seals

Triassic Dolomites

The Triassic deposits consist of alternating carbonates, clays, and evaporites. Carbonates correspond to massive, black, fractured dolomites that are proven as a good reservoir. CO_2 and gas discoveries are recorded in the Utique-1 well (northeast onshore Tunisia).

In fact, the Utique-1 well, drilled in 1995 (northeast onshore Tunisia), tested gas (mainly CO_2) from the Triassic dolomites with a porosity as much as 23% and with high permeability, which is mainly related to an intensive fracturing and karstification process.

The seal is assumed by Triassic evaporites, shales, and/or the overlying impermeable formations.

Early Cretaceous Sandstones (M'Cherga Formation)

In Tunisia, the first discovery of hydrocarbon was made in the Early Cretaceous lenticular sandstones (1948) in the Cape Bon area (northeast onshore Tunisia),

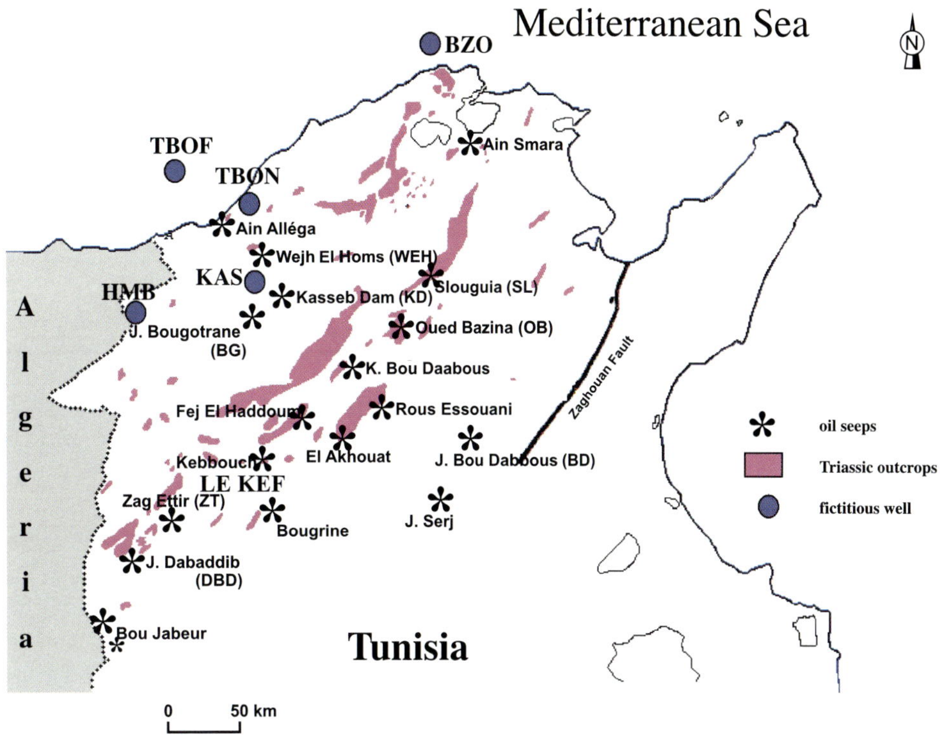

FIGURE 10. Distribution map of oil seeps in northern Tunisia. KAS = Kasseb; HMB = Hammam Bourguiba; BZO = Bizerte; TBON = Tabarka onshore; TBOF = Tabarka offshore.

which has produced gas since the 1950s (Jbel Abderrahmane field).

These sandstones are beige, brown, very fine to fine grained, lenticular in shape, and showing cross-stratification associated with sedimentary and burrowing features.

The hectometric lenticular fractured sandstones of Valanginian–Hauterivian age can reach 27 m (89 ft) of thickness (Kef Lemsene section, Ghardimaou area) and are well capped by intraformational shales.

The porosity ranges between 8 and 16%, and the permeability is as much as 30 md. Reservoir quality can be enhanced by dissolution and karstification processes.

Middle Campanian–Early Maastrichtian (Abiod Formation)

In northern Tunisia, the middle Campanian–early Maastrichtian time interval developed a thick limestone section, generally deposited in a platform domain, and commonly known as the Abiod Formation, which produced oil and gas in several fields in northeastern and central Tunisia (Maamoura, Zinnia, and Sidi El Kilani fields). The depositional facies includes sequences of both homogenous and rhythmic and thick massive beddings.

Within this area, the middle Campanian to early Maastrichtian Abiod Formation is composed of two limy members separated by an alternating marl and limestone member (Fourati and Hamouda, 1996).

The lower limy member (middle Campanian) consists of highly fractured limestones locally chalky and showing some thin intercalations of marls in the uppermost part, where frequent slump marks are recorded. It is made of wackestones to packstones, rich in planktonic foraminifera, associated with some benthic forams; debris of bivalves and echinoids are also present. The matrix is mainly micritic. The lower member thickness ranges between 40 m (130 ft) in the south and 130 m (430 ft) in the north. This lower limy member is well sealed by a relatively thick (80–300 m; 260–1000 ft) middle shaly member.

The upper limy member (early Maastrichtian) is made up of rhythmic to massive beds of fractured limestones, including sparse macrofauna composed of inocerams, echinoids, and stemmed crinoids. Limestones consist of bioclastic wackestones-packstones rich in planktonic forams (Globigerinae, Globotruncana) and common benthonic forams associated with pelecypods debris. The matrix is micritic and contains microsparitic to sparitic cement. The thickness of upper limy member ranges from 40 m (130 ft) in the south to as much as 180 m (590 ft) in the north.

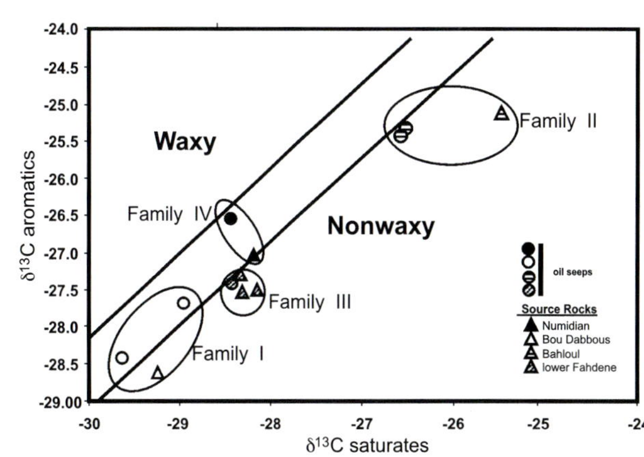

FIGURE 11. $\delta^{13}C$ isotopic composition of saturate and aromatic hydrocarbon fractions of oil seep and source rock samples from northern Tunisia shown on the Sofer (1984) type plot.

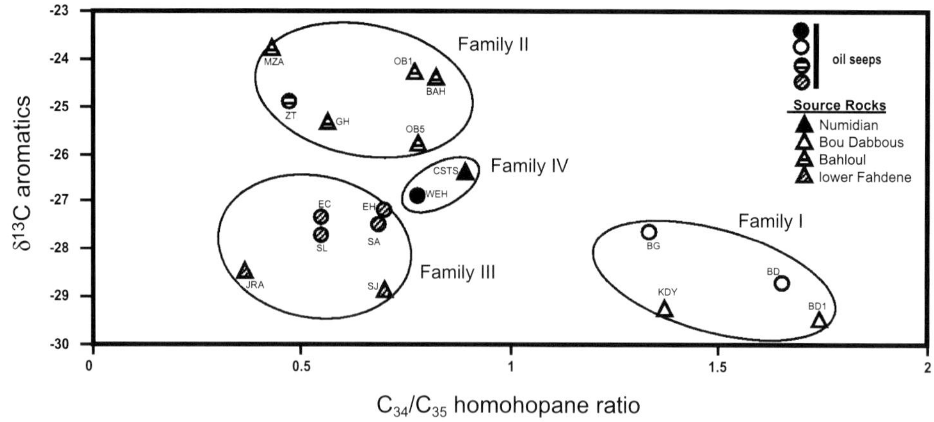

FIGURE 12. $\delta^{13}C$ of aromatics vs. C_{35}/C_{34} homohopane ratio plot of oil seeps and source rocks. MZA = Maiza; ZT = Zag Ettir; OB (1&5) = Oued Bazina; BAH = Oued Bahloul; EC = Echeid; SL = Slouguia; EH = El Haroun; SA = Sidi Ameur; JRA = Jebel Rihane; SJ = Jebel Serj; WEH = Wejh El Homs; CSTS = Cap Serrat; BG = Bougotrane; KDY = Kasseb Dam Ypresian.

The sealing of this upper member is assumed by the Paleocene shales of El Haria Formation to as much as 700 m (2300 ft).

The main diagenetic processes that can influence reservoir properties are represented by dissolution, cementation, and compaction. However, the secondary porosities (intragranular, vuggy, and moldic) are later partly filled by sparitic and dolosparitic cement. The total estimated porosity is as much as 10%, and the measured porosity ranges from 12 to 17%.

The permeability is highly improved by fracturing. Multidirectional fractures are observed (north–south, east–west, northeast–southwest, and northwest–southeast trends). Theses fractures were generated essentially by the folding as shear fractures. Additionally, fault-related fractures are also common. The latest north–south stress field generated northwest–southeast to north–south open fractures. The qualitative permeability indicates that the north–south direction is favorable for fluid migration.

Ypresian (Bou Dabbous and El Gueria Formation)

In northern Tunisia, the Ypresian time interval is characterized mostly by carbonate deposits, except the northernmost area, where shales, breccia, and cherts are common.

From the northwest to the southeast, numerous facies are displayed (Figure 14). These lateral distributions could be related to the paleotopography and paleobathymetry that are closely controlled by the Late Cretaceous–Paleogene tectonics. The main event consists of the detachment and displacement toward the southeast of the inner crystalline units (Kabylyan equivalent), in front of which the foredeep is created. This basin is related to the southeast by the forebulge with associated normal faulting. This architecture had controlled the facies and thickness distribution of the Ypresian sequences. Hence, two major depositional environments are defined:

- The foredeep basin, where reworked Ypresian deposits made of shales, breccias, and cherts accumulated, is situated in front of the crystalline nappes. The northern slope of the foredeep sequence included in the Addissa and Ain Draham tectonic units (Rouvier, 1977) issued from the erosion of the inner nappes, whereas the main depocenter of the foredeep basin comprises more shales. Progressively toward the southeast, frequent carbonate levels interbedded with shales (Melloula facies) appear in the Ypresian sequence and pass laterally to black laminated limestones with Globigerinae and locally echinoids and cherts (Bou Dabbous facies). The Bou Dabbous facies included in the Kasseb and Ed Diss units is organized in

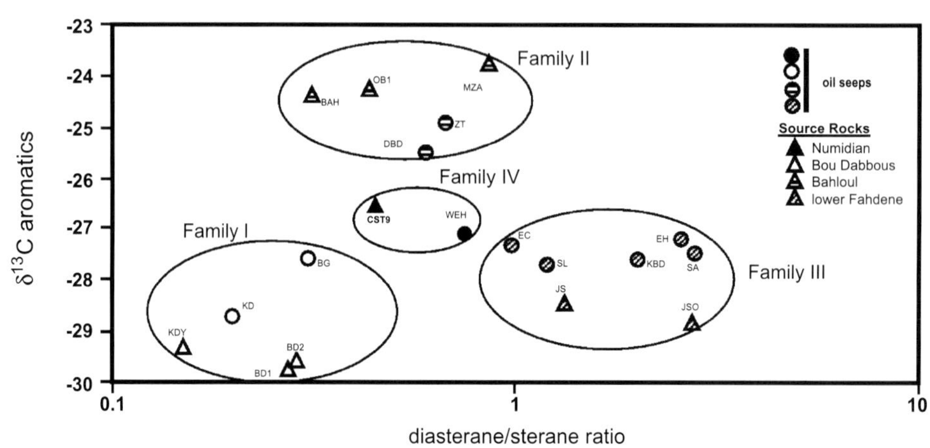

FIGURE 13. $\delta^{13}C$ of aromatics vs. diasterane/sterane ratio plot of oil seeps and source rocks. See Figure 12 caption for abbreviation definitions.

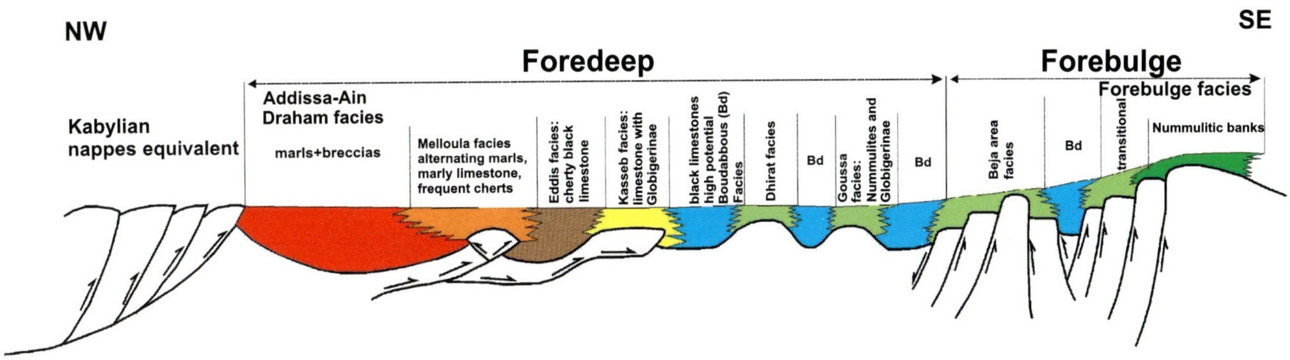

FIGURE 14. Foredeep-forebulge architecture with related Ypresian sequences.

ramp folds, on the top of which are localized bioaccumulations and reworked Ypresian sequences composed of limestones with corals, polyps, nummulitic and bivalves debris, coccolites, Globigerinae, and breccias (transitional facies). This facies localized particularly on the southern edge of the basin.

- Forebulge (platform). This domain, in relation to the southeast in the foredeep basin, corresponds to the platform and is mostly affected by simultaneous normal faults that closely controlled the lateral distribution and thickness of the Ypresian facies. Three different facies are mainly present in this area:
  - Nummulitic banks in the inner shelf, on the top of uplifted blocks (El Gueria facies)
  - Reworked sequences with Globigerinae, nummulitic and bivalves debris, corals, breccias, and breccias in the slopes of the tilted blocks (mixed facies)
  - Laminated black limestones with Globigerinae filling the half grabens and grabens (Bou Dabbous facies).

Before the 1990s, El Gueria nummulitic facies was considered as the main proven reservoir level, producing oil in several fields in the central-eastern Tunisia and in the Gulf of Gabes. Since 1991 and following the discovery of the Belli oil field in northeastern Tunisia, the Bou Dabbous Globigerinae fractured limestones, known as a good source rock, represent one of the main targets of exploration in northern Tunisia thrust belt. These limestones are generally overlain by thick middle to late Eocene Souar shales and principally by the allochthonous Numidian turbiditic sequence that acted as an efficient seal and obviously as the effective overburden for maturity.

Based on petrographic thin-section studies and petrophysical analysis, several diagenetic events have been identified in the Ypresian limestones. Cementation, compaction, and dissolution are the most important features controlling the reservoir qualities.

In addition to significant intragranular and intergranular primary porosities, the late stage of the dissolution process leads to the development of secondary moldic, vuggy, and microfracture porosities.

These primary and secondary porosities are partially filled by microsparitic to sparitic and dolomicrosparitic cement.

The remaining estimated porosity ranges from 8 to 16%, and the permeability is expected to be low, but fracturing seems to be the major process, improving the petrophysical characteristics of these limestones.

The results of fracture modeling conducted on the outcrop sections show the following:

- The two-dimensional modeling indicates at least two interesting reopened fracture directions trending north–south and northwest–southeast.
- The three-dimensional simulation shows that the ellipsoid of permeability is subcircular in shape elongated N30°. This indicates that all directions are favorable for fluid migration with a discrete N30° trend, probably resulting from conjugate faults induced by the Numidian thrusting.

Oligocene–Early Miocene (Numidian Turbidites and Its Equivalent)

Numidian Turbidites

The Numidian flysch sequence shows a variety of reworked material deposited during the main Oligocene–Miocene thrusting in a foredeep basin. It is composed of turbiditic sequences made of gravity sediments interbedded with thin autochthonous shales (El Maherssi, 1992).

Petrophysical analysis of the sandy-conglomeratic levels from outcropping Numidian sections indicates moderate to locally good porosities, ranging from 11 to 23% and moderate to good nitrogen permeabilities, ranging from 9 to 595 md.

The best porosity-permeability values are related generally to the coarse fuxoturbiditic-debris-grain flow

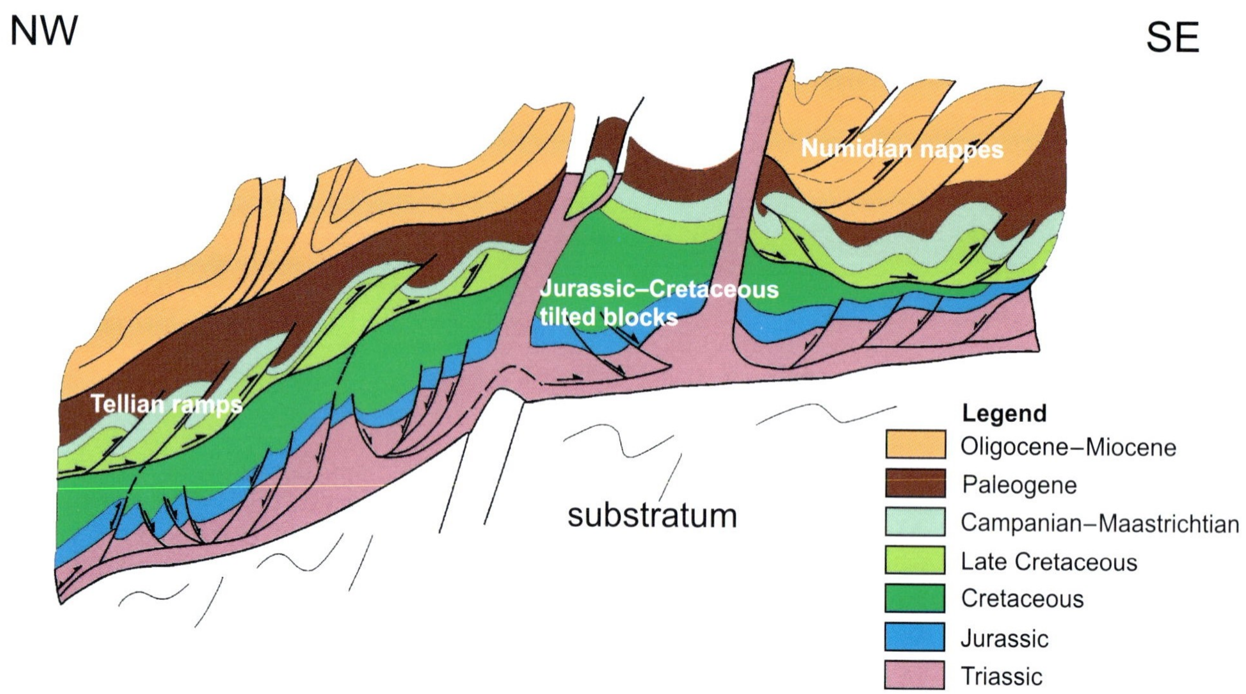

FIGURE 15. Synthetic geological section showing different potential traps in northern Tunisia onshore (no scale).

levels frequent in the Numidian sequence. Fracturing enhanced the petrophysical characteristics of the fine turbiditic facies.

Compaction and quartz cementation are the most important diagenetic events identified from thin-section petrography and scanning electron microscopy analysis. They have affected the Numidian sandy-conglomeratic levels by modifying porosity, permeability, and locally, the original mineralogy.

Sealing is assumed by the interbedded Cretaceous–Eocene allochthonous units, the intraformational shales, and the late Miocene–Pliocene sequences (shales, marls, and evaporites). The allochthonous Triassic salt that remobilized from older diapirism at the base of the thrust sheets could be considered as a seal.

Autochthonous Numidian Equivalent

The autochthonous Oligocene–early Miocene Bejaoua Group represents the equivalent of the Numidian flysch. It is made up of siliciclastic shallow-marine sequence and consists of conglomerates, sandstones, and glauconitic shales. It rests unconformably on the early deformed substratum and is diachronous.

These sequences could be considered as a potential reservoir in this area, especially when they unconformably rest on the Cretaceous and Eocene source rocks or are overlain by the Numidian sheets.

The total thickness ranges from 150 to 300 m (500 to 1000 ft), and sandy horizons are as much as 12 m (40 ft) thick. The measured porosity is as much as 17%, and it is mainly represented by secondary intragranular, moldic, and vuggy porosities, associated with intergranular primary porosity.

Traps

In northern Tunisia, the polyphasic tectonics induced a vertical succession of three structural levels (Figure 15; Table 1).

- The lowest one is obviously characterized by (listric) normal faulting, associated locally to a Triassic salt flowage. It affected the post-Triassic to the Late Cretaceous series. If well preserved, these tilted blocks represent potential traps that could be sourced either by Late Cretaceous or Early Cretaceous source rocks.
- The middle deformational zone, squeezed between two major decollement levels, respectively, the Santonian shales at the base and the lower Numidian shales at the top, is represented by the Late Cretaceous–Paleogene thrust anticlines and detached folds as flexural-slip or flexural-flow folds. Such structures are assumed to be potential traps mainly in the onshore area, where primary targets are the Maastrichtian and Ypresian fractured limestones and the autochthonous Oligocene–early Miocene sandy lenses.
- The upper structural style is represented by the independent Numidian nappes, which are disconnected from the lower levels and characterized by

Table 1. Tectonic events and associated potential traps in northern Tunisia thrust belt.

Event	*Trap*	*Reservoir*	*Seal*
Triassic–Jurassic rifting and margin breakup	tilted blocks	Jurassic carbonates	Cretaceous marls
Cretaceous extension	tilted blocks	Jurassic-Cretaceous carbonates	Cretaceous marls
	hanging walls	Cretaceous sandy lenses	
	rollovers		
Pyrenean compression	compressional faulted blocks	Campanian–Maastrichtian limestones	Paleocene shales
	anticlines	Ypresian limestones	Lutetian shales
	ramps	Oligocene–Miocene sandy lenses	Oligocene–Miocene marls
	slip folds		
Alpine thrusting	thrusted anticlines	Numidian sandy lenses	late Miocene shales
	ramps	Cretaceous and Eocene limestones	Numidian shales
	triangle zones		Paleocene and Lutetian shales
	pop-ups		Cretaceous shales
			Triassic salt

flat to high-angle thrusts, ramps, overturned anticline, backthrusts, and triangle zones. The uppermost horizon is locally represented by the successor late Miocene–Pliocene basins that have also been inverted by the early Quaternary shortening. Within the offshore, the Numidian sandstones are targeted, such as in Sicily, where gas is trapped in ramps, triangle zones, and thrust anticlines.

BASIN MODELING STUDY AND TIMING OF THE PETROLEUM SYSTEMS

A one-dimensional basin modeling of burial reconstruction, conductive heat transfer, and petroleum generation has been performed on northern Tunisia (offshore and onshore areas), using 1-D Genex® software (GenTec module) to quantify subsidence, maturation, generation, and expulsion histories.

On the basis of the regional geological framework defined in this study, some fictitious wells have been modeled. These fictitious wells, namely, Kasseb, Hammam Bourguiba, Bizerte, Tabarka onshore (TBON), and Tabarka offshore (TBOF) (Figure 10), have been constructed using only interpretive structural models and measured geological sections (based on field work) because of the lack of subsurface information on the onshore area and the bad quality of the available seismic data in the offshore. These fictitious wells are made of several normal tectonic units. An example of the sedimentation history from TBON fictitious well is given in Table 2.

Thermal History

Geochemical parameters, such as T_{max} and HI from source rock levels outcropping close to the different fictitious wells, have been used as optimization and control parameters for the determination of the thermal histories in the simulation of various fictitious wells. The best match corresponds to a constant heat flow (during geological time) ranging between 69 and 71 mW/m^2.

The simulation of generation and expulsion processes has been focused on several source rocks from the autochthonous (lower Fahdene, Bahloul, and Bou Dabbous), the para-autochthonous (Bou Dabbous formation), and the allochthonous (Numidian and Bou Dabbous) series.

Source Rocks, Generation, and Expulsion Histories

Source Rocks

Geochemical evaluation of the studied area suggests that the Bahloul and the Bou Dabbous levels are characterized by type II kerogen, with mean total organic carbon values of 2 and 1.8%, respectively. The lower Fahdene formation and the Numidian shales total organic

TABLE 2. Sedimentation history of the TBON fictitious well.

Formations	*Type**	*Period (Ma)*		*Depth (m)*		*Amount (m)*	
		End	*Start*	*Top*	*Bottom*	*Eroded*	*Missing*
Unit 1	NU						
Lower Pliocene/Quaternary	F	0	3	0	250		
Unconformity	U	3	5			50	
Lower Medjerda	M	5	5.5				50
Base Medjerda	F	5.5	6	250	600		
Babouch	F	6	21	600	1200		
Khroumir 1	F	21	22	1200	1700		
Khroumir 2	F	22	23	1700	1950		
Khroumir 3	F	23	24	1950	2450		
Base Numidian 2	F	24	28	2450	2800		
Base Numidian 1	F	28	32	2800	3300		
Fault 1	RF	14.1					
Unit 2	NU						
Triassic	F	14.2	14.3	3300	3400		
Fault 2	RF	14.15					
Ain Draham unit	NU						
Fortuna	F	21	33	3400	3550		
Unconformity	U	33	36			100	
Lower Souar	M	36	40				100
Base Souar	F	40	50	3550	4150		
Bou Dabbous	F	50	55	4150	4280		
El Haria	F	55	60	4280	4680		
Fault 3	RF	14.2					
Ed Diss unit	NU						
Fortuna	F	20	33	4680	4830		
Souar	F	33	50	4830	5530		
Bou Dabbous	F	50	55	5530	5660		
El Haria	F	55	66	5660	5810		
Fault 4	RF	40					
Unit 3	NU						
El Haria	F	55	66	5810	6060		
Abiod	F	66	82.5	6060	6410		
Aleg	F	82.5	88	6410	6610		
Fault 5	RF	41					
Unit 4	NU						
Aleg	F	89	91	6610	6810		
Bahloul	F	91	92	6810	6825		
Upper Fahdene	F	92	97.5	6825	6925		
Lower Fahdene	F	97.5	108	6925	7225		

**Nu = normal unit; F = formation; U = unconformity; M = missing; RF = reverse fault.*

carbon contents, at present time, are relatively low, ranging from 1.00 to 1.3% and from 0.5 to 1.0%, respectively. Their organic matter could be dominated by type II kerogen with some type III kerogen contribution, probably related to poor anoxic conditions at sea bottom and to an imperfect preservation of the organic matter (this is confirmed by petrographic study) (Figure 16a).

Generation and Expulsion Histories

Examples of the hydrocarbon generation and expulsion histories (TBON) of the source rocks mentioned above are shown in Figures 16 and 17.

These histories are not synchronous for each source rock and differ from one locality to another.

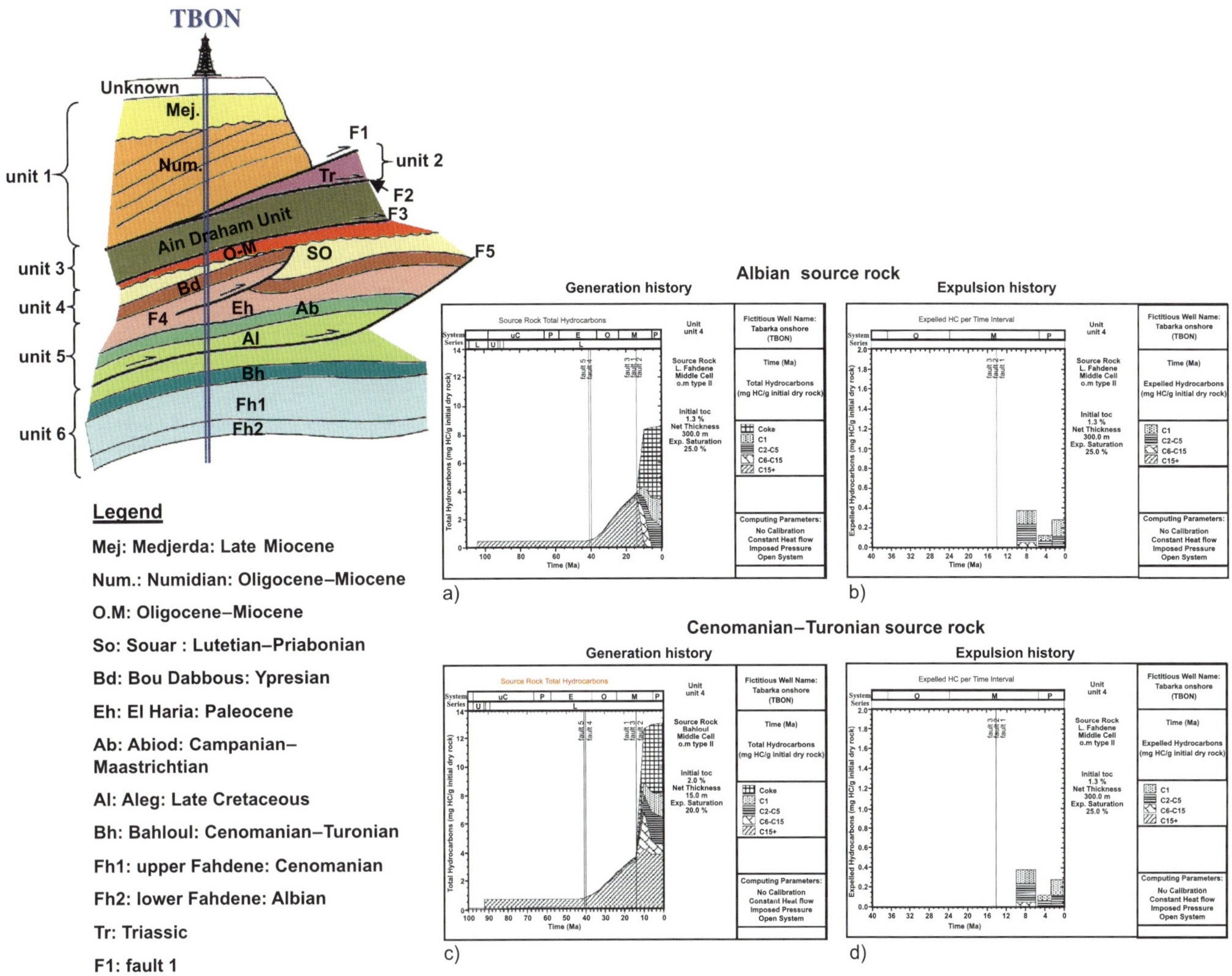

FIGURE 16. Generation and expulsion histories of the Albian and the Cenomanian–Turonian source rocks penetrated by Tabarka onshore fictitious well (TBON).

The lower Fahdene source rock penetrated by the TBON pseudowells started to generate oil as early as the Late Cretaceous (Figure 16a). The peak oil generation was reached at different times, ranging from the late Serravalian to the late Miocene. In general, the maximum expulsion phase of generated hydrocarbon occurred between 14 and 5 Ma. This source rock expelled mainly gas and condensate (Figure 16b).

The onset of the oil generation from the Bahloul source rock corresponds to the Late Cretaceous–Paleocene. The peak oil generation was reached during the Tortonian (Figure 17a), followed immediately by the expulsion of C_{15+} oil and, recently, by the expulsion of condensate and gas (Figure 17b).

The autochthonous Bou Dabbous formation started to generate oil from the Oligocene up to the Serravalian. The main oil window was entered at Serravalian, and the main expulsion phase occurred during the last 10 m.y. (Figure 17c, d).

The generation history of the allochthonous Bou Dabbous source rock is more complex and different from one locality to another. In general, the main oil window was entered since the Serravalian, and the peak oil generation was entered during the Tortonian or later. The Numidian shales generated small to moderate quantities of hydrocarbon without reaching the peak oil (Figure 17c).

CONCLUSIONS

The tectonic history of northern Tunisia is made of two major cycles:

- the opening cycle, corresponding to the Tethyan rifting and margin development
- the shortening cycle, represented by the Pyrenean–Alpine orogenies.

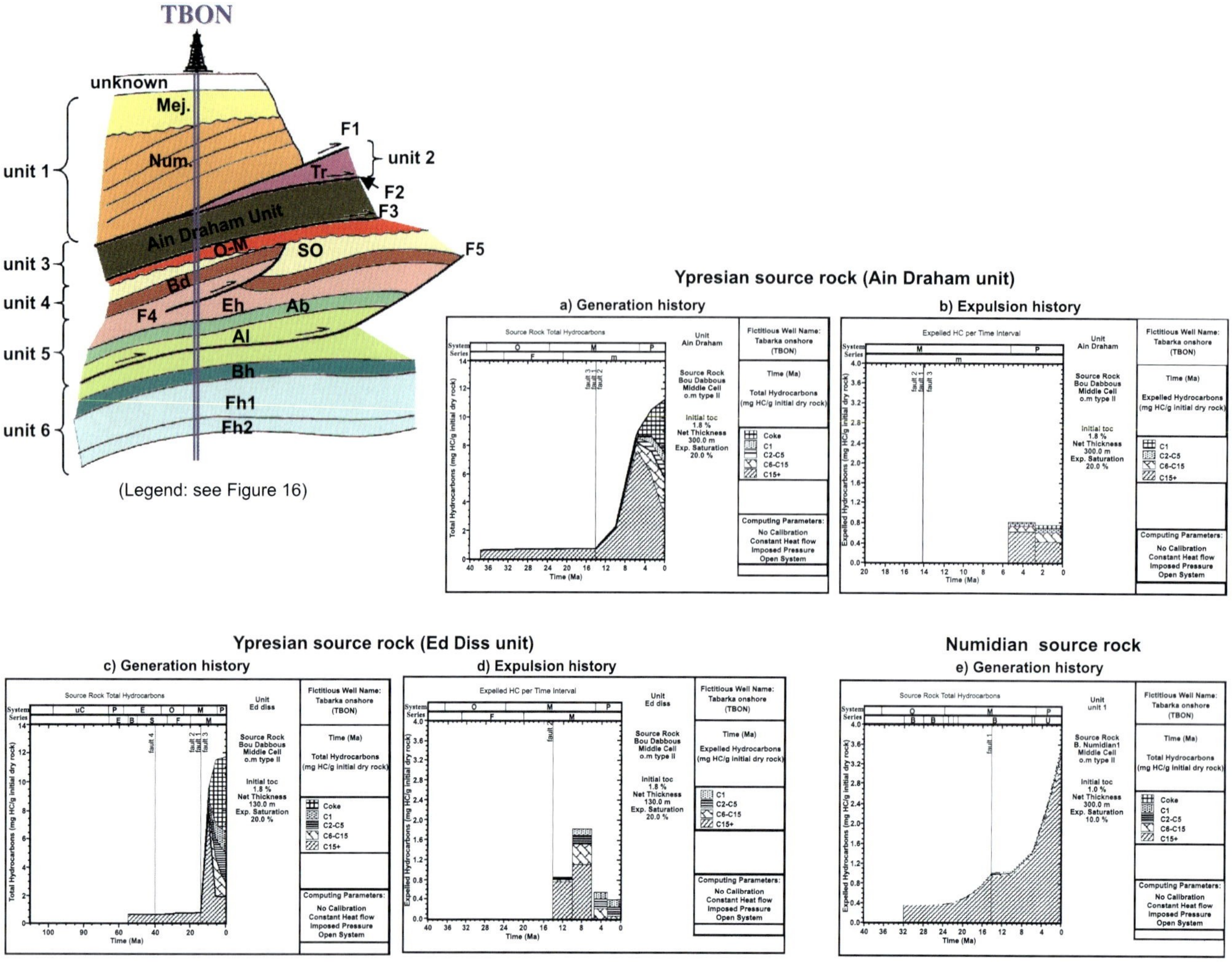

FIGURE 17. Generation and expulsion histories of the Ypresian and Numidian source rocks penetrated by the Tabarka fictitious well (TBON).

The inversion occurred at the end of the Cretaceous and led to important changes of the depositional environments and related sedimentary regimes.

Four reservoirs are described: the Maastrichtian and Ypresian fractured limestones and the Oligocene early allochthonous turbiditic and its autochthonous sandy equivalent. These reservoirs that were associated with four source rocks (Albian, Cenomanian–Turonian, Ypresian, and Oligocene–early Miocene) led to at least three active petroleum systems.

The main geological feature affecting the hydrocarbon generation is the Serravalian–Tortonian thrusting.

If traps formed during Serravalian–Tortonian thrusting period are preserved, these should contain mainly oil and gas and condensate. Older traps, when preserved, are likely to contain oil sourced by Cretaceous source rocks.

The timing of hydrocarbon expulsion constitutes a favorable parameter for exploration, because it is generally late Miocene–Pliocene, postdating the formation of major traps. Different structures, such as ramps, thrusted anticlines, triangle zones, pop-ups, and undergone well-conserved faulted blocks, should be attractive traps for significant oil and gas accumulations.

Unconformities, thrust surfaces, high-angle reverse faults, and normal faults are thought to be migration pathways. Vertical migration is also considered, particularly in the Numidian reservoir source rock couple.

Hence, this fold and thrust belt is one of the highest prospective areas in Tunisia despite its early stage of exploration, which could be considered as an encouraging factor. The main targets are represented by the Ypresian and the Maastrichtian in the onshore and the Numidian and Ypresian in the offshore area (Table 3).

ACKNOWLEDGMENTS

Authors acknowledge ETAP management for its support and permission to publish this chapter. We would like to thank ETAP geoscientists, particularly A. M'rabet, J. Friha, R. Ghenima, as well as our colleagues

TABLE 3. Main targets in northern Tunisia thrust belt.

	Onshore Targets		*Offshore Targets*	
	Ypresian	***Maastrichtian***	***Numidian***	***Ypresian***
Source Rock	Ypresian: oil and gas prone	Albian: oil and gas prone	Cenomanian–Turonian: gas prone	Ypresian: gas and oil prone
		Cenomanian–Turonian: oil and gas prone	Numidian: gas and oil prone	
Reservoir	Fractured limestones:	Chalky fractured limestones:	Oligocene–Miocene turbidites:	Fractured limestones:
	Bou Dabbous formation	Maastrichtian–Campanian	Numidian sandy formation	Bou Dubbous Formation
	transitional facies	Abiod Formation		transitional facies
	Nummilitic banks			
Seal	Lutetian shales:	Paleocene shales	Cretaceous-Paleogene allochthonous shales	Lutetian shales:
	Oligocene–Miocene shales	Oligocene–Miocene shales	Numidian shales	Oligocene–Miocene shales
	Numidian shales	Numidian shales	Miocene–Pliocene shales and evaporates	Numidian shales
	Miocene–Pliocene shales and evaporates			Miocene–Pliocene shales and evaporates
Trap Type	ramps	ramps	ramps	ramps
	faulted anticlines	faulted anticlines	faulted anticlines	faulted anticlines
	pop-ups	pop-ups	pop-ups	pop-ups
	tilted blocks	tilted blocks	triangle zones	tilted blocks
			pinch-outs	

J. Ben Yaagoub, S. Manaa, and N. Issaoui (Eni-Agip Tunisia); S. Merlini, A. Riva, S. Torichelli, M. Comesso, V. Arnera, G. Davoli, C. Rizzi, F. Cassini, C. Rizzi, A. Quadrelli, and M. Pajola (Eni-Agip Milano) and also P. Tremolieres and P. F. Burollet (France).

Some aspects of this study have been achieved in collaboration with oil companies (Agip, PREUSSAG Energy, Turkish Petroleum national company, Societe de recherche et d'exploration petroliere de Tunisie, SEREPT) and academic institutions (Institut Français du Pétrole, France, Tunis University, Oklahoma University). We acknowledge them as well.

REFERENCES CITED

Ben Ferjani, A., P. F. Burollet, and F. Mejri, 1990, Petroleum geology of Tunisia: ETAP Memoir 1, 194 p.

Caron, M., 1985, Cretaceous planktonic foraminifera, *in* H. M. Boli, J. B. Sanders, and K. Perchnielon, eds., Plankton stratigraphy: University of Cambridge Press, p. 17–86.

Cohen, C., S. Shamel, and P. Boyd-Kaygi, 1980, Neogene deformation in northern Tunisia: Geological Society of America Bulletin, v. 91, p. 225–237.

Crampon, N., 1971, Etude geologique de la bordure des Mogods, du pays de Bizerte et du Nord des Hedhils (Tunisie Septentrionale): Ph.D. thesis, University of Nancy, France, v. 2, 549 p.

El Euchi, H., A. Gaaya, F. Hamouda, and F. Mejri, 1997a, Northern Tunisia — A promising area for petroleum exploration, *in* 59th European Association of Geoscientists and Engineers Petroleum Conference, Geneva, p. 256.

El Euchi, H., A. Gaaya, R. Ghenima, and M. Saidi, 1997b, Eocene structures under the Numidian thrust belt (abs.): European Association of Geoscientists and Engineers-Society of Exploration Geophysicists Conference, September 1997, Moscow.

El Euchi, H., A. Ouahchi, and A. Mrabet, 2000, Tunisian Aptian–Albian carbonate platform: Dynamics and petroleum systems (abs.): 62th European Association of Geoscientists and Engineers Annual Conference, Glasgow, June 2000.

El Maherssi, C., 1992, Dynamique de dépôt du flysh numidien de Tunisie (Oligocène–Miocène): Ph.D. thesis. Ecole des Mines de Paris, Memoir 15, 246 p.

Fourati, L., and F. Hamouda, 1996, Abiod limestone reservoir characterisation in Beja block (northern Tunisia), *in* Proceedings of 5th Tunisian Petroleum Conference: ETAP Memoir 10, p. 363–380.

Gottis, C. H., 1953, Stratigraphie et tectonique du flysh numidien en Tunisie Septentrionale: Comptes Rendus de l'Academie des Sciences, Paris, v. 236, no. 10, p. 1059–1061.

Grantham, P. J., 1986a, The occurrence of unusual C_{27}

and C_{29} sterane predominances in two types of Oman crude oil: Organic Geochemistry, v. 9, p. 1–10.

Grantham, P. J., 1986b, Sterane isomerization and moretane/hopane ratios in crude oils derived from Tertiary source rocks: Organic Geochemistry, v. 9, p. 293–304.

Kujawski, H., 1969, Contribution à l'étude géologique de la région des Hedils et Bejaoua Orientale: Ph.D. thesis: Annales des Mines et de la Geologie (Tunis), v. 24, 281 p.

Moldowan, J. M., C. Y. Lee, P. Sundaraman, T. Salvatori, A. Alajbeg, B. Gjukic, and C. J. Demaison, 1989, Source correlation and maturity assessment of selected oil and rocks from the central Adriatic basin (Italy and Yugoslavia): American Chemical Society of Petrological Chemistry Preprints, v. 34, no. 1, p. 112–121.

Rouvier, H., 1977, Géologie de l'extrème Nord Tunisien: tectoniques et Paléogéographies superposées à l'extrémité Nord-orientale de la chaine maghrébin: Thèse d'etat de l'université de Paris VI, France, 703 p.

Saidi, M., 1993, Etude géologique et géochimique des roches mères albo-vraconiennes dans le domaine de la Tunisie centro-septentrionale: Thèse troisieme cycle, Université de Tunis, Faculté des sciences de Tunis, Tunisia, 140 p.

Saidi, M., and R. P. Philip, 1997, Geochemical characterization and relationship of oil-source rock from northern Tunisia (abs.): AAPG Annual Meeting Program, v. 6, p. A-100.

Sofer, Z., 1984, Stable carbon isotope compositions of crude oils: Application to source rock depositional environments and petroleum alteration: AAPG Bulletin, v. 68, p. 31–49.

Soussi, M., 2002, Le Jurassique de la Tunisie Atlasique Stratigraphie, dynamique, sédimentaire, paléogéoraphie et intérêt pétrolier. Documents STU Lyon no. 157-2002, UFR des Sciences de la Terre, Université Claude Bernard Lyon 1 France. 363 p., 156 figures, 45 planches. Publié avec le concours du Ministére des affaires étrangères françaises, Projet CMCU 99F1008. ISBN 2-85454-114-6. ISSN 0750-6635.

Talbi, 1991, Etude géologique et géochimique des facies riches en matière organique d'age albien du bassin de Bir M'cherga (Nord-Est de la Tunisie). Determinisme de leur genese et ineret petrolier de la région: Thèse de troisieme cycle, Université de Tunis, Faculté des Sciences de Tunis, 110 p.

Ten Haven, H. L., M. Rohmer, J. Rullkotter, and P. Bisseret, 1989, Tetrahymanol, the most likely precursor of gammacerane occurs ubiquitously in marine sediments: Geochimica et Cosmochimica Acta, v. 53, p. 3073–3079.

Turki, M. M., 1985, Poly-cinématique et contrôle sédimentaire associé sur la cicatrice Zaghouan-Nebhana: Ph.D. thesis, University of Tunis, Tunisia, 252 p.

20

Granath, J. W., and P. Casero, 2004, Tectonic setting of the petroleum systems of Sicily, *in* R. Swennen, F. Roure, and J. W. Granath, eds., Deformation, fluid flow, and reservoir appraisal in foreland fold and thrust belts: AAPG Hedberg Series, no. 1, p. 391–411.

Tectonic Setting of the Petroleum Systems of Sicily

James W. Granath
Granath and Associates, Houston, Texas, U.S.A.

Piero Casero
Consultant, Rome, Italy

ABSTRACT

Petroleum systems in Sicily are divided between the Neogene fold and thrust belt of the Sicilian accretionary prism and its foreland. By far, more production has been established in the foreland on the Ragusa platform, where generally heavy oil from Triassic and Jurassic source rocks has been trapped in similarly aged fractured carbonate mound reservoirs. Traps are structures recurrently rejuvenated since the Cretaceous. Although models of source maturation and migration history emphasize Pliocene–Pleistocene charge related to subsidence in the foreland, occurrences of light oil and exploration data suggest intermittent periods of charge since the Mesozoic. Hydrocarbon deposits in the immediate vicinity of the thrust front, in the narrow foredeep depression, are also heavy and have characteristics in common with those more remote in the foreland from the front; they do not appear to have been significantly influenced by or mixed with oils from within the fold belt, although variations in their chemistry have yet to be investigated in terms of source correlation. Thrust sheets in the fold and thrust belt carry similar Mesozoic stratigraphy to the foreland, namely, a series of carbonate platforms and basins with potentially similar source rocks, and can reasonably be expected to have undergone a similar pre-Neogene history.

Commercial hydrocarbons in the fold belt, however, are limited to thermo- and biogenic gas-charged Tertiary reservoirs and questionably Tertiary-sourced oil offshore to the west. These all currently are viable exploration plays. Numerous seeps and tar deposits indicate that at least one petroleum system was active in the fold belt, but only one tar occurrence links that system to the Mesozoic carbonate section. Recent structural studies have indicated that a two-phase history of thrust emplacement and its redeformation by duplexing on lower detachments characterizes much of the Sicilian fold and thrust belt and affords exploration targets that have yet to be satisfactorily tested. Furthermore, structural petrological and fluid-inclusion studies in

DOI:10.1306/1025702H1854

Neogene-thrusted rocks of the lower portions of the frontal fold belt indicate that a light hydrocarbon system used Neogene detachment surfaces as a preferential migration route in at least part of the accretionary prism.

Insofar as the foreland may give an insight into the early history of the fold belt platforms and basins, a general sense of the hydrocarbon migration history in the fold belt might include the following: (1) early maturation and migration of a variety of oils into carbonate mound reservoirs during a long period of time in the latter half of the Mesozoic and Paleogene (like in the foreland plays), bolstered by (2) a charge of immature oil generated during rapid depression in the foredeep immediately adjacent to the moving thrust front (yet to be demonstrated), and (3) generation of higher maturity oil and gas in the thrust wedge during stacking of thrust sheets and remigration and mixing with the earlier generated fluids (an untested play).

INTRODUCTION

Sicily lies astride a segment of the Apennine–Maghreb Mountain chain (Figure 1) that stretches from the Alps in Europe to the Atlas Mountains in North Africa and records most of the tectonic events involved in the complex interaction of Europe and Africa prior to and during the opening and closing of western Tethys. The island is surrounded by three seas (Figure 1) that frame the geology of Sicily. To the south, the western Mediterranean Sea is underlain by the Pelagian platform that includes onshore Sicily as well as Malta. Paleomagnetic data show no difference in apparent polar wandering paths between southeast Sicily and North Africa (Meloni et al., 1997, and references therein; Speranza et al., 2000, and references therein), thus firmly tying the geology of Sicily to that of North Africa, a situation that has persisted since the Permian (Stampfli et al., 2001). Mesozoic carbonates dominate the geology of the Pelagian platform and are built on stretched North African continental crust. A useful analogy to visualize the pre-Alpine situation might be a setting similar to and comparable in size to the modern Bahama Banks (adjusted of course for the biotic assemblage) adjacent to Tunisia and Libya. The Moho under the Pelagian platform is as shallow as 20 km (13 mi), and the Moho under Sicily ranges from 25 km (16 mi) in depth along the southern coast outside the fold belt to 30 km (19 mi) under the fold belt and 35 km (22 mi) along the northern edge of the island (Dal Piaz and Nicolich, 1991; Nicolich, 2001). Given something like 10 km (6 mi) for the thickness of the thrust stack (Catalano et al., 1993) and an allowance for pre-Permian sediments, it is evident that the continental crust of Sicily has undergone considerable thinning.

The Ionian Basin lies to the southeast of Sicily. It is the last remnant of Mesozoic oceanic crust in the central Mediterranean (Catalano et al., 2001), where it is subducting under the Calabrian Arc (Figure 1). The abyssal plain of the Ionian Sea (Finetti, 1981, 1985; Morelli, 1985) is separated from Sicily by a continent-ocean transition at the Malta escarpment. To the north, the Tyrrhenian Sea is composed of Neogene oceanic crust developed in a back-arc environment (Kastens et al., 1988).

The Pelagian crust, with its carbonate-dominated cover, has been colliding with a north-dipping subduction zone since the Miocene, and its northern fringe is deformed in the Sicilian thrust belt. Thus, although Sicily lies in the Apennine–Maghreb accretionary prism, its tectonic character is that of a foreland fold and thrust belt. It is composed of a complex stack of thrust sheets involving the Mesozoic platform and elements of Alpine crust, of the precollision forearc region, and of various syntectonic deposits. As a result, the prism is much more similar with retroarc fold and thrust belts around the world in terms of its architecture than it is to oceanic accretionary prisms, but its hydrocarbon systems are heavily influenced by details peculiar to Sicily's setting, particularly factors that isolate the foreland area from the fold belt proper.

In recent years, major discoveries have been made at the Val d'Agri fields in Basilicata on mainland Italy (La Bella et al., 1996), said to total nearly 1 billion bbl of oil recoverable (Holton, 1999), in environments that are similar to the fold and thrust belt in Sicily. Obviously, the sparsely explored fold and thrust belts of Italy may harbor major hydrocarbon potential, but the geology is difficult. Here, we review that geological history of Sicily, with a special emphasis on how it relates to hydrocarbon occurrences and thereby more generally to the behavior of fluids in the system.

HISTORY OF HYDROCARBON EXPLORATION

Hydrocarbons have been known in Sicily for many centuries from a variety of surface manifestations. For example, bitumen-impregnated limestones were used for building purposes in southeastern Sicily, largely because

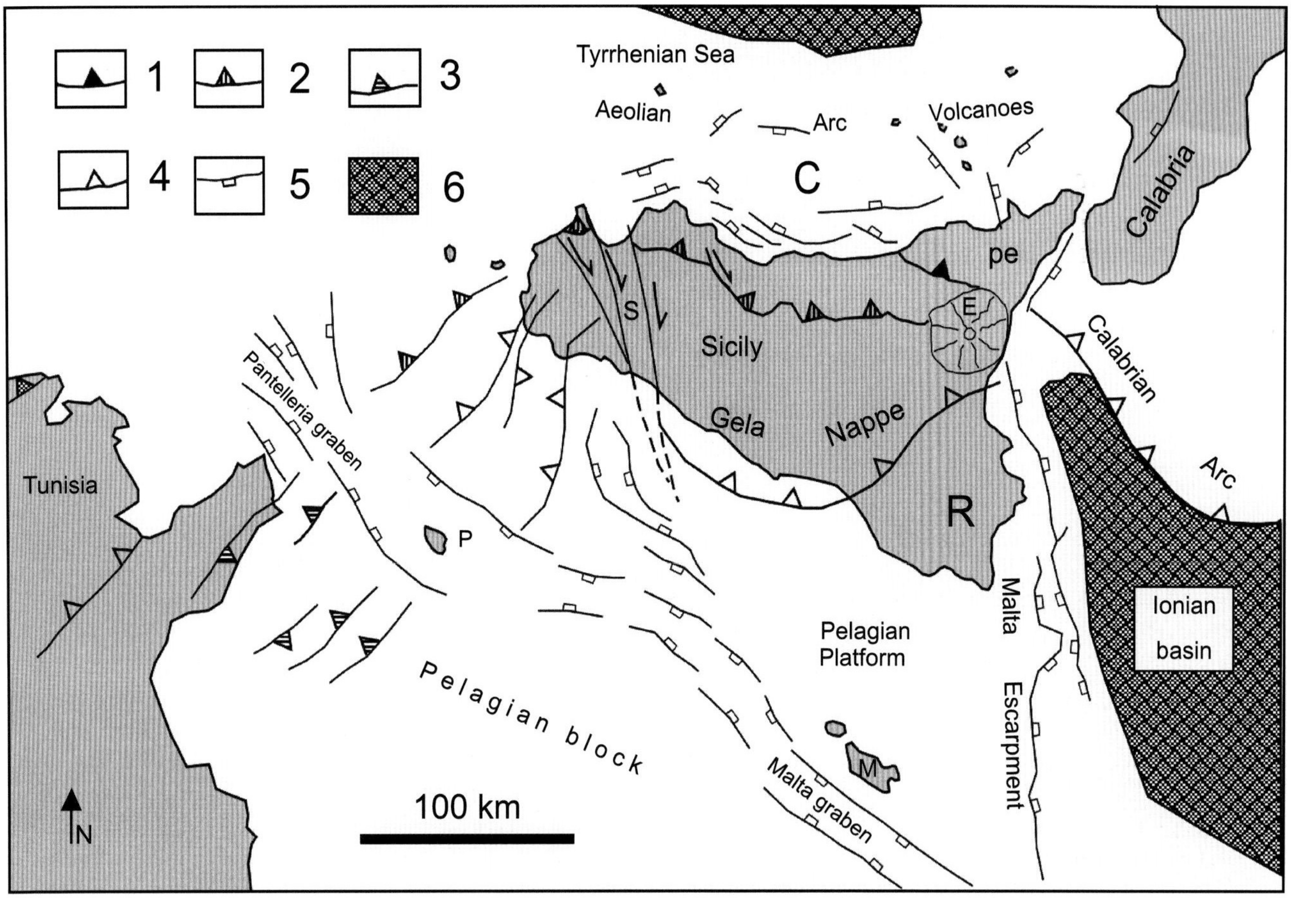

FIGURE 1. Tectonic sketch map of the Sicilian region. S = Segesta fault zone; C = Cefalù Basin; R = Ragusa plateau; pe = Peloritani nappe; E = Mt. Etna; P = Pantelleria Island; M = Malta; 1 = basal thrust carrying Peloritani Mountain sheet, including fundamentally European basement into the Sicilian accretionary prism; 2 = nappes associated with Neotethyan and northern platformal sections; 3 = foreland uplifts; 4 = deformation front of the Sicilian and Calabrian accretionary prisms and frontal front in Tunisia; 5 = extensional fault systems; 6 = oceanic crust in the Tyrrhenian and Ionian basins. Modified after Pepe et al. (2000) and Casero and Roure (1994).

of their ease of workability, and as a medium to fashion tubing for aqueducts, because of their impermeability (Kafka and Kirkbride, 1959; Mattavelli et al., 1993). Surficial asphalt deposits were exploited during the 19th and early 20th centuries, and several shallow wells were drilled before World War II. J. E. Thomas, a founder and the first president of the AAPG, recognized the southeastern dip of the Ragusa plateau away from its western-bounding normal faults (Figure 2) and the possibility that the structural culmination there afforded a focus for migrating hydrocarbons (A.W. Bally, 2003, personal communication). He initiated surface geological studies in the areas in the 1940s that predated the first modern exploration permits, which were signed beginning in the 1950s. Those studies eventually led to the discovery of Ragusa field in 1953 at the Ragusa-1 well (Figure 3). This was the first major oil discovery in Mesozoic reservoirs. Other discoveries followed both in Tertiary rocks and in the Mesozoic: Gela in 1956, Cammarata-Pozzillo in 1959, and the first offshore discovery in Gela-21, also in 1959. Onshore exploration peaked in 1959 and tapered off in 1970, in response to the heavy quality of oil discovered, difficult geology and marginal seismic quality, low oil prices, and the passage of a stricter Sicilian hydrocarbon law.

After 1969, part of the Sicilian offshore was opened to exploration, with the discovery of light oil at Nilde in 1972 and at Mila in 1978. The price boom of the late 1970s and the advent of digital seismic acquisition in Sicily spurred activity, and heavy crude was discovered at Perla in 1979, Vega in 1980, and Prezioso in 1983. A modest onshore resurgence of activity in the 1980s followed adjustments to the hydrocarbon law and better seismic acquisition and processing. Several small discoveries were made. Other parts of the offshore were opened at that time, but little success was realized. Since 1992, activity has been dormant both onshore and offshore, although the hydrocarbon law has once again been modified (this time to be similar to Italian law). Currently, several permits are held by several operators, but activity is minimal.

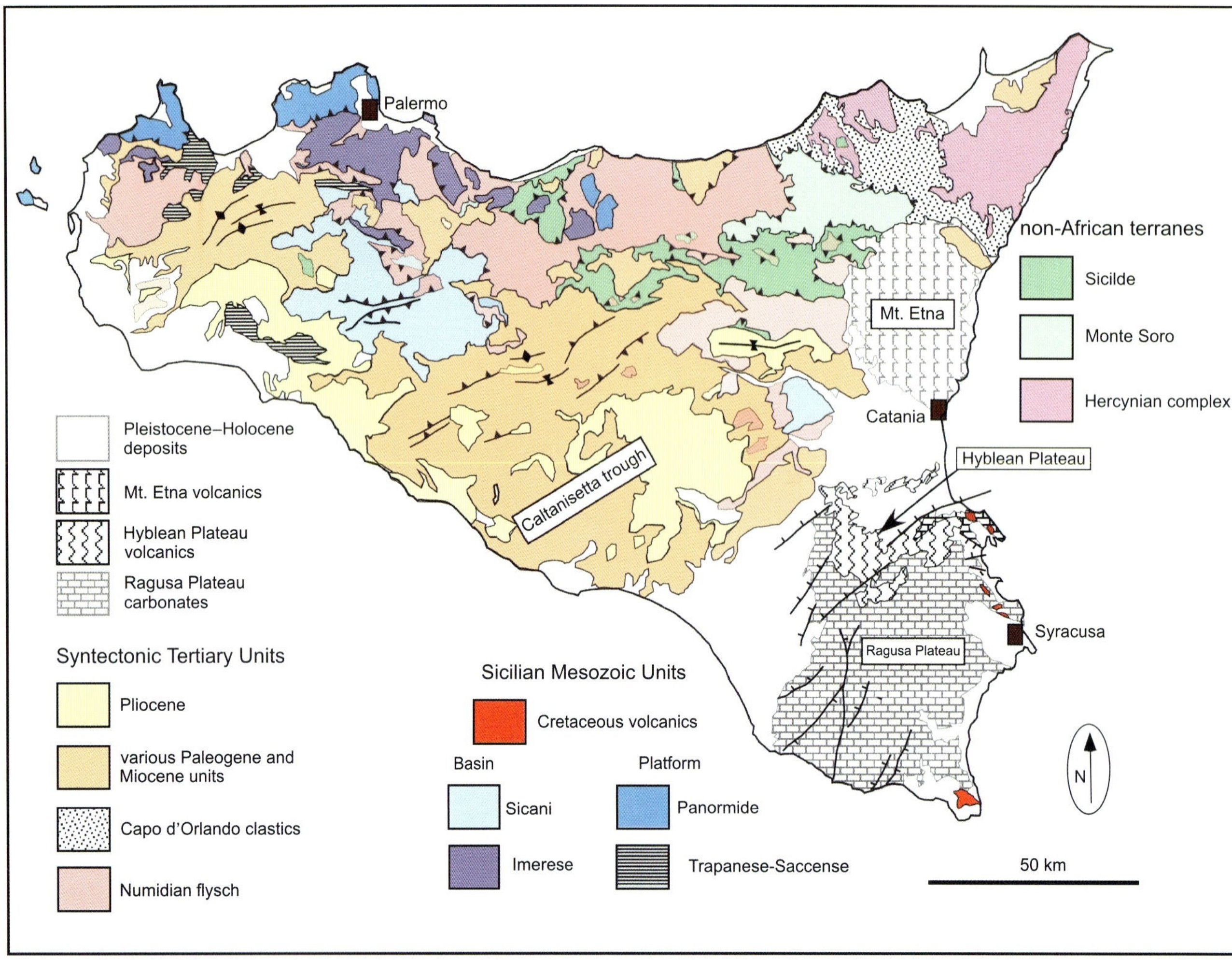

FIGURE 2. Geological map of Sicily generalized from Bigi et al. (1983).

Altogether, some 350 wells have been drilled, of which only 21 are offshore. Reserves are some 500 MMBO, with about half already produced. Gas totals about 500 bcf, of which just under two thirds has been produced. The principal fields are listed in Table 1, and the geological aspects of the various plays linked to those discoveries are tabulated in the Appendix in this chapter.

REGIONAL GEOLOGICAL FRAMEWORK

Tectonic Setting of Sicily in the Apennine–Calabrian–Maghreb System

The entire island of Sicily north of the Gela nappe front (Figure 1) is a strongly arcuate Neogene fold and thrust belt in the active accretionary prism that stretches continuously from the Apennine fold belt and Calabrian Arc in Europe to the Maghreb system of North Africa. It lies above a north-dipping seismogenic (Gasparini et al., 1985) subduction zone, along which most of the remnants of Tethys have been consumed and into which African lithosphere is currently being subducted (Amato and Cimini, 2001). The volcanic arc associated with the Sicilian sector is located to the north of the island itself in the Aeolian Islands, beyond the extensional Cefalù Basin. Synorogenic extension in the Cefalù Basin began synchronously with shortening in the accretionary prism, initially as a subaerial wedge-top basin, but it was much amplified later by extension and back-arc spreading in the Tyrrhenian Sea (Pepe et al., 2000). Extension in the Tyrrhenian Sea dates from rifting and clockwise rotation of Sardinia in the Tortonian to back-arc oceanic spreading in the late Pliocene to Quaternary.

Tectonic Elements in the Pelagian Platform

The Sicilian accretionary prism is thrust upon the Ragusa plateau, exposed in southeast Sicily, and bordered on its north by the chiefly volcanic Hyblean Plateau (Figures 2, 3). The Ragusa plateau is the major subaerial exposure of the Pelagian platform (Figure 1)

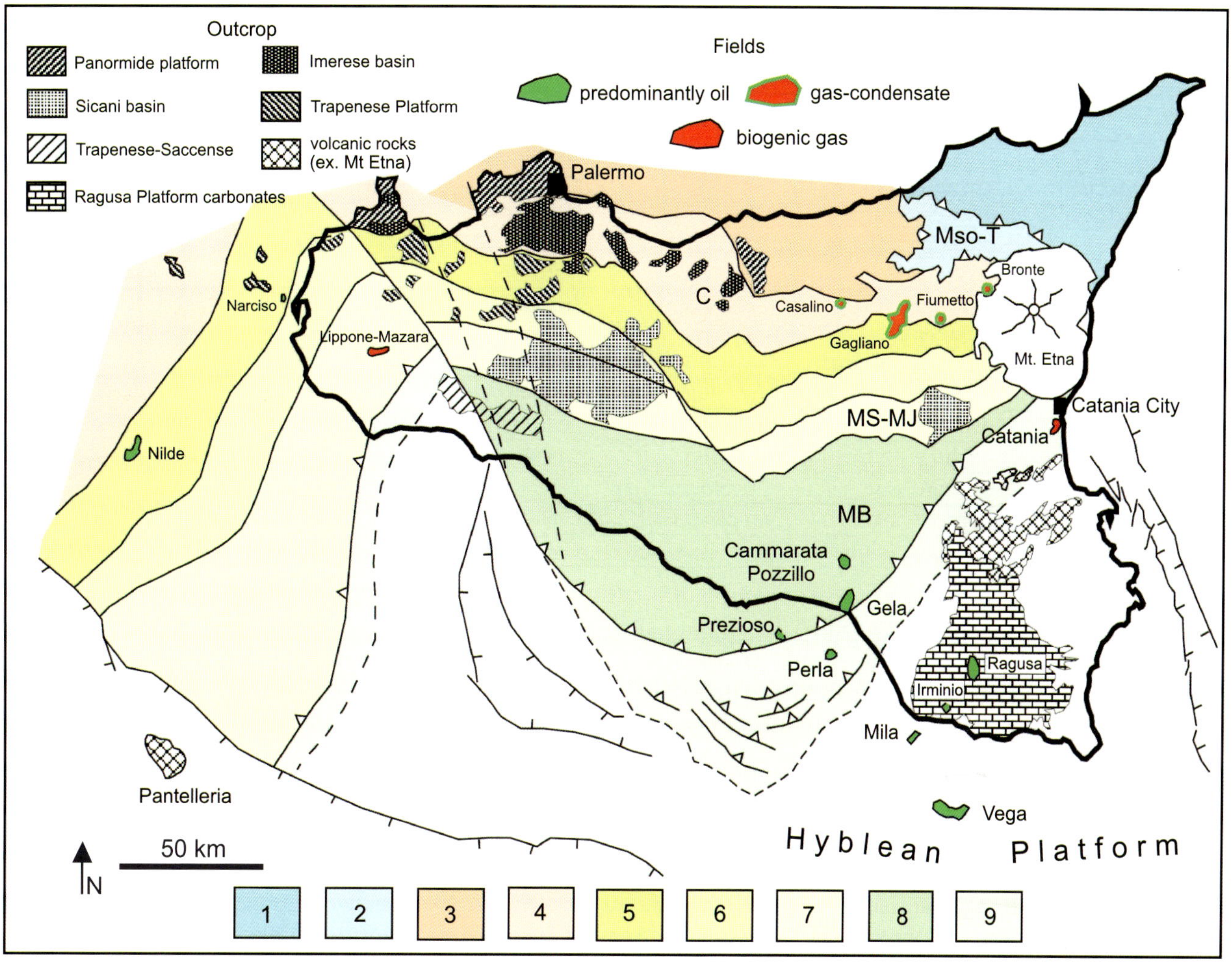

FIGURE 3. Map showing the sequential southward progression of foredeep (flysch) basins that envelope the Mesozoic hydrocarbon-bearing sequences. All units north of the Gela nappe front are allochthonous. Outcrop of the Mesozoic rocks and the hydrocarbon fields are shown in patterns. 1 = Peloritan crystalline terrain of European origin; 2 = Mt. Soro flysch, Neotethyan seafloor section; 3 = inner Numidian flysch and related Panormide platformal rocks; 4 = outer Numidian flysch and related Imerese domain basinal deposits; 5 = inner Messinian foredeep; 6 = central Messinian foredeep; 7 = outer Messinian foredeep; 8 = Gela nappe; 9 = Gela foredeep and the Carla trough (Tunisian trough). Areas mentioned in the text: C = Caltavuturo; MS–MJ = Monte Scalpello–Monte Judica; MB = Monte Bubonia; Mso-T = Monte Soro and Troïna. Modified after Roure et al. (2002, their figure 2).

outside of the island of Malta. Several of the important hydrocarbon deposits in Sicily are trapped in structures in the Ragusa plateau, i.e., Vega, Mila, Ragusa, and Irminio fields (Figure 3; Table 1).

Since about 0.5 Ma, the northern part of the Pelagian platform, including Malta, has been decoupled from the rest of Italy, from the Pelagian block of North Africa, and the Ionian Basin by recurrent tectonic activity of two major tectonic zones. This triangular-shaped area is bounded on the east by the Malta escarpment and the southwest by the Pantelleria and Malta graben (Figure 1), otherwise known as the Medina wrench system (Jongsma et al., 1987).

At the east end of the island, just south of Mount Etna, the thrust front intersects the northern extent of the Malta escarpment, locally known as the Siracusa escarpment, to connect to the Calabrian Arc at a major inflection in the Neogene thrust front (Figure 1). The Malta escarpment extends to the south-southeast for some 250 km (150 mi) to the Malta graben, separating the Pelagian platform from the Ionian abyssal plain to the east. The ocean-continent transition at the escarpment dates from pre-Alpine time and could have suffered inversion in the Alpine collision. Late Miocene to Quaternary block faulting is documented on the escarpment (Groupe Escarmed, 1983). Contemporary volcanism is concentrated at the junction of the Siracusa escarpment and the thrust belt in the Mount Etna area. Pre-Etnan basaltic volcanism began at about 5 Ma and was later superceded by the trachyandesites and

Table 1. Principal oil and gas fields of Sicily.

	Location	*API (°)*	*Reservoir*	*Age*	*Depth (m)*
Nilde	offshore; west	39	Nilde limestone	middle Miocene	1560
Narciso	offshore; west	29	Fortuna limestone	Oligocene	1920
Lippone-Mazara	onshore; west	gas	Terravecchia sandstone	Tortonian	630
Prezioso	offshore; Hyblean	19.5	Gela limestone	Triassic	4950
Gela	onshore; Hyblean	10	Noto-Gela	Triassic	3200
Cammarata-Pozillo	onshore; Gela nappe	7–10	Siracusa limestone	Liassic	3200
Ragusa	onshore; Hyblean	19.3	Gela dolomite	Triassic	1750
Mila	offshore; Hyblean	36	Noto limestone	Triassic	3400
Irminio	onshore; Hyblean	31	Noto limestone	Triassic	2500
Vega	offshore; Hyblean	16	Siracusa limestone	Liassic	2500
Catania	onshore; east	biogas	Ribera sandstone	Pleistocene	500
Gagliano-Fiumetto	onshore; fold belt	gas-condensate	Collesano sandstone	early Miocene	1600–2200
Bronte	onshore; fold belt	gas-condensate	Collesano sandstone	early Miocene	1250

trachytes of Etna itself (Behncke, 2001). The bimodal volcanic progression suggests that rifting, in conjunction with the arc setting of Etna, is fundamentally important (Serri et al., 2001), but the details of the magma genesis are controversial. Flexure-related reactivation at the ocean-continent transition may take the form of a tear or a rift in the subducting lithosphere that permits differential flexure or rollback of the slab and access of magma generated in the surrounding mantle and downgoing slab to the surface (Di Geronimo et al., 1978; Gillot et al., 1994; Continisio et al., 1997; McGuire et al., 1997; Monaco et al., 1997; Gvirtzman and Nur, 1999; Doglioni et al., 2001).

Transecting the Pelagian platform south of Sicily are the Pantelleria and Malta graben (Figure 1), which intersect the Malta escarpment southeast of Malta itself and eventually are masked by the accretionary prism at the Mediterranean Ridge some 200 km (120 mi) to the east. This is a complicated mosaic of thrust, normal, inversion, and transcurrent faults, generally forming a right-lateral system whose activity dates from the late Aquitanian to Pliocene in Malta (Dart et al., 1993). Quaternary sedimentary infill and associated volcanic rocks (Illies, 1981; Groupe Escarmed, 1983; Cello, 1987; Calanchi et al., 1989) indicate that this zone has recently been active. Jongsma et al. (1987) have suggested on plate kinematic grounds that this broad zone of deformation forms a transform plate boundary marking as much as 100 km (62 mi) of right-lateral motion between Africa and a Hyblean microplate (Mazzoli and Helman, 1994). The relative offset in the internal nappes of the Maghrebs vs. Sicily (Figure 1) is of comparable size, but the motion may be overestimated. The offset may represent a lateral ramp, however, as the structures may not be perfect piercing points. Indeed, Casero and Roure (1994) have traced facies patterns across the zone with little offset.

Possibly more important to Sicily itself but poorly documented in the literature is the Segesta lineament, a north-northeast trend bisecting the major fold belt syntaxis in western Sicily (Figures 1, 3). This feature juxtaposes differing facies and structures across strands of a (presumably) tear fault in the allochthonous units. Within the allochthons, the Segesta zone probably functions as a lateral ramp to the thrust sheets and aids the internal kinematics of the fold belt, which involves significant rotations (see below). Its projection into the foreland on the Pelagian platform is subtle; it displaces pre-Tertiary sedimentary facies (Casero and Roure, 1994), presumably through reactivating older crustal-scale structures.

Heat Flow, Volcanism, and Recent Uplift

Heat flow is generally elevated across Sicily (Figure 4), especially so along the Malta and Pantelleria graben and in the Aeolian Islands, but also in a trend across the Ragusa and Hyblean plateaus north through Mount Etna. Mount Etna has been mentioned in connection with its relationship to the Malta escarpment (Figure 1), but all three of these areas are characterized by young volcanism, generally basaltic in character (Serri et al., 2001, and references therein). Della Vedova et al. (2001) emphasize water circulation in connection with the southeastern Sicily anomalies, but the area has experienced recurrent volcanism since the Jurassic (Figure 5). Basalts and tuffs occur in the Jurassic Streppenosa, Modica, and Buccheri formations and prominently in the Late Cretaceous Amerillo Formation (Patacca et al., 1979; Scandone et al., 1981, Rocchi et al., 1996). Scandone et al. (1981) report similar rocks from dredge sites to the east on the Malta escarpment. The same area of southeastern Sicily also shows high and variable magnetic anomalies (Chiappini et al., 2000), which, if caused

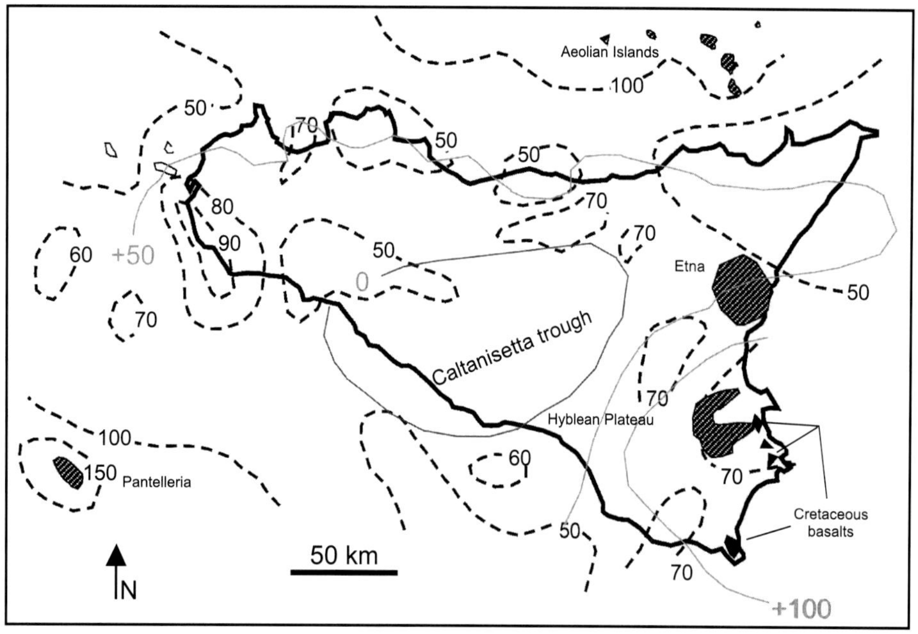

FIGURE 4. Heat flow and Bouguer anomaly sketch map of Sicily, also showing distribution of in-situ volcanic rocks, mostly basalts, and Mt. Etna. Heat flow in milliwatts per square meter after Della Vedova et al. (2001). Bouguer anomaly in milligal after Carrozzo et al. (1986). Locations of volcanics after Serri et al. (2001).

by volcanics distributed in the section, suggest that a very large area of southeastern Sicily experienced volcanism for a major part of the Mesozoic. Insofar as that volcanism may have been associated with periodic extension, heat flow is likely to have been at least mildly elevated for the majority of the post-Triassic history of the area.

Elsewhere in Sicily, heat-flow anomalies occur mainly in areas where carbonates occur at or near the surface and where structural studies suggest that stacking of carbonates characterizes the fold belt. The coincidence of a large area of uniform normal heat flow with a normal Bouguer signature in the Caltanisetta trough in central Sicily (Figure 4), where carbonate stacking is minimal, may indicate that the high heat flow is a ramification of the near-surface lithologies and associated water circulation.

GEOLOGY OF SICILY ITSELF

Stratigraphy of Sicily

The stratigraphy of Sicily has been approached from many vantage points. The most convenient, in view of the complexity of the physical stratigraphy, is to divide the coverage into that of the Ragusa plateau outside the fold belt (columns 7 and 8 in Figure 5) and that in the fold belt (columns 1–6). Figure 5 is offered here to summarize the general distribution of units in Sicily and to link them to the occurrences of hydrocarbons. In the southeast, on the Ragusa plateau, Patacca et al. (1979) remains the standard reference and is used here in column 7 of Figure 5. It supercedes the nomenclature developed by Rigo and Barbieri (1959) and Schmidt Di Friedberg (1965), which used formational names from disparate and tectonically separated localities. Nevertheless, this older nomenclature has persisted in the petroleum industry and is shown in column 8 of Figure 5 for reference only.

The Sicilian terranes of African origin (columns 3–8 in Figure 5) are broadly similar to sequences elsewhere in Italy and are composed of thick, platformal sequences of shallow-water carbonates with intervening deep-water, basinal sections of cherty limestones and radiolarian cherts, both superceded by open-marine and pelagic limestones and argillites. This tectonostratigraphic arrangement developed during and after Triassic and Early Jurassic crustal thinning and intracontinental rifting along the northern African margin. The platform-basin system persisted through the Jurassic and the Cretaceous into the Paleogene, with the filling of the basinal region and the development of shoaling indicators such as bauxitic horizons and rudist reefs on the platforms. In the southeasternmost parts of Sicily, the Ragusa plateau represents the exposed portion of the autochthonous Pelagian platform (Patacca et al., 1979), separated from the thrust front by the narrow Gela foredeep of middle to upper Pliocene and Pleistocene sediments (Figure 2). From Oligocene onward, the stratigraphy is dominated by clastics and marls that herald the impending deformation, molasse deposited in the foredeep of the southward-advancing thrust belt, and suprawedge piggyback basins.

Tectonostratigraphic Components of the Sicilian Accretionary Prism

Sicily is composed of several tectonostratigraphic provinces telescoped together during the Neogene orogeny between the Serravelian and the present (Figure 6). The bulk of the island has clear affinities with African geology, but in the northeastern corner of Sicily, the Peloritani Mountains are composed of fundamentally

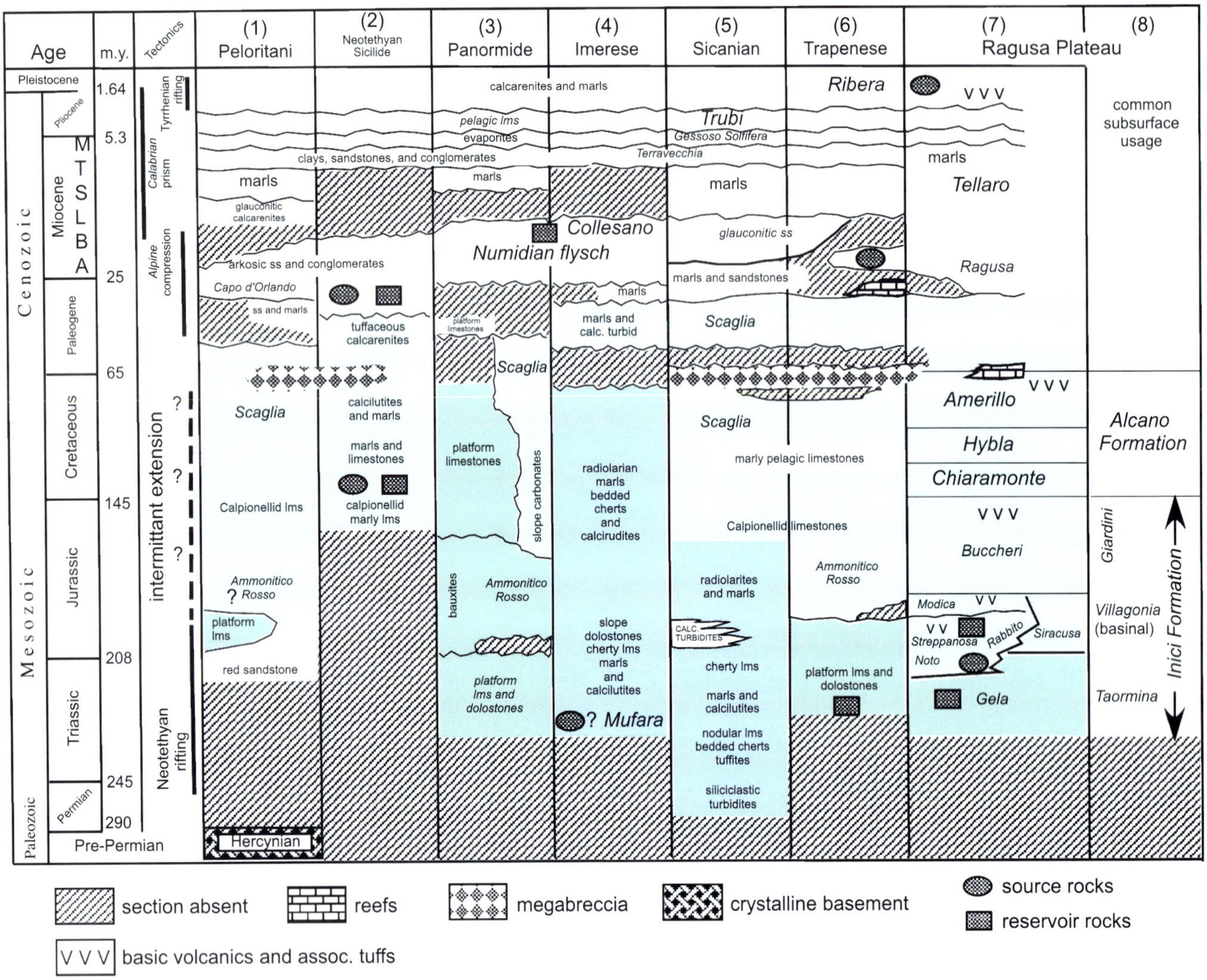

FIGURE 5. Stratigraphy of Sicily and associated source and reservoir rocks. Columns correspond to the various lithostratigraphic domains. Formation names in italics, rock types in standard font. Columns 1–6 for the fold belt after Catalano et al. (1993, 2000) and Ronchi et al. (2000). Usage of formation names for the Ragusa platform (column 7) from Patacca et al. (1979) and in column 8 as is commonly used in the subsurface, e.g., in well logs, after Rigo and Barbieri (1959) and Schmidt Di Friedberg (1965). Shades of blue denote various important carbonate environments: dark blue = basinal facies, medium blue = platforms, and light blue = various open-marine and pelagic facies, including the Streppenosa basin. Miocene stages: M = Messinian, T = Tortonian, S = Serravalian, L = Langhian, B = Burdigalian, A = Aquitanian. Timescale from Harland et al. (1989). lms = limestones.

European Hercynian basement and tectonically juxtaposed Mesozoic sedimentary rocks, both overlain by late Oligocene and younger coarse terrigenous strata, Messinian evaporitic, marly, and calcarenitic strata (Maccarrone et al., 2000). The Peloritani Mountains lie in thrust contact above a Late Jurassic, Cretaceous, and Paleogene sequence (Figures 2, 3, 6) of flyschlike turbidites (especially well exposed in the Monte Soro and Troïna areas) (Figure 3) that, by reason of a major late Eocene unconformity, is commonly interpreted to have been deformed during late Alpine events (Roure et al., 1990, and references therein). The Monte Soro and Peloritani lie in thrust contact over a Cretaceous to lower Miocene assemblage of flysch units, the Sicilide nappe, apparently deposited in the Tethyan ocean and thrust onto the fundamentally African Sicilian section. Hence, the northeast appears to be Tethyan and European in origin and is foreign to the African terranes over which it has been thrust.

Within the fold and thrust belt, the Mesozoic is represented by several platformal and intervening basinal units that are now interdigitated in the allochthons (Figure 6). The Panormide platformal unit outcrops in the north along the coast (Figures 2, 3), where it lies structurally above all other African terranes and below the Tethyan underpinnings of the Peloritani. In western Sicily, to the south and west of Palermo, the Trapanese–Saccense platformal units lie structurally below the Panormide and intervening basinal facies, where they

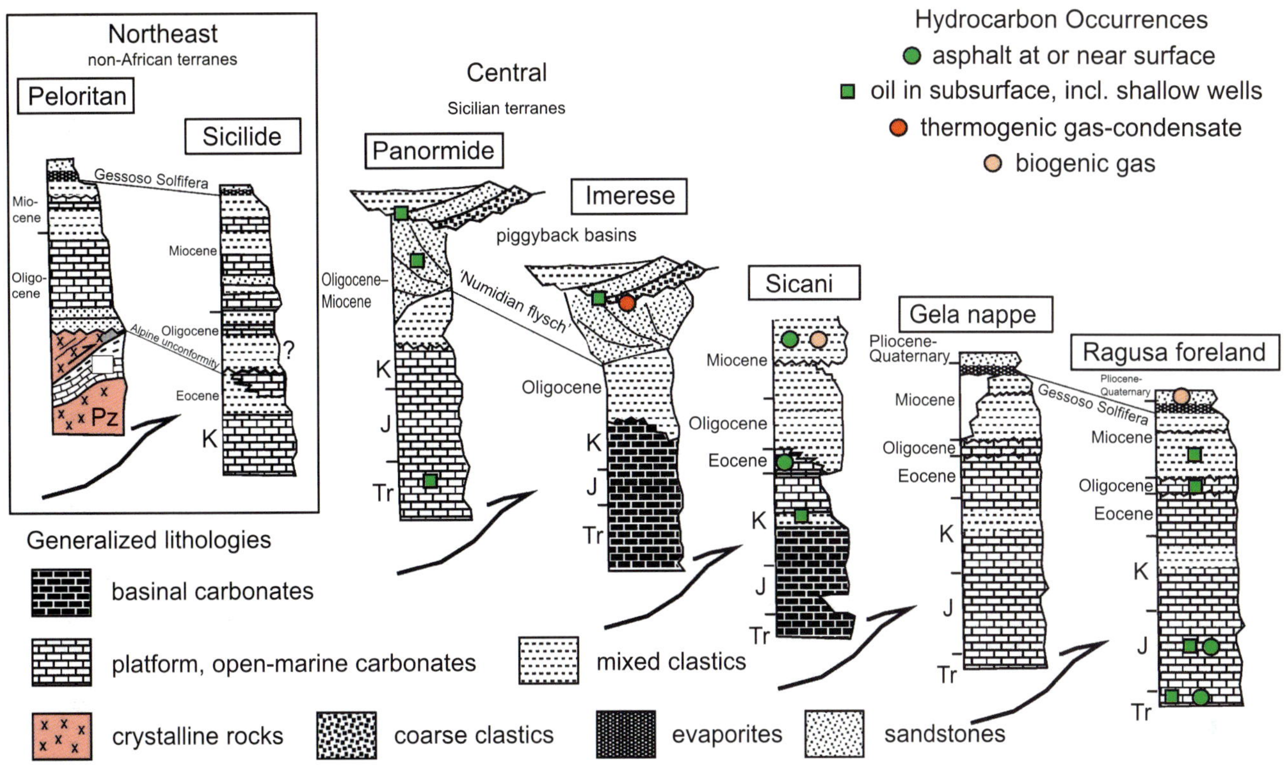

FIGURE 6. Tectonostratigraphic relationships between the various allochthonous units in Sicily and the Ragusa foreland and occurrences of hydrocarbons. See Figure 5 for more conventional stratigraphic detail. Inset shows relationship in the northeast: Peloritan (European nappe emplaced over Tethyan Sicilide, which, in turn, are emplaced over Sicani terrane). In central Sicily, the Panormide and Imerese are thrust over the Sicani. Emplacement ages especially of the more internal units are a matter of controversy (see De Capoa et al., 2000, and references therein). Emplacement of the Peloritani and Sicilide units over the Panormide, for example, is commonly considered Aquitanian, but De Capoa et al. (2000) argue for an age no older than Burdigalian. The more external units were emplaced in the Messinian, and the Gela nappe is commonly considered Pliocene–Pleistocene in age. Elements of figure after Roure et al. (1990).

are imbricated and thrust above their foreland equivalents in the subsurface (Catalano et al., 2000).

Two major basinal units are involved in the thrust belt. The Imerese unit is the major Mesozoic basinal unit southwest of Palermo to the south of the city of Cefalù (Figure 2) in structural windows through overlying thrust sheets containing the Panormide unit, Numidian flysch, and Neogene synorogenic sediments. A similar unit occurs south of the Imerese and in east-central Sicily 25 km (16 mi) southwest of Mount Etna, the Sicani unit. Both are composed of deep-water carbonate sections with important occurrences of alkali basaltic rocks that are related to the Triassic–Jurassic fragmentation of the carbonate platform (Lucido et al., 1978; Guarnieri et al., 2000).

Associated with the Panormide and Imerese units in the north is a nappe of Numidian flysch, which represents distal offshore African Neotethyan deposition that was caught up in the accretionary prism in the early phases of deformation in Sicily. It lies in thrust contact over both the platform and basinal units and in apparent depositional relationship to upper Miocene units in the fold belt. Thus, it represents a sedimentary unit deposited into and tectonically emplaced onto the northern side of the Sicilian passive margin before Neogene imbrication of the Mesozoic section itself. Presumably, it was derived from the Tethyan ocean floor, accretionary complex, or the trench immediately north of Sicily prior to the arrival of the northern reaches of the Pelagian platform.

Paralleling the north-to-south progression of Neogene deformation is a continuous sequence of foredeep deposits ranging in age from Aquitanian in the north to middle Pliocene in the Gela foredeep (Figure 3). Presently, they are either involved in the thrust sheets with their substrate or form piggyback basins carried southward with the progression of the thrust front. They form a large part of the outcrop in Sicily (Figure 2). These units record

- the systematic deepening of deposition as the Pelagian platform and its northern equivalents entered the foredeep and approached the leading edge of the Sicilian accretionary prism and
- synorogenic wedge-top deposition in piggyback basins during the southward progression of the thrust belt.

Emplacement of the Thrust Sheets of Sicily

Outside the Ragusa plateau in the southeast, all the platforms and basinal units alike are entirely allochthonous (Figure 3), being detached along surfaces in the Triassic and locally in the Permian (Roure et al., 1990) much like the mainland Apennine chain (Bally et al., 1986). Highly deformed red and green argillites in the Cretaceous and Paleogene section mark the base of many of the thrust sheets. Although previously thought to be olistostromes (Rigo de Righi, 1957; Broquet et al., 1984), these "argille varicolore" are more probably tectonic in origin and are the sole representatives of melange in the Sicilian accretionary prism (Roure et al., 1990).

Relationships among the thrust sheets are complex. Paleomagnetic data clearly indicate that the sheets have undergone large differential rotation with respect to the foreland and North Africa, generally a clockwise rotation (Channell et al., 1980, 1990; Oldow et al., 1990; Speranza et al., 1999), thus contributing to the strong arcuate character of the Gela front. Also contributing to the rotational history of sheet emplacement and the asymmetry of the Gela salient are several strike-slip faults, functioning primarily as lateral ramps during thrust emplacement, but also locally facilitating escape blocks from west to east (Monaco et al., 1997) and linking extensional elements near the north coast (Abate et al., 1998; Renda et al., 2000). The allochthonous structures onshore are pervasively overprinted by normal faults that presumably contributed to the maintenance of a critical taper as the fold belt has developed and underwent uplift (Roure et al., 1990).

STRUCTURAL STYLES AND TRAPS

General Statement

The surface geology of Sicily is exceedingly well mapped, particularly by programs at the universities in Palermo and Catania [e.g., in the Ragusa plateau by Lentini (1984) and Grasso (1999) and in key areas of the fold belt, such as those by Di Stefano and Vitale (1993) and Catalano et al. (1978)]. *The Structural Model of Italy* (Bigi et al., 1983) summarized those efforts up to 1983 and stands as a standard small-scale reference. What is clear from the mapping are both the complexity of the structural geology and the shallow level of exposure, especially in the fold belt. This is a subaerial fold belt with much of its young carapace intact. Neogene units dominate exposure in Sicily (Figure 2), and although they are intimately involved in the deformation, in many cases, they obscure the deeper structure. The lateral discontinuity of the thrust systems in Figure 2 illustrates the point. Note also that there are several cases in Figure 2 where thrust sheets that are stacked in standard older-over-younger relationships are cut by thrust faults that emplace younger units over older units. Seismic imaging is consequently difficult and further hampered by the rough terrain and the dominantly carbonate character of much of the section.

Structural Styles and Traps in the Foreland

Structures in the southeastern Sicily harbor the overwhelming bulk of the discovered hydrocarbon deposits in Sicily, especially the liquid ones. The discoveries on the Ragusa plateau are located on a north-northeast-trending high bounded by normal faults (Figure 2) that attracted Thomas' attention to Sicily in the 1940s. Local dip reversals on the otherwise southeast-dipping plateau that have survived from the Oligocene to the present seem to afford the best traps.

The traps throughout the southeast are intimately related to the stratigraphic architecture of the Triassic and Jurassic carbonate section in the Streppenosa basin (Figures 3, 5), isolated and modified by recurrent normal faulting that initiated during the Tethyan regional extension. Most of the fields are related in some way to different facies of carbonate buildups in the otherwise subsiding Streppenosa intraplatformal basin (Patacca et al., 1979). Ragusa, Gela, Irminio, Mila, and possibly Perla fields sit on reef mound highs in the basin. Cammarata field appears to be developed in resedimented carbonate derived marginally from a nearby high, and Vega appears to be near a particularly sharp edge of a carbonate buildup. Insofar as these are Mesozoic, ramp-style biohermal mounds (as opposed to true reefs) in the carbonate platform, Pedley (1990) has argued that the shelf-edge fields are localized and sharpened by syndepositional faulting. Indeed, Ragusa and Gela both have undergone repeated rejuvenation, evidenced through the many unconformities and thinned section across the structure both in the Early Jurassic and intermittently since (Kafka and Kirkbride, 1959; Rocco, 1959). Lindsey et al. (2002) described the structural setting of Irminio in terms of a reverse fault that separates the field from Ragusa, presumably a Neogene reactivation of an earlier normal fault.

No commercial deposits of oil and gas have been discovered outside the Streppenosa basin in the southeastern Sicily foreland (Figure 3), that is, elsewhere on the more stable platform in areas characterized by the development of Siracusa formation (see column 7; Figure 5). Only gas and oil shows have been encountered in the handful of wells involved, which has led Pedley (1990) to surmise that cap rocks have not developed to seal traps and/or that the traps are isolated from hydrocarbon sources by low-permeability facies, thus affording no carrier horizons to charge the structures.

Architecture in the Fold and Thrust Belt

A complete review of the structure of the Sicilian fold and thrust belt is beyond the scope of this chapter, but several points are germane to a discussion of the petroleum systems for two major reasons. The details of the structure, of course, dictate the traps in the petroleum systems, but in addition, palinspastic restorations of the Neogene deformation would define the distribution of potentially source-bearing rocks in basinal sections relative to traps as the system evolved. Insofar as the Streppenosa basin is the major source region for the Pelagian platform hydrocarbon occurrences (see below), similar potential source intervals are of major interest in the thrust belt.

Block rotations aside, the Sicilian fold and thrust belt conforms to the forward-vergent, piggyback style of deformation familiar from fold belts worldwide. Its architecture is defined by several dominant detachment levels. Ignoring the intracrystalline detachment under the Peloritani, horizons near the base of the Mesozoic carbonate section form the basal detachment of the whole system (Figure 6) as in the Apennines. The decollement typically ramps across the Mesozoic carbonate section into one or more horizons in the Cretaceous and Paleogene argillites or into the Miocene and Pliocene evaporates and shales, the latter especially near the thrust front. The thrust front is commonly a triangle zone that developed above the detachment localized in the Miocene or Pliocene.

Tectonic inversion of the older Mesozoic normal faults is evident in the Gela trough, and their preservation into the Neogene is an important trapping mechanism. Within the fold and thrust belt itself, pre-Neogene inversion structures have probably been a significant factor in migration and trapping, but the Neogene overprint has, for the most part, precluded their recognition. Major unconformities in the fold belt, however, where Paleogene sections lie over older Jurassic basinal facies as in the Monte Scalpello–Monte Judica area (Broquet, 1968; Pion-Leflaive et al., 1990; Larroque, 1993) and in the Imerese domain near Caltavuturo (Figure 3), may in fact represent Alpine inversion features (Casero and Roure, 1994).

As evident in Figure 2, there is considerable complication in the thrust sheets because of both duplexing of the Mesozoic and numerous minor detachment levels, which give rise to the most obvious exploration targets. The Gagliano gas-condensate field (Figure 3; Table 1), for example, with approximately 700 bcf initially in place, is the largest and most important onshore field in the fold and thrust belt. Production is from the Miocene Collesano formation (Figure 5), in thrust slices composed of the section from the Mesozoic carbonates to Oligocene–Miocene (Numidian) flysch, yet the field lies below minor folds in the flysch associated with intraformational detachments.

The map pattern of the Sicilian fold and thrust belt (Figure 2) is a patchwork of klippen, piggyback basins, breached anticlines, and tectonic windows, suggesting that deeper structure has caused redeformation of the now-exposed nappes and had a strong influence on outcrop pattern. Bello et al. (2000) have combined two- and three-dimensional seismic data, well information, and gravity and magnetic data to develop a comprehensive picture of the deformational history of the fold belt in eastern Sicily (Figure 7A). A two-phase picture of accretionary wedge development resulted from their study, with an initial emplacement of allochthonous units over the now-buried northern reaches of the Pelagian platform in the latest Oligocene and Miocene. Forward progression of the wedge on sub- and intra-Mesozoic detachments cut southward redeforming higher thrust sheets and duplicating the Mesozoic section at depth as far south as the Numidian flysch belt in the Gagliano area (Figures 3, 7). Windows occur where the shallower, older thrusts have been exposed in folds above buried ramps that branch from the sub-Mesozoic detachment (Figure 7). Pliocene–Pleistocene piggyback deposition has been preserved south of the tip of the deeper second-generation Mesozoic wedge, where back-thrusts have facilitated its emplacement, and north of the younger south-vergent ramps. Because the syntectonic basins appear to occupy flats in the younger deformational system, they date the southward progression of the basal detachment. This younger phase of deformation presumably facilitated thickening of the thrust wedge and its southward advance through a process of tectonic underplating of younger para-autochthonous thrust wedges under older, higher thrust sheets. The outcrop of Sicani basinal section (Figures 2, 3) in east-central Sicily is particularly important in that it implies the exposure of the deeper structure containing potential source in the fold belt.

Similar structural relationships occur in the west, where Catalano et al. (2000) report a superstructure of basinal carbonate facies Mesozoic rocks (the Imerese and Sicani domains of Figures 2, 3) along with Numidian flysch, which was emplaced during the Miocene and redeformed above younger thrust sheets that are composed of the platformal Trapanese and Saccense carbonates during the latest Miocene to the Pleistocene (Figure 7B).

The deeper parts of Figure 7B differ from their interpretation, which deserves some comment. Their plates I and II show three major detachments. The highest emplaces the basinal Imerese–Sicani section over the platformal Trapanese–Saccense with an undetermined

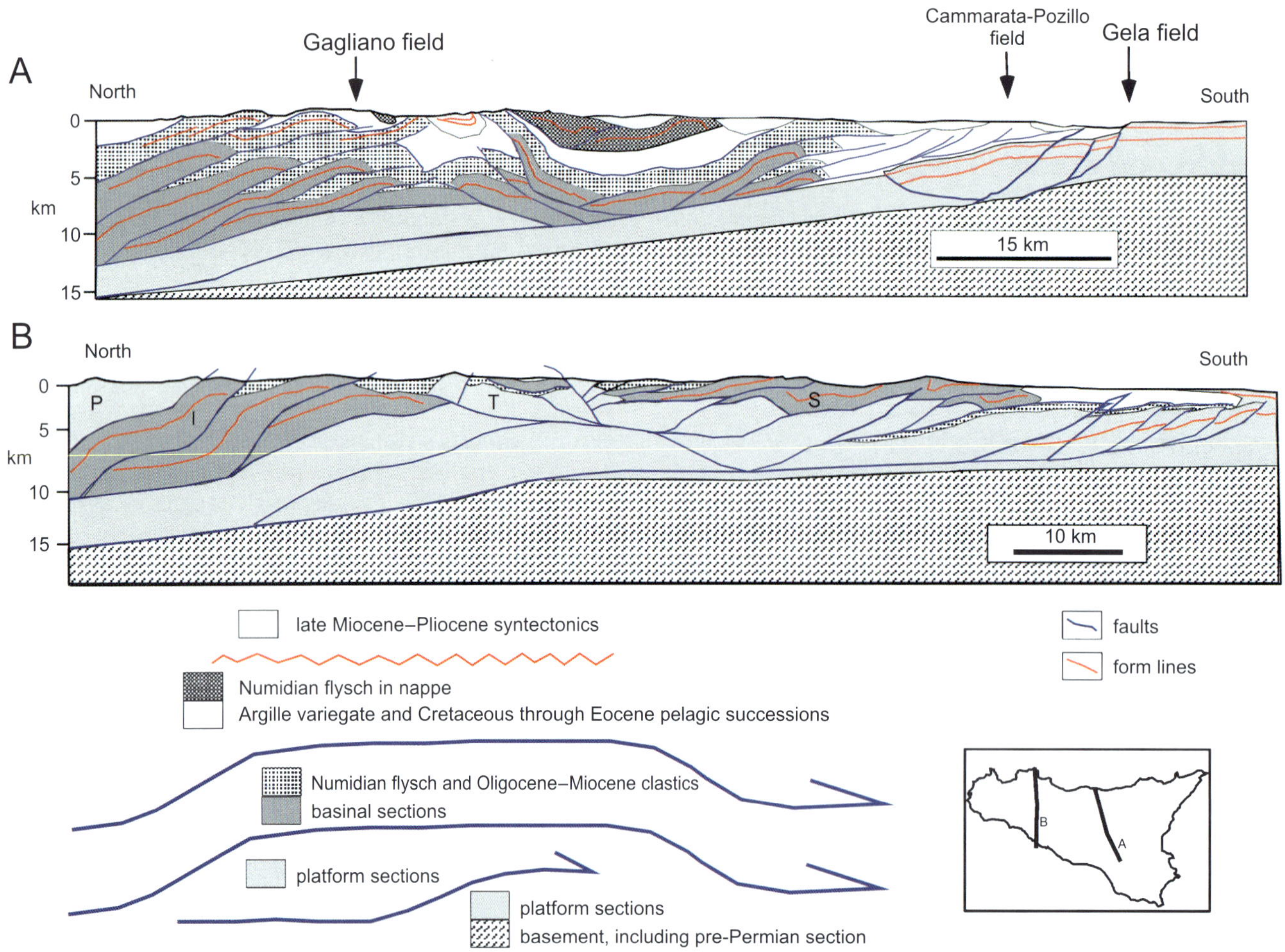

FIGURE 7. Generalized crustal-scale structure sections across Sicily. The major oil and gas fields indicated in the figure are shown in analogous positions to their true locations. Key indicates the emplacement sequence: para-autochthonous sheets of platform rocks of the Hyblean domain refold higher sheets during emplacement. Section A redrawn and modified from Bello et al. (2000). Section B reinterpreted as described in the text from Catalano et al. (1996, 2000). Section B is also extended into the offshore to the south and to the north where the Panormide domain is thrust on Imerese. Late normal faulting overprint is neglected. P = Panormide, I = Imerese, T = Trapanese, and S = Sicani domains (terranes).

distance. The underlying Trapanese–Saccense, in turn, is thrust some 50 km (30 mi) to the south on a detachment overlying Oligocene–Miocene terrigenous rocks, carrying the basinal section piggyback style. Finally, the deepest detachment redeforms these two upper nappes on a sequence of para-autochthonous thrust slices that stretches to the south coast. Hence, the overall shortening of Catalano et al. (2000) telescopes the section to about a third of its original length and implies that the basinal section originally lay outboard (to the north) of both slabs of Trapanese–Saccense platform. Figure 7B accepts the shallow seismic and well-controlled interpretation from Catalano et al. (2000) but takes a more conservative point of view on the shortening, similar to the sections in Catalano et al. (1996). The shallow deformation is attributed to the duplexing of the platformal underpinnings of the allochthonous basinal section in a similar fashion to Bello et al. (2000), which is generalized here in Figure 7A. The net effect is to more intimately associate the basinal sections with the platforms, thus also taking a more optimistic viewpoint on the exploration potential of the west, especially in light of the many hydrocarbon occurrences there (Figure 8).

The younger phase of the fold-belt development generated the most attractive exploration play in the Sicilian fold belt. Duplexes, ramp anticlines, and refolded higher sheets can generate closures in the buried structure that constitute an as-yet unrealized play in both the carbonate Mesozoic and its overlying siliciclastic cover. As Bello et al. (2000) demonstrate, newer, better seismic data are necessary over much of Sicily to test this play.

SOURCE ROCKS AND OIL-TO-SOURCE CORRELATION

Source Distribution

Proven thermogenic source rocks throughout the central Mediterranean are associated with Mesozoic basinal facies and the margins of those basins. The best developed occur in Triassic to Jurassic anoxic intraplatformal basins and troughs (Zappaterra, 1994) that developed on the margins or wholly in platformal areas, where relatively straightforward source-to-reservoir migration routes are available to charge traps. In Sicily, the Streppenosa basin (Figure 8 and column 7 of Figure 5) is an archetypical example, where the bulk of Sicily's oil fields lies (Figure 3).

Tertiary source facies have been established in the fold belt (Figure 5). However, the distribution of Mesozoic source facies is problematic. Although the Sicani and Imerese basinal facies outcrop in the fold belt, their generative potential is yet to be established. The Mufara Formation is a putative source interval but yet to be typed to any hydrocarbon. The subsurface distribution of basinal facies under late orogenic cover is unmapped. For example, Catalano et al. (2000) infer a clear physical connection between these basinal units and the Streppenosa basin itself, whereas Zappaterra (1994) draws a distinction in the tectonic evolution of the two and an intervening platformal barrier between them. Better seismic imaging north of the Gela nappe front is clearly needed here to constrain the subsurface structural geometry well enough to allow realistic palinspastic restorations of the fold belt and, in turn, specific stratigraphic relationships between the various Mesozoic facies belts.

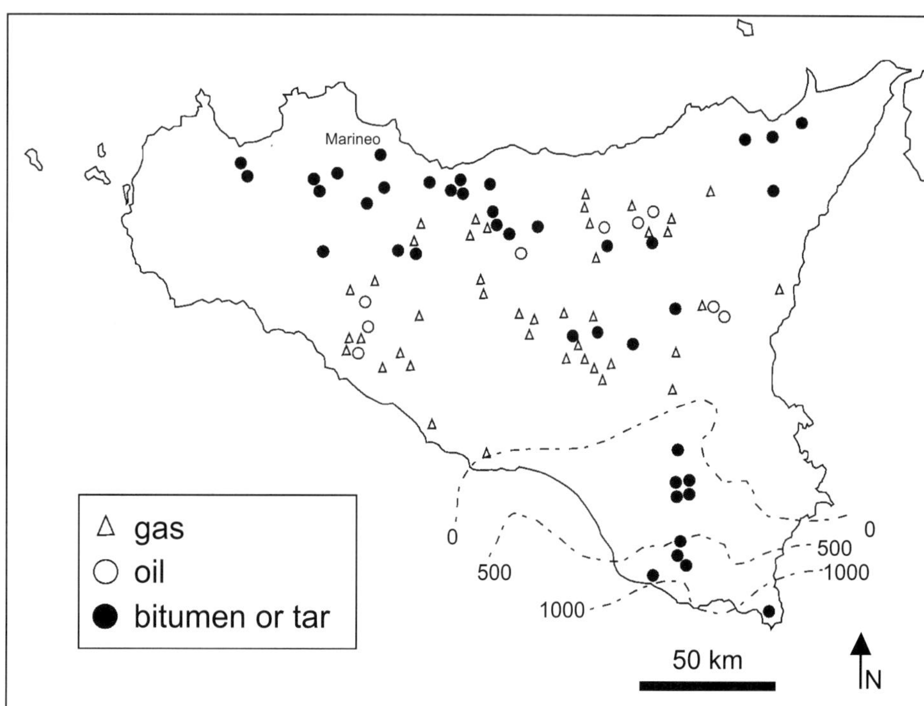

FIGURE 8. Seeps and asphalt occurrences in Sicily. Dashed lines are isopachs (in meters) on Streppenosa Formation, outlining the Streppenosa (source rock) Basin. Isopachs after Patacca et al. (1979); seep and asphalt locations after Rigo de Righi (1957) and Colombo and Sironi (1959). A major issue in source rock distribution for Sicily is if the Streppenosa basin closes off as shown or is connected in some way with the basinal sections to the north (Figures 2, 3).

Ragusa Foreland

By far, the greatest focus of geochemical studies in Sicily has been on the Mesozoic systems of the Hyblean Plateau in southeastern Sicily, primarily on the Noto and Streppenosa formations. The oils of this area are largely heavy (API < 20°), with high sulfur contents of 3–9%, except for two notable accumulations at Mila and Irminio (Figure 3), which have API values in excess of 30°. The heavy oils are generally water undersaturated, with varying salinities. At Cammarata–Pozzillo and Ponte Dirillo fields, for example, salinities range from 80,000 to 90,000 ppm, and at Ragusa, salinities are 100,000 ppm, whereas at Gela, they vary considerably more, from 20,000 to 100,000 in different horizons (Dalla Casa et al., 1981). Gas is associated at Gela and Ragusa, with as much as 80% of the gas being CO_2.

Pieri and Mattavelli (1986) linked the detailed chemistry of the heavy oils to early, immature generation instead of biodegradation, as did Novelli et al. (1988), who typed them to sources primarily in the Noto formation (Rhaetian in age). The Noto contains primarily types I and II kerogen that are sometimes in excess of 1% by volume, whereas the Streppenosa is considerably leaner and dominated by type III. Brosse et al. (1988) studied kerogen both in the Noto and Streppenosa formations distributed between black shales and clayey laminites and, alternatively, in a variety of laminated limestones, micrites, wackestones, packstones, dolomitic breccias, and their recrystallized equivalents. They found that the best oil potential was in clay-rich facies in the Noto formation, and that the kerogens have a variety of kinetic characteristics. However, all appeared to have early generative capabilities, so that the occurrence of light oil in the Ragusa foreland appears to be a ramification of the geological history that is as yet unexplained.

The maturation modeling studies of Novelli et al. (1988) emphasized rapid Pliocene–Pleistocene burial and heating during the Gela foreland subsidence in connection with the generation of the heavy oils, a conclusion corroborated by Brosse et al. (1989) specifically for the Gela area. Brosse et al. (1989), however, contrasted Ragusa with Gela and found that variations in the pre-Tertiary subsidence history and in the kerogen composition from different facies could lead to maturation, hydrocarbon generation, and migration as early as the Jurassic in the deepest depocenters of the Streppenosa basin (Figure 8). Ocampo et al. (1993) found that biomarkers uniquely typed a Noto source rock to oils of the small Ponte Dirillo field (very near Gela), but that significant differences in Gela oils suggested a different generation and/or source history.

Exploration data would also suggest that multiple periods of hydrocarbon trapping have occurred. Multiple migration events are evident in the reservoir in Irminio with the first migration as early as the middle Cretaceous (Lindsey et al., 2002). Isopachs on various stratigraphic levels of the Ragusa field show a thin section above the reservoir (Kafka and Kirkbride, 1959), indicating recurrent rejuvenation of the trap well before the depression in the Gela foredeep. Furthermore, flattening exercises on various horizons in connection with exploration studies seem to suggest that Oligocene structural highs that have survived through the foreland basin flexure and associated faulting to remaining highs today are the most successful exploration targets, implying that a major period of petroleum migration occurred prior to the emplacement of at least most of the fold and thrust belt. Numerous dry holes have been drilled on Pliocene–Pleistocene traps in the Ragusa foreland. Although there may be many reasons for the dry holes, the occurrences of traps charged with light oil, as well as the structural history of successful traps, would suggest that a more complicated maturation-migration history is involved throughout the Ragusa plateau. In any case, the simple model attributing maturation and migration to subsidence in the fold-belt foreland is insufficient to account for the distribution of oil fields.

Fold-belt Oils and Tars

Although many hydrocarbon seeps and asphalts occur in the fold belt (Figure 8), no modern geochemical studies have been made to type them to sources. No discoveries in the Mesozoic sequence have been recorded in the onshore fold belt, although one tar south of Palermo at Marineo (Figure 8) occurs in Triassic dolomite. Most of the occurrences seem to be similar to the heavy oils in the foreland, but several have more paraffinic characteristics. Heavy-metal concentrations vary widely. The source of these hydrocarbons is uncertain but is likely to be similar in depositional and tectonic setting to the source in the Streppenosa basin. Because rich Mesozoic source rocks, such as the Noto formation, are of limited distribution outside the Streppenosa basin (Figure 5), other candidates have been proposed. The Triassic Mufara Formation is common in the north, but of limited total organic carbon. Likewise, younger formations have some carbon content, but their total organic carbon and maturation levels have received little study. A systematic study of the chemistry of these occurrences in connection with their structural setting would be of value in assessing the petroleum potential of the fold and thrust belt. Again, better seismic imaging would contribute much to the geographic aspects of such a study.

Melange, Mud Diapirs, and Fluid Flux through the Fold and Thrust Belt

Connections between overpressure generation, dewatering processes, and deformation in accretionary prisms and fold and thrust belts are documented in a rich literature (e.g., Langseth and Moore, 1990; Shipley et al., 1995) and are intimately related to hydrocarbon migration. Prisms with at least partially subaerial histories, such as Sicily, are similar to many retroarc fold and thrust belts and differ from intraoceanic prisms in that meteoric water factors into the system, providing hydrodynamic recharge otherwise unavailable and an additional diagenetic influence over potential reservoir rocks. Both detachment-related melange and mud diapirs indicate that Sicily is no departure from the norm, but also that meteoric recharge, overpressure, and deformation in the wedge are linked to at least one period of petroleum migration.

Guilhaumou and Larroque (1993), Larroque (1993), and Guilhaumou et al. (1994) have studied the argille varicolore that are associated with detachment surfaces and mud diapirs in the Monte Scalpello and Monte Bubonia areas (Figure 2). Detachment surfaces in the duplex at Monte Scalpello that carry the Tertiary section are characterized by a sheared Late Cretaceous and Tertiary argillaceous matrix containing blocks of argillite. Calcite veins in the blocks have low-salinity fluid inclusions with low homogenization temperatures (<100°C). Basal detachment surfaces, i.e., those carrying the Mesozoic carbonate succession, are marked by mélange of sheared argillitic matrix and brecciated blocks of mudstones, sandstones, and radiolarian cherts carrying Lower Cretaceous fossils. The blocks contain several generations of quartz and calcite veins with two types of fluid inclusions. One set containing low-salinity water exists in a range of veins, from quartz veins with inclusions that have homogenization temperatures of 150°C to overprinted veins of calcite with inclusions homogenized at more than 200°C. Significantly, the calcite veins also have high-temperature pyrobitumen inclusions. A second set of water-depleted inclusions in the calcite veins carry light oil with no gas. The surrounding melange is in thermal disequilibrium with the veins, recording only a lower thermal overprint in the form of clay transformation. The authors concluded that the detachment channelized several periods of fluid flux through the melange. Overpressure resulted in naturally hydrofractured rocks with passing hot fluids that, in at least one case, were associated with hydrocarbons from a deeper source.

Similar lithologies to the upper detachment melange at Monte Scalpello occur in a body that crosscuts stratigraphy at Monte Bubonia (Figure 3) and breaches the core of an anticline in the Gela nappe (Larroque, 1993). The matrix is sheared red and green argillites, with blocks of *Nummulites*-bearing Eocene calcarenites, Numidian flysch, and various elements of the Mesozoic and Paleogene section. Importantly, carbonate blocks containing superimposed crack-and-seal calcite veins are also present. Isotopic studies indicate that cements in the carbonates are normal-marine cements, but the veins relate to hydrothermal fluids. No oils were encountered in these upper detachment rocks, suggesting the possibility that, at least locally, the hydrocarbon source rocks lay deeper in the section than that accessed by the upper fluid pathways. Presumably, the source rocks were in the Triassic–Jurassic part of the section, which was represented in blocks of the lower detachment and not of the upper detachment.

Offshore Oils

Tertiary-reservoired oils in the offshore west of Sicily at Nilde and Narciso (Figure 3) likewise have not been conclusively typed to sources but are assumed to be derived from associated Oligocene–Miocene source rocks.

Gas and Gas-condensate

Biogenic gas has been produced at two fields in the fold belt. At the Lippone-Mazara field (Figure 3) in western Sicily, gas that was produced from the Tortonian Terravecchia sandstone (Figure 5) is presumed to derive from surrounding shales, whereas at Catania, south of Mount Etna, Pleistocene Ribera sands reservoir gas again sourced from the surrounding shales. Thermogenic gas and associated condensate derived from the surrounding Miocene Collesano formation charge the Gagliano field.

RESERVOIRS

Sicilian reservoirs include a variety of complex fractured carbonates in the Triassic and Jurassic of the Hyblean platform. They have complex episodes of charge and diagenetic alteration. Lindsey et al. (2002) describe a sequence of events in Irminio field involving the emergence of the reservoir in the Lias, a period of oil migration in middle Cretaceous time, followed by an influx of hot water and bitumen baking. A second period of charge began in the Paleogene and continued well into the Neogene, when a period of stylolitization and dedolomitization also occurred.

The distribution of these reservoirs is controlled by the Mesozoic paleogeography, in that the best reservoirs are concentrated in and fringe the carbonate mound buildups in the Streppenosa basin, where cyclic effects of dolomitization, karstification and subaerial dissolution, emersion, burial, and fracturing that promote reservoir quality are coupled with periods of fine-grained deposition that afford suitable seals. Presumably, similar reservoir rocks occur elsewhere in Sicily, where the Neotethyan extended terranes afford similar histories. At locations in the Gela nappe, for example, Mesozoic carbonates exhibit cataclastic fabrics with marine dolomitic cements.

Cretaceous units similar to those elsewhere in Italy (Mattavelli et al., 1993) are potential reservoirs, such as debris flows and turbiditic units in the basinal facies, but these have yet to be tested in Sicily. Tertiary terrigenous sandstones that are related to the time-transgressive advance of the fold and thrust belt provide the other major play in Sicily, as at Nilde or Catania (Figure 3). These are units deposited in the migrating foredeep basin or in the piggyback basins. Reservoir quality depends on cementation history, of course, and on the provenance of the sands, either carbonate or siliciclastic.

SUMMARY AND CONCLUSIONS

Hydrocarbons occur in two main megasystems in Sicily: in the Ragusa plateau and in the fold and thrust belt. The Ragusa plateau is separated from the fold and thrust belt by a very young and narrow foredeep (Figure 2). Currently, the plateau carbonates dip southeastward away from a strong normal fault system along its western edge, thus setting up a high along the main Ragusa–Irminio–Mila oil field trend. The plateau is partially underlain by the intraplatformal Streppenosa basin of Triassic and Jurassic age, which extends to the south offshore well away from the fold belt. It contains the well-established Noto and Streppenosa source rocks, which have been typed to the predominantly (but not exclusively) heavy-liquid hydrocarbons in most of the fields. Several lines of evidence suggest that a multiphase maturation and migration was involved in charging the fields:

- multiple ages of mild deformation, as reflected in thinning isopachs over structures
- long history of potentially elevated heat flow
- petrographic studies of the diagenetic history of reservoir rocks
- observations that Oligocene structures that have survived to the present are more likely to contain substantial hydrocarbons
- variations among the oils involved

Although there would seem to potentially be some charge into heavy-oil fields in the narrow foredeep, notably Gela and Perla, from petroleum systems in the frontal parts of the fold and thrust belt, oil-bearing fluid inclusions from basal detachment-related melange are much lighter and do not suggest that heavy oils were a component of the flux. Other fluid inclusions suggest that fresher water would also be generated and potentially flush foredeep reservoirs as easily as they may be charged. Consequently, the bulk of the oils appear to have been sourced from within the foreland itself, and the Ragusa plateau has served as its own kitchen, transport, and trap system using the large fetch area of the Streppenosa source basin.

Light oils occur in Irminio and Mila, but it is unlikely that these light oils originated outside the foreland for the reasons above. All of these deposits are located in the zero isopach of the Streppenosa basin, so it appears that the foreland oils are sourced from the deeper parts of the Mesozoic dominantly carbonate section of the intraplatformal basin during multiple phases of generation and migration. Modeling studies suggests that maturation and migration were certainly aided by flexure of the foreland and Neogene burial, but the deeper parts of the Streppenosa basin may have generated hydrocarbons as early as the Jurassic. Cretaceous charge has also been proposed for Irminio (Lindsey et al., 2002).

Unsuccessful tests are common in the Ragusa plateau and its offshore. Many of them are located on young structural highs and poorly positioned to test older closures. Structures in the foreland have undergone several stages of rejuvenation, with the distinct possibility that many of them have been either ruptured or spilled either during Alpine structural influences on the area or during the emplacement of the Sicilian fold and thrust belt. Hence, the hydrocarbon system in the foreland is a complex one but appears to be unlinked, for the most part, to the fold belt, except as a potential analog for the early prefold belt phases of hydrocarbon generation elsewhere in Sicily.

Hydrocarbons in the fold and thrust belt are limited to biogenic and thermogenic gas onshore, numerous seeps and outcropping tar deposits, and oil of questionable source in offshore fields west of Sicily. All of these deposits are confined to Tertiary reservoirs (with the exception of one occurrence of tar in Triassic rocks) largely because few valid tests of pre-Tertiary traps have been performed in the fold belt. Nevertheless, several plays are possible, as outlined in the Appendix. That a live petroleum system was active at one time is indicated by the fluid-inclusion studies in the Monte Scalpello area and numerous seeps and tars in the outcrop; but whether liquids have survived in commercial quantities remains a question. Relatively new concepts of the structural architecture in the fold belt offer the possibility that future tests of buried closures would confirm or deny the potential of the fold belt. Establishing sedimentological similarities and/or physical connections between the rich Streppenosa intraplatformal basin and the Jurassic basinal sections in the fold belt would be a great step forward in demonstrating an effective Mesozoic source in the fold belt. Better seismic imaging in a systematic survey would also be a step forward.

ACKNOWLEDGMENTS

J. W. Granath thanks Forest Oil International for providing the opportunity to become familiar with the geology of Sicily, Luigi Albanesi for his mentorship in Italian petroleum exploration, and A. W. Bally for his thoughtful review of an early form of the manuscript. Both of us thank the conveners of the conference *Deformation, Fluid Flow, and Reservoir Appraisal in Foreland Fold and Thrust Belts* held in Palermo in May 2002 for requesting this contribution to the proceedings volume.

APPENDIX: SICILIAN PETROLEUM PLAYS

Sicilian petroleum systems can be classified according to their structural and geodynamic province (Figure 2), starting from the hinterland toward the foreland. Efficient petroleum systems, as indicated by the main oil and gas fields, are characterized first, followed by other, underexplored and potential plays. Only plays that have been successful in discovering hydrocarbons have been listed. The Mesozoic carbonates in the fold and thrust belt have been disappointing to date. This should be a major play, but a realistic test of structures in the thrust stack has not yet been made. Focus has largely been on the shallower, surface anticlines and not those involved in the deeper, younger sheets described by Bello et al. (2000) and Catalano et al. (2000).

Petroleum Systems in the Fold Belt

Petroleum System of the Outer Numidian Domain, Exemplified by the Casalini, Bronte-San Nicola, and Gagliano and Fiumetto Fields

1) Reservoir: Argillaceous sandstone of the Collesano formation [Tortonian(?)]; porosity with microfractures 6–15%; low permeability
2) Trap: Imbricated thrust folds that are low amplitude, south verging, and involving well-bedded carbonate substratum (basinal Imerese units); structural closure by bedding dips on three flanks and by fault toward the south; vertical closure exceeding 200 m (660 ft)
3) Seal: Argillites from the paleo-oceanic Sicilides units
4) Source rock: Argillites from the Collesano formation or siliceous shales of the Crisanti Formation (Liassic)
5) Maturation: At regional scale, these two formations become increasingly buried toward the north and laterally enter the gas window
6) Migration: Lateral and updip from the north
7) Hydrocarbon type: Gas (90% methane) and condensate (62° API) with gas-oil ratio averaging 7000

Petroleum System of the Inverted Inner Messinian Foredeep, Exemplified by Nilde, Norma, and Narciso Fields

1) Reservoir: Fractured and most likely karstified bioclastic Serravalian–lower Tortonian(?) carbonates at the top of former passive-margin sequence (Nilde limestone or lateral equivalent of the Fortuna Formation)
2) Trap: Steeply dipping, imbricated thrust folds verging toward the southeast
3) Seal: Shales at the base of the upper Tortonian Terravecchia Formation
4) Source rock: Carbonate source according to oil analyses made by Agip; no potential source rock has yet been evidenced in the former passive-margin sequence at Nilde, but a lateral contribution of outer Numidian sources is also possible
5) Maturation: During Messinian times, at the onset of inversion episodes
6) Migration: Lateral and updip from the west, where Numidian series become increasingly thick, both stratigraphically and by tectonic duplication
7) Hydrocarbon types: Light oil in Nilde (38° API), becoming progressively heavier toward the east (21–25° API in Narciso); farther east, away from the oil kitchen, only heavy oil and CO_2 occur (Nora Nl, Orlando 1 and 2)

Petroleum System of the Outer Messinian Foredeep, Exemplified by Lippone-Mazara Fields

1) Reservoir: Argillaceous turbiditic sandstones of the central part of the lower Messinian Terravecchia Formation
2) Traps: East-trending thrust folds and pop-up structures of the Sicani domain resulting from inversion of high-angle structures of the former African passive margin
3) Seal: Argillites from the Terravecchia Formation
4) Source rock: The same argillites
5) Maturation: Biogenic gas
6) Migration: Synchronous with folding (i.e., Messinian)
7) Hydrocarbon type: Gas/methane = 99%

Petroleum Systems of the Hyblean Foreland

Gela-type Play: Gela, Ragusa, Ponte Dirillo, and Prezioso Fields

1) Reservoir: Porous and fractured dolomites of the Gela Formation (Norian–Rhaetian)
2) Traps: Multistage faulted anticlines, with late-stage transpersonal reactivation during Apennine compressional episodes (upper Pliocene)
3) Seal: Black shales of the Streppenosa Formation
4) Source rocks: Black shales and bituminous limestones of the Noto formation (Rhaetian) and Streppenosa Formation (Hettangian)
5) Maturation: Deep burial already affected source rocks during the Mesozoic southwestward of the Hyblean Plateau, resulting in early oil generation and expulsion
6) Migration: The Noto and Streppenosa formations constitute in the south lateral basinal equivalents of shallow-water platform dolomites of the Gela Formation. This global architecture of the foreland allows a direct connection between very thick source rocks and reservoir sequences. During the Late Jurassic, Cretaceous, and Cenozoic, continuous subsidence led to the burial of source rocks into the oil window. Since the Messinian, the progressive uplift of the Hyblean Plateau has resulted in a regional tilting of the margin, thus inducing long-range migration pathways from the oil kitchen toward the traps. Vertical migration is also locally controlled by high-angle faults, thus allowing a remigration of hydrocarbons from Mesozoic toward Miocene

reservoirs (i.e., in the Ragusa area, where bituminous calcarenites are cropping out and mined).

7) Hydrocarbon types: Low-maturity heavy oil with sulfur; oil gravity increases with the distance of migration
8) Special cases: Mila and Irminio fields are members of this same petroleum system, but they are located near the facies transition between the Streppenosa–Noto and Gela formations; for instance, the Mila reservoirs are rather reefal and thus distinct from those of the nearby platform; hydrocarbons are light (30° API), and the reserves appear to be the product of multiple periods of migration (Lindsey et al., 2002)

Perla-type Play: Perla and Vega Fields

1) Reservoir: Partially dolomitized and fractured oolitic limestones and algal pelsparites from open-platform environment (Siracusa and Inici formations); intergranular and intercrystalline and vuggy porosity (12–16%)
2) Traps: Mixed, stratigraphic, and structural; a facies transition from platform carbonates of the Siracusa formation and shales from the Villagonia (Modica) formation occurs toward the northeast; this facies boundary has been reactivated by numerous north-verging reverse faults during Late Cretaceous inversion episodes, thus increasing further the stratigraphic closure
3) Seal: Pelagic marly limestones of the Buccheri (Giardini) formation
4) Source rock: Mainly black shales of the Noto and Streppenosa formations; because of regional migration trends, a contribution of lateral equivalents of the Siracusa formation in a back-reef position should be also considered
5) Maturation: In case the source rock belongs to the deep Noto–Streppenosa basin, maturation would be rather old, as in the Gela play; instead, if source rocks are located in the back-reef area, they would have entered the oil window more recently, i.e., during the deposition of the Pliocene–Quaternary flexural sequence
6) Migration: Vertical along high-angle fractures (in case the source is located in the Noto–Streppenosa basin) or lateral and updip from a hypothetical oil kitchen located toward the southwest
7) Hydrocarbon type: Moderately mature and not biodegraded heavy oil (12–16° API) with sulfur

Catania and Cisina Gas Fields in the Pliocene–Pleistocene Foredeep of Gela

1) Reservoir: Multilayered Pleistocene sandstones
2) Traps: Trapping anticlines and/or recent transpressional reactivation of fractures
3) Seals: Interbedded argillites
4) Source rocks: Interbedded argillites
5) Maturation: Biogenic gas
6) Migration: From the argillites toward nearby sandstones

REFERENCES CITED

Abate, B., A. Incandella, F. Nigro, and P. Renda, 1998, Pliocene–Pleistocene strike slip tectonics in the Trapani Mountains (NW Sicily): Bollettino della Societa Geologica Italiana, v. 117, p. 555–567.

Amato, A., and G. B. Cimini, 2001, Deep structure from seismic tomography, *in* G. B. Vai and I. P. Martini, eds., Anatomy of an orogen: The Apennines and adjacent Mediterranean basins: Dordrecht, Kluwer Academic Publishers, p. 33–45.

Bally, A. W., L. Burbi, C. Cooper, and R. Ghelardoni, 1986, Balanced sections and seismic reflection profiles across the central Apennines: Memorie della Societa Geologica Italiana, v. 35, p. 257–310.

Behncke, B., 2001, Volcanism in the southern Apennines and Sicily, *in* G. B. Vai and I. P. Martini, eds., Anatomy of an orogen: The Apennines and adjacent Mediterranean basins: Dordrecht, Kluwer Academic Publishers, p. 105–120.

Bello, M., A. Franchino, and S. Merlini, 2000, Structural model of eastern Sicily: Memorie della Societa Geologica Italiana, v. 55, p. 61–70.

Bigi, G., D. Cosentino, M. Parotto, R. Sartori, and P. Scandone, 1983, Structural model of Italy: Consiglio Nazionale delle Ricerche Progetto Finalizzato Geodinamic, scale 1:500,000, 3 sheets.

Broquet, P. 1968, Etude géologique de la region des Madonies (Sicile): Ph.D. thesis, University of Lille, Lille, France, 797 p.

Broquet, P., G. Duée, G. Mascle, and R. Truillet, 1984, Evolution structurale alpine récente de la Sicile et sa signification géodynamique: Revue de Géographie Physique et de Géologie Dynamique, v. 25, p. 75–85.

Brosse, E., J. P. Loreau, A. Y. Huc, A. Frixa, L. Martellini, and A. Riva, 1988, The organic matter of interlayered carbonates and clays sediments— Trias/Lias, Sicily: Organic Geochemistry, v. 13, p. 433–443.

Brosse, E., A. Riva, S. Santucci, M. Bernon, J. P. Loreau, A. Frixa, and F. Laggoun-Défarg, 1989, Some sedimentological and geochemical characters of the Late Triassic Noto formation, source rock in the Ragusa Basin (Sicily): Organic Geochemistry, v. 16, p. 715–734.

Calanchi, N., P. Colantoni, P. L. Rossi, M. Saitta, and G. Serri, 1989, The Strait of Sicily continental rift systems: Physiography and petrochemistry of the submarine volcanic centres: Marine Geology, v. 87, p. 55–83.

Carrozzo, M. T., D. Luzio, C. Margiotta, and T. Quarta, 1986, Gravity map of Italy: Consiglio Nazionale delle Ricerche, Progretto Finalizzato Geodinamica, scale 1:500,000, 2 sheets.

Casero, P., and F. Roure, 1994, Neogene deformations at the Sicilian–North African plate boundary, *in* F. Roure,

ed., Peri-Tethyan platforms: Paris, Éditions Technip, p. 27–45.
Catalano, R., B. Abate, and P. Renda, 1978, Carta Geologica dei Monti di Palermo: Istituto di Geologia della Università di Palermo, scale 1:50,000, 1 sheet.
Catalano, R., P. Di Stefano, F. Nigro, and F. P. Vitale, 1993, Sicily mainland and its offshore: A structural comparison, *in* M. D. Max and P. Colantoni, eds., Geological development of the Sicilian–Tunisian platform: United Nations Educational, Scientific, and Cultural Organization Report in Marine Science, v. 58, p. 19–24.
Catalano, R., P. Di Stefano, A. Sulli, and P. Vitale, 1996, Peleogeography and structure of the central Mediterranean: Sicily and its offshore area: Tectonophysics, v. 260, p. 291–323.
Catalano, R., A. Franchino, S. Merlini, and A. Sulli, 2000, Central western Sicily structural setting interpreted from seismic reflection profiles: Memorie della Societa Geologica Italiana, v. 55, p. 5–16.
Catalano, R., C. Doglioni, and S. Merlini, 2001, On the Mesozoic Ionian Basin: Geophysical Journal International, v. 144, p. 49–64.
Cello, G., 1987, Structure and deformation processes in the Strait of Sicily rift zone: Tectonophysics, v. 141, p. 237–241.
Channell, J. E. T., R. Catalano, and B. d'Argenio, 1980, Paleomagnetism and the deformation of the Mesozoic continental margin in Sicily: Tectonophysics, v. 61, p. 391–407.
Channell, J. E. T., J. S. Oldow, R. Catalano, and B. d'Argenio, 1990, Paleomagnetically determined rotations in the western Sicilian fold and thrust belt: Tectonics, v. 9, p. 641–660.
Chiappini, M., A. Meloni, E. Boschi, O. Faggioni, N. Beverini, C. Carmisciano, and I. Marson, 2000, Shaded relief magnetic anomaly map of Italy and surrounding marine seas: Annali di Geofisica, v. 43, p. 983–989.
Colombo, U., and G. Sironi, 1959, Geochemical analyses of Italian oils and asphalts: Proceedings of the 5th World Petroleum Congress, v. 5, p. 177–206.
Continisio, R., F. Ferrucci, G. Gaudiosi, D. Lo Bascio, and G. Ventura, 1997, Malta escarpment and Mt. Etna: Early stages of an asymmetric rifting process? Evidences from geophysical and geological data: Acta Vulcanologica, v. 9, p. 45–53.
Dalla Casa, G., E. Henking, M. Medici, M. Perego, P. Rossi, M. Sattanino, and A. Vitaliani, 1981, The occurrence of heavy crude and tar sands in Italy, *in* R. F. Meyer and C. T. Steele, eds., The future of heavy crude and tar sands: New York, McGraw-Hill, p. 97–106.
Dal Piaz, G. V., and R. Nicolich, 1991, Carta della Moho e lineamenti tettonici, in Structural Model of Italy: Consiglio Nazionale delle Ricerche Geodinamica, scale 1:500,000, 2 sheets.
Dart, C. J., D. W. J. Bosence, and K. R. McClay, 1993, Stratigraphy and structure of the Maltese graben system: Journal of the Geological Society (London), v. 150, p. 1153–1166.
De Capoa, P., F. Guerrera, V. Perrone, F. Serrano, and M. Tramontana, 2000, The onset of syn-orogenic sedimentation in the flysch basin of the Sicilian Maghrebids: State of the art and new biostratigraphic constraints: Eclogae Geologicae Helvetiae, v. 93, p. 65–79.
Della Vedova, B., S. Bellani, G. Pellis, and P. Squarci, 2001, Deep temperatures and surface heat-flow distribution, *in* G. B. Vai and I. P. Martini, eds., Anatomy of an orogen, the Apennines and adjacent Mediterranean basins: Dordrecht, Kluwer Academic Publishers, p. 65–76.
Di Geronimo, I., F. Ghisetti, F. Lentini, and L. Vezzani, 1978, Lineamenti neotettonici della Sicilia orientale: Memorie della Società Geologica Italiana, v. 19, p. 543–549.
Di Stefano, P., and F. Vitale, 1993, Carta Geologica dei Monti Sicani Occidentali: Palermo, Università Degli Studi di Palermo, Dipartimento dei Geologia e Gedesia, scale 1:50,000, 1 sheet.
Doglioni, C., F. Innocenti, and G. Mariotti, 2001, Why Mt. Etna? Terra Nova, v. 13, p. 25–31.
Finetti, I., 1981, Geophysical study on the evolution of the Ionian Sea, *in* F. C. Wezel, ed., Sedimentary basins of Mediterranean margins: Consiglio Nazionale delle Ricerche, p. 465–484.
Finetti, I., 1985, Structure and evolution of the central Mediterranean (Pelagian and Ionian) Sea, *in* D. J. Stanley and F. C. Wezel, eds., Geological evolution of the Mediterranean basins: New York, Springer-Verlag, p. 215-230.
Gasparini, C., G. Iannacone, P. Scandone, and R. Scarpa, 1985, Seismotectonics of the Calabrian Arc: Tectonophysics, v. 84, p. 267–286.
Gillot, P. Y., G. Kieffer, and R. Romano, 1994, The evolution of Mount Etna in the light of potassium-argon dating: Acta Vulcanologica, v. 5, p. 81–87.
Grasso, M., 1999, Carta Geologica del Settore Centro-Meridionale dell'Altopiano Ibleo (Provincia di Ragusa, Sicilia Sud-Orientale): Instituto di Geologia e Geofisica, Università di Catania: Firenze, scale 1:50,000, 1 sheet.
Groupe Escarmed, 1983, Exemples de sédimentation condensée sur les escarpements de la Mer Ionienne (Méditerranée orientale): Obervations à partir du submersible "Cyana.": Revue Institut Français du Pétrole, v. 38, p. 427–438.
Guarnieri, P., D. Puglisi, and K. Balogh, 2000, Age and provenance of the pillow basalt exotics within the Eocene Imerese succession (north-western Sicily): Memorie della Società Geologica Italiana, v. 55, p. 235–241.
Guilhaumou, N., and C. Larroque, 1993, Organic matter associated to warm fluid circulation in hydraulic breccias (abs.): Terra Abstracts, v. 5, p. 458.
Guilhaumou, N., C. Larroque, E. Nicot, F. Roure, and J.-F. Stephan, 1994, Mineralized veins resulting from fluid flow in decollement zones of the Sicilian prism: Evidence from fluid inclusions: Bulletin de la Société Géologique de France, v. 5, p. 425–436.
Gvirtzman, Z., and A. Nur, 1999, The formation of Mount Etna as the consequence of slab rollback: Nature, v. 401, p. 782–785.
Harland, W. B., R. L. Armstrong, A. V. Cox, L. G. Craig,

A. G. Smith, and D. G. Smith, 1989, A geologic time scale: Cambridge, Cambridge University Press, 263 p.

Holton, J., 1999, Southern Apennines success bodes well for potential off southern Italy: Oil & Gas Journal, v. 97 (December 6, 1999), no. 48, p. 65–68.

Illies, J. H., 1981, Graben formation: The Maltese Islands, a case history: Tectonophysics, v. 73, p. 151–168.

Jongsma, D., J. M. Woodside, G. C. P. King, and J. E. van Hinte, 1987, The Medina wrench: A key to the kinematics of the central and eastern Mediterranean over the past 5 Ma: Earth and Planetary Science Letters, v. 82, p. 87–106.

Kafka, F. T., and R. K. Kirkbride, 1959, The Ragusa oil field, Sicily: Proceedings of the 5th World Petroleum Congress, v. 5, p. 233–257.

Kastens, K., et al., 1988, ODP Leg 107 in the Tyrrhenian Sea: Insights into passive margin and back-arc basin evolution: Geological Society of America Bulletin, v. 100, p. 1140–1156.

La Bella, G., L. Bertelli, and L. Savini, 1996, Monte Alpi 3D, a challenging 3-D survey in the Apennine range, southern Italy: First Break, v. 14, p. 285–294.

Langseth, M. G., and J. C. Moore, conveners, 1990, Role of fluids in sediment accretion, deformation, diagenesis, and metamorphism in subduction zones: Journal of Geophysical Research, v. 95, p. 8737–9147.

Larroque, C., 1993, L'accrétion tectonique de sediments: Modalités géométriques et cinétiques, paléo-circulations de fluides: Ph.D. thesis, University of Nice-Sophia Antipolis, Nice, France, 286 p.

Lentini, F., 1984, Garta geologica della Sicilia sud-orientale: Instituto di scienze della Terra, Università di Catania: Firenze, scale 1:100,000, 1 sheet.

Lindsey, D. D., J. R. Kirkpatrick, and P. W. Choquette, 2002, A very unusual platform-edge carbonate mound reservoir— Irminio oil field, SE Sicily (abs): Paper F07, 64th European Association of Geoscientists and Engineers Conference and Exposition, Florence, Italy, May 27–30, 2002, CD-ROM.

Lucido, G., P. M. Nuccio, M. Valenza, and G. Gunta, 1978, Magmatism in the Sicano basin (Sicily) related to Meso-Cenozoic tectonics of the North-African palaeomargin: Mineralogica et Petrographica Acta, v. 22, p. 55–69.

Maccarrone, M., L. Montanari, and P. Pino, 2000, Stratigraphic characters of the Miocene in the Peloritani Mountains (NE Sicily): Memorie della Società Geologica Italiana, v. 55, p. 243–249.

Mattavelli, L., M. Pieri, and G. Groppi, 1993, Petroleum exploration in Italy: A review: Marine and Petroleum Geology, v. 10, p. 410–425.

Mazzoli, S., and M. Helman, 1994, Neogene patterns of relative plate motion for Africa–Europe: Some implications for recent central Mediterranean tectonics: Geologische Rundschau, v. 83, p. 464–468.

McGuire, W. J., I. S. Stewart, and S. J. Saunders, 1997, Intra-volcanic rifting at Mount Etna in the context of regional tectonics: Acta Vulcanologica, v. 9, p. 147–156.

Meloni, A., L. Alfonsi, F. Florindo, L. Sagnotti, F. Speranza, and A. Winkler, 1997, Neogene and Quaternary geodynamic evolution of the Italian peninsula: The contribution of paleomagnetic data: Annali di Geofisica, v. XL, p. 705–727.

Monaco, C., P. Tapponnier, L. Tortorici, and P. Y. Gillot, 1997, Late Quaternary slip rates on the Acireale–Piedimonte normal faults and tectonic origin of Mt. Etna (Sicily): Earth and Planetary Science Letters, v. 147, p. 125–139.

Morelli, C., 1985, Geophysical contributions to knowledge of the Mediterranean crust, *in* D. J. Stanley and F. C. Wezel, eds., Geological Evolution of the Mediterranean basins: New York, Springer-Verlag, p. 65-82.

Nicolich, R., 2001, Deep seismic transects, *in* G. B. Vai and I. P. Martini, eds., Anatomy of an orogen, the Apennines and adjacent Mediterranean basins: Dordrecht, Kluwer Academic Publishers, p. 47–52.

Novelli, L., D. H. Welte, L. Mattavelli, M. N. Yalçin, D. Cinelli, and K. J. Schmitt, 1988, Hydrocarbon generation in southern Sicily. A three-dimensional computer aided basin modeling study: Organic Geochemistry, v. 13, p. 153–164.

Ocampo, R., A. Riva, J. M. Trendel, J. Riolo, H. J. Callot, and P. Albrecht, 1993, Petroporphyrins as biomarkers in oil-oil and oil-source rock correlations: Energy and Fuels, v. 7, p. 191–193.

Oldow, J. S., J. E. T. Channell, R. Catalano, and B. d'Argenio, 1990, Contemporaneous thrusting and large-scale rotations in the western Sicilian fold and thrust belt: Tectonics, v. 9, p. 661–681.

Patacca, E., P. Scandone, G. Giunta, and V. Liquori, 1979, Mesozoic paleotectonic evolution of the Ragusa zone (southeastern Sicily): Geologica Romana, v. 18, p. 331–369.

Pedley, H. M., 1990, Syndepositional tectonics affecting Cenozoic and Mesozoic deposition in the Malta and SE Sicily areas (central Mediterranean) and their bearing on Mesozoic reservoir development in the N Malta offshore region: Marine and Petroleum Geology, v. 7, p. 171–180.

Pepe, F., G. Bertoni, F. Cella, and E. Marsella, 2000, Rifted margin formation in the south Tyrrhenian Sea: A high-resolution seismic profile across the north Sicily passive continental margin: Tectonics, v. 19, p. 241–257.

Pieri, H. M., and L. Mattavelli, 1986, The geological framework of Italian petroleum resources: AAPG Bulletin, v. 70, p. 103–130.

Pion-Leflaive, E., J. Rozemond, and A. Torrez, 1990, Déformation et sedimentation associée au prisme d'accrétion sicilien: Institut Français du Pétrole, Ecole Nationale Superieure de Physique de Marseille internal report 37764, 80 p.

Renda, P., E. Tavarnelli, M. Tramutoli, and E. Gueguen, 2000, Neogene deformations of northern Sicily, and their implications for the geodynamics of the southern Tyrrhenian Sea margin: Memorie della Società Geologica Italiana, v. 55, p. 53–59.

Rigo de Righi, F., 1957, Olistostromi neogenici in Sicilia: Bollettino della Societa Geologica Italiana, v. 75, p. 185–214.

Rigo, M., and F. Barbieri, 1959, Stratigrafia pratica applicata in Sicilia: Bollettino del Servizio Geologico d'Italia, v. 80, p. 351-441.

Rocchi, S., G. Langaretti, L. Ferrari, and D. Carniolo, 1996, Evoluzione del magmetismo nel sottosuolo della Sicilia sud-orientale: Dati sul chimismo dei clinpirosseni: Memorie della Società Geologica Italiana, v. 51, p. 1101–1113.

Rocco, P., 1959, Gela in Sicilia, un singolore campo petrolifero: Rivisita Mineraria Siciliana, v. 10, p. 167–188.

Ronchi, P., F. Lottaroli, and T. Ricchiuto, 2000, Sedimentary and diagenetic aspects of the Liassic Inici Fm. and its stratigraphic context (Sicily Channel, Italy): Memorie della Società Geologica Italiana, v. 55, p. 261–269.

Roure, F., D. G. Howell, C. Müller, and I. Moretti, 1990, Late Cenozoic subduction complex of Sicily: Journal of Structural Geology, v. 12, p. 259–266.

Roure, F, R. Swennen, and P. Casero, 2002, Tectonics and petroleum systems of the Sicilian foreland fold and thrust belt: Field Trip Guidebook, American Association of Petroleum Geologists–Institut Français du Pétrole Hedberg Research Conference on Deformation, Fluid Flow and Reservoir Appraisal in Foreland Fold and Thrust Belts, Palermo, May 2002.

Scandone, P., E. Patacca, R. Radoicic, W. B. Ryan, M. B. Cita, M. Rawaon, H. Chezar, E. Miller, J. McKenzie, and S. Rossi, 1981, Mesozoic and Cenozoic rocks from Malta escarpment (Central Mediterranean): AAPG Bulletin, v. 65, p. 1299–1319.

Schmidt di Friedberg, P., 1965, Lithostratigrafia petrolifera della Sicilia: Rivista Mineraria Sicilia, v. 15, p. 3–43.

Serri, G., F. Innocenti, and P. Manetti, 2001, Magmatism from Mesozoic to Present: Petrogenesis, time-space distribution and geodynamic implications, *in* G. B. Vai and I. P. Martini, eds., Anatomy of an orogen, the Apennines and adjacent Mediterranean basins: Dordrecht, Kluwer Academic Publishers, p. 77–103.

Shipley, T. H., et al., 1995, Proceedings of the Ocean Drilling Program, Initial Reports, p. 156.

Speranza, F., R. Maniscalco, M. Mattei, A. Di Stefano, R. W. H. Butler, and R. Funiciello, 1999, Timing and magnitude of rotations in the frontal thrust systems of southwestern Sicily: Tectonics, v. 18, p. 1178–1197.

Speranza, F., R. Maniscalco, M. Mattei, and R. Funiciello, 2000, Paleomagnetism in the Sicilian Maghrebides: A review of the data and implications for the tectonic styles and shortening estimates: Memorie della Società Geologica Italiana, v. 55, p. 243–249.

Stampfli, G. M., J. Mosar, P. Favre, A. Pillevuit, and J.-C. Vannay, 2001, Permo-Mesozoic evolution of the western Tethys realm: The Neotethys east-Mediterranean basin connection, *in* P. Ziegler, W. Cavazza, A. Robertson, and S. Crasquin-Soleau, eds., Peri-Tethyan rift/wrench basins and passive margins: Peri-Tethys Memoir 6: Mémoire Muséum National d'Historie Naturelle, v. 186, p. 51–108.

Zappaterra, E., 1994, Source-rock distribution model of the peri-Adriatic region: AAPG Bulletin, v. 78, p. 333–354.

21

Kirkwood, D., M. Lavoie, and J. S. Marcil, 2004, Structural style and hydrocarbon potential in the Acadian foreland thrust and fold belt, Gaspé Appalachians, Canada, *in* R. Swennen, F. Roure, and J. W. Granath, eds., Deformation, fluid flow, and reservoir appraisal in foreland fold and thrust belts: AAPG Hedberg Series, no. 1, p. 412–430.

Structural Style and Hydrocarbon Potential in the Acadian Foreland Thrust and Fold Belt, Gaspé Appalachians, Canada

D. Kirkwood
Département de Géologie et de Génie Géologique, Université Laval, Québec, Canada

M. Lavoie
Département de Géologie et de Génie Géologique, Université Laval, Québec, Canada

J. S. Marcil
Junex Inc., Québec, Canada

ABSTRACT

For more than 150 yr, the Gaspé Peninsula in the northern Appalachians has generated interest for oil and gas exploration. Although oil shows and seeps are abundant, oil production has been minor. The hydrocarbon potential is restricted to Late Ordovician to Early Devonian rocks of the Gaspé belt, and different play concepts have been explored over the years. The recent discovery of the Junex-Galt natural gas field has renewed interest for onshore hydrocarbon exploration in this area of eastern Canada.

We propose a new structural model for the Middle Devonian Acadian orogeny involving development of an early foreland thrust and fold belt with deformation accommodated by folding, considerable tectonic wedging, blind north-verging thrusts, south-verging backthrusts, and a possible triangle zone followed by strike-slip faulting that partially dissected the thrust and fold belt. Previous interpretations of the tectonic evolution of the Gaspé belt during the Acadian orogeny was that of a tranpressional orogen basically characterized by folds and strike-slip faults compatible with a wrench tectonic model. Instead, new field work and seismic data that were acquired by the Ministère des Richesses Naturelles du Québec (2000–2001) and presented here suggest enough shortening during the initial compressive phase to produce south-verging folds and south-directed motion along north-dipping reverse faults.

DOI:10.1306/1025703H13127

The fold and thrust belt structural style proposed in this chapter offers a promising geological setting for potential structural traps and new hydrocarbon plays in the entire Gaspé Peninsula. Highly fractured zones are expected in domal structures above important blind thrusts and at thrust tips in areas of fault-propagation folding. Hydrocarbon migration has been documented through fractures in the Upper Ordovician to Lower Devonian Gaspé belt succession at different stages of its tectonic history. Major anticlines in the northeastern part of the Connecticut Valley–Gaspé synclinorium, i.e., Mississippi, Bald Mountain, and Holland anticlines, could be fault propagation located above blind thrusts. The Junex-Galt natural gas field is a good example of such a tectonic setting. Gas is trapped in a folded and fractured Devonian limestone that is located between two regional strike-slip faults at a depth of about 2250 m (7400 ft). Geochemical analyses from recovered oils in the area indicate a Devonian as well as an Ordovician origin very similar to oils from western Newfoundland. A large Ordovician anticline, which was imaged on old and newly acquired seismic lines, underlie the Galt area and represent deeper plays located above blind thrusts deep within the Acadian fold and thrust belt.

INTRODUCTION

Since the mid-1800s, eastern Gaspé Peninsula has generated interest for oil and gas exploration (see Lavoie and Bourque, 2001, for a more detailed account of hydrocarbon exploration in the Gaspé). It is noteworthy to mention that the work of William Logan in the early 1840s in the Gaspé Peninsula (Logan, 1844) was instrumental in focusing hydrocarbon exploration on structural anticlines. Howell (1932) mentioned the importance of Logan's early work in the Gaspé in his classic paper on the historical development of the structural theory of accumulation of oil and gas. According to Howell (1932), following the discovery of oil in the Drake well in Pennsylvania, both T. S. Hunt and H. D Rogers recalled Logan's observations that seepages and sinks were located on anticlines in northeastern Gaspé (Rogers, 1860; Hunt 1861). This idea inspired both of them to formulate the anticlinal theory for hydrocarbon accumulation in the early 1860s (see details in Howell, 1932). Although more than 100 oil shows and seeps have been identified and almost 80 wells have been drilled, eastern Gaspé still remains an underexplored area because of its remote setting and faulted host rocks that complicate both exploration and production. Hydrocarbon production has been minor, and no major oil discoveries have been made so far.

The first wells drilled in Canada were located just a few kilometers west of the town of Gaspé, in the folded, faulted, and fractured Devonian Gaspé sandstones. From 1850 to 1950, approximately 70 shallow holes were drilled in eastern Gaspé, and oil was recovered from some of these. From the 1960s onward, hydrocarbon exploration efforts in the Gaspé were dramatically reduced because of the exploration boom in Western Canada. Nonetheless, detailed onshore seismic surveys in eastern Gaspé and Anticosti Island, as well as offshore surveys in the Gulf of St. Lawrence, were carried out in the 1970s by the Société Québecoise d'Initiative Pétrolière and various industry partners. As a result of the 427-km (265-mi) onshore seismic reflection survey, six deep wells were drilled to test different hydrocarbon play concepts in eastern Gaspé, and a limited amount of oil was produced. In the mid-1990s, renewed interest for hydrocarbon exploration in the Gaspé Peninsula followed a significant oil discovery in western Newfoundland in the Port-au-Port area. Many oil companies and junior exploration companies acquired exploration rights in the Gaspé Peninsula.

In the early 1990s, Jaltin Inc. (now Junex Inc.) worked their Galt prospect in northeastern Gaspé, and in 1993, they reentered the Galt 1 well. Their efforts led to the development of the Galt field, a 500-mmcf/80-ac (32-ha) gas field in production since 2001. Following the discovery of notable oil shows in open fractures in rocks of the Gaspé belt in western Gaspé in the late 1990s, the Québec government (Secteur Énergie of the Ministère des Richesses Naturelles du Québec) conducted a regional seismic survey over a 3-yr time span. Almost 400 km (250 mi) of new seismic reflection data were acquired from October 2000 to July 2002, providing excellent transects that covered the entire peninsula from north to south (Figure 1). The Québec government regional seismic survey and the discovery of the Junex-Galt natural gas field stimulated the renewed interest in onshore hydrocarbon exploration in this area of eastern Canada.

The hydrocarbon potential in the Gaspé Peninsula is restricted to Late Ordovician to Early Devonian rocks of the Gaspé belt, and various play concepts have been tested over the years (Lavoie and Bourque, 2001). Rocks

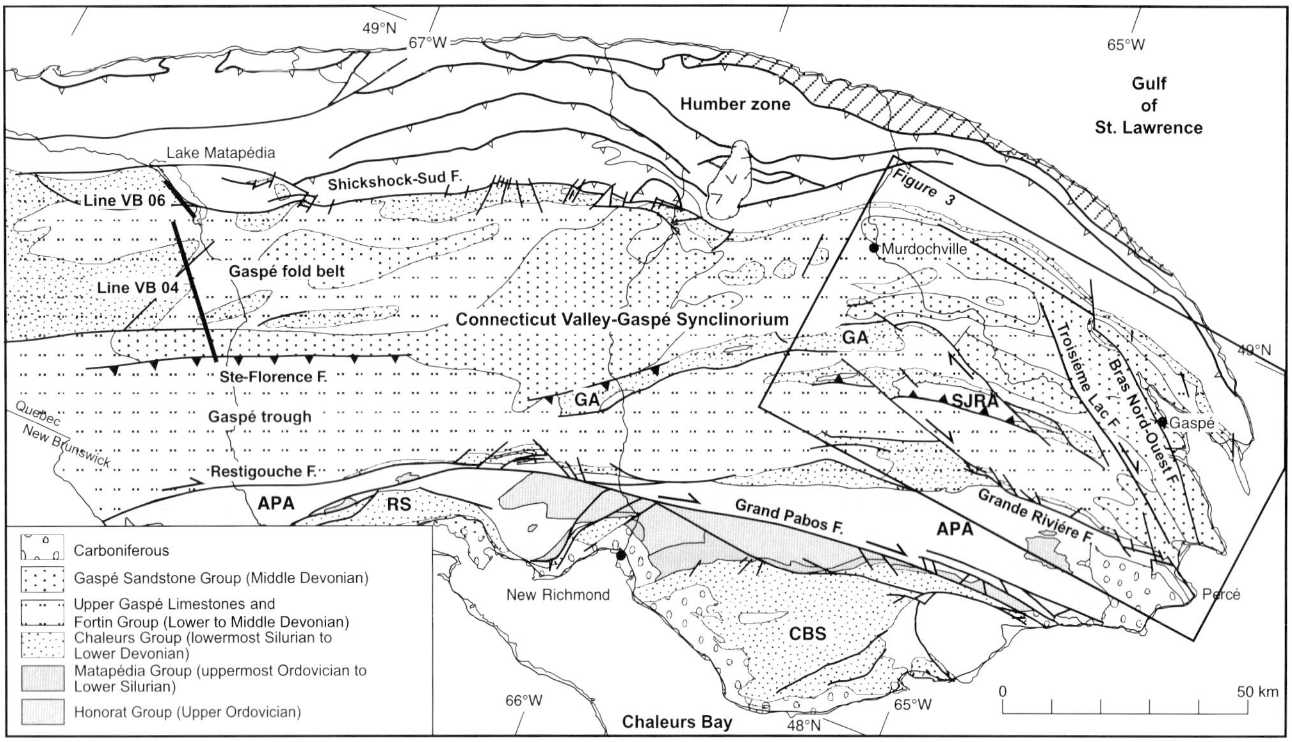

FIGURE 1. Simplified geological map of the Gaspé Peninsula, northern Appalachians, highlighting the distribution of Upper Ordovician to Upper Devonian rocks. Cambrian to Ordovician Humber zone allochthons are in white. ▨ Middle Ordovician syntectonic flysch of the Humber zone. Taconian thrusts are identified with white triangles, Acadian thrusts with black filled triangles. Updated from Bourque et al. (2001, their figure 2). Note location of MRNQ (Ministère des ressources naturelles du Québec) government seismic lines VB 06 and VB 04 in western Gaspé. APA = Aroostook-Percé anticlinorium; CBS = Chaleurs Bay synclinorium; SJRA = Saint-Jean River anticline; RS = Ristigouche syncline, GA = Gastonguay anticline.

of the Gaspé belt were deformed mainly during the Devonian Acadian orogeny, which was dominated by dextral transpression in the Gaspé (Malo and Kirkwood, 1995). Compression in the transpressive belt was initially manifested by widespread folding and the development of reverse faults. Continued compression resulted in dextral transcurrent faulting along steeply dipping strike-slip faults (Kirkwood, 1999). In this chapter, we present new insights on the structural style of the Gaspé belt and propose a new structural model for the early Acadian orogeny involving the development of a foreland thrust and fold belt. Our interpretation differs from that of previously published models primarily because of our detailed structural mapping in northeastern Gaspé, our use of recently released public-domain seismic reflection data from the Québec government in western Gaspé, and unpublished industry three-dimensional seismic data. The new model offers a promising geological setting for potential structural traps and new hydrocarbon plays in the entire Gaspé Peninsula. New regional and detailed field work in the Saint-Jean River anticline in eastern Gaspé helped define the structural geometry of the area (Lavoie and Kirkwood, 2002). This regional south-facing anticlinorium is bounded to the south by a north-dipping backthrust that is linked to an underlying north-directed blind thrust. This tectonic wedging geometry is typical of a thrust and fold belt style of deformation. Structural relationships show that strike-slip faulting postdates the development of the fold and thrust belt. Strike-slip faults compartmentalize the large-scale fold structures.

The Junex-Galt natural gas field is a good example of such a structural setting. Gas is trapped in a folded and fractured Devonian limestone that is located between two regional strike-slip faults at a depth of about 2250 m (7400 ft). A large, anticlinorial structure located above blind thrusts that imaged on old and newly acquired seismic lines underlies the Galt area at a depth of more than 3000 m (10,000 ft) and represents more than 2000 m (6600 ft) of folded, faulted, and fractured rocks.

REGIONAL CONTEXT

The Appalachian orogen is made up of deformed Paleozoic rocks that have been broadly divided into five Cambrian–Ordovician tectonostratigraphic zones defined by Williams (1979) and Williams et al. (1988)

and post-Taconian deformed belts (see Williams, 1995). In the Gaspé Peninsula, the Humber zone, the westernmost of the tectonostratigraphic zones, mainly crops out in the northern part of the peninsula. The Humber zone consists of a folded belt tectonically emplaced over a Cambrian–Ordovician passive-margin sequence during the Taconian orogeny (St-Julien and Hubert, 1975). The fold and thrust belt includes allochthonous Cambrian–Ordovician slope and rise deposits of the Laurentian continental margin divided into four main thrust sheets and an overlying syntectonic Middle to Late Ordovician foreland basin succession (Figure 2). The passive-margin succession is exposed on land farther to the northeast on Anticosti Island (Figure 2), where the Romaine Formation, the Upper Cambrian to Lower Ordovician carbonate shelf, rests on a southward-dipping Grenvillian-type crust and is overlain by a shallow-marine Middle Ordovician to Lower Silurian succession (Sanford, 1993). This succession comprises carbonates and argillaceous limestones (the Mingan, Vauréal, Black River, Trenton, Ellis Bay, Becsie, Gun River, Jupiter, and Chicotte formations) as well as the Caradocian black shales of the Macasty Formation (Figure 2). The Taconian orogenic event occurred during the Upper Ordovician and was recently dated as having occurred from 459 to 446 Ma in the Gaspé Peninsula (Pincivy et al., 2003). Logan's line crops out in northeastern Humber zone and represents the northernmost limit of tectonically transported thrust sheets. The Cambrian–Ordovician rocks of the Humber zone were subsequently deformed by Middle Devonian folds and faults, although their effect is still regarded as minor compared to the Ordovician structures (Lynch and Arsenault, 1997). Fault slices and slivers of Cambrian–Ordovician ophiolitic rocks, melanges, arc, and backarc volcanics and forearc basin deposits of the oceanic Dunnage tectonostratigraphic zone occur to the south of the Humber zone along the Shickshock fault.

Post-Taconian rocks of the Gaspé Peninsula occur in the Gaspé belt, a deformed successor basin (Bourque et al., 1995) that overlies the exposed southern part of the Humber zone. This basin developed following the Taconian orogeny and comprises Late Ordovician to Middle Devonian clastic and carbonate rocks (Bourque et al., 2000). In northeastern and northwestern Gaspé, Silurian strata of the Gaspé belt unconformably overlie the Ordovician folded belt (Malo and Bourque, 1993). In north-central Gaspé, the contact between both zones occurs along the Shickshock Sud fault, a regional fault with a long-lived, complex faulting history from thrusting during the Ordovician Taconian orogeny, to normal faulting during the Silurian to Devonian time span, to oblique strike-slip faulting during the Middle Devonian Acadian orogeny (Malo and Kirkwood, 1995). Flat-lying Carboniferous rocks unconformably overlie the Silurian–Devonian succession in southern Gaspé Peninsula (Figure 1).

The Gaspé belt comprises four broad temporal and lithological rock assemblages (Bourque et al., 2000) (Figure 2). (1) The oldest assemblage consists of Upper Ordovician–lowermost Silurian deep-water fine-grained siliciclastic and carbonate facies of the Honorat and Matapedia Groups (Malo, 1988). (2) The following assemblage is characterized by shallow- to deep-shelf facies of the Chaleurs Group deposited during the Silurian to earliest Devonian time interval (Bourque et al., 1993). (3) The next assemblage consists of Lower Devonian mixed siliciclastic and carbonate fine-grained outer shelf to basin facies of the Upper Gaspé limestones and Fortin Groups (Lavoie, 1992). (4) Finally, upper Lower to Middle Devonian nearshore to terrestrial coarse-grained facies of the Gaspé sandstones make up the youngest assemblage. During the Early Devonian, the successor basin evolved into a foreland basin that developed in front of the Acadian orogenic wedge (Malo, 2001). The Devonian lithostratigraphic succession of the Upper Gaspé Limestone, Fortin Group, and Gaspé Sandstone can be viewed as a foreland basin sequence comprised of a Pragian lower argillaceous limestone-dominated unit, the Forillon, Shiphead, and Indian Cove formations, partially coeval with or followed by a lower Emsian turbidite-dominated unit, the Fortin Group, and the York Lake and York River formations, capped by a middle to late Emsian molasse made up of coarser grained sandstones and conglomerates of the Battery Point and Malbaie formations.

STRUCTURAL EVENTS IN THE GASPÉ BELT

Three orogenic events affected the rocks of the Gaspé Peninsula: the Taconian orogeny (St-Julien and Hubert, 1975), the Salinic disturbance (Boucot, 1962), and the Acadian orogeny (Malo and Bourque, 1993). Rocks of the Gaspé belt were mainly deformed during the Acadian orogeny, although they also record slight instabilities attributed to the Salinic disturbance (Bourque et al., 2000). The Gaspé belt is separated into three structural domains that are characterized by differing stratigraphic and structural features. These are, from north to south, the Connecticut Valley–Gaspé synclinorium, the Aroostook-Percé anticlinorium, and the Chaleurs Bay synclinorium (Figure 1). The structure of the belt is dominated by regional northeast-trending Acadian folds, with fold wavelengths and interlimb angles that differ in each domain and a cleavage that also varies in intensity in the different domains (Kirkwood, 1999). Regional faults either border the domains (e.g., the Shickshock Sud fault to the north) or cut across them (e.g., the Grand Pabos fault in southern Gaspé) (Figure 1).

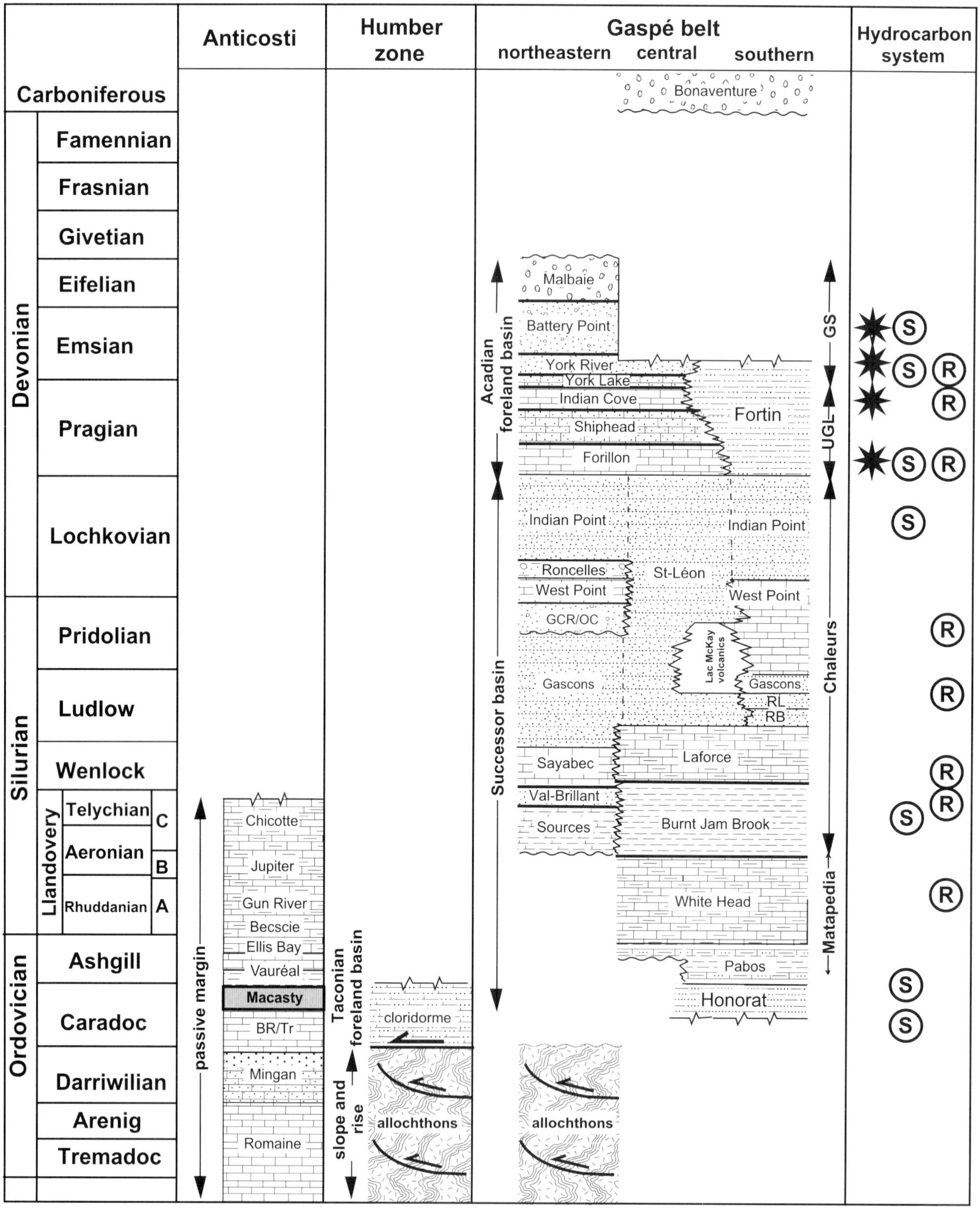

FIGURE 2. Generalized stratigraphic columns for northeastern Québec Appalachians showing the passive-margin sequence (Anticosti Island) and folded belts (Humber zone and Gaspé belt). Hydrocarbon system is indicated with documented reservoir rocks (R), oil occurrences (✷), and source rocks (S). Updated from Bourque et al. (2000, their figure 3), Slivitzky et al. (1991), and Sandford (1993). GCR/OC = Griffon Cove River/Owl Capes Mb.; RL = Ruisseau Louis Mb.; RB = Ruisseau Bleu Mb.; UGL = Upper Gaspé Limestone; BR/Tr = Black River/Trenton formations; GS = Gaspé Sandstone.

The Salinic event is best recorded in the Gaspé belt succession of northeastern Gaspé (Bourque et al., 2000). It induced normal faulting, minor folding, and a resulting unconformity that, although it is only locally observed, can be very important, cutting unequally into the Late Ordovician and Silurian sequence, in places indicating almost 1500 m (500 ft) of differential uplift (Bourque et al., 1993). The unconformity is clearly dated as Late Silurian (late Ludlow to Pridolian) (Bourque et al., 1993; Malo and Bourque, 1993). The proposed model for the Salinic event in the Gaspé Peninsula is based on sedimentological analysis of the Gaspé belt rocks in the northeastern part of the peninsula (Bourque, 1977; Lavoie, 1992) and relates the Salinic unconformity to the interplay between normal faulting and sea level changes (Bourque, 2001). Topographically high blocks were eroded, and erosional surfaces occurred in marginal basin facies of the Chaleurs succession of northern Gaspé (Bourque, 2002). Bourque (2001) and Malo (2001) propose that the Late Silurian to Early Devonian period corresponded to an episode of regional extension in Gaspé Peninsula between two compressive events. It was followed by a second tectonic pulse recognized as the Middle Devonian Acadian orogeny (Malo and Kirkwood, 1995). In the northern Appalachians, some workers propose that both phases belong to the same tectonic event, the Acadian orogeny, and that a zone of active deformation migrated across the Laurentian margin from the Upper Silurian (early Ludlow, 423 Ma) to the Upper Devonian (Givetian, 387 Ma) (Bradley et al., 1998). Nevertheless, because of their far-foreland position in the Québec reentrant with respect to the evolving orogenic wedge, rocks of the Gaspé Peninsula have recorded two spatially and temporally distinct deformation phases (Malo, 2001). Our model offers an alternative explanation (see below) and proposes instead that normal faulting occurred locally in the far foreland in front of an advancing orogenic wedge, whereas folding and thrusting developed coevally in southern Gaspé in the Late Silurian to Early Devonian (Kirkwood et al., 2002).

Although the Salinic event locally affected the Gaspé belt rocks, regional deformational features, such as folds, faults, and cleavage development in the Gaspé belt rocks are attributed to the Middle Devonian Acadian orogeny. Deformation of the Gaspé belt during the Acadian orogeny evolved from an essentially pure shear to dominantly simple shear shortening in a transpressive setting, arising from oblique convergence between the composite Laurentia and Gondwana supercontinents (Kirkwood et al., 1995; Malo and Kirkwood, 1995). The deformation in the transpressive belt was initially accommodated and distributed over the entire area through the development of contractional structures, such as folds, cleavage, and reverse faults (Kirkwood, 1995). Continued deformation brought about further flattening of the folds and resulted in transcurrent faulting along steeply dipping east- to east-northeast-striking brittle-ductile shear zones (e.g., Grand Pabos, Grande Rivière, and Shickshock Sud faults). Most of these regional faults are dextral strike-slip faults (Malo et al., 1992). An exception is the Sainte-Florence fault, a reverse fault that separates the Connecticut Valley–Gaspé synclinorium into the northern Gaspé folded belt and the southern Gaspé trough (Figure 1). The upper limit for the time of displacement along the strike-slip faults is the uppermost Devonian, given by plutons in northern New Brunswick and western Gaspé that are dated at 364 and 368 ± 2 Ma and have metamorphic aureoles that are not disturbed by movement along the Restigouche dextral strike-slip fault (J. Carroll, 2002, personal communication). In eastern Gaspé, two distinct sets of dextral strike-slip faults are present: east-trending dextral strike-slip faults, such as the Grand Pabos and Grande Rivière faults, and northwest-trending dextral strike-slip faults, such as the Bras-Nord Ouest and Troisième Lac faults (Figures 1, 3). Chronological relationships in this area indicate that the east-trending system, e.g., the Grande Rivière fault, is crosscut and displaced by the northwest-trending system, e.g., the Ruisseau Blanc fault, a subsidiary of and a subparallel fault to the Troisième Lac fault (Kirkwood, 1989).

STRUCTURAL STYLE OF THE SAINT-JEAN RIVER ANTICLINE

Structural Features

Regional and detailed field work during the 1995, 1996, 2001, and 2002 field seasons in the Saint-Jean River area in northeastern Gaspé documented the structural style of the lower part of the Gaspé belt succession (Kirkwood et al., 2001; M. Lavoie, 2003, personal communication). The Saint-Jean River anticline, located 30 km (20 mi) southwest of Gaspé in the Gaspé Peninsula, is a 200-km^2 (49,400-ac) structure that exposes the Upper Ordovician to lowermost Silurian White Head Formation and the Lower Silurian to Lower Devonian Chaleurs Group (Figure 3). The anticlinorial structure trends roughly east–west, with its eastern tip originating just west of the Troisième Lac fault and extends 50 km (30 mi) toward the west. It is dissected by a set of southeast-trending dextral strike-slip faults (Figure 3). Detailed mapping has outlined the geometry of the 6–8-km (4–5-mi)-wide anticline, partially truncated along its southern limb by a major east-trending, south-directed reverse fault, the Saint-Jean River fault (Figure 4) (Lavoie and Kirkwood, 2002). The northern limb of the regional structure dips gently to the north, whereas the southern limb dips more steeply to the south. In the core of the anticlinorial structure, argillaceous limestones of

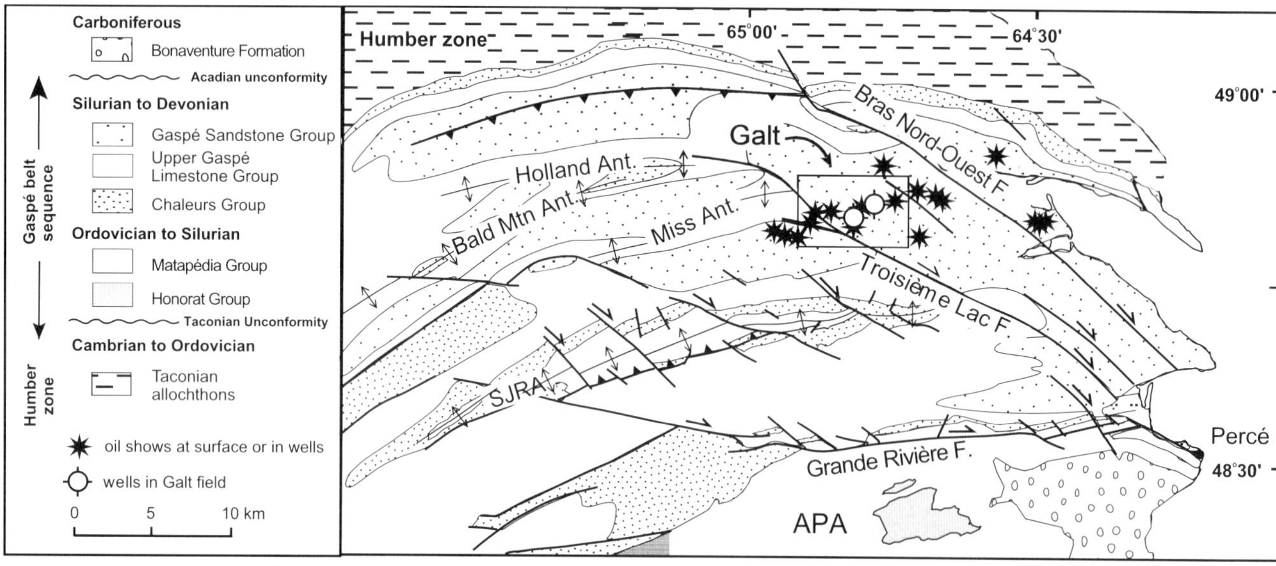

FIGURE 3. Geological map of the Gaspé belt in northeastern Gaspé showing distribution of major anticlines. Location of Galt natural gas field with oil seeps and wells is highlighted in squared area. See location in Figure 1. APA = Aroostook-Percé anticlinorium, SJRA = Saint-Jean River anticline.

the White Head Formation are tightly folded and affected by a pervasive pressure-solution cleavage that strikes N260° and dips 85° to the north.

South-dipping faults are common in the White Head limestones in the core of the anticline. These faults are oriented east–west, with dips ranging between 15 and 80° to the south. The fault planes are discrete, thin surfaces coated with calcite slickenfibers, indicating oblique to north-directed slip. Calcite crack-and-seal slip foliations on the slip surfaces also indicate downdip, bedding-parallel movement. A widely spaced, curvilinear fracture cleavage occurs in the hanging wall of the bedding-parallel faults and disappears in 10–15 cm (4–6 in.). The physical aspect of the fracture cleavage is different from the regional cleavage, a tightly spaced, fold-axial planar dissolution cleavage (Kirkwood et al., 2001). The fault surfaces commonly follow the bedding plane but locally cut up section through the limestones in a pattern indicative of a ramp and flat thrust geometry.

North-dipping faults that trend east–west and dip from 50 to 70° are located along the southern limb of the regional anticline. Mesoscopic fault-related folding, dragging of bedding along the fault planes, and calcite slickenfibers indicate reverse slip along the fault surfaces. The south-directed reverse faults define a deformed zone that extends for 30 km (20 mi) laterally and varies in thickness and in geometry. In the central part of the structure, the zone consists of a single fault plane that cuts up section and juxtaposes Upper Ordovician limestones of the White Head Formation in the hanging wall, with Middle Devonian turbidites of the Fortin Group in the footwall (Figure 4). Dip separation reaches a maximum of 1500 m (5000 ft). Toward the east, multiple fault planes and folds affect the southern limb of the structure (Figure 4), dip separation along the faults decreases, and eventually, deformation is entirely accommodated by folding in the eastern nose of the anticline.

Major northwest-trending dextral strike-slip faults cut across and displace lithological contacts, the axial trace of the anticline, as well as the southern limb reverse faults (Figure 4).The strike-slip faults are parallel to the Troisième Lac fault system. At the outcrop scale, the faulted zones define 100–300-m (330–1000-ft)-wide corridors characterized by numerous mesoscopic brittle-ductile faults and Riedel shear fractures with centimetric to decimetric displacements, an important volume of calcite, and associated veins. "CS-type" fabrics, sigmoidal dragging of bedding, calcite slickenfibers, and calcite steps indicate horizontal to gently plunging slip. Most strike-slip faults display a dextral sense of shear, although some north-northeast-trending faults display a sinistral sense of shear. This geometry is typical of the low-strain zones of Acadian strike-slip faults (Kirkwood and Malo, 1993).

Interpretation

The fold and fault geometry of the Saint-Jean River anticline is that of a south-verging anticline in the hanging wall of a backthrust, itself located above stacked Upper Ordovician to Lower Silurian limestones of the White Head Formation (Figure 4). The south-facing anticline is peculiar, because most folds in eastern Gaspé are either upright or north verging. Likewise, south-directed transport along the reverse faults on the southern limb of the anticline seems to be anomalous. Previous geological interpretations proposed that this

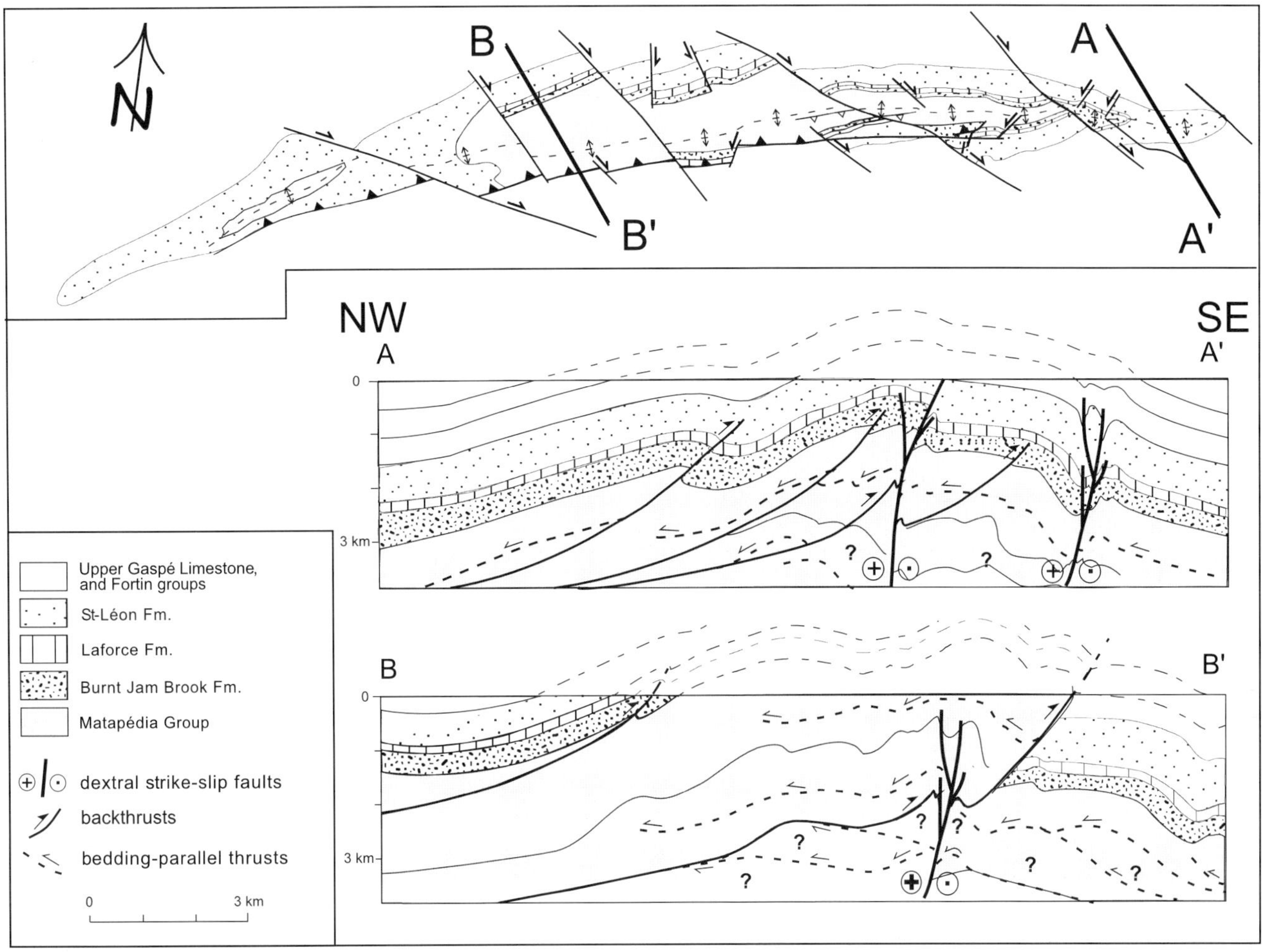

FIGURE 4. Schematic structural cross sections through the Saint-Jean River anticline illustrating structural and chronological relationships between folds, bedding-parallel north-directed faults, backthrusts, and strike-slip faults described in text. Inset: simplified geology of the Saint-Jean River anticline and location of cross sections. ⊙ = toward the viewer; ⊕ = away from the viewer; **?** = basement unknown.

fault was either a subvertical normal fault or an east-trending dextral strike-slip fault (Brisebois, 1981; Simard, 1987, 1988), which developed during Acadian dextral transpression.

The north-directed thrusts in the Saint-Jean River area are very similar to low-angle thrusts documented by Kirkwood et al. (2001) in the Percé area, some 50 km (30 mi) to the south. However, good-quality outcrops in the core of the Saint-Jean River anticline help decipher structural relationships between the different fault systems. The south-dipping fault surfaces in the core of the anticline are clearly folded together with the bedding planes and are crosscut by the strike-slip faults; they are thus pre- to synfolding and precede strike-slip faulting. We interpret these faults as early north-directed, low-angle thrusts that we relate to stacking in the basal part of the Gaspé belt succession (Figure 4). As for the north-dipping reverse faults along the southern limb of the anticline, they are also clearly crosscut by the strike-slip faults. We tentatively suggest that the reverse faults are backthrusts related to tectonic wedging at depth and possibly form the upper structural contact of a relict triangle zone instead of simple backthrusts located at the frontal tip of the fold and thrust belt. Blind thrusting at depth farther to the north imaged on industry seismic lines (Roksandic and Granger, 1981) favors this interpretation. Acadian deformation in the Saint-Jean River anticline area can be thus be separated into two distinct stages: (1) early low-angle thrusting, folding, and backthrusting and (2) dextral strike-slip faulting (Lavoie and Kirkwood, 2002).

EARLY ACADIAN FORELAND FOLD AND THRUST BELT

Based on the structural relationships between folds, faults, and cleavage planes in the Saint-Jean River anticline and on the structural geometry imaged at depth on seismic profiles (see below), we propose a revised structural model involving the development of a foreland fold and thrust belt during an early phase of the Acadian orogeny that preceded the strike-slip deformation.

Newly released seismic profiles from the Québec government's regional survey provide excellent images of the subsurface geometry of the Silurian–Devonian rocks of the Gaspé belt and the Acadian fold and thrust belt. Profiles shot in western Gaspé in 2000 show several reflectors located at depths of 1500–5000 m (5000–16,000 ft) that display structures involving the Early Silurian Val-Brillant sandstone and the Sayabec carbonate ramp that are typical of thin-skinned deformation (Figure 5), including imbricate thrust faults, thrust-propagation folds, duplexes, and a possible triangle zone (Beausoleil et al., 2002; Morin and Laliberté, 2002a). Some of these structural features could be geometrical analogs to the north-directed thrusts and south-directed backthrusts that occur in the basal part of the Gaspé belt succession of the Saint-Jean River anticline (Figure 4). Preliminary interpretation of seismic data along the central Gaspé transect confirms the same structural style as in the western Gaspé belt, i.e., thin-skinned thrusting in the Silurian rocks (Morin and Laliberté, 2002b). The structures imaged on the seismic profiles record significant horizontal shortening in rocks of the Gaspé belt at depth and indicate that broad open folds mapped at the surface in the Connecticut Valley–Gaspé synclinorium formed above blind structures that are located in the deeper units. This structural geometry is much more compatible with a compressional fold and thrust belt than with the transpressive tectonic regime that has previously been invoked for the Acadian orogeny (Malo and Béland, 1989; Malo and Bourque, 1993; Malo and Kirkwood, 1995). The shallow, south-dipping reflector at the northwestern end of the profile, located at the contact between the Gaspé belt succession and the Humber zone and interpreted as an Acadian detachment surface (Figure 5), possibly emerges further north in the Humber zone. Similar south-dipping reflectors can also be interpreted below the Gaspé belt succession, in what is interpreted as the underlying Humber and Dunnage zones. These reflectors could be detachment surfaces in the Taconian fold and thrust belt or, alternatively, Acadian detachments that cut through and restructure the Taconian belt.

Kirkwood et al. (2002) concluded that deformation related to folding and cleavage development during the initial phase of the Acadian orogeny was probably more important than previously reported. The structural geometry of the Saint-Jean River anticline and the occurrence of east-trending thrusts and backthrusts is not compatible with a wrench tectonic model driven by regional east-trending strike-slip faults. As already mentioned, the first stage of the Acadian progressive deformation history of the Gaspé belt is characterized by the sequential development of folds, cleavage, and apparently minor reverse faulting (Kirkwood et al., 1995). The pure-shear component during this first stage of the Acadian deformation accounts for 70% of the total strain (Kirkwood, 1999), and the thrust and fold belt model proposed here offers a more realistic explanation for such a considerable amount of shortening.

The structural model is also supported by microstructural, isotopic, thermal, and fluid-inclusion data (Kirkwood et al., 2001), as well as regional quantitative

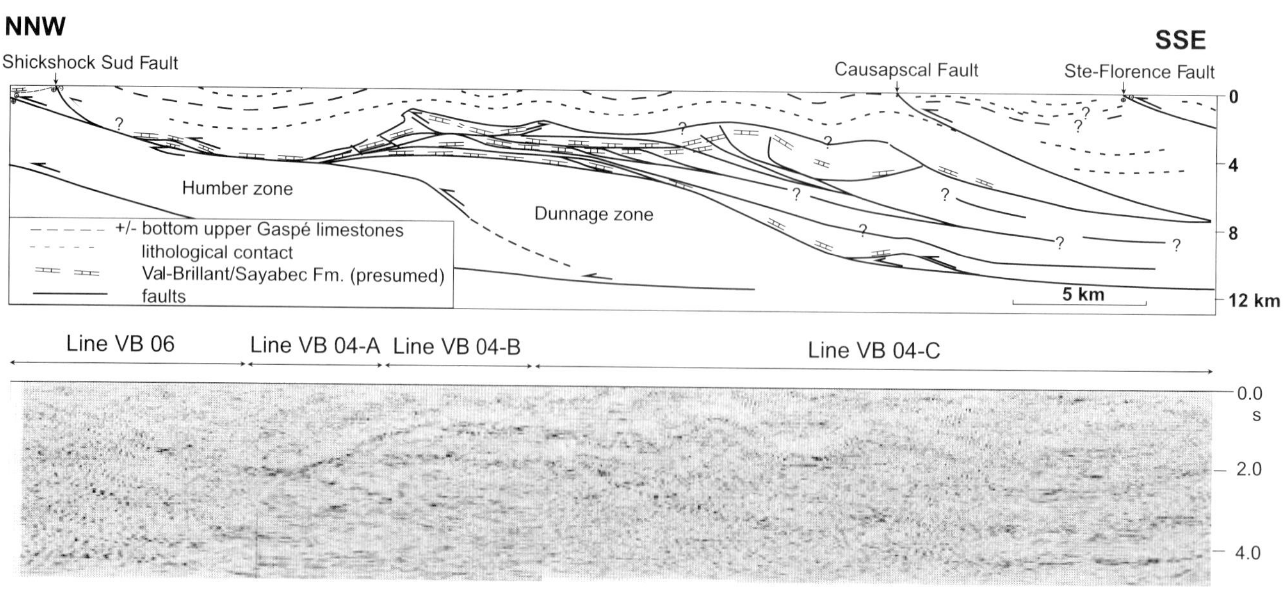

FIGURE 5. Uninterpreted and interpreted Ministère des Ressources Naturelles du Québec seismic lines VB 06, VB 04-A, VB 04-B, and VB 04-C. Prestack two-way traveltimes down to 5 s. See location of lines in Figure 1. Interpretation by C. Morin and J.-Y. Laliberté (2001). Published with permission from the Ministère des Ressources Naturelles du Québec.

strain analysis data (Kirkwood, 1995, 1999). A detailed tectonodiagenetic study in the White Head Formation of northeastern Gaspé, including some samples from the Saint-Jean River anticline, concluded that the Upper Ordovician to Lower Silurian limestones were subjected to shallow to moderate burial at maximum depths of approximately 1500 m (5000 ft), followed by an episode of fracture development, during which the rocks were uplifted approximately 800 m (2600 ft). The uplift was followed by deeper burial, fracturing, folding, and strike-slip faulting, with maximum depths reaching 3800 m (12,500 ft) (Kirkwood et al., 2001). The microthermometric and isotope data point toward an influx of marinelike water at shallow-burial conditions during uplift. Kirkwood et al. (2001) related the uplift to normal faulting during the Salinic event. However, based on the structural features documented in the Saint-Jean River anticline area and features imaged on public and industry seismic lines, we now believe that this episode of fracturing and uplift is related to thrusting and structural stacking during the early Acadian. This is a more realistic mechanism for at least 1 km (0.6 mi) of relative uplift of the Upper Ordovician to Lower Silurian White Head limestones.

TECTONIC EVOLUTION OF THE ACADIAN FORELAND FOLD AND THRUST BELT

Figure 6 illustrates our preferred model for the structural evolution of the Gaspé belt during the Early to Middle Devonian, with emphasis on the basic geometry and sequence of events. This model is consistent with a two-stage history of north-directed thrusting followed by dextral strike-slip faulting. More complete discussions on the tectonosedimentary development of the Gaspé belt can be found in Bourque et al. (2000) and Malo (2001).

During the Late Silurian to Early Devonian, an active compressional orogenic system formed south of the present-day location of the Gaspé Peninsula (Figure 6A) (Malo and Kirkwood, 1995; Bradley et al., 1998) and slightly outboard of northwestern Newfoundland (Waldron and Stockmal, 1994; Stockmal et al., 1998). The Gaspé Basin was a marine successor basin that evolved into a foreland basin located in front of the growing Acadian orogenic wedge (Malo, 2001). Palinspastic reconstruction of the basin in pre-Acadian times shows that the basin extended southeast as far as the present-day location of Cape Breton Island (see figures in Kirkwood, 1999). A transgression at the beginning of the Lochkovian resulted in a deepening-upward sequence represented by the fine-grained sediments and turbidites of the upper part of the Chaleurs Group (Indian Point Formation) and the fine-grained deep-water limestones of the Upper Gaspé Limestone (Bourque et al., 2000). In the southern part of the basin, numerous depocenters with orogen-derived gravels (Harrison Member) formed coevally with reef limestone bodies and terrestrial red beds (Plage Woodmans Member) of the West Point Formation, indicating locally uplifted areas in the late Lochkovian to Pridolian (Bourque et al., 2000). In the northern part of the Gaspé Basin, continent-derived siliciclastics of the St-Léon and Indian Point formations were deposited in the distal foreland and far foreland, thus recording accelerated subsidence of the basin. Deposition was coincident with well-documented drowning of the shelf that continued from the Wenlockian to the Pragian in the Gaspé (see Bourque et al., 2000, their figure 11). We relate the subsidence of the basin to the encroachment of Acadian allochthons farther south, an event that has been dated to range from as early as Upper Ordovician to Upper Silurian in the Newfoundland Appalachians (Stockmal et al., 1998). We also propose that normal faulting in the far foreland of the Gaspé Basin (Figure 6A) was controlled by the flexure of the Laurentian margin because of the tectonic loading typical of foredeeps and foreland basins of all ages (Bradley and Kidd, 1991). Thrusting during growth of the orogenic wedge would have created a high topography to the south, and this could explain the deposition of coarse-grained clastic deposits and the creation of local uplifts. In such tectonically active settings, fault scarps may produce various types of mass gravitational flows in response to the slope instabilities that are characteristic of submarine environments (Martinsen, 1994). However, observations by Bourque et al. (1993) in the upper Chaleurs Group conglomerate bodies indicate that they did not form along a major fault scarp at the front of the wedge but by erosion of local paleotopographic highs instead (Bourque et al., 2000). Thus, the orogenic wedge must have been a considerable distance away from the southern Gaspé Basin.

In Pragian time, two distinct but laterally equivalent assemblages were deposited in the Gaspé Basin. The fine-grained carbonates of the Upper Gaspé Limestone formed on a northern outer shelf in the far foreland, whereas the siliciclastic turbidites of the Fortin Group were deposited as flysch to the south in the deep foreland basin (Bourque et al., 2000). Migration of the orogenic wedge toward the continent was recorded by abundant slumping in the Fortin siliciclastics in the Gaspé Basin. The deformation front probably advanced as far as the southernmost part of the Gaspé Basin, causing imbrication of the Matapedia and Chaleurs Group rocks coeval to sedimentation in the foreland basin (Figure 6B). As suggested by Malo (2001), the tectonically active southern margin of the basin was most probably a topographic high, because there is no record for deposition in this area. Different lines of evidence

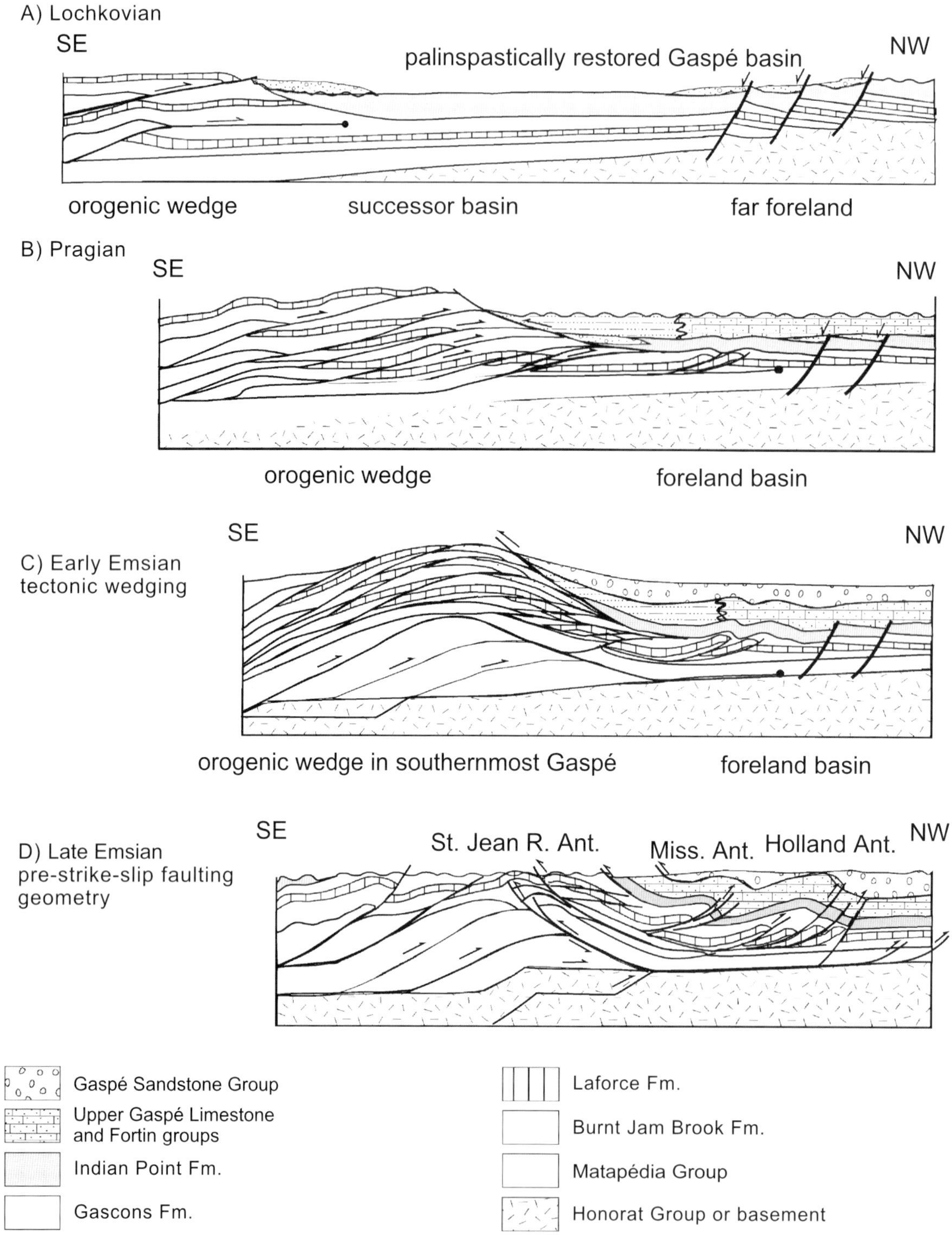

FIGURE 6. Schematic interpretation of sequential development of the Acadian foreland fold and thrust belt, previous to strike-slip faulting. (A) During the Lochkovian, deposition of the fine-grained sediments and turbidites of the upper part of the Chaleurs Group in the successor basin with interbedded orogen-derived gravels to the south. Normal faulting in the far-foreland accelerating basin subsidence to the north. (B) In Pragian times, the growing orogenic wedge migrated toward the continent and approached the southern limit of the Gaspé Basin. Deposition of the Fortin flyschs occurred in the southern part of the basin coeval with the Upper Gaspé Limestone to the north. Imbrication of the lower part of the Gaspé belt sequence occurred. (C) In the early Emsian, tectonic imbrication and stacking occurred in the Upper Ordovician to Silurian successions in the southern Gaspé, with possible development of an anticlinal stack. Backthrusts were developed on the northwestern side of the stack in response to tectonic wedging. (D) Late Emsian folding of foreland basin succession above blind thrusts in the northern Gaspé belt. Detachment surfaces propagate and extend beneath the Humber zone to the north.

point to tectonic instability and possible exhumation in the southernmost part of the basin. The Lower Devonian Archibald Settlement subarial volcanic rocks in southwestern Gaspé indicate deposition in a shallow basin. In the Ristigouche syncline (see Figure 1 for location), Bourque and Lachambre (1980) mapped the Salinic unconformity as an angular unconformity, implying mild folding of the Upper Ordovician to Silurian rocks before deposition of the late Pridolian Chaleurs Group rocks. Bourque et al. (2001) mentioned that erosional surfaces apparently cut as deep as the Matapedia Group, suggesting as much as 1500 m (5000 ft) of differential uplift. Thermal maturation data for the Chaleurs Group rocks in the Chaleurs Bay synclinorium indicate a very low-grade burial metamorphism (Hesse and Dalton, 1991), also suggesting shallow pressure and temperature conditions for the southern Gaspé Basin.

The structural features described in the Saint-Jean River anticline depict stacking in the basal part of the Gaspé belt succession, and tectonic wedging at depth is imaged on government seismic profiles in northern Gaspé (Figure 5). Continued convergence during the early Emsian could have been responsible for tectonic imbrication, stacking, and possible development of duplex structures in the Upper Ordovician to Silurian successions in the southern Gaspé Basin (Figure 6C). The deformation front migrated even farther north as blind thrusts in a tectonic wedge, involving the deeper units of the foreland basin. In the northern and eastern parts of the foreland basin, deep-water carbonates were replaced by shallow-marine to terrestrial sediments of the Gaspé Sandstone Group, while siliciclastic sediments of the Fortin Group were still depositing in the deep foreland basin to the west (Bourque et al., 2001). Deltaic, distal-braided stream, meandering river, and proximal braid-plain deposits that infilled the Gaspé foreland basin imply that the migrating orogenic wedge had advanced well within the Québec reentrant and was located just south of the present-day position of the Gaspé Peninsula. The Emsian thrust and fold belt in southern Gaspé evolved as a dominantly piggyback sequence, with thrusting being older in the south than in the north, as indicated by Lower Devonian deformation in the Ristigouche syncline that is evidenced by the angular unconformity between the Upper Silurian and lowermost Devonian rocks of the Chaleurs Group (Figure 6C).

By late Emsian (Figure 6D), the deformation front had migrated northward beneath the foreland basin at least as far as the northern limit of the Gaspé belt, i.e., the Shickshock Sud fault. It probably extended even farther north in the Humber zone and under the Gulf of St. Lawrence, as suggested by Rocksandic and Granger (1981). Syndepositional normal faults in the far foreland were most probably reactivated as contractional structures at this stage of development of the fold and thrust belt. The structural geometry of the Connecticut Valley–Gaspé synclinorium was probably established during the Emsian with its characteristic thin-skinned structural style, including blind thrusts beneath outcropping anticlines and fault-propagation folds at depth (Figure 6C), as imaged on government seismic lines farther to the west (Figure 5). Deposition of the middle to late Emsian clastic wedge, comprising molassic coarse-grained sandstones and conglomerates of the Battery Point and Malbaie formations, indicates that erosion of the orogen occurred close by. Strike-slip faulting must have occurred in post-Emsian times, because strike-slip faults displace stratigraphic contacts between the upper Emsian Battery Point Formation and lower Eifelian Malbaie formation in northeastern Gaspé (Kirkwood, 1989).

The structure of the Saint-Jean River anticline indicates that older, north-directed thrusts affecting the deeper Matapedia Group limestones were deformed and passively folded by younger, south-directed thrusts and folds (Figure 4). A triangle zone could have been initiated in the early Emsian as transported thrust slices wedged beneath the upper part of the Gaspé belt sequence, creating an anticlinal thrust sheet stack caused by tectonic accretion of younger and deeper thrust units. In such a scenario, the backthrusts documented in the Saint-Jean River anticline can be interpreted as foreland-dipping thrusts that splay upward, at the tip of the thrust wedge, from the tip of a blind thrust, creating a tectonic wedge geometry or a triangle zone. The model provides for deformation that is accommodated mainly by in-sequence thrusting in the Matapedia and Honorat rocks at the base of the Gaspé belt sequence and possibly even by thrusting of underlying basement rocks. Tectonic wedging provides a mechanism for bringing rocks of the Matapedia Group close to the surface in the Saint-Jean River anticline, as required by isotopic and fluid-inclusion data. However, documented slip along the hinterland-directed backthrusts in the Saint-Jean River anticline does not exceed 1500 m (5000 ft). The amount of shortening in the Gaspé belt sequence implied by cross sections (Figures 4, 6) and seismic profiles (Figure 5) must be accommodated by slip along detachment surfaces farther north at depth in the foreland and possibly even along detachment surfaces that imbricate the Humber zone. No evidence exists as of yet to assume both thin-skinned and thick-skinned thrusting in the Gaspé belt as documented in the Appalachian structural front in western Newfoundland (Stockmal et al., 1998).

The structural development proposed here implies that the southern and northern Gaspé belt represent two distinct tectonostratigraphic domains with contrasting stratigraphies and deformation styles. In Figure 6, it is clear that the schematic structural geometry of southern Gaspé is that of a transported domain, although the amount of displacement is uncertain. The present-day structure of the Chaleurs Bay synclinorium, which is characterized by tight folds and thrust faults, is probably a result of compression during thrusting and folding. Horizontal shortening of 30–45% has been estimated in rocks of the Chaleurs Bay synclinorium by cross section restoration of regional folds and faults and by strain computed for cleavage development (Kirkwood, 1999). A detailed strain study of Gaspé belt also supports the contrasting deformation in both domains (Kirkwood, 1999). Based on regional and local three-dimensional strain analysis, shortening values range from 20% in the Gaspé folded belt to as much as 77% south of the Sainte-Florence fault in the Gaspé trough and 65% in the Aroostook-Percé anticlinorium. Stratigraphic evidence for the Chaleurs Group rocks (Bourque et al., 1993) and the Upper Gaspé limestones (Lavoie, 1992) also indicates different paleogeographic settings between northeastern and southern Gaspé, although the boundary between both areas was located along the Grande Rivière fault in eastern Gaspé and its northwestern extension (see Figure 1). Thickening of the Burnt Jam Brook and Saint-Léon formations west of the Grande Rivière fault was interpreted by Bourque et al. (2000) as suggesting higher average subsidence and sedimentation rates with deep-shelf facies of the Chaleurs Group, basinal facies of the Upper Gaspé Limestone, and turbiditic siliciclastics of the Fortin Group depositing west of the Grande Rivière fault. We suggest instead that the boundary between the two domains is located just south of the Saint-Jean River backthrust, where outcropping rocks of the Gaspé belt form a series of transported thrust sheets. We also suggest that the western extension of this structural boundary be located just south of the Gastonguay anticline in central Gaspé and along the Sainte-Florence fault in western Gaspé (see Figure 1).

OVERVIEW OF THE GASPÉ BELT HYDROCARBON SYSTEM

In this part of the Canadian Appalachians, the hydrocarbon potential is restricted to Upper Ordovician to Early Devonian rocks of the Gaspé belt. Most surface seeps occur in northeastern Gaspé and have been spatially associated with northwest-trending Acadian strike-slip faults, namely, the Troisième Lac and Bras Nord-Ouest faults. The general east–west trend of surface seeps in the Galt area, however, suggests that east-trending folds and faults are important structural features for trapping hydrocarbons (Figure 3). Several Devonian units host the seeps (Figure 2): the Indian Cove and Forillon formations of the Upper Gaspé Limestone, the York River and Battery Point formations of the Gaspé sandstones, and the rocks near the contact between the Upper Gaspé Limestone and the Gaspé Sandstone (Lavoie and Bourque, 2001). Oils recovered from wells and seeps include dark, slightly biodegraded oil, light oil-condensates, and retrograde condensates (Idiz et al., 1997). Over the years, different play concepts have been explored: depositional plays in porous sandstones and reef facies, diagenetic plays, structural traps in anticlinorial structures (Rocksandic and Granger, 1981), and fractured reservoir plays (Aguilera and Lavoie, 2002). The hydrocarbon potential of the Silurian and Devonian rocks of the Gaspé Peninsula is addressed in 11 articles in the June 2001 issue of the *Canadian Society of Petroleum Geology Bulletin*. The collection of papers is a result of a collaborative research project initiated in 1996 by Shell Canada to evaluate the hydrocarbon potential of this area.

Most of the Silurian–Devonian rock succession in the northeastern part of the Gaspé Peninsula is within the oil or gas window. The main parameters that characterize the hydrocarbon system, as presented by Lavoie and Bourque (2001; see their figure 1), are outlined in Figure 2. Potential reservoir rock units include clastic rocks of the Val-Brillant and Gascons formations and the Gaspé Sandstone, the Sayabec and West Point reef complexes (Bourque et al., 2001; Lavoie and Chi, 2001), as well as fractured limestones of the White Head, Forillon, and Indian Cove formations (Kirkwood et al., 2001; Lavoie et al., 2001). Although total organic carbon values in Silurian–Devonian rocks of the Gaspé belt are generally very low, ranging from 0.1 to 0.3%, good source rock intervals are present locally, with total organic carbon values above 1% (Bertrand and Malo, 2001). Based on total organic carbon and hydrogen index values, the Burnt Jam Brook, Indian Point, Forillon, York River, and Battery Point formations represent fair to good source rocks. The basal part of the York River contains as much as 5% total organic carbon and hydrogen index values close to 600 in outcrop (Bertrand and Malo, 2001). Other probable source rocks are the Ordovician units underlying the Silurian–Devonian succession, because geochemical analyses indicate that recovered Gaspé oils belong to the same family of oils as Ordovician oil recovered from western Newfoundland in the Port-au-Port Peninsula (Idiz et al., 1997; Rodgers et al., 1998). Rocks of the Cambrian–Ordovician Québec Supergroup that represent the basement to the Silurian–Devonian succession between the Troisième Lac and Bras Nord-Ouest faults have been suggested as possible source rocks in northeastern Gaspé on the basis of their total organic carbon content (Bertrand and Malo, 2001). The basal part of the Silurian–Devonian succession in the southern Gaspé Peninsula might also contain potential source rocks; however, no total organic carbon values are available in this part of the succession (Kirkwood et al., 2001). It consists of the Upper Ordovician Honorat Group, which contains a black shale unit of the same age as the Utica and Macasty shales, identified as excellent source rocks in the St. Lawrence Lowlands and Anticosti Island (Bertrand, 1990).

THE GALT NATURAL GAS FIELD

The Junex-Galt natural gas field is located 15 km (9 mi) west of the town of Gaspé and approximately 15 km (9 mi) north of the Saint-Jean River anticline, in the northern part of the Connecticut Valley–Gaspé synclinorium (Figure 3). The geometry in the Galt area, a large anticlinorial structure located above blind thrusts in the Gaspé belt sequence, is an excellent example of the structural style north of the Saint-Jean River anticline. It provides additional evidence supporting a fold and thrust belt tectonic setting. The production of the Galt field started in January 2001. During the first 3 months of production, 100,000 m^3 (3.5×10^6 ft^3) of high-Btu natural gas was produced for a copper smelter located near the new gas field. The result of this first production confirmed the hydrocarbon potential of the structural trap and the availability of the resource for future customers. The estimated ultimate gas recovery is projected at 6.6 bcf (Aguilera and Lavoie, 2002).

The Galt anticline is a regional, east-trending anticlinal structure that is situated between two regional southwest-trending strike-slip faults, the Troisième Lac and a subsidiary fault to the Bras Nord-Ouest faults. The host rocks of the reservoir are fractured limestones of the Lower Devonian Upper Gaspé Limestone. The Galt 1 well, drilled to a depth of 2500 m (8000 ft) in the mid-1980s, encountered oil- and gas-bearing Devonian limestones from 900 to 2000 m (3000 to 6600 ft). In 1993 and 1994, Junex, Inc. (formerly Jaltin, Inc.) reentered the well that tested for oil from the Indian Cove Formation and gas from the underlying Forillon Formation at 1500–2500-m (5000–8200-ft) depth. The Galt anticline is a complex structure formed mainly by a fault-related fold in the upper 2500 m (8000 ft) and a larger dome structure located at a depth of 3000 m (10,000 ft) (Aguilera and Lavoie, 2002). The structural geometry of the east-trending anticline allows for a four-way closure, with northwest-trending strike-slip faults bordering the fold to the west and east, thus forming a 10 × 8-km (6 × 5-mi) domal structure. Numerous north-directed and south-directed thrusts cut the main structure, producing excellent fracture porosity (Figure 7). Enhanced fracture density may also have been caused by tectonic inversion of syndepositional normal faults occurring in the Upper Silurian to Lower Devonian succession. Regionally, the Galt dome is a 100-km^2 (40-mi^2) structural culmination that has focused the migration and accumulation of hydrocarbon.

The Galt field petroleum system developed in the Gaspé belt depositional sequence. It involves four distinct reservoirs (Figure 7). In the upper 500 m (1600 ft) of the structure, heavy oil occurs in sandstones of the York River Formation. Two reservoirs containing light oil have been defined in the partially recrystalized and dolomitized calcareous mudstone of the Indian Cove Formation at depths of 500–1500 m (1600–5000 ft). Core samples of the Indian Cove contain many brecciated zones. The deeper reservoir is the gas-productive Forillon Formation between 1500 and 2500 m (5000 and 8000 ft). It is made up of three subunits that are, from bottom to top, a silty micrite-dolomicrite, a cherty micrite-dolomicrite with fossil fragments and bioturbation, and a cherty micrite-dolomicrite interlayered with crinoidal calcarenite (Amyot, 1984). The noncherty section has the best gas production potential (Aguilera and Lavoie, 2002). In outcrop, the Forillon

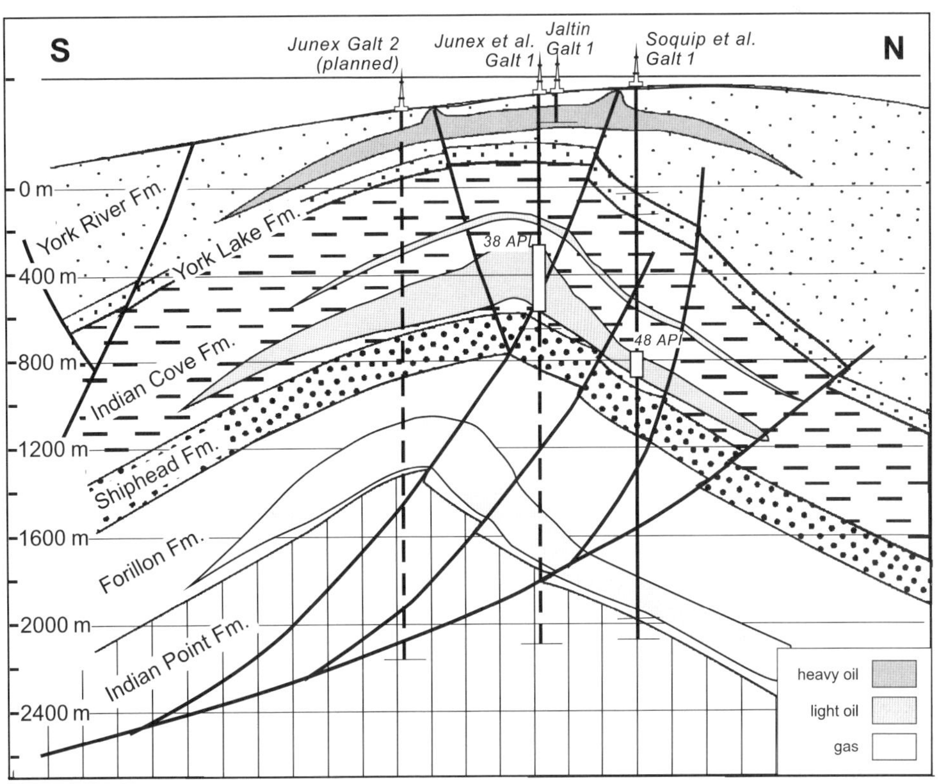

FIGURE 7. Schematic structural cross section through the Galt natural gas field showing the Galt anticline and the four distinct reservoirs hosted in the Upper Gaspé Limestone.

Formation shows intense fracturing spatially related to small-scale folds or strike-slip faults. Evaluation of fractured limestones of the Forillon Formation using fracture completion log shows porosities of greater than 6% and a partitioning coefficient in the order of 5%, which represents the fraction of total porosity that is made out of fractures (Aguilera and Lavoie, 2002). Recent seismic surveys and a detailed structural study of a surface analog in the area suggest that the highly fractured zone of the lower Forillon occurs at the tip of a small thrust fault network originating at depth in the Silurian rocks (Figure 7). The seal of the Forillon gas trap is formed by the argilaceous limestones of the Shiphead Formation. Oil, gas, and condensates also occur at depth in silty limestones of the Indian Point Formation of the Chaleurs Group and possibly as deep as in the Upper Ordovician part of the succession.

Two possible source rocks exist for the hydrocarbon-bearing Indian Cove and Forillon formations. The Lower to Middle Devonian Chaleurs Group and Upper Gaspé Limestone have been identified as containing good source rock intervals, with total organic carbon values locally above 1% in the Indian Point and Forillon formations (Bertrand and Malo, 2001). Other possible source rocks may occur in the deeper Ordovician rocks. However, the hydrocarbon potential of the Upper Ordovician carbonate units of the basal part of the Gaspé belt succession remains to be investigated. These units are chronostratigraphically correlated with the Ordovician platformal sequence on Anticosti Island and may represent distal equivalents of the platform deposited in the Gaspé Basin. Our detailed structural analysis in the Saint-Jean River anticline highlights the presence of contractional structures in this part of the sequence, suggesting that imbricated distal platformal sequences might also be present beneath the Galt field at depth.

POTENTIAL STRUCTURAL TRAPS IN THE GASPÉ

In the Saint-Jean River anticline, five distinct fracture sets have been identified and can be related either to regional folding or to the strike-slip faulting event (Lavoie, 2002; Lavoie and Kirkwood, 2002). Fractures related to the strike-slip faulting event locally enhance connectivity by connecting strata-bound fold-related fractures and might be an explanation for the spatial association between oil seeps and strike-slip faults in northeastern Gaspé. Recent studies have documented hydrocarbon migration through the Upper Ordovician to Lower Devonian Gaspé belt succession at different stages of its tectonic history. The development of fractures related to Lochkovian to Pragian synsedimentary normal faulting, Pragian to early Emsian folding and thrusting, and post-Emsian strike-slip faulting contributed effectively to porosity development and enhanced permeability in the entire Gaspé belt succession. Petrography of fluid inclusions in veins of the Upper Ordovician White Head limestones indicates that hydrocarbon-rich fluids did indeed migrate through the fracture network during the Late Silurian to Middle Devonian (Kirkwood et al., 2001). The presence of liquid hydrocarbon inclusions in the oldest vein set indicates that the development of these fractures occurred while the source rocks were well within the oil window and the host rock, the White Head limestones, either within or above the oil window. Methane inclusions in the younger vein sets indicate that development of late Acadian fractures occurred when source rocks and limestones of the White Head Formation were at depths greater than the oil window. The oil and methane inclusions also indicate that source rocks were present below the White Head in the Saint-Jean River anticline and Percé areas (Kirkwood et al., 2001). The Lower Devonian Upper Gaspé Limestone also shows significant open-fracture porosity and are characterized by repeated fracture development and dissolution during burial and associated hydrocarbon migration (Lavoie et al., 2001). Late hydrocarbon migration is also documented in the Upper Silurian–lowermost Devonian West Point reef limestone in eastern Gaspé (Bourque et al., 2001), with primary pores remaining partially open during significant burial depth.

The fold and thrust belt structural style proposed in this chapter offers a promising geological setting for potential structural traps and new hydrocarbon plays in the entire Gaspé Peninsula. Highly fractured zones are expected to be located in domal structures above important blind thrusts and at thrust tips in areas of fault-propagation folding. This type of play can also be found in many areas in the northern part of the Connecticut Valley–Gaspé synclinorium. Many fault-propagation folds have been identified based on geological mapping and recent seismic surveying, such as the Galt, Mississippi, Bald Mountain, and Holland anticlines (see Figure 3). Larger structures located above south-directed thrusts, such as the Saint-Jean River, Castonguay, and Joncas anticlines, are also very attractive targets for the exploration of fractured traps in limestones. We believe that economically viable hydrocarbon production can be found at depths of more than 1 km (0.6 mi), with both oil and gas present in the Lower Devonian, fractured, and dolomitized Upper Gaspé Limestone, the Silurian West Point reefs and Val-Brillant sandstones, and the Upper Ordovician. The Upper Ordovician fractured limestones of the White Head Formation, imbricated at depth, represent an interesting play, because they are located stratigraphically above the Upper Ordovician Honorat Group, which contains a possible source rock, a black shale unit of the same age as the Utica and Macasty shales.

CONCLUSION

Recent field work in the Upper Ordovician to Silurian rocks of the northeastern Gaspé belt and the structural geometry of the belt imaged at depth on seismic profiles support a new structural model of the Gaspé belt. The revised model involves development of a foreland fold and thrust belt during an early phase of the Acadian orogeny that preceded the Middle Devonian Acadian strike-slip deformation. The fold and fault geometry of the Saint-Jean River anticline is that of a south-verging anticline in the hanging wall of a south-directed backthrust, itself located above stacked Upper Ordovician to Lower Silurian limestones of the White Head Formation. The backthrusting is related to tectonic wedging at depth, a geometry typical of a fold-and-thrust-belt style of deformation. Public and industry seismic lines image similar north-directed thrust faults mapped at the surface as well as blind faults located beneath anticlines in the northern part of the Gaspé belt. These north-directed thrusts flatten out in the subsurface. Reflectors are mostly located at depths of 1500–5000 m (5000–16,000 ft) and imbricate the Silurian part of the Gaspé belt succession. Structural features, such as imbricate thrust faults, thrust-propagation folds, and duplexes, define the thin-skinned structural style at depth and record significant horizontal shortening in rocks of the Gaspé belt. Structural relationships in the eastern Gaspé show that strike-slip faulting postdates the development of the fold and thrust belt. Geometrically, strike-slip faults compartmentalize the large-scale fold and fault structures. The structural geometry of the Saint-Jean River anticline, as well as in the subsurface Gaspé belt, is much more compatible with a compressional fold and thrust belt than with the transpressive tectonic regime that has previously been invoked for the Acadian orogeny.

In the northern part of the Gaspé belt, the contact between the Gaspé belt succession and the Humber zone in subsurface corresponds to a shallow, south-dipping reflector that we interpret as an Acadian detachment surface. The detachment surface extends farther north into the Humber zone, i.e., the Taconian fold and thrust belt, where it most probably emerges at the surface. Similar south-dipping reflectors are interpreted in the Humber and Dunnage zones underlying the Gaspé belt. These reflectors could be detachment surfaces in the Taconian fold and thrust belt or, alternatively, Acadian detachments that cut through and restructure the Taconian belt.

The Junex-Galt natural gas field in the northeastern part of the Gaspé belt is a good example of a fractured reservoir located in fractured limestones in a fault-related fold structural trap. Geometrically, the Galt structure is a large, east-trending anticline located above a blind thrust in the upper part of the Gaspé belt sequence. To the west and east, the structure is limited by two regional strike-slip faults that produce a dome geometry and provide lateral closure for the structure. In northeastern Gaspé, fractures related to the strike-slip faulting event locally enhance connectivity by connecting strata-bound fold-related fractures. Hydrocarbon migration through the Upper Ordovician to Lower Devonian Gaspé belt succession is documented at different stages of its tectonic history. Highly fractured zones can be expected in doubly plunging anticlines above important blind thrusts and at thrust tips in areas of fault-propagation folding.

ACKNOWLEDGMENTS

The model presented in this chapter is the result of detailed field work conducted by the coauthors in northeastern Gaspé. Our ideas evolved because of numerous discussions with colleagues who have worked for many years in the Gaspé and have seen almost every outcrop. Special thanks go to Michel Malo, Pierre-André Bourque, Daniel Brisebois, and Rudolf Bertrand, who unselfishly shared all of their unpublished results. We would like to acknowledge C. Morin and J.-Y. Laliberté from the Ministère des Ressources Naturelles du Québec for permission to publish their seismic line. We are also indebted to Junex, Inc., for their continued financial and logistical support of our research projects in eastern Gaspé, and particularly to Jean-Yves Lavoie, for his relentless and aggressive exploration of the Gaspé Peninsula over the years and his support of new ideas. Thanks to the Geological Survey of Canada, Québec Office (National Mapping Project [NATMAP]–Geological Bridges of eastern Canada) for financial and logistical support during the 2001 field season. Additional field work and scholarships to M. Lavoie were funded by a Natural Sciences and Engineering Research Council individual research grant to D. Kirkwood. M. Lavoie acknowledges financial support from the Fonds pour la Formation de Chercheurs et l'Aide á la recherche. This chapter was substantially improved because of *AAPG Bulletin* reviewers, A. Bally, R. Price, and G. Stockmal. Their comments helped improve the organization of the chapter and made the final version more easily accessible to a sophisticated worldwide audience.

REFERENCES CITED

Aguilera, R., and J.-Y. Lavoie, 2002, Trucking of compressed natural gas (CNG) from the naturally fractured Galt field in the Gaspé Area of Québec: Calgary, Alberta, Society of Petroleum Engineers Gas Technology Symposium, April 30– May, 2002, Paper 75673.

Amyot, G., 1984, Lithostratigraphie de sous-surface de l'est de la Gaspésie: Ministère de l'Énergie et des ressources du Québec, ET-83-11, 75 p.

Beausoleil, C., M. Malo, C. Morin, J.-Y. Lalibeté, and D. Brisebois, 2002, Contrasting Taconian and Acadian structural styles along the new geophysical seismic reflection profiles in western Gaspé Appalachians, Matapedia Valley (abs.): Canadian Society of Petroleum Geology Diamond Jubilee Convention, Calgary, Program and Abstracts, p. 67.

Bertrand, R., 1990, Maturation thermique et histoire de l'enfouissement et de la génération des hydrocarbures du bassin de l'archipel de Mingan et de l'Ile d'Anticosti: Canadian Journal of Earth Sciences, v. 27, p. 731–741.

Bertrand, R., and M. Malo, 2001, Source rock analysis, thermal maturation and hydrocarbon generation histories in the Siluro-Devonian rocks of the Gaspé belt basin: Bulletin of Canadian Petroleum Geology, v. 49, p. 238–261.

Boucot, A. J., 1962, Appalachian Silurian–Devonian, *in* K. Coe, ed., Some aspects of the Variscan fold belt: Manchester, Manchester University Press, p. 155–163.

Bourque, P.-A., 1977, Silurian and basal Devonian of northeastern Gaspé Peninsula: Québec Ministry of Natural Resources Special Paper ES-29, 323 p.

Bourque, P.-A., 2001, Sea-level fluctuations, syn-sedimentary tectonics, reef settlement and development, and implications for hydrocarbon exploration in Silurian–lowermost Devonian Gaspé belt, Québec Appalachians: Bulletin of Canadian Petroleum Geology, v. 49, p. 217–237.

Bourque, P.-A., 2002, Reservoir potential in the southern Gaspé belt, Québec Appalachians (abs.): Canadian Society of Petroleum Geology Diamond Jubilee Convention, June 3–7, 2002, Calgary, Alberta, Program with Abstracts: http://www.cspgconvention.org/2002abstract_author.htm (accessed September 2004).

Bourque, P.-A., and G. Lachambre, 1980, Stratigraphie du Silurien et du Dévonien basal du Sud de la Gaspésie: Québec Ministry of Energy and Resources Special Paper ES-30, 123 p.

Bourque, P.-A., C. Gosselin, D. Kirkwood, M. Malo, and P. St-Julien, 1993, Le Silurien du segment appalachien Gaspésie-Matapédia-Témiscouata, Québec: Stratigraphie, géologie structurale et paléogéographie: Québec Ministry of Energy and Resources, MB 93-25, 115 p. + 23 maps.

Bourque, P.-A., D. Brisebois, and M. Malo, 1995, Gaspé belt, *in* H. Williams, ed., Geology of the Appalachian/Caledonian orogen in Canada and Greenland: Geological Survey of Canada, Geology of Canada, v. 6, chapter 4, p. 316–351 (also in Geological Society of America, The geology of North America, v. F-1, p. 316–351).

Bourque, P.-A., M. Malo, and D. Kirkwood, 2000, Sedimentary history and paleogeography of a shelf segment at the NW margin of Iapetus during Silurian to Early Devonian time: Geological Society of America Bulletin, v.112, p. 4–20.

Bourque, P.-A., M. M. Savard, G. Chi, and P. Dansereau, 2001, Diagenetic history and porosity evolution of the Upper Silurian–lowermost Devonian West Point reef limestones, Gaspé belt, Québec Appalachians: Bulletin of Canadian Petroleum Geology, v. 49, p. 299–326.

Bradley, D. C., and W. S. F. Kidd, 1991, Flexural extension of the upper continental crust in collisional foredeeps: Bulletin of the Geological Society of America, v. 103, 1416–1438.

Bradley D. C., R. D. Tucker, D. R. Lux, A. G. Harris, and D. C. McGregor, 1998, Migration of the Acadian orogen and foreland basin across the northern Appalachians: U.S. Geological Survey Open-file Report 98-770, 79 p.

Brisebois, D., 1981, Géologie de la région de Gaspé: Québec, Ministère de l'Énergie et des Ressources Naturelles, DPV-824.

Hesse, R., and E. Dalton, 1991, Diagenetic and low-grade metamorphism terranes of the Gaspé Peninsula related to the geological structures of the Taconian and Acadian orogenic belts, Québec Appalachians: Journal of Metamorphic Geology, v. 9, p. 775–790.

Howell, J. V., 1932, Historical development of the structural theory of accumulation of oil and gas (reprinted 1998), *in* N. H. Foster and E. A. Beaumont, eds., Traps and seals: I. Structural/fault-seal and hydrodynamics traps: AAPG Treatise of Petroleum Geology Reprint Series, no. 6, p. 3–26.

Hunt, T. S., 1861, Notes on the history of petroleum or rock oil: Canadian Naturalist, v. 6, p. 241–255.

Idiz, E., M. G. Fowler, and R. Bertrand, 1997, Oils and source rocks from the Gaspé Peninsula, Québec; geochemical characteristics and provenance (abs.): Canadian Society of Petroleum Geology–Society of Sedimentary Geology Joint Convention, June 1–6, 1997 Calgary, Alberta, Program with Abstracts, p. 139.

Kirkwood, D., 1989, Géologie structurale de la région de Percé: Québec Ministry of Energy and Resources, ET 87-17, 42 p + 2 maps.

Kirkwood, D., 1995, Progressive deformation history of a transpressive belt, northern Appalachians: Tectonophysics, v. 241, p. 15–34.

Kirkwood, D., 1999, A palinspastic restoration of a post-Taconian successor basin deformed within a transpressive setting, northern Appalachians: Tectonics, v. 18, p. 1027–1040.

Kirkwood, D., and M. Malo, 1993, Across strike geometry of the Grand Pabos fault zone: Evidence for Devonian dextral transpression in the Québec Appalachians: Canadian Journal of Earth Sciences, v. 30, p. 1363–1373.

Kirkwood, D., M. Malo, P. St-Julien, and P. Therrien, 1995, Progressive deformation history of a folded slate belt in a transpressive setting, northern Appalachians: Journal of Structural Geology, v. 17, p. 329–344.

Kirkwood, D., M. M. Savard, and G. Chi, 2001, Microstructural analysis and cement characterization of pre-Acadian and Acadian deformation stages of the White Head Formation— Evaluation of the HC reservoir potential in northeastern Gaspé: Bulletin of Canadian Petroleum Geology, v. 49, p. 262–281.

Kirkwood, D., M. Lavoie, V. Lavoie, and J.-S. Marcil, 2002, Acadian tectonic wedging, stacking, and triangle zone in northeastern Gaspé Appalachians? (abs.): Canadian Society of Petroleum Geology Diamond Jubilee Convention, Calgary, Program with Abstracts, p. 186.

Lavoie, D., 1992, Carbonate sedimentation in an extensional tectonic regime: The Lower Devonian Upper Gaspé limestones, Québec Appalachians: Canadian Journal of Earth Sciences, v. 29, p. 118–128.

Lavoie, V., 2002, Étude détaillée de la fracturation à l'anticlinal de la Rivière Saint-Jean: B.Sc. thesis, Université Laval, Québec, 47 p.

Lavoie, D., and P.-A. Bourque, 2001, The history of hydrocarbon exploration in the Silurian–Devonian Gaspé belt: 100 years of modest success: Bulletin of Canadian Petroleum Geology, v. 49, p. 180–185.

Lavoie, D., and G. Chi, 2001, The Lower Silurian Sayabec Formation in northern Gaspé: Carbonate diagenesis and reservoir potential: Bulletin of Canadian Petroleum Geology, v. 49, p. 282–298.

Lavoie, D., G. Chi, and M. Fowler, 2001, The Lower Devonian upper Gaspé Limestones in eastern Gaspé: Carbonate diagenesis and reservoir potential. Bulletin of Canadian Petroleum Geology, v. 49, p. 346–365.

Lavoie, M., and D. Kirkwood, 2002, Structural Analysis of the St.-Jean River anticline, eastern Gaspé, northern Appalachians (abs.): Canadian Society of Petroleum Geology Diamond Jubilee Convention, Calgary, Program with Abstracts, p. 195.

Lavoie, J.-Y., J.-S. Marcil, and R. Aguillera, 2002, Exploration efforts in the Galt area: First economic hydrocarbon discovery in the eastern Gaspé Peninsula (abs.): Canadian Society of Petroleum Geology Diamond Jubilee Convention, Calgary, Program with abstracts, p. 194.

Logan, Sr., W. R., 1844, Canada Geological Survey Report of Progress, p. 41.

Lynch, G., and O. Arsenault, 1997, Stratigraphy and deformation of the Humber zone in Gaspésie Québec, *in* Current research 1997-D: Geological Survey of Canada, p. 1–8.

Malo, M., 1988, Stratigraphy of the Aroostook-Percé anticlinorium in the Gaspé Peninsula, Québec: Canadian Journal of Earth Sciences, v. 25, p. 893–908.

Malo, M., 2001, Late Silurian–Early Devonian tectonosedimentary history of the Gaspé belt in the Gaspé Peninsula: From a transtensional Salinic basin to an Acadian foreland basin: Bulletin of Canadian Petroleum Geology, v. 49, p. 202–216.

Malo, M., and J. Béland, 1989, Acadian strike-slip tectonics in the Gaspé region, Québec Appalachians: Canadian Journal of Earth Sciences, v. 26, p. 1764–1777.

Malo, M., and P.-A. Bourque, 1993, Timing of the deformation events from Late Ordovician to mid-Devonian in the Gaspé Peninsula, *in* D. C. Roy, and S. J. Skehan, eds., The Acadian orogeny: Recent studies in New England, maritime Canada, and the autochthonous foreland: Geological Society of America Special Paper 275, p. 101–122.

Malo, M., and D. Kirkwood, 1995, Faulting and progressive deformation history of the Gaspé belt, northern Appalachians: A review, *in* J. Hibbard, C. van Staal, and P. Cawood, eds., Current perspectives in Appalachian–Caledonian geology, Geological Association of Canada Special Paper 41, p. 267–282.

Malo, M., D. Kirkwood, G. DeBroucker, and P. St-Julien, 1992, A re-evaluation of the position of the Baie Verte-Brompton line in the Québec Appalachians: The influence of Middle Devonian strike-slip faulting in Gaspé Peninsula: Canadian Journal of Earth Sciences, v. 29, p. 1265–1273.

Martinsen, O., 1994, Mass movements, *in* A. Maltmann, ed., The geological deformation of sediments: London, Chapman & Hall, p. 127–165.

Morin, C., and J.-Y. Laliberté, 2001, Interprétation du transect de sismique réflexion St-Cléophas-Ronceaux, programme géophysique Val-Brillant: Ministère des Ressources Naturelles, Québec, Secteur de l'énergie, Direction du gaz et du pétrole, 22 ième Séminaire de la recherche géologique, Château Fontenac, Novembre 28 et 29, 2001.

Morin, C., and J.-Y. Laliberté, 2002a, The unexpected Silurian–Devonian structural style in western Gaspé—New insight for promising hydrocarbon plays (abs.): Canadian Society of Petroleum Geology Diamond Jubilee Convention, Calgary, Program and Abstracts: http://cspgconvention.org/2002abstracts_author.htm (accessed September 2004).

Morin, C., and J.-Y. Laliberté, 2002b, Overview of the very first mega-transect seismic survey throughout the Appalachian Gaspé belt (abs.): Canadian Society of Petroleum Geology Diamond Jubilee Convention, Calgary, Program and Abstracts: http://cspgconvention.org/2002abstracts_author.htm (accessed September 2004).

Pincivy, A., M. Malo, G. Ruffet, A. Tremblay, and P. E. Sacks, 2003, Regional metamorphism of the Appalachian Humber zone of Gaspé Peninsula: $^{40}Ar/^{39}Ar$ evidence for crustal thickening during the Taconian orogeny: Canadian Journal of Earth Sciences, v. 40, p. 301–315.

Rodgers, K. M., M. M. Savard, M. Fowler, and R. Betrand, 1998, A geochemical study of oils and possible source rocks from the Gaspé Peninsula, Québec (abs.): Geological Association of Canada-Mineralogical Association of Canada-Association des Professionals Geologies, et Geophysiciens du Québec 1998, abstract volume, p. A-158.

Rogers, H. D., 1860, On the distribution and probable origin of the Petroleum or Rock Oil of Western Pennsylvania, New York and Ohio: Proceedings Philosophical Society of Glasgow, v. 4, p. 355–359.

Roksandic, M. M., and B. Granger, 1981, Structural styles of Anticosti Island, Gaspé Passage, and eastern Gaspé Peninsula inferred from reflection seismic data, *in* Field meeting, Anticosti-Gaspé, 1981, *in* P. J. Lespérance, ed., Stratigraphy and paleontology v. II: International Union of Geological Sciences Subcommission on Silurian Stratigraphy and Ordovician–Silurian Boundary Working Group: Département de géologie, Université de Montréal, p. 211–221.

Sanford, B. V., 1993, St. Lawrence Platform— Geology, *in* D. F. Stott, G. G. Cook, and J. D. Aitken, eds., Sedimentary cover of the craton in Canada, Geological Survey of Canada, Geology of Canada, no. 5, p. 723–786.

Simard, M., 1987, Géologie de la région de Dufour: Québec, Ministère de l'Énergie et des Ressources Naturelles, ET-87-12.

Simard, M., 1988, Géologie de la région de Joncas-Power: Québec, Ministère de l'Énergie et des Ressources Naturelles, ET-86-10.

Slivitzky, A., P. St-Julien, and G. Lachambre, 1991, Synthèse géologique du Cambro-Ordovicien du Nord de la Gaspésie: Québec, Ministère de l'Énergie et des Ressources, ET 88-14.

St-Julien, P., and C. Hubert, 1975, Evolution of the Taconian orogen in the Québec Appalachians: American Journal of Science, v. 275-A, p. 337–362.

Stockmal, G. S., A. Slingsby, and J. W. F. Waldron, 1998, Deformation styles at the Appalachian structural front, western Newfoundland: Implications of new industry seismic reflection data: Canadian Journal of Earth Sciences, v. 35, p. 1288–1306.

Waldron, J. W. F., and G. S. Stockmal, 1994, Structural and tectonic evolution of the Humber zone, western Newfoundland: 2. A regional model for Acadian thrust tectonics: Tectonics, v. 13, p. 1498–1513.

Williams, H., 1979, Appalachian orogen in Canada: Canadian Journal of Earth Sciences, v. 16, p. 792–807.

Williams, H., S. P. Colman-Sadd, and H. S. Swinden, 1988, Tectonic-stratigraphic subdivisions of central Newfoundland, *in* Current Research, part B: Geological Survey of Canada, Paper 88-1B, p. 91–98.

Williams, H. R., ed., 1995. Geology of the Appalachian–Caledonian orogen in Canada and Greenland: Geological Survey of Canada, The geology of Canada, v. 6, 943 p.